Handbook of Epistemic Logic

Handbook of Epistemic Logic

Edited by

Hans van Ditmarsch

Joseph Y. Halpern

Wiebe van der Hoek

Barteld Kooi

© Individual authors and College Publications 2015. All rights reserved.

ISBN 978-1-84890-158-2

College Publications
Scientific Director: Dov Gabbay
Managing Director: Jane Spurr

http://www.collegepublications.co.uk

Cover produced by Laraine Welch
Printed by Lightning Source, Milton Keynes, UK

All rights reserved. No part of this publication may be reproduced, stored in a retrieval system or transmitted in any form, or by any means, electronic, mechanical, photocopying, recording or otherwise without prior permission, in writing, from the publisher.

Contents

Preface

Work on epistemic logic (reasoning about knowledge) originated with philosophers in the early 1960s. Since then, it has played a significant role not only in philosophy, but also in computer science and economics. This handbook reports significant progress in a field that has now really matured, but is still very active. We hope this books makes it easier for new researchers to enter the field, while giving experts a chance to appreciate work in related areas.

The first chapter in the handbook gives a brief introduction to epistemic logic, as well as providing an overview of the rest of the book. What follows are then 11 chapters, making up three parts of the book. Each of those chapters focuses on a particular topic; they can be read independently, in any order. Each episode ends with a section of notes that provides some historical background, references, and, occasionally, more details on points not covered in the chapter. Because we expect that chapters will often be read independently, each chapter has its own bibliography.

Many people were involved in the production of this book. First and foremost, we thank the authors for their contributions. Some of the authors also acted as readers; in addition, we also had additional external readers who provided authors with useful comments. These include Philippe Balbiani, Thomas Bolander, Jan Broersen, Masoud Koleini, Emiliano Lorini, Cláudia Nalon, Ji Ruan, Elias Tsakas, and Boèna Woźna Szcześniak. Finally, we thank College Publications for their flexibility and patience, especially Jane Spurr.

Hans van Ditmarsch, Joe Halpern, Wiebe van der Hoek, and Barteld Kooi

Contributors

Thomas Ågotnes
Department of Information Science and Media Studies, University of Bergen, Norway
Email: thomas.agotnes@infomedia.uib.no.

Johan van Benthem
Institute of Logic, Language and Computation, University of Amsterdam, the Netherlands, and Department of Philosophy, Stanford University, USA
Email: johan.vanbenthem@uva.nl.

Giacomo Bonanno
Department of Economics, University of California, Davis, USA
Email: gfbonanno@ucdavis.edu.

Jan Broersen
Department of Information and Computing Sciences, Utrecht University, the Netherlands
Email: J.M.Broersen@uu.nl.

Lorenz Demey
Institute of Philosophy, KU Leuven, Belgium
Email: lorenz.demey@hiw.kuleuven.be.

Hans van Ditmarsch
LORIA, CRNS, University of Lorraine, France
Email: hans.van-ditmarsch@loria.fr.

Clare Dixon
Department of Computer Science, University of Liverpool, UK
Email: cldixon@liverpool.ac.uk.

Valentin Goranko
Department of Philosophy, Stockholm University, Sweden
Email: valentin.goranko@philosophy.su.se.

Joseph Y. Halpern
Computer Science Department, Cornell University, USA
Email: halpern@cs.cornell.edu.

Andreas Herzig
IRIT-LILaC, University of Toulouse and CNRS, France
Email: herzig@irit.fr.

Wiebe van der Hoek
Department of Computer Science, University of Liverpool, UK
Email: wiebe@liverpool.ac.uk.

Wojciech Jamroga
Institute of Computer Science, Polish Academy of Sciences, Poland
Email: w.jamroga@ipipan.waw.pl.

Barteld Kooi
Philosophy, University of Groningen, the Netherlands
Email: B.P.Kooi@rug.nl.

Gerhard Lakemeyer
Department of Computer Science, RWTH Aachen University, Germany
Email: gerhard@informatik.rwth-aachen.de.

Hector J. Levesque
Department of Computer Science, University of Toronto, Canada
Email: hector@cs.toronto.edu.

Alessio Lomuscio
Department of Computing, Imperial College London, UK
Email: a.lomuscio@imperial.ac.uk.

John-Jules Ch. Meyer
Department of Information and Computing Sciences, Utrecht University, and the Alan Turing Institute Almere, the Netherlands
Email: jj@cs.uu.nl.

Lawrence S. Moss
Mathematics Department, Indiana University, Bloomington USA
Email: lsm@cs.indiana.edu.

Cláudia Nalon
Department of Computer Science, University of Brasília, Brazil
Email: nalon@unb.br.

Wojciech Penczek
Institute of Computer Science, Polish Academy of Sciences, Poland
Email: penczek@ipipan.waw.pl.

Riccardo Pucella
Forrester Research, Inc., USA
Email: riccardo@acm.org.

R. Ramanujam
Theoretical Computer Science, The Institute of Mathematical Sciences, India
Email: jam@imsc.res.in.

Joshua Sack
Institute of Logic, Language and Computation, University of Amsterdam, the Netherlands
Email: joshua.sack@gmail.com.

Burkhard C. Schipper
Department of Economics, University of California, Davis, USA
Email: bcschipper@ucdavis.edu.

Sonja Smets
Institute of Logic, Language and Computation, University of Amsterdam, the Netherlands
Email: S.J.L.Smets@uva.nl.

Michael Wooldridge
Department of Computer Science, University of Oxford, UK
Email: mjw@cs.ox.ac.uk.

Chapter 1

An Introduction to Logics of Knowledge and Belief

Hans van Ditmarsch
Joseph Y. Halpern
Wiebe van der Hoek
Barteld Kooi

Contents

Abstract This chapter provides an introduction to some basic concepts of epistemic logic, basic formal languages, their semantics, and proof systems. It also contains an overview of the handbook, and a brief history of epistemic logic and pointers to the literature.

1.1 Introduction to the Book

This introductory chapter has four goals:

1. an informal introduction to some basic concepts of epistemic logic;

2. basic formal languages, their semantics, and proof systems;

Chapter 1 of the *Handbook of Epistemic Logic*, H. van Ditmarsch, J.Y. Halpern, W. van der Hoek and B. Kooi (eds), College Publications, 2015, pp. 1–51.

3. an overview of the handbook; and

4. a brief history of epistemic logic and pointers to the literature.

In Section 1.2, we deal with the first two items. We provide examples that should help to connect the informal concepts with the formal definitions. Although the informal meaning of the concepts that we discuss may vary from author to author in this book (and, indeed, from reader to reader), the formal definitions and notation provide a framework for the discussion in the remainder of the book.

In Section 1.3, we outline how the basic concepts from this chapter are further developed in subsequent chapters, and how those chapters relate to each other. This chapter, like all others, concludes with a section of notes, which gives all the relevant references and some historical background, and a bibliography.

1.2 Basic Concepts and Tools

As the title suggests, this book uses a formal tool, *logic*, to study the notion of *knowledge* ("episteme" in Greek, hence *epistemic logic*) and belief, and, in a wider sense, the notion of *information*.

Logic is the study of reasoning, formalising the way in which certain conclusions can be reached, given certain premises. This can be done by showing that the conclusion can be *derived* using some deductive system (like the axiom systems we present in Section 1.2.5), or by arguing that the *truth* of the conclusion must follow from the truth of the premises (truth is the concern of the semantical approach of Section 1.2.2). However, first of all, the premises and conclusions need to be presented in some formal *language*, which is the topic of Section 1.2.1. Such a language allows us to specify and verify properties of complex systems of interest.

Reasoning about knowledge and belief, which is the focus of this book, has subtleties beyond those that arise in propositional or predicate logic. Take, for instance, the law of excluded middle in classical logic, which says that for any proposition p, either p or $\neg p$ (the negation of p) must hold; formally, $p \vee \neg p$ is valid. In the language of epistemic logic, we write $K_a p$ for 'agent a knows that p is the case'. Even this simple addition to the language allows us to ask many more questions. For example, which of the following formulas should be valid, and how are they related? What kind of 'situations' do the formulas describe?

- $K_a p \vee \neg K_a p$
- $K_a p \vee K_a \neg p$

- $K_a(p \vee \neg p)$
- $K_a p \vee \neg K_a \neg p$

It turns out that, given the semantics of interest to us, only the first and third formulas above are valid. Moreover as we will see below, $K_a p$ logically implies $\neg K_a \neg p$, so the last formula is equivalent to $\neg K_a \neg p$, and says 'agent a considers p possible'. This is incomparable to the second formula, which says agent a knows whether p is true'.

One of the appealing features of epistemic logic is that it goes beyond the 'factual knowledge' that the agents have. Knowledge can be about knowledge, so we can write expressions like $K_a(K_a p \rightarrow K_a q)$ (a knows that if he knows that p, he also knows that q). More interestingly, we can model knowledge about other's knowledge, which is important when we reason about communication protocols. Suppose *Ann* knows some fact m ('we meet for dinner the first Sunday of August'). So we have $K_a m$. Now suppose Ann e-mails this message to Bob at Monday 31st of July, and Bob reads it that evening. We then have $K_b m \wedge K_b K_a m$. Do we have $K_a K_b m$? Unless Ann has information that Bob has actually read the message, she cannot assume that he did, so we have $(K_a m \wedge \neg K_a K_b m \wedge \neg K_a \neg K_b m)$.

We also have $K_a K_b \neg K_a K_b m$. To see this, we already noted that $\neg K_a K_b m$, since Bob might not have read the message yet. But if *we* can deduce that, then Bob can as well (we implicitly assume that all agents can do perfect reasoning), and, moreover, Ann can deduce *that*. Being a gentleman, Bob should resolve the situation in which $\neg K_a K_b m$ holds, which he could try to do by replying to Ann's message. Suppose that Bob indeed replies on Tuesday morning, and Ann reads this on Tuesday evening. Then, on that evening, we indeed have $K_a K_b K_a m$. But of course, Bob cannot assume Ann read the acknowledgement, so we have $\neg K_b K_a K_b K_a m$. It is obvious that if Ann and Bob do not want any ignorance about knowledge of m, they better pick up the phone and verify m. Using the phone is a good protocol that guarantees $K_a m \wedge K_b m \wedge K_a K_b m \wedge K_b K_a m \wedge K_a K_b K_a m \wedge \ldots$, a notion that we call *common knowledge*; see Section 1.2.2.

The point here is that our formal language helps clarify the effect of a (communication) protocol on the information of the participating agents. This is the focus of Chapter 12. It is important to note that requirements of protocols can involve both knowledge and ignorance: in the above example for instance, where Charlie is a roommate of Bob, a goal (of Bob) for the protocol might be that he knows that Charlie does *not* know the message ($K_b \neg K_c m$), while a goal of Charlie might even be $K_c K_b \neg m$. Actually, in the latter case, it may be more reasonable to write $K_c B_b \neg m$: Charlie knows that Bob believes that there is no dinner on Sunday. A temporal progression from $K_b m \wedge \neg K_a K_b m$ to $K_b K_a m$ can be viewed as learning. This raises interesting questions in the study of epistemic protocols: given an initial and final specification of information, can we find a sequence of messages that take us from the former to the latter? Are there optimal such sequences? These questions are addressed in Chapter 5, specifically Sections 5.7 and 5.9.

Here is an example of a scenario where the question is to derive a sequence of messages from an initial and final specification of information. It is taken from Chapter 12, and it demonstrates that security protocols that aim to ensure that certain agents stay ignorant cannot (and do not) always rely on the fact that some messages are kept secret or hidden.

> Alice and Betty each draw three cards from a pack of seven cards, and Eve (the eavesdropper) gets the remaining card. Can players Alice and Betty learn each other's cards without revealing that information to Eve? The restriction is that Alice and Betty can make only public announcements that Eve can hear.

We assume that (it is common knowledge that) initially, all three agents know the composition of the pack of cards, and each agent knows which cards she holds. At the end of the protocol, we want Alice and Betty to know which cards each of them holds, while Eve should know only which cards she (Eve) holds. Moreover, messages can only be public announcements (these are formally described in Chapter 6), which in this setting just means that Alice and Betty can talk to each other, but it is common knowledge that Eve hears them. Perhaps surprisingly, such a protocol exists, and, hopefully less surprisingly by now, epistemic logic allows us to formulate precise epistemic conditions, and the kind of announcements that should be allowed. For instance, no agent is allowed to lie, and agents can announce only what they know. Dropping the second condition would allow Alice to immediately announce Eve's card, for instance. Note there is an important distinction here: although Alice knows that there is an announcement that she can make that would bring about the desired state of knowledge (namely, announcing Eve's card), there is not something that Alice knows that she can announce that would bring about the desired state of knowledge (since does not in fact know Eve's card). This distinction has be called the *de dicto/de re* distinction in the literature. The connections between knowledge and strategic ability are the topic of Chapter 11.

Epistemic reasoning is also important in distributed computing. As argued in Chapter 5, processes or programs in a distributed environment often have only a limited view of the global system initially; they gradually come to know more about the system. Ensuring that each process has the appropriate knowledge needed in order to act is the main issue here.

The chapter mentions a number of problems in distributed systems where epistemic tools are helpful, like agreement problems (the dinner example of Ann and Bob above would be a simple example) and the problem of mutual exclusion, where processes sharing a resource must ensure that only one process uses the resource at a time. An instance of the latter is provided in Chapter 8, where epistemic logic is used to specify a correctness property of the *Railroad Crossing System.* Here, the agents Train, Gate and Controller must ensure, based on the type of signals that they send, that the train is never at the crossing while the gate is 'up'. Chapter 8 is on model checking; it provides techniques to automatically verify that such properties (specified in an epistemic temporal language; cf. Chapter 5) hold. Epistemic tools to deal with the problem of mutual exclusion are also discussed in Chapter 11, in the context of dealing with *shared file updates.*

Reasoning about knowing what others know (about your knowledge) is also typical in strategic situations, where one needs to make a decision based on how others will act (where the others, in turn, are basing their decision on their reasoning about you). This kind of scenario is the focus of game theory. *Epistemic* game theory studies game theory using notions from epistemic logic. (Epistemic game theory is the subject of Chapter 9 in this book.) Here, we give a simplified example of one of the main ideas. Consider the game in Figure 1.1.

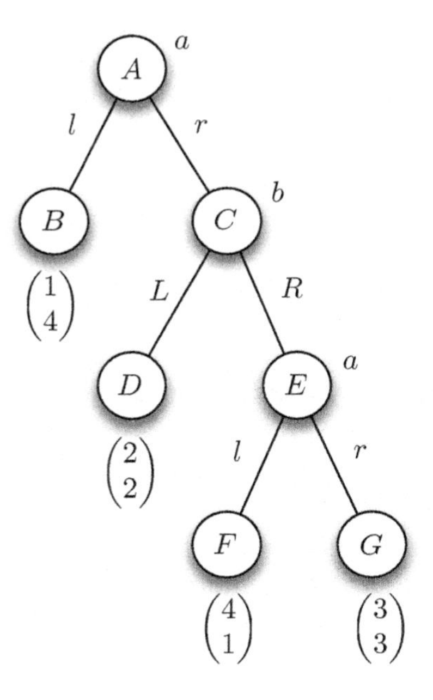

Figure 1.1: A simple extensive form game.

This model represents a situation where two players, a and b, take turns, with a starting at the top node A. If a plays l ('left') in this node, the game ends in node B and the payoff for a is 1 and that for b is 4. If a, however, plays r in A, the game proceeds to node C, where it is b's turn. Player b has a choice between playing L and R (note that we use upper case to distinguish b's moves from a's moves). The game continues until a terminal node is reached. We assume that both players are *rational*; that is, each prefers a higher outcome for themselves over a lower one. What will a play in the start node A?

One way to determine what will happen in this game is to use backward. Consider node E. If that node is reached, given that a is rational (denoted rat_a), a will play l here, since she prefers the outcome 4 over 3 (which she would get by playing r). Now consider node C. Since b knows that a is rational, he knows that his payoff when playing R at C is 1. Since b is rational, and playing L in C gives him 2, he will play L. The only thing needed to conclude this is $(rat_b \wedge K_b rat_a)$. Finally, consider node A. Player a can reason as we just did, so a knows that she has a choice between the payoff of 2 she would obtain by playing r and the payoff of 1 she would obtain by playing l. Since a is rational, she plays r at A. Summarising, the condition that justifies a playing r at A and b playing L at B is

$$rat_a \wedge K_a rat_b \wedge K_a K_b rat_a \wedge rat_b \wedge K_b rat_a$$

This analysis predicts that the game will end in node D. Although this analysis used only 'depth-two' knowledge (a knows that b knows), to perform a similar analysis for longer variants of this game requires deeper and deeper knowledge of rationality. In fact, in many epistemic analyses in game theory, common knowledge of rationality is assumed. The contribution of epistemic logic to game theory is discussed in more detail in Chapter 9.

1.2.1 Language

Most if not all systems presented in this book extend propositional logic. The language of propositional logic assumes a set At of primitive (or atomic) propositions, typically denoted $p, q, \ldots$, possibly with subscripts. They typically refer to statements that are considered basic; that is, they lack logical structure, like 'it is raining', or 'the window is closed'. Classical logic then uses Boolean operators, such as $\neg$ ('not'), $\wedge$ ('and'), $\vee$, ('or'), $\rightarrow$ ('implies'), and $\leftrightarrow$ ('if and only if'), to build more complex formulas. Since all those operators can be defined in terms of $\wedge$ and $\neg$ (see Definition 1.2), the formal definition of the language often uses only these two connectives. Formulas are denoted with Greek letters: $\varphi, \psi, \alpha, \ldots$. So, for instance, while $(p \wedge q)$ is the conjunction of two primitive propositions, the formula $(\varphi \wedge \psi)$ is a conjunction of two arbitrary formulas, each of which may have further structure.

When reasoning about knowledge and belief, we need to be able to refer to the subject, that is, the agent whose knowledge or belief we are talking about. To do this, we assume a finite set Ag of agents. Agents are typically denoted $a, b, \ldots, i, j, \ldots$, or, in specific examples, *Alice*, *Bob*, ... To reason about knowledge, we add operators K_a to the language of classical logic, where $K_a\varphi$ denotes 'agent a knows (or believes) φ'. We typically let the context determine whether K_a represents knowledge or belief. If it is necessary to reason knowledge and belief simultaneously, we use operators K_a for knowledge and B_a for belief. Logics for reasoning about knowledge are sometimes called *epistemic* logics, while logics for reasoning about belief are called *doxastic* logics, from the Greek words for knowledge and belief. The operators K_a and B_a are examples of *modal* operators. We sometimes use $\Box$ or $\Box_a$ to denote a generic modal operator, when we want to discuss general properties of modal operators.

Definition 1.1 (An Assemblage of Modal Languages)
Let At be a set of primitive propositions, Op a set of modal operators, and Ag a set of agent symbols. Then we define the language $\mathsf{L}(\mathsf{At}, \mathsf{Op}, \mathsf{Ag})$ by the following BNF:

$$\varphi := p \mid \neg\varphi \mid (\varphi \wedge \varphi) \mid \Box\varphi,$$

where $p \in \mathsf{At}$ and $\Box \in \mathsf{Op}$. ⊣

Typically, the set Op depends on Ag. For instance, the language for multi-agent epistemic logic is $\mathsf{L}(\mathsf{At}, \mathsf{Op}, \mathsf{Ag})$, with $\mathsf{Op} = \{K_a \mid a \in \mathsf{Ag}\}$, that is, we have a knowledge operator for every agent. To study interactions between knowledge and belief, we would have $\mathsf{Op} = \{K_a, B_a \mid a \in \mathsf{Ag}\}$. The language of propositional logic, which does not involve modal operators, is denoted $\mathsf{L}(\mathsf{At})$; *propositional formulas* are, by definition, formulas in $\mathsf{L}(\mathsf{At})$.

Definition 1.2 (Abbreviations in the Language)
As usual, parentheses are omitted if that does not lead to ambiguity. The following abbreviations are also standard (in the last one, $A \subseteq \mathsf{Ag}$).

description/name	*definiendum*	*definiens*
false	$\bot$	$p \wedge \neg p$
true	$\top$	$\neg\bot$
disjunction	$\varphi \vee \psi$	$\neg(\neg\varphi \wedge \neg\psi)$
implication	$\varphi \rightarrow \psi$	$\neg\varphi \vee \psi$
dual of K	$M_a\varphi$ or $\hat{K}_a\varphi$	$\neg K_a \neg\varphi$
everyone in A knows	$E_A\varphi$	$\bigwedge_{a \in A} K_a\varphi$

Note that $M_a\varphi$, which say 'agent a does not know $\neg\varphi$', can also be read 'agent a considers φ possible'. ⊣

Let $\Box$ be a modal operator, either one in Op or one defined as an abbreviation. We define the nth iterated application of $\Box$, written $\Box^n$, as follows:

$$\Box^0\varphi = \varphi \text{ and } \Box^{n+1}\varphi = \Box\Box^n\varphi.$$

We are typically interested in iterating the E_A operator, so that we can talk about 'everyone in A knows', 'everyone in A knows that everyone in A knows', and so on.

Finally, we define two measures on formulas.

Definition 1.3 (Length and modal depth)
The *length* $|\varphi|$ and the *modal depth* $d(\varphi)$ of a formula φ are both defined inductively as follows:

$$\begin{array}{lllllll}
|p| & = & 1 & \text{and} & d(p) & = & 0 \\
|\neg\varphi| & = & |\varphi|+1 & \text{and} & d(\neg\varphi) & = & d(\varphi) \\
|(\varphi\wedge\psi)| & = & |\varphi|+|\psi|+1 & \text{and} & d(\varphi\wedge\psi) & = & max\{d(\varphi),d(\psi)\} \\
|\Box_a\varphi| & = & |\varphi|+1 & \text{and} & d(\Box\varphi) & = & 1+d(\varphi).
\end{array}$$

In the last clause, $\Box_a$ is a modal operator corresponding to a single agent. Sometimes, if $A \subseteq \mathsf{Ag}$ is a group of agents and $\Box_A$ is a group operator (like E_A, D_A or C_A), $|\Box_A\varphi|$ depends not only on φ, but also on the cardinality of A. ⊣

So, $|\Box_a(q \wedge \Box_b p)| = 5$ and $d(\Box_a(q \wedge \Box_b p)) = 2$. Likewise, $|\Box_a q \wedge \Box_b p| = 5$ while $d(\Box_a q \wedge \Box_b p) = 1$.

1.2.2 Semantics

We now define a way to systematically determine the *truth value* of a formula. In propositional logic, whether p is true or not 'depends on the situation'. The relevant situations are formalised using *valuations*, where a valuation

$$V : \mathsf{At} \to \{true, false\}$$

determines the truth of primitive propositions. A valuation can be extended so as to determine the truth of all formulas, using a straightforward inductive definition: $\varphi \wedge \psi$ is true given V iff each of φ and ψ is true given V, and $\neg\varphi$ is true given V iff φ is false given V. The truth conditions of disjunctions, implications, and bi-implications follow directly from these two clauses and Definition 1.2. To model knowledge and belief, we use ideas that go back to Hintikka. We think of an agent a as considering possible a number of different situations that are consistent with the information that the agent has. Agent a is said to know (or believe) φ, if φ is true in all the situations that a considers possible. Thus, rather than using a single situation to give meaning to modal formulas, we use a *set* of such situations; moreover, in each situation, we consider, for each agent, what other situations he or she considers possible. The following example demonstrates how this is done.

Example 1.1
Bob is invited for a job interview with Alice. They have agreed that it will take place in a coffeehouse downtown at noon, but the traffic is quite unpredictable, so it is not guaranteed that either Alice or Bob will arrive on time. However, the coffeehouse is only a 15-minute walk from the bus stop where Alice plans to go, and a 10-minute walk from the metro station where Bob plans to go. So, 10 minutes before the interview, both Alice and Bob will know whether they themselves will arrive on time. Alice and Bob have never met before. A Kripke model describing this situation is given in Figure 1.2.

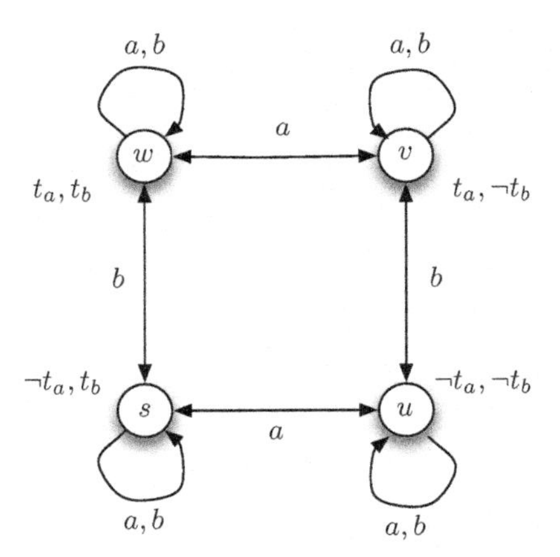

Figure 1.2: The Kripke model for Example 1.1.

Suppose that at 11:50, both Alice and Bob have just arrived at their respective stations. Taking t_a and t_b to represent that Alice (resp., Bob) arrive on time, this is a situation (denoted w in Figure 1.2) where both t_a and t_b are true. Alice knows that t_a is true (so in w we have $K_a t_a$), but she does not know whether t_b is true; in particular, Alice considers possible the situation denoted v in Figure 1.2, where $t_a \wedge \neg t_b$ holds. Similarly, in w, Bob considers it possible that the actual situation is s, where Alice is running late but Bob will make it on time, so that $\neg t_a \wedge t_b$ holds. Of course, in s, Alice knows that she is late; that is, $K_a \neg t_a$ holds. Since the only situations that Bob considers possible at world w are w and s, he knows that he will be on time ($K_b t_b$), and knows that Alice knows whether or not she is on time ($K_b(K_a t_a \vee K_a \neg t_a)$). Note that the latter fact follows since $K_a t_a$ holds in world w and $K_a \neg t_a$ holds in world s, so $K_a t_a \vee K_a \neg t_a$ holds in both worlds that Bob considers possible. ⊣

This, in a nutshell, explains what the models for epistemic and doxastic look like: they contain a number of situations, typically called *states* or *(possible) worlds*, and binary relations on states for each agent, typically called *accessibility relations*. A pair (v, w) is in the relation for agent a if, in world v, agent a considers state w possible. Finally, in every state, we need to specify which primitive propositions are true.

Definition 1.4 (Kripke frame, Kripke model)
Given a set At of primitive propositions and a set Ag of agents, a *Kripke model* is a structure $M = \langle S, R^{\mathsf{Ag}}, V^{\mathsf{At}} \rangle$, where

- $S \neq \emptyset$ is a set of states, sometimes called the *domain* of M, and denoted $\mathcal{D}(M)$;
- R^{Ag} is a function, yielding an accessibility relation $R_a \subseteq S \times S$ for each agent $a \in \mathsf{Ag}$;
- $V^{\mathsf{At}} : S \rightarrow (\mathsf{At} \rightarrow \{true, false\})$ is a function that, for all $p \in \mathsf{At}$ and $s \in S$, determines what the truth value $V^{\mathsf{At}}(s)(p)$ of p is in state s (so $V^{\mathsf{At}}(s)$ is a propositional valuation for each $s \in S$).

We often suppress explicit reference to the sets At and Ag, and write $M = \langle S, R, V \rangle$, without upper indices. Further, we sometimes write sR_at or R_ast rather than $(s,t) \in R_a$, and use $R_a(s)$ or R_as to denote the set $\{t \in S \mid R_ast\}$. Finally, we sometimes abuse terminology and refer to V as a valuation as well.

The class of all Kripke models is denoted $\mathcal{K}$. We use $\mathcal{K}_m$ to denote the class of Kripke models where $|\mathsf{Ag}| = m$. A *Kripke frame* $F = \langle S, R \rangle$ focuses on the graph underlying a model, without regard for the valuation. ⊣

More generally, given a modal logic with a set Op of modal operators, the corresponding Kripke model has the form $M = \langle S, R^{\mathsf{Op}}, V^{\mathsf{At}} \rangle$, where there is a binary relation $R_\Box$ for every operator $\Box \in \mathsf{Op}$. Op may, for example, consist of a knowledge operator for each agent in some set Ag and a belief operator for each agent in Ag.

Given Example 1.1 and Definition 1.4, it should now be clear how the truth of a formula is determined given a model M and a state s. A pair (M, s) is called a *pointed model*; we sometimes drop the parentheses and write M, s.

Definition 1.5 (Truth in a Kripke Model)
Given a model $M = \langle S, R^{\mathsf{Ag}}, V^{\mathsf{At}}\rangle$, we define what it means for a formula φ to be true in (M, s), written $M, s \models \varphi$, inductively as follows:

$$\begin{array}{lll} M, s \models p & \text{iff} & V(s)(p) = \textit{true} \text{ for } p \in \mathsf{At} \\ M, s \models \varphi \wedge \psi & \text{iff} & M, s \models \varphi \text{ and } M, s \models \psi \\ M, s \models \neg\varphi & \text{iff} & \text{not } M, s \models \varphi \ \ (\text{often written } M, s \not\models \varphi) \\ M, s \models K_a\varphi & \text{iff} & M, t \models \varphi \text{ for all } t \text{ such that } R_a st. \end{array}$$

More generally, if $M = \langle S, R^{\mathsf{Op}}, V^{\mathsf{At}}\rangle$, then for all $\Box \in \mathsf{Op}$:

$$M, s \models \Box\varphi \text{ iff } (M, t) \models \varphi \text{ for all } t \text{ such that } R_\Box st.$$

Recall that M_a is the dual of K_a; it easily follows from the definitions that

$$M, s \models M_a\varphi \text{ iff there exists some } t \text{ such that } R_a st \text{ and } M, t \models \varphi.$$

We write $M \models \varphi$ if $M, s \models \varphi$ for all $s \in S$. ⊣

Example 1.2
Consider the model of Figure 1.2. Note that $K_a p \vee K_a\neg p$ represents the fact that agent a knows *whether* p is true. Likewise, $M_a p \wedge M_a \neg p$ is equivalent to $\neg K_a\neg p \wedge \neg K_a p$: agent a is ignorant about p. We have the following (in the final items we write E_{ab} instead of $E_{\{a,b\}}$):

1. $(M, s) \models t_b$: truth of a primitive proposition in s.

2. $M, s \models (\neg t_a \wedge K_a\neg t_a \wedge \neg K_b\neg t_a) \wedge (t_b \wedge \neg K_a t_b \wedge K_b t_b)$: at s, a knows that t_a is false, but b does not; similarly, b knows that t_b is true, but a does not.

3. $M \models K_a(K_b t_b \vee K_b\neg t_b) \wedge K_b(K_a t_a \vee K_a\neg t_a)$: in all states of M, agent a knows that b knows whether t_b is true, and b knows that a knows whether t_a is true.

4. $M \models K_a(M_b t_b \wedge M_b\neg t_b) \wedge K_b(M_a t_a \wedge M_a\neg t_a)$ in all states of M, agent a knows that b does not know whether t_a is true, and b knows that a does not know whether t_b is true.

5. $M \models E_{ab}((K_a t_a \vee K_a\neg t_a) \wedge (M_a t_b \wedge M_a\neg t_b))$: in all states, everyone knows that a knows whether t_a is true, but a does not know whether t_b is true.

6. $M \models E_{ab}E_{ab}((K_a t_a \vee K_a\neg t_a) \wedge (M_a t_b \wedge M_a\neg t_b))$: in all states, everyone knows what we stated in the previous item.

This shows that the model M of Figure 1.2 is not just a model for a situation where a knows t_a but not t_b and agent b knows t_b but not t_a; it represents much more information. ⊣

As the following example shows, in order to model certain situations, it may be necessary that some propositional valuations occur in more than one state in the model.

Example 1.3
Recall the scenario of the interview between Alice and Bob, as presented in Example 1.1. Suppose that we now add the information that in fact Alice will arrive on time, but Bob is not going to be on time. Although Bob does not know Alice, he knows that his friend Carol is an old friend of Alice. Bob calls Carol, leaving a message on her machine to ask her to inform Alice about Bob's late arrival as soon as she is able to do so. Unfortunately for Bob, Carol does not get his message on time. This situation can be represented in state M, v of the model of Figure 1.3.

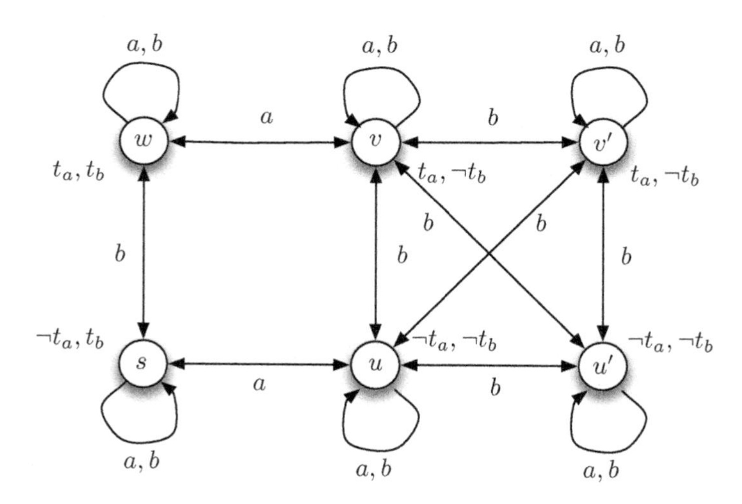

Figure 1.3: The Kripke model for Example 1.3.

Note that in (M, v), we have $\neg K_a \neg t_b$ (Alice does not know that Bob is late), but also $M_b(K_a \neg t_b)$ (Bob considers it possible that Alice knows that Bob is late). So, although the propositional valuations in v and v' are the same, those two states represent different situations: in v agent a is uncertain whether $\neg t_b$ holds, while in v' she knows $\neg t_b$. Also, in M, v, Bob considers it possible that both of them will be late, and that Alice knows this: this is because $R_b v u'$ holds in the model, and $M, u' \models K_a(\neg t_a \wedge \neg t_b)$.⊣

We often impose restrictions on the accessibility relation. For example, we may want to require that if, in world v, agent a considers world w possible, then in w, agent a should consider v possible. This requirement would make R_a symmetric. Similarly, we might require that, in each world w, a considers w itself possible. This would make R_a reflexive. More generally, we are interested in certain subclasses of models (typically characterized by properties of the accessibility relations).

Definition 1.6 (Classes of models, validity, satisfiability)
Let $\mathcal{X}$ be a class of models, that is, $\mathcal{X} \subseteq \mathcal{K}$. If $M \models \varphi$ for all models M in $\mathcal{X}$, we say that φ *is valid in* $\mathcal{X}$, and write $\mathcal{X} \models \varphi$. For example, for validity in the class of all Kripke models $\mathcal{K}$, we write $\mathcal{K} \models \varphi$. We write $\mathcal{X} \not\models \varphi$ when it is not the case that $\mathcal{X} \models \varphi$. So $\mathcal{X} \not\models \varphi$ holds if, for some model $M \in \mathcal{X}$ and some $s \in \mathcal{D}(M)$, we have $M, s \models \neg\varphi$. If there exists a model $M \in \mathcal{X}$ and a state $s \in \mathcal{D}(M)$ such that $M, s \models \varphi$, we say that φ is *satisfiable in* $\mathcal{X}$. ⊣

We now define a number of classes of models in terms of properties of the relations R_a in those models. Since they depend only on the accessibility relation, we could have defined them for the underlying frames; indeed, the properties are sometimes called *frame properties.*

Definition 1.7 (Frame properties)
Let R be an accessibility relation on a domain of states S.

1. R is *serial* if for all s there is a t such that Rst. The class of serial Kripke models, that is, $\{M = \langle S, R, V \rangle \mid \text{every } R_a \text{ is serial}\}$ is denoted $\mathcal{KD}$.

2. R is *reflexive* if for all s, Rss. The class of reflexive Kripke models is denoted $\mathcal{KT}$.

3. R is *transitive* if for all s, t, u, if Rst and Rtu then Rsu. The class of transitive Kripke models is denoted $\mathcal{K}4$.

4. R is *Euclidean* if for all s, t, and u, if Rst and Rsu then Rtu. The class of Euclidean Kripke models is denoted $\mathcal{K}5$

5. R is *symmetric* if for all s, t, if Rst then Rts. The class of symmetric Kripke models is denoted $\mathcal{KB}$

6. We can combine properties of relations:

   (a) The class of reflexive transitive models is denoted $\mathcal{S}4$.

   (b) The class of transitive Euclidean models is denoted $\mathcal{K}45$.

   (c) The class of serial transitive Euclidean models is denoted $\mathcal{KD}45$.

(d) R is an *equivalence relation* if R is reflexive, symmetric, and transitive. It not hard to show that R is an equivalence relation if R is reflexive and Euclidean. The class of models where the relations are equivalence relations is denoted $\mathcal{S}5$.

As we did for $\mathcal{K}_m$, we sometimes use the subscript m to denote the number of agents, so $\mathcal{S}5_m$, for instance, is the class of Kripke models with $|\,\mathsf{Ag}\,| = m$.⊣

Of special interest in this book is the class $\mathcal{S}5$. In this case, the accessibility relations are equivalence classes. This makes sense if we think of $R_a st$ holding if s and t are indistinguishable by agent a based on the information that a has received. $\mathcal{S}5$ has typically been used to model knowledge. In an $\mathcal{S}5$ model, write $s \sim_a t$ rather than $R_a st$, to emphasize the fact that R_a is an equivalence relation. When it is clear that $M \in \mathcal{S}5$, when drawing the model, we omit reflexive arrows, and since the relations are symmetric, we connect states by a line, rather than using two-way arrows. Finally, we leave out lines that can be deduced to exist using transitivity. We call this the *S5 representation* of a Kripke model. Figure 1.4 shows the S5 representation of the Kripke model of Figure 1.3.

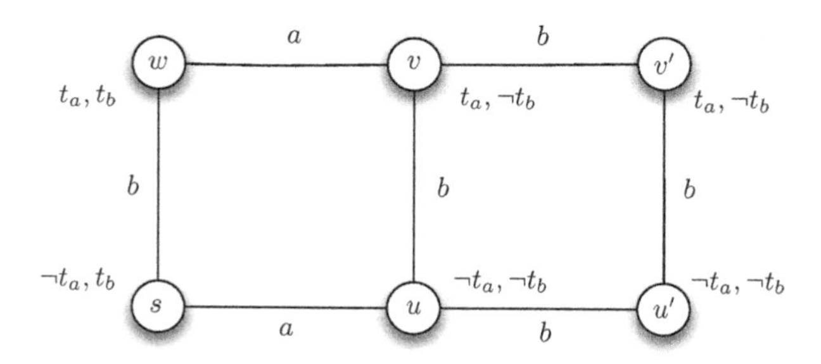

Figure 1.4: The S5 representation of the Kripke model in Figure 1.3.

When we restrict the classes of models considered, we get some interesting additional valid formulas.

Theorem 1.1 (Valid Formulas)
Parts (c)–(i) below are valid formulas, where α is a substitution instance of a propositional tautology (see below), φ and ψ are arbitrary formulas, and $\mathcal{X}$ is one of the classes of models defined in Definition 1.7; parts (a), (b), and (j) show that we can infer some valid formulas from others.

(a) If $\mathcal{X} \models \varphi \rightarrow \psi$ and $\mathcal{X} \models \varphi$, then $\mathcal{X} \models \psi$.

(b) If $\mathcal{X} \models \varphi$ then $\mathcal{X} \models K\varphi$.

(c) $\mathcal{X} \models \alpha$.

(d) $\mathcal{X} \models K(\varphi \to \psi) \to (K\varphi \to \psi)$.

(e) $\mathcal{KD} \models K\varphi \to M\varphi$.

(f) $\mathcal{T} \models K\varphi \to \varphi$.

(g) $\mathcal{K}4 \models K\varphi \to KK\varphi$.

(h) $\mathcal{K}5 \models \neg K\varphi \to K\neg K\varphi$.

(i) $\mathcal{KB} \models \varphi \to KM\varphi$.

(j) If $\mathcal{X} \subseteq \mathcal{Y}$ then $\mathcal{Y} \models \varphi$ implies that $\mathcal{X} \models \varphi$. $\dashv$

Since $\mathcal{S}5$ is the smallest of the classes of models considered in Definition 1.7, it easily follows that all the formulas and inference rules above are valid in $\mathcal{S}5$. To the extent that we view $\mathcal{S}5$ as the class of models appropriate for reasoning about knowledge, Theorem 1.1 can be viewed as describing properties of knowledge. As we shall see, many of these properties apply to the standard interpretation of belief as well.

Parts (a) and (c) emphasise that we represent knowledge in a logical framework: modus ponens is valid as a reasoning rule, and we take all propositional tautologies for granted. In part (c), α is a *substitution instance* of a propositional tautology. For example, since $p \vee \neg p$ and $p \to (q \to p)$ are propositional tautologies, α could be $Kp \vee \neg Kp$ or $K(p \vee q) \to (Kr \to K(p \vee q))$. That is, we can substitute an arbitrary formula (uniformly) for a primitive proposition in a propositional tautology. Part (b) says that agents know all valid formulas, and part (d) says that an agent is able to apply modus ponens to his own knowledge. Part (e) is equivalent to $K\varphi \to \neg K\neg\varphi$; an agent cannot at the same time know a proposition and its negation. Part (f) is even stronger: it says that what an agent knows must be true. Parts (g) and (h) represent what has been called *positive* and *negative introspection*, respectively: an agent knows what he knows and what he does not know. Part (i) can be shown to follow from the other valid formulas; it says that if something is true, the agent knows that he considers it possible.

Notions of Group Knowledge

So far, all properties that we have encountered are properties of an individual agent's knowledge. such as E_A, defined above. In this section we introduce two other notions of group knowledge, *common knowledge* C_A and *distributed knowledge* D_A, and investigate their properties.

Example 1.4 (Everyone knows and distributed knowledge)
Alice and Betty each has a daughter; their children can each either be at the playground (denoted p_a and p_b, respectively) or at the library ($\neg p_a$, and $\neg p_b$, respectively). Each child has been carefully instructed that, if she ends up being on the playground without the other child, she should call her mother to inform her. Consider the situation described by the model M in Figure 1.5.

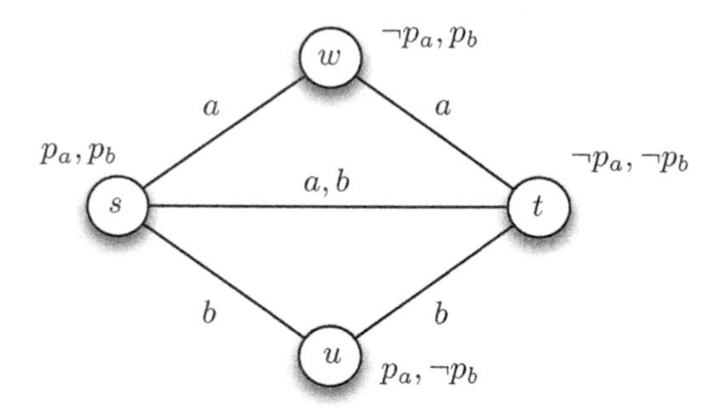

Figure 1.5: The (S5 representation of the) model for Example 1.4.

We have

$$M \models ((\neg p_a \wedge p_b) \leftrightarrow K_a(\neg p_a \wedge p_b)) \wedge ((p_a \wedge \neg p_b) \leftrightarrow K_b(p_a \wedge \neg p_b)).$$

This models the agreement each mother made with her daughter. Now consider the situation at state s. We have $M, s \models K_a \neg(p_a \wedge \neg p_b)$, that is, Alice knows that it is not the case that her daughter is alone at the playground (otherwise her daughter would have informed her). What does each agent know at s? If we consider only propositional facts, it is easy to see that Alice knows $p_a \to p_b$ and Betty knows $p_b \to p_a$. What does everyone know at s? The following sequence of equivalences is immediate from the definitions:

$$\begin{array}{ll} & M, s \models E_{\{a,b\}}\varphi \\ \text{iff} & M, s \models K_a\varphi \wedge K_b\varphi \\ \text{iff} & \forall x(R_a sx \Rightarrow M, x \models \varphi) \text{ and } \forall y(R_b sy \Rightarrow M, y \models \varphi) \\ \text{iff} & \forall x \in \{s, w, t\}\,(M, x \models \varphi) \text{ and } \forall y \in \{s, u, t\}\,(M, y \models \varphi) \\ \text{iff} & M \models \varphi. \end{array}$$

Thus, in this model, what is known by everyone are just the formulas valid in the model. Of course, this is not true in general.

Now suppose that Alice and Betty an opportunity to talk to each other. Would they gain any new knowledge? They would indeed. Since $M, s \models$

$K_a(p_a \rightarrow p_b) \wedge K_b(p_b \rightarrow p_a)$, they would come to know that $p_a \leftrightarrow p_b$ holds; that is, they would learn that their children are at least together, which is certainly not valid in the model. The knowledge that would emerge if the agents in a group A were allowed to communicate is called *distributed knowledge in* A, and denoted by the operator D_A. In our example, we have $M, s \models D_{\{a,b\}}(p_a \leftrightarrow p_b)$, although $M, s \models \neg K_a(p_a \leftrightarrow p_b) \wedge \neg K_b(p_a \leftrightarrow p_b)$. In other words, distributed knowledge is generally *stronger* than any individual's knowledge, and we therefore cannot define $D_A\varphi$ as $\bigvee_{i \in A} K_i\varphi$, the dual of general knowledge that we may have expected; that would be weaker than any individual agent's knowledge. In terms of the model, what would happen if Alice and Betty could communicate is that Alice could tell Betty that he should not consider state u possible, while Betty could tell Alice that she should not consider state w possible. So, after communication, the only states considered possible by both agents at state s are s and t. This argument suggests that we should interpret D_A as the necessity operator ($\Box$-type modal operator) of the relation $\bigcap_{a \in A} R_a$. By way of contrast, it follows easily from the definitions that E_A can be interpreted as the necessity operator of the relation $\bigcup_{a \in A} R_a$. ⊣

The following example illustrates common knowledge.

Example 1.5 (Common knowledge)
This time we have two agents: a sender (s) and a receiver (r). If a message is sent, it is delivered either immediately or with a one-second delay. The sender sends a message at time t_0. The receiver does not know that the sender was planning to send the message. What is each agent's state of knowledge regarding the message?

To reason about this, let s_z (for $z \in \mathbb{Z}$) denote that the message was sent at time $t_0 + z$, and, likewise, let d_z denote that the message was delivered at time $t = z$. Note that we allow z to be negative. To see why, consider the world $w_{0,0}$ where the message arrives immediately (at time t_0). (In general, in the subscript (i, j) of a world $w_{i,j}$, i denotes the time that the message was sent, and j denotes the time it was received.) In world $w_{0,0}$, the receiver considers it possible that the message was sent at time $t_0 - 1$. That is, the receiver considers possible the world $w_{-1,0}$ where the message was sent at $t_0 - 1$ and took one second to arrive. In world $w_{-1,0}$, the sender considers possible the world $w_{-1,-1}$ where the message was sent at time $t_0 - 1$ and arrived immediately. And in world $w_{-1,-1}$, the receiver considers possible a world $w_{-2,-1}$ where the message as sent at time $t_0 - 2$. (In general, in world $w_{n,m}$, the message is sent at time $t_0 + n$ and received at time $t_0 + m$.) In addition, in world $w_{0,0}$, the sender considers possible world $w_{0,1}$, where the message is received at time $t_0 + 1$. The situation is described in the following model M.

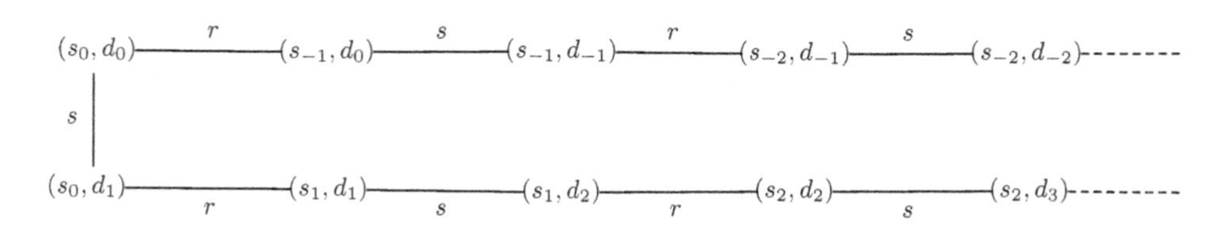

Figure 1.6: The (S5 representation of the) model for Example 1.5.

Writing E for 'the sender and receiver both know', it easily follows that

$$M, w_{0,0} \models s_0 \wedge d_0 \wedge \neg E \neg s_{-1} \wedge \neg E \neg d_1 \wedge \neg E^3 \neg s_{-2}.$$

The notion of φ being *common knowledge* among group A, denoted $C_A\varphi$, is meant to capture the idea that, for all n, $E^n\varphi$ is true. Thus, φ is *not* common among A if someone in A considers it possible that someone in A considers it possible that ... someone in A considers it possible that φ is false. This is formalised below, but the reader should already be convinced that in our scenario, even if it is common knowledge among the agents that messages will have either no delay or a one-second delay, it is *not* common knowledge that the message was sent at or after time $t_0 - m$ for any value of m! ⊣

Definition 1.8 (Semantics of three notions of group knowledge) Let $A \subseteq \mathsf{Ag}$ be a group of agents. Let $R_{E_A} = \cup_{a \in A} R_a$. As we observed above,

$$(M, s) \models E_A\varphi \text{ iff for all } t \text{ such that } R_{E_A}st, \text{ we have } (M, t) \models \varphi.$$

Similarly, taking $R_{D_A} = \cap_{a \in A} R_a$, we have

$$(M, s) \models D_A\varphi \text{ iff for all } t \text{ such that } R_{D_A}st, \text{ we have } (M, t) \models \varphi.$$

Finally, recall that the *transitive closure* of a relation R is the smallest relation R^+ such that $R \subseteq R^+$, and such that, for all x, y, and z, if R^+xy and R^+yz then R^+xz. We define R_{C_A} as $R^+_{E_A} = (\bigcup_{a \in A} R_a)^+$. Note that, in Figure 1.6, *every* pair of states is in the relation $R^+_{C_{\{r,s\}}}$. In general, we have $R_{C_A}st$ iff there is some path $s = s_0, s_1, \ldots, s_n = t$ from s to t such that $n \geq 1$ and, for all $i < n$, there is some agent $a \in A$ for which $R_a s_i s_{i+1}$. Define

$$(M, s) \models C_A\varphi \text{ iff for all } t \text{ such that } R_{C_A}st, (M, t) \models \varphi.$$

It is almost immediate from the definitions that, for $a \in A$, we have

$$\mathcal{K} \models (C_A\varphi \to E_A\varphi) \wedge (E_A\varphi \to K_a\varphi) \wedge (K_a\varphi \to D_A\varphi). \tag{1.1}$$

Moreover, for $\mathcal{T}$ (and hence also for $\mathcal{S}4$ and $\mathcal{S}5$), we have

$$\mathcal{T} \models D_a\varphi \to \varphi.$$

The relative strengths shown in (1.1) are strict in the sense that none of the converse implications are valid (assuming that $A \neq \{a\}$).

We conclude this section by defining some languages that are used later in this chapter. Fixing At and Ag, we write L_X for the language $\mathsf{L}(\mathsf{At}, \mathsf{Op}, \mathsf{Ag})$, where

$$\begin{array}{ll} X = K & \text{if } \mathsf{Op} = \{K_a \mid a \in \mathsf{Ag}\} \\ X = CK & \text{if } \mathsf{Op} = \{K_a, C_A \mid a \in \mathsf{Ag}, A \subseteq \mathsf{Ag}\} \\ X = DK & \text{if } \mathsf{Op} = \{K_a, D_A \mid a \in \mathsf{Ag}, A \subseteq \mathsf{Ag}\} \\ X = CDK & \text{if } \mathsf{Op} = \{K_a, C_A, D_A \mid a \in \mathsf{Ag}, A \subseteq \mathsf{Ag}\} \\ X = EK & \text{if } \mathsf{Op} = \{K_a, E_A \mid a \in \mathsf{Ag}, A \subseteq \mathsf{Ag}\}. \end{array}$$

Bisimulation

It may well be that two models (M, s) and (M', s') 'appear different', but still satisfy the same formulas. For example, consider the models (M, s), (M', s'), and (N, s_1) in Figure 1.7. As we now show, they satisfy the same formulas. We actually prove something even stronger. We show that all of (M, s), (M, t), (M', s'), (N, s_1), (M, s_2), and (N, s_3) satisfy the same formulas, as do all of (M, u), (M, w), (M', w'), (N, w_1), and (N, w_2). For the purposes of the proof, call the models in the first group *green*, and the models in the second group *red*. We now show, by induction on the structure of formulas, that all green models satisfy the same formulas, as do all red models. For primitive propositions, this is immediate. And if two models of the same colour agree on two formulas, they also agree on their negations and their conjunctions. The other formulas we need to consider are knowledge formulas. Informally, the argument is this. Every agent considers, in every pointed model, both green and red models possible. So his knowledge in each pointed model is the same. We now formalise this reasoning.

Definition 1.9 (Bisimulation)
Given models $M = (S, R, V)$ and $M' = (S', R', V')$, a non-empty relation $\mathfrak{R} \subseteq S \times S'$ is a *bisimulation between* M *and* M' iff for all $s \in S$ and $s' \in S'$ with $(s, s') \in \mathfrak{R}$:

- $V(s)(p) = V'(s')(p)$ for all $p \in \mathsf{At}$;
- for all $a \in \mathsf{Ag}$ and all $t \in S$, if $R_a st$, then there is a $t' \in S'$ such that $R'_a s't'$ and $(t, t') \in \mathfrak{R}$;

- for all $a \in \mathsf{Ag}$ and all $t' \in S'$, if $R'_a s' t'$, then there is a $t \in S$ such that $R_a st$ and $(t, t') \in \mathfrak{R}$.

We write $(M, s) \leftrightarroweq (M', s')$ iff there is a bisimulation between M and M' linking s and s'. If so, we call (M, s) and (M', s') *bisimilar.* ⊣

Figure 1.7 illustrates some bisimilar models. In terms of the models

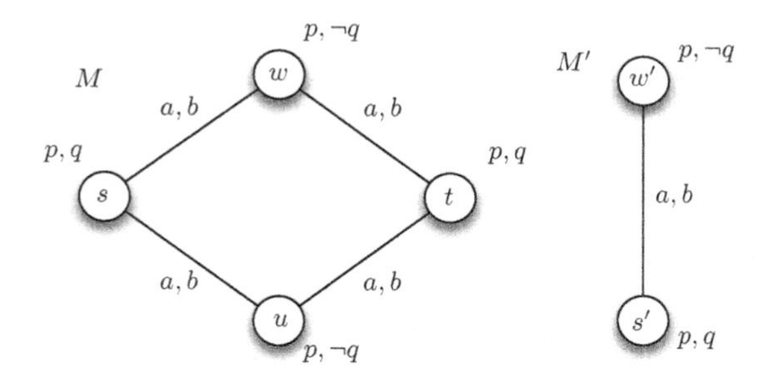

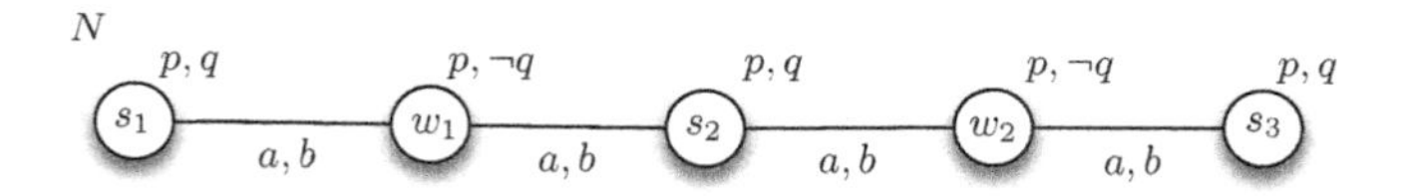

Figure 1.7: Bisimilar models.

of Figure 1.7, we have $M, s \leftrightarroweq M', s'$, $M, s \leftrightarroweq N, s_1$, etc. We are interested in bisimilarity because, as the following theorem shows, bisimilar models satisfy the same formulas involving the operators K_a and C_A.

Theorem 1.2 (Preservation under bisimulation)
Suppose that $(M, s) \leftrightarroweq (M', s')$. Then, for all formulas $\varphi \in \mathsf{L}_{CK}$, we have

$$M, s \models \varphi \Leftrightarrow M', s' \models \varphi. \qquad \dashv$$

The proof of the theorem proceeds by induction on the structure of formulas, much as in our example. We leave the details to the reader.

Note that Theorem 1.2 does not claim that distributed knowledge is preserved under bisimulation, and indeed, it is not, i.e., Theorem 1.2 does not hold for a language with D_A as an operator. Figure 1.8 provides a witness for this. We leave it to the reader to check that although $(M, s) \leftrightarroweq (N, s_1)$ for the two pointed models of Figure 1.8, we nevertheless have $(M, s) \models \neg D_{\{a,b\}} p$ and $(N, s_1) \models D_{\{a,b\}} p$.

We can, however, generalise the notion of bisimulation to that of a *group bisimulation* and 'recover' the preservation theorem, as follows. If $A \subseteq \mathsf{Ag}$,

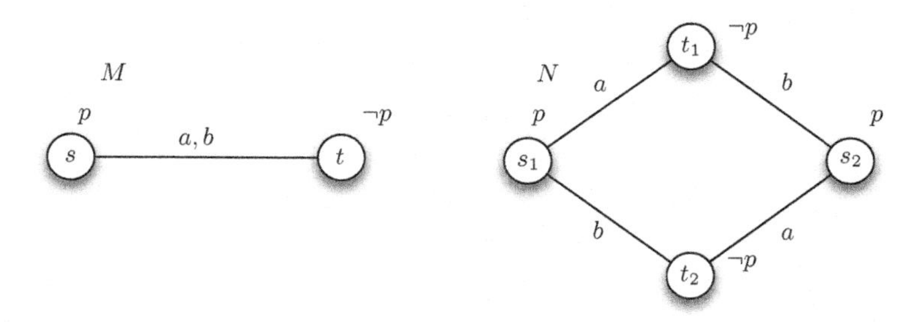

Figure 1.8: Two bisimilar models that do not preserve distributed knowledge.

s and t are states, then we write $R_A st$ if $A = \{a \mid R_a st\}$. That is, $R_A st$ holds if the set of agents a for which s and t are a-connected is exactly A. (M, s) and (M', s') are *group bisimilar*, written $(M, s) \leftrightarroweq_{group} (M', s')$, if the conditions of Definition 1.9 are met when every occurrence of an individual agent a is replaced by the group A. Obviously, being group bisimilar implies being bisimilar. Note that the models (M, s) and (N, s_1) of Figure 1.8 are bisimilar, but not group bisimilar. The proof of Theorem 1.3 is analogous to that of Theorem 1.2.

Theorem 1.3 (Preservation under bisimulation)
Suppose that $(M, s) \leftrightarroweq_{group} (M', s')$. Then, for all formulas $\varphi \in \mathsf{L}_{CDK}$, we have

$$M, s \models \varphi \Leftrightarrow M', s' \models \varphi. \qquad \dashv$$

1.2.3 Expressivity and Succinctness

If a number of formal languages can be used to model similar phenomena, a natural question to ask is which language is 'best'. Of course, the answer depends on how 'best' is measured. In the next section, we compare various languages in terms of the computational complexity of some reasoning problems. Here, we consider the notions of *expressivity* (what can be expressed in the language?) and *succinctness* (how economically can one say it?).

Expressivity

To give an example of expressivity and the tools that are used to study it, we start by showing that finiteness of models cannot be expressed in epistemic logic, even if the language includes operators for common knowledge and distributed knowledge.

Theorem 1.4
There is no formula $\varphi \in \mathsf{L}_{CDK}$ such that, for all $\mathcal{S}5$-models $M = \langle S, R, V \rangle$,

$$M \models \varphi \text{ iff } S \text{ is finite} \qquad \dashv$$

Proof Consider the two models M and M' of Figure 1.9. Obviously,

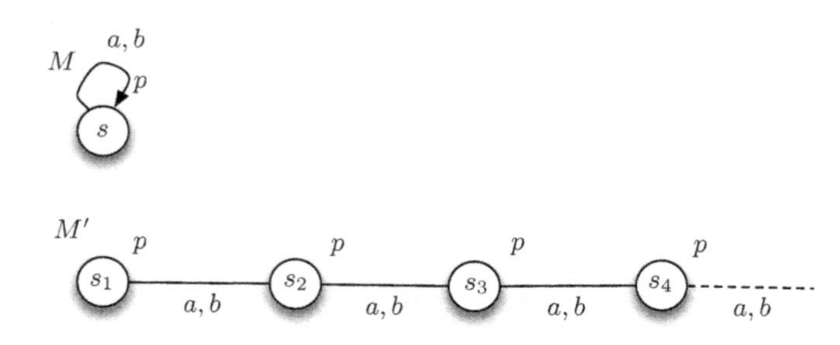

Figure 1.9: A finite and an infinite model where the same formulas are valid.

M is finite and M' is not. Nevertheless, the two models are easily seen to be group bisimilar, so they cannot be distinguished by epistemic formulas. More precisely, for all formulas $\varphi \in \mathsf{L}_{CDK}$, we have $M, s \models \varphi$ iff $M', s_1 \models \varphi$ iff $M', s_2 \models \varphi$ iff $M', s_n \models \varphi$ for some $n \in \mathbb{N}$, and hence $M \models \varphi$ iff $M' \models \varphi$. $\dashv$

It follows immediately from Theorem 1.4 that finiteness cannot be expressed in the language L_{CDK} in a class $\mathcal{X}$ of models containing $\mathcal{S}5$.

We next prove some results that let us compare the expressivity of two different languages. We first need some definitions.

Definition 1.10
Given a class $\mathcal{X}$ of models, formulas φ_1 and φ_2 are *equivalent on* $\mathcal{X}$, written $\varphi_1 \equiv_{\mathcal{X}} \varphi_2$, if, for all $(M, s) \in \mathcal{X}$, we have that $M, s \models \varphi_1$ iff $M, s \models \varphi_2$. Language L_2 is *at least as expressive as* L_1 on $\mathcal{X}$, written $\mathsf{L}_1 \sqsubseteq_{\mathcal{X}} \mathsf{L}_2$ if, for every formula $\varphi_1 \in \mathsf{L}_1$, there is a formula $\varphi_2 \in \mathsf{L}_2$ such that $\varphi_1 \equiv_{\mathcal{X}} \varphi_2$. L_1 and L_2 are *equally expressive* on $\mathcal{X}$ if $\mathsf{L}_1 \sqsubseteq_{\mathcal{X}} \mathsf{L}_2$ and $\mathsf{L}_2 \sqsubseteq_{\mathcal{X}} \mathsf{L}_1$. If $\mathsf{L}_1 \sqsubseteq_{\mathcal{X}} \mathsf{L}_2$ but $\mathsf{L}_2 \not\sqsubseteq_{\mathcal{X}} \mathsf{L}_1$, then L_2 is *more expressive than* L_1 on $\mathcal{X}$, written $\mathsf{L}_1 \sqsubset_{\mathcal{X}} \mathsf{L}_2$.$\dashv$

Note that if $\mathcal{Y} \subseteq \mathcal{X}$, then $\mathsf{L}_1 \sqsubseteq_{\mathcal{X}} \mathsf{L}_2$ implies $\mathsf{L}_1 \sqsubseteq_{\mathcal{Y}} \mathsf{L}_2$, while $\mathsf{L}_1 \not\sqsubseteq_{\mathcal{Y}} \mathsf{L}_2$ implies $\mathsf{L}_1 \not\sqsubseteq_{\mathcal{X}} \mathsf{L}_2$. Thus, the strongest results that we can show for the classes of models of interest to us are $\mathsf{L}_1 \sqsubseteq_{\mathcal{K}} \mathsf{L}_2$ and $\mathsf{L}_1 \not\sqsubseteq_{\mathcal{S}5} \mathsf{L}_2$

With these definitions in hand, we can now make precise that common knowledge 'really adds' something to epistemic logic.

Theorem 1.5
$\mathsf{L}_K \sqsubseteq_{\mathcal{K}} \mathsf{L}_{CK}$ and $\mathsf{L}_K \not\sqsubseteq_{\mathcal{S}5} \mathsf{L}_{CK}$. $\dashv$

Proof Since $\mathsf{L}_K \subseteq \mathsf{L}_{CK}$, it is obvious that $\mathsf{L}_K \sqsubseteq_{\mathcal{K}} \mathsf{L}_{CK}$. To show that $\mathsf{L}_{CK} \not\sqsubseteq_{S5} \mathsf{L}_K$, consider the sets of pointed models $\mathcal{M} = \{(M_n, s_1) \mid n \in \mathbb{N}\}$ and $\mathcal{N} = \{(N_n, t_1) \mid n \in \mathbb{N}\}$ shown in Figure 1.10. The two models M_n and N_n differ only in (M_n, s_{n+1}) (where p is false) and (N_n, t_{n+1}) (where p is true). In particular, the first $n-1$ states of (M_n, s_1) and (N_n, t_1) are the same. As a consequence, it is easy to show that,

$$\text{for all } n \in \mathbb{N} \text{ and } \varphi \in \mathsf{L}_K \text{ with } d(\varphi) < n, \ (M_n, s_1) \models \varphi \text{ iff } (N_n, t_1) \models \varphi. \tag{1.2}$$

Clearly $\mathcal{M} \models C_{\{a,b\}}\neg p$ while $\mathcal{N} \models \neg C_{\{a,b\}}\neg p$. If there were a formula $\varphi \in \mathsf{L}_K$ equivalent to $C_{\{a,b\}}\neg p$, then we would have $\mathcal{M} \models \varphi$ while $\mathcal{N} \models \neg\varphi$. Let $d = d(\varphi)$, and consider the pointed models (M_{d+1}, s_1) and (N_{d+1}, t_1). Since the first is a member of $\mathcal{M}$ and the second of $\mathcal{N}$, the pointed models disagree on $C_{\{a,b\}}\neg p$; however, by (1.2), they agree on φ. This is obviously a contradiction, therefore a formula $\varphi \in \mathsf{L}$ that is equivalent to $C_{\{a,b\}}\neg p$ does not exist.

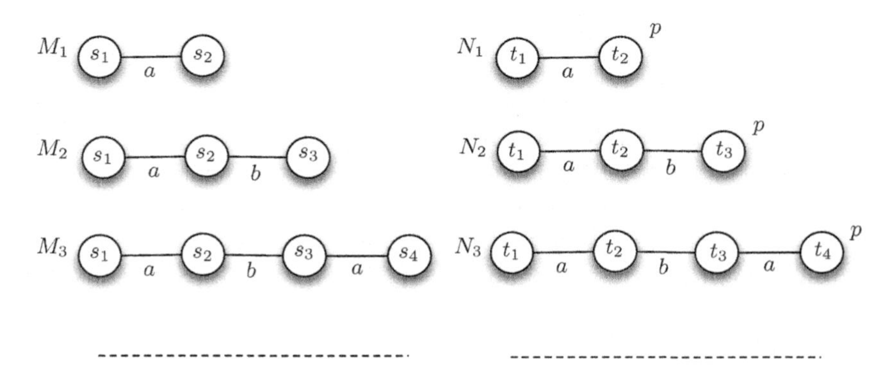

Figure 1.10: Models M_n and N_n. The atom p is only true in the pointed models (N_n, s_{n+1}).

⊣

The next result shows, roughly speaking, that distributed knowledge is not expressible using knowledge and common knowledge, and that common knowledge is not expressible using knowledge and distributed knowledge.

Theorem 1.6

(a) $\mathsf{L}_K \sqsubseteq_{\mathcal{K}} \mathsf{L}_{DK}$ and $\mathsf{L}_K \not\sqsubseteq_{S5} \mathsf{L}_{DK}$;

(b) $\mathsf{L}_{CK} \not\sqsubseteq_{S5} \mathsf{L}_{DK}$;

(c) $\mathsf{L}_{DK} \not\sqsubseteq_{S5} \mathsf{L}_{CK}$;

(d) $\mathsf{L}_{CK} \sqsubseteq_{\mathcal{K}} \mathsf{L}_{CDK}$ and $\mathsf{L}_{CDK} \not\sqsubseteq_{\mathcal{S}5} \mathsf{L}_{CK}$;

(e) $\mathsf{L}_{DK} \sqsubseteq_{\mathcal{K}} \mathsf{L}_{CDK}$ and $\mathsf{L}_{CDK} \not\sqsubseteq_{\mathcal{S}5} \mathsf{L}_{DK}$. ⊣

Proof For part (a), $\sqsubseteq_{\mathcal{K}}$ holds trivially. We use the models in Figure 1.8 to show that $\mathsf{L}_{DK} \not\sqsubseteq_{\mathcal{S}5} \mathsf{L}_{K}$. Since $(M, s) \underline{\leftrightarrow} (N, s_1)$, the models verify the same L-formulas. However, L_{DK} discriminates them: we have $(M, s) \models \neg D_{\{a,b\}}p$, while $(N, s_1) \models D_{\{a,b\}}p$. Since (M, s) and (N, s_1) also verify the same L_{CK}-formulas, part (3) also follows.

For part (b), observe that (1.2) is also true for all formulas $\varphi \in \mathsf{L}_{DK}$, so the formula $C_{\{a,b\}}\neg p \in \mathsf{L}_{CK}$ is not equivalent to a formula in L_{DK}.

Part (c) is proved using exactly the same models and argument as part (a).

For part (d), $\sqsubseteq$ is obvious. To show that $\mathsf{L}_{CDK} \not\sqsubseteq_{\mathcal{S}5} \mathsf{L}_{DK}$, we can use the models and argument of part (b). Similarly, for part (e), $\sqsubseteq$ is obvious. To show that $\mathsf{L}_{CDK} \not\sqsubseteq_{\mathcal{S}5} \mathsf{L}_{DK}$, we can use the models and argument of part (a). ⊣

We conclude this discussion with a remark about distributed knowledge. We informally described distributed knowledge in a group as the knowledge that would obtain were the agents in that group able to communicate. However, Figure 1.8 shows that this intuition is not quite right. First, observe that both a and b know the same formulas in (M, s) and (N, s_1); they even know the same formulas in (M, s) and (N, s_1). That is, for all $\varphi \in \mathsf{L}_K$, we have

$$(M, s) \models K_a\varphi \text{ iff } (M, s) \models K_b\varphi \text{ iff } (N, s_1) \models K_a\varphi \text{ iff } (N, s_1) \models K_b\varphi$$

But if both agents possess the same knowledge in (N, s_1), how can communication help them in any way, that is, how can it be that there is distributed knowledge (of p) that no individual agent has? Similarly, if a has the same knowledge in (M, s) in (N, s_1), and so does b, why would communication in one model (N) lead them to know p, while in the other, it does not? Semantically, one could argue that in s_1 agent a could 'tell' agent b that t_2 'is not possible', and b could 'tell' a that t_1 'is not possible'. But how would verify the same formulas? This observation has led some researchers to require that distributed knowledge be interpreted in what are called *bisimulation contracted models* (see the notes at the end of the chapter for references). Roughly, a model is bisimulation contracted if it does not contain two points that are bisimilar. Model M of Figure 1.8 is bisimulation contracted, model N is not.

Succinctness

Now suppose that two languages L_1 and L_2 are equally expressive on $\mathcal{X}$, and also that their computational complexity of the reasoning problems for them is equally good, or equally bad. Could we still prefer one language over the other? *Representational succinctness* may provide an answer here: it may be the case that the description of some properties is much shorter in one language than in the other.

But what does 'much shorter' mean? The fact that there is a formula L_1 whose length is 100 characters less than the shortest equivalent formula in L_2 (with respect to some class $\mathcal{X}$ of models) does not by itself make L_1 much more succinct that L_2.

We want to capture the idea that L_1 is exponentially more succinct than L_2. We cannot do this by looking at just one formula. Rather, we need a sequence of formulas $\alpha_1, \alpha_2, \alpha_3, \ldots$ in L_1, where the gap in size between α_n and the shortest formula equivalent to α_n in L_2 grows exponentially in n. This is formalised in the next definition.

Definition 1.11 (Exponentially more succinct)
Given a class $\mathcal{X}$ of models, L_1 is *exponentially more succinct* than L_2 on $\mathcal{X}$ if the following conditions hold:

(a) for every formula $\beta \in \mathsf{L}_2$, there is a formula $\alpha \in \mathsf{L}_1$ such that $\alpha \equiv_{\mathcal{X}} \beta$ and $|\alpha| \leq |\beta|$.

(b) there exist $k_1, k_2 > 0$, a sequence $\alpha_1, \alpha_2, \ldots$ of formulas in L_1, and a sequence $\beta_1, \beta_2, \ldots$ of formulas in L_2 such that, for all n, we have:

 (i) $|\alpha_n| \leq k_1 n$;

 (ii) $|\beta_n| \geq 2^{k_2 n}$;

 (iii) β_n is the shortest formula in L_2 that is equivalent to α_n on $\mathcal{X}$. $\dashv$

In words, L_1 is exponentially more succinct than L_2 if, for every formula $\beta \in \mathsf{L}_2$, there is a formula in L_1 that is equivalent and no longer than β, but there is a sequence $\alpha_1, \alpha_2, \ldots$ of formulas in L_1 whose length increases at most linearly, but there is no sequence $\beta_1, \beta_2, \ldots$ of formulas in L_2 such that β_n is the equivalent to α_n and the length of the formulas in the latter sequence is increasing better than exponentially.

We give one example of succinctness results here. Consider the language L_{EK}. Of course, E_A can be defined using the modal operators K_i for $i \in A$. But, as we now show, having the modal operators E_A in the language makes the language exponentially more succinct.

Theorem 1.7
The language L_{EK} is exponentially more succinct than L_K on $\mathcal{X}$, for all X between $\mathcal{K}$ and $\mathcal{S}5$. ⊣

Proof Clearly, for every formula α in $(L)_K$, there is an equivalent formula in L_{EK} that is no longer than α, namely, α itself. Now consider the following two sequences of formulas:

$$\alpha_n = \neg E^n_{\{a,b\}} \neg p$$

and

$$\beta_1 = \neg(K_a \neg p \wedge K_b \neg p), \text{ and } \beta_n = \neg(K_a \neg \beta_{n-1} \wedge K_b \neg \beta_{n-1}).$$

If we take $| E_A \varphi | = | A | + | \varphi |$, then it is easy to see that $| \alpha_n | = 2n + 3$, so $| \alpha_n |$ is increasing linearly in n. On the other hand, since $| \beta_n | > 2 \, | \beta_{n-1} |$, we have $| \beta | \geq 2^n$. It is also immediate from the definition of $E_{\{a,b\}}$ that β_n is equivalent to α_n for all classes $\mathcal{X}$ between $\mathcal{K}$ and $\mathcal{S}5$. To complete the proof, we must show that there is no formula shorter than β_n in $\mathcal{L}_\mathcal{K}$ that is equivalent to α_n. This argument is beyond the scope of this book; see the notes for references. ⊣

1.2.4 Reasoning problems

Given the machinery developed so far, we can state some basic reasoning problems in semantic terms. They concern *satisfiability* and *model checking.* Most of those problems are typically considered with a specific class of models and a specific language in mind. So let $\mathcal{X}$ be some class of models, and let L be a language.

Decidability Problems

A decidability problem checks some input for some property, and returns 'yes' or 'no'.

Definition 1.12 (Satisfiability)
The satisfiability problem for $\mathcal{X}$ is the following reasoning problem.

PROBLEM:	satisfiability in $\mathcal{X}$, denoted $\text{SAT}_\mathcal{X}$.
INPUT:	a formula $\varphi \in \mathsf{L}$.
QUESTION:	does there exist a model $M \in \mathcal{X}$ and a state $s \in \mathcal{D}(M)$ such that $M, s \models \varphi$?
OUTPUT:	'yes' or 'no'.

Obviously, there may well be formulas that are satisfiable in some Kripke model (or generally, in a class $\mathcal{Y}$), but not in $\mathcal{S}5$ models. Satisfiability in $\mathcal{X}$ is closely related to the problem of *validity* in $\mathcal{X}$, due to the following equivalence: φ is valid in $\mathcal{X}$ iff $\neg\varphi$ is not satisfiable in $\mathcal{X}$.

PROBLEM:	validity in $\mathcal{X}$, denoted $\text{VAL}_{\mathcal{X}}$.
INPUT:	a formula $\varphi \in \mathsf{L}$.
QUESTION:	is it the case that $\mathcal{X} \models \varphi$?
OUTPUT:	'yes' or 'no'.

The next decision problem is computationally and conceptually simpler than the previous two, since rather than quantifying over a set of models, a specific model is given as input (together with a formula).

Definition 1.13 (Model checking)
The model checking problem for $\mathcal{X}$ is the following reasoning problem:

PROBLEM:	Model checking in $\mathcal{X}$, denoted $\text{MODCHECK}_{\mathcal{X}}$.
INPUT:	a formula $\varphi \in \mathsf{L}$ and a pointed model (M, s) with $M \in \mathcal{X}$ and $s \in \mathcal{D}(M)$.
QUESTION:	is it the case that $M, s \models \varphi$?
OUTPUT::	'yes' or 'no'.

The field of *computational complexity* is concerned with the question of how much of a resource is needed to solve a specific problem. The resources of most interest are *computation time* and *space*. Computational complexity then asks questions of the following form: if my input were to increase in size, how much more space and/or time would be needed to compute the answer? Phrasing the question this way already assumes that the problem at hand can be solved in finite time using an algorithm, that is, that the problem is *decidable*. Fortunately, this is the case for the problems of interest to us.

Proposition 1.1 (Decidability of SAT and MODCHECK)
If $\mathcal{X}$ is one of the model classes defined in Definition 1.7, $(M, s) \in \mathcal{X}$, and φ is a formula in one of the languages defined in Definition 1.1, then both $\text{SAT}_{\mathcal{X}}(\varphi)$ and $\text{MODCHECK}_{\mathcal{X}}((M, s), \varphi)$ are decidable. ⊣

In order to say anything sensible about the additional resources that an algorithm needs to compute the answer when the input increases in size, we need to define a notion of size for inputs, which in our case are formulas and models. Formulas are by definition finite objects, but models can in principle be infinite (see, for instance, Figure 1.6). The following fact is the key to proving Fact 1.1. For a class of models $\mathcal{X}$, let $\mathcal{F}in(\mathcal{X}) \subseteq \mathcal{X}$ be the set of models in $\mathcal{X}$ that are finite.

Proposition 1.2 (Finite model property)
For all classes of models in Definition 1.7 and languages L in Definition 1.1, we have, for all $\varphi \in \mathsf{L}$,

$$\mathcal{X} \models \varphi \text{ iff } \mathcal{F}in(\mathcal{X}) \models \varphi. \qquad \dashv$$

Fact 1.2 does not say that the models in $\mathcal{X}$ and the finite models in $\mathcal{X}$ are the same in any meaningful sense; rather, it says that we do not gain valid formulas if we restrict ourselves to finite models. It implies that a formula is satisfiable in a model in $\mathcal{X}$ iff it is satisfiable in a finite model in $\mathcal{X}$. It follows that in the languages we have considered so far, 'having a finite domain' is not expressible (for if there were a formula φ that were true only of models with finite domains, then φ would be a counterexample to Fact 1.2).

Definition 1.14 (Size of Models)
For a finite model $M = \langle S, ^{\mathsf{Ag}}, V^{\mathsf{At}} \rangle$, the size of M, denoted $\|M\|$, is the sum of the number of states ($| S |$, for which we also write $| M |$) and the number of pairs in the accessibility relation ($| R_a |$) for each agent $a \in \mathsf{Ag}$.$\dashv$

We can now strengthen Fact 1.2 as follows.

Proposition 1.3
For all classes of models in Definition 1.7 and languages L in Definition 1.1, we have, for all $\varphi \in \mathsf{L}$, φ is satisfiable in $\mathcal{X}$ iff there is a model $M \in \mathcal{X}$ such that $| \mathcal{D}(M) | \leq 2^{|\varphi|}$ and φ is satisfiable in M. $\dashv$

The idea behind the proof of Proposition 1.3 is that states that 'agree' on all subformulas of φ can be 'identified'. Since there are only $| \varphi |$ subformulas of φ, and $2^{|\varphi|}$ truth assignments to these formulas, the result follows. Of course, work needs to done to verify this intuition, and to show that an appropriate model can be constructed in the right class $\mathcal{X}$.

To reason about the complexity of a computation performed by an algorithm, we distinguish various complexity classes. If a deterministic algorithm can solve a problem in time polynomial in the size of the input, the problem is said to be in P. An example of a problem in P is to decide, given two finite Kripke models M_1 and M_2, whether there exists a bisimulation between them. Model checking for the basic multi-modal language is also in P; see Proposition 1.4.

In a *nondeterministic* computation, an algorithm is allowed to 'guess' which of a finite number of steps to take next. A nondeterministic algorithm for a decision problem says 'yes' or *accepts the input* if the algorithm says 'yes' to an appropriate sequence of guesses. So a nondeterministic algorithm can be seen as generating different branches at each computation step, and the answer of the nondeterministic algorithm is 'yes' iff one of the branches results in a 'yes' answer.

The class NP is the class of problems that are solvable by a nondeterministic algorithm in polynomial time. Satisfiability of propositional logic is an example of a problem in NP: an algorithm for satisfiability first guesses an appropriate truth assignment to the primitive propositions, and then verifies that the formula is in fact true under this truth assignment.

A problem that is at least as hard as any problem in NP is called NP-*hard.* An NP-hard problem has the property that any problem in NP can be reduced to it using a polynomial-time reduction. A problem is NP-*complete* if it is both in NP and NP-hard; satisfiability for propositional logic is well known to be NP-complete. For an arbitrary complexity class C, notions of C-hardness and C-completeness can be similarly defined.

Many other complexity classes have been defined. We mention a few of them here. An algorithm that runs in space polynomial in the size of the input it is in PSPACE. Clearly if an algorithm needs only polynomial time then it is in polynomial space; that is P $\subseteq$ PSPACE. In fact, we also have NP $\subseteq$ PSPACE. If an algorithm is in NP, we can run it in polynomial space by systematically trying all the possible guesses, erasing the space used after each guess, until we eventually find one that is the 'right' guess. EXPTIME consists of all algorithms that run in time exponential in the size of the input; NEXPTIME is its nondeterministic analogue. We have P $\subseteq$ NP $\subseteq$ PSPACE $\subseteq$ EXPTIME $\subseteq$ NEXPTIME. One of the most important open problems in computer science is the question whether P $=$ NP. The conjecture is that the two classes are different, but this has not yet been proved; it is possible that a polynomial-time algorithm will be found for an NP-hard problem. What is known is that P $\neq$ EXPTIME and NP $\neq$ NEXPTIME.

The complement $\bar{P}$ of a problem P is the problem in which all the 'yes' and 'no' answers are reversed. Given a complexity class C, the class co-C is the set of problems for which the complement is in C. For every deterministic class C, we have co-C $=$ C. For nondeterministic classes, a class and its complement are, in general, believed to be incomparable. Consider, for example, the satisfiability problem for propositional logic, which, as we noted above, is NP-complete. Since a formula φ is valid if and only if $\neg\varphi$ is not satisfiable, it easily follows that the validity problem for propositional logic is co-NP-complete. The class of NP-complete and co-NP-complete problems are believed to be distinct.

We start our summary of complexity results for decision problems in modal logic with model checking.

Proposition 1.4
Model checking formulas in $\mathsf{L}(\mathsf{At}, \mathsf{Op}, \mathsf{Ag})$, with $\mathsf{Op} = \{K_a \mid a \in \mathsf{Ag}\}$, in finite models is in P. ⊣

Proof We now describe an algorithm that, given a model $M = \langle S, R^{\mathsf{Ag}}, V^{\mathsf{At}}\rangle$ and a formula $\varphi \in \mathsf{L}$, determines in time polynomial in $|\varphi|$ and $\|M\|$ whether $M, s \models \varphi$. Given φ, order the subformulas $\varphi_1, \ldots \varphi_m$ of φ in such a way that, if φ_i is a subformula of φ_j, then $i < j$. Note that $m \leq |\varphi|$. We claim that

> (*) for every $k \leq m$, we can label each state s in M with either φ_j (if φ_j if true at s) or $\neg\varphi_j$ (otherwise), for every $j \leq k$, in $k\|M\|$ steps.

We prove (*) by induction on m. If $k = 1$, φ_m must be a primitive proposition, and obviously we need only $|M| \leq \|M\|$ steps to label all states as required. Now suppose (*) holds for some $k < m$, and consider the case $k+1$. If φ_{k+1} is a primitive proposition, we reason as before. If φ_{k+1} is a negation, then it must be $\neg\varphi_j$ for some $j \leq k$. Using our assumption, we know that the collection of formulas $\varphi_1, \ldots, \varphi_k$ can be labeled in M in $k\|M\|$ steps. Obviously, if we include $\varphi_{k+1} = \neg\varphi_j$ in the collection of formulas, we can do the labelling in k more steps: just use the opposite label for φ_{k+1} as used for φ_i. So the collection $\varphi_1, \ldots, \varphi_{k+1}$ can be labelled in M in at $(k+1)\|M\|$ steps, are required. Similarly, if $\varphi_{k+1} = \varphi_i \wedge \varphi_j$, with $i, j \leq k$, a labelling for the collection $\varphi_1, \ldots, \varphi_{k+1}$ needs only $(k+1)\|M\|$ steps: for the last formula, in each state s of M, the labelling can be completed using the labellings for φ_i and φ_j. Finally, suppose φ_{k+1} is of the form $K_a\varphi_j$ with $j \leq k$. In this case, we label a state s with $K_a\varphi_j$ iff each state t with $R_a st$ is labelled φ_j. Assuming the labels φ_j and $\neg\varphi_j$ are already in place, this can be done in $|R_a(s)| \leq \|M\|$ steps. ⊣

Proposition 1.4 should be interpreted with care. While having a polynomial-time procedure seems attractive, we are talking about computation time polynomial *in the size of the input.* To model an interesting scenario or system often requires 'big models'. Even for one agent and n primitive propositions, a model might consist of 2^n states. Moreover, the procedure does not check properties of the model either, for instance whether it belongs to a given class $\mathcal{X}$.

We now formulate results for satisfiability checking. The results depend on two parameters: the class of models considered (we focus on

$\mathcal{K}, \mathcal{T}, \mathcal{S}4, \mathcal{KD}45$ and $\mathcal{S}5$) and the language. Let $\mathsf{Ag}_{=1}$ consist of only one agent, let $\mathsf{Ag}_{\geq 1} \neq \emptyset$ be an arbitrary set of agents, and let $\mathsf{Ag}_{\geq 2}$ be a set of at least two agents. Finally, let $\mathsf{Op} = \{K_a \mid a \in \mathsf{Ag}\}$.

Theorem 1.8 (Satisfiability)
The complexity of the satisfiability problem is

1. NP-complete if $\mathcal{X} \in \{\mathcal{KD}45, \mathcal{S}5\}$ and $\mathsf{L} = \mathsf{L}(\mathsf{At}, \mathsf{Op}, \mathsf{Ag}_{=1})$;
2. PSPACE-complete if
   (a) $\mathcal{X} \in \{\mathcal{K}, \mathcal{T}, \mathcal{S}4\}$ and $\mathsf{L} = \mathsf{L}(\mathsf{At}, \mathsf{Op}, \mathsf{Ag}_{\geq 1})$, or
   (b) $\mathcal{X} \in \{\mathcal{KD}45, \mathcal{S}5\}$ and $\mathsf{L} = \mathsf{L}(\mathsf{At}, \mathsf{Op}, \mathsf{Ag}_{\geq 2})$;
3. EXPTIME-complete if
   (a) $\mathcal{X} \in \{\mathcal{K}, \mathcal{T}$ and $\mathsf{L} = \mathsf{L}(\mathsf{At}, \mathsf{Op} \cup \{C\}, \mathsf{Ag}_{\geq 1})$, or
   (b) $\mathcal{X} \in \{\mathcal{S}4, \mathcal{KD}45, \mathcal{S}5\}$ and $\mathsf{L} = \mathsf{L}(\mathsf{At}, \mathsf{Op} \cup \{C\}, \mathsf{Ag}_{\geq 2})$. $\dashv$

From the results in Theorem 1.8, it follows that the satisfiability problem for logics of knowledge and belief for one agent, $\mathcal{S}5$ and $\mathcal{KD}45$, is exactly as hard as the satisfiability problem for propositional logic. If we do not allow for common knowledge, satisfiability for the general case is PSPACE-complete, and with common knowledge it is EXPTIME-complete. (Of course, common knowledge does not add anything for the case of one agent.)

For validity, the consequences of Theorem 1.8 are as follows. We remarked earlier that if satisfiability (in $\mathcal{X}$) is in some class C, then validity is in co-C. Hence, checking validity for the cases in item 1 is co-NP-complete. Since co-PSPACE = PSPACE, the validity problem for the cases in item 2 is PSPACE-complete, and, finally, since co-EXPTIME = EXPTIME, the validity problem for the cases in item 3 is EXPTIME-complete. What these results on satisfiability and validity mean in practice? Historically, problems that were not in P were viewed as too hard to deal with in practice. However, recently, major advances have been made in finding algorithms that deal well with many NP-complete problems, although no generic approaches have been found for dealing with problems that are co-NP-complete, to say nothing of problems that are PSPACE-complete and beyond. Nevertheless, even for problems in these complexity classes, algorithms with humans in the loop seem to provide useful insights. So, while these complexity results suggest that it is unlikely that we will be able to find tools that do automated satisfiability or validity checking and are guaranteed to always give correct results for the logics that we focus on in this book, this should not be taken to say that we cannot write algorithms for satisfiability, validity, or model checking that are useful for the problems of practical interest. Indeed, there is much work focused on just that.

1.2.5 Axiomatisation

In the previous section, the formalisation of reasoning was defined around the notion of *truth*: $\mathcal{X} \models \varphi$ meant that φ is true in all models in $\mathcal{X}$. In this section, we discuss a form of reasoning where a conclusion is inferred purely based on its syntactic form. Although there are several ways to do this, in epistemic logic, the most popular way to define deductive inference is by defining a *Hilbert-style axiom system*. Such systems provide a very simple notion of formal proofs. Some formulas are valid merely because they have a certain syntactic form. These are the axioms of the system. The rules of the system say that one can conclude that some formula is valid due to other formulas being valid. A formal proof or *derivation* is a list of formulas, where each formula is either an axiom of the system or can be obtained by applying an inference rule of the system to formulas that occur earlier in the list. A proof or derivation of φ is a derivation whose last formula is φ.

Basic system

Our first definition of such a system will make the notion more concrete. We give our definitions for a language where the modal operators are K_a for the agents in some set Ag, although many of the ideas generalise to a setting with arbitrary modal operators.

Definition 1.15 (System K)
Let $\mathsf{L} = \mathsf{L}(\mathsf{At}, \mathsf{Op}, \mathsf{Ag})$, with $\mathsf{Op} = \{K_a \mid a \in \mathsf{Ag}\}$. The axiom system **K** consists of the following axioms and rules of inference:

1	All substitution instances of propositional tautologies.
K	$K_a(\varphi \to \psi) \to (K_a\varphi \to K_a\psi)$ for all $a \in \mathsf{Ag}$.
MP	From φ and $\varphi \to \psi$ infer ψ.
Nec	From φ infer $K_a\varphi$.

$\dashv$

Here, formulas in the axioms **1** and **K** have to be interpreted as *axiom schemes*: axiom **K** for instance denotes all formulas $\{K_a(\varphi \to \psi) \to (K_a\varphi \to K_a\psi) \mid \varphi, \psi \in \mathsf{L}\}$. The rule **MP** is also called *modus ponens*; **Nec** is called *necessitation*. Note that the notation for axiom **K** and the axiom system **K** are the same: the context should make clear which is intended.

To see how an axiom system is actually used, we need to define the notion of *derivation*.

Definition 1.16 (Derivation)
Given a logical language L, let **X** be an axiom system with axioms $\mathbf{Ax_1}, \ldots$, $\mathbf{Ax_n}$ and rules $\mathbf{Ru_1}, \ldots \mathbf{Ru_k}$. A *derivation* of φ in **X** is a finite sequence $\varphi_1, \ldots, \varphi_m$ of formulas such that: (a) $\varphi_m = \varphi$, and (b) every φ_i in the sequence is either an instance of an axiom or else the result of applying a rule to formulas in the sequence prior to φ_i. For the rules **MP** and **Nec**, this means the following:

MP $\varphi_h = \varphi_j \to \varphi_i$, for some $h, j < i$.

That is, both φ_j and $\varphi_j \to \varphi_i$ occur in th sequence before φ_i.

Nec $\varphi_i = K_a\varphi_j$, for some $j < i$;

If there is a derivation for φ in **X** we write $\mathbf{X} \vdash \varphi$, or $\vdash_{\mathbf{X}} \varphi$, or, if the system **X** is clear from the context, we just write $\vdash \varphi$. We then also say that φ is a *theorem* of **X**, or that **X** *proves* φ. The sequence $\varphi_1, \ldots, \varphi_m$ is then also called a *proof of* φ *in* **X**. $\dashv$

Example 1.6 (Derivation in K)
We first show that

$$\mathbf{K} \vdash K_a(\varphi \wedge \psi) \to (K_a\varphi \wedge K_a\psi). \tag{1.3}$$

We present the proof as a sequence of numbered steps (so that the formula φ_i in the derivation is given number i). This allows us to justify each step in the proof by describing which axioms, rules of inference, and previous steps in the proof it follows from.

1. $(\varphi \wedge \psi) \to \varphi$ **1**
2. $K_a((\varphi \wedge \psi) \to \varphi)$ **Nec**, 1
3. $K_a((\varphi \wedge \psi) \to \varphi) \to (K_a(\varphi \wedge \psi) \to K_a\varphi)$ **K**
4. $K_a(\varphi \wedge \psi) \to K_a\varphi$ **MP**, 2, 3
5. $(\varphi \wedge \psi) \to \psi$ **1**
6. $K_a((\varphi \wedge \psi) \to \psi)$ **Nec**, 5
7. $K_a((\varphi \wedge \psi) \to \psi) \to (K_a(\varphi \wedge \psi) \to K_a\psi)$ **K**
8. $K_a(\varphi \wedge \psi) \to K_a\psi$ **MP**, 6, 7
9. $(K_a(\varphi \wedge \psi) \to K_a\varphi) \to$
   $((K_a(\varphi \wedge \psi) \to K_a\psi) \to (K_a(\varphi \wedge \psi) \to (K_a\varphi \wedge K_a\psi)))$ **1**
10. $(K_a(\varphi \wedge \psi) \to K_a\psi) \to (K_a(\varphi \wedge \psi) \to (K_a\varphi \wedge K_a\psi))$ **MP**, 4, 9
11. $K_a(\varphi \wedge \psi) \to (K_a\varphi \wedge K_a\psi)$ **MP**, 8, 10

Lines 1, 5, and 9 are instances of propositional tautologies (this can be checked using a truth table). Note that the tautology on line 9 is of the form $(\alpha \to \beta) \to ((\alpha \to \gamma) \to (\alpha \to (\beta \wedge \gamma)))$. A proof like that above may look cumbersome, but it does show what can be done using only the axioms and rules of **K**. It is convenient to give names to properties that are derived, and so build a library of theorems. We have, for instance that $\mathbf{K} \vdash \mathbf{KCD}$, where **KCD** ('$K$-over-conjunction-distribution') is

$$\mathbf{KCD} \quad K_a(\alpha \wedge \beta) \to K_a\alpha \text{ and } K_a(\alpha \wedge \beta) \to K_a\beta.$$

The proof of this follows steps 1 - 4 and steps 5 - 8, respectively, of the proof above. We can also derive new rules; for example, the following rule: **CC** ('combine conclusions') is derivable in **K**:

$$\mathbf{CC} \quad \text{from } \alpha \to \beta \text{ and } \alpha \to \gamma \text{ infer } \alpha \to (\beta \wedge \gamma).$$

The proof is immediate from the tautology on line 9 above, to which we can, given the assumptions, apply modus ponens twice. We can give a more compact proof of $K_a(\varphi \wedge \psi) \to (K_a\varphi \wedge K_a\psi)$ using this library:

1.	$K_a(\varphi \wedge \psi) \to K_a\varphi$	**KCD**
2.	$K_a(\varphi \wedge \psi) \to K_a\psi$	**KCD**
3.	$K_a(\varphi \wedge \psi) \to (K_a\varphi \wedge K_a\psi)$	**CC**, 1, 2

⊣

For every class $\mathcal{X}$ of models introduced in the previous section, we want to have an inference system **X** such that derivability in **X** and validity in $\mathcal{X}$ coincide:

Definition 1.17 (Soundness and Completeness)
Let L be a language, let $\mathcal{X}$ be a class of models, and let **X** be an axiom system. The axiom system is said to be

1. *sound* for $\mathcal{X}$ and the language L if, for all formulas $\varphi \in \mathsf{L}$, $\mathbf{X} \vdash \varphi$ implies $\mathcal{X} \models \varphi$; and

2. *complete* for $\mathcal{X}$ and the language L if, for all formulas $\varphi \in \mathsf{L}$, $\mathcal{X} \models \varphi$ implies $\mathbf{X} \vdash \varphi$.

We now provide axioms that characterize some of the subclasses of models that were introduced in Definition 1.7.

Definition 1.18 (More axiom systems)
Consider the following axioms, which apply for all agents $a \in \mathsf{Ag}$:

T.	$K_a\varphi \to \varphi$
D.	$M_a\top$
B.	$\varphi \to K_aM_a\varphi$
4.	$K_a\varphi \to K_aK_a\varphi$
5.	$\neg K_a\varphi \to K_a\neg K_a\varphi$

A simple way to denote axiom systems is just to add the axioms that are included together with the name **K**. Thus, **KD** is the axiom system that has all the axioms and rules of the system **K** (**1**, **K**, and rules **MP** and **Nec**) together with **D**. Similarly, **KD45** extends **K** by adding the axioms **D**, **4** and **5**. System **S4** is the more common way of denoting **KT4**, while **S5** is the more common way of denoting **KT45**. If it is necessary to make explicit that there are m agents in Ag, we write $\mathbf{K}_m$, $\mathbf{KD}_m$, and so on. ⊣

Using S5 to model knowledge

The system **S5** is an extension of **K** with the so-called 'properties of knowledge'. Likewise, **KD45** has been viewed as characterizing the 'properties of belief'. The axiom **T** expresses that knowledge is *veridical*: whatever one knows, must be true. (It is sometimes called the *truth axiom*.) The other two axioms specify so-called *introspective agents*: **4** says that an agent knows what he knows (positive introspection), while **5** says that he knows what he does not know (negative introspection). As a side remark, we mention that axiom **4** is superfluous in **S5**; it can be deduced from the other axioms.

All of these axioms are idealisations, and indeed, logicians do not claim that they hold for all possible interpretations of knowledge. It is only human to claim one day that you know a certain fact, only to find yourself admitting the next day that you were wrong, which undercuts the axiom **T**. Philosophers use such examples to challenge the notion of knowledge in the first place (see the notes at the end of the chapter for references to the literature on logical properties of knowledge). Positive introspection has also been viewed as problematic. For example, consider a pupil who is asked a question φ to which he does not know the answer. It may well be that, by asking more questions, the pupil becomes able to answer that φ is true. Apparently, the pupil knew φ, but was not aware he knew, so did not know that he knew φ.

The most debatable among the axioms is that of negative introspection. Quite possibly, a reader of this chapter does not know (yet) what Moore's paradox is (see Chapter 6), but did she know before picking up this book that she did not know that?

Such examples suggest that a reason for ignorance can be lack of *awareness*. Awareness is the subject of Chapter 3 in this book. Chapter 2 also has an interesting link to negative introspection: this chapter tries to capture what it means to claim 'All I know is φ'; in other words, it tries to give an account of 'minimal knowledge states'. This is a tricky concept in the presence of axiom **5**, since all ignorance immediately leads to knowledge!

One might argue that 'problematic' axioms for knowledge should just be omitted, or perhaps weakened, to obtain an appropriate system for knowledge, but what about the basic principles of modal logic: the axiom **K** and the rule of inference **Nec**. How acceptable are they for knowledge? As one might expect, we should not take anything for granted. **K** assumes perfect reasoners, who can infer logical consequences of their knowledge. It implies, for instance, that under some mild assumptions, an agent will know what day of the week July 26, 5018 will be. All that it takes to answer this question is that (1) the agent knows today's date and what day of the week it is today, (2) she knows the rules for assigning dates, computing leap years, and so on (all of which can be encoded as axioms in an epistemic logic with the appropriate set of primitive propositions). By applying K to this collection of facts, it follows that the agent must know what day of the week it will be on July 26, 5018. Necessitation assumes agents can infer all **S5** theorems: agent a, for instance, would know that $K_b(K_b q \land \neg K_b(p \to \neg K_b q))$ is equivalent to $(K_b q \land M_b p)$. Since even telling whether a formula is propositionally valid is co-NP-complete, this does not seem so plausible.

The idealisations mentioned in this paragraph are often summarised as *logical omniscience*: our **S5** agent would know everything that is logically deducible. Other manifestations of logical omniscience are the equivalence of $K(\varphi \land \psi)$ and $K\varphi \land K\psi$, and the derivable rule in **K** that allows one to infer $K\varphi \to K\psi$ from $\varphi \to \psi$ (this says that agents knows all logical consequences of their knowledge).

The fact that, in reality, agents are *not* ideal reasoners, and not logically omniscient, is sometimes a feature exploited by computational systems. Cryptography for instance is useful because artificial or human intruders are, due to their limited capacities, not able to compute the prime factors of a large number in a reasonable amount of time. Knowledge, security, and cryptographic protocols are discussed in Chapter 12

Despite these problems, the **S5** properties are a useful idealisation of knowledge for many applications in distributed computing and economics, and have been shown to give insight into a number of problems. The **S5** properties are reasonable for many of the examples that we have already given; here is one more. Suppose that we have two processors, a and b, and that they are involved in computations of three variables, x, y, and z. For simplicity, assume that the variables are Boolean, so that they are either 0 or 1. Processor a can read the value of x and of y, and b can read y and z. To model this, we use, for instance, 010 as the state where $x = 0 = z$, and $y = 1$. Given our assumptions regarding what agents can see, we then have $x_1y_1z_1 \sim_a x_2y_2z_2$ iff $x_1 = x_2$ and $y_1 = y_2$. This is a simple manifestation of an *interpreted system*, where the accessibility relation is based on what an agent can see in a state. Such a relation is an equivalence relation. Thus, an interpreted system satisfies all the knowledge axioms. (This is formalised in Theorem 1.9(1) below.)

While **T** has traditionally been considered an appropriate axiom for knowledge, it has not been considered appropriate for *belief*. To reason about belief, **T** is typically replaced by the weaker axiom **D**: $\neg B_a \bot$, which says that the agent does not believe a contradiction; that is, the agent's beliefs are consistent. This gives us the axiom system **KD45**. We can replace **D** by the following axiom **D′** to get an equivalent axiomatisation of belief:

$$\mathbf{D}': \quad K_a\varphi \to \neg K_a \neg\varphi.$$

This axioms says that the agent cannot know (or believe) both a fact and its negation. Logical systems that have operators for both knowledge and belief often include the axiom $K_a\varphi \to B_a\varphi$, saying that knowledge entails belief.

Axiom systems for group knowledge

If we are interested in formalising the knowledge of just one agent a, the language $\mathsf{L}(\mathsf{At}, \{K_a\}, \mathsf{Ag})$ is arguably too rich. In the logic $\mathbf{S5}_1$ it can be shown that every formula is equivalent to a *depth-one* formula, which has no nested occurrences of K_a. This follows from the following equivalences, all of which are valid in $\mathcal{S}5$ as well as being theorems of **S5**: $KK\varphi \leftrightarrow K\varphi$; $K\neg K\varphi \leftrightarrow \neg K\varphi$; $K(K\varphi \vee \psi) \leftrightarrow (K\varphi \vee K\psi)$; and $K(\neg K\varphi \vee \psi) \leftrightarrow \neg K\varphi \vee K\psi$. From a logical perspective things become more interesting in the multi-agent setting.

We now consider axiom systems for the notions of group knowledge that were defined earlier. Not surprisingly, we need some additional axioms.

Definition 1.19 (Logic of common knowledge)
The following axiom and rule capture common knowledge.

Fix.	$C_A\varphi \to E_A(\varphi \wedge C_A\varphi)$.
Ind.	From $\varphi \to E_A(\psi \wedge \varphi)$ infer $\varphi \to C_A\psi$.

For each axiom system **X** considered earlier, let **XC** be the result of adding **Fix** and **Ind** to **X**. ⊣

The *fixed point axiom* **Fix** says that common knowledge can be viewed as the fixed point of an equation: common knowledge of φ holds if everyone knows both that φ holds and that φ is common knowledge. **Ind** is called the *induction rule*; it can be used to derive common knowledge 'inductively'. If it is the case that φ is 'self-evident', in the sense that if it is true, then everyone knows it, and, in addition, if φ is true, then everyone knows ψ, we can show by induction that if φ is true, then so is $E_A^k(\psi \wedge \varphi)$ for all k. It follows that $C_A\varphi$ is true as well. Although common knowledge was defined as an 'infinitary' operator, somewhat surprisingly, these axioms completely characterize it.

For distributed knowledge, we consider the following axioms for all $A \subseteq$ Ag:

W.	$K_a\varphi \to D_A\varphi$ if $a \in A$.
$\mathbf{K_D}$.	$D_A(\varphi \to \psi) \to (D_A\varphi \to D_A\psi)$.
$\mathbf{T_D}$.	$D_A\varphi \to \varphi$.
$\mathbf{D_D}$.	$\neg D_A\neg\top$.
$\mathbf{B_D}$.	$\varphi \to D_A\neg D_A\neg\varphi$.
$\mathbf{4_D}$.	$D_A\varphi \to D_AD_A\varphi$.
$\mathbf{5_D}$.	$\neg D_A\varphi \to D_A\neg D_A\varphi$.

These axioms have to be understood as follows. It may help to think about distributed knowledge in a group A as the knowledge of a wise man, who has been told, by every member of A, what each of them knows. This is captured by axiom **W**. The other axioms indicate that the wise man has at least the same reasoning abilities as distributed knowledge to the system $\mathbf{S5}_m$, we add the axioms $\mathbf{W}, \mathbf{K_D}, \mathbf{T_D}, \mathbf{4_D}$, and $\mathbf{5_D}$ to the axiom system. For $\mathbf{K}_m$, we add only **W** and $\mathbf{K_D}$.

Proving Completeness

We want to prove that the axiom systems that we have defined are sound and complete for the corresponding semantics; that is, that **K** is sound and complete with respect to $\mathcal{K}$, **S5** is sound and complete with respect to $\mathcal{S}5$, and so on. Proving soundness is straightforward: we prove by induction on k that any formula proved using a derivation of length k is valid. Proving completeness is somewhat harder. There are different approaches, but the common one involves to show that if a formula is not derivable, then there is a model in which it is false. There is a special model called the *canonical model* that simultaneously shows this for all formulas. We now sketch the construction of the canonical model.

The states in the canonical model correspond to *maximal consistent sets of formulas*, a notion that we define next. These sets provide the bridge between the syntactic and semantic approach to validity.

Definition 1.20 (Maximal consistent set)
A formula φ is *consistent with axiom system* **X** if we cannot derive $\neg\varphi$ in **X**. A finite set $\{\varphi_1, \ldots, \varphi_n\}$ of formulas is consistent with **X** if the conjunction $\varphi_1 \wedge \ldots \wedge \varphi_n$ is consistent with **X**. An infinite set Γ of formulas is consistent with **X** if each finite subset of Γ is consistent with **X**. Given a language L

and an axiom system $\mathbf{X}$, a *maximal consistent set* for $\mathbf{X}$ and L is a set Γ of formulas in L that is consistent and *maximal*, in the sense that every strict superset Γ' of Γ is inconsistent. ⊣

We can show that a maximal consistent set Γ has the property that, for every formula $\varphi \in \mathsf{L}$, exactly one of φ and $\neg\varphi$ is in Γ. If both were in Γ, then Γ would be inconsistent; if neither were in Γ, then Γ would not be maximal. A maximal consistent set is much like a state in a Kripke model, in that every formula is either true or false (but not both) at a state. In fact, as we suggested above, the states in the canonical model can be identified with maximal consistent sets.

Definition 1.21 (Canonical model)
The *canonical model for* L *and* $\mathbf{X}$ is the Kripke model $M = \langle S, R, V\rangle$ defined as follows:

- S is the set of all maximal consistent sets for $\mathbf{X}$ and L;
- $\Gamma R_a \Delta$ iff $\Gamma|K_a \subseteq \Delta$, where $\Gamma|K_a = \{\varphi \mid K_a\varphi \in \Gamma\}$;
- $V(\Gamma)(p) = \mathit{true}$ iff $p \in \Gamma$. ⊣

The intuition for the definition of R_a and V is easy to explain. Our goal is to show that the canonical model satisfies what is called the *Truth Lemma*: a formula φ is true at a state Γ in the canonical model iff $\varphi \in \Gamma$. (Here we use the fact that the states in the canonical model are actually sets of formulas—indeed, maximal consistent sets.) We would hope to prove this by induction. The definition of V ensures that the Truth Lemma holds for primitive propositions. The definition of R_a provides a necessary condition for the Truth Lemma to hold for formulas of the form $K_a\varphi$. If $K_a\varphi$ holds at a state (maximal consistent set) Γ in the canonical model, then φ must hold at all states Δ that are accessible from Γ. This will be the case if $\Gamma|K_a \subseteq \Delta$ for all states Δ that are accessible from Γ (and the Truth Lemma applies to φ and Δ).

The Truth Lemma can be shown to hold for the canonical model, as long as we consider a language that does not involve common knowledge or distributed knowledge. (The hard part comes in showing that if $\neg K_a\varphi$ holds at a state Γ, then there is an accessible state Δ such that $\neg\varphi \in \Delta$. That is, we must show that the R_a relation has 'enough' pairs.) In addition to the Truth Lemma, we can also show that the canonical model for axiom system $\mathbf{X}$ is a model in the corresponding class of models; for example, the canonical model for $\mathbf{S5}$ is in $\mathcal{S}5$.

Completeness follows relatively easily once these two facts are established. If a formula $\varphi \in \mathsf{L}$ cannot be derived in $\mathbf{X}$ then $\neg\varphi$ must be consistent with $\mathbf{X}$, and thus can be shown to be an element of a maximal consistent set, say Γ. Γ is a state in the canonical model for **X** and L. By the Truth Lemma, $\neg\varphi$ is true at Γ, so there is a model where φ is false, proving the completeness of **X**.

This argument fails if the language includes the common knowledge operator. The problem is that with the common knowledge operator in the language, the logic is not *compact*: there is a set of formulas such that all its finite subsets are satisfiable, yet the whole set is not satisfiable. Consider the set $\{E_A^n p \mid n \in \mathbb{N}\} \cup \{\neg C_A p\}$, where $A \subseteq \mathsf{Ag}$ is a group with at least two agents. Each finite subset of this set is easily seen to be satisfiable in a model in $\mathcal{S}5$ (and hence in a model in any of the other classes we have considered), but the whole set of formulas is not satisfiable in any Kripke model. Similarly, each finite subset of this set can be shown to be consistent with **S5C**. Hence, by definition, the whole set is consistent with **S5C** (and hence all other axiom systems we have considered). This means that this set must be a subset of a maximal consistent set. But, as we have observed, there is no Kripke model where this set of formulas is satisfied.

This means that a different proof technique is necessary to prove completeness. Rather than constructing one large canonical model for all formulas, for each formula φ, we construct a finite canonical model tailored to φ. And rather than considering maximal consistent subsets to the set of all formulas in the language, we consider maximal consistent sets of the set of subformulas of φ.

The canonical model $M_\varphi = \langle S_\varphi, R, V\rangle$ for φ and **KC** is defined as follows:

- S_φ is the set of all maximal consistent sets of subformulas of φ for **KC**;
- $\Gamma R_a \Delta$ iff $(\Gamma|K_a) \cup \{C_A\psi \mid C_A\psi \in \Gamma \text{ and } a \in A\} \subseteq \Delta$.
- $V(\Gamma)(p) = \mathit{true}$ iff $p \in \Gamma$.

The intuition for the modification to the definition of R_a is the following: Again, for the Truth Lemma to hold, we must have $\Gamma|K_a \subseteq \Delta$, since if $K_a\psi \in \Gamma$, we want ψ to hold in all states accessible from Γ. By the fixed point axiom, if $C_A\psi$ is true at a state s, so is $E_AC_A\psi$; moreover, if $a \in A$, then $K_aC_A\psi$ is also true at s. Thus, if $C_A\psi$ is true at Γ, $C_A\psi$ must also be true at all states accessible from Γ, so we must have $\{C_A\psi \mid C_A\psi \in \Gamma$ and $a \in A\} \subseteq \Delta$. Again, we can show that the Truth Lemma holds for the canonical model for φ and **KC** for subformulas of φ; that is, if ψ is a subformula of φ, then ψ is true at a state Γ in the canonical model for φ and **KC** iff $\varphi \in \Gamma$.

We must modify this construction somewhat for axiom systems that contain the axiom **4** and/or **5**. For axiom systems that contain **4**, we redefine

R_a so that $\Gamma R_a \Delta$ iff $(\Gamma \mid K_a) \cup \{C_A\psi \mid C_A\psi \in \Gamma \text{ and } a \in A\} \cup \{K_a\psi \mid K_a\psi \in \Gamma\} \subseteq \Delta$. The reason that we want $\{K_a\varphi \mid K_a\varphi \in \Gamma\} \subseteq \Delta$ is that if $K_a\psi$ is true at the state Γ, so is $K_aK_a\psi$, so $K_a\psi$ must be true at all worlds accessible from Γ. An obvious question to ask is why we did not make this requirement in our original canonical model construction. If both $K_a\psi$ and $K_aK_a\psi$ are subformulas of φ, then the requirement is in fact not necessary. For if $K_a\psi \in \Gamma$, then consistency will guarantee that $K_aK_a\psi$ is as well, so the requirement that $\Gamma|K_a \subseteq \Delta$ guarantees that $K_a\psi \in \Delta$. However, if $K_a\psi$ is a subformula of φ but $K_aK_a\psi$ is not, this argument fails.

For systems that contain **5**, there are further subtleties. We illustrate this for the case of **S5**. In this case, we require that $\Gamma R_a \Delta$ iff $\{K_a\psi \mid K_a\psi \in \Gamma\} = \{K_a\psi \mid K_a\psi \in \Delta\}$ and $\{C_A\psi \mid C_A\psi \in \Gamma \text{ and } a \in A\} = \{C_A\psi \mid C_A\psi \in \Delta \text{ and } a \in A\}$. Notice that the fact that $\{K_a\psi \mid K_a\psi \in \Gamma\} = \{K_a\psi \mid K_a\psi \in \Delta\}$ implies that $\Gamma|K_a = \Delta|K_a$. We have already argued that having **4** in the system means that we should have $\{K_a\psi \mid K_a\psi \in \Gamma\} \subseteq \{K_a\psi \mid K_a\psi \in \Delta\}$. For the opposite inclusion, note that if $K_a\psi \notin \Gamma$, then $\neg K_a\psi$ should be true at the state Γ in the canonical model, so (by **5**) $K_a\neg K_a\psi$ is true at Γ, and $\neg K_a\psi$ is true at Δ if $\Gamma R_a \Delta$. But this means that $K_a\psi \notin \Delta$ (assuming that the Truth Lemma applies). Similar considerations show that we must have $\{C_A\psi \mid C_A\psi \in \Gamma \text{ and } a \in A\} = \{C_A\psi \mid C_A\psi \in \Delta \text{ and } a \in A\}$, using the fact that $\neg C_A\psi \to E_A\neg C_A\psi$ is provable in **S5C**.

Getting a complete axiomatisation for languages involving distributed knowledge requires yet more work; we omit details here.

We summarise the main results regarding completeness of epistemic logics in the following theorem. Recall that, for an axiom system **X**, the axiom system **XC** is the result of adding the axioms **Fix** and **Ind** to **X**. Similarly, **XD** is the result of adding the 'appropriate' distributed knowledge axioms to **X**; specifically, it includes the axiom **W**, together with every axiom $\mathbf{Y_D}$ for which **Y** is an axiom of **X**. So, for example, **S5D** has the axioms of **S5** together with **W**, $\mathbf{K_D}$, $\mathbf{T_D}$, $\mathbf{4_D}$, and $\mathbf{5_D}$.

Theorem 1.9
If $(\mathsf{At}, \mathsf{Op}, \mathsf{Ag})$, **X** is an axiom systems that includes all the axioms and rules of **K** and some (possibly empty) subset of $\{\mathbf{T}, \mathbf{4}, \mathbf{5}, \mathbf{D}\}$, and $\mathcal{X}$ is the corresponding class of Kripke models, then the following hold:

1. if $\mathsf{Op} = \{K_a \mid a \in \mathsf{Ag}\}$, then **X** is sound and complete for $\mathcal{X}$ and L;

2. if $\mathsf{Op} = \{K_a \mid a \in \mathsf{Ag}\} \cup \{C_A \mid A \subseteq \mathsf{Ag}\}$, then **XC** is sound and complete for $\mathcal{X}$ and L;

3. if $\mathsf{Op} = \{K_a \mid a \in \mathsf{Ag}\} \cup \{D_A \mid A \subseteq \mathsf{Ag}\}$, then **XD** is sound and complete for $\mathcal{X}$ and L;

4. if $\mathsf{Op} = \{K_a \mid a \in \mathsf{Ag}\} \cup \{C_A \mid A \subseteq \mathsf{Ag}\} \cup \{D_A \mid A \subseteq \mathsf{Ag}\}$, then **XCD** is sound and complete for $\mathcal{X}$ and L. ⊣

1.3 Overview of the Book

The book is divided into three parts: informational attitudes, dynamics, and applications. Part I, informational attitudes, considers ways that basic epistemic logic can be extended with other modalities related to knowledge and belief, such as "only knowing", "awareness", and probability. There are three chapters in Part I:

Only Knowing Chapter 2, on only knowing, is authored by Gerhard Lakemeyer and Hector J. Levesque. What do we mean by 'only knowing'? When we say that an agent knows p, we usually mean that the agent knows *at least* p, but possibly more. In particular, knowing p does not allow us to conclude that q is not known. Contrast this with the situation of a knowledge-based agent, whose knowledge base consists of p, and nothing else. Here we would very much like to conclude that this agent does not know q, but to do so requires us to assume that p is all that the agent knows or, as one can say, the agent only knows p. In this chapter, the logic of only knowing for both single and multiple agents is considered, from both the semantic and proof-theoretic perspective. It is shown that only knowing can be used to capture a certain form of honesty, and that it relates to a form of non-monotonic reasoning.

Awareness Chapter 3, on logics where knowledge and awareness interact, is authored by Burkhard Schipper. Roughly speaking, an agent is unaware of a formula φ if φ is not on his radar screen (as opposed to just having no information about φ, or being uncertain as to the truth of φ). The chapter discusses various approaches to modelling (un)awareness. While the focus is on axiomatisations of structures capable of modelling knowledge and awareness, structures for modelling probabilistic beliefs and awareness, are also discussed, as well as structures for awareness of unawareness.

Epistemic Probabilistic Logic Chapter 4, authored by Lorenz Demey and Joshua Sack, provides an overview of systems that combine probability theory, which describes quantitative uncertainty, with epistemic logic, which describes qualitative uncertainty. By combining knowledge and probability, one obtains a very powerful account of information and information flow. Three types of systems are investigated: systems that describe uncertainty of agents at a single moment in time, systems where the uncertainty changes over time, and systems that describe the actions that cause these changes.

Part II on dynamics of informational attitudes considers aspects of how knowledge and belief change over time. It consists of three chapters:

Knowledge and Time Chapter 5, on knowledge and time, is authored by Clare Dixon, Cláudia Nalon, and Ram Ramanujam. It discusses the dynamic aspects of knowledge, which can be characterized by a combination of temporal and epistemic logics. The chapter presents the language and axiomatisation for such a combination, and discusses complexity and expressivity issues. It presents two different proof methods (which apply quite broadly): *resolution* and *tableaux*. Levels of knowledge and the relation between knowledge and communication in distributed protocols are also discussed, and an automata-theoretic characterisation of the knowledge of finite-state agents is provided. The chapter concludes with a brief survey on applications.

Dynamic Epistemic Logic Chapter 6, on dynamic epistemic logic, is authored by Lawrence Moss. Dynamic Epistemic Logic (**DEL**) extends epistemic logic with operators corresponding to *epistemic actions*. The most basic epistemic action is a public announcement of a given sentence to all agents. In the first part of the chapter, a logic called **PAL** (*public announcement logic*), which includes announcement operators, is introduced. Four different axiomatisations for **PAL** are given and compared. It turns out that **PAL** without common knowledge is reducible to standard epistemic logic: the announcement operators may be translated away. However, this changes once we include common knowledge operators in the language. The second part of Chapter 6 is devoted to more general epistemic actions, such as private announcements.

Dynamic Logics of Belief Change Chapter 7, on belief change, is authored by Johan van Benthem and Sonja Smets. The chapter gives an overview of current dynamic logics that describe belief update and revision. This involves a combination of ideas from belief revision theory and dynamic epistemic logic. The chapter describes various types of belief change, depending on whether the information received is 'hard' or 'soft'. The chapter continues with three topics that naturally complement the setting of single steps of belief change: connections with probabilistic approaches to belief change, long-term temporal process structure including links with formal learning theory, and multi-agent scenarios of information flow and belief revision in games and social networks. It ends with a discussion of alternative approaches, further directions, and windows to the broader literature.

Part III considers applications of epistemic logic in various areas. It consists of five chapters:

Model Checking Temporal Epistemic Logic Chapter 8, authored by Alessio Lomuscio and Wojciech Penczek, surveys work on model checking systems against temporal-epistemic specifications. The focus is on two approaches to verification: approaches based on *ordered binary decision diagrams* (OBDDs) and approaches based on translating specifications to propositional logic, and then applying propositional satisfiability checkers (these are called *SAT-based* approaches). OBDDs provide a compact representation for propositional formulas; they provide powerful techniques for efficient mode checking; SAT-based model checking is the basis for many recent symbolic approach to verification. The chapter also discusses some more advanced techniques for model checking.

Epistemic Foundations of Game Theory Chapter 9, authored by Giacomo Bonanno, provides an overview of the epistemic approach to game theory. Traditionally, game theory focuses on interaction among intelligent, sophisticated and rational individuals. The epistemic approach attempts to characterize, using epistemic notions, the behavior of rational and intelligent players who know the structure of the game and the preferences of their opponents and who recognize each other's rationality and reasoning abilities. The focus of the analysis is on the implications of common belief of rationality in strategic-form games and on dynamic games with perfect information.

BDI Logics Chapter 10, on logics of beliefs, desires, and intentions (BDI), is authored by John-Jules Ch. Meyer, Jan Broersen and Andreas Herzig. Various formalisations of BDI in logic are considered, such as the approach of Cohen and Levesque (recast in dynamic logic), Rao and Georgeff's influential BDI logic based on the branching-time temporal logic CTL*, the KARO framework, in which action together with knowledge (or belief) is the primary concept on which other agent notions are built, and BDI logics based on STIT (seeing to it that) logics, such as XSTIT.

Knowledge and Ability Chapter 11, authored by Thomas Ågotnes, Valentin Goranko, Wojciech Jamroga and Michael Wooldridge, relates epistemic logics to various logics for *strategic abilities*. It starts by discussing approaches from philosophy and artificial intelligence to modelling the interaction of agents knowledge and abilities, and then focuses on concurrent game models and the alternating-time temporal logic ATL. The authors discuss how ATL enables reasoning about agents' coalitional abilities to achieve qualitative objectives in concurrent game models, first assuming complete information and then under incomplete information and uncertainty about the structure of the game model. Finally, extensions of ATL that allow explicit reasoning about the interaction of knowledge and strategic abilities are considered; this leads to the notion of *constructive knowledge*.

Knowledge and Security Chapter 12, on knowledge and security, is authored by Riccardo Pucella. A persistent intuition in the field of computer security says that epistemic logic, and more generally epistemic concepts, are relevant to the formalisation of security properties. What grounds this intuition is that much work in the field is based on epistemic concepts. Confidentiality, integrity, authentication, anonymity, non-repudiation, all can be expressed as epistemic properties. This survey illustrates the use of epistemic concepts and epistemic logic to formalise a specific security property, *confidentiality*. Confidentiality is a prime example of the use of knowledge to make a security property precise. It is explored in two large domains of application: cryptographic protocol analysis and multi-level security systems.

1.4 Notes

The seminal work of the philosopher Jaakko Hintikka (1962) is typically taken as the starting point of modern epistemic logic. Two texts on epistemic logic by computer scientists were published in 1995: one by Fagin, Halpern, Moses, and Vardi (1995) and the other by Meyer and van der Hoek (1995). Another influential text on epistemic logic, which focuses more on philosophical aspects, is by Rescher (2005). Formal treatments of the notion of knowledge in artificial intelligence, in particular for reasoning about action, go back to the work of Moore (1977). In the mid-1980s, the conference on Theoretical Aspects of Reasoning About Knowledge (TARK), later renamed to "Theoretical Aspects of *Rationality* and Knowledge, was started (1986); in the mid-1990s, the Conference on Logic and Foundations of Game and Decision Theory (LOFT) (1996) began. These two conferences continue to this day, bringing together computer scientists, economists, and philosophers.

Our chapter is far from the first introduction to epistemic logic. The textbooks by Fagin et al. (1995) and by Meyer and van der Hoek (1995) each come with an introductory chapter; more recent surveys and introductions can be found in the book by van Ditmarsch, van der Hoek, and Kooi (2007, Chapter 2), in a paper on epistemic logic and epistemology by Holliday (2014), in the chapter by Bezhanishvili and van der Hoek (2014), which provides a survey of semantics for epistemic notions, and in online resources (Hendricks and Symons 2014, Wikipedia).

Halpern (1987) provides an introduction to applications of knowledge in distributed computing; the early chapters of the book by Perea (2012) give an introduction to the use of epistemic logic in game theory. As we already said, more discussion of the examples in Section 1.1 can be found in the relevant chapters. Public announcements are considered in Chapter 6; protocols are studied in Chapter 12 and, to some extent, in Chapter 5; strategic ability is the main topic of Chapter 11; epistemic foundations of game theory are considered in Chapter 9; distributed computing is touched on in Chapter 5, while examples of model checking distributed protocols are given in Chapter 8.

The use of Kripke models puts our approach to epistemic logic firmly in the tradition of modal logic, of which Kripke is one of the founders (see Kripke (1963)). Modal logic has become *the* framework to reason not only about notions as knowledge and belief, but also about agent attitudes such as desires and intentions (Rao and Georgeff, 1991), and about notions like time (Emerson, 1990), action (Harel, 1984), programs (Fischer and Ladner, 1979), reasoning about obligation and permission (von Wright, 1951), and combinations of them. Modern references to modal logic include the textbook by Blackburn, de Rijke, and Venema (2001) and the handbook edited by Blackburn, van Benthem, and Wolter (2006).

Using modal logic to formalise knowledge and belief suggests that one has an idealised version of these notions in mind. The discussion in Section 1.2.5 is only the tip of the iceberg. Further discussion of logical omniscience can be found in (Stalnaker, 1991; Sim, 1997) and in (Fagin et al., 1995, Chapter 9). There is a wealth of discussion in the philosophy and psychology literature of the axioms and their reasonableness (Koriat, 1993; Larsson, 2004; Zangwill, 2013). Perhaps the most controversial axiom of knowledge is **5**; which was dismissed in the famous claim by Donald Rumsfeld that there are 'unknown unknowns' (see `http://en.wikipedia.org/wiki/There_are_known_knowns`). Some approaches for dealing with lack of knowledge using awareness avoid this axiom (and, indeed, all the others); see Chapter 3.

Broadly speaking, philosophers usually distinguish between the *truth* of a claim, our *belief* in it, and the *justification for the claim*. These are often considered the three key elements of knowledge. Indeed, there are papers that define knowledge as justified true belief. There has been much debate of this definition, going back to Gettier's (1963) *Is justified true belief knowledge?*. Halpern, Samet, and Segev (2009) provide a recent perspective on these issues.

The notion of *common knowledge* is often traced back to the philosopher David Lewis's (1969) independently developed by the sociologist Morris Friedell (1969). Work on common knowledge in economics was initiated by Robert Aumann (1976); John McCarthy's (1990) work involving common knowledge had a significant impact in the field of artificial intelligence. Good starting points for further reading on the topic of common knowledge are by Vanderschraaf and Sillari (2014) and by Fagin et al. (1995, Chapter 6). Section 9.5 compares the notions of common knowledge with that of common belief.

Distributed knowledge was discussed first, in an informal way, by Hayek (1945), and then, in a more formal way, by Hilpinen (1977). It was rediscovered and popularized by Halpern and Moses (1990), who originally called it *implicit knowledge.*

The notion of bisimulation is a central notion in modal logic, providing an answer to the question when two models are 'the same' and is discussed in standard modal logic texts (Blackburn et al., 2001, 2006). Bisimulation arises quite often in this book, including in Chapters 5, 6, and 7.

We mentioned below Theorem 1.8, when discussing complexity of validity, that some recent advances make NP-complete problems seem more tractable: for this we refer to work by Gomes, Kautz, Sabharwal, and Selman (2008).

We end this brief discussion of the background literature by providing the pointers to the technical results mentioned in our chapter. Theorem 1.1 gives some standard valid formulas for several classes of models (see Fagin et al. (1995, Chapter 2.4) for a textbook treatment). Theorem 1.2 is a folk theorem in modal logic: for a proof and discussion, see Blackburn et al. (2006, Chapter 2.3). Proposition 1.3 is proved by Fagin et al. (1995) as Theorem 3.2.2 (for the case $\mathcal{X} = \mathcal{K}$) and Theorem 3.2.4 (for $\mathcal{X} = \mathcal{T}, \mathcal{S}4, \mathcal{KD}45$, and $\mathcal{S}5$). Proposition 1.4 is Proposition 3.2.1 by Fagin et al. (1995). Theorem 1.8 is proved by Halpern and Moses (1992).

Although the first proofs of completeness for multi-agent versions of axiom systems of the form $\mathbf{X}_m$ and $\mathbf{XC}_m$ are by Halpern and Moses (1992), the ideas go back much earlier. In particular, the basic canonical model construction goes back to Makinson (1966) (see Blackburn et al. (2001, Chapter 4) for a discussion), while the idea for completeness of axiom systems of the form **XC** is already in the proof of Kozen and Parikh (1981) for proving completeness of dynamic logic. Completeness for axiom systems of the form **XD** was proved by Fagin, Halpern, and Vardi (1992) and by van der Hoek and Meyer (1992). A novel proof is provided by Wang (2013, Chapter 3). Theorem 1.6 is part of logical folklore. A proof of Theorem 1.7 was given by French, van der Hoek, Iliev, and Kooi (2013).

Acknowledgements The authors are indebted to Cláudia Nalon for a careful reading. Hans van Ditmarsch is also affiliated to IMSc, Chennai, as associated researcher, and he acknowledges support from European Research Council grant EPS 313360. Joseph Y. Halpern was supported in part by NSF grants IIS-0911036 and CCF-1214844, by AFOSR grant FA9550-09-1-0266, by ARO grants W911NF-09-1-0281 and W911NF-14-1-0017, and by the Multidisciplinary University Research Initiative (MURI) program administered by the AFOSR under grant FA9550-12-1-0040.

References

Aumann, R. J. (1976). Agreeing to disagree. *Annals of Statistics 4*(6), 1236–1239.

Bezhanishvili, N. and W. van der Hoek (2014). Structures for epistemic logic. In A. Baltag and S. Smets (Eds.), *Logical and Informational Dynamics, a volume in honour of Johan van Benthem*, pp. 339–381. Springer.

Blackburn, P., M. de Rijke, and Y. Venema (2001). *Modal Logic.* Cambridge University Press: Cambridge, England.

Blackburn, P., J. van Benthem, and F. Wolter (Eds.) (2006). *Handbook of Modal Logic.* Elsevier Science Publishers B.V.: Amsterdam, The Netherlands.

van Ditmarsch, H., W. van der Hoek, and B. Kooi (2007). *Dynamic Epistemic Logic.* Berlin: Springer.

Emerson, E. A. (1990). Temporal and modal logic. In J. van Leeuwen (Ed.), *Handbook of Theoretical Computer Science Volume B: Formal Models and Semantics*, pp. 996–1072. Elsevier Science Publishers B.V.: Amsterdam, The Netherlands.

Fagin, R., J. Y. Halpern, Y. Moses, and M. Y. Vardi (1995). *Reasoning About Knowledge.* The MIT Press: Cambridge, MA.

Fagin, R., J. Y. Halpern, and M. Y. Vardi (1992). What can machines know? on the properties of knowledge in distributed systems. *Journal of the ACM 39*(2), 328–376.

Fischer, M. and R. Ladner (1979). Propositional dynamic logic of regular programs. *Journal of Computer and System Sciences 18*, 194–211.

French, T., W. van der Hoek, P. Iliev, and B. Kooi (2013). On the succinctness of some modal logics. *Artificial Intelligence 197*, 56–85.

Friedell, M. (1969). On the structure of shared awareness. *Behavioral Science 14*(1), 28–39. A working paper with the same title was published in 1967 by the Center for Research on Social Organization, University of Michigan.

Gettier, E. (1963). Is justified true belief knowledge? *Analysis 23*, 121Ð123.

Gomes, C. P., H. Kautz, A. Sabharwal, and B. Selman (2008). Satisfiability solvers. In *Handbook of Knowledge Representation*, pp. 89–133.

Halpern, J. Y. (1987). Using reasoning about knowledge to analyze distributed systems. In J. F. Traub, B. J. Grosz, B. W. Lampson, and N. J. Nilsson (Eds.), *Annual Review of Computer Science, Volume 2*, pp. 37–68. Palo Alto, Calif.: Annual Reviews Inc.

Halpern, J. Y. and Y. Moses (1990). Knowledge and common knowledge in a distributed environment. *Journal of the ACM 37*(3), 549–587. A preliminary version appeared in *Proc. 3rd ACM Symposium on Principles of Distributed Computing*, 1984.

Halpern, J. Y. and Y. Moses (1992). A guide to completeness and complexity for modal logics of knowledge and belief. *Artificial Intelligence 54*, 319–379.

Halpern, J. Y., D. Samet, and E. Segev (2009). Defining knowledge in terms of belief: The modal logic perspective. *The Review of Symbolic Logic 2*(3), 469–487.

Harel, D. (1984). Dynamic logic. In D. Gabbay and F. Guenther (Eds.), *Handbook of Philosophical Logic Volume II — Extensions of Classical Logic*, pp. 497–604. D. Reidel Publishing Company: Dordrecht, The Netherlands. (Synthese library Volume 164).

Hayek, F. (1945). The use of knowledge in society. *American Economic Review 35*, 519–530.

Hendricks, V. and J. Symons (retrieved 2014). Epistemic logic. In E. N. Zalta (Ed.), *The Stanford Encyclopedia of Philosophy*. `http://plato.stanford.edu/archives/spr2014/entries/logic-epistemic/`.

Hilpinen, R. (1977). Remarks on personal and impersonal knowledge. *Canadian Journal of Philosophy 7*, 1–9.

Hintikka, J. (1962). *Knowledge and Belief.* Cornell University Press: Ithaca, NY. Reprint: 'Knowledge and Belief', in: Texts in Philosophy, Vol. 1, Kings College Publications, 2005.

van der Hoek, W. and J.-J. Meyer (1992). Making some issues of implicit knowledge explicit. *International Journal of Foundations of Computer Science 3*(2), 193–224.

Holliday, W. (2014). Epistemic logic and epistemology. to appear, preliminary version at `http://philosophy.berkeley.edu/file/814/el_episteme.pdf`.

Koriat, A. (1993). How do we know that we know? the accessibility model of the feeling of knowing. *Psychological review 100*, 609Ð639.

Kozen, D. and R. Parikh (1981). An elementary proof of the completeness of PDL. *Theoretical Computer Science 14*(1), 113–118.

Kripke, S. (1963). Semantical analysis of modal logic. *Zeitschrift für Mathematische Logik und Grundlagen der Mathematik 9*, 67–96.

Larsson, S. (2004). The magic of negative introspection.

Lewis, D. (1969). *Convention — A Philosophical Study.* Harvard University Press: Cambridge, MA.

LOFT (since 1996). Logic and the foundations of game and decision theory. `http://www.econ.ucdavis.edu/faculty/bonanno/loft.html`.

Makinson, D. (1966). On some completeness theorems in modal logic. *Zeitschrift für Mathematische Logik und Grundlagen der Mathematik 12*, 379–384.

McCarthy, J. (1990). Formalization of two puzzles involving knowledge. In V. Lifschitz (Ed.), *Formalizing Common Sense : Papers by John McCarthy*, Ablex Series in Artificial Intelligence. Norwood, N.J.: Ablex Publishing Corporation. original manuscript dated 1978–1981.

Meyer, J.-J. C. and W. van der Hoek (1995). *Epistemic Logic for AI and Computer Science.* Cambridge University Press: Cambridge, England.

Moore, R. C. (1977). Reasoning about knowledge and action. In *Proceedings of the Fifth International Joint Conference on Artificial Intelligence (IJCAI-77)*, Cambridge, MA.

Perea, A. (2012). *Epistemic Game Theory.* Cambridge, U.K.: Cambridge University Press.

Rao, A. S. and M. P. Georgeff (1991, April). Modeling rational agents within a BDI-architecture. In R. Fikes and E. Sandewall (Eds.), *Proceedings of Knowledge Representation and Reasoning (KR&R-91)*, pp. 473–484. Morgan Kaufmann Publishers: San Mateo, CA.

Rescher, N. (2005). *Epistemic Logic: A Survey of the Logic of Knowledge.* University of Pittsburgh Press.

Sim, K. M. (1997). Epistemic logic and logical omniscience: A survey. *International Journal of Intelligent Systems 12*(1), 57–81.

Stalnaker, R. (1991). The problem of logical omniscience, I. *Synthese 89*(3), 425–440.

TARK (since 1986). Theoretical aspects of rationality and knowledge. `http://www.tark.org`.

Vanderschraaf, P. and G. Sillari (retrieved 2014). Common knowledge. In E. N. Zalta (Ed.), *The Stanford Encyclopedia of Philosophy.* `http://plato.stanford.edu/archives/spr2014/entries/common-knowledge/`.

Wang, Y. (2013). *Logical Dynamics of Group Knowledge and Subset Spaces.* Ph. D. thesis, University of Bergen.

Wikipedia. Epistemic modal logic. `http://en.wikipedia.org/wiki/Epistemic_modal_logic`.

von Wright, G. H. (1951). Deontic logic. *Mind 60*(237), 1–15.

Zangwill, N. (2013). Does knowledge depend on truth? *Acta Analytica 28*(2), 139–144.

Part I

Informational Attitudes

Chapter 2

Only Knowing

Gerhard Lakemeyer and Hector J. Levesque

Contents

Abstract When we say that an agent knows p, we usually mean that the agent knows *at least* p, but possibly more. In particular, knowing p does not allow us to conclude that q is not known. Contrast this with the situation of a knowledge-based agent, whose knowledge base consists of p, and nothing else. Here we would very much like to conclude that this agent does not know q, but to do so requires us to assume that p is all that the agent knows or, as we will say, the agent only knows p. In this chapter, we consider the logic of only knowing for both single and multiple agents and from both the semantic and proof-theoretic perspective. Among other things, we will see how only knowing can be used to capture a certain form of honesty and how it relates to a particular form of nonmonotonic reasoning.

2.1 Introduction

The topic of this chapter is perhaps best motivated in terms of what has been called a *knowledge-based* agent or system. The assumption here is that

Chapter 2 of the *Handbook of Epistemic Logic*, H. van Ditmarsch, J.Y. Halpern, W. van der Hoek and B. Kooi (eds), College Publications, 2015, pp. 55–75.

the knowledge of the system will be represented explicitly by a collection of formulas in some logical language, called its *knowledge base* or KB. So the KB can be thought of as providing the explicit beliefs of the system. We can then ask what the system's *implicit* beliefs would be given that the KB represents *all* that it knows, or as we will say, that it *only knows* the KB. In case the KB consists, say, of sentences in propositional logic and we are interested only in beliefs from the same language, then the answer is in principle quite straightforward: simply define the implicit beliefs to be those formulas that are logically entailed by the KB.[1]

The notion of only knowing becomes more complex, and more interesting, if we add an explicit notion of belief to the language and assume that the system is able to *introspect* on its own beliefs. For that we first endow the language with an epistemic operator K for knowledge or belief. Now suppose KB consists of a single atomic proposition p, that is, KB only knows p. Clearly, we would expect that the KB knows $(p \vee q)$ yet it does not know q and, by introspection, knows that it does not know q. While we can capture the first in any normal modal logic in terms of $\mathrm{K}p$ logically entailing $\mathrm{K}(p \vee q)$, there is, surprisingly, no such equivalent for the other two. In particular, neither $\neg\mathrm{K}q$ nor $\mathrm{K}\neg\mathrm{K}q$ are entailed by $\mathrm{K}p$. This is because the meaning of $\mathrm{K}p$ is that *at least* p is known, but perhaps more, including q. In other words, classical modal logic as introduced in Chapter 1 is not able to fully capture (in terms of suitable entailments) what a system or an agent knows and does not know.

And this is where the story of only knowing begins. The idea is to introduce a new modal operator O for only knowing and define its truth conditions such that $\mathrm{O}p$ logically entails all of the above: $\mathrm{K}(p \vee q)$, $\neg\mathrm{K}q$, $\mathrm{K}\neg\mathrm{K}q$, and much more.

In the rest of the chapter, we study only knowing both semantically and proof-theoretically. To get us started, we first concentrate on the single-agent case, and later move on to only knowing for many agents.

2.2 Single-Agent Only Knowing

While there are a number of ways to give meaning to only knowing, we begin with one that has a particularly simple semantics. To simplify matters even further, we also limit ourselves to the propositional case, even though the original logic was developed for a full first-order language with equality. Besides the semantics we will, in addition, look at its proof the-

[1]Things become more complicated already when we are concerned about computational complexity, in which case classical entailment may be too strong, an issue we largely ignore in this chapter.

ory, connections to a form of nonmonotonic reasoning, and consider some variants.

2.2.1 Syntax and Semantics

Let At be a countably infinite set of atomic propositions. We define the language OL by the following BNF:

$$\varphi := p \mid \neg\varphi \mid (\varphi \wedge \varphi) \mid \mathrm{K}\varphi \mid \mathrm{O}\varphi,$$

where $p \in \mathsf{At}$ and K and O are modal operators for knowing and only knowing, respectively.

As usual, other logical connectives such as $\vee$, $\rightarrow$, or $\leftrightarrow$ are understood as syntactic abbreviations and will be used freely throughout this chapter. $\top$ and $\bot$ will be used as shorthand for $(p \vee \neg p)$ and $(p \wedge \neg p)$, respectively. A formula is called *objective* if it does not mention any modal operator; it is called *subjective* if every atom is within the scope of a modal operator.

To give meaning to K and O, we use what is perhaps the simplest of all possible-world models: sets of worlds. In other words, an agent knows or believes[2] a sentence φ just in case φ is true in all worlds the agent considers possible. As this set will be the same for every world, there is no need to introduce an explicit accessibility relation. In fact, worlds themselves can be defined as valuations, that is, a world w is a mapping $w : \mathsf{At} \longrightarrow \{\mathsf{true}, \mathsf{false}\}$. We call a set of worlds e the agent's epistemic state. We denote the set of all worlds as e_0.

As expected, knowing a sentence φ with respect to an epistemic state e simply means that φ is true in all $w \in e$. What should be the meaning of only knowing? As we discussed earlier, the intuition behind only knowing φ is that, while φ and all its consequences are known, sentences that do not follow from φ are not known. In other words we need to minimize what is known given φ. One way to achieve this is to maximize e by requiring that any world that satisfies φ must also a member of e.

Given a world w and an arbitrary set of worlds e, the formal semantics is then simply this:

$$\begin{array}{lll}
e, w \models p & \text{iff} & w(p) = \mathit{true} \qquad p \in \mathsf{At} \\
e, w \models (\varphi \wedge \psi) & \text{iff} & e, w \models \varphi \text{ and } e, w \models \psi \\
e, w \models \neg\varphi & \text{iff} & e, w \not\models \varphi \\
e, w \models \mathrm{K}\varphi & \text{iff} & \text{for all } w', \text{ if } w' \in e, \text{ then } e, w' \models \varphi \\
e, w \models \mathrm{O}\varphi & \text{iff} & \text{for all } w', \; w' \in e \text{ iff } e, w' \models \varphi
\end{array}$$

[2]While we allow for the possibility that agents can have false beliefs, we still use the terms belief and knowledge interchangeably.

A set of formulas Γ logically implies a formula φ ($\Gamma \models \varphi$) if for all w and e, if $e, w \models \gamma$ for all $\gamma \in \Gamma$, then $e, w \models \varphi$. A formula φ is valid ($\models \varphi$) if $\{\} \models \varphi$. φ is satisfiable if $\not\models \neg\varphi$.

If φ is objective, we often write $w \models \varphi$ as φ does not depend on e. Similarly, if φ is subjective, we write $e \models \varphi$ as w does not matter in this case.

2.2.2 Some properties of OL

As far as K is concerned, it is easy to see that OL behaves like the modal logic *K45*, that is, the following formulas are valid:

K: $\mathrm{K}\varphi \wedge \mathrm{K}(\varphi \rightarrow \psi) \rightarrow \mathrm{K}\psi$

4: $\mathrm{K}\varphi \rightarrow \mathrm{KK}\varphi$

5: $\neg\mathrm{K}\varphi \rightarrow \mathrm{K}\neg\mathrm{K}\varphi$

In particular, note that positive and negative introspection obtains because the worlds in e are globally accessible from every world.

Let us now focus on the properties of O. When we compare the semantic rules for K and O, then the only difference is that O uses an "iff" instead of "then." An equivalent formulation for the O-rule would be:

$$e, w \models \mathrm{O}\varphi \quad \text{iff} \quad e, w \models \mathrm{K}\varphi \text{ and for all } w', \text{if } e, w' \models \varphi, \text{ then } w' \in e$$

From this it follows immediately that $(\mathrm{O}\varphi \rightarrow \mathrm{K}\varphi)$ is valid. The converse does not hold. For example, let $e = \{w \mid w \models p \wedge q\}$. Then $e \models \mathrm{K}p$ but $e \not\models \mathrm{O}p$, since there are worlds that satisfy p that are not in e. Intuitively, e does not only know p because it knows more, namely q.

In case φ is objective, there is a unique epistemic state that only knows it, namely $\{w \mid w \models \varphi\}$. Hence we have

Proposition 2.1
For any objective φ and arbitrary ψ, either $\models \mathrm{O}\varphi \rightarrow \mathrm{K}\psi$ or $\models \mathrm{O}\varphi \rightarrow \neg\mathrm{K}\psi$. $\dashv$

Note that ψ can be an arbitrary formula. For example, we get $\models \mathrm{O}p \rightarrow \mathrm{K}(p \vee q)$ and $\models \mathrm{O}p \rightarrow \neg\mathrm{K}q$, but also $\models \mathrm{O}p \rightarrow \mathrm{KK}(p \vee q)$ and $\models \mathrm{O}p \rightarrow \mathrm{K}\neg\mathrm{K}q$. Regarding the objective beliefs when only knowing an objective formula, it is easy to see that these are precisely the ones that can be obtained by ordinary propositional reasoning:

Proposition 2.2
For any objective formulas φ and ψ, $\models \mathrm{O}\varphi \rightarrow \mathrm{K}\psi$ iff $\models \varphi \rightarrow \psi$. $\dashv$

In general, only knowing may not determine a unique epistemic state. For example, $\models \mathrm{O}(\mathrm{K}p \rightarrow p) \equiv (\mathrm{O}\top \vee \mathrm{O}p)$, that is, there are exactly two epistemic states that only know $(\mathrm{K}p \rightarrow p)$, the set of all worlds and the set of worlds that satisfy p. There are also formulas φ that cannot be only known at all even if φ is knowable. For example, $\neg\mathrm{OK}p$ is valid. To see why, consider $e = \{w \mid w \models p\}$, the only plausible candidate. Clearly $e \models \mathrm{K}p$ and hence $e \models \mathrm{KK}p$. But now consider a w^* such that $w^* \models \neg p$. Then $e, w^* \models \mathrm{K}p$ but $w^* \notin e$. Therefore, $e \not\models \mathrm{OK}p$.

Reducing only knowing to a disjunction of only knowing objective formulas as in the above example of $\mathrm{O}(\mathrm{K}p \rightarrow p)$ can be generalized:

Theorem 2.1
Let φ be an arbitrary formula such that $\not\models \neg\mathrm{O}\varphi$. Then there are objective formulas $\psi_1, \ldots, \psi_n$ such that $\models \mathrm{O}\varphi \equiv \mathrm{O}\psi_1 \vee \ldots \vee \mathrm{O}\psi_n$. ⊣

How difficult is it to decide whether a formula in OL is satisfiable? Let us call a formula *basic* if it does not mention O. If we restrict our attention to basic formulas, meaning $K45$, it is well known that the decision problem is NP-complete. It turns out that it is most likely harder for all of OL. More precisely, satisfiability in OL is Σ_2^p-complete, where Σ_2^p refers to the class of problems that can be decided by a non-deterministic Turing machine that uses an NP-oracle. As we will see later, this result is not too surprising as OL captures a certain form of nonmonotonic reasoning of the same complexity class.

2.2.3 Proof Theory

Perhaps the most concise way to characterize the valid sentences of OL is by providing an axiom system. As we will see, this is surprisingly simple if we view O not as a primitive notion but instead defined in terms of K and a new operator N in the following way. One way to read $\mathrm{O}\varphi$ is to say that φ is believed and nothing more, whereas $\mathrm{K}\varphi$ says that φ is believed, and perhaps more. In other words, $\mathrm{K}\varphi$ means that φ *at least* is believed to be true. A natural dual to this is to say that φ *at most* is believed to be false, which we write $\mathrm{N}\varphi$. The idea behind introducing this operator is that $\mathrm{O}\varphi$ would then be *definable* as $(\mathrm{K}\varphi \wedge \mathrm{N}\neg\varphi)$, that is, at least φ is believed and at most φ is believed. So, *exactly* φ is believed. In other words, we are taking K to specify a lower bound on what is believed (since there may be other beliefs) and N to specify an upper bound on beliefs (since there may be fewer beliefs). What is actually believed must lie between these two bounds. These bounds can be seen most clearly when talking about objective sentences.

Given an epistemic state as specified by a set of world states e, to say that $\mathrm{K}\varphi$ is true with respect to e is to say that e is a subset of the states where φ is true. By symmetry then, $\mathrm{N}\neg\varphi$ will be true when the set of states satisfying φ are a subset of e. The fact that e must contain all of these states means that nothing else can be believed that would eliminate any of them. This is the sense in which no more than φ is known. Finally, as before, $\mathrm{O}\varphi$ is true iff both conditions hold and the two sets coincide.

This leads us to the precise definition of $\mathrm{N}\varphi$:

$$e, w \models \mathrm{N}\varphi \quad \text{iff} \quad \text{for every } w', \text{ if } e, w' \not\models \varphi \text{ then } w' \in e.$$

from which the original constraint on $\mathrm{O}\varphi$ follows trivially.

So what are the properties of believing at most that φ is false? It is very easy to show that if φ is valid, then $\mathrm{N}\varphi$ will be valid too, if $\mathrm{N}\varphi$ and $\mathrm{N}(\varphi\rightarrow\psi)$ are both true, then so is $\mathrm{N}\psi$, and if some subjective σ is true, then so is $\mathrm{N}\sigma$. In other words, remarkably enough, N behaves like an ordinary belief operator: it is closed under logical implication and exhibits perfect introspection. This is most clearly seen by rephrasing very slightly the definition of N and comparing it to that of K:

$$e, w \models \mathrm{N}\varphi \quad \text{iff} \quad \text{for every } w' \notin e,\ e, w' \models \varphi.$$

Letting $\overline{e}$ stand for the set of states not in e, we have

$$e, w \models \mathrm{N}\varphi \quad \text{iff} \quad \text{for every } w' \in \overline{e},\ e, w' \models \varphi.$$

So N is like a belief operator with one important difference: we use the complement of e. In possible-world terms, we range over the *inaccessible* possible world states. In other words, K and N give us two belief-like operators: one, with respect to e, and one with respect to $\overline{e}$.

In the following, a *subjective* sentence is understood as before except that any occurrence of K (but not necessarily all) can be replaced by N. The axioms for OL are then given as follows:[3]

1. Axioms of propositional logic.

2. $\mathrm{L}(\varphi\rightarrow\psi) \rightarrow \mathrm{L}\varphi\rightarrow\mathrm{L}\psi$.

3. $\sigma \rightarrow \mathrm{L}\sigma$, where σ is subjective.

4. The N vs. K axiom:
   $(\mathrm{N}\varphi\rightarrow\neg\mathrm{K}\varphi)$, where $\neg\varphi$ is any consistent objective formula.

[3]In the following axioms, L is used as a modal operator standing for either K or N. Multiple occurrences of L in the same axiom should be uniformly replaced by K or by N. For example, $\mathrm{N}(\varphi\rightarrow\psi) \rightarrow \mathrm{N}\varphi\rightarrow\mathrm{N}\psi$ is an instance of Axiom 2.

5. The definition of O: $\mathrm{O}\varphi \leftrightarrow (\mathrm{K}\varphi \wedge \mathrm{N}\neg\varphi)$.

6. The inference rule of Modus ponens: From φ and $(\varphi \rightarrow \psi)$, infer ψ.

7. The inference rule of necessitation:
   From φ infer $\mathrm{K}\varphi$ and $\mathrm{N}\varphi$.

The notion of derivability of a formula φ with no premises ($\vdash \varphi$) is then defined in the usual way.

The first thing to notice is that N, taken by itself, has precisely the same properties as K, that is, K and N can both be thought of as ordinary belief operators except, of course, that there is a strong connection between the two. For one, Axiom 3 expresses that both K and N are perfectly and mutually introspective. For example, $(\mathrm{K}\varphi \rightarrow \mathrm{NK}\varphi)$ is an instance of 3. If we think of K and N as two agents, then this says that each agent has perfect knowledge of what the other knows. The other and more interesting connection between the two is, of course, Axiom 4, which is valid because K and N together range over all possible world states.

Proving that the axioms are sound is straightforward. While completeness can be proved using the standard Henkin-style technique of showing that every maximally consistent set of formulas is satisfiable, there is a subtlety. Recall that K and N are defined semantically with respect to complementary sets of worlds e and $\overline{e}$. While Axiom 4 guarantees that the two sets are exhaustive, that is, cover all possible worlds, it is not sufficient to enforce that the two sets are disjoint. More precisely, there are maximally consistent sets of formulas that are satisfiable only if we allow the sets of worlds for K and N to overlap. One solution then is to first consider a variant of the semantics, where K and N are indeed interpreted with respect to 2 overlapping and exhaustive sets of worlds, show that the axioms are still sound with respect to this variant and that the valid formulas under this variant are the same as before, and finally show completeness in a Henkin-style fashion.

Theorem 2.2
For any φ of OL, $\models \varphi$ iff $\vdash \varphi$. $\dashv$

It is perhaps worth pointing out that the assumption that the language comes with an infinite set of atoms is essential for the completeness proof. To see why, let us assume, to the contrary, that there are only finitely many atoms. For simplicity let us assume there is only a single atom p. Then it is easy to see that $\mathrm{K}p \wedge \neg\mathrm{K}\bot \rightarrow \mathrm{O}p$ would be valid. In words, if I know p and my beliefs are consistent then I only know p. This is because there is only a single world where p is true, and this world must be in the epistemic state to satisfy the antecedent of the implication.

2.2.4 Only Knowing and Autoepistemic Logic

It turns out that only knowing is closely related to autoepistemic logic (AEL), which was introduced by Robert Moore to capture a form of default reasoning. The idea is, roughly, that one should be able to draw plausible conclusions solely on the basis of one's own knowledge and ignorance. In terms of the classical birds-fly example, if I know that Tweety is a bird and I do not know that Tweety does not fly then I am entitled to conclude that Tweety indeed flies. Such a default can be expressed as

$$\delta = \mathrm{K}Bird(tweety) \wedge \neg\mathrm{K}\neg Fly(tweety) \rightarrow Fly(tweety),$$

where $Bird(tweety)$ and $Fly(tweety)$ are treated as atomic propositions. If we let $\mathrm{KB} = Bird(tweety)$, then

$$\mathrm{O}(\mathrm{KB} \wedge \delta) \rightarrow \mathrm{K}Fly(tweety) \text{ is valid,}$$

that is, the desired default conclusion indeed follows from the assumptions. Notice that we need only knowing here to conclude that Tweety flies. In particular, $\mathrm{K}(Bird(tweety) \wedge \delta) \rightarrow \mathrm{K}Fly(tweety)$ is not valid because there certainly are epistemic states where $Bird(tweety) \wedge \delta$ is known as well as $\neg Fly(tweety)$. In contrast, when $Bird(tweety) \wedge \delta$ is all that is known, it follows logically that $\neg Fly(tweety)$ is not known, and hence the default conclusion obtains.

While this can be proved easily with a semantic argument, we give a proof using the axioms. To simplify matters, we will freely use derivations which involve only properties of propositional logic (PL) or the logic K45 and mark them as such:

1. $\mathrm{O}(\mathrm{KB} \wedge \delta) \rightarrow \mathrm{K}(\mathrm{KB} \wedge \delta)$ Ax. 5; PL.
2. $\mathrm{K}(\mathrm{KB} \wedge \delta) \rightarrow (\mathrm{K}\neg\mathrm{K}\neg Fly(tweety) \rightarrow \mathrm{K}Fly(tweety))$ K45.
3. $(\mathrm{K}\neg\mathrm{K}\neg Fly(tweety) \rightarrow \mathrm{K}Fly(tweety)) \rightarrow (\neg\mathrm{K}\neg Fly(tweety) \rightarrow \mathrm{K}Fly(tweety))$ K45.
4. $\mathrm{O}(\mathrm{KB} \wedge \delta) \rightarrow (\neg\mathrm{K}\neg Fly(tweety) \rightarrow \mathrm{K}Fly(tweety))$ 1–3; PL.
5. $\mathrm{O}(\mathrm{KB} \wedge \delta) \rightarrow \mathrm{N}\neg(\mathrm{KB} \wedge \delta)$ Ax. 5; PL.
6. $\mathrm{N}\neg(\mathrm{KB} \wedge \delta) \rightarrow \mathrm{N}(\mathrm{KB} \rightarrow \neg\delta)$ K45.
7. $\mathrm{N}(\mathrm{KB} \rightarrow \neg\delta) \rightarrow \mathrm{N}(\mathrm{KB} \rightarrow \neg Fly(tweety))$ K45.
8. $\mathrm{N}(\mathrm{KB} \rightarrow \neg Fly(tweety)) \rightarrow \neg\mathrm{K}(\mathrm{KB} \rightarrow \neg Fly(tweety))$ Ax. 4.
9. $\neg\mathrm{K}(\mathrm{KB} \rightarrow \neg Fly(tweety)) \rightarrow \neg\mathrm{K}\neg Fly(tweety)$ K45.

10. $\mathrm{O}(\mathrm{KB} \wedge \delta) \rightarrow \neg\mathrm{K}\neg \mathit{Fly}(\mathit{tweety})$ 5–9; PL.

11. $\mathrm{O}(\mathrm{KB} \wedge \delta) \rightarrow \mathrm{K}\, \mathit{Fly}(\mathit{tweety})$ 4, 10; PL.

The fact that only knowing captures default conclusions such as "Tweety flies unless known otherwise" is no accident. It can be shown that there is an exact correspondence between only knowing and AEL. For that we need to briefly recall the notion of a *stable expansion*, which is central to AEL.

A set of basic formulas E is called a stable expansion of a formula φ iff

$$\gamma \in E \quad \text{iff} \quad \{\varphi\} \cup \{\mathrm{K}\psi \mid \psi \in E\} \cup \{\neg\mathrm{K}\beta \mid \beta \notin E\} \models_{pl} \gamma,$$

where $\models_{pl}$ means entailment in propositional logic. Without going into details, stable expansions have the property that they minimize what is known while still believing φ. Most importantly, this minimization of what is believed allows default inferences to be modeled. For example, $\mathit{Bird}(\mathit{tweety}) \wedge \delta$ has a unique stable expansion, and it contains $\mathit{Fly}(\mathit{tweety})$. As the following theorem tells us, there is a one-to-one correspondence between stable sets and the beliefs that result from only knowing:

Theorem 2.3
E is a stable expansion of a basic formula φ iff for some e such that $e \models \mathrm{O}\varphi$, $E = \{\text{basic } \alpha \mid e \models \mathrm{K}\alpha\}$. ⊣

AEL has been criticized for sometimes sanctioning expansions that are not grounded in facts. Perhaps the simplest example is $\varphi = (\mathrm{K}p \rightarrow p)$. There are two stable expansions, one that contains p and another that does not. Intuitively, we would prefer the latter as there really is no good reason to believe p based on φ. Given the theorem above, the same criticism applies to only knowing, of course. In the nonmonotonic reasoning community, a number of alternatives to AEL have been proposed to address the problem of ungroundedness. It turns out that many of them can also be captured by modifying the semantics of only knowing. For example, the notion of *moderately grounded stable expansions* introduced by Konolige can be formalized in terms of only knowing, using a new modal operator O^{K}, as follows:

$$e, w \models \mathrm{O}^{\mathrm{K}}\varphi \quad \text{iff} \quad e, w \models \mathrm{O}\varphi \text{ and for all } e' \supsetneq e, \quad e', w \not\models \mathrm{O}\varphi.$$

We now have that $\mathrm{O}^{\mathrm{K}}(\mathrm{K}p \rightarrow p)$ is uniquely satisfied by the set of all worlds and hence $\mathrm{O}^{\mathrm{K}}(\mathrm{K}p \rightarrow p) \rightarrow \neg\mathrm{K}p$. Unfortunately, there are still cases, where this strengthened form of only knowing is still admits ungrounded epistemic states. For example, $\mathrm{O}^{\mathrm{K}}[(\mathrm{K}p \rightarrow p) \wedge (\neg\mathrm{K}p \rightarrow q)]$ is satisfied by two epistemic states, one where only p is known and another where only q is known. Again, believing p does not seem reasonable. In order to rule out the unwanted epistemic state in such cases, even more restricted forms of only knowing are needed. While this can be done, a discussion would lead is too far afield; we refer instead to Section 2.4 for references to the relevant literature.

2.2.5 Honesty

A fundamental principle behind the idea of only knowing φ has been to minimize what is known. Semantically this amounts to maximizing the set of worlds considered possible. When φ is objective, practically everyone agrees that the corresponding (unique) e in this case should be the set of all worlds satisfying φ. When φ is not objective, the picture is less clear, as the discussion about grounded vs. ungrounded stable expansions or epistemic states illustrates.

A somewhat more radical view would be to require from the start that in order to only know φ, there must be a *unique* epistemic state which minimizes what is believed. Only knowing as defined so far clearly does not meet this requirement. Halpern and Moses (HM) were the first to characterize such a form of only knowing, which we denote by a new modal operator O^{HM}.[4] The semantics is again surprisingly simple:

$$e, w \models \mathrm{O}^{\mathrm{HM}}\varphi \quad \text{iff} \quad e, w \models \mathrm{K}\varphi \text{ and for all } e', \text{ if } e', w \models \mathrm{K}\varphi \text{ then } e' \subseteq e.$$

From this it follows immediately that for e to only know φ, the formula must be known and no superset of e knows it. So clearly O^{HM} applies a form of minimization to what is believed, but as we will see below, it differs from the principles we have seen so far. Interestingly, there is at most one e that satisfies $\mathrm{O}^{\mathrm{HM}}\varphi$: for if $e \models \mathrm{O}^{\mathrm{HM}}\varphi$ and $e' \models \mathrm{O}^{\mathrm{HM}}\varphi$, then the definition requires e and e' to be subsets of each other and hence to be equal.

HM call a formula φ *honest* if $\mathrm{O}^{\mathrm{HM}}\varphi$ is satisfiable and *dishonest* otherwise. An important property of honest formulas is the following:

Proposition 2.3
φ is honest iff for all e and e', if $e \models \mathrm{K}\varphi$ and $e' \models \mathrm{K}\varphi$, then $e \cup e' \models \mathrm{K}\varphi$. $\dashv$

Clearly, all objective formulas are honest and also many non-objective ones, including $\mathrm{K}p \rightarrow p$, which is only known according to HM by e_0. A typical dishonest formula is $\varphi = \mathrm{K}p \vee \mathrm{K}q$. (Let $e_p = \{w \mid w \models p\}$ for any atomic proposition p.) Then $e_p \models \mathrm{K}\varphi$ and $e_q \models \mathrm{K}\varphi$, yet $e_p \cup e_q \not\models \mathrm{K}\varphi$.

O^{HM} differs significantly from both O (and O^{K}), even in the case where O is uniquely satisfied:

[4] While Halpern and Moses introduced their version of only knowing merely meta-theoretically, we use a modal operator for consistency.

> Let $\varphi = \neg\mathrm{K}p \rightarrow q$. Then $e_q \models \mathrm{O}\varphi$, and the same for O^{K}. But φ is dishonest according to HM, which follows immediately after observing that $\models \mathrm{K}\varphi \equiv (\mathrm{K}p \vee \mathrm{K}q)$.

Note that $\neg\mathrm{K}p \rightarrow q$ can be understood as the default "assume q unless p is known." Hence O^{HM} does not seem suitable for default reasoning.

Another example where O^{HM} and O differ is Kp, which is honest with $e_p \models \mathrm{O}^{\mathrm{HM}}\mathrm{K}p$. On the other hand, OK$p$ is unsatisfiable, as seen before.

With $K45$ as the base logic, calling Kp honest may seem counter-intuitive, as believing Kp is not grounded in any (objective) facts. For that reason, it may be more appropriate to use $S5$ as the base logic for O^{HM}, where the real world is assumed to be a member of e and hence $\mathrm{K}p \rightarrow p$ is valid, which provides the desired grounding.

The complexity of deciding whether a basic formula is honest is $\Delta_2^{p,log(n)}$-complete. Here $\Delta_2^{p,log(n)}$ refers to the class of problems that can be decided by a deterministic Turing machine that is allowed up to $log(n)$ calls to an NP-oracle, where n is the size of the input. Likewise, deciding whether $\mathrm{O}^{\mathrm{HM}}\varphi \rightarrow \mathrm{K}\psi$ is valid for honest and basic φ and basic ψ is also $\Delta_2^{p,log(n)}$-complete.

2.3 Multi-Agent Only Knowing

Let us now turn to the multi-agent case. We will begin by extending the version of only knowing introduced initially and later discuss other variants.

When considering n agents, extending the language OL to multi-agent OL_n is straightforward: we simply introduce modal operators K_a, N_a, and O_a, for each agent a. As in Section 2.2.3, O_a will be definable in terms of K_a and N_a. Sometimes we will use L_i to mean any of the operators K_a and N_a.

Adapting the semantics of only knowing from one to many agents complicates matters substantially. A major issue is that the simple notion of a set of possible worlds to represent an agent's epistemic state no longer suffices, even under the assumption that agents continue to be fully introspective. For suppose there are two agents, Alice and Bob. Then Alice may consider it possible that p is true, that Bob believes that p is false and that Bob does not believe that Alice believes p. In other words, among Alice's possibilities there should be one where $\varphi = p \wedge \mathrm{K}_b\neg p \wedge \neg\mathrm{K}_b\mathrm{K}_a p$ comes out true. Thus it is not enough to consider which atomic propositions are true at a certain state of affairs, but also what is and what is not believed by other agents. Moreover, we need to make sure that models allow for enough possibilities, since only knowing needs to take into account what is considered impossible (see the discussion of the N-operator in the single agent case). For example, if φ is all Alice knows, then she should consider a state of affairs where $\mathrm{K}_b\mathrm{K}_a p$ holds impossible.

In the following we will consider one approach to the semantics that is based on Kripke models and, in particular, canonical models. While it does have shortcomings, as will be discussed later, the advantage is that it uses familiar notions from modal logic, which were introduced in Chapter 1.

2.3.1 The Canonical-Model Approach

Recall that a Kripke model M is a triple $\langle S, R, V\rangle$, where S is a nonempty set of states (or worlds), R is a mapping from agents into $S \times S$, and V is a mapping from states and atoms into $\{\mathsf{true}, \mathsf{false}\}$. As in the single-agent case, we want agents to be fully introspective with respect to their own beliefs. It is well known that this can be achieved by restricting R to be transitive and Euclidean for every agent. We will not repeat the semantics for atoms and the Boolean connectives here. The meaning of $\mathrm{K}_a\varphi$ is also as defined in Chapter 1:

$$M, s \models \mathrm{K}_a\varphi \quad \text{iff} \quad M, t \models \varphi \text{ for all } t \text{ such that } sR_a t.$$

Turning to N_a, we need to take care of the issue of mutual introspection between K_a and N_a, which we got "for free" in the single agent case. More precisely, when we consider worlds t which are not R_a-accessible from s, only those t are of interest which themselves have the same accessible worlds as s. Let us write $s \approx_a t$ if $R_a(s) = R_a(t)$. Then

$$M, s \models \mathrm{N}_a\varphi \quad \text{iff} \quad M, t \models \varphi \text{ for all } t \text{ such that not } sR_a t \text{ and } s \approx_a t.$$

O_a can then be defined as expected:

$$M, s \models \mathrm{O}_a\varphi \quad \text{iff} \quad M, t \models \mathrm{K}_a\varphi \wedge \mathrm{N}_a\neg\varphi.$$

We still need to address the issue that not all of the models include enough conceivable states of affairs, because otherwise we cannot draw the proper conclusions about what an agent does not know based on what she only knows. For example, consider $M = \langle S, R, V\rangle$ with $S = \{s\}$, $R_a(s) = s$ and $V(s)(p) = V(s)(q) = \mathsf{true}$. Then clearly $M, s \models \mathrm{K}_a p$. We also have $M, s \models \mathrm{N}_a\neg p$, because there are no non-accessible worlds, and hence $M, s \models \mathrm{O}_a p$, yet $M, s \not\models \neg\mathrm{K}_a q$.

So we need to find models that by design offer enough diversity to account for every conceivable state of affairs. One such candidate is the canonical model for the logic $K45_n$. The general idea of a canonical model was introduced already in Chapter 1. Here we specialize the definition for our purposes. As before, let us call a formula *basic* if it does not mention N_a and O_a for any a.

$M^c = \langle S^c, R^c, V^c \rangle$ is called the canonical $K45_n$-model if

- S is the set of all maximal consistent sets of basic formulas for $K45_n$;
- $\Gamma R_a \Delta$ iff $\{\varphi \mid K_a\varphi \in \Gamma\} \subseteq \Delta$;
- $V(\Gamma)(p) = \mathit{true}$ iff $p \in \Gamma$.

We define validity in $\mathsf{OL_n}$ only with respect to the canonical model, that is, a formula φ of $\mathsf{OL_n}$ is valid iff for all $s \in S^c$, $M^c, s \models \varphi$.

Before we turn to some of the properties of $\mathsf{OL_n}$, let us first consider in what sense M^c has enough possibilities. Let us call a formula φ *a-objective* if for all modal operators $\mathrm{K}_b, \mathrm{N}_b, \mathrm{O}_b$ in φ that do not occur in the scope of another modal operator, $b \neq a$. For example, $p \wedge \mathrm{K}_b\neg p \wedge \neg \mathrm{K}_b\mathrm{K}_a p$ is a-objective but $p \wedge \mathrm{K}_a\mathrm{K}_b\neg p \wedge \neg\mathrm{K}_b\mathrm{K}_a p$ is not. Intuitively, a-objective generalizes objective in the sense that, from agent's a point of view, the beliefs of other agents are also treated as objective facts.

Likewise, a formula φ is *a-subjective* if for all modal operators $\mathrm{K}_b, \mathrm{N}_b, \mathrm{O}_b$ in φ that do not occur in the scope of another modal operator, $b = a$. For example, $\mathrm{N}_a\mathrm{K}_b p \vee \mathrm{K}_a\mathrm{K}_a q$ is a-subjective, but $\mathrm{N}_a\mathrm{K}_b p \vee \mathrm{K}_b\mathrm{K}_a q$ is not.

For an arbitrary Kripke model M and world s, let

$$\mathrm{obj}_a(M, s) = \{\varphi \,|\, \varphi \text{ basic and } a\text{-objective s.t. } M, s \models \varphi\}.$$

A conceivable state of affairs for agent a, called *a-set,* is then any set of formulas Γ such that $\Gamma = \mathrm{obj}_a(M, s)$ for some M and s. The following theorem says that the set of conceivable states of affairs for agent a is both exhaustive and the same at all worlds of the canonical model.

Theorem 2.4
Let $s \in S^c$. Then for every a-set Γ there is exactly one world t in S^c such that $\mathrm{obj}_a(M^c, t) = \Gamma$ and $s \approx_a t$. $\dashv$

As far as only believing basic formulas is concerned, the canonical-model approach has all the right properties in the sense that the properties of the single-agent case generalize in a natural way. For example, for any a-objective basic formula φ, $\mathrm{O}_a\varphi$ is uniquely satisfied in the sense that if $M^c, s \models \mathrm{O}_a\varphi$, then $\{\mathrm{obj}_a(M^c, t) \,|\, t \in R_a(s)\}$ is the set of all a-sets containing φ.

For example, if $M^c, s \models \mathrm{O}_a(p \wedge \mathrm{K}_b\neg p \wedge \neg\mathrm{K}_b\mathrm{K}_i p)$ then the following formulas are also satisfied at s:

1. $\mathrm{K}_a\mathrm{K}_b\neg p$;

2. $\neg K_a q$;

3. $\neg K_a K_b q$;

4. $\neg K_a \neg K_b q$.

O_a also captures default reasoning the same way O does in the single-agent case. For example, consider $\neg K_a K_b p \rightarrow \neg K_b p$,, which says that unless a believes that b believes p, a assumes that b does not believe p. Then $O_a(\neg K_a K_b p \rightarrow \neg K_b p) \rightarrow K_a \neg K_b p$ is valid.

2.3.2 Proof Theory

The following proof theory is a natural generalization of the proof theory for a single agent.

1. Axioms of propositional logic.

2. $L_a(\varphi \rightarrow \psi) \rightarrow L_a\varphi \rightarrow L_a\psi$.

3. $\sigma \rightarrow L_a\sigma$, where σ is a-subjective.

4. The N vs. K axiom:
   $(N_a\varphi \rightarrow \neg K_a\varphi)$, where $\neg\varphi$ is a $K45$-consistent a-objective basic formula.

5. The definition of O: $O_a\varphi \leftrightarrow (K_a\varphi \wedge N_a\neg\varphi)$.

6. The inference rule of Modus ponens:
   From φ and $(\varphi \rightarrow \psi)$, infer ψ.

7. The inference rule of necessitation:
   From φ infer $K_a\varphi$ and $N_a\varphi$.

When comparing the proof theory to the one in Section 2.2.3, the similarities are obvious. In fact, the only real differences are that objective and subjective are replaced by a-objective and a-subjective, respectively, and that Axiom 4 refers to $K45$-consistency instead of consistency in propositional logic. It is not hard to show that the axioms are sound for $\mathsf{OL_n}$:

Theorem 2.5
For every formula φ in $\mathsf{OL_n}$, if $\vdash \varphi$ then $\models \varphi$. ⊣

While intuitively appealing, the proof theory turns out to be complete only for a subset of the language, where only knowing is essentially applied to basic sentences only. Formally, let $\mathsf{OL_n^-}$ consist of all formulas φ in $\mathsf{OL_n}$ such that, in φ, no N_b and O_b may occur within the scope of a K_a, N_a, O_a for $a \neq b$.

Theorem 2.6
For every formula φ in OL_n^-, if $\models \varphi$ then $\vdash \varphi$. ⊣

The reason why the axioms are incomplete for all of OL_n lies in Axiom 4. More precisely, notice that the axiom refers to $K45$-consistency of basic formulas only. However, $\mathrm{N}_a\mathrm{O}_b p \rightarrow \neg \mathrm{K}_a\mathrm{O}_b p$ is also valid in OL_n, but we cannot use Axiom 4 to prove it.

Apart from this technical deficiency of the axioms, there is actually a much deeper problem with the semantics of OL_n, which leads to undesirable properties of only knowing non-basic formulas. For example, $\mathrm{O}_a\neg\mathrm{O}_b p$ is unsatisfiable, that is, it is impossible for Alice to only know that Bob does not only know p, which seems counter-intuitive. The root cause of this undesirable property is a fundamental weakness of the canonical model in that not every set of a-sets (or conceivable states of affairs) is possible. In other words, there are sets of a-sets Γ for which there is no world in the canonical model where the R_a-accessible worlds cover exactly Γ. Note that this is different from the semantics for single agents, where any arbitrary set of worlds can be an epistemic state.

Consequently, while the canonical-model approach seems to do have all the right properties when restricted to OL_n^-, serious problems arise beyond OL_n^-. There are ways to address these problems, but the technical apparatus needed is considerable and would take us too far afield (see Section 2.4 for pointers to the literature).

As for complexity, it can be shown that deciding whether a formula φ in OL_n^- can be inferred from the axioms or, equivalently, whether φ is valid is PSPACE-complete. In other words, the decision problem is no harder than for $K45_n$. We remark that other logics of multi-agent only knowing alluded to in the previous paragraph remain in the same complexity class.

2.3.3 Honesty Revisited

In the final technical section of this chapter on only knowing, we return to the notion of honesty, this time for many agents, following a proposal by Halpern. We will depart from the presentation and assumptions made so far in two aspects. For one, this time we will dispense with the use of a modal operator to capture honesty, as not much would be gained by it. For another, we also drop the assumption that agents are fully introspective. One reason for doing so is to show that the principles behind honesty do not necessarily depend on introspection. Another is because the technical treatment is actually simpler if we do not insist on introspection. To simplify matters even further we also assume that there are only finitely many atomic propositions, although with some extra work almost all the results mentioned in this section carry over to infinitely many atoms.

In the following, we will be concerned with the logics K_n, T_n, and $S4_n$. The language is $\mathsf{OL_n}$ restricted to basic formulas and finitely many atoms. For any $\mathcal{S} \in \{K_n, T_n, S4_n\}$ and φ, we call a pointed model (M, s) of logic $\mathcal{S}$ a pointed $\mathcal{S}$-model and we write $\models_{\mathcal{S}} \varphi$ to mean that φ is valid in $\mathcal{S}$.

In order to see how honesty can be generalized from the single to the multi-agent case, let us go back and revisit the underlying principles for one agent:

1. An epistemic state is characterized by a set of *possibilities*, which for a single agent is just a set of worlds.

2. For a formula φ to be honest there must be a maximum set of possibilities where φ is known in the sense that any other epistemic state where φ is known is a subset of this maximum set.

The idea will be to come up with an appropriate notion of possibilities for many agents in the context of Kripke models and then essentially lift the definition of honesty from the single-agent case. From the discussion at the beginning of Section 2.3, it is clear that possibilities can no longer be just worlds, as other agents' beliefs also come into play when agents consider what is possible. Rather than defining possibilities in terms of a canonical model as in the previous subsection,we will derive them from arbitrary Kripke models and represent them in a uniform way.

First we need to introduce the notion of an ω-*tree*, which will serve as a canonical representation of a pointed model in the sense that any two pointed models that agree on all sentences have identical corresponding ω-trees.

An ω-tree T is an infinite sequence $\{T_0, T_1, \ldots\}$ where each T_k is a k-*tree*. A k-tree is a labelled tree of depth k defined inductively as follows: a 0-tree consists of a single node labelled by a truth assignment to the atomic propositions; a $(k+1)$-tree consists of a root labelled by a truth assignment with a (possibly empty) set of outgoing edges labelled by agents leading to the roots of distinct k-trees.

Every k-tree T_k can be turned into a corresponding Kripke model $M(T_k)$ in the obvious way: the worlds are the nodes of T_k, the accessibility relation for agent a is constructed from the edges labelled by a, and the valuation function is constructed from the truth assignments labelling the nodes of T_k.

The more interesting direction is to construct, for a given pointed model (M, s), an ω-tree $T_{M,s} = \{T_{M,s,0}, T_{M,s,1}, T_{M,s,2}, \ldots\}$ that can be said to uniquely represent the pointed model. The construction proceeds inductively roughly as follows:

1. $T_{M,s,0}$ consists of a single node labelled by the truth assignment at s.

2. The root of $T_{M,s,k+1}$ is a node r labelled by the truth assignment at world s. For each world $s' \in R_a(s)$, add an edge labelled a from the root r to the root of the k-tree $T_{M,s',k}$ for the pointed model (M, s'), avoiding duplicates.

Let the depth of a formula φ be the maximal nesting of modal operators within φ. Then we have the following:

Theorem 2.7

1. For any pointed model (M, s), the k-tree $T_{M,s,k}$ is unique for all k;
2. (M, s) and $(M(T_{M,s,k}), s)$ agree on the truth value of all formulas of depth k;
3. if (M, s) and (M', s') agree on the truth value of all formulas, then their corresponding ω-trees are identical. $\dashv$

In essence, the theorem is the justification that $T_{M,s}$ can be taken as the canonical representation of (M, s).

Given a pointed $\mathcal{S}$-model (M, s), we can now represent a's possibilities in s as $Poss_a^{\mathcal{S}}(M, s) = \{T_{M,t} \mid t \in R_a(s)\}$. A formula φ is then called $\mathcal{S}$-a-honest if there is a pointed $\mathcal{S}$-model (M, s) such that $M, s \models \varphi$, and for all pointed $\mathcal{S}$-models (M', s'), if $M', s' \models \mathrm{K}_a\varphi$ then $Poss_a^{\mathcal{S}}(M', s') \subseteq Poss_a^{\mathcal{S}}(M, s)$.

If the context is clear or not important, we will write honest instead of $\mathcal{S}$-a-honest and dishonest instead of not $\mathcal{S}$-a-honest. Note how the definition of honesty generalizes the version for single agents (Section 2.2.5) in a fairly natural way, with $Poss_a^{\mathcal{S}}(M, s)$ and $Poss_a^{\mathcal{S}}(M', s')$ taking the place of e and e'.

In the following, we will see two equivalent characterizations of $\mathcal{S}$-a-honesty, which suggest that the definition is not unreasonable.

Theorem 2.8

A formula φ is $\mathcal{S}$-a-honest iff (a) $\mathrm{K}_a\varphi$ is $\mathcal{S}$-consistent and (b) for all formulas $\psi_1, \ldots, \psi_k$, if $\models_{\mathcal{S}} \mathrm{K}_a\varphi \rightarrow (\mathrm{K}_a\psi \vee \ldots \vee \mathrm{K}_a\psi_k)$, then $\models_{\mathcal{S}} \mathrm{K}_a\varphi \rightarrow \mathrm{K}_a\psi_i$ for some i. $\dashv$

From this theorem it follows immediately that the formula $\varphi = K_a p \vee \mathrm{K}_a q$ (for atoms p and q) is dishonest in T_n and $S4_n$ because $\mathrm{K}_a\varphi \models (\mathrm{K}_a p \vee \mathrm{K}_a q)$ is valid in T_n and $S4_n$, yet neither $\mathrm{K}_a\varphi \rightarrow \mathrm{K}_a p$ nor $\mathrm{K}_a\varphi \rightarrow \mathrm{K}_a p$ are. Note that the validity of $\mathrm{K}_a\varphi \models (\mathrm{K}_a p \vee \mathrm{K}_a q)$ depends on the validity $\mathrm{K}_a\varphi \rightarrow \varphi$, which holds in both T_n and $S4_n$, but not in K_n. In fact, $K_a p \vee \mathrm{K}_a q$ is honest in K_n, indeed:

Theorem 2.9
All formulas are K_n-a-honest. ⊣

Let us call a set Γ an $\mathcal{S}$-a-belief set if $\Gamma = \{\varphi \mid M, s \models \mathrm{K}_a\varphi\}$ for some pointed $\mathcal{S}$-model (M, s).

Theorem 2.10
A formula φ is $\mathcal{S}$-a-honest iff there is an $\mathcal{S}$-a-belief set Γ containing φ that is a subset of every $\mathcal{S}$-a-belief set containing φ. ⊣

Given the notion of a-objective formulas introduced in Section 2.3.1, it is not difficult to generalize these results to the logics $K45_n$ and $KD45_n$. Roughly, instead of ω-trees we need to introduce a-objective ω-trees, where each k-tree is such that the root has no a-successors. A version of Theorem 2.8 then also holds for $\mathcal{S} \in \{K45_n, KD45_n\}$ by restricting the ψ_i to a-objective formulas. Similarly, a version of Theorem 2.10 holds by restricting the assertion about subsets to the a-objective formulas contained in the respective $\mathcal{S}$-a-belief sets. The logic $S5_n$ poses another complication because what is true in the actual world must be considered possible. Again, this can be addressed by another modification of ω-trees, but we will not go into the details here.

Finally, let us remark that deciding whether a formula is honest is PSPACE-complete for all the logics considered in this section.

2.4 Notes

The logic of only knowing reported on in Section 2.2 (except 2.2.5) is the propositional fragment of a much richer first-order epistemic logic, which first appeared in (Levesque, 1990) and later in (Levesque and Lakemeyer, 2001). The completeness proof, which uses an alternative overlapping semantics for K and N is due to Halpern and Lakemeyer (2001), and the computational complexity of single-agent only knowing was examined by Rosati (2000). Pratt-Hartmann (2000) introduced a logic of *total knowledge*, which is also based on sets of possible worlds and which agrees with O on objective formulas. Interestingly, total knowledge of an arbitrary formula always reduces to total knowledge of an objective formula. The notion of inaccessible worlds (as modelled by N) was independently investigated by Humberstone (1983) in the context of arbitrary Kripke structures.

As we saw in Section 2.2.4, this version of only knowing is closely connected to the autoepistemic logic by Moore (1985). In (Lakemeyer and Levesque, 2005, 2006) we developed variants which capture the modification of autoepistemic logic by Konolige (1988) and the default logic by Reiter (1980), respectively. The latter was further explored by Lian and

Waaler (2008). Recently, we presented yet another variant (Lakemeyer and Levesque, 2012), which lines up with a nonmonotonic logic in the style proposed by McDermott and Doyle (1980). A connection between only knowing and default reasoning with preferences was investigated by Engan, Langholm, Lian, and Waaler (2005).

The canonical-model approach to multi-agent only knowing appeared in Lakemeyer (1993). Independently, Halpern (1993) proposed an alternative semantics, resulting in a slightly different logic. As we saw, both had shortcomings, which were overcome in another proposal (Halpern and Lakemeyer, 2001) extending the canonical-model approach. An alternative approach using an extended form of Kripke structures is discussed by Waaler (2004) and by Waaler and Solhaug (2005). Recently, Belle and Lakemeyer (2010) developed a new semantics of only knowing for many agents, which is much closer to Levesque's original possible-world account and where the underlying language is first-order. Moreover, the propositional fragment of the logic was shown to be equivalent to the one by Halpern and Lakemeyer (2001).

The work on honesty began with a paper by Halpern and Moses (1984), which Section 2.2.5 is based on. The multi-agent extension discussed in Section 2.3.3 was developed by Halpern (1997). The ω-trees, which are key in defining *possibilities* for many agents in a uniform way, are closely related to knowledge structures proposed by Fagin, Halpern, and Vardi (1991). Other approaches to honesty, or, more generally, the problem of minimizing knowledge for various modal logics, can be found in (Jaspars, 1990; Schwarz and Truszczynski, 1994; van der Hoek, Jaspars, and Thijsse, 2004). The last also surveys only knowing, yet with a stronger emphasis on honesty, compared to this chapter.

Acknowledgements Gerhard Lakemeyer is also affiliated with the Department of Computer Science at the University of Toronto.

References

Belle, V. and G. Lakemeyer (2010). Multi-agent only-knowing revisited. In F. Lin, U. Sattler, and M. Truszczynski (Eds.), *KR*, pp. 49–59. AAAI Press.

Engan, I., T. Langholm, E. H. Lian, and A. Waaler (2005). Default reasoning with preference within only knowing logic. In C. Baral, G. Greco, N. Leone, and G. Terracina (Eds.), *LPNMR*, Volume 3662 of *Lecture Notes in Computer Science*, pp. 304–316. Springer.

Fagin, R., J. Y. Halpern, and M. Y. Vardi (1991). A model-theoretic analysis of knowledge. *J. ACM 38*(2), 382–428.

Halpern, J. Y. (1993). Reasoning about only knowing with many agents. In *AAAI*, pp. 655–661.

Halpern, J. Y. (1997). A theory of knowledge and ignorance for many agents. *J. Log. Comput. 7*(1), 79–108.

Halpern, J. Y. and G. Lakemeyer (2001). Multi-agent only knowing. *J. Log. Comput. 11*(1), 41–70.

Halpern, J. Y. and Y. Moses (1984). Towards a theory of knowledge and ignorance: Preliminary report. In *Proceedings of the Non-Monotonic Reasoning Workshop, New Paltz, NY 12561 October 17-19*, pp. 125–143.

van der Hoek, W., J. Jaspars, and E. Thijsse (2004). Theories of knowledge and ignorance. In S. Rahman, J. Symons, D. M. Gabbay, and J. P. van Bendegem (Eds.), *Logic, Epistemology, and the Unity of Science*, Volume 1 of *Logic, Epistemology, And The Unity Of Science*, pp. 381–418. Springer Netherlands.

Humberstone, I. L. (1983). Inaccessible worlds. *Notre Dame J. Formal Logic 24*(3), 346–352.

Jaspars, J. (1990). A generalization of stability and its application to circumscription of positive introspective knowledge. In E. Börger, H. K. Büning, M. M. Richter, and W. Schönfeld (Eds.), *CSL*, Volume 533 of *Lecture Notes in Computer Science*, pp. 289–299. Springer.

Konolige, K. (1988). On the relation between default and autoepistemic logic. *Artif. Intell. 35*(3), 343–382.

Lakemeyer, G. (1993). All they know: A study in multi-agent autoepistemic reasoning. In *IJCAI-93*.

Lakemeyer, G. and H. J. Levesque (2005). Only-knowing: Taking it beyond autoepistemic reasoning. In M. M. Veloso and S. Kambhampati (Eds.), *AAAI*, pp. 633–638. AAAI Press / The MIT Press.

Lakemeyer, G. and H. J. Levesque (2006). Towards an axiom system for default logic. In *AAAI*, pp. 263–268. AAAI Press.

Lakemeyer, G. and H. J. Levesque (2012). Only-knowing meets nonmonotonic modal logic. In G. Brewka, T. Eiter, and S. A. McIlraith (Eds.), *KR*, pp. 350–357. AAAI Press.

Levesque, H. J. (1990). All I know: a study in autoepistemic logic. *Artif. Intell. 42*(2-3), 263–309.

Levesque, H. J. and G. Lakemeyer (2001). *The Logic of Knowledge Bases.* The MIT Press.

Lian, E. H. and A. Waaler (2008). Computing default extensions by reductions on OR. In G. Brewka and J. Lang (Eds.), *KR*, pp. 496–506. AAAI Press.

McDermott, D. V. and J. Doyle (1980). Non-monotonic logic I. *Artif. Intell. 13*(1-2), 41–72.

Moore, R. C. (1985). Semantical considerations on nonmonotonic logic. *Artif. Intell. 25*(1), 75–94.

Pratt-Hartmann, I. (2000). Total knowledge. In H. A. Kautz and B. W. Porter (Eds.), *AAAI/IAAI*, pp. 423–428. AAAI Press / The MIT Press.

Reiter, R. (1980). A logic for default reasoning. *Artif. Intell. 13*(1-2), 81–132.

Rosati, R. (2000). On the decidability and complexity of reasoning about only knowing. *Artif. Intell. 116*(1-2), 193–215.

Schwarz, G. and M. Truszczynski (1994). Minimal knowledge problem: A new approach. *Artif. Intell. 67*(1), 113–141.

Waaler, A. (2004). Consistency proofs for systems of multi-agent only knowing. In R. A. Schmidt, I. Pratt-Hartmann, M. Reynolds, and H. Wansing (Eds.), *Advances in Modal Logic*, pp. 347–366. King's College Publications.

Waaler, A. and B. Solhaug (2005). Semantics for multi-agent only knowing: extended abstract. In R. van der Meyden (Ed.), *TARK*, pp. 109–125. National University of Singapore.

Chapter 3

Awareness

Burkhard C. Schipper

Contents

Abstract Unawareness refers to the lack of conception rather than the lack of information. This chapter discusses various epistemic approaches to modeling (un)awareness from computer science and economics that have been developed over the last 25 years. While the focus is on axiomatizations of structures capable of modeling knowledge and propositionally determined awareness, we also discuss structures for modeling probabilistic beliefs and awareness as well as structures for awareness of unawareness. Further topics, such as dynamic awareness, games with unawareness, the theory of decisions under unawareness, and applications are just briefly mentioned.

3.1 Introduction

Formalized notions of awareness have been studied both in computer science and economics. In computer science the original motivation was mainly the

Chapter 3 of the *Handbook of Epistemic Logic*, H. van Ditmarsch, J.Y. Halpern, W. van der Hoek and B. Kooi (eds), College Publications, 2015, pp. 77–146.

modeling of agents who suffer from different forms of logical non-omniscience. The aim was to introduce logics that are more suitable than traditional logics for modeling beliefs of humans or machines with limited reasoning capabilities. In economics the motivation is similar but perhaps less ambitious. The goal is to model agents who may not only lack information but also conception. Intuitively, there is a fundamental difference between not knowing that an event obtained and not being able to conceive of that event. Despite such a lack of conception, agents in economics are still assumed to be fully rational in the sense of not making any errors in information processing such as violating introspection of beliefs for events they can conceive. By letting unawareness stand for lack of conception, economists seem to have aimed for a narrow notion of awareness, while computer scientists are more agnostic about the appropriate notion of awareness. Economists and computer scientists seem to have slightly different tastes over formalisms too. While computer scientists are clearly inspired by Kripke structures but formalize awareness syntactically, economists seem to prefer purely event-based approaches similar to Aumann structures or Kripke frames as well as Harsanyi type-spaces. This may be due to different uses of those models. Central to economists are applications to game theory in which players in principle may use the model to reason about other players, to reason about other players' reasoning, reasoning about that etc. Consequently, states can be interpreted as "subjective" descriptions by players. In contrast, states in awareness structures by computer scientists are best understood as an outside analyst's description of agents' reasoning. They are typically not "accessible" to the agent themselves. Because of the different emphasis it is perhaps somewhat surprising that some approaches to awareness in computer science turn out to be equivalent in terms of expressivity to the approach taken in economics, especially with the focus on the notion of awareness that has been called "awareness generated by primitive propositions". At a first glance, the name of this notion suggests that it is essentially syntactic because it refers to primitive propositions or atomic formulae, thus presupposing a syntactic formalism. Yet, it also makes clear that syntactic constructions (such as the order of formulae in a conjunction of formulae) do not play a role in determining this notion of awareness. Hence, this notion should be well-suited to be captured with event-based approaches that economists have focused on. Consequently, the literature on awareness can be viewed from at least two angles. On one hand, the structures proposed by economists are equivalent to a subclass of structures proposed by computer scientists. On the other hand, the different modeling approaches pursued by economists make some of their structures directly more amenable to applications in game theory and allow us to isolate without further assumptions the effect of unawareness from other forms of logical non-omniscience. Throughout the chapter, I comment on the slightly different perspectives.

We find it useful to repeatedly refer as a backdrop to the following simple example:

Example 3.1 (Speculative trade)
There are two agents, an owner o of a firm, and a potential buyer b of the firm. The status quo value of the firm is \$100 per share. The owner of the firm is aware of a potential lawsuit that reduces the value of the firm by \$20 per share. The owner does not know whether the lawsuit will occur. The buyer is unaware of the lawsuit and the owner knows that the buyer is unaware of the lawsuit. The buyer, however, is aware of a potential innovation that increases the value of the firm by \$20 per share. The buyer does not know whether the innovation will occur. The owner is unaware of the innovation and the buyer knows that the owner is unaware of the innovation.

A question of interest to economists is whether speculative trade between the owner and buyer is possible. Speculative trade is trade purely due to differences in information/awareness. In this example, we may phrase the question as follows: Suppose that the buyer offers to purchase the firm from the owner for \$100 per share. Is the owner going to sell to her? $\dashv$

The purpose of the verbal description of the example is threefold: First, the "real-life" application should serve as a motivation to studying unawareness in multi-agent settings. We will provide an answer to what extent speculative trade is possible under unawareness in Section 3.4.2. Second, we will formalize the example using different approaches to awareness. This will allow us to illustrate the main features of those approaches and make them easily accessible with the help of simple graphs. Third, an answer to this question is relevant for showing how results on "No-speculative-trade" or "No-agreeing-to-disagree" under unawareness differ from standard results without unawareness.

3.2 Preliminary discussion

3.2.1 Some properties of awareness in natural languages

The words "aware" and "unaware" are used in many contexts with many different connotations. Sometimes "aware" is used in place of "knowing" like in the sentence "I was aware of the red traffic light." On the other hand we interpret "aware" to mean "generally taking into account", "being present in mind", "thinking about" or "paying attention to" like in the sentence "Be aware of sexually transmitted diseases!" In fact, the last sentence resonates closely with the etymology of "aware" since it has its roots in the old English "gewær" (which itself has roots in the German "gewahr") emphasizing to be "wary". In psychiatry, lack of self-awareness means that a patient is oblivious to aspects of an illness that is obvious to social contacts. This is arguably closer to "not knowing". But it also implies that the patient lacks introspection of her/his lack of knowledge of the illness. It turns out that lack of negative introspection will play a crucial role in modeling unawareness (see Section 3.2.2). In neuroscience, being aware is taken as making/having/enjoying some experience and being able to specify the content of consciousness. While the precise connotations of all those uses of awareness are different, they have in common that the agent is able to conceive something. Being unaware means then that he lacks conception of something.

Describing properties of awareness and unawareness informally with words like "knowing", "not knowing", "lack of conception", "not thinking about it" etc. does not make awareness amenable to formal analysis. We turn now to formal approaches.

3.2.2 Some properties of awareness in a formal language

One attempt to avoid ambiguities of natural languages is to use a formal language. Given a nonempty set of agents $\mathsf{Ag} = \{1, \dots, n\}$ indexed by a and a nonempty set of atomic formulae At (also called primitive propositions) as well as the special formula $\top$, the formulae φ of the language $\mathsf{L}_n^{K,A}(\mathsf{At})$ are defined by the following grammar

$$\varphi ::= p \mid \neg\varphi \mid \varphi \wedge \psi \mid K_a\varphi \mid A_a\varphi,$$

where $p \in \mathsf{At}$ and $a \in \mathsf{Ag}$. As usual, the propositional connectives $\neg$ and $\wedge$ denote negation and conjunction, respectively. The epistemic modal operators K_a and A_a are named knowledge and awareness, respectively. For instance, $A_a\varphi$ is read as "agent a is aware of φ". Atomic formulae represent propositions such as "penicillium rubens has antibiotic properties" that are not themselves formed of other propositions. As usual, disjunction $\vee$, implication $\rightarrow$, and bi-implication $\leftrightarrow$ are abbreviations, defined by $\varphi \vee \psi := \neg(\neg\varphi \wedge \neg\psi)$, $(\varphi \rightarrow \psi) := (\neg\varphi \vee \psi)$, and $(\varphi \leftrightarrow \psi) := (\varphi \rightarrow \psi) \wedge (\psi \rightarrow \varphi)$, respectively.

One immediate property of awareness is implicitly assumed with the introduction of just one modal operator A_a: there are no "degrees" of awareness. That is, one does not have statements like "agent a is more aware of φ than she is of ψ but less aware of φ than she is of χ". I believe this is justified. The notion of awareness is essentially dichotomous. An agent is either aware of φ or unaware of φ.

With some abuse of notation, let $\mathsf{At}(\varphi)$ denote the set of primitive propositions that appear in φ defined inductively as follows:

- $\mathsf{At}(\top) := \emptyset$,
- $\mathsf{At}(p) := p$, for $p \in \mathsf{At}$,
- $\mathsf{At}(\neg\varphi) := \mathsf{At}(\varphi)$,
- $\mathsf{At}(\varphi \wedge \psi) := \mathsf{At}(\varphi) \cup \mathsf{At}(\psi)$,
- $\mathsf{At}(K_a\varphi) := \mathsf{At}(\varphi) =: \mathsf{At}(A_a\varphi)$.

With this notation on hand, we can formalize the property of "awareness generated by primitive propositions":

AGPP. $A_a\varphi \leftrightarrow \bigwedge_{p\in\mathsf{At}(\varphi)} A_a p$

An agent a is aware of φ if and only if she is aware of every primitive proposition that appears in φ. This property is not completely innocent. If an agent can think about every primitive proposition $p \in \mathsf{At}(\varphi)$, can she also think about all those primitive propositions joint together in some potentially very complicated formula φ? Isn't one feature of unawareness that an agent is sometimes unable to "connect various thoughts"? This property differentiates unawareness generated by primitive propositions from other forms of logical non-omniscience.

Using this syntactic approach, we can state easily further properties of awareness that have been considered in the literature:

KA. $K_a\varphi \rightarrow A_a\varphi$ (Knowledge implies Awareness)
AS. $A_a\neg\varphi \leftrightarrow A_a\varphi$ (Symmetry)
AKR. $A_a\varphi \leftrightarrow A_aK_a\varphi$ (Awareness Knowledge Reflection)
AR. $A_a\varphi \leftrightarrow A_aA_a\varphi$ (Awareness Reflection)
AI. $A_a\varphi \rightarrow K_aA_a\varphi$ (Awareness Introspection)

KA relates awareness to knowledge. Quite naturally, knowing φ implies being aware of φ. Symmetry is natural if we take the idea of awareness of φ to mean "being able to think about φ". For example, if an agent can think about that penicillium rubens could have antibiotic properties then she can also think about that penicillium rubens does not have antibiotic properties. Symmetry makes clear that awareness is different from notions of knowledge as knowledge does not satisfy symmetry. Awareness Knowledge Reflection is also natural: if an agent can think about some particular proposition, then she can also think about her knowledge of that proposition and vice versa. Similarly, if an agent is aware of a proposition, then she knows that she is aware of this proposition (Awareness Introspection). Awareness Reflection states that an agent can reflect on her awareness.[1]

[1] We use the term "introspection" when the agent reasons about knowledge of knowledge or awareness. We use the term "reflection" when the agent reasons about awareness of knowledge or awareness.

Modica and Rustichini defined awareness in terms of knowledge by

$$A_a\varphi := K_a\varphi \vee (\neg K_a\varphi \wedge K_a\neg K_a\varphi), \quad \text{(MR)}$$

which in propositional logic is equivalent to

$$A_a\varphi = K_a\varphi \vee K_a\neg K_a\varphi,$$

a definition of awareness used frequently in the literature especially in economics. The MR definition of awareness and some of the properties discussed above make use of the knowledge modality. This just begs the question about the properties of knowledge. One property often assumed is

5. $\neg K_a\varphi \rightarrow K_a\neg K_a\varphi$ (Negative Introspection)

This property is implicitly used (together with others) in most economic applications to model agents who are free from "mistakes in information processing." It is immediate that the MR definition of awareness is equivalent to Negative Introspection. Thus, if Negative Introspection is a valid formula, then awareness as defined by MR must be trivial in the sense that the agent is aware of every formula. The discussion then begs the questions about which formulae are valid. Answers to this question are given with various structures discussed in sequel.

3.3 Awareness and knowledge

3.3.1 Awareness structures

Awareness structures are the first structures in the literature for modeling awareness. One way to model awareness is to augment Kripke structures with a syntactic awareness correspondence. More generally, this turns out to be a rather flexible approach for modeling logical non-omniscience. It has its roots in the logic of implicit and explicit beliefs. That's why also an epistemic modal operator L_a interpreted as "implicit knowledge" is considered. We denote the resulting language $\mathsf{L}_n^{L,K,A}(\mathsf{At})$.

An *awareness structure* is a tuple $M = (S, (R_a)_{a\in\mathsf{Ag}}, (\mathcal{A}_a)_{a\in\mathsf{Ag}}, V)$ where $(S, (R_a)_{a\in\mathsf{Ag}}, V)$ is a Kripke structure and $\mathcal{A}_a : \Omega \longrightarrow 2^{\mathsf{L}_n^{L,K,A}(\mathsf{At})}$ is agent a's awareness correspondence[2] that associates with each state $s \in S$ the set of formulae $\mathcal{A}_a(s) \subseteq \mathsf{L}_n^{L,K,A}(\mathsf{At})$ of which agent a is aware at state s. A Kripke structure $(S, (R_a)_{a\in\mathsf{Ag}}, V)$ consists of a nonempty set of states S, and for each agent $a \in \mathsf{Ag}$ a binary relation $R_a \subseteq S \times S$, the accessibility relation.[3] Intuitively, $(s, t) \in R_a$ is interpreted as 'at state s agent a considers the

[2]The name "correspondence" refers to a set-valued function.

[3]See Chapter 1 of this handbook.

state t possible'. Of particular interest is the case when R_a is an equivalence relation, i.e., when R_a is a relation that is reflexive (i.e., $(s, s) \in R_a$, for all $s \in S$), transitive (i.e., for all $s_1, s_2, s_3 \in S$, if $(s_1, s_2) \in R_a$, and $(s_2, s_3) \in R_a$, then $(s_1, s_3) \in R_a$) and Euclidean (i.e., for all $s_1, s_2, s_3 \in S$, if $(s_1, s_2) \in R_a$ and $(s_1, s_3) \in R_a$, then $(s_2, s_3) \in R_a$). In this case (and only in this case), R_a forms a partition of the state space S, i.e., a collection of disjoint subsets such that the union covers S. We call an awareness structure in which each agent's accessibility relation forms a partition a *partitional awareness structure*. Our exposition focuses on partitional awareness structures.[4] The valuation function $V : S \times \mathsf{At} \longrightarrow \{true, false\}$ assigns to each state and atomic formula a truth value.

The awareness correspondence offers a very flexible approach to modeling various notions of awareness, some of which have yet to be explored. Restrictions depend on which interpretation of awareness is desired in applications. For instance, economists became interested in framing affects including presentation order effects. Presentation order effects may be relevant when information is acquired with the help of online search engines, which typically present lists of search results. Rather than going through all of them, we usually stop when we found a satisfactory result. There may be search results further down the list of which we remain unaware but which we would find much more relevant if we were aware of them. Our awareness of search results depends crucially on the order in which they are presented and on our search aim. If we consider lists as conjunctions of propositions in which the order matters (i.e., for which we do not assume commutativity), then we may be able to model presentation order effects with awareness correspondences because at some state $s \in S$ we may have $\varphi \wedge \psi \in \mathcal{A}_a(s)$ but $\psi \wedge \varphi \notin \mathcal{A}_a(s)$ while in another state $s' \neq s$, $\psi \wedge \varphi \in \mathcal{A}_a(s')$. Such "realistic" approaches have not been explored in the awareness literature. But especially for such kind of applications, the syntactic awareness correspondence may have an advantage over event-based approaches that we will introduce later.

Note that by definition, the awareness correspondence imposes a dichotomous notion of awareness because at a state s a formula can be either in $\mathcal{A}_a(s)$ or not in $\mathcal{A}_a(s)$. Thus, at state s, agent a is either aware or unaware

[4]The aim of our exposition is not necessarily to present the most general setting. Nor do we believe that partitional information structures describe best how humans reason in real life. Rather, partitional information structures serve as a "rational" benchmark. It allows us to demonstrate that one can model both, a "strong" notion of knowledge *and* unawareness. Thus, we can isolate unawareness from errors of information processing associated with lack of introspection. Moreover, as it turns out, the interpretation of propositionally generated unawareness as "lack of conception" is most transparent in partitional awareness structures.

of that formula.

Given how the literature has developed so far, two restrictions on the awareness correspondence that are jointly called *propositionally determined awareness* are of particular interest:

- *Awareness is generated by primitive propositions* if for all $a \in \mathsf{Ag}$ and $s \in S$, $\varphi \in \mathcal{A}_a(s)$ if and only if $\mathsf{At}(\varphi) \subseteq \mathcal{A}_a(s)$.
- *Agents know what they are aware of* if for all $a \in \mathsf{Ag}$ and $s, s' \in S$, $(s, s') \in R_a$ implies $\mathcal{A}_a(s') = \mathcal{A}_a(s)$.

A satisfaction relation specifies for each awareness structure and state which formulae are true. We denote by $M, s \models \varphi$ that φ is true (or satisfied) at state s in the awareness structure M, and define inductively on the structure of formulae in $\mathsf{L}_n^{L,K,A}(\mathsf{At})$,

$M, s \models \top$, for all $s \in S$,
$M, s \models p$, for $p \in \mathsf{At}$ if and only if $V(s, p) = true$,
$M, s \models \varphi \wedge \psi$ if and only if both $M, s \models \varphi$ and $M, s \models \psi$,
$M, s \models \neg\varphi$ if and only if $M, s \not\models \varphi$,
$M, s \models L_a\varphi$ if and only if $M, t \models \varphi$ for all $t \in S$ such that $(s, t) \in R_a$,
$M, s \models A_a\varphi$ if and only if $\varphi \in \mathcal{A}_a(s)$,
$M, s \models K_a\varphi$ if and only if $M, s \models A_a\varphi$ and $M, s \models L_a\varphi$.

The satisfaction relation is standard except for awareness ($A_a\varphi$), which is new, and for knowledge ($K_a\varphi$), which is defined by $L_a\varphi \wedge A_a\varphi$. In the presence of an implicit knowledge modality L_a, the knowledge modality K_a is called explicit knowledge. Agent a explicitly knows φ if she is aware of φ and implicitly knows φ.

At this point, it may be helpful to illustrate awareness structures with our example.

Example 3.2 (Speculative trade (continued))
Denote by ℓ the atomic formula "the lawsuit is brought against the firm" and by n "the novel use of the firm's product is discovered". Figure 3.1 depicts an awareness structure that models the speculative trade example from the Introduction.[5] There are four states. For simplicity, we name each state by the atomic formulae that are true or false at that state. For instance the upper right state $(n, \neg\ell)$, n is true and ℓ is false. The awareness correspondences are indicated by clouds, one for each player. For each state, the blue solid cloud represents the awareness set of the owner while the red intermitted cloud represents the awareness set of the potential buyer. For

[5]We choose "clouds" to depict the awareness sets so as to suggest the interpretation of "thinking about".

Figure 3.1: An Awareness Structure for the Speculative Trade Example

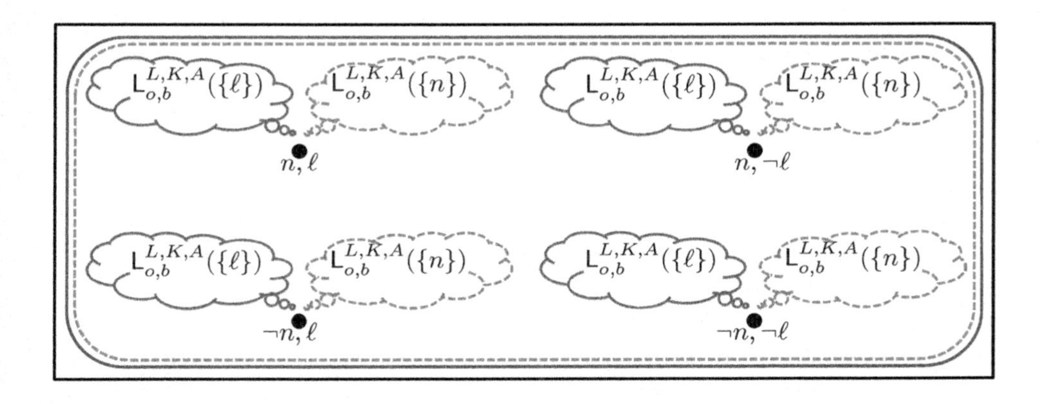

graphical simplicity, we represent the accessibility relations of agents by possibility sets rather than arrows, a practice common in game theory. The blue solid-lined possibility set belongs to the owner while the one with the red intermitted line is the buyer's. Each agent's information is trivial as neither can distinguish between any states.

This simple figure models the story that we outlined in the Introduction. In any state, the owner is unaware of the potential innovation but aware of the lawsuit because his awareness set never contains formulae involving n but only formulae involving ℓ. Similarly, at any state the buyer is unaware of the potential lawsuit but aware of the innovation because his awareness set never contains formulae involving ℓ but only formulae involving n. The accessibility relations show us that the owner does not know whether the lawsuit obtains because he cannot distinguish between states in which the lawsuit obtains and states where it doesn't. Analogously, the buyer does not know whether the innovation obtains because she cannot distinguish between states in which the innovation obtains from states where it doesn't obtain. But the owner knows implicitly that the buyer is unaware of the lawsuit because at every state of his possibility set, the buyer's awareness set does not contain formulae involving ℓ. Moreover, he also explicitly knows that the buyer is unaware of the lawsuit because he implicitly knows it and his own awareness set contains formulae involving ℓ. (An analogous statement holds for the buyer n.)

This example is very special because each agent's accessibility relation is trivial and each agent's awareness correspondence is constant across states. Nevertheless, the example illustrates some particular features and properties of awareness structures. One thing to note is that the accessibility relation models implicit knowledge and not necessarily explicit knowledge. For instance, at every state the owner implicitly knows that he is unaware of the innovation although he does not explicitly know it because his awareness set never contains formulae that involve n. This is actually a general property (see axiom UI_L below), which we find hard to interpret. ⊣

The discussion of the example just begs the question about what are the general properties of awareness and knowledge in awareness structures? What are *all* properties of awareness and knowledge in awareness structures? To answer these questions, we characterize the properties of awareness and knowledge in terms of formulae that are valid in awareness structures. To set the stage for such an axiomatization, we need to introduce the following standard notions (see Chapter 1 of this handbook): An *axiom* is a formula assumed. An *inference rule* infers a formula (i.e., a conclusion) from a collection of formulae (i.e., the hypothesis). An *axiom system* consists of a collection of axioms and inferences rules. A *proof* in an axiom system consists of a sequence of formulae, where each formula is either an axiom in the axiom system or follows by an application of an inference rule. A proof is a proof of a formula φ if the last formula in the proof is φ. A formula φ is *provable* in an axiom system if there is a proof of φ in the axiom system. The set of *theorems* of an axiom system is the smallest set of formulae that contains all axioms and that is closed under inference rules of the axiom system.

Given a class $\mathcal{M}^{FH}$ of awareness structures, a formula φ is *valid* in $\mathcal{M}^{FH}$ if $M, s \models \varphi$ for every awareness structure $M \in \mathcal{M}^{FH}$ and state s in M. An axiom system is said to be *sound* for a language L with respect to a class $\mathcal{M}^{FH}$ of awareness structures if every formula in L that is provable in the axiom system is valid with respect to every awareness structure in $\mathcal{M}^{FH}$. An axiom system is *complete* for a language L with respect to a class of awareness structures $\mathcal{M}^{FH}$ if every formula in L that is valid in $\mathcal{M}^{FH}$ is provable in the axiom system. A sound and complete axiomatization for a class of awareness structures characterizes these awareness structures in terms of properties of knowledge and awareness as codified in the axiom system.

Consider the following axiom system:[6]

[6]Some of the axioms involving awareness have been introduced in the literature under various different names. Here we attempt to assign them intuitive acronyms. "R" stands for "reflection" and "I" for introspection.

Prop. All substitutions instances of tautologies of propositional logic, including the formula $\top$.

KL. $K_a\varphi \leftrightarrow L_a\varphi \wedge A_a\varphi$ (Explicit Knowledge is Implicit Knowledge and Awareness)

AS. $A_a\neg\varphi \leftrightarrow A_a\varphi$ (Symmetry)

AC. $A_a(\varphi \wedge \psi) \leftrightarrow A_a\varphi \wedge A_a\psi$ (Awareness Conjunction)

AKR. $A_a\varphi \leftrightarrow A_aK_a\varphi$ (Awareness Explicit Knowledge Reflection)

ALR. $A_a\varphi \leftrightarrow A_aL_a\varphi$ (Awareness Implicit Knowledge Reflection)

AR. $A_a\varphi \leftrightarrow A_aA_a\varphi$ (Awareness Reflection)

AI_L. $A_a\varphi \rightarrow L_aA_a\varphi$ (Awareness Introspection)

UI_L. $\neg A_a\varphi \rightarrow L_a\neg A_a\varphi$ ()

K. $(L_a\varphi \wedge L_a(\varphi \rightarrow \psi)) \rightarrow L_a\psi$ (Distribution Axiom)

T. $L_a\varphi \rightarrow \varphi$ (Implicit Knowledge Truth Axiom)

4. $L_a\varphi \rightarrow L_aL_a\varphi$ (Implicit Positive Introspection Axiom)

5. $\neg L_a\varphi \rightarrow L_a\neg L_a\varphi$ (Implicit Negative Introspection Axiom)

MP. From φ and $\varphi \rightarrow \psi$ infer ψ (modus ponens)

Gen. From φ infer $L_a\varphi$ (Implicit Knowledge Generalization)

Note that each of the axioms and inference rules is an instance of a *scheme*; it defines an infinite collection of axioms (inference rules, respectively), one for each choice of formulae.

Axioms AS, AKR, and AR were motivated in Section 3.2.2. Awareness Conjunction (AC) has a similar flavor as the property "awareness generated by primitive propositions" (AGPP) introduced in Section 3.2.2. Axioms ALR and AI_L are similar to axioms AKR and AI, respectively, but with explicit knowledge replaced by implicit knowledge. Axioms and inferences rules Prop., K, T, 4, 5, MP, and Gen. together make up the well-known axiom system $\mathsf{S5}$ but are stated here with implicit knowledge modalities. Axiom KL links explicit knowledge and implicit knowledge via awareness. Explicit knowledge is implicit knowledge and awareness. This resonates well with the interpretation of awareness as "being present in mind". Explicit knowledge, i.e., knowledge that one is aware of, is knowledge that is "present in mind". The notion of knowledge usually considered in economics corresponds to explicit knowledge despite the fact that standard properties of knowledge are now imposed on implicit knowledge! The axiom Unawareness Introspection (UI_L) is hard to interpret since it has no analog in which implicit knowledge is replaced by explicit knowledge. We denote the above axiom system by $\mathsf{S5}_n^{L,K,A}$ because it is analogous to $\mathsf{S5}$.

Theorem 3.1
For the language $\mathsf{L}_n^{L,K,A}(\mathsf{At})$, the axiom system $\mathsf{S5}_n^{L,K,A}$ is a sound and complete axiomatization with respect to partitional awareness structures in which awareness is determined by propositions. $\dashv$

The theorem is proved by modifying the proof for the well-known result that S5 is a sound and complete axiomatization with respect to partitional Kripke structures for the language $\mathsf{L}^L(\mathsf{At})$.

While we focus our exposition on the strong notion of knowledge as encapsulated in axiom systems analogous to S5, the literature considers weaker notions of knowledge in non-partitional awareness structures. There are also versions of awareness structures in which the Kripke relations are replaced by a neighborhood correspondences $\mathcal{N}_a : S \longrightarrow 2^{2^S}$ in the spirit of neighborhood semantics. Intuitively, $\mathcal{N}_a(s)$ is the list of events that agent a knows at state s. The satisfaction relation for the case of implicit knowledge of a formula is then modified accordingly to $M, s \models L_a\varphi$ if and only if $\{t \in S : M, t \models \varphi\} \in \mathcal{N}_a(s)$. Modelers interested in studying various forms of logical non-omniscience will welcome the additional flexibility that awareness neighborhood structures provide over awareness (Kripke) structures. But this additional generality is less helpful when being interested in isolating the effect of unawareness per se in the presence of a strong notion of knowledge as the one encapsulated in S5.[7]

The axiomatization of Theorem 3.1 is somewhat dissatisfactory as most of the properties are stated in terms of implicit knowledge, a notion that we find very hard to interpret in the context of propositionally determined awareness. (In fact, since explicit knowledge is defined in terms of implicit knowledge and awareness, the expressivity of language $\mathsf{L}_n^{L,A}(\mathsf{At})$ is equal to $\mathsf{L}_n^{L,K,A}(\mathsf{At})$.) Implicit knowledge that is not explicit knowledge is as if Isaac Newton would say "I know the theory of relativity but unfortunately I am not aware of it". In economics, we are only interested in knowledge that the agent is aware of, that can guide her decisions, and that in principle could be tested with choice experiments. While an outsider may be able to reason about the implicit knowledge of an agent, it is hard to see how the agent herself could reason about her implicit knowledge that is not explicit knowledge as well. Some authors in the awareness literature interpret implicit knowledge as "knowledge that the agent would have if she were aware of it". But this interpretation is flawed because if she really becomes aware of it, then maybe her explicit knowledge would not correspond anymore to her earlier implicit knowledge because her state of mind would have changed from the one she was in when she had this implicit knowledge and was unaware.

[7]It should be feasible to model awareness using a hybrid of a Kripke structure and a neighborhood structure where, in place of a syntactic awareness correspondence of awareness structures, the neighborhood correspondence lists for each state the set of events that the agent is aware of at that state while knowledge continues to be modeled by the accessibility relation as in Kripke structures. To our knowledge, nobody has explored such a syntax-free approach.

Fortunately, it is possible to axiomatize awareness structures without an implicit knowledge modality using the language $\mathsf{L}_n^{K,A}(\mathsf{At})$. The following axiom system that we may denote by $\mathsf{S5}_n^{K,A}$ has been in part already motivated in Section 3.2.2:

Prop. All substitutions instances of tautologies of propositional logic, including the formula $\top$.
KA. $K_a\varphi \to A_a\varphi$ (Knowledge implies Awareness)
AS. $A_a\neg\varphi \leftrightarrow A_a\varphi$ (Symmetry)
AC. $A_a(\varphi \wedge \psi) \leftrightarrow A_a\varphi \wedge A_a\psi$ (Awareness Conjunction)
AKR. $A_a\varphi \leftrightarrow A_aK_a\varphi$ (Awareness Knowledge Reflection)
AR. $A_a\varphi \leftrightarrow A_aA_a\varphi$ (Awareness Reflection)
AI. $A_a\varphi \to K_aA_a\varphi$ (Awareness Introspection)
K. $(K_a\varphi \wedge K_a(\varphi \to \psi)) \to K_a\psi$ (Distribution Axiom)
T. $K_a\varphi \to \varphi$ (Axiom of Truth)
4. $K_a\varphi \to K_aK_a\varphi$ (Positive Introspection Axiom)
5_A. $\neg K_a\varphi \wedge A_a\varphi \to K_a\neg K_a\varphi$ (Weak Negative Introspection Axiom)
MP. From φ and $\varphi \to \psi$ infer ψ (modus ponens)
Gen_A. From φ and $A_a\varphi$ infer $K_a\varphi$ (Modified Knowledge Generalization)

Note that we now require awareness for the formalization of the negative introspection axiom 5_A and knowledge generalization Gen_A. This is quite intuitive. If an agent does not know φ and is also not aware of φ, how could she know that she doesn't know φ? Moreover, how could she infer knowledge of φ from φ if she isn't aware of φ?

The definition of the satisfaction relation must be modified with respect to knowledge as follows:

> $M, s \models K_a\varphi$ if and only if $M, s \models A_a\varphi$ and $M, t \models \varphi$ for all $t \in S$ such that $(s,t) \in R_a$.

Theorem 3.2
For the language $\mathsf{L}_n^{K,A}(\mathsf{At})$, the axiom system $\mathsf{S5}_n^{K,A}$ is a sound and complete axiomatization with respect to partitional awareness structures in which awareness is determined by propositions. $\dashv$

Axiomatizing awareness structures for the language $\mathsf{L}_n^{K,A}(\mathsf{At})$ does not completely avoid the issue of implicit knowledge because implicit knowledge and not explicit knowledge is represented by the relation R_a. This is quite in contrast to the interpretation of information partitions in economics and game theory despite the fact that R_a is here assumed to be partitional. Note further that the issue with implicit knowledge is potentially much less severe if we are not interested in propositionally determined awareness. With other forms of logical non-omniscience, implicit knowledge may be to a certain extent present in the mind and the objections raised here against the notion of implicit knowledge may be misguided when different notions of awareness are considered.

Recall from Section 3.2.2 the MR definition of awareness in terms knowledge, $A_a\varphi = K_a\varphi \vee K_a\neg K_a\varphi$.

Lemma 3.1
For any partitional awareness structure in which awareness is generated by primitive propositions, the formula $A_a\varphi \leftrightarrow K_a\varphi \vee K_a\neg K_a\varphi$ is valid. ⊣

The proof relies mainly on weak negative introspection and awareness generated by primitive propositions.

That is, in the class of partitional awareness structures we can define awareness in terms of knowledge and the expressive power of the language $\mathsf{L}_n^{K,A}(\mathsf{At})$ is equivalent to $\mathsf{L}_n^{K}(\mathsf{At})$. Note that the lemma does not hold in general for awareness structures which are not necessarily partitional.

In conclusion, we like to emphasize that the strength of awareness structures is their flexibility. Potentially they can be used to model many interesting notions of awareness and logical non-omniscience. But this flexibility comes also at a cost. Because of the syntactic awareness correspondence, the semantics of awareness structures is not completely syntax free. All applications that appeared so far in the literature avoided specifying any syntax and just work with states instead. Generally, modeling approaches in economics using syntax are extremely rare. Another issue is that awareness structures model reasoning about awareness and knowledge of agents from an outside analyst's point of view. While an outsider can easily use an awareness structure to analyze agents' interactive reasoning, it is hard to imagine that agents' themselves could also use the awareness structure to analyze their reasoning. For instance, if the buyer in the speculative trade example is presented with the awareness structure in Figure 3.1, then presumably she must become aware of the lawsuit. Writing down an awareness structure from the buyer's point of view, would require us to erase everything involving ℓ (and analogously for the owner). The states in an awareness structure are "objective" descriptions of situations in the eyes of an analyst but not necessarily the agents themselves. This may be problematic in game theoretic models where we are interested in modeling strategic situations from each player's perspective. In the following sections we will introduce structures for modeling awareness that feature "subjective" states, have a syntax-free semantics, and in some cases avoid the notion of implicit knowledge altogether.

3.3.2 Impossibility of unawareness in Kripke frames

We have seen that awareness structures are not syntax-free since they involve a syntactic awareness correspondence. This has been criticized early on. Others conjectured the impossibility of a purely semantic approach to awareness. Why would it be difficult to devise a syntax-free semantics for modeling awareness? Suppose we would "erase" all syntactic components of awareness structures; could we still model non-trivial awareness? When erasing each agent's awareness correspondence and the valuation from awareness structures, we are left with what is known as an Aumann structure or Kripke frame.

Let S be a nonempty space of states and consider a set of events which we may take for simplicity to be the set of all subsets 2^S. A natural occurrence like "penicillium rubens has antibiotic properties" is represented by an event, which is simply a subset of states $E \in 2^S$. That is, E is the set of states in which "penicillium rubens has antibiotic properties". Instead modal operators on propositions, we model epistemic notions of knowledge and unawareness by operators on events. The knowledge operator of agent a is denoted by $\mathbf{K}_a : 2^S \longrightarrow 2^S$. For the event $E \in 2^S$, the set $\mathbf{K}_a(E)$ represents the event that agent a knows the event E. Yet, to interpret the operator $\mathbf{K}_a$ as knowledge, we should impose properties that reasonable notions knowledge should satisfy. Here we just require the knowledge operator to satisfy one extremely basic property: Agent a always knows the state space, i.e.,

$$\mathbf{K}_a(S) = S \text{ (Necessitation)}.$$

Note that all notions of knowledge or belief used in economics, game theory or decision theory satisfy this property, including the knowledge operator defined from an accessibility relation[8] or possibility correspondence, the probability p-belief operator[9] as well as belief operators for ambiguous beliefs.[10] Yet, this property is not as innocent as it may appear at the first

[8] I.e., $\mathbf{K}_a(E) := \{s \in S : (s, s') \in R_a \text{ implies } s' \in E\}$.

[9] Let $t_a : S \longrightarrow \Delta(S)$ be a type mapping that assigns to each state in s a probability measure on S, where $\Delta(S)$ denotes the set of all probability measures on S. For $p \in [0, 1]$, the probability-of-at-least-p-belief operator is defined on events by $\mathbf{B}_a^p(E) := \{s \in S : t_a(\omega)(E) \geq p\}$.

[10] Let $C(S)$ denote the set of capacities on S, i.e., the set of set functions $\nu : 2^S \longrightarrow [0, 1]$ satisfying monotonicity (for all $E, F \subseteq S$, if $E \subseteq F$ then $\nu(E) \leq \nu(F)$) and normalization ($\nu(\emptyset) = 0$, $\nu(S) = 1$). Capacities are like probability "measures" except that they do not necessarily satisfy additivity. Capacities have been used extensively in decision theory to model Knightian uncertainty, ambiguous beliefs, or lack of confidence in one's probabilistic beliefs. Typically, Knightian uncertainty is distinguished from risk in economics. Risk refers to situations in which the agent reasons probabilistically while Knightian uncertainty refer to a situation in which the agent is unable to form probability

glance. The state space is the universal event; it always obtains. We may interpret S as a tautology. Thus, necessitation can be interpreted as knowing tautologies. Such a basic property may be violated by agents who lack logical omniscience and face potentially very complicated tautologies. For instance, mathematicians work hard to "discover new" theorems. They obviously don't know all theorems beforehand. There may be at least two reasons for why it is hard to know all tautologies. It may be due to logical non-omniscience in the sense that the agent does not realize all implications of her knowledge. Alternatively, it may be because the agent is unaware of some concepts referred to in the tautology.

To see how Necessitation may conflict with lack of awareness, we need a formal notion of unawareness in this purely event-based setting. Denote the unawareness operator of agent a by $\mathbf{U}_a : 2^S \longrightarrow 2^S$. For the event $E \in 2^S$, the set $\mathbf{U}_a(E)$ represents the event that agent a is unaware of the event E. In our verbal discussion of the use of the term awareness in psychiatry in Section 3.2.1, we noted already that lack of awareness may imply lack of negative introspection. Formally we can state

$$\mathbf{U}_a(E) \subseteq \neg\mathbf{K}_a(E) \cap \neg\mathbf{K}_a\neg\mathbf{K}_a(E) \text{ (Plausibility)}$$

This property is implied by the Modica-Rustichini definition of awareness discussed earlier. Being unaware of an event implies not knowing the event and not knowing that you don't know the event. The negation of an event is here defined by the relative complement of that event with respect to S, i.e., $\neg E := S \setminus E$ is the event that the event E does not occur. Conjunction of events is given by the intersection of events. Thus, $E \cap F$ denotes the event that the event E and the event F occurs. Implication of events is given by the subset relation; $E \subseteq F$ denotes that the event E implies the event F.

The next property states that an agent lacks positive knowledge of her unawareness. That is, she never knows that she is unaware of an event.

$$\mathbf{K}_a\mathbf{U}_a(E) \subseteq \emptyset \text{ (KU Introspection)}$$

While we may know that, in principle, there could exist some events that we are unaware of (an issue we will turn to in Section 3.5), we cannot know that we are unaware of a specific event E. In the same vein we may also require that if an agent is aware that she is unaware of an event, then she should be aware of the event. Stated in the contrapositive,

$$\mathbf{U}_a(E) \subseteq \mathbf{U}_a\mathbf{U}_a(E) \text{ (AU Reflection).}$$

judgements. Let $t_a : S \longrightarrow C(S)$ be a type mapping that assigns to each state a capacity on S. The capacity-of-at-least-p-belief operator is now defined analogously to the probability-of-at-least-p-belief operator.

In order to interpret $\mathbf{K}_a$ as knowledge we should certainly impose further properties but we can already show a very simple but conceptually important impossibility result according to which the above notion of unawareness is inconsistent with any notion of knowledge satisfying Necessitation. Since we did not even assume that the knowledge operator is derived from an accessibility relation (or a possibility correspondence), Theorem 3.3 below will apply more generally to any state-space model with knowledge satisfying necessitation, and not just to Kripke frames or Aumann structures.

Theorem 3.3
If a state-space model satisfies Plausibility, KU-introspection, AU-reflection, and Necessitation, then $\mathbf{U}_a(E) = \emptyset$, for any event $E \in 2^S$. ⊣

Proof. $\mathbf{U}_a(E) \overset{AU-Refl.}{\subseteq} \mathbf{U}_a(\mathbf{U}_a(E)) \overset{Plaus.}{\subseteq} \neg\mathbf{K}_a(\neg\mathbf{K}_a(\mathbf{U}_a(E))) \overset{KU-Intro.}{=} \neg\mathbf{K}_a(S) \overset{Nec.}{=} \emptyset$. □

This shows that the ("standard") state-space approach is incapable of modeling unawareness. Thus, we need more structure for modeling nontrivial unawareness than what Kripke structures have to offer.

Our brief discussion of Necessitation already suggests that more careful descriptions of states (i.e., syntactic approaches) are useful for modeling awareness. Tautologies are descriptions that are true in every state. Knowing tautologies seems to imply that at every state the agent is able to reason with a language that is as expressive as the most complicated tautology. But if she can use this rich language at every state, then she should be able to describe and reason about any event expressible in this language and thus it may not come as a surprise that she must be aware of all events. The syntactic approach is nicely fine-grained as the "internal structure" of states can be made explicit. This allows us to write formally properties like "awareness generated by primitive propositions" (AGPP), $A_a\varphi \leftrightarrow \bigwedge_{p\in\mathsf{At}(\varphi)} A_a p$, a property that we discussed already in Section 3.2.2.

It may be worthwhile to ask how awareness structures circumvent impossibility results like the one discussed in this section. The literature identified two assumptions that are implicitly satisfied in every event-based approach like Aumann structures or Kripke frames. One of them is *event-sufficiency.* It says that if two formulae are true in exactly the same subset of states, then (1) the subset of states in which the agent knows one formula must coincide with the subset of states in which the agent knows the other formula and (2) the subset of states in which the agent is unaware of one formula must coincide with the subset of states in which the agent is unaware of the other formula. Clearly, awareness structures do not satisfy event-sufficiency since two formulae may be true exactly at the same subset of states but the awareness correspondence may be such that the agent is aware of one but not the other in some states. It is the awareness correspondence that allow awareness structures to overcome the impossibility result.

3.3.3 Unawareness frames

Inspired by Aumann structures, we now introduce an event-based approach to unawareness, that is, a syntax-free semantics for multi-agent unawareness. To circumvent the impossibility results of Theorem 3.3, we work with a lattice of state spaces rather than a single state-space.

Let $\mathcal{S} = (\{S_\alpha\}_{\alpha \in \mathcal{A}}, \succeq)$ be a complete lattice of disjoint *state-spaces*, with the partial order $\succeq$ on $\mathcal{S}$.[11] A complete lattice is a partially ordered set in which each subset has a least upper bound (i.e., supremum) and a greatest lower bound (i.e., infimum). If S_α and S_β are such that $S_\alpha \succeq S_\beta$ we say that "S_α is more expressive than S_β." Intuitively, states of S_α "describe situations with a richer vocabulary" than states of S_β". Denote by $\Omega = \bigcup_{\alpha \in \mathcal{A}} S_\alpha$ the union of these spaces. This is by definition of $\{S_\alpha\}_{\alpha \in \mathcal{A}}$ a disjoint union.

For every S and S' such that $S' \succeq S$, there is a surjective projection $r_S^{S'} : S' \longrightarrow S$, where r_S^S is the identity.[12]("$r_S^{S'}(\omega)$ is the restriction of the description ω to the more limited vocabulary of S.") Note that the cardinality of S is smaller than or equal to the cardinality of S'. We require the projections to commute: If $S'' \succeq S' \succeq S$ then $r_S^{S''} = r_S^{S'} \circ r_{S'}^{S''}$. If $\omega \in S'$, denote $\omega_S = r_S^{S'}(\omega)$. If $D \subseteq S'$, denote $D_S = \{\omega_S : \omega \in D\}$. Intuitively, projections "translate" states from "more expressive" spaces to states in "less expressive" spaces by "erasing" facts that can not be expressed in a lower space.

For $D \subseteq S$, denote $D^\uparrow = \bigcup_{S' \in \{S' : S' \succeq S\}} \left(r_S^{S'}\right)^{-1}(D)$. ("All the extensions of descriptions in D to at least as expressive vocabularies.") This is the union of inverse images of D in weakly higher spaces.

An *event* is a pair (E, S), where $E = D^\uparrow$ with $D \subseteq S$, where $S \in \mathcal{S}$. D is called the *base* and S the *base-space* of (E, S), denoted by $S(E)$. If $E \neq \emptyset$, then S is uniquely determined by E and, abusing notation, we write E for (E, S). Otherwise, we write $\emptyset^S$ for $(\emptyset, S)$. Note that not every subset of Ω is an event. Intuitively, some fact may obtain in a subset of a space. Then this fact should be also "expressible" in "more expressive" spaces. Therefore the event contains not only the particular subset but also its inverse images in "more expressive" spaces.

---

[11]Recall that a binary relation is a partial order if it is reflexive, antisymmetric, and transitive.

[12]Recall that a function $f : X \longrightarrow Y$ is surjective (or called onto) if for every $y \in Y$ there is some $x \in X$ such that $f(x) = y$.

Let Σ be the set of *events* of Ω, i.e., sets $D^{\uparrow}$ such that $D \subseteq S$, for some state space $S \in \mathcal{S}$. Note that unless $\mathcal{S}$ is a singleton, Σ is not an algebra because it contains distinct $\emptyset^S$ for all $S \in \mathcal{S}$. The event $\emptyset^S$ should be interpreted as a "logical contradiction phrased with the expressive power available in S". It is quite natural to have distinct vacuous events since "contradictions can be phrased with differing expressive powers".

If $(D^{\uparrow}, S)$ is an event where $D \subseteq S$, the negation $\neg(D^{\uparrow}, S)$ of $(D^{\uparrow}, S)$ is defined by $\neg(D^{\uparrow}, S) := ((S \setminus D)^{\uparrow}, S)$. Note that, by this definition, the negation of an event is an event. Abusing notation, we write $\neg D^{\uparrow} := \neg(D^{\uparrow}, S)$. By our notational convention, we have $\neg S^{\uparrow} = \emptyset^S$ and $\neg\emptyset^S = S^{\uparrow}$, for each space $S \in \mathcal{S}$. $\neg D^{\uparrow}$ is typically a proper subset of the complement $\Omega \setminus D^{\uparrow}$, that is, $(S \setminus D)^{\uparrow} \subsetneqq \Omega \setminus D^{\uparrow}$. Intuitively, there may be states in which the description of an event $D^{\uparrow}$ is both expressible and valid – these are the states in $D^{\uparrow}$; there may be states in which its description is expressible but invalid – these are the states in $\neg D^{\uparrow}$; and there may be states in which neither its description nor its negation are expressible – these are the states in $\Omega \setminus \left(D^{\uparrow} \cup \neg D^{\uparrow}\right) = \Omega \setminus S\left(D^{\uparrow}\right)^{\uparrow}$. Thus unawareness structures are not standard state-space models because the definition of negation prevents them from satisfying what are called *real states*.

If $\left\{\left(D_\lambda^{\uparrow}, S_\lambda\right)\right\}_{\lambda \in L}$ is a collection of events (with $D_\lambda \subseteq S_\lambda$, for $\lambda \in L$), their conjunction $\bigwedge_{\lambda \in L}\left(D_\lambda^{\uparrow}, S_\lambda\right)$ is defined by $\bigwedge_{\lambda \in L}\left(D_\lambda^{\uparrow}, S_\lambda\right) := \left(\left(\bigcap_{\lambda \in L} D_\lambda^{\uparrow}\right), \sup_{\lambda \in L} S_\lambda\right)$. Note, that since $\mathcal{S}$ is a *complete* lattice, $\sup_{\lambda \in L} S_\lambda$ exists. If $S = \sup_{\lambda \in L} S_\lambda$, then we have $\left(\bigcap_{\lambda \in L} D_\lambda^{\uparrow}\right) = \left(\bigcap_{\lambda \in L}\left(\left(r_{S_\lambda}^S\right)^{-1}(D_\lambda)\right)\right)^{\uparrow}$. Again, abusing notation, we write $\bigwedge_{\lambda \in L} D_\lambda^{\uparrow} := \bigcap_{\lambda \in L} D_\lambda^{\uparrow}$ (we will therefore use the conjunction symbol $\wedge$ and the intersection symbol $\cap$ interchangeably).

Intuitively, to take the intersection of events $(D_\lambda^{\uparrow}, S_\lambda)_{\lambda \in L}$, we express them "most economically in the smallest language" in which they are all expressible $S = \sup_{\lambda \in L} S_\lambda$, take the intersection, and then the union of inverse images obtaining the event $\left(\bigcap_{\lambda \in L}((r_{S_\lambda}^S)^{-1}(D_\lambda))\right)^{\uparrow}$ that is based in S.

We define the relation $\subseteq$ between events (E, S) and (F, S'), by $(E, S) \subseteq (F, S')$ if and only if $E \subseteq F$ as sets *and* $S' \preceq S$. If $E \neq \emptyset$, we have that $(E, S) \subseteq (F, S')$ if and only if $E \subseteq F$ as sets. Note however that for $E = \emptyset^S$ we have $(E, S) \subseteq (F, S')$ if and only if $S' \preceq S$. Hence we can write $E \subseteq F$ instead of $(E, S) \subseteq (F, S')$ as long as we keep in mind that in the case of $E = \emptyset^S$ we have $\emptyset^S \subseteq F$ if and only if $S \succeq S(F)$. It follows from these definitions that for events E and F, $E \subseteq F$ is equivalent to $\neg F \subseteq \neg E$ only when E and F have the same base, i.e., $S(E) = S(F)$.

Intuitively, in order to say "E implies F" we must be able to express F in the "language" used to express E. Hence, it must be that $S(F) \preceq S(E)$. The inclusion is then just $E \cap S(E) \subseteq F \cap S(E)$.

The disjunction of $\left\{D_\lambda^\uparrow\right\}_{\lambda\in L}$ is defined by the de Morgan law $\bigvee_{\lambda\in L} D_\lambda^\uparrow = \neg\left(\bigwedge_{\lambda\in L} \neg\left(D_\lambda^\uparrow\right)\right)$. Typically $\bigvee_{\lambda\in L} D_\lambda^\uparrow \subsetneqq \bigcup_{\lambda\in L} D_\lambda^\uparrow$, and if all D_λ are nonempty we have that $\bigvee_{\lambda\in L} D_\lambda^\uparrow = \bigcup_{\lambda\in L} D_\lambda^\uparrow$ holds if and only if all the $D_\lambda^\uparrow$ have the same base-space.

So far, we have just described an event structure. To formalize the state of mind of an agent, a possibility correspondence is introduced analogous to the one in standard game theory. For each agent $a \in \mathsf{Ag}$ there is a *possibility correspondence* $\Pi_a : \Omega \to 2^\Omega$ with the following properties:

> *Confinement:* If $\omega \in S$ then $\Pi_a(\omega) \subseteq S'$ for some $S' \preceq S$.
> *Generalized Reflexivity:* $\omega \in \Pi_a^\uparrow(\omega)$ for every $\omega \in \Omega$.
> *Stationarity:* $\omega' \in \Pi_a(\omega)$ implies $\Pi_a(\omega') = \Pi_a(\omega)$.
> *Projections Preserve Ignorance:* If $\omega \in S'$ and $S \preceq S'$ then $\Pi_a^\uparrow(\omega) \subseteq \Pi_a^\uparrow(\omega_S)$.
> *Projections Preserve Knowledge:* If $S \preceq S' \preceq S''$, $\omega \in S''$ and $\Pi_a(\omega) \subseteq S'$ then $(\Pi_a(\omega))_S = \Pi_a(\omega_S)$.

Note that Generalized Reflexivity implies that if $S' \preceq S$, $\omega \in S$ and $\Pi_a(\omega) \subseteq S'$, then $r_{S'}^S(\omega) \in \Pi_a(\omega)$. Additionally, we have the possibility correspondence is serial, i.e., $\Pi_a(\omega) \neq \emptyset$, for all $\omega \in \Omega$.

The possibility correspondence is the analogue to the accessibility relations in Kripke structures. Generalized Reflexivity and Stationarity are the analogues of the partitional properties of the possibility correspondence in partitional Aumann structures or Kripke structures. In particular, Generalized Reflexivity yields the truth property; Stationarity will guarantee the introspection properties (see Proposition 3.1). It captures both transitivity and Euclideaness.

The properties Projections Preserve Ignorance and Projections Preserve Knowledge guarantee the coherence of knowledge and awareness of individuals down the lattice structure. They compare the possibility sets of an individual in a state ω and its projection ω_S. The properties guarantee that, first, at the projected state ω_S the individual knows nothing she does not know at ω, and second, at the projected state ω_S the individual is not aware of anything she is unaware of at ω (Projections Preserve Ignorance). Third, at the projected state ω_S the individual knows every event she knows at ω, provided that this event is based in a space lower than or equal to S (Projections Preserve Knowledge). These properties also imply that at the projected state ω_S the individual is aware of every event she is aware of at ω, provided that this event is based in a space lower than or equal to S.[13]

The *knowledge operator* of agent a on events E is defined, as usual in Aumann structures, by

$$\mathbf{K}_a(E) := \{\omega \in \Omega : \Pi_a(\omega) \subseteq E\},$$

if there is a state ω such that $\Pi_a(\omega) \subseteq E$, and by $\mathbf{K}_a(E) := \emptyset^{S(E)}$ otherwise.

The *awareness operator* of agent a on events E can be defined by

$$\mathbf{A}_a(E) := \left\{\omega \in \Omega : \Pi_a(\omega) \subseteq S(E)^{\uparrow}\right\},$$

if there is a state ω such that $\Pi_a(\omega) \subseteq S(E)^{\uparrow}$, and by $\mathbf{A}_a(E) := \emptyset^{S(E)}$ otherwise. Thus, an agent is aware of an event if she considers possible states in which this event is "expressible".

Both, the knowledge and awareness operators are well-defined and easy to work with:

Lemma 3.2
If E is an event, then $\mathbf{K}_a(E)$ and $\mathbf{A}_a(E)$ are $S(E)$-based events. ⊣

The proof of the lemma makes use of the properties imposed on the possibility correspondence as does the proof of the following proposition.

The *unawareness operator* on events is defined as the negation of awareness, $\mathbf{U}_a(E) := \neg\mathbf{A}_a(E)$.

Proposition 3.1
The Knowledge and Awareness operators satisfy following properties:

Necessitation: $\mathbf{K}_a(\Omega) = \Omega$,
Distribution: $\mathbf{K}_a\left(\bigcap_{\lambda \in L} E_\lambda\right) = \bigcap_{\lambda \in L} \mathbf{K}_a(E_\lambda)$,
Monotonicity: $E \subseteq F$ implies $\mathbf{K}_a(E) \subseteq \mathbf{K}_a(F)$,
Truth: $\mathbf{K}_a(E) \subseteq E$,
Positive Introspection: $\mathbf{K}_a(E) \subseteq \mathbf{K}_a\mathbf{K}_a(E)$,
Negative Non-Introspection: $\neg\mathbf{K}_a(E) \cap \neg\mathbf{K}_a\neg\mathbf{K}_a(E) \subseteq \neg\mathbf{K}_a\neg\mathbf{K}_a\neg\mathbf{K}_a(E)$,
Weak Negative Introspection: $\neg\mathbf{K}_a(E) \cap \mathbf{A}_a\neg\mathbf{K}_a(E) = \mathbf{K}_a\neg\mathbf{K}_a(E)$,
MR Awareness: $\mathbf{A}_a(E) = \mathbf{K}_a(E) \cup \mathbf{K}_a\neg\mathbf{K}_a(E)$,
Strong Plausibility: $\mathbf{U}_a(E) = \bigcap_{n=1}^{\infty} (\neg\mathbf{K}_a)^n(E)$,
KU Introspection: $\mathbf{K}_a\mathbf{U}_a(E) = \emptyset^{S(E)}$,
AU Reflection: $\mathbf{U}_a(E) = \mathbf{U}_a\mathbf{U}_a(E)$,
Symmetry: $\mathbf{A}_a(E) = \mathbf{A}_a(\neg E)$,
A-Conjunction: $\bigcap_{\lambda \in L} \mathbf{A}_a(E_\lambda) = \mathbf{A}_a\left(\bigcap_{\lambda \in L} E_\lambda\right)$,

[13] The literature also mentions another property, called Projections Preserve Awareness, which is known to follow from other properties.

AK-Reflection: $\mathbf{A}_a(E) = \mathbf{A}_a\mathbf{K}_a(E)$,
A-Reflection: $\mathbf{A}_a(E) = \mathbf{A}_a\mathbf{A}_a(E)$,
A-Introspection: $\mathbf{A}_a(E) = \mathbf{K}_a\mathbf{A}_a(E)$. ⊣

The event-based approach lends itself well to study *interactive* reasoning about knowledge and awareness. Common knowledge can be defined in the usual way. The mutual knowledge operator on events is defined by

$$\mathbf{K}(E) := \bigcap_{a \in \mathsf{Ag}} \mathbf{K}_a(E).$$

The common knowledge operator on events is defined by

$$\mathbf{CK}(E) := \bigcap_{n=1}^{\infty} \mathbf{K}^n(E).$$

Analogously we can define mutual and common awareness. The mutual awareness operator on events is defined by

$$\mathbf{A}(E) = \bigcap_{a \in \mathsf{Ag}} \mathbf{A}_a(E),$$

and the common awareness operator by

$$\mathbf{CA}(E) = \bigcap_{n=1}^{\infty} (\mathbf{A})^n (E).$$

Proposition 3.2
The following multi-agent properties obtain: For all events E and agents $a, b \in \mathsf{Ag}$,

1. $\mathbf{A}_a(E) = \mathbf{A}_a\mathbf{A}_b(E)$,
2. $\mathbf{A}_a(E) = \mathbf{A}_a\mathbf{K}_b(E)$,
3. $\mathbf{K}_a(E) \subseteq \mathbf{A}_a\mathbf{K}_b(E)$,
4. $\mathbf{A}(E) = \mathbf{K}(S(E)^{\uparrow})$,
5. $\mathbf{A}(E) = \mathbf{CA}(E)$,
6. $\mathbf{K}(E) \subseteq \mathbf{A}(E)$,
7. $\mathbf{CK}(E) \subseteq \mathbf{CA}(E)$,
8. $\mathbf{CK}(S(E)^{\uparrow}) \subseteq \mathbf{CA}(E)$. ⊣

At this point, it may be useful to illustrate unawareness frames with our speculative trade example:

Example 3.3 (Speculative trade (continued))
Consider the unawareness structure depicted in Figure 3.2. There are four spaces. Space $S_{\{n,\ell\}}$ is the richest space in which both the lawsuit and the innovation are expressible. Both spaces, $S_{\{n\}}$ and $S_{\{\ell\}}$, are less expressive than $S_{\{n,\ell\}}$. $S_{\{n\}}$ is the space in which only the innovation is expressible while $S_{\{\ell\}}$ is the space in which only the lawsuit is expressible. Finally,

neither the innovation nor the lawsuit are expressible in the lowest space, $S_\emptyset$. We let $\succeq$ be defined by $S_{\{n,\ell\}} \succeq S_{\{n\}} \succeq S_\emptyset$ and $S_{\{n,\ell\}} \succeq S_{\{\ell\}} \succeq S_\emptyset$. Projections from higher to lower spaces are indicated by the grey dotted lines. For instance, state $(\neg n, \neg\ell) \in S_{\{n,\ell\}}$ projects to $(\neg n) \in S_{\{n\}}$. It also projects to $(\neg\ell) \in S_{\{\ell\}}$. Both $(\neg n) \in S_{\{n\}}$ and $(\neg\ell) \in S_{\{\ell\}}$ project to $(\top) \in S_\emptyset$. The possibility correspondence is given by the blue solid and red intermitted arrows and soft-edged rectangles for the owner and the potential buyer, respectively. At any state in $S_{\{n,\ell\}}$ the owner's possibility set is at $S_{\{\ell\}}$. Thus, he is unaware of the innovation but aware of the lawsuit. Further, the owner's possibility set includes all states in $S_{\{\ell\}}$, which means that he does not know whether the lawsuit obtains or not. Since at every state in $S_{\{\ell\}}$ the buyer's possibility set is on S, in any state in $S_{\{\ell\}}$ the buyer is unaware of the lawsuit and the owner knows that. At any state in $S_{\{n,\ell\}}$ the buyer's possibility set is at $S_{\{n\}}$. Thus, he is unaware of the lawsuit but aware of the innovation. Further, the buyer's possibility includes all states in $S_{\{n\}}$, which means that she does not know whether the lawsuit obtains or not. Since at every state in $S_{\{n\}}$ the owner's possibility set is on S, in any state in $S_{\{n\}}$ the owner is unaware of the innovation and the buyer knows that. Thus, the unawareness frame of Figure 3.2 models the speculative trade example. In comparison to awareness structures, we observe that the

Figure 3.2: An Unawareness Frame for the Speculative Trade Example

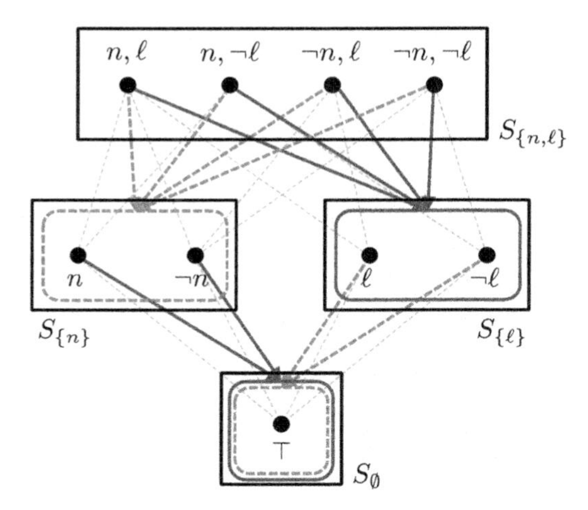

possibility correspondences model explicit knowledge.[14] In fact, together

[14]This is not to say that one couldn't define implicit knowledge in unawareness frames. An "implicit" possibility correspondence could be defined from the possibility correspondence by taking the inverse images of possibility sets in the upmost space.

with the lattice structure, they also model awareness determined by the space in which the possibility set lies. Although we used suggestive labels such as (n, ℓ) etc., unawareness frames are syntax-free. Finally, while the entire unawareness frame is the analysts model of the situation, it contains directly the "submodels" of agents. For instance, the sublattice consisting of the two spaces, $S_{\{\ell\}}$ and $S_{\emptyset}$, corresponds to the model of the owner while the sublattice consisting of the two spaces, $S_{\{n\}}$ and $S_{\emptyset}$, is the buyer's model. Moreover, the sublattice $S_{\emptyset}$ is the model that both agents attribute to each other. The states in all those spaces can be interpreted as subjective descriptions of situations in the respective agent's mind. $\dashv$

More general versions of unawareness frames have been studied in which the possibility correspondence does not necessarily satisfy generalized reflexivity and stationarity. In such a case of course, Truth, Positive Introspection, Negative Non-Introspection, and Weak Negative Introspection may fail, and agents may make mistakes in information processing on top of being unaware. More interestingly in such structures, KU-Introspection fails as well. This suggests that unawareness may not only be consistent with a strong notion of knowledge like the one embodied in the properties of S5 of unawareness structures but that it may even be enhanced by it.

There is also variant of unawareness frames in the literature in which the property Projections Preserve Knowledge of the possibility correspondence is dropped. The motivation is to study to what extent unawareness can constrain an agent's knowledge and can impair her reasoning about what other agents know.

Finally, unawareness frames can be augmented with decision theoretic primitives like preference relation over acts, i.e., functions from states to real numbers. One is then able to characterize properties of the possibility correspondence by corresponding properties of a decision maker's preference relation. This allows, at least in principle, to design choice experiments that reveal knowledge and awareness of the agent.

3.3.4 Unawareness structures

While the event-based approach of unawareness frames is a tractable approach to modeling nontrivial reasoning about knowledge and awareness among multiple agents, it leaves many questions open. For instance, when introducing the event-based approach we often alluded to intuitive explanations typeset in quotation marks that referred to "expressibility" etc. What justifies such an interpretation? Is it possible to link formally the "expressibility" of state spaces to the expressivity of languages? What does the expressivity of languages has to do with the notion of awareness used in unawareness frames? We also saw that an event may obtain in some states, its negation may obtain in others, and yet in others this event or its negation may not even be defined. This suggests that implicitly a three-valued logic is lurking behind the approach. Again, can we make this explicit? Moreover, Proposition 3.1 presents properties that awareness and knowledge satisfy in the event-based approach. But are these *all* the properties? That is, can we axiomatize the event-based approach in terms of all the properties of awareness and knowledge? Moreover, can we guarantee that the event-based approach is comprehensive enough so that we can model with it all situations with such properties. These questions can be addressed by introducing a logical apparatus and constructing a canonical unawareness structure.

Consider the language $\mathsf{L}_n^{K,A}(\mathsf{At})$ and define, as in Section 3.2.2, awareness in terms of knowledge by

$$A_a\varphi := K_a\varphi \vee K_a\neg K_a\varphi.$$

With this definition, we consider the following axiom system that we call $\widetilde{\mathsf{S5}}_n^{K,A}$:

Prop. All substitutions instances of tautologies of propositional logic, including the formula $\top$.

AS. $A_a\neg\varphi \leftrightarrow A_a\varphi$ (Symmetry)

AC. $A_a(\varphi \wedge \psi) \leftrightarrow A_a\varphi \wedge A_a\psi$ (Awareness Conjunction)

$\mathrm{A}_a\mathrm{K}_b\mathrm{R}$. $A_a\varphi \leftrightarrow A_aK_b\varphi$, for all $b \in \mathsf{Ag}$ (Awareness Knowledge Reflection)

T. $K_a\varphi \to \varphi$ (Axiom of Truth)

4. $K_a\varphi \to K_aK_a\varphi$ (Positive Introspection Axiom)

MP. From φ and $\varphi \to \psi$ infer ψ (modus ponens)

RK. For all natural numbers $n \geq 1$, if $\varphi_1, \varphi_2, \ldots, \varphi_n$ and φ are such that $\mathsf{At}(\varphi) \subseteq \bigcup_{i=1}^n \mathsf{At}(\varphi_i)$, then $\varphi_1 \wedge \varphi_2 \wedge \cdots \wedge \varphi_n \to \varphi$ implies $K_a\varphi_1 \wedge K_a\varphi_2 \wedge \cdots \wedge K_a\varphi_n \to K_a\varphi$. (RK-Inference)

We also define the modality U_a by $U_a\varphi := \neg A_a\varphi$ read as "agent a is unaware of φ".

Remark 3.4

The Modica and Rustichini definition of awareness and axiom system $\widetilde{\mathsf{S5}}_n^{K,A}$ implies:

K. $K_a\varphi \wedge K_a(\varphi \to \psi) \to K_a\varphi$

$K_a\varphi \wedge K_a\psi \to K_a(\varphi \wedge \psi)$

NNI. $U_a\varphi \to \neg K_a\neg K_a\neg K_a\varphi$

AI. $A_a\varphi \to K_aA_a\varphi$

AGPP. $A_a\varphi \leftrightarrow \bigwedge_{p\in\mathsf{At}(\varphi)} A_ap$

Gen_A. If φ is a theorem, then $A_a\varphi \to K_a\varphi$ is a theorem.

For every $\mathsf{At}' \subseteq \mathsf{At}$, let $S_{\mathsf{At}'}$ be the set of maximally consistent sets $\omega_{\mathsf{At}'}$ of formulae in the sublanguage $\mathsf{L}_n^{K,A}(\mathsf{At}')$. Given a language $\mathsf{L}_n^{K,A}(\mathsf{At}')$, a set of formulae Γ is consistent with respect to an axiom system if and only if there is no formula φ such that both φ and $\neg\varphi$ are provable from Γ. ω_{At} is maximally consistent if it is consistent and for any formula $\varphi \in \mathsf{L}_n^{K,A}(\mathsf{At}') \setminus \omega_{\mathsf{At}'}$, the set $\omega_{\mathsf{At}'} \cup \{\varphi\}$ is not consistent. By standard arguments one can show that every consistent subset of $\mathsf{L}_n^{K,A}(\mathsf{At}')$ can be extended to a maximally consistent subset $\omega_{\mathsf{At}'}$ of $\mathsf{L}_n^{K,A}(\mathsf{At}')$. Moreover, $\Gamma \subseteq \mathsf{L}_n^{K,A}(\mathsf{At}')$ is a maximally consistent subset of $\mathsf{L}_n^{K,A}(\mathsf{At}')$ if and only if Γ is consistent and for every $\varphi \in \mathsf{L}_n^{K,A}(\mathsf{At}')$, $\varphi \in \Gamma$ or $\neg\varphi \in \Gamma$.

Clearly, $\{S_{\mathsf{At}'}\}_{\mathsf{At}' \subseteq \mathsf{At}}$ is a complete lattice of disjoint spaces by set inclusion defined on the set of atomic formulae. Define the partial order on $\{S_{\mathsf{At}'}\}_{\mathsf{At}' \subseteq \mathsf{At}}$ by $S_{\mathsf{At}_1} \succeq S_{\mathsf{At}_2}$ if and only if $\mathsf{At}_1 \supseteq \mathsf{At}_2$. Let $\Omega := \bigcup_{\mathsf{At}' \subseteq \mathsf{At}} S_{\mathsf{At}'}$. For any $S_{\mathsf{At}_1} \succeq S_{\mathsf{At}_2}$, surjective projections $r^{\mathsf{At}_1}_{\mathsf{At}_2} : S_{\mathsf{At}_1} \longrightarrow S_{\mathsf{At}_2}$ are defined by $r^{\mathsf{At}_1}_{\mathsf{At}_2}(\omega) := \omega \cap \mathsf{L}_n^{K,A}(\mathsf{At}_2)$.

Theorem 3.5
For every ω and $a \in \mathsf{Ag}$, the possibility correspondence defined by

$$\Pi_a(\omega) := \left\{ \omega' \in \Omega : \text{ For every formula } \varphi, \begin{array}{l} \text{(i) } K_a\varphi \text{ implies } \varphi \in \omega', \text{ and} \\ \text{(ii) } A_a\varphi \in \omega \text{ if and only if} \\ \qquad \varphi \in \omega \text{ or } \neg\varphi \in \omega \end{array} \right\}$$

satisfies Confinement, Generalized Reflexivity, Projections Preserve Ignorance, and Projections Preserve Knowledge. Moreover, for every formula φ, the set of states $[\varphi] := \{\omega \in \Omega : \varphi \in \omega\}$ is a $S_{\mathsf{At}(\varphi)}$-based event, and $[\neg\varphi] = \neg[\varphi]$, $[\varphi \wedge \psi] = [\varphi] \cap [\psi]$, $[K_a\varphi] = \mathbf{K}_a[\varphi]$, $[A_a\varphi] = \mathbf{A}_a[\varphi]$, and $[U_a\varphi] = \mathbf{U}_a[\varphi]$. $\dashv$

The canonical unawareness structure is constructed such that states are consistent and comprehensive descriptions, and the "internal" descriptions of the states is reflected by operations on the events.

We can extend unawareness frames to unawareness structures by adding a valuation. An unawareness structure $M = \Big(\mathcal{S}, (r^{S'}_{S})_{S' \succeq S; S, S' \in \mathcal{S}}, (\Pi_a)_{a \in \mathsf{Ag}}, V\Big)$ is an unawareness frame $\Big(\mathcal{S}, (r^{S'}_{S})_{S' \succeq S; S, S' \in \mathcal{S}}, (\Pi_a)_{a \in \mathsf{Ag}}\Big)$ and a valuation $V : \mathsf{At} \longrightarrow \Sigma$ that assigns to each atomic formula in At an event in Σ. The set $V(p)$ is the event in which p obtains.

The satisfaction relation is defined inductively on the structure of formulae in $\mathsf{L}_n^{K,A}(\mathsf{At})$

$M, \omega \models \top$, for all $\omega \in \Omega$,
$M, \omega \models p$ if and only if $\omega \in V(p)$,

$M, \omega \models \varphi \wedge \psi$ if and only if $[\varphi] \cap [\psi]$,
$M, \omega \models \neg\varphi$ if and only if $\omega \in [\neg\varphi]$,
$M, \omega \models K_a\varphi$ if and only if $\omega \in \mathsf{K}_a[\varphi]$,

where $[\varphi] := \{\omega' \in \Omega : M, \omega' \models \varphi\}$, for every formula φ. Note that $[\varphi]$ is an event in the unawareness structure. Recall that $A_a\varphi := K_a\varphi \vee K_a\neg K_a\varphi$. Thus, indeed the satisfaction relation is defined for formulae in $\mathsf{L}_n^{K,A}(\mathsf{At})$.

Our aim is to state a characterization of unawareness structures in terms of properties. More precisely, we seek a complete and sound axiomatization of unawareness structures. To this extent we need to define first the notion of validity. Recall that a formula is said to be valid in a Kripke structure if it is true in *every* state. Yet, in unawareness structures, a nontrivial formula is not even defined in all states. Thus, the definition of validity for Kripke structures is not directly applicable to unawareness structures. But the remedy is straightforward. We say that *φ is defined in state ω* in M if $\omega \in \bigcap_{p \in \mathsf{At}(\varphi)} (V(p) \cup \neg V(p))$. Now, we say φ is valid in M if $M, \omega \models \varphi$ for all ω in which φ is defined. φ is valid if it is valid in all M. Note that this generalized definition of validity is identical to the notion of validity for Kripke structures if $\mathcal{S} = \{S\}$, i.e., if the lattice of spaces of the unawareness structure is a singleton and thus the unawareness structure is a Kripke structure. The notions of soundness and completeness are now defined analogous to Kripke structures but using the generalized definition of validity.

Theorem 3.6
For the language $\mathsf{L}_n^{K,A}(\mathsf{At})$, the axiom system $\widetilde{\mathsf{S5}}_n^{K,A}$ is a sound and complete axiomatization with respect to unawareness structures. ⊣

An alternative axiomatization of unawareness structures extends the language by adding a non-standard implication operator. Recall that in an unawareness structure a formula may not be defined at every state. Implicitly, the non-standard implication operator combines standard implication with an "at least as defined" relation on formulae. That is, formula φ implies (non-standardly) ψ is valid only if ψ is true whenever φ is true and ψ is "at least as defined as" φ. Then, define a satisfaction relation such that every formula in the extended language is defined in every state *across all* spaces. In such a setting, one can apply directly the definition of validity as for Kripke structures. In this approach, propositional logic is axiomatized with respect to the non-standard implication operator. Unawareness structures are axiomatized by an axiom system that is similar to S5 but makes use of the non-standard implication operator. Yet, there are also analogous axiomatizations of unawareness structures in the literature in which the possibility correspondence does not necessarily satisfy generalized reflexivity or stationarity.

How are unawareness structures related to awareness structures introduced earlier? It turns out that despite the differences in motivation, their semantics are equivalent in terms of expressibility. That is, everything that can be described about awareness and knowledge in one structure can be described in the other structure and vice versa. More formally, let $\mathcal{M}^{HMS}(\mathsf{At})$ be the class of unawareness structures over At.

Theorem 3.7
For any partitional awareness structure $M = (\Omega, (R_a)_{a\in\mathsf{Ag}}, (\mathcal{A}_a)_{a\in\mathsf{Ag}}, V) \in \mathcal{M}^{FH}(\mathsf{At})$ in which awareness is propositionally determined, there exists an unawareness structure $M' = (\Omega', (\Pi_a)_{a\in\mathsf{Ag}}, (r^{\mathsf{At}_1}_{\mathsf{At}_2})_{\mathsf{At}_1,\mathsf{At}_2\subseteq\mathsf{At}}, V') \in \mathcal{M}^{HMS}(\mathsf{At})$ such that $\Omega' := \Omega \times 2^{\mathsf{At}}$, $S_{\mathsf{At}_1} = \Omega \times \{\mathsf{At}_1\}$ for all $\mathsf{At}_1 \subseteq \mathsf{At}$, and for all $\varphi \in \mathsf{L}^K_n(\mathsf{At})$, if $\mathsf{At}(\varphi) \subseteq \mathsf{At}_1$, then $M, \omega \models \varphi$ if and only if $M', (\omega, \mathsf{At}_1) \models \varphi$.

Conversely, for every unawareness structure M with $M = (\Omega, (\Pi_a)_{a\in\mathsf{Ag}}, (r^{\mathsf{At}_1}_{\mathsf{At}_2})_{\mathsf{At}_1,\ \mathsf{At}_2\subseteq\mathsf{At}}, V) \in \mathcal{M}^{HMS}(\mathsf{At})$, there exists a partitional awareness structure $M' = (\Omega, (R_a)_{a\in\mathsf{Ag}}, (\mathcal{A}_a)_{a\in\mathsf{Ag}}, V') \in \mathcal{M}^{FH}(\mathsf{At})$ in which awareness is propositionally determined such that for all $\varphi \in \mathsf{L}^K_n(\mathsf{At})$, if $\omega \in S_{\mathsf{At}_1}$ and $\mathsf{At}(\varphi) \subseteq \mathsf{At}_1$, then $(M, \omega) \models \varphi$ if and only if $(M', \omega) \models \varphi$. $\dashv$

The proof is by induction on the structure of φ. There are also versions of this result in the literature in which the possibility correspondences of unawareness structures (and accessibility relations of awareness structures, respectively) do not necessarily satisfy generalized reflexivity or stationarity (reflexivity, transitivity, and Euclideaness, respectively).

The previous result implies alternative axiomatizations:

Corollary 3.1
For the language $\mathsf{L}^{K,A}_n(\mathsf{At})$, the axiom system $\mathsf{S5}^{K,A}_n$ is a sound and complete axiomatization with respect to unawareness structures. For the language $\mathsf{L}^{K,A}_n(\mathsf{At})$, the axiom system $\widetilde{\mathsf{S5}}^{K,A}_n$ is a sound and complete axiomatization with respect to partitional awareness structures in which awareness is propositionally determined. $\dashv$

As mentioned previously, there are variants of unawareness structures in the literature in which the property Projections Preserve Knowledge is dropped. These can be axiomatized with a multiple knowledge modalities, one for each sub-language.

3.3.5 Generalized standard models

In economics, the first structures introduced for modeling unawareness were generalized standard models. They are confined to a single agent. In retrospect, unawareness structures introduced in the previous section can be understood as multi-agent generalizations of generalized standard models.

A *generalized standard model* $M = (S, \Omega, \rho, \Pi, V)$ over At consists of a space of objective states S and a collection of nonempty disjoint subjective state spaces $\{\tilde{S}_{\mathsf{At}'}\}_{\mathsf{At}' \subseteq \mathsf{At}}$ with $\Omega := \bigcup_{\mathsf{At}' \subseteq \mathsf{At}} \tilde{S}_{\mathsf{At}'}$. Ω and S are disjoint. Further, there is a surjective projection $\rho : S \longrightarrow \Omega$. Moreover, the agent has a generalized $\Pi : S \longrightarrow 2^{\Omega}$ that satisfies:

> *Generalized Reflexivity:* if $s \in S$, then $\Pi(s) \subseteq \tilde{S}_{\mathsf{At}'}$ for some $\mathsf{At}' \subseteq \mathsf{At}$,
> *Stationarity:* $\rho(s) = \rho(t)$, then $\Pi(s) = \Pi(t)$.

Finally, there is a valuation $V : \mathsf{At} \longrightarrow 2^S$ such that if $\rho(s) = \rho(t) \in \tilde{S}_{\mathsf{At}'}$ then for all $p \in \mathsf{At}'$ either $s, t \in V(p)$ or $s, t \notin V(p)$.

Intuitively, the states in the subjective state-space $\tilde{S}_{\mathsf{At}'}$ describe situations conceivable by an agent who is aware of atomic formulae in At' only. Generalized reflexivity confines in each objective situation the perception of the agent to subjective situations that are all described with same "vocabulary". Stationarity means that the agents' perception depends only on her subjective states and summarizes transitivity and Euclideaness of the possibility correspondence.

Two caveats are to note: First, it is a single-agent structure. It is not immediate how to extent generalized standard models to a multi-agent setting. If we add additional possibility correspondences, one for each agent, then agents could reason about each other's knowledge but presumably only at the same awareness level. At state $s \in S$, agent a may know that agent b does not know the event $E \subseteq \tilde{S}_{\mathsf{At}'}$. But since at every state in $\tilde{S}_{\mathsf{At}'}$, agent b's possibility set must be a subset of $\tilde{S}_{\mathsf{At}'}$ as well, agent a is forced to know that b is aware of E. This is avoided in unawareness frames of Section 3.3.3 where at $\tilde{s} \in \tilde{S}_{\mathsf{At}'}$, agent b's possibility set may be a subset of states in yet a lower space $\tilde{S}_{\mathsf{At}''}$ with $\mathsf{At}'' \subsetneqq \mathsf{At}'$. Generalized standard models are limited to the single-agent case. Yet, unawareness is especially interesting in interactive settings, where different agents may have different awareness and knowledge, and reason about each others awareness and knowledge.

Second, the condition on the valuation, if $\rho(s) = \rho(t) \in \tilde{S}_{\mathsf{At}'}$ then for all $p \in \mathsf{At}'$ either $s, t \in V(p)$ or $s, t \notin V(p)$, is also a condition on the projection ρ. Deleting the valuation does not yield straightforwardly an event-based approach or frame similar to Aumann structures, but one would need to add instead conditions on the projections.

We extend Π to a correspondence defined on the domain $S \cup \Omega$ by if $\tilde{s} \in \Omega$ and $\tilde{s} = \rho(s)$, then define $\Pi(\tilde{s}) = \Pi(s)$. This extension is well-defined by stationarity. A generalized standard model is said to be partitional if Π restricted to Ω is partitional. We also extend V to a valuation having the range $S \cup \Omega$ by defining $\tilde{V}(p) = V(p) \cup \bigcup_{\mathsf{At}' \subseteq \mathsf{At}} \left\{\tilde{s} \in \tilde{S}_{\mathsf{At}'} : p \in \mathsf{At}', \rho^{-1}(\tilde{s}) \subseteq V(p)\right\}$.

For $\omega \in S \cup \Omega$, we define inductively on the structure of formulae in $\mathsf{L}_1^{K,A}(\mathsf{At})$ the satisfaction relation

$M, \omega \models \top$, for all $\omega \in \Omega$,
$M, \omega \models p$ if and only if $\omega \in \tilde{V}(p)$,
$M, \omega \models \varphi \wedge \psi$ if and only if both $M, \omega \models \varphi$ and $M, \omega \models \psi$,
$M, \omega \models \neg\varphi$ if and only if $M, \omega \not\models \varphi$ and either $\omega \in S$ or $\omega \in \tilde{S}_{\mathsf{At}'}$ and $\varphi \in \mathsf{L}_1^{K,A}(\mathsf{At}')$,
$M, \omega \models K\varphi$ if and only if $M, \omega' \models \varphi$ for all $\omega' \in \Pi(\omega)$.

Recall the Modica and Rustichini definition of awareness in terms of knowledge, $A\varphi := K\varphi \vee K\neg K\varphi$. Thus, indeed the satisfaction relation is defined for formulae in $\mathsf{L}_1^{K,A}(\mathsf{At})$.

At this point, it may be useful to illustrate generalized standard models with our speculative trade example. Yet, since generalized standard models are defined for a single agent only, we cannot model the speculative trade example. While we could construct a separate generalized standard model for each of the agents, these models could not model the agent's reasoning about the other agent's awareness and knowledge. For instance, the sublattice consisting of the two spaces $S_{\{n,\ell\}}$ and $S_{\{\ell\}}$ in Figure 3.2 can be viewed as a generalized standard model of the owner.

To prove a characterization of generalized standard models in terms of properties of knowledge and awareness, we need to define validity. Recall that a formula is said to be valid in a Kripke structure if it is true in *every* state. In generalized standard models, the notion of validity is restricted to objective states in S only. We say φ is *objectively valid* in M if $M, \omega \models \varphi$ for all $\omega \in S$. The notions of soundness and completeness are now defined analogous to Kripke structures but using the notion of objective validity.

Consider the following axiom system called U.

Prop. All substitutions instances of tautologies of propositional logic, including the formula $\top$.
AS. $A\neg\varphi \leftrightarrow A\varphi$ (Symmetry)
AC. $A(\varphi \wedge \psi) \rightarrow A\varphi \wedge A\psi$
T. $K\varphi \rightarrow \varphi$ (Axiom of Truth)
4. $K\varphi \rightarrow KK\varphi$ (Positive Introspection Axiom)
MP. From φ and $\varphi \rightarrow \psi$ infer ψ (modus ponens)
M, C. $K(\varphi \wedge \psi) \leftrightarrow K\varphi \wedge K\psi$ (Distribution)
N. $K\top$
RK_{sa}. From $\varphi \leftrightarrow \psi$ infer $K\varphi \leftrightarrow K\psi$, where φ and ψ are such that $\mathsf{At}(\varphi) = \mathsf{At}(\psi)$.

Theorem 3.8
The axiom system U is a complete and sound axiomatization of objective validity for the language $\mathsf{L}_1^{K,A}(\mathsf{At})$ with respect to partitional generalized standard models. $\dashv$

When we restrict partitional awareness structures that are propositionally determined to a single-agent, then those awareness structures and generalized standard structures are equally expressive. Everything that can be described about awareness and knowledge in a generalized standard model can be described in partitional awareness structures in which awareness is propositionally determined. Let $\mathcal{M}^{MR}(\mathsf{At})$ be the class of generalized standard models over At.

Theorem 3.9
For any partitional awareness structure $M = (S, R, \mathcal{A}, V) \in \mathcal{M}^{FH}(\mathsf{At})$ in which awareness is propositionally determined, there exists a generalized standard model $M' = (S, \Omega, \Pi, \rho, V') \in \mathcal{M}^{MR}(\mathsf{At})$ such that for all formulae $\varphi \in \mathsf{L}_1^K(\mathsf{At})$, $M, s \models \varphi$ if and only if $M', s \models \varphi$.

Conversely, for every generalized standard model $M = (S, \Omega, \Pi, \rho, V) \in \mathcal{M}^{MR}(\mathsf{At})$, there exists a partitional awareness structure $M' = (S, R, \mathcal{A}, V')$ $\in \mathcal{M}^{FH}(\mathsf{At})$ in which awareness is propositionally determined such that for all $\varphi \in \mathsf{L}_1^K(\mathsf{At})$, $M, s \models \varphi$ if and only if $M', s \models \varphi$. $\dashv$

In the literature there are also analogous results for generalized standard models (awareness structures, respectively) for which generalized reflexivity and stationarity (reflexivity, transitivity, and Euclideaness) may fail.

3.3.6 Product models

In the economics literature, product models have been introduced around the same time as unawareness structures. The starting point of product models is a subset of questions about the relevant aspects of the world that can be answered either in the affirmative or negative. Awareness then differs by the subset of questions the agent has in mind. Such an approach to awareness is quite natural since lacking conception of some aspects of the world implies that one is not even able to ask questions about these aspects.

Originally, product models were confined to a single agent only. The extension to the multi-agent setting is non-trivial. We focus on the single-agent case but will consider the multi-agent case in the speculative trade example below. The primitives of the product model $M = (Q^*, \Omega^*, \mathcal{A}, P)$ are a set of questions Q^* and a space of objective states $\Omega^* := \prod_{q \in Q^*} \{1_q, 0_q\}$. Each state is a profile of zeros and ones; each component corresponding to a question in Q^*. The question is answered in the affirmative if the component corresponding to it is one. Further, an awareness correspondence $\mathcal{A} : \Omega^* \longrightarrow 2^{Q^*}$ assigns to each state a subset of questions that the agent is aware of at that state. Finally, a possibility correspondence $P : \Omega^* \longrightarrow 2^{\Omega^*} \setminus \{\emptyset\}$ assigns to each state a subset of states that the agent implicitly considers possible. We assume

Reflexivity: $\omega^* \in P(\omega^*)$ for all $\omega^* \in \Omega^*$.
Stationarity: For all $\omega_1^*, \omega_2^* \in \Omega^*$, $\omega_1^* \in P(\omega_2^*)$ implies both $P(\omega_1^*) = P(\omega_2^*)$ and $\mathcal{A}(\omega_1^*) = \mathcal{A}(\omega_2^*)$.

For $Q \subseteq Q^*$, let $\Omega_Q = \prod_{q \in Q}\{1_q, 0_q\}$ denote the state space in which states contain answers to questions in the subset Q only. Using the awareness correspondence, the subjective state-space of the agent at $\omega^* \in \Omega^*$ is $\Omega_{\mathcal{A}(\omega^*)}$.

Let $? : \bigcup_{Q \subseteq Q^*} 2^{\Omega_Q} \longrightarrow 2^{Q^*}$ denote the correspondence that assigns to each subset of each state space the set of questions that define the space. That is, if $E \subseteq \Omega_Q$, then $?(E) = Q$.

For $Q \subseteq Q^*$, denote by $r_Q : \Omega^* \longrightarrow \Omega_Q$ the surjective projection. If $E \subseteq \Omega_Q$, denote by $E^* = r_Q^{-1}(E)$ the set of objective states in Ω^* that project to E, i.e., the inverse image of E in Ω^*. An event is a pair $E = (E^*, ?(E))$. We define $\neg E := (\neg E^*, ?(E))$, $E_1 \wedge E_2 := (E_1^* \cap E_2^*, ?(E_1) \cup ?(E_2))$. Disjunction is defined by the De Morgan law using negation and conjunction as just defined. Note that there are many vacuous events $\emptyset_Q := \neg\Omega_Q$, $Q \subseteq Q^*$, one for each subset of questions.

The unawareness operator is defined by

$$\mathbf{U}(E) = \{\omega^* \in \Omega^* : ?(E) \nsubseteq \mathcal{A}(\omega^*)\}.$$

The awareness operator is defined by $\mathbf{A}(E) := \neg\mathbf{U}(E)$. Note that for every event E, the $\mathbf{U}(E)$ and $\mathbf{A}(E)$ are subsets of the objective space. Thus, they capture an agent's reasoning about awareness from an outside modeler's point of view.

There are two knowledge operators. As our notation suggests, "objective knowledge" is best understood as implicit knowledge

$$\mathbf{L}(E) := \{\omega^* \in \Omega^* : P(\omega^*) \subseteq E^*\}.$$

The second knowledge operator refers to "subjective knowledge from the modeler's perspective". It is equivalent to explicit knowledge

$$\mathbf{K}(E) := \mathbf{L}(E) \cap \mathbf{A}(E).$$

Note however, that both $\mathbf{L}(E)$ and $\mathbf{K}(E)$ are subsets of the objective state-space and therefore not necessarily "accessible" to the agent. As remedy, one can define a subjective possibility correspondence and use it to define a knowledge operator reflecting "subjective knowledge from the agent's perspective". We focus on implicit and explicit knowledge only but will illustrate also the subjective versions in the speculative trade example below.

Proposition 3.3
For the product model, the following properties obtain for any events E, E_1, E_2 and $\omega^* \in \Omega^*$,

Subjective Necessitation: $\omega^* \in \mathbf{K}(\Omega(\omega^*))$,
Distribution: $\mathbf{K}(E_1) \cap \mathbf{K}(E_2) = \mathbf{K}(E_1 \wedge E_2)$,
Monotonicity: $E_1^* \subseteq E_2^*$ and $?(E_1) \supseteq ?(E_2)$ implies $\mathbf{K}(E_1) \subseteq \mathbf{K}(E_2)$,
Truth: $\mathbf{K}(E) \subseteq E^*$,
Positive Introspection: $\mathbf{K}(E) \subseteq \mathbf{KK}(E)$,
Weak Negative Introspection: $\neg\mathbf{K}(E) \cap \mathbf{A}(E) = \mathbf{K}\neg\mathbf{K}(E)$,
MR Awareness: $\mathbf{A}(E) = \mathbf{K}(E) \cup \mathbf{K}\neg\mathbf{K}(E)$,
Strong Plausibility: $\mathbf{U}(E) = \bigcap_{n=1}^{\infty}(\neg\mathbf{K})^n(E)$,
KU Introspection: $\mathbf{KU}(E) = \emptyset_{Q^*}$,
AU Reflection: $\mathbf{U}(E) = \mathbf{UU}(E)$,
Symmetry: $\mathbf{A}(E) = \mathbf{A}(\neg E)$. ⊣

The proof follows from definitions and properties of the possibility correspondence.

At this point, it may be instructive to consider as an illustration the speculative trade example.

Example 3.4 (Speculative trade (continued))
The originally product model is confined to a single agent only. Thus, we cannot use it to model the speculative trade example. We will illustrate a multi-agent extension of the product model with the speculative trade example. The set of questions is $\{n, \ell\}$, where we let n and ℓ stand for the questions "Is the innovation true?" and "Is the lawsuit true?", respectively. The objective state space is $\Omega^* = \Omega_{\{n,\ell\}} = \{1_n, 0_n\} \times \{1_\ell, 0_\ell\}$; the upmost space in Figure 3.3. For instance, at the state $(1_n, 0_\ell)$ the question "Is the innovation true?" is answered in the affirmative, "The innovation is true.", while the question "Is the lawsuit true?" is answered in the negative, "The lawsuit is not true." The awareness correspondences are indicated in Figure 3.3 by "speech bubbles" above each state. The solid blue speech bubbles belong to the owner, while the intermitted red speech bubbles are the buyer's. Both awareness correspondences are very special as they are constant on Ω^*. At every state Ω^*, the owner is aware only of questions involving the lawsuit while the buyer is only aware of questions involving the innovation. The possibility correspondences are indicated in Figure 3.3 by the solid blue and intermitted red soft-edged rectangles for the owner and buyer, respectively. Again, the possibility correspondences are very special in this example as no agent can distinguish any objective states in Ω^*.

Given the awareness correspondences defined on the set of objective states Ω^*, we can construct the subjective state spaces of both agents by considering for each agent only the questions of which he is aware. At every state in Ω^*, the buyer's subjective state space is the space to the left, $\Omega_b(\Omega^*)$, while the owner's subjective state space is the space to the right, $\Omega_o(\Omega^*)$.

Figure 3.3: A Product Model for the Speculative Trade Example

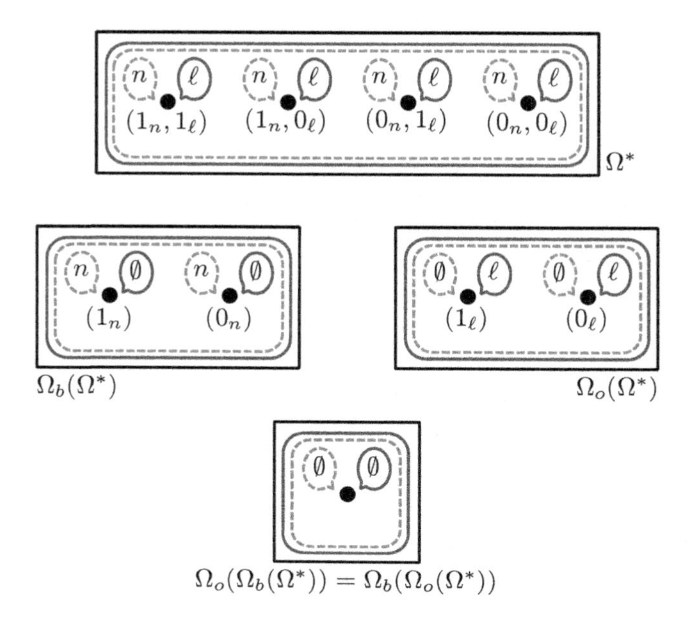

So far, these are all the primitives of the product model. One can also define subjective versions of the awareness and possibility correspondences that do not play a role in Proposition 3.3, but that are useful for modeling the example. To define the subjective awareness correspondence on $\Omega_b(\Omega^*)$, we extend the objective awareness correspondence on Ω^* to the subjective states in $\Omega_b(\Omega^*)$ and $\Omega_o(\Omega^*)$ by restricting the awareness sets at the subjective states to questions available at those subjective states, respectively.[15] Similarly, we can extend the objective possibility correspondences defined on Ω^* to subjective states by taking the projections to $\Omega_b(\Omega^*)$ and $\Omega_o(\Omega^*)$, respectively.

The subjective awareness correspondences allow us to defined another subjective state space shown as the lowest space in Figure 3.3. At every state in $\Omega_b(\Omega^*)$, the owner is unaware of the innovation (and the lawsuit), hence his subjective state space (in the eyes of the buyer) is $\Omega_o(\Omega_b(\Omega^*))$. Similarly, at every state in the owner's subjective state space $\Omega_o(\Omega^*)$, the buyer is unaware of the lawsuit (and the innovation); thus his subjective state space (in the eyes of the owner) is $\Omega_b(\Omega_o(\Omega^*)$. Both spaces are defined

[15]Note that, for instance, the owner's awareness correspondence on $\Omega_b(\Omega^*)$ cannot really be interpreted as the owner's subjective awareness correspondence but rather as the owner's awareness correspondence as perceived by the buyer.

from an empty set of questions. They are identical and singleton. Again, we can extend the awareness and possibility correspondences to the lowest space as outlined above.

This example illustrates the multi-agent product model. As it should be clear by now, it models the introductory example of speculative trade. It also suggests that product models bear features both of awareness structures and of unawareness frames. First, with awareness structures it has in common that awareness is modeled with an awareness correspondence although questions are used as primitive instead atomic formulae. Since one can define a one-to-one relation between questions and atomic formulae, the upmost space is analogous to the awareness structure depicted in Figure 3.1. We will use this relationship more generally in the discussion below. Second, we see clearly that the possibility correspondence models implicit knowledge and not necessarily explicit knowledge. For instance, the possibility sets on the objective space Ω^* and the owner's (the buyer's, resp.) possibility sets on $\Omega_b(\Omega^*)$ (on $\Omega_o(\Omega^*)$, respectively) can be understood only in terms of implicit knowledge. With unawareness frames it has in common the idea of subjective states and the lattice structure. ⊣

The following discussion is confined to the single-agent product model. The product model is analogous to a frame but with an additional set of questions as a primitive. In order to extend it to a structure, we need to relate questions to formulae and introduce a valuation. Let $b : Q^* \longrightarrow \mathsf{At}$ be a bijection. For every question $q \in Q^*$ there is exactly one primitive proposition p such that $b(q) = p$. The bijection is interpreted as assigning to each question $q \in Q^*$ exactly one primitive proposition $p \in \mathsf{At}$ that stands for "q is answered affirmatively" and q stands for "Is p true?". We can consider now the language $\mathsf{L}_1^{L,K,A}(b(Q^*))$.

A valuation $V : b(Q^*) \longrightarrow 2^{\Omega^*}$ is defined by $V(p) = \{\omega^* \in \Omega^* : r_{\{b^{-1}(p)\}}(\omega^*) = 1\}$. The corresponding subjective event is $[p] = (V(p), \{b^{-1}(p)\})$. Note $[p]^* = V(p)$. The satisfaction relation is defined by induction on the structure of formulae:

$M, \omega^* \models p$ if and only if $\omega^* \in V(p)$,
$M, \omega^* \models \neg\varphi$ if and only if $\omega^* \in \neg[\varphi]^*$,
$M, \omega^* \models \varphi \wedge \psi$ if and only if $\omega^* \in [\varphi]^* \wedge [\psi]^*$,
$M, \omega^* \models A\varphi$ if and only if $\omega^* \in \mathbf{A}[\varphi]$,
$M, \omega^* \models L\varphi$ if and only if $\omega^* \in \mathbf{L}[\varphi]$,
$M, \omega^* \models K\varphi$ if and only if $\omega^* \in \mathbf{K}[\varphi]$.

Note that the satisfaction relation is defined only for objective states in Ω^*. Thus, similar to product models discussed in the previous section, the setting so far allows for a notion of objective validity only. We can relate product models to structures discussed in the previous sections. Let $\mathcal{M}^{Li}$ denote the class of product models over the set of questions Q^*.

Theorem 3.10
For any partitional awareness structure $M = (S, R, \mathcal{A}, V) \in \mathcal{M}^{FH}(\mathsf{At})$ in which awareness is generated by primitive propositions there is a product model $M = (b^{-1}(\mathsf{At}), S, P, \mathcal{A}', V') \in \mathcal{M}^{Li}$ with $\mathcal{A}' := b^{-1} \circ \mathcal{A}$ such that for any $\varphi \in \mathsf{L}_1^{L,K,A}(\mathsf{At})$, $M, s \models \varphi$ if and only if $M', s \models \varphi$.

Conversely, for every product model $M = (Q^*, \Omega^*, P, \mathcal{A}) \in \mathcal{M}^{Li}$ there is a partitional awareness structure $M' = (\Omega^*, R, \mathcal{A}', V') \in \mathcal{M}^{FH}(b(Q^*))$ in which awareness is generated by primitive propositions such that for any $\varphi \in \mathsf{L}_1^{L,K,A}(b(Q^*))$, $M, \omega^* \models \varphi$ if and only if $M', \omega^* \models \varphi$. ⊣

3.4 Awareness and probabilistic belief

3.4.1 Type spaces with unawareness

Models of unawareness are mostly applied in strategic contexts when agents are players in a game and have to take decisions that are rational/optimal with respect to their state of mind. In such situations, it is extremely helpful for players to judge uncertain events probabilistically. To model such agents, we need to replace the qualitative notions of knowledge or belief discussed until now by the quantitative notion of probabilistic beliefs. In standard game theory with incomplete information, this is done with Harsanyi type spaces. Type spaces do not just model in a parsimonious way a player's belief about some basic uncertain events but also their beliefs about other players' beliefs, beliefs about that, and so on. That is, they model infinite hierarchies of beliefs. Under unawareness, the problem is complicated by the fact that agents may also be unaware of different events and may form beliefs about other players' unawareness, their belief about other players' beliefs about unawareness, etc. Combining ideas from unawareness frames and Harsanyi type spaces, define an unawareness type space $\left(\mathcal{S}, (r_S^{S'})_{S' \succeq S; S, S' \in \mathcal{S}}, (t_a)_{a \in \mathsf{Ag}}\right)$ by a complete lattice of disjoint measurable spaces $\mathcal{S} = \{S_\alpha\}_{\alpha \in \mathcal{A}}$, each with a σ-field $\mathcal{F}_S$, and measurable surjective projections $(r_S^{S'})_{S' \succeq S; S, S' \in \mathcal{S}}$. Let $\Delta(S)$ be the set of probability measures on $(S, \mathcal{F}_S)$. We consider this set itself as a measurable space endowed with the σ-field $\mathcal{F}_{\Delta(S)}$ generated by the sets $\{\mu \in \Delta(S) : \mu(D) \geq p\}$, where $D \in \mathcal{F}_S$ and $p \in [0, 1]$.

For a probability measure $\mu \in \Delta(S')$, the marginal $\mu_{|S}$ of μ on $S \preceq S'$ is defined by

$$\mu_{|S}(D) := \mu\left(\left(r_S^{S'}\right)^{-1}(D)\right), \quad D \in \mathcal{F}_S.$$

Let S_μ be the space on which μ is a probability measure. Whenever $S_\mu \succeq S(E)$, we abuse notation slightly and write

$$\mu(E) = \mu\left(E \cap S_\mu\right).$$

If $S(E) \npreceq S_\mu$, then we say that $\mu(E)$ is undefined.

For each agent $a \in \mathsf{Ag}$, there is a *type mapping* $t_a : \Omega \longrightarrow \bigcup_{\alpha\in\mathcal{A}} \Delta\left(S_\alpha\right)$, which is measurable in the sense that for every $S \in \mathcal{S}$ and $Q \in \mathcal{F}_{\Delta(S)}$ we require $t_a^{-1}(Q) \cap S \in \mathcal{F}_S$. Analogous to properties of the possibility correspondence in unawareness frames, the type mapping t_a should satisfy the following properties:

> *Confinement:* If $\omega \in S'$ then $t_a(\omega) \in \triangle(S)$ for some $S \preceq S'$.
> (2) If $S'' \succeq S' \succeq S$, $\omega \in S''$, and $t_a(\omega) \in \triangle(S')$ then $t_a(\omega_S) = t_a(\omega)_{|S}$.
> (3) If $S'' \succeq S' \succeq S$, $\omega \in S''$, and $t_a(\omega_{S'}) \in \triangle(S)$ then $S_{t_a(\omega)} \succeq S$.

$t_a(\omega)$ represents agent a's belief at state ω. The properties guarantee the consistent fit of beliefs and awareness at different state spaces. *Confinement* means that at any given state $\omega \in \Omega$ an agent's belief is concentrated on states that are all described with the same "vocabulary" - the "vocabulary" available to the agent at ω. This "vocabulary" may be less expressive than the "vocabulary" used to describe statements in the state ω.

Properties (2) and (3) compare the types of an agent in a state $\omega \in S'$ and its projection to ω_S, for some $S \preceq S'$. Property (2) means that at the projected state ω_S the agent believes everything she believes at ω given that she is aware of it at ω_S. Property (3) means that at ω an agent cannot be unaware of an event that she is aware of at the projected state $\omega_{S'}$.

Define the set of states at which agent a's type or the marginal thereof coincides with her type at ω by $Ben_a(\omega) := \left\{\omega' \in \Omega : t_a(\omega')_{|S_{t_a(\omega)}} = t_a(\omega)\right\}$. This is an event of the unawareness-belief frame although it may not be a measurable event (even in a standard type-space). It is assumed that if $Ben_a(\omega) \subseteq E$, for an event E, then $t_a(\omega)(E) = 1$. This assumption implies introspection with respect to beliefs.

For agent $a \in \mathsf{Ag}$ and an (not necessarily measurable) event E, define the *awareness operator* by

$$\mathbf{A}_a(E) := \{\omega \in \Omega : t_a(\omega) \in \Delta(S), S \succeq S(E)\}$$

if there is a state ω such that $t_a(\omega) \in \Delta(S)$ with $S \succeq S(E)$, and by $\mathbf{A}_a(E) := \emptyset^{S(E)}$ otherwise. This is analogous to awareness in unawareness frames.

For each agent $a \in \mathsf{Ag}$, $p \in [0,1]$, and measurable event E, the *probability-of-at-least-p-belief operator* is defined as usual by

$$\mathbf{B}_a^p(E) := \{\omega \in \Omega : t_a(\omega)(E) \geq p\},$$

if there is a state ω such that $t_a(\omega)(E) \geq p$, and by $\mathbf{B}_a^p(E) := \emptyset^{S(E)}$ otherwise.

Lemma 3.3
If E is an event, then both $\mathbf{A}_a(E)$ and $\mathbf{B}_a^p(E)$ are $S(E)$-based events. $\dashv$

The proof follows from the properties of the type mapping and the definitions.

The *unawareness operator* is defined by $\mathbf{U}_a(E) := \neg\mathbf{A}_a(E)$.

Let Ag be an at most countable set of agents. Interactive beliefs are defined as usual. The *mutual p-belief operator* $\mathbf{B}^p$ is defined analogously to the mutual knowledge operator in Section 3.3.3 with $\mathbf{K}_a$ replaced by $\mathbf{B}_a$. The *common certainty operator* $\mathbf{CB}^1$ is defined analogously to the common knowledge operator but with $\mathbf{K}$ replaced $\mathbf{B}^1$.

Proposition 3.4
Let E and F be events, $\{E_l\}_{l=1,2,\ldots}$ be an at most countable collection of events, and $p, q \in [0,1]$. The following properties of belief obtain:

(o) $\mathbf{B}_a^p(E) \subseteq \mathbf{B}_a^q(E)$, for $q \leq p$,
(i) Necessitation: $\mathbf{B}_a^1(\Omega) = \Omega$,
(ii) Additivity: $\mathbf{B}_a^p(E) \subseteq \neg\mathbf{B}_a^q(\neg E)$, for $p + q > 1$,
(iiia) $\mathbf{B}_a^p\left(\bigcap_{l=1}^{\infty} E_l\right) \subseteq \bigcap_{l=1}^{\infty} \mathbf{B}_a^p(E_l)$,
(iiib) for any decreasing sequence of events $\{E_l\}_{l=1}^{\infty}$, $\mathbf{B}_a^p\left(\bigcap_{l=1}^{\infty} E_l\right) = \bigcap_{l=1}^{\infty} \mathbf{B}_a^p(E_l)$,
(iiic) $\mathbf{B}_a^1\left(\bigcap_{l=1}^{\infty} E_l\right) = \bigcap_{l=1}^{\infty} \mathbf{B}_a^1(E_l)$,
(iv)] Monotonicity: $E \subseteq F$ implies $\mathbf{B}_a^p(E) \subseteq \mathbf{B}_a^p(F)$,
(va) Introspection: $\mathbf{B}_a^p(E) \subseteq \mathbf{B}_a^1\mathbf{B}_a^p(E)$,
(vb) Introspection II: $\mathbf{B}_a^p\mathbf{B}_a^q(E) \subseteq \mathbf{B}_a^q(E)$, for $p > 0$.

Proposition 3.5
Let E be an event and $p, q \in [0,1]$. The following properties of awareness and belief obtain:

1. Plausibility: $\mathbf{U}_a(E) \subseteq \neg\mathbf{B}_a^p(E) \cap \neg\mathbf{B}_a^p\neg\mathbf{B}_a^p(E)$,
2. Strong Plausibility: $\mathbf{U}_a(E) \subseteq \bigcap_{n=1}^{\infty} \left(\neg\mathbf{B}_a^p\right)^n(E)$,
3. B^p U Introspection: $\mathbf{B}_a^p\mathbf{U}_a(E) = \emptyset^{S(E)}$ for $p \in (0,1]$ and $\mathbf{B}_a^0\mathbf{U}_a(E) = \mathbf{A}_a(E)$,
4. AU Reflection: $\mathbf{U}_a(E) = \mathbf{U}_a\mathbf{U}_a(E)$,
5. Weak Necessitation: $\mathbf{A}_a(E) = \mathbf{B}_a^1\left(S(E)^{\uparrow}\right)$,
6. $\mathbf{B}_a^p(E) \subseteq \mathbf{A}_a(E)$ and $\mathbf{B}_a^0(E) = \mathbf{A}_a(E)$,
7. $\mathbf{B}_a^p(E) \subseteq \mathbf{A}_a\mathbf{B}_a^q(E)$,
8. Symmetry: $\mathbf{A}_a(E) = \mathbf{A}_a(\neg E)$,
9. A Conjunction: $\bigcap_{\lambda \in L} \mathbf{A}_a(E_\lambda) = \mathbf{A}_a\left(\bigcap_{\lambda \in L} E_\lambda\right)$,

10. AB^p Reflection: $\mathbf{A}_a \mathbf{B}_a^p(E) = \mathbf{A}_a(E)$,
11. Awareness Reflection: $\mathbf{A}_a \mathbf{A}_a(E) = \mathbf{A}_a(E)$, and
12. $\mathbf{B}_a^p \mathbf{A}_a(E) = \mathbf{A}_a(E)$. ⊣

Proposition 3.6
Let E be an event and $p, q \in [0,1]$. The following multi-person properties obtain:

1. $\mathbf{A}_a(E) = \mathbf{A}_a \mathbf{A}_b(E)$,
2. $\mathbf{A}_a(E) = \mathbf{A}_a \mathbf{B}_b^p(E)$,
3. $\mathbf{B}_a^p(E) \subseteq \mathbf{A}_a \mathbf{B}_b^q(E)$,
4. $\mathbf{B}_a^p(E) \subseteq \mathbf{A}_a \mathbf{A}_b(E)$,
5. $\mathbf{CA}(E) = \mathbf{A}(E)$,
6. $\mathbf{CB}^1(E) \subseteq \mathbf{CA}(E)$,
7. $\mathbf{B}^p(E) \subseteq \mathbf{CA}(E)$, $\mathbf{B}^0(E) = \mathbf{CA}(E)$,
8. $\mathbf{B}^p(E) \subseteq \mathbf{A}(E)$, $\mathbf{B}^0(E) = \mathbf{A}(E)$,
9. $\mathbf{A}(E) = \mathbf{B}^1(S(E)^{\uparrow})$,
10. $\mathbf{CA}(E) = \mathbf{B}^1(S(E)^{\uparrow})$,
11. $\mathbf{CB}^1(S(E)^{\uparrow}) \subseteq \mathbf{A}(E)$,
12. $\mathbf{CB}^1(S(E)^{\uparrow}) \subseteq \mathbf{CA}(E)$, ⊣

We view unawareness type spaces are the probabilistic analogue to unawareness frames.

Unawareness type spaces capture unawareness and beliefs, beliefs about beliefs (including beliefs about unawareness), beliefs about that and so on in a parsimonious way familiar from standard type spaces. That is, hierarchies of beliefs are captured implicitly by states and type mappings. This begs two questions: First, can we construct unawareness type spaces from explicit hierarchies of beliefs? Such a construction, if possible, is complicated by the multiple awareness levels involved. Player 1 with a certain awareness level may believe that player 2 has a lower awareness level. Moreover, he may believe that player 2 believes that player 1 has yet an even lower awareness level, etc. The second question that arises is whether there exists a universal unawareness type space in the sense that *every* hierarchy of beliefs is represented in it. In using such a universal type space for an application, the modeler ensures that she can analyze any hierarchy of beliefs. These questions have been answered in the literature; see the notes section at the end of this chapter.

The presentation of unawareness type spaces is somewhat divorced from awareness structures and unawareness structures presented earlier. Those structures we could axiomatize. We could describe in minute detail knowledge and awareness of all agents in each state. While the hierarchical construction of unawareness type spaces retains the flavor of explicit descriptions of beliefs, it begs the question of whether unawareness type spaces could be axiomatized using a logic with modal operators p_a^{μ} interpreted as "agent a assigns probability at least μ", for rational numbers $\mu \in [0,1]$. That is, can we axiomatize the probabilistic analogue of awareness structures or unawareness structures? An satisfactory extension to multi-agent unawareness is still open; see again the notes section at the end of this chapter for comments on the existing literature.

3.4.2 Speculative trade and agreement

In this section we revisit the speculative trade example discussed earlier. Unawareness type spaces allow us now to provide an answer to the question posed in the introduction, namely whether at a price of $100 per share the owner is going to sell to the buyer. If this question is answered in the affirmative, then under unawareness we have a counterexample to the "No-speculative-trade" theorem for standard structures, thus illustrating that asymmetric awareness may have different implications from asymmetric (standard) information. In standard structures, if there is a common prior probability (i.e., common among agents), then common certainty of willingness to trade implies that agents are indifferent to trade. To address how our example fits to the "No-speculative-trade" theorems, we need to recast it into an unawareness type space with a common prior. This is illustrated in Figure 3.4.

Figure 3.4: An Unawareness Type Space with a Common Prior for the Speculative Trade Example

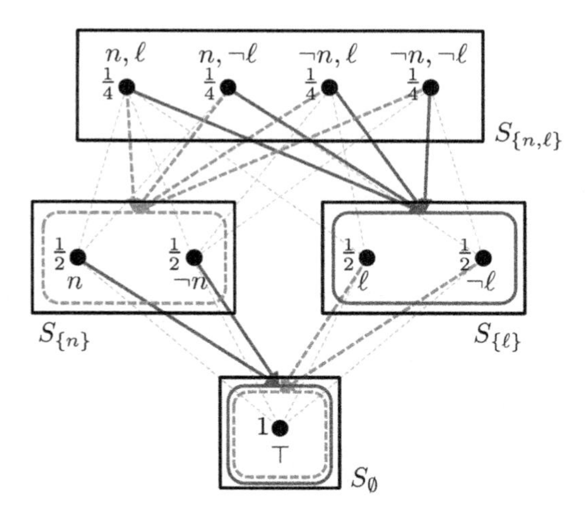

The type-mappings are represented in Figure 3.4 as follows. At any state in the upmost space $S_{\{n,\ell\}}$, the buyer's belief has full support on the left space $S_{\{n\}}$ given by the red intermitted soft-edged rectangle and the owner's

belief has full support on $S_{\{\ell\}}$ given by the solid blue soft-edge rectangle. At any state in $S_{\{n\}}$ the owner's belief has full support on the lowest space $S_{\emptyset}$. Analogously, the owner is certain that the buyer is unaware of the law suit since at any state in $S_{\{\ell\}}$ the belief of the buyer has full support on the space $S_{\emptyset}$. This example is analogous to Figure 3.2 except that the supports of types are displayed rather the possibility sets and we write to the left of each state its common prior probability as well. For instance, state (n, ℓ) has common prior probability $\frac{1}{4}$. We see that, for instance, the common prior on S_n is the marginal of the common prior on $S_{\{n,\ell\}}$. Indeed, the common prior in unawareness type spaces generally constitutes a projective system of probability measures. Both agents' beliefs are consistent with the common prior. Of course, referring to a "prior" is misleading terminology under unawareness as it is nonsensical to think of a prior stage at which all agents are aware of all states while at the interim stage, after they received their type, they are unaware of some events. Rather than understanding the prior as a primitive of the model, it should be considered as derived from the types of players. As in standard structures, it is a convex combination of types.

Say that the buyer prefers to buy at price x if his expected value of the firm is at least x, while the owner prefers to sell at price x if her expected value is at most x. The buyer strictly prefers to buy at price x if his expected value of the firm is strictly above x, while the owner strictly prefers to sell at price x if her expected value is strictly below x. Note that at any state in $S_{\{n,\ell\}}$, the owner's beliefs are concentrated on $S_{\{\ell\}}$ and thus his expected value of a firm's share is \$90. Similarly, at any state in $S_{\{n,\ell\}}$, the buyer's beliefs are concentrated on $S_{\{n\}}$ and thus her expected value of a firm's share is \$110. Thus the owner strictly prefers to sell at the price \$100 while the buyer strictly prefers to buy at the price of \$100. Moreover, at the price \$100 it is common certainty that each agent prefers to trade because each agent strictly prefers to trade at \$100 and is certain that the other agent is indifferent between trading or not at \$100. Hence, we have a common prior, common certainty of willingness to trade but each agent has a strict preference to trade. We conclude that speculative trade is possible under asymmetric awareness while it is ruled out under symmetric awareness by the standard "No-speculative-trade" theorems.

At a second glance, we realize that speculative trade is a knife-edge case in this example. Suppose that there are some transaction costs. For instance, the government may require the buyer to pay a tax of \$1 per share. Then the owner knows that the buyer is not just indifferent between buying or not but must have a strict preference to trade as well (similar for the buyer). This leads to the question of whether the common prior assumption rules out common certainty of strict preference to trade. One can prove for unawareness type spaces that a non-degenerate common prior rules out common certainty of strict preference to trade. Thus, arbitrary small transaction costs such as the famous Tobin tax on transactions rule out speculation under unawareness. The "No-speculative-trade" result under unawareness is also relevant for the following reason: one may casually conjecture that any behavior is possible when awareness is allowed to vary among agents and thus behavior under unawareness may have no testable predictions. The "No-speculative-trade" result under unawareness shows that this is not the case.

3.5 Awareness of unawareness

According to KU introspection, an agent never knows or believes that she is unaware of a specific event. This does not mean that she couldn't know that she is unaware of *something.* There is a difference between knowing (or not knowing) that you are unaware of *the proposition* φ and knowing (or not knowing) that *there exists some proposition* that you are unaware of. A primary care physician may be unaware of a specific disease and may not even realize that she is unaware of this specific disease. Nevertheless she may refer a patient to a specialist because she believes that the specialist is aware of some diseases that she doesn't even think or have heard about. One can argue that agents should generally induce from prior experience and experience with other agents that they may be unaware of something. All previous approaches outlined so far are silent on awareness of unawareness of something. As the discussion suggests, we could model awareness of unawareness with an existential quantifier like "I am uncertain about whether *there exists* a proposition that I am unaware of". In this section, we will present several alternative approaches.

3.5.1 Propositional quantifiers and extended awareness structures

Given a nonempty set of agents $\mathsf{Ag} = \{1, \ldots, n\}$ indexed by a, a countable infinite set of primitive propositions At as well as a countable infinite set of variables X, the languages are $\mathsf{L}_n^{\forall,K,A}(\mathsf{At}', \mathsf{X})$, $\emptyset \neq \mathsf{At}' \subseteq \mathsf{At}$. Different from $\mathsf{L}_n^{K,A}(\mathsf{At})$ introduced earlier, we allow for quantification with domain $\mathsf{L}_n^{K,A}(\mathsf{At})$: If φ is a formula in $\mathsf{L}_n^{K,A}(\mathsf{At}')$, then $\forall x \varphi$ is a formula in $\mathsf{L}_n^{\forall,K,A}(\mathsf{At}', \mathsf{X})$. That is, the domain of quantification are just quantifier-free formulae. As usual, we define $\exists x \varphi$ by $\neg \forall x \neg \varphi$.

An occurrence of a variable x is free in a formula φ if it is not bound by a quantifier. More formally, define inductively: If φ does not contain a quantifier, then every occurrence of x is free in φ. An occurrence of the variable x is free in $\neg\varphi$, (in $K_a\varphi$ and $A_a\varphi$, respectively) if and only if its corresponding occurrence is free in φ. An occurrence of the variable x is free in $\varphi \wedge \psi$ if and only if the corresponding occurrence of x in φ or ψ is free. An occurrence of x is free in $\forall y \varphi$ if and only if the corresponding occurrence of x is free in φ and x is different from y. A formula that contains no free variables is a *sentence*.

If ψ is a formula, we denote by $\varphi[x/\psi]$ the formula that results in replacing all free occurrences of the variable x in φ with ψ.

An *extended awareness structure* is a tuple $M = (S, (R_a)_{a\in \mathsf{Ag}}, (\mathcal{A}_a)_{a\in \mathsf{Ag}}, V, \mathcal{A}t)$, where $(S, (R_a)_{a\in \mathsf{Ag}}(\mathcal{A}_a)_{a\in \mathsf{Ag}}, V)$ is an awareness structure as introduced in Section 3.3.1 and $\mathcal{A}t : S \longrightarrow 2^{\mathsf{At}} \setminus \{\emptyset\}$ is a correspondence that assigns to each state s in S a nonempty subset of primitive propositions in At. We require that at each state every agent is only aware of sentences that are in the language of this state. That is, $\mathcal{A}_a(s) \subseteq \mathsf{L}_n^{\forall,K,A}(\mathcal{A}t(s), \mathsf{X})$. Moreover, the properties "awareness generated by primitive propositions" and "agents know what they are aware of" take the following form: for all $a \in \mathsf{Ag}$ and $s, s' \in S$, if $(s, s') \in R_a$ then $\mathcal{A}_a(s) \subseteq \mathsf{L}_n^{\forall,K,A}(\mathcal{A}t(s'), \mathsf{X})$.

The satisfaction relation is defined inductively on the structure of formulae in $\mathsf{L}_n^{\forall,K,A}(\mathsf{At}, \mathsf{X})$ as follows:

$M, s \models p$ if and only if $p \in \mathcal{A}t(s)$ and $V(s, p) = \mathit{true}$,
$M, s \models \varphi \wedge \psi$ if and only if $\varphi, \psi \in \mathsf{L}_n^{\forall,K,A}(\mathcal{A}t(s), \mathsf{X})$ and both $M, s \models \varphi$ and $M, s \models \psi$,
$M, s \models \neg\varphi$ if and only if $\varphi \in \mathsf{L}_n^{\forall,K,A}(\mathcal{A}t(s), \mathsf{X})$ and $M, s \not\models \varphi$,
$M, s \models A_a\varphi$ if and only if $\varphi \in \mathsf{L}_n^{\forall,K,A}(\mathcal{A}t(s), \mathsf{X})$ and $\varphi \in \mathcal{A}_a(s)$,
$M, s \models K_a\varphi$ if and only if $\varphi \in \mathsf{L}_n^{\forall,K,A}(\mathcal{A}t(s), \mathsf{X})$ and both $M, s \models A_a\varphi$ and $M, t \models \varphi$ for all $t \in S$ such that $(s, t) \in R_a$,
$M, s \models \forall x\varphi$ if and only if $\varphi \in \mathsf{L}_n^{\forall,K,A}(\mathcal{A}t(s), \mathsf{X})$ and $M, s \models \varphi[x/\psi]$ for all $\psi \in \mathsf{L}_n^{K,A}(\mathsf{At})$.

Note that for a formula φ to be true at a world s, we require now also that $\varphi \in \mathsf{L}_n^{\forall,K,A}(\mathcal{A}t(s), \mathsf{X})$. We say that a formula φ is defined at s if (with the obvious abuse of notation) $\mathsf{At}(\varphi) \subseteq \mathcal{A}t(s)$. Otherwise, φ is undefined at s. Like in unawareness structures, a formula may not be defined in all states in an extended awareness structure. This requires us to define validity analogous to unawareness structures, that is, a formula φ is valid in an extended awareness structure M if $M, s \models \varphi$ for all s in which φ is defined. The notions of soundness and completeness are as defined previously.

Consider the following axiom system that we call $\mathsf{S5}_n^{\forall,K,A}$:

Prop. All substitution instances of valid formulae of propositional logic.

AGPP. $A_a\varphi \leftrightarrow \bigwedge_{p\in \mathsf{At}(\varphi)} A_a p$

AI. $A_a\varphi \rightarrow K_a A_a\varphi$

KA. $K_a\varphi \rightarrow A_i\varphi$

K. $(K_a\varphi \wedge K_a(\varphi \rightarrow \psi)) \rightarrow K_a\psi$

T. $K_a\varphi \rightarrow \varphi$

4. $K_a\varphi \rightarrow K_a K_a\varphi$

5_A. $\neg K_a\varphi \wedge A_a\varphi \rightarrow K_a\neg K_a\varphi$

$1_\forall$. $\forall x\varphi \rightarrow \varphi[x/\psi]$ if ψ is a quantifier-free sentence

$6_\forall$. $\forall x(\varphi \rightarrow \psi) \rightarrow (\forall x\varphi \rightarrow \forall x\psi)$

$N_\forall$. $\varphi \rightarrow \forall x\varphi$ if x is not free in φ

FA. $\forall x U_a x \rightarrow K_a \forall x U_a x$

Barcan^*_A. $(A_a(\forall x\varphi) \wedge \forall x(A_a x \rightarrow K_a\varphi)) \rightarrow K_a(\forall x A_a x \rightarrow \forall x\varphi)$

MP. From φ and $\varphi \rightarrow \psi$ infer ψ.

Gen_A. From φ infer $A_a\varphi \rightarrow K_a\varphi$.

$Gen_\forall$. If $q \in \mathsf{At}$, then from φ infer $\forall x\varphi[q/x]$.

All axioms and inference rules that do not involve quantification were discussed earlier. $1_\forall$ means that if a universally quantified formula is true then so is every instance of it. FA and Barcan^*_A are more difficult to interpret. FA says that if an agent is unaware of everything then she knows that she is unaware of everything. It is hard to judge the reasonableness of this axiom as the hypothesis of being unaware of everything is extreme. At a first glance, it may even appear paradoxical: If she knows that she is unaware of everything then by KA she is aware that she is unaware of everything. But if she is aware that she is unaware of everything, how can she be unaware of everything? Recall that quantification is just over quantifier-free sentences. Thus, the agent may be unaware of every quantifier-free sentence but still be aware that she is unaware of every quantifier-free sentence.

Barcan^*_A should be contrasted with the "standard" Barcan axiom $\forall x K_a\varphi \rightarrow K_a\forall x\varphi$: If the agent knows $\varphi[x/\psi]$ for every quantifier-free sentence ψ, then she knows $\forall x\varphi$. Barcan^*_A. also connects knowledge with quantification but it requires awareness. In the antecedent it requires that the agent is aware of the formula $\forall x\varphi$. Moreover, the agent is required to know $\varphi[x/\psi]$ only if she is aware of ψ. In the conclusions, $\forall x\varphi$ is true only if the agent is aware of all (quantifier-free) formulae.

Theorem 3.11
For the language $\mathsf{L}_n^{\forall,K,A}(\mathsf{At},\mathsf{X})$, the axiom system $\mathsf{S5}_n^{\forall,K,A}$ is a sound and complete axiomatization with respect to extended awareness structures. $\dashv$

The completeness part of the proof is by constructing a canonical model but dealing appropriately with complications arising from quantification.

The soundness parts for modus ponens, Barcan$^*_A$, and Gen$_\forall$ are nonstandard.

Extended awareness structures merge features of both awareness structures and unawareness structures introduced in Sections 3.3.1 and 3.3.4, respectively. Recall that in awareness structures the awareness correspondence associates potentially different subsets of formulae with different states but all formulae are defined at each state, while in unawareness structures potentially different subsets of formulae are defined at different states. This difference between the formalisms are immaterial as long as we are "just" interested in modeling reasoning about knowledge and propositionally determined awareness and do not care about the important conceptual issue of whether structures can be viewed from an agent's subjective perspective. In extended awareness structures, the difference between formulae defined at a state and the formulae that an agent is aware of at that state is of conceptual significance for a second reason. Roughly these are the labels that the agent is aware that he is unaware of.

3.5.2 First-order logic with unawareness of objects

We now present a first-order modal logic with unawareness in order to model awareness of unawareness. Different from extended awareness structures, the quantification is over objects rather than over quantifier-free formulae.

Given a nonempty set of agents $\mathsf{Ag} = \{1, \ldots, n\}$, a countable infinite set of variables X, and k-ary predicates P for every $k = 1, 2, \ldots$, the set of atomic formulae At is generated by $P(x_1, \ldots, x_k)$ where $x_1, \ldots, x_k \in \mathsf{X}$. We require that there is a unary predicate E. The intended interpretation of $E(x)$ is "x is real". The language we consider is $\mathsf{L}_n^{\forall,L,K,A}(\mathsf{At}, \mathsf{X})$. We allow for quantification: If φ is a formula in $\mathsf{L}_n^{\forall,L,K,A}(\mathsf{At}, \mathsf{X})$ and $x \in \mathsf{X}$, then $\forall x \varphi \in \mathsf{L}_n^{\forall,L,K,A}(\mathsf{At}, \mathsf{X})$. We define a variable to be free in a formula as in Section 3.5.1. Moreover, if φ is a formula, we denote by $\varphi[x/y]$ the formula that results from replacing all free occurrences of x with y.

An *object-based unawareness structure* is a tuple $M = (S, D, \{D(s)\}_{s \in S}, (\Pi_a)_{a \in \mathsf{Ag}}, (\mathcal{A}_a)_{a \in \mathsf{Ag}}, \pi)$, where S is a nonempty set of states, D is a nonempty set of objects, $D(s)$ is a nonempty subset of D containing objects that are "real" in s, and $\Pi_a : S \longrightarrow 2^S$ is a possibility correspondence of agent $a \in \mathsf{Ag}$. We focus here on the case in which, for each agent $a \in \mathsf{Ag}$, the possibility correspondence forms a partition of the state-space. That is, we assume that it satisfies

Reflexivity: $s \in \Pi_a(s)$ for all $s \in S$,
Stationarity: $s' \in \Pi_a(s)$ implies $\Pi_a(s') = \Pi_a(s)$, for all $s, s' \in S$.

$\mathcal{A}_a : S \longrightarrow 2^D$ is the awareness correspondence of agent $a \in \mathsf{Ag}$. Different from awareness structures or the product model discussed earlier, the awareness correspondence in object-based unawareness structures assigns subsets of objects to states. We focus here on the case, in which for each agent $a \in \mathsf{Ag}$, the possibility correspondence and the awareness correspondence satisfy jointly

$$s' \in \Pi_a(s) \text{ implies } \mathcal{A}_a(s') = \mathcal{A}_a(s).$$

Thus analogous to the corresponding property in awareness structures, agents know what they are aware of.

π is a state-dependent assignment of a k-ary relation $\pi(s)(P) \subseteq D^k$ to each k-ary predicate P. Intuitively, the assignment π ascribes in each state and to each property the subset of objects satisfying this property at that state. It is sometimes called a classical first-order interpretation function.

A valuation $V : \mathsf{X} \longrightarrow D$ assigns to each variable an object. Intuitively, $V(x)$ denotes the object referred to by variable x, provided that x is free in a given formula. Call V' is an x-alternative valuation of V if, for every variable y except possibly x, $V'(y) = V(y)$.

Since the truth value of a formula depends on the valuation, on the left-hand side of $\models$ we have a model, a state in the model, and a *valuation*. The satisfaction relation is defined inductively on the structure of formulae in $\mathsf{L}_n^{\forall,L,K,A}(\mathsf{At}, \mathsf{X})$ as follows:

$M, s, V \models E(x)$ if and only if $V(x) \in D(s)$,
$M, s, V \models P(x_1, \ldots, x_k)$ if and only if $(V(x_1), \ldots, V(x_k)) \in \pi(s)(P)$,
$M, s, V \models \neg\varphi$ if and only if $M, s, V \not\models \varphi$,
$M, s, V \models \varphi \wedge \psi$ if and only if $M, s, V \models \varphi$ and $M, s, V \models \psi$,
$M, s, V \models \forall x\varphi$ if and only if $M, s, V' \models \varphi$ and $V'(x) \in D(s)$ for every x-alternative valuation V',
$M, s, V \models A_a\varphi$ if and only if $V(x) \in \mathcal{A}_a(s)$ for every x that is free in φ,
$M, s, V \models L_a\varphi$ if $M, s', V \models \varphi$ for all $s' \in \Pi_a(s)$,
$M, s, V \models K_a\varphi$ if and only if $M, s, V \models A_a\varphi$ and $M, s, V \models L_a\varphi$.

A formula φ is valid in the object-based unawareness structure M under the valuation V if $M, s, V \models \varphi$ for all $s \in S$. The notions of soundness and completeness are now the standard notions.

The unawareness operator is defined, as usual, as the negation of awareness, that is, $U_a\varphi := \neg A_a\varphi$.

Consider the following axiom system that we call $\mathsf{S5}_n^{\forall,L,K,A}$:

Prop. All substitution instances of valid formulae of propositional logic.
A. $A_a\varphi$ if there is no free variable in φ;
AC$^-$. $A_a\varphi \wedge A_a\psi \to A_a(\varphi \wedge \psi)$
A3. If every variable free in ψ is also free in φ, then $A_a\varphi \to A_a\psi$.
K. $L_a(\varphi \to \psi) \to (L_a\varphi \to L_a\psi)$
T. $L_a\varphi \to \varphi$
4. $L_a\varphi \to L_aL_a\varphi$
5. $\neg L_a\varphi \to L_a\neg L_a\varphi$
KL. $K_a\varphi \leftrightarrow A_a\varphi \wedge L_a\varphi$.
AI$_L$. $A_a\varphi \to L_aA_a\varphi$.
UI$_L$. $U_a\varphi \to L_aU_a\varphi$.
E. $\forall x E(x)$.
$1_{\forall,E}$. $\forall x\varphi \to (E(y) \to \varphi[x/y])$.
$6_\forall$. $\forall x(\varphi \to \psi) \to (\forall x\varphi \to \forall x\psi)$
$N^*_\forall$. $\varphi \leftrightarrow \forall x\varphi$ if x is not free in φ
MP. From φ and $\varphi \to \psi$ infer ψ.
Gen$_L$. From φ infer $L_a\varphi$.
Gen$_{\forall,L}$. For all natural numbers $n \geq 1$, from $\varphi \to L_a(\varphi_1 \to \cdots \to L_a(\varphi_n \to L_a\psi)\cdots)$ infer $\varphi \to L_a(\varphi_1 \to \cdots \to L_a(\varphi_n \to L_a\forall x\psi)\cdots)$, provided that x is not free in $\varphi, \varphi_1, \ldots, \varphi_n$.
Gen$^*_\forall$. From φ infer $\forall x\varphi$.

Theorem 3.12
For the language $\mathsf{L}_n^{\forall,L,K,A}(\mathsf{At},\mathsf{X})$, the axiom system $\mathsf{S5}_n^{\forall,L,K,A}$ is a sound and complete axiomatization with respect to object-based unawareness structures. ⊣

For object-based unawareness structures, the literature considered also the analogue of "frames". This approach is not purely event-based, though, as it requires the modeler to consider for each event also the set of objects referred to in the event. Formally, an *event* in an object-based unawareness frame is a pair (E, O), where $E \in 2^S$ is a subset of states and $O \in 2^D$ is a subset of objects. (E is now a set of states, not the existence predicate introduced earlier.) We let $states(E, O) := E$, and $objects(E, O) := O$. Negation and conjunction of events are defined by

$$\begin{aligned} \neg(E, O) &:= (S \setminus E, O), \\ \bigwedge_i (E_i, O_i) &:= \left(\bigcap_i E_i, \bigcup_i O_i\right). \end{aligned}$$

The negation pertains to the set of states in which E does not obtain but refers to the same set of objects. The conjunctions of events is the set of states in which all these events obtain and the union of objects referred to by those events. Conjunction is defined by the De-Morgan law by

$$\bigvee_i (E_i, O_i) = \neg\left(\bigwedge_i \neg(E_i, O_i)\right) = \left(\bigcup_i E_i, \bigcup_i O_i\right).$$

We let Σ denote the set of all events.

For each agent $a \in \mathsf{Ag}$, the awareness operator is defined on events by

$$\mathsf{A}_a(E, O) \quad := \quad (\{s \in S : O \subseteq \mathcal{A}_a(s)\}, O).$$

As before, the unawareness operator is defined as the negation of the awareness, $\mathsf{U}_a(E, O) := \neg\mathsf{A}_a(E, O)$.

The implicit knowledge operator is defined on events by

$$\mathsf{L}_a(E, O) \quad := \quad (\{s \in S : \Pi_a(s) \subseteq E\}, O).$$

Explicit knowledge is defined, as in awareness structures, by the conjunction of awareness and implicit knowledge, i.e.,

$$\mathsf{K}_a(E, O) \quad := \quad \mathsf{A}_a(E, O) \wedge \mathsf{L}_a(E, O).$$

Awareness, implicit knowledge, and explicit knowledge of an event with a given subset of objects are events, respectively, with the same subset of objects.

Properties are defined as functions $p : D \longrightarrow \Sigma$ such that $p(o) = (E_o^p, O^p \cup \{o\})$ for some $E_o^p \in 2^S$ and $O^p \in 2^D$. E_o^p is the set of states in which object o possesses property p and O^p is the set of objects referred to in that property. For instance, a property could be "... has as many legs as horses." If object o is a unicorn, then E_o^p is the set of states in which this unicorn has as many legs as horses and O^p is the set of horses.

Object-based unawareness structures allow for quantification over objects. In the object-based unawareness frame, we will consider quantified events. We focus on an actualist quantifier that ranges over objects that "actually exist". Formally, first define the property

$$e(o) = (\{s \in S : o \in D(s)\}, \{o\}).$$

That is, $e(o)$ is the event that object o exists. For any property p, the event that all (actually existing) objects satisfy property p is defined by

$$Allp = \left(\bigcap_{o \in D} E_o^{e \to p}, O^p\right).$$

Allp obtains if all existing objects possess property p. Quantified events satisfy the following properties:

(i) $All(\wedge_i p_i) = \wedge_i(All p_i)$
(ii) If $s \in E_o^p$ for every $o \in D$, then $s \in states(All p)$.
(iii) If $E_o^p = E_o^q$ for every $o \in D$, then $states(All p) = states(All q)$.

At this point, it may be helpful to consider our simple example.

Example 3.5 (Speculative trade (continued))
Let ℓ denote the object "lawsuit" and n the object "innovation". Figure 3.5 presents a simply objective-based unawareness frame. There are four states. Below each state we indicate which atomic formulas are true or false. The picture is analogous to the corresponding picture for awareness structures (Figure 3.1) except that the awareness correspondence is now indicated by rectangular text bubbles above states in which we indicate the set of objects the agent is aware of that state. The blue solid-lined rectangular text bubbles belong to the owner while the red intermitted-lined are the buyer's. Each agent's awareness correspondence is very special in this example because it is constant across states. As in Figure 3.1, the soft-edged rectangles indicate the possibility correspondences. The blue solid-lined possibility set belongs to the owner, while the red intermitted-lined is the buyer's. Both agents consider all states possible. This simple figure models the story out-

Figure 3.5: An Object-Based Unawareness Frame for the Speculative Trade Example

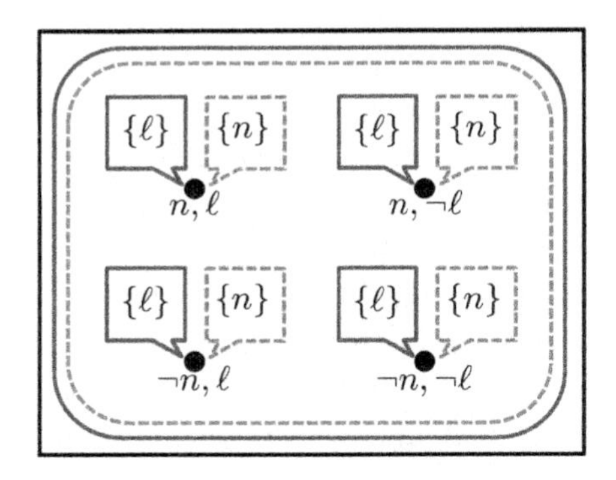

lined in the introduction. The awareness correspondences shows that at any state, the owner is aware of the lawsuit and unaware of the innovation, while the buyer is aware of the innovation and unaware of the lawsuit. The possibility correspondences model implicit knowledge. The owner does not implicitly know whether the law suit obtains, and implicitly knows that the buyer is unaware of the lawsuit. But he also explicitly knows that because he is aware of the lawsuit. The buyer does not implicitly know whether the innovation obtains, and implicitly knows that the owner is unaware of the innovation. But she also explicitly knows that because she is aware of the innovation. Both agents also implicitly know what they are unaware of. This is hard to interpret. Similar to awareness structures discussed earlier, object-based unawareness structures are best understood from an outside modeler's point of view. The same structure can generally not be used as an analytical device by the agent herself to reason about her and other agents' knowledge and awareness.

We should mention that the introductory speculative trade example does not do full justice to objective-based unawareness structures as the example does not make use of quantification over objects. ⊣

The object-based unawareness frame allows us to easily compare this approach to unawareness frames introduced in Section 3.3.3. Already on an informal level, some differences become obvious. First, quantification is not explicitly considered in unawareness frames. Second, similar to awareness structures, the set of states in object-based unawareness structures are "objective" descriptions that should be interpreted as given to an outside observer. This is different from "subjective" descriptions in unawareness structures. Third, there is no axiomatization for a language with explicit knowledge and awareness only; the possibility correspondences model implicit knowledge. Arguably, explicit knowledge is what is of ultimate interest in applications and this is the notion focused on in unawareness structures. Finally, object-based unawareness frames model unawareness about propositions where the unawareness arises from unawareness of objects referred to in the propositions, and no unawareness of properties is considered. In contrast, unawareness structures model unawareness of abstract propositions. Yet, the more fine-grained distinction between objects and properties in object-based unawareness structures may yield an advantage in some applications where this distinction may be necessary.

More formally, common to both frames is that there is a set of events Σ, a negation operator $\neg$, a conjunction operator $\wedge$, and for each agent $a \in \mathsf{Ag}$ a knowledge operator K_a and an awareness operator A_a defined on events in Σ. We say that a frame $(\Sigma, \neg, \wedge, (\mathsf{K}_a, \mathsf{A}_a)_{a \in \mathsf{Ag}})$ can be embedded into a frame $(\Sigma', \neg, \wedge, (\mathsf{K}'_a, \mathsf{A}'_a)_{a \in \mathsf{Ag}})$ if there is an injective function $f : \Sigma \longrightarrow \Sigma'$ with the following properties: For any events $E, F \in \Sigma$,

Negation-Preserving: $f(\neg E) = \neg f(E)$
Conjunction-Preserving: $f(E \wedge F) = f(E) \wedge f(F)$
Knowledge-Preserving: $f(\mathsf{K}_a(E)) = \mathsf{K}'_a(f(E))$
Awareness-Preserving: $f(\mathsf{A}_a(E)) = \mathsf{A}'_a(f(E))$

Theorem 3.13
Every object-based unawareness frame can be embedded into some unawareness frame. Conversely, every unawareness frame can be embedded into an object-based unawareness frame that does not necessarily satisfy that agents know what they are aware of. ⊣

The proof is constructive in that one can construct an embedding and show that it "works". Moreover, one can show that the property of 'agents know what they are aware of' is required for any embedding of object-based unawareness frames into unawareness frames. One can also show that if generalized reflexivity (i.e., the truth axiom) and stationarity (i.e, the introspection properties of knowledge) are dropped from unawareness frames, then any object-based unawareness frame (not necessarily satisfying reflexivity, stationarity, or 'agents know what they are aware of') can be embedded into an unawareness frame and vice versa. While this result is mathematically more general and "cleaner" than Theorem 3.13 (because it has the full converse), it is of less interest because we want to know how even strong notions of knowledge are embedded into various frames. So far, it is open whether with strong notions of knowledge a full converse can be obtained with some different embedding than the one used in the proof.

3.5.3 Neighborhood semantics and first-order logic with awareness

In the literature, there is also a first-order logic of awareness with a semantics based on awareness neighborhood frames mentioned already in Section 3.3.1. The language is as in Section 3.5.2. That is, $\mathsf{L}_n^{\forall,L,K,A}(\mathsf{At},\mathsf{X})$ consists of well-formed formulae having the following syntax

$$\varphi ::= P(x_1,\ldots,x_k) \mid A!_a(x) \mid \neg\varphi \mid \varphi\wedge\psi \mid L_a\varphi \mid K_a\varphi \mid A_a\varphi \mid \forall x\varphi$$

$A!_a(x)$ denotes the awareness predicate of agent a, which is somewhat similar to the agent-independent existence predicate $E(x)$ in Section 3.5.2.

An awareness neighborhood structure is $M = (S, (\mathcal{N}_a)_{a\in\mathsf{Ag}}, (\mathcal{A}_a)_{a\in\mathsf{Ag}}, D, (D_a)_{a\in\mathsf{Ag}}, \pi, (\pi_a)_{a\in\mathsf{Ag}})$ in which S is a space of states and $\mathcal{N}_a : S \longrightarrow 2^{2^S}$ is the neighborhood correspondence of agent a that assigns to each state the set of events that the agent knows at this state. As in awareness structures, $\mathcal{A}_a : S \longrightarrow 2^{L_n^{\forall,L,K,A}(\mathsf{At},\mathsf{X})}$ is the awareness correspondence of agent a. D is a nonempty set called the domain. $D_a : S \longrightarrow 2^D$ is a correspondence of agent a assigning to each state s a subjective domain $D_a(s)$ of objects. Intuitively, $D_a(s)$ represents the objects that agent a is aware of at state s. As in the previous section, π is a state-dependent assignment of a k-ary relation $\pi(s)(P) \subseteq D^k$ to each k-ary predicate P. Finally, for each agent $a \in \mathsf{Ag}$,

π_a is a state-dependent assignment of a k-ary relation $\pi_a(s)(P) \subseteq D_a^k(s)$ to each k-ary predicate P that may possibly agree partially with π. This "agent-based" assignment is motivated by the desire to model also awareness of properties.

Recall that in first-order logic an atomic formula takes the form $P(x_1, \ldots, x_k)$ where P is a k-ary predicate and $x_1, \ldots, x_k \in \mathsf{X}$. The notion of 'awareness generated by atomic formulae' is analogous to awareness structures, i.e., for all $s \in S$, $\varphi \in \mathcal{A}_a(s)$ if and only if $\mathsf{At}(\varphi) \subseteq \mathcal{A}_a(s)$. We require that $P(x_1, \ldots, x_k) \in \mathcal{A}_a(s)$ if and only if (i) $V(x_\ell) \in D_a(s)$ for $\ell = 1, \ldots, k$, and (ii) $(V(x_1), \ldots, V(x_k)) \in \pi_a(s)(P)$. As before, $V : \mathsf{X} \longrightarrow D$ is a valuation or substitution. By Property (i), if at a state an agent is aware of an atomic formula then at that state she must be aware of any object referred to in the atomic formula. This formalizes the idea of awareness of objects. Property (ii) is interpreted as formalizing the idea of awareness of properties of objects. In order for an agent to be aware of a given atomic formula, she needs to be also aware of the property mentioned in the formula, i.e., she needs to be aware that the objects in the formula enjoy a given property. This is different from Section 3.5.2 where only awareness of objects is considered.

The satisfaction relation is defined by

$M, s, V \models A!_a(x)$ if and only if $V(x) \in D_a(s)$,
$M, s, V \models P(x_1, \ldots, x_k)$ if and only if $(V(x_1), \ldots, V(x_k)) \in \pi(s)(P)$
$M, s, V \models \neg\varphi$ if and only if $M, s, V \not\models \varphi$,
$M, s, V \models \varphi \wedge \psi$ if and only if $M, s, V \models \varphi$ and $M, s, V \models \psi$,
$M, s, V \models \forall x \varphi(x)$ if and only if $M, s, V' \models \varphi$ for every x-alternative valuation V' for which $V'(y) = V(y)$ for all $y \neq x$,
$M, s, V \models L_a\varphi$ if and only if $\{t \in S : M, t, V \models \varphi\} \in \mathcal{N}_a(s)$,
$M, s, V \models A_a\varphi$ if and only if $\varphi \in \mathcal{A}_a(s)$,
$M, s, V \models K_a\varphi$ if and only if $M, s, V \models A_a\varphi$ and $M, s, V \models L_a\varphi$.

In addition to some axioms on awareness discussed already in Section 3.3.1, we present two axioms that relate awareness and quantifiers. Recall that $\exists x\varphi$ stands for $\neg\forall x \neg\varphi$. The first axiom is

A$\exists$. $A_a\varphi[x/y] \rightarrow A_a\exists x\varphi(x)$.

If agent a is aware that y, which substitutes a free x, has property φ, then she is aware that there exists an x with property φ. Second,

A$\forall$. $A_a\forall x\varphi(x) \rightarrow (A!_a(y) \rightarrow A_a\varphi[x/y])$.

If agent a is aware that any x has property φ, then, provided that a is aware of y, she is aware that y, which substitutes a free x, has property φ. This axiom has a similar flavor as the axiom $1_{\forall,E}$ in the previous section except that it now involves awareness.

3.5.4 Awareness of unawareness without quantifiers

In this section we model awareness of unawareness by propositional constants such as "agent a is aware of everything" and "agent b is aware of everything that agent a is aware of" rather than with quantifiers. We will present a two-stage semantics in order to allow an agent also to be uncertain about her awareness of unawareness.

Let $\mathsf{L}_n^{L,K,A,F,R}(\mathsf{At})$ consists of well-formed formulae having the following syntax

$$\varphi ::= p \mid F_a \mid R_{ab} \mid \varphi \mid \neg\varphi \mid \varphi \wedge \psi \mid L_a\varphi \mid K_a\varphi \mid A_a\varphi$$

The propositional constants, F_a and R_{ab} for $a, b \in \mathsf{Ag}$, are new. Formula F_a stands for "agent a is aware of everything" (i.e., "full" awareness) while formula R_{ab} reads "agent b is aware of everything that agent a is aware of" (i.e., "relative" awareness). (Note that, different from previous sections, R_{ab} does not denote the accessibility relation but a propositional constant.)

A *modified awareness structure* $M = (S, (\Pi_a)_{a\in\mathsf{Ag}}, (\mathcal{A}_a)_{a\in\mathsf{Ag}}, (\trianglerighteq_a^s)_{a\in\mathsf{Ag}, s\in S}, V)$ consists of a space of states S and for each agent $a \in \mathsf{Ag}$ a possibility correspondence $\Pi_a : S \longrightarrow 2^S$. As in the previous sections, we will focus on the case where Π_a forms a partition of S. That is, for all $a \in \mathsf{Ag}$ and $s \in S$, we require

Reflexivity: $s \in \Pi_a(s)$, and
Stationarity: $s' \in \Pi_a(s)$ implies $\Pi_a(s') = \Pi_a(s)$.

Modified awareness structures in which every Π_a forms a partition are called *partitional modified awareness structures.*

The awareness correspondence $\mathcal{A}_a : S \longrightarrow 2^{\mathsf{At}}$ assigns to each state a subset of atomic formulae. Note that different from awareness structures introduced in Section 3.3.1, the co-domain of the awareness correspondence is restricted to the set of all subsets of atomic formulae only (instead allowing for the entire language). We focus on the case in which *agents know what they are aware of,* that is, for all $a \in \mathsf{Ag}$ and $s \in S$,

$$s' \in \Pi_a(s) \text{ implies } \mathcal{A}_a(s') = \mathcal{A}_a(s).$$

The next ingredient is new: For each state $s \in S$, $\trianglerighteq_a^s$ is a preorder (i.e., a reflexive and transitive binary relation) on $\mathsf{Ag} \cup \{\mathsf{At}\}$ with $\mathsf{At} \in \max_{\trianglerighteq_a^s}\{\mathsf{Ag} \cup \{\mathsf{At}\}\}$. The preorder $\trianglerighteq_a^s$ describes agent a's conjecture about the relative extent of all agent's awareness at state s. $b \trianglerighteq_a^s c$ means that agent a conjectures in state s that agent b's awareness is more extensive than agent c's awareness. $b \trianglerighteq_a^s \mathsf{At}$ means that agent a conjectures agent b to be aware of everything. $\mathsf{At} \trianglerighteq_a^s b$ means that agent a conjectures agent b to be not more than aware of everything (which does not imply that agent

a is aware of everything). We focus on the case in which for $s \in S$ and agents $a, b, c \in \mathsf{Ag}$, the awareness correspondences and the preorders jointly satisfy the condition that we may dub *coherent relative awareness*:

$$\mathcal{A}_a(s) \cap \mathcal{A}_b(s) \not\supseteq \mathcal{A}_a(s) \cap \mathcal{A}_c(s) \text{ implies } b \not\trianglerighteq_a^s c.$$

This condition may be interpreted as saying that agent a's conjecture at state s about agent b's awareness relative to agent c's awareness is based on those agents' actual awareness conditional on agent a's awareness at that state.

The last component of the modified awareness structure is the valuation function $V : S \times \mathsf{At} \longrightarrow \{true, false\}$.

There is a two-stage semantics. At the first stage, an "individualized preliminary" truth value is assigned to every formula at every state. At the second stage, the final truth value is assigned. We denote the individualized preliminary satisfaction relation of agent a by $\models_a^1$ and the final satisfaction relation by $\models$. The individualized preliminary satisfaction relation is defined inductively on the structure of formulae in $\mathsf{L}_n^{L,K,A,F,R}(\mathsf{At})$ as follows:

$M, s \models_a^1 p$ if and only if $V(s, p) = true$,
$M, s \models_a^1 F_b$ if and only if $b \trianglerighteq_a^s \mathsf{At}$,
$M, s \models_a^1 R_{bc}$ if and only if $c \trianglerighteq_a^s b$,
$M, s \models_a^1 \neg\varphi$ if and only if $M, s \not\models_a^1 \varphi$,
$M, s \models_a^1 \varphi \wedge \psi$ if and only if both $M, s \models_a^1 \varphi$ and $M, s \models_a^1 \psi$,
$M, s \models_a^1 L_b\varphi$ if and only if $M, s' \models_b^1 \varphi$ for all $s' \in \Pi_b(s)$,
$M, s \models_a^1 A_b\varphi$ if and only if $\mathsf{At}(\varphi) \subseteq \mathcal{A}_b(s)$,
$M, s \models_a^1 K_b\varphi$ if and only if both $M, s \models_a^1 A_b\varphi$ and $M, s \models_a^1 L_b\varphi$.

All clauses with the exception of the second and third clause are familiar from the satisfaction relation defined for awareness structures. In the modified awareness structure M, formula F_b is preliminarily true for agent a at state s if and only if at state s agent a conjectures that agent b is aware of everything. Similarly, in the modified awareness structure M, formula R_{bc} is preliminarily true for agent a at state s if and only if at state s agent a conjectures that agent c's awareness is more extensive than agent b's awareness. Note that although the preliminary satisfaction relation is individualized, it is hard to interpret it as a subjective notion because states in an awareness structure should be interpreted as "objective" descriptions from an modeler's point of view and not necessarily from the agent's point of view.

The final satisfaction relation is defined inductively on the structure of formulae in $\mathsf{L}_n^{L,K,A,F,R}(\mathsf{At})$ and makes use of the individualized preliminary satisfaction relation as follows:

$M, s \models p$ if and only if $V(s,p) = true$,
$M, s \models F_a$ if and only if $\mathcal{A}_a(s) = \mathsf{At}$,
$M, s \models R_{ab}$ if and only if $\mathcal{A}_b(s) \supseteq \mathcal{A}_a(s)$,
$M, s \models \neg\varphi$ if and only if $M, s \not\models \varphi$,
$M, s \models \varphi \wedge \psi$ if and only if both $M, s \models \varphi$ and $M, s \models \psi$,
$M, s \models L_a\varphi$ if and only if $M, s' \models_a^1 \varphi$ for all $s' \in \Pi_a(s)$,
$M, s \models A_a\varphi$ if and only if $\mathsf{At}(\varphi) \subseteq \mathcal{A}_a(s)$,
$M, s \models K_b\varphi$ if and only if both $M, s \models A_b\varphi$ and $M, s \models L_b\varphi$.

The second and third clauses use now the awareness correspondences instead individual conjectures captured by the preorders. In the modified awareness structure M, formula F_a is true at state s if and only if at state s agent a is aware of everything. Similarly, in the modified awareness structure M, formula R_{ab} is true at state s if and only if at state s agent b's awareness as given by his awareness set is more extensive than agent a's awareness. Most important is the clause giving semantics to implicit knowledge, which refers to the preliminary satisfaction relation of agent a. In the modified awareness structure M, agent a implicitly knows formula φ at state s if φ is *preliminary true* for agent a at every state that he considers possible at s. Thus, whether or not an agent implicitly knows a formula depends on his preliminary satisfaction relation at states that he considers possible. This can be different from the final satisfaction relation for formulas involving F_a and R_{ab}.

The notion of validity is as in Kripke or awareness structures using the final satisfaction relation $\models$ just defined.

The aim is to characterize modified awareness structures in terms of properties of knowledge and awareness. To state the axiom system, it will be helpful to define the sublanguage $\mathsf{L}^- \subseteq \mathsf{L}_n^{L,K,A,F,R}(\mathsf{At})$ that consists exactly of the set of formulae whose final truth values in any state and any modified awareness structure coincides with the individualized preliminary truth values. Define L^- inductively as follows:

$p \in \mathsf{At}$ implies $p \in \mathsf{L}^-$,
$A_a\varphi, L_a\varphi, K_a\varphi \in \mathsf{L}^-$ for any $\varphi \in \mathsf{L}_n^{L,K,A,F,R}(\mathsf{At})$,
$\varphi \in \mathsf{L}^-$ implies $\neg\varphi \in \mathsf{L}^-$,
$\varphi, \varphi \in \mathsf{L}^-$ implies $\varphi \wedge \varphi \in \mathsf{L}^-$.

Note that $\varphi \in \mathsf{L}^-$ does not imply that F_a or R_{ab} for some $a, b \in \mathsf{Ag}$ could not be a subformula of φ. With this definition on hand, we consider the following axiom system that we call $\mathsf{S5}_n^{L,K,A,F,R}$:

Prop. All substitution instances of tautologies of propositional logic.

KL. $K_a\varphi \leftrightarrow L_a\varphi \wedge A_a\varphi$ (Explicit Knowledge is Implicit Knowledge and Awareness)

AS. $A_a\neg\varphi \leftrightarrow A_a\varphi$ (Symmetry)

AC. $A_a(\varphi \wedge \psi) \leftrightarrow A_a\varphi \wedge A_a\psi$ (Awareness Conjunction)

AKR. $A_a\varphi \leftrightarrow A_aK_a\varphi$ (Awareness Explicit Knowledge Reflection)

ALR. $A_a\varphi \leftrightarrow A_aL_a\varphi$ (Awareness Implicit Knowledge Reflection)

AR. $A_a\varphi \leftrightarrow A_aA_a\varphi$ (Awareness Reflection)

AI. $A_a\varphi \rightarrow K_aA_a\varphi$ (Awareness Introspection)

F0. A_aF_b

R0. A_aR_{bc}

F1. $F_a \rightarrow A_a\varphi$

F2. $F_a \rightarrow R_{ba}$

F3. $F_a \wedge R_{ab} \rightarrow F_b$

R1. $R_{ab} \rightarrow (A_a\varphi \rightarrow A_b\varphi)$

R2. $R_{ab} \wedge R_{bc} \rightarrow R_{ac}$ (Transitivity of Relative Awareness)

R3. R_{aa} (Reflexivity of Relative Awareness)

R4. $A_a\varphi \rightarrow K_a((A_b\varphi \wedge \neg A_c\varphi) \rightarrow \neg R_{bc})$

K. $(L_a\varphi \wedge L_a(\varphi \rightarrow \psi)) \rightarrow L_a\psi$ (Distribution Axiom)

T$^-$. $L_a\varphi \rightarrow \varphi$ for any $\varphi \in \mathsf{L}^-$ (Modified Implicit Knowledge Truth Axiom)

4. $L_a\varphi \rightarrow L_aL_a\varphi$ (Implicit Positive Introspection Axiom)

5. $\neg L_a\varphi \rightarrow L_a\neg L_a\varphi$ (Implicit Negative Introspection Axiom)

MP. From φ and $\varphi \rightarrow \psi$ infer ψ (modus ponens)

Gen$^-$. From φ proved without application of F1 or R1 infer $L_a\varphi$ (Modified Implicit Knowledge Generalization)

Most axiom schemes are familiar from previous sections. F0 means that the agent is always aware that an agent is aware of everything. R0 means that the agent is always aware that an agent's awareness is as extensive as another (or the same) agent's awareness. F1 states that full awareness implies awareness of any particular formula. Of course, these properties do not necessarily imply that an agent is aware of everything or is as aware as other agents. F2 means that full awareness implies relative awareness with respect to any agent.

F3 says that relative awareness of one agent with respect to a second agent implies full awareness of the first agent in the case that the second agent is fully aware. R1 states that if agent b's awareness is as extensive as agent a's awareness, then agent a being aware of a formula implies that agent b must be aware of it as well. R2 encapsulates the idea that relative awareness is transitive among agents. If agent b is as aware as agent a and agent c is as aware as agent b, then also agent c must be as aware as agent a. R3 states that relative awareness is reflexive in the sense that every agent is aware of everything that he is aware of. Finally, R4 says that when agent a is aware of a formula then he knows that if agent b is aware of it and agent c is not, then c's awareness is not as extensive as b's awareness. It is closely connected to the condition "coherent relative awareness".

The set of formulae for which the individualized preliminary truth values and the final truth values agree, L^-, play a role in the statement of the truth axiom T^-, which is restricted to just these formulae. This means that the agent may be delusional with respect to reasoning about full awareness of an agent or the awareness of an agent relative to another. Similarly, implicit knowledge generalization, Gen^-, has been weakened as it applies only to theorems that are deduced without use of axioms F1 or R1. We note that most of the axioms are stated in terms of implicit knowledge. As mentioned previously, axioms and inference rules that involve implicit knowledge are hard to interpret as it is not necessarily the knowledge that is "present in the agent's mind".

Theorem 3.14
For the language $\mathsf{L}_n^{L,K,A,F,R}(\mathsf{At})$, the axiom system $\mathsf{S5}_n^{L,K,A,F,R}$ is a sound and complete axiomatization with respect to partitional modified awareness structures in which agents know what they are aware of and relative awareness conjectures are coherent. ⊣

Modeling awareness of unawareness with propositional constants rather than quantification yields a language that is less expressive than the approaches introduced in Sections 3.5.1 to 3.5.3. For instance, we cannot express that an agent a knows that there is no more than one proposition that agent b is aware of but agent c is not. Nevertheless, the approach allows for modeling awareness of unawareness in relevant examples such as the doctor example mentioned earlier.

3.6 Notes

We present here detailed references for this chapter. This is followed by an overview over the different approaches, and finally, we mention a number of omitted approaches to awareness. For a comprehensive bibliography, see `http://www.econ.ucdavis.edu/faculty/schipper/unaw.htm`.

The seminal paper is by Fagin and Halpern (1988), who were the first to present a formal approach of modeling awareness. Their starting point was the Logic of Implicit and Explicit Belief by Levesque (1984). Awareness structures of Section 3.3.1 have been introduced by Fagin and Halpern (1988). Theorem 3.1 is proved by Halpern (2001), who proves also axiomatizations for non-partitional awareness structures. Huang and Kwast

(1991) study variants of awareness structures and discuss various notions of negation and implications in the context of logical non-omniscience. Sillari (2008a,b) uses a version of awareness structures in which he replaces the Kripke relations by a neighborhood correspondences in the spirit of neighborhood semantics originally introduced by Montague (1970) and by Scott (1970) (see Section 3.5.3). Theorem 3.2 is proved by Halpern (2001) with an additional inference rule that Halpern and Rêgo (2008) show indirectly to be unnecessary. We like to mention that Halpern (2001) and van Ditmarsch, French, Velázquez-Quesada, and Wáng (2013) prove also axiomatizations for awareness structures that are not necessarily partitional. The latter paper also introduces another notion of knowledge that they dub "speculative knowledge". This notion is similar to explicit knowledge except that an agent always speculatively knows tautologies even though these tautologies may involve primitive propositions that she is unaware of. Lemma 3.1 is proved in (Halpern, 2001). The definition of awareness in terms of knowledge as $K_a\varphi \vee (\neg K_a\varphi \wedge K_a \neg K_a\varphi)$ is by Modica and Rustichini (1994).

The syntactic notion of awareness of awareness structures has been criticized early on by Konolige (1986). The impossibility of a purely semantic approach to awareness has been conjectured, for instance, by Thijsse (1991) who writes that "...a purely semantic and fully recursive approach would be preferable, but I believe it is intrinsically impossible, due to the psychological nature of awareness." Theorem 3.3 is due to Dekel, Lipman, and Rustichini (1998b), who present also impossibility results in which necessitation is weakened or replaced by monotonicity. Chen, Ely, and Luo (2012) and Montiel Olea (2012) provide further elaborations of those impossibility results.

The event-based approach of unawareness frames of Section 3.3.3 is due to Heifetz, Meier, and Schipper (2006). Board, Chung, and Schipper (2011) study properties of unawareness frames in which the possibility correspondence does not necessarily satisfy generalized reflexivity and stationarity. Galanis (2013a) studies a variant of unawareness frames in which he drops the property Projections Preserve Knowledge of the possibility correspondence. Schipper (2014) complements unawareness frames with decision theoretic primitives to characterize properties of the possibility correspondence by corresponding properties of a decision maker's preference relation. This extends the approach by Morris (1996, 1997) for standard states-spaces to unawareness frames.

The canonical unawareness structure of Section 3.3.4 was introduced by Heifetz, Meier, and Schipper (2008). Remark 3.4 and Theorems 3.5 and 3.6 are proved by Heifetz et al. (2008). Theorem 3.7 is due to Halpern and Rêgo (2008), who prove also analogous axiomatizations of unawareness structures in which the possibility correspondence does not necessarily satisfy generalized reflexivity and stationarity. Galanis (2011) axiomatizes his variant of unawareness structures with multiple knowledge modalities, one for each sub-language.

Modica and Rustichini (1999) were the first economists to present a semantics for propositionally determined awareness and partitional knowledge in the case of a single agent. The exposition of generalized standard models in Section 3.3.5 follows mostly Halpern (2001). The proof of Theorem 3.8 is by Modica and Rustichini (1999). Theorem 3.9 is by Halpern (2001), who also proves analogous results for generalized standard models (awareness structures, respectively) for which generalized reflexivity and stationarity (reflexivity, transitivity, and Euclideaness, respectively) may fail.

The product models of Section 3.3.6 were introduced by Li (2009) and the multi-agent extension is found in (Li, 2008a). Proposition 3.3 is proved in (Li, 2009). She does not present an axiomatization of product models. Theorem 3.10 is proved by Heinsalu (2012) for product models (awareness structures, respectively) for which reflexivity and stationarity (reflexivity, transitivity, and Euclideaness) may fail. But it is straightforward to extend it to any corresponding subset of those properties. The relationship between the multi-agent extension of the product model by Li (2008a) and the rest of the literature is still open.

Type spaces with unawareness in Section 3.4.1 were proposed by Heifetz, Meier, and Schipper (2013b). They are inspired by Harsanyi type spaces without unawareness introduced by Harsanyi (1967) (see also Mertens and Zamir (1985)). The proofs of the results in this section are by Heifetz et al. (2013b). Heifetz, Meier, and Schipper (2012) present the hierarchical construction and show the existence of a universal unawareness type space analogous to Mertens and Zamir (1985) when the space of underlying uncertainties is compact Hausdorff. Heinsalu (2014) proved independently the measurable case in an non-constructive approach (see also Pinter and Udvari (2012)).

Sadzik (2007) presents extensions of both awareness and unawareness structures to the probabilistic cases and provides axiomatizations. In a recent paper, Cozic (2012) also extends (Heifetz and Mongin, 2001) to the case of unawareness of a single-agent analogous to generalized standard structures of Modica and Rustichini (1999). Meier (2012) devises an infinitary axiom system that he shows to be strongly sound and strongly complete for standard type spaces without unawareness. An extension to unawareness and to multi-agent settings with unawareness is still open.

Speculative trade is closely related to agreeing-to-disagree. Aumann's famous "No-agreeing-to-disagree" result says that if agents share a common prior probability measure, then it cannot be common knowledge that their posteriors disagree (Aumann, 1976). The "No-speculate-trade" result says that if agents share a common prior probability measure, then if the expectations of a random variable (e.g., the price of a stock) are common certainty among agents, then these expectations must agree (Milgrom and Stokey, 1982). Both results apply for standard structures without unawareness. The speculative trade Example 3.1 *with unawareness* has been considered first by Heifetz et al. (2006). Heifetz et al. (2013b) prove a generalization of the "No-speculative-trade" result mentioned in Section 3.4.2 according to which a non-degenerate common prior rules out common certainty of strict preference to trade under unawareness. They also prove a generalized of the "No-agreeing-to-disagree" result under unawareness. The infinite case of the "No-speculative-trade" result under unawareness is proved by Meier and Schipper (2014b). For the definition and discussion of the common prior under unawareness, see Heifetz et al. (2013b). While the common prior assumption is a sufficient condition for the "No-speculative-trade" result, it is not necessary under unawareness, which is in contrast to standard state-spaces. See Heifetz et al. (2013b) for a counterexample.

Extended awareness structures with propositional quantifiers of Section 3.5.1 are due to Halpern and Rêgo (2013). They improve upon an earlier approach by Halpern and Rêgo (2009) in which they also extend the syntax to allow for quantification over quantifier-free formulae. Unfortunately, in this earlier approach formulae expressing that an agent considers it possible that she is aware of all formulae and also considers it possible that she is not aware of all formulae are not satisfiable in any of their extended awareness structures. This is a serious limitation for applications as it is very natural to consider agents who may be uncertain about whether they are aware of everything or not. As remedy, Halpern and Rêgo (2013) allow different languages to be defined at different states, which is very much in the spirit of Modica and Rustichini (1999) and Heifetz et al. (2008). The proof of Theorem 3.11 is by Halpern and Rêgo (2013). Halpern and Rêgo (2013) also explore the connection between awareness and unawareness structures by showing that quantifier-free fragment of their logic is characterized by exactly the same axioms as the logic of Heifetz et al. (2008). Moreover, they show that under minimal assumptions they can dispense with the syntactic notion of awareness by Fagin and Halpern (1988), as this notion of awareness is essentially equivalent to the one used in (Heifetz et al., 2006, 2008; Li, 2009; Modica and Rustichini, 1999).

The extension of first-order modal logic to unawareness of Section 3.5.2 is by Board and Chung (2011a). The proof of Theorem 3.12 by Board and Chung (2011a) is a version not imposing the assumption of a partitional possibility correspondence. Theorem 3.13 is proved by Board et al. (2011).

A first-order logic of awareness but with a semantics based on awareness neighborhood frames in Section 3.5.3 has been introduced by Sillari

(2008a,b). He does not show soundness and completeness but Sillari (2008a) suggests that results by Arló-Costa and Pacuit (2006) could be extended to awareness. Sillari (2008b) proves two theorems. First, he shows that awareness neighborhood structures (without restrictions on the awareness correspondences and neighborhood correspondences) are equally expressive to impossible possible worlds structures introduced by Rantala (1982a,b) and by Hintikka (1975). This complements results on equal expressivity of awareness (Kripke) structures and impossible possible worlds structures by Wansing (1990). Second, he shows an analogous result for quantified impossible possible worlds structures and quantified awareness neighborhood structures. This implies that one should be able to model awareness also with impossible possible worlds structures. Yet, without knowing how exactly various restrictions on awareness and belief translate into impossible possible worlds, it is not clear how tractable it would be to model awareness with these structures.

The idea of modeling awareness of unawareness by propositional constants in Section 3.5.4 is originally due to Ågotnes and Alechina (2007). The two-stage semantics is by Walker (2014). Axioms F1 to F3 and R1 to R3 appear also in (Ågotnes and Alechina, 2007). The proof of Theorem 3.14 is by Walker (2014).

Figure 3.6 provides an overview over approaches discussed in this chapter. For lack of space, we excluded in this picture probabilistic approaches to unawareness (Section 3.4). While the upper part of the figure lists single-agent structures, the middle part shows multi-agent structures. Finally, the lower part presents structures with awareness of unawareness. Roughly, we indicate generalizations by an arrow and equivalence by a bi-directional arrow. Beside the arrows, we sometimes list articles that show the connection between the approaches. Often these results imply further relationships. The interested reader should consult the original papers for the precise notions of equivalence. Figure 3.6 also shows a connection to the impossible worlds approach by Rantala (1982a,b). Wansing (1990) shows that it is equally expressive to awareness structures of Fagin and Halpern (1988). Thijsse (1991), Thijsse and Wansing (1996), and Sillari (2008a,b) present further results along those lines.

There are several approaches to awareness mentioned in Figure 3.6 that we haven't discussed so far. Around the same time as Modica and Rustichini (1994) published the first paper on awareness in economics, Pires (1994) finished her doctoral dissertation in economics at MIT with an unpublished chapter on awareness that unfortunately has been ignored in the literature so far. She presents a model of non-trivial awareness for a single-agent that essentially captures awareness generated by primitive propositions. She al-

Figure 3.6: Partial overview over the literature

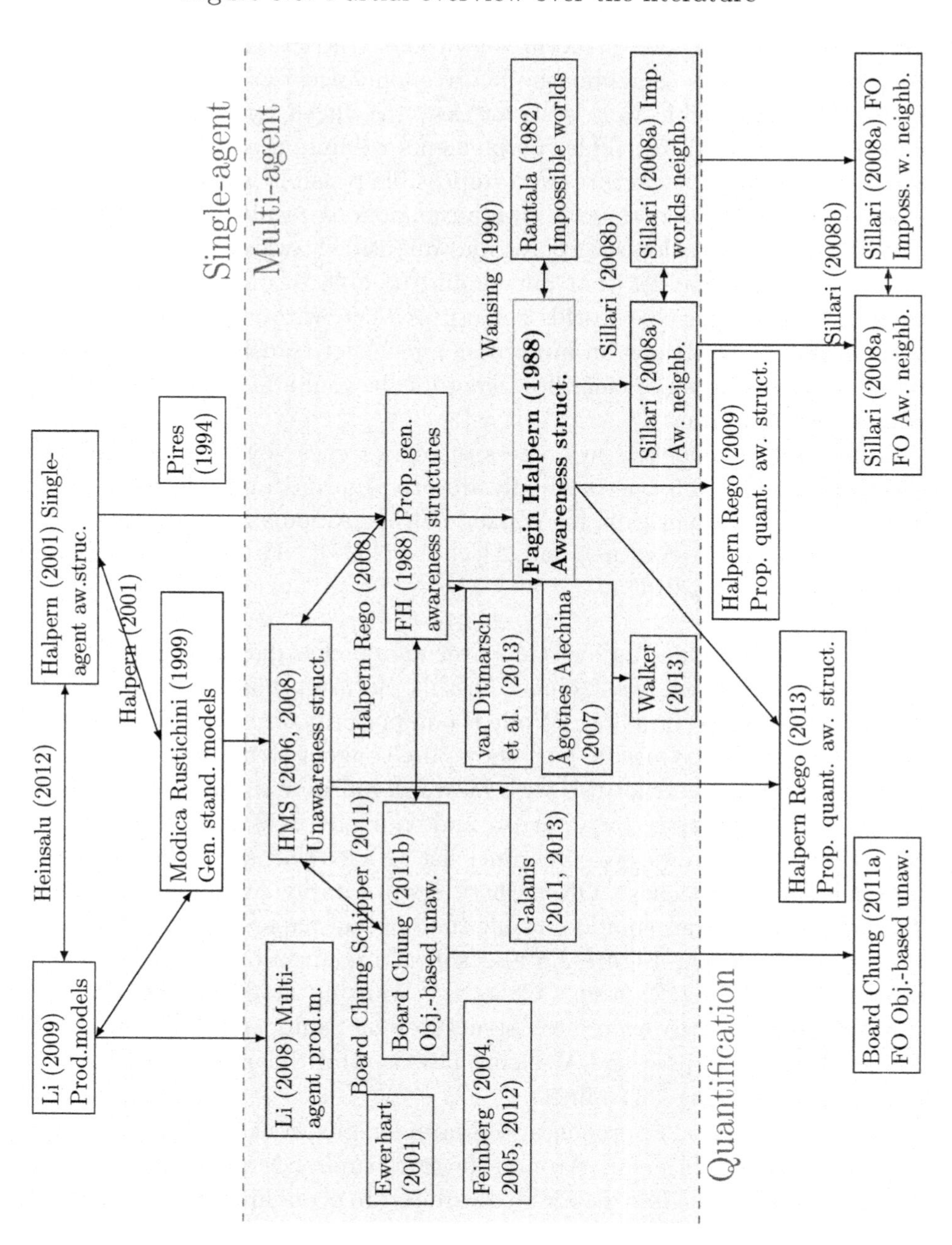

ready considers weak negative introspection, weak necessitation, and plausibility as properties of awareness and knowledge. Although she introduces both a logic and a state-space semantics, she shows no soundness or completeness results. She also anticipates modeling awareness of unawareness very much in the spirit of later works by Ågotnes and Alechina (2007) and Walker (2014). Finally, she studies updating of awareness as refinements of conceivable states. Ewerhart (2001) introduces a state-space model in which at each state an agent may only be aware of a subset of states. But since no special event structure is assumed, an agent may be aware of an event but unaware of its complement or vise versa, thus violating symmetry. Ewerhart (2001) considers both implicit and explicit knowledge and the model satisfies, for instance, weak negative introspection with respect to explicit knowledge but not weak necessitation. Under an additional richness assumption, it satisfies KU-introspection and AU-introspection, strong plausibility with "$\subseteq$" but not the Modica-Rustichini definition of awareness unless unawareness is trivial. He proves a generalization of Aumann's "No-agreeing-to-disagree" theorem for his models with unawareness. Feinberg (2004, 2005, 2012) provides different versions of an approach that models interactive awareness of components of games by explicit unbounded sequences of mutual views of players. Among the properties imposed on awareness is that (1) if a player is aware of what an opponent is aware of, then the player herself is also aware of it, and (2), if a player ("she") is aware that an opponent ("he") is aware of something, then she is also aware that the opponent is aware that he is aware of it. These two properties are satisfied also by the notion of propositionally determined awareness discussed earlier. Yet, the precise connection between Feinberg's approach and the rest of the literature is still open.

Our review leaves out many topics. For instance, some of the discussed papers also contain results on the *complexity of deciding the satisfiability* of formulas, e.g., (Ågotnes and Alechina, 2007; van Ditmarsch and French, 2011a; Fagin and Halpern, 1988; Halpern and Rêgo, 2009). Complexity may also be related to awareness on a conceptual level. Already Fagin and Halpern (1988) suggested that one may want to consider a computational-based notion of awareness of agents who may lack the computational ability to deduce all logical consequences of their knowledge. Fagin, Halpern, Moses, and Vardi (1995) (Chapter 10.2.2) discuss the connection between algorithmic knowledge and awareness. One may also conceive of a computational-based notion of awareness of an object that roughly corresponds to the amount of time needed to generate that object within a certain environment. Such an approach is pursued by Devanur and Fortnow (2009) using Kolmogorov complexity.

For lack of space, we are not able to cover *dynamics of awareness*. In a series of papers, van Ditmarsch et al. (2009; 2011a; 2011b; 2012; 2013) study changes of awareness and information, introduce the notions of awareness bisimulation and epistemic awareness action models. Van Benthem et al. (2010), Grossi and Velázquez-Quesada (2009), and Hill (2010) also analyze logics for changes of awareness. Changes of awareness have been studied as a stochastic process by Modica (2008). It is also a topic of active research in decision theory (Karni and Vierø, 2013b,a; Li, 2008b) and comes up naturally in *dynamic games with unawareness* (Feinberg, 2004, 2012; Grant and Quiggin, 2013; Halpern and Rêgo, 2014; Heifetz, Meier, and Schipper, 2011, 2013a; Meier and Schipper, 2012; Nielsen and Sebald, 2012; Ozbay, 2008; Rêgo and Halpern, 2012). Since most applications of awareness are in a multi-agent setting, extending games to unawareness has been an important recent conceptual innovation in game theory. Beside the aforementioned works on dynamic games with unawareness, there has been also work on static games with incomplete information and unawareness, see Feinberg (2012), Meier and Schipper (2014a) and Sadzik (2007).

Decision theoretic approaches to unawareness are pursued in (Karni and Vierø, 2013b,a; Li, 2008b; Schipper, 2013, 2014). There is a growing literature to unforeseen contingencies in decision theory. Here is not the space to give an adequate review and the interested reader may want to consult Dekel, Lipman, and Rustichini (1998a) for an early review of some of the approaches. Most work on unforeseen contingencies is best understood in terms of awareness of unawareness discussed in Section 3.5.

We expect many applications of unawareness to emerge. Besides the study of speculative trade discussed in Section 3.4.2, it has already been applied to the strategic *disclosure of verifiable information* (and awareness) (Heifetz et al., 2011; Schipper and Woo, 2014; Li, Peitz, and Zhao, 2014), *political awareness* and modern electoral campaigning (Schipper and Woo, 2014), the *value of information* under unawareness (Galanis, 2013b), *incomplete contracting* (Auster, 2013; Filiz-Ozbay, 2012; Grant, Kline, and Quiggin, 2012; Lee, 2008; Thadden and Zhao, 2012a,b), the study of *legal doctrines* (Board and Chung, 2011b), and *fair disclosure* in financial markets (Liu, 2008).

Acknowledgements I thank Hans van Ditmarsch, Joe Halpern, and Fernando R. Velázquez Quesada for comments on an earlier draft. Moreover, I thank Aviad Heifetz, Sander Heinsalu, Li Jing, Salvatore Modica, and Giacomo Sillari for helpful discussions. Financial support through NSF SES-0647811 is gratefully acknowledged. A part of the paper has been written when the author visited New York University.

References

Ågotnes, T. and N. Alechina (2007). Full and relative awareness: A decidable logic for reasoning about knowledge of unawareness. In D. Samet (Ed.), *Theoretical Aspects of Rationality and Knowledge, Proceedings of the 11th Conference (TARK 2007)*, pp. 6–14. Presses Universitaires De Louvain.

Arló-Costa, H. and E. Pacuit (2006). First-order classical modal logic. *Studia Logica 84*, 171–210.

Aumann, R. (1976). Agreeing to disagree. *Annals of Statistics 4*, 1236–1239.

Auster, S. (2013). Asymmetric awareness and moral hazard. *Games and Economic Behavior 82*, 503–521.

van Benthem, J. and F. Velázquez-Quesada (2010). The dynamic of awareness. *Synthese 177*, 5–27.

Board, O. and K. Chung (2011a). Object-based unawareness: Axioms. University of Minnesota.

Board, O. and K. Chung (2011b). Object-based unawareness: Theory and applications. University of Minnesota.

Board, O., K. Chung, and B. Schipper (2011). Two models of unawareness: Comparing the object-based and subjective-state-space approaches. *Synthese 179*, 13–34.

Chen, Y., J. Ely, and X. Luo (2012). Note on unawareness: Negative introspection versus au introspection (and ku introspection). *International Journal of Game Theory 41*, 325–329.

Cozic, M. (2012). Probabilistic unawareness. Université Paris-Est Créteil Va-de-Marne.

Dekel, E., B. Lipman, and A. Rustichini (1998a). Recent developments in modeling unforeseen contingencies. *European Economic Review 42*, 523–542.

Dekel, E., B. Lipman, and A. Rustichini (1998b). Standard state-space models preclude unawareness. *Econometrica 66*, 159–173.

Devanur, N. and L. Fortnow (2009). A computational theory of awareness and decision making. In A. Heifetz (Ed.), *Theoretical Aspects of Rationality and Knowledge, Proceedings of the 12th Conference (TARK 2009)*, pp. 99–107. ACM.

van Ditmarsch, H. and T. French (2009). Awareness and forgetting of facts and agents. In P. Boldi, G. Vizzari, G. Pasi, and R. Baeza-Yates (Eds.), *Proceedings of WI-IAT Workshops, Workshop WLIAMAS*, pp. 478–483. IEEE Press.

van Ditmarsch, H. and T. French (2011a). Becoming aware of propositional variables. In M. Banerjee and A. Seth (Eds.), *ICLA 2011*, LNAI 6521, pp. 204–218. Springer.

van Ditmarsch, H. and T. French (2011b). On the dynamics of awareness and certainty. In D. Wang and M. Reynolds (Eds.), *Proceedings of 24th Australasian Joint Conference on Artificial Intelligence (AI 11)*, LNAI 7106, pp. 727–738. Springer.

van Ditmarsch, H., T. French, and F. Velázquez-Quesada (2012). Action models for knowledge and awareness. In *Proceedings of the 11th International Conference on Autonomous Agents and Miltiagent Systems (AAMAS 2012)*.

van Ditmarsch, H., T. French, F. Velázquez-Quesada, and Y. Wáng (2013). Knowledge, awareness, and bisimulation. In B. Schipper (Ed.), *Proceedings of the 14. Conference on Theoretical Aspects of Rationality and Knowledge (TARK 2013)*, pp. 61–70. University of California, Davis.

Ewerhart, C. (2001). Heterogeneous awareness and the possibility of agreement. Universität Mannheim.

Fagin, R. and J. Y. Halpern (1988). Belief, awareness, and limited reasoning. *Artificial Intelligence 34*, 39–76.

Fagin, R., J. Y. Halpern, Y. Moses, and M. Vardi (1995). *Reasoning about knowledge*. MIT Press.

Feinberg, Y. (2004). Subjective reasoning - games with unawareness. Stanford University.

Feinberg, Y. (2005). Games with incomplete awareness. Stanford University.

Feinberg, Y. (2012). Games with unawareness. Stanford University.

Filiz-Ozbay, E. (2012). Incorporating unawareness into contract theory. *Games and Economic Behavior 76*, 181–194.

Galanis, S. (2011). Syntactic foundation for unawareness of theorems. *Theory and Decision 71*, 593–614.

Galanis, S. (2013a). Unawareness of theorems. *Economic Theory 52*, 41–73.

Galanis, S. (2013b). The value of information under unawareness. University of Southampton.

Grant, S., J. Kline, and J. Quiggin (2012). Differential awareness and incomplete contracts: A model of contractual disputes. *Journal of Economic Behavior and Organization 82*, 494–504.

Grant, S. and J. Quiggin (2013). Inductive reasoning about unawareness. *Economic Theory 54*, 717–755.

Grossi, D. and F. Velázquez-Quesada (2009). Twelve angry men: A study on the fine-grain of announcements. In X. He, J. Horty, and E. Pacuit (Eds.), *LORI 2009*, LNAI 5834, pp. 147–160. Springer.

Halpern, J. Y. (2001). Alternative semantics for unawareness. *Games and Economic Behavior 37*, 321–339.

Halpern, J. Y. and L. Rêgo (2009). Reasoning about knowledge of unawareness. *Games and Economic Behavior 67*, 503–525.

Halpern, J. Y. and L. Rêgo (2013). Reasoning about knowledge of unawareness revisited. *Mathematical Social Sciences 65*, 73–84.

Halpern, J. Y. and L. Rêgo (2014). Extensive games with possibly unaware players. *Mathematical Social Sciences 70*, 42–58.

Halpern, J. Y. and L. C. Rêgo (2008). Interactive unawareness revisited. *Games and Economic Behavior 61*, 232–262.

Harsanyi, J. (1967). Games with incomplete information played 'bayesian' players, part i, ii, and ii. *Management Science 14*, 159–182, 320–334, 486–502.

Heifetz, A., M. Meier, and B. Schipper (2006). Interactive unawareness. *Journal of Economic Theory 130*, 78–94.

Heifetz, A., M. Meier, and B. Schipper (2008). A canonical model for interactive unawareness. *Games and Economic Behavior 62*, 304–324.

Heifetz, A., M. Meier, and B. Schipper (2011). Prudent rationalizability in generalized extensive-form games. University of California, Davis.

Heifetz, A., M. Meier, and B. Schipper (2012). Unawareness, beliefs, and speculative trade. University of California, Davis.

Heifetz, A., M. Meier, and B. Schipper (2013a). Dynamic unawareness and rationalizable behavior. *Games and Economic Behavior 81*, 50–68.

Heifetz, A., M. Meier, and B. Schipper (2013b). Unawareness, beliefs, and speculative trade. *Games and Economic Behavior 77*, 100–121.

Heifetz, A. and P. Mongin (2001). Probability logic for type spaces. *Games and Economic Behavior 35*, 31–53.

Heinsalu, S. (2012). Equivalence of the information structure with unawareness to the logic of awareness. *Journal of Economic Theory 147*, 2453–2468.

Heinsalu, S. (2014). Universal type spaces with unawareness. *Games and Economic Behavior 83*(C), 255–266.

Hill, B. (2010). Awareness dynamics. *Journal of Philosophical Logic 39*, 113–137.

Hintikka, J. (1975). Impossible possible worlds vindicated. *Journal of Philosophical Logic 4*, 475–484.

Huang, Z. S. and K. Kwast (1991). Awareness, negation and logical omniscience. In J. van Eijck (Ed.), *Logics in AI*, Volume 478 of *Lectures Notes in Computer Science*, pp. 282–300. Springer.

Karni, E. and M. Vierø (2013a). Probabilistic sophistication and reverse bayesianism. Queen's University.

Karni, E. and M. Vierø (2013b). "Reverse Bayesianism": A choice-based theory of growing awareness. *American Economic Review 103*, 2790–2810.

Konolige, K. (1986). What awareness insn't: A sentenial view of implicit and explicit belief. In J. Y. Halpern (Ed.), *Proceedings of the First Conference on Theoretical Aspects of Reasoning and Knowledge (TARK I)*, pp. 241–250. Morgan Kaufmann.

Lee, J. (2008). Unforeseen contingencies and renegotiation with asymmetric information. *Economic Journal 118*, 678–694.

Levesque, H. (1984). A logic of implicit and explicit belief. In *AAAI-84 Proceedings*, pp. 198–220.

Li, J. (2008a). Interactive knowledge with unawareness. University of Pennsylvania.

Li, J. (2008b). A note on unawareness and zero probability. University of Pennsylvania.

Li, J. (2009). Information structures with unawareness. *Journal of Economic Theory 144*, 977–993.

Li, S., M. Peitz, and X. Zhao (2014). Vertically differentiated duopoly with unaware consumers. *Mathematical Social Sciences 70*, 59–67.

Liu, Z. (2008). Fair disclosure and investor asymmetric awareness in stock markets. University of Buffalo.

Meier, M. (2012). An infinitary probability logic for type spaces. *Israel Journal of Mathematics 192*, 1–58.

Meier, M. and B. Schipper (2012). Conditional dominance in games with unawareness. University of California, Davis.

Meier, M. and B. Schipper (2014a). Bayesian games with unawareness and unawareness perfection. *Economic Theory 56*, 219–249.

Meier, M. and B. Schipper (2014b). Speculative trade under unawareness: The infinite case. *Economic Theory Bulletin 2*(2), 147–160.

Mertens, J. F. and S. Zamir (1985). Formulation of bayesian analysis for games with incomplete information. *International Journal of Game Theory 14*, 1–29.

Milgrom, P. and N. Stokey (1982). Information, trade and common knowledge. *Journal of Economic Theory 26*, 17–27.

Modica, S. (2008). Unawareness, priors and posteriors. *Decisions in Economics and Finance 31*, 81–94.

Modica, S. and A. Rustichini (1994). Awareness and partitional information structures. *Theory and Decision 37*, 107–124.

Modica, S. and A. Rustichini (1999). Unawareness and partitional information structures. *Games and Economic Behavior 27*, 265–298.

Montague, R. (1970). Universal grammar. *Theoria 36*, 373–398.

Montiel Olea, J. (2012). A simple characterization of trivial unawareness. Harvard University.

Morris, S. (1996). The logic of belief and belief change: A decision theoretic approach. *Journal of Economic Theory 69*, 1–23.

Morris, S. (1997). Alternative notions of knowledge. In M. Bacharach, L.-A. Gérad-Varet, P. Mongin, and H. Shin (Eds.), *Epistemic logic and the theory of games and decisions*, pp. 217–234. Kluwer Academic Press.

Nielsen, C. and A. Sebald (2012). Unawareness in dynamic psychological games. University of Copenhagen.

Ozbay, E. (2008). Unawareness and strategic announcements in games with uncertainty. University of Maryland.

Pinter, M. and Z. Udvari (2012). Generalized type-spaces. Corvinus University Budapest.

Pires, C. (1994). Do i know ω? an axiomatic model of awareness and knowledge. Universidade Nova de Lisboa.

Rantala, V. (1982a). Impossible world semantics and logical omniscience. *Acta Philosophica Fennica 35*, 106–115.

Rantala, V. (1982b). Quantified modal logic: non-normal worlds and propositional attitudes. *Studia Logica 41*, 41–65.

Rêgo, L. and J. Y. Halpern (2012). Generalized solution concepts in games with possibly unaware players. *International Journal of Game Theory 41*, 131–155.

Sadzik, T. (2007). Knowledge, awareness and probabilistic beliefs. Stanford University.

Schipper, B. C. (2013). Awareness-dependent subjective expected utility theory. *International Journal of Game Theory 42*, 725–753.

Schipper, B. C. (2014). Preference-based unawareness. *Mathematical Social Sciences 70*, 34–41.

Schipper, B. C. and H. Y. Woo (2014). Political awareness, microtargeting of voters, and negative electoral campaigning. University of California, Davis.

Scott, D. (1970). Advice in modal logic. In K. Lambert (Ed.), *Philosophical problems in logic*, pp. 143–173. Reidel.

Sillari, G. (2008a). Models of awareness. In G. Bonanno, W. van der Hoek, and M. Wooldridge (Eds.), *Logic and the foundations of games and decisions*, pp. 209–240. University of Amsterdam.

Sillari, G. (2008b). Quantified logic of awareness and impossible possible worlds. *Review of Symbolic Logic 1*, 514–529.

Thadden, E. v. and X. Zhao (2012a). Incentives for unaware agents. *Review of Economic Studies 79*, 1151–1174.

Thadden, E. v. and X. Zhao (2012b). Multitask agency with unawareness. Hong Kong University of Science and Technology.

Thijsse, E. (1991). On total awareness logic. with special attention to monotonicity constraints and flexibility. In M. de Rijke (Ed.), *Diamond and defaults: Studies in pure and applied intensional logic*, pp. 309–347. Kluwer.

Thijsse, E. and H. Wansing (1996). A fugue on the themes of awareness logic and correspondence. *Journal of Applied Non-Classical Logics 6*, 127–136.

Walker, O. (2014). Unawareness with "possible" possible worlds. *Mathematical Social Sciences 70*, 23–33.

Wansing, H. (1990). A general possible worlds framework for reasoning about knowledge and belief. *Studia Logica 49*, 523–539.

Chapter 4

Epistemic Probabilistic Logic

Lorenz Demey and Joshua Sack

Contents

Abstract This chapter provides an overview of systems that combine probability theory, which describes quantitative uncertainty, with epistemic logic, which describes qualitative uncertainty. Both the quantitative and qualitative uncertainty in the systems we discuss are from the viewpoint of agents. At first, we investigate systems that describe uncertainty of agents at a single moment in time, and later we look at systems where the uncertainty changes over time, as well as systems that describe the actions that cause these changes.

Chapter 4 of the *Handbook of Epistemic Logic*, H. van Ditmarsch, J.Y. Halpern, W. van der Hoek and B. Kooi (eds), College Publications, 2015, pp. 147–201.

4.1 Introduction

This chapter provides an overview of systems that combine epistemic logic with probability theory. Combining these two frameworks is a natural move, since they can both be seen as formal accounts of the information of one or several agents. Furthermore, both frameworks deal not only with the agents' information at a single moment in time, but also with how it changes over time, as new information becomes available. Each framework has its own distinctive features, and by combining them, we obtain very powerful accounts of information and information flow.

Epistemic logic takes a *qualitative* perspective on information, and works with a modal operator K. Formulas such as $K\varphi$ can be interpreted as 'the agent knows that φ', 'the agent believes that φ', or, more generally speaking, 'φ follows from the agent's current information'. Probability theory, on the other hand, takes a *quantitative* perspective on information, and works with numerical probability functions P. Formulas such as $P(\varphi) = k$ can be interpreted as 'the probability of φ is k'. In the present context, probabilities will usually be interpreted subjectively, and can thus be taken to represent the agents' degrees of belief or credences.

As a direct consequence of its qualitative nature, epistemic logic provides a relatively *coarse-grained* perspective on information. With respect to one and the same formula φ, epistemic logic distinguishes only three epistemic attitudes: knowing its truth ($K\varphi$), knowing its falsity ($K\neg\varphi$), and being ignorant about its truth value ($\neg K\varphi \wedge \neg K\neg\varphi$). Probability theory, on the other hand, is much more *fine-grained*: it distinguishes infinitely many epistemic attitudes with respect to φ, viz. assigning it probability k ($P(\varphi) = k$), for every number $k \in [0, 1]$.

While epistemic logic thus is a coarser account of information, it is typically used with a wider *scope*. From its very origins, epistemic logic has not only been concerned with knowledge about 'the world', but also with knowledge about knowledge, i.e. with *higher-order information*. Typical discussions focus on principles such as positive introspection ($K\varphi \to KK\varphi$). On the other hand, probability theory rarely talks about principles involving higher-order probabilities, such as $P(\varphi) = 1 \to P(P(\varphi) = 1) = 1$.[1] This issue becomes even more pressing in multi-agent scenarios. Natural examples might involve an agent a not having any information about a proposition

[1] A notable exception is 'Miller's principle', which states that $P_1(\varphi \mid P_2(\varphi) = b) = b$. The probability functions P_1 and P_2 can have various interpretations, such as the probabilities of two agents, subjective probability (credence) and objective probability (chance), or the probabilities of one agent at different moments in time—in the last two cases, the principle is also called the 'principal principle' or the 'principle of reflection', respectively.

φ, while being certain that another agent, b, does have this information. In epistemic logic this is naturally formalized as

$$\neg K_a\varphi \wedge \neg K_a\neg\varphi \wedge K_a(K_b\varphi \vee K_b\neg\varphi).$$

A formalization in probability theory might look as follows:

$$P_a(\varphi) = 0.5 \wedge P_a(P_b(\varphi) = 1 \vee P_b(\varphi) = 0) = 1.$$

However, because such statements make use of 'nested' probabilities, they are rarely used in standard treatments of probability theory.

These differences with respect to higher-order information resurface in the treatment of dynamics, i.e. *information change*. As new information becomes available, the agents' information states should change accordingly. Probability theory typically uses Bayesian updating to represent such changes (but other, more complicated update mechanisms are available as well). Dynamic epistemic logic, on the other hand, interprets new information as changing the epistemic model, and then uses the new, updated model to represent the agents' updated information states.

Finally, both epistemic logic and probability theory are mathematically well-understood. Since epistemic logic can draw upon the extensive toolbox of modal logic, there is a plethora of results regarding its meta-theoretical properties (sound and complete axiomatizations, complexity, etc.). Similarly, probability theory, as axiomatized by Kolmogorov, has firm measure-theoretic foundations.

For all these reasons, the study of systems that combine epistemic logic and probability theory has proven to be very fruitful. Such combined systems inherit the fine-grained perspective on information from probability theory, and the representation of higher-order information from epistemic logic. Their dynamic versions provide a unified perspective on changes in first- and higher-order information. By incorporating the complementary perspectives of (dynamic) epistemic logic and probability theory, they thus yield very powerful and mathematically well-understood accounts of information and information flow. Throughout the chapter, we will illustrate the expressive and deductive power of these combined systems by showing how they can be used to analyze various scenarios in which both knowledge and probability play a role.

The remainder of this chapter is organized as follows. Section 4.2 introduces a basic system of probabilistic propositional logic. This system actually remains very close to probability theory proper, but it will allow us to make some technical remarks in their most general form. Section 4.3 introduces the first full-fledged combination of epistemic logic and probability theory, viz. probabilistic epistemic logic. This system represents the agents' knowledge and probabilities at a single point in time, but not yet their dynamics. Section 4.4 shows how interpreted systems (a particular sub-field of epistemic logic that has been applied extensively in computer science) can be extended with probabilities. Sections 4.5 and 4.6 describe dynamic extensions of the system of probabilistic epistemic logic introduced in Section 4.3. Section 4.5 focuses on a particularly simple type of dynamics, viz. public announcements. It describes a probabilistic version of the well-known system of public announcement logic, and extensively discusses the role of higher-order information in public announcements and Bayesian conditionalization. In Section 4.6, we introduce a more general update mechanism, which can deal with more intricate types of dynamics. This is a probabilistic version of the 'product update' mechanism in dynamic epistemic logic. Section 4.7 briefly indicates some applications and open questions for the systems discussed in this chapter. Section 4.8 contains concluding remarks, and Section 4.9 provides pointers for further reading: as in all chapters in this book, the details of all bibliographical references are postponed until this final section.

Sections 4.2–4.6 all have the same goal (viz. introducing some system of epistemic logic with probabilities), and therefore these sections are structured in roughly the same way. First, the models are introduced. Then, we discuss the logic's language and its semantics (and in Sections 4.2 and 4.3 also its expressivity). Next, we present a sound and complete proof system, and we conclude by discussing its decidability and complexity.

Finally, it should be emphasized that throughout this chapter, we will distinguish between 'fully general' kinds of models and their 'more elementary' versions: in Section 4.2 we distinguish between probabilistic models and *discrete* probabilistic models (Definition 4.2), and in Sections 4.3–4.6 between probabilistic relational models and *simplified* probabilistic relational models (Definition 4.5). These distinctions are essentially based on the same criterion, viz. whether the models' probabilistic components are *probability spaces* in the general measure-theoretic sense of the word, or *discrete probabilistic structures*, which are simpler, but less general (see Definition 4.1).

4.2 Probabilistic Propositional Logic

This section discusses a basic system of probabilistic propositional logic, which remains very close to probability theory proper: it does not have an epistemic component, and, more importantly, it does not allow reasoning about higher-order probabilities. This will allow us to make some technical remarks in their most general form (which will also apply to the more complex systems that are introduced later). Subsection 4.2.1 introduces probability spaces and probabilistic models. Subsection 4.2.2 defines the language that is interpreted on these models, and Subsection 4.2.3 discusses its expressivity. Subsection 4.2.4 provides a complete axiomatization. Finally, Subsection 4.2.5 discusses decidability and complexity properties.

4.2.1 Probabilistic Models

There are two ways of adding probabilistic information to a set of states S. The first is to define probabilities directly on the states of S. The second is to define probabilities on (a subcollection of) the subsets of S. These two approaches are formalized in the notions of *discrete probability structure* and *probability space*, respectively.

Definition 4.1
A *discrete probability structure* is a tuple $\langle S, p\rangle$, where S is an arbitrary set, whose elements will usually be called 'states' or 'possible worlds', and $p\colon S \to [0,1]$ is a function with countable support (i.e. the set $\{s \in S \mid p(s) > 0\}$ is countable), such that $\sum_{s\in S} p(s) = 1$.

A *probability space* is a tuple $\langle S, \mathcal{A}, \mu\rangle$, where S is an arbitrary (potentially uncountable) set called the *sample space*, $\mathcal{A} \subseteq \wp(S)$ is a σ-*algebra* over S (i.e. $S \in \mathcal{A}$, and $\mathcal{A}$ is closed under complements and countable unions), and $\mu\colon \mathcal{A} \to [0,1]$ is a *probability measure* (a countably additive function[2] such that $\mu(S) = 1$). The structure $(S, \mathcal{A})$ is called a *measurable space*, and the elements of $\mathcal{A}$ are called the *measurable sets* of the space.

The function p on individual states in a discrete probability structure is naturally extended to a function p^+ on sets of states, by putting

$$p^+ : \wp(S) \to [0,1] : X \mapsto p^+(X) \stackrel{\text{def}}{=} \sum_{x\in X} p(x).$$

Note that it follows immediately that $p^+(S) = 1$. Given this construction, it is easy to check that every discrete probability structure gives rise to a probability space, by taking the σ-algebra of the space to consist of *all* subsets: if $\langle S, p\rangle$ is a discrete probability structure, then $\langle S, \wp(S), p^+\rangle$ is a probability space. Hence, probability spaces can be seen as a generalization of discrete probability structures—or vice versa, discrete probability structures can be seen as a special case of probability spaces, viz. those whose σ-algebra is the powerset of their sample space.

[2] A real-valued set function is *countably additive* iff for any countable collection of sets A_i that are pairwise disjoint ($A_i \neq A_j$ for each $i \neq j$), it holds that $f(\bigcup_{i=1}^{\infty} A_i) = \sum_{i=1}^{\infty} f(A_i)$.

Discrete probability structures, defined by point-functions, have the advantage of simplicity, as well as having numerous countable settings for examples and applications. The probability spaces, with their set-functions defined on σ-algebras, have the advantage of generality. The purpose of the σ-algebra is to restrict the domain of the probability set-function from the entire powerset of the sample space to a smaller σ-algebra (cf. *supra*). This is unavoidable, for example, when we wish to define a uniform probability distribution over an infinite set: one cannot assign equal probability to all singletons while maintaining countable additivity (since the sum of all probabilities must be at most 1). Throughout the chapter, we will make use of both discrete probability structures and probability spaces.

Let AtProp be a countable set of *atomic propositions*, which we assume as given throughout this chapter. Probabilistic models and discrete probabilistic models are defined as follows:

Definition 4.2
A *discrete probabilistic model* is a tuple $\mathbb{M} = \langle S, p, V\rangle$, where $\langle S, p\rangle$ is a discrete probability structure, and $V\colon \mathsf{AtProp} \to \wp(S)$ is a valuation. The class of all discrete probabilistic models will be denoted $\mathcal{C}_d^{PPL}$.

A *probabilistic model* is a tuple $\mathbb{M} = \langle S, \mathcal{A}, \mu, V\rangle$, where $\langle S, \mathcal{A}, \mu\rangle$ is a probability space, and $V\colon \mathsf{AtProp} \to \mathcal{A}$ is a valuation. The class of all probabilistic models will be denoted $\mathcal{C}^{PPL}$. ⊣

The details of discrete probability structures and probability spaces have already been discussed above. The valuation V determines which atomic propositions are true at which states; intuitively, $s \in V(p)$ means that p is true at the state s. Note that in the definition of probabilistic model, the valuation is required to map atomic propositions to the σ-algebra $\mathcal{A}$, instead of to the full powerset $\wp(S)$ (as is usually done in modal settings). This restriction ensures that every Boolean formula has a probability,[3] which will be used in the definition of the formal semantics.

4.2.2 Language and Semantics

The basic probability language is defined in layers. For a given set of atomic propositions AtProp, let $\mathcal{L}_{BL}(\mathsf{AtProp})$ be the set of propositional (Boolean) formulas, given by the grammar:

$$\varphi ::= p \mid \neg\varphi \mid (\varphi \wedge \varphi)$$

[3]The σ-algebra of measurable sets thus plays a role similar to that of the modal algebra of admissible sets in a generalized relational (Kripke) model: both are meant to restrict the range of the valuation function, to prevent certain 'problematic' sets from becoming truth sets of formulas.

where $p \in \mathsf{AtProp}$. Then let $\mathcal{T}(\mathsf{AtProp})$ be a set of terms, given by the grammar

$$t ::= aP(\varphi) \mid t + t$$

where $a \in \mathbb{Q}$ is a rational number, and $\varphi \in \mathcal{L}_{BL}(\mathsf{AtProp})$. Then we let $\mathcal{L}_{PPL}(\mathsf{AtProp})$ be the set of probability formulas, given by the grammar:

$$f ::= t \geq a \mid \neg f \mid (f \wedge f)$$

where $a \in \mathbb{Q}$ and $t \in \mathcal{T}(\mathsf{AtProp})$. We restrict the numbers a to being rational in order to ensure that the language is countable, and to ensure that each number is easy to represent.

Formulas of the form $t \geq a$ are called atomic probability formulas. We allow for linear combinations in atomic probability formulas, because this additional expressivity is useful when looking for a complete axiomatization, and because it allows us to express comparative judgments such as 'φ is at least twice as probable as ψ': $P(\varphi) \geq 2P(\psi)$. We will discuss this in more detail in Subsection 4.2.3.

The formula $P(\varphi) \geq 2P(\psi)$ is actually an abbreviation for $P(\varphi) - 2P(\psi) \geq 0$. In general, we introduce the following abbreviations:

$$\begin{array}{lll}
\sum_{\ell=1}^{n} a_\ell P(\varphi_\ell) \geq b & \text{for} & a_1P(\varphi_1) + \cdots + a_nP(\varphi_n) \geq b, \\
a_1P_i(\varphi_1) \geq a_2P_i(\varphi_2) & \text{for} & a_1P(\varphi_1) + (-a_2)P(\varphi_2) \geq 0, \\
\sum_{\ell=1}^{n} a_\ell P(\varphi_\ell) \leq b & \text{for} & \sum_{\ell=1}^{n} (-a_\ell)P(\varphi_\ell) \geq -b, \\
\sum_{\ell=1}^{n} a_\ell P(\varphi_\ell) < b & \text{for} & \neg(\sum_{\ell=1}^{n} a_\ell P(\varphi_\ell) \geq b), \\
\sum_{\ell=1}^{n} a_\ell P(\varphi_\ell) > b & \text{for} & \neg(\sum_{\ell=1}^{n} a_\ell P(\varphi_\ell) \leq b), \\
\sum_{\ell=1}^{n} a_\ell P(\varphi_\ell) = b & \text{for} & \sum_{\ell=1}^{n} a_\ell P(\varphi_\ell) \geq b \wedge \sum_{\ell=1}^{n} a_\ell P(\varphi_\ell) \leq b.
\end{array}$$

The formal semantics is defined in layers, just like the language itself. Given a probabilistic model $\mathbb{M}$ (discrete or otherwise) with valuation V, we first define a function $[\![\cdot]\!]^{\mathbb{M}} \colon \mathcal{L}_{BL} \to \wp(S)$ by putting

$$\begin{array}{lll}
[\![p]\!]^{\mathbb{M}} & = & V(p), \\
[\![\neg\varphi]\!]^{\mathbb{M}} & = & S - [\![\varphi]\!]^{\mathbb{M}}, \\
[\![\varphi \wedge \psi]\!]^{\mathbb{M}} & = & [\![\varphi]\!]^{\mathbb{M}} \cap [\![\psi]\!]^{\mathbb{M}}.
\end{array}$$

It is easy to check that in the case of $\mathbb{M}$ being a non-discrete probabilistic model, the set $[\![\varphi]\!]^{\mathbb{M}}$ is measurable for all $\varphi \in \mathcal{L}_{BL}$. We omit the superscript $\mathbb{M}$ when it is understood by context. We then define a relation $\models$ between probabilistic models $\mathbb{M} = \langle S, \mathcal{A}, \mu, V\rangle$ and formulas $f \in \mathcal{L}_{PPL}$ by

$$\begin{array}{lll}
\mathbb{M} \models a_1P(\varphi_1) + \cdots + a_nP(\varphi_n) \geq a & \text{iff} & a_1\mu([\![\varphi_1]\!]^{\mathbb{M}}) + \cdots \\
 & & +a_n\mu([\![\varphi_n]\!]^{\mathbb{M}}) \geq a, \\
\mathbb{M} \models \neg f & \text{iff} & \mathbb{M} \not\models f, \\
\mathbb{M} \models f_1 \wedge f_2 & \text{iff} & \mathbb{M} \models f_1 \text{ and } \mathbb{M} \models f_2.
\end{array}$$

If $\mathbb{M}$ is a *discrete* probabilistic model, i.e. $\mathbb{M} = \langle S, p, V \rangle$, then the first semantic clause makes use of the additive lifting p^+ of the probability function p:

$$\mathbb{M} \models a_1 P(\varphi_1) + \cdots + a_n P(\varphi_n) \geq a \text{ iff } a_1 p^+(\llbracket \varphi_1 \rrbracket^{\mathbb{M}}) + \cdots + a_n p^+(\llbracket \varphi_n \rrbracket^{\mathbb{M}}) \geq a.$$

It should be emphasized that $\models$ is only defined for $\mathcal{L}_{PPL}$-formulas. Hence, for a propositional atom p and a probabilistic model $\mathbb{M}$, it makes sense to ask whether $\mathbb{M} \models P(p) \geq 0.6$, but not whether $\mathbb{M} \models p$. More importantly, note that probability formulas are of the form $P(\varphi) \geq k$, where φ is a Boolean combination of propositional atoms. In other words, higher-order probabilities cannot be expressed in $\mathcal{L}_{PPL}$: formulas such as $P(P(q) \geq 0.7) \geq 0.6$ are not well-formed.

We will finish this subsection by showing that despite its limitation to first-order probabilities, this framework is quite powerful, and can be used to naturally formalize rather intricate scenarios.

Example 4.1

Three indistinguishable balls are simultaneously dropped down a tube. Inside the tube, each ball gets stuck (independently of the other balls) with some small probability ε. In other words, each ball rolls out at the other end of the tube with probability $1 - \varepsilon$, and does not roll out with probability ε. Because the three balls are indistinguishable, we cannot know *which* ball(s) got stuck; we can only count the *number* of balls that roll out of the tube. (Of course, if we count 3 (resp. 0) balls rolling out, then we do know that no (resp. all) balls have gotten stuck.) What is the probability that exactly 2 balls will roll out?

We define the sample space as $S \stackrel{\text{def}}{=} \{(s_1, s_2, s_3) \mid s_i \in \{0, 1\}\}$, where

$$s_i \stackrel{\text{def}}{=} \begin{cases} 1 & \text{if ball } i \text{ rolls out,} \\ 0 & \text{if ball } i \text{ does not roll out.} \end{cases} \qquad \dashv$$

We consider the propositional atoms $\mathsf{rollout}_n$, which is to be read as 'exactly n balls roll out'. Obviously, we put $V(\mathsf{rollout}_n) \stackrel{\text{def}}{=} \{(s_1, s_2, s_3) \in S \mid s_1 + s_2 + s_3 = n\}$. Let $\mathcal{A}$ be the σ-algebra generated by $\{V(\mathsf{rollout}_n) \mid 0 \leq n \leq 3\}$. This reflects the fact that we can only observe the *number* of balls rolling out of the tube; for example, the singleton set $\{(1, 1, 0)\}$ (which contains the information that b_3 got stuck, and b_1 and b_2 roll out) is not in $\mathcal{A}$. Finally, we define a probability measure μ by putting $\mu(V(\mathsf{rollout}_n)) \stackrel{\text{def}}{=} \binom{3}{n}(1-\varepsilon)^n \varepsilon^{3-n}$ (i.e. a binomial distribution).[4] The probabilistic model $\mathbb{M} \stackrel{\text{def}}{=} \langle S, \mathcal{A}, \mu, V \rangle$ fully captures the scenario. For example, if $\varepsilon = 0.1$, one can check that $\mathbb{M} \models P(\mathsf{rollout}_2) = 0.243$.

[4]The binomial coefficient $\binom{3}{n}$ is defined as $\frac{3!}{n!(3-n)!}$.

4.2.3 The Expressivity of Linear Combinations

The set of terms $\mathcal{T}(\mathsf{AtProp})$ introduced in the previous subsection contains not only terms of the form $P(\varphi)$ (with $\varphi \in \mathcal{L}_{BL}(\mathsf{AtProp})$), but also linear combinations: $a_1 P(\varphi_1) + \cdots + a_n P(\varphi_n)$. Besides having technical motivations, this leads to the language $\mathcal{L}_{PPL}(\mathsf{AtProp})$ being highly expressive; for example, (i) it can express *comparative* probability judgments (of the form $P(\varphi) \geq P(\psi)$), and (ii) the $\geq$-comparison suffices to define all others (cf. the abbreviations stated above for $>$, $\leq$, etc.).

However, this expressivity gain should not be exaggerated. For example, if we restrict ourselves to 'single' probabilities—i.e. terms of the form $P(\varphi)$, and thus probability formulas of the form $P(\varphi) \geq b$—, all the comparisons are already definable. Let's first consider the case of $P(\varphi) \leq b$. With the abbreviations mentioned above in mind, this can be defined as $-P(\varphi) \geq -b$. However, this involves scalar multiplication (with -1) of $P(\varphi)$, and thus already takes us outside the narrow realm of 'single' probabilities. There exists, however, an alternative definition that stays inside this realm, viz. $P(\neg\varphi) \geq 1 - b$ (note that if $\varphi \in \mathcal{L}_{BL}(\mathsf{AtProp})$, then $\neg\varphi \in \mathcal{L}_{BL}(\mathsf{AtProp})$ as well, and thus $\neg\varphi$ is perfectly allowed inside $P(\cdot)$). Defining the other comparisons is now straightforward:

$$\begin{array}{lll} P(\varphi) \leq b & \text{for} & P(\neg\varphi) \geq 1 - b, \\ P(\varphi) < b & \text{for} & \neg\,(P(\varphi) \geq b), \\ P(\varphi) > b & \text{for} & \neg\,(P(\varphi) \leq b), \\ P(\varphi) = b & \text{for} & P(\varphi) \geq b \wedge P(\varphi) \leq b. \end{array}$$

Next, consider the (finite)[5] additivity property of probability measures: $\mu(X \cup Y) = \mu(X) + \mu(Y)$ for all disjoint $X, Y \in \mathcal{A}$. Using linear combinations of probabilities, this can be expressed almost 'literally': $P(\varphi \vee \psi) = P(\varphi) + P(\psi)$ whenever $\neg(\varphi \wedge \psi)$ is a tautology. The axiomatization that is introduced in the next subsection uses another, equivalent expression, which also involves linear combinations: $P(\varphi \wedge \psi) + P(\varphi \wedge \neg\psi) = P(\varphi)$. This suggests an alternative way of expressing additivity which does *not* make use of linear combinations:

$$\big(P(\varphi \wedge \psi) = a \wedge P(\varphi \wedge \neg\psi) = b\big) \rightarrow P(\varphi) = a + b.$$

There exist alternative axiomatizations of probabilistic propositional logic along these lines. Note that the original formula (with a linear combination) does not explicitly contain the numbers a and b, and is therefore able

[5]Probability measures actually satisfy the stronger *countable* additivity requirement. Since this requirement cannot easily be captured in a finitary logic (although there exist systems in which it is captured by means of an infinitary rule in a finitary logic), it is customary to focus on formulas for *finite* additivity.

to express additivity in a *single* formula (for given φ, ψ, of course). The alternative formulation (without linear combinations) should be seen as a *scheme*: it corresponds to the (countable) set of formulas

$$\{\Big((P(\varphi \wedge \psi) = a \wedge P(\varphi \wedge \neg\psi) = b) \rightarrow P(\varphi) = a + b\Big) \in \mathcal{L}_{PPL}(\mathsf{AtProp}) \mid a, b \in [0,1] \cap \mathbb{Q}\}.$$

Finally, note that linear combinations do not make the language more powerful when *distinguishing* between models. However, they *do* make the language more powerful when *characterizing* classes of models. For any language $\mathcal{L}$ and models $\mathbb{M}_1, \mathbb{M}_2$ (on which $\mathcal{L}$ is interpretable), we define:

$$\mathbb{M}_1 \text{ and } \mathbb{M}_2 \text{ are } \mathcal{L}\text{-}equivalent \quad \text{iff} \quad \forall \varphi \in \mathcal{L}\colon \mathbb{M}_1 \models \varphi \Leftrightarrow \mathbb{M}_2 \models \varphi.$$

Informally, two models are $\mathcal{L}$-equivalent if $\mathcal{L}$ cannot distinguish them. Furthermore, for any class $\mathcal{C}$ of models (on which $\mathcal{L}$ is interpretable) and formula $\varphi \in \mathcal{L}$, we define:

$$\mathcal{C} \text{ is } characterized \text{ by } \varphi \quad \text{iff} \quad \text{for all } \mathbb{M}\colon \mathbb{M} \in \mathcal{C} \Leftrightarrow \mathbb{M} \models \varphi.$$

Let $\mathcal{L}^*_{PPL}(\mathsf{AtProp})$ be the language that is obtained from $\mathcal{L}_{PPL}(\mathsf{AtProp})$ by only allowing atomic probability formulas of the form $P(\varphi) \geq a$ (in other words, linear combinations of probability terms are not allowed). Then one can show the following:

Lemma 4.1
Consider arbitrary probabilistic models $\mathbb{M}_1 = \langle S_1, \mathcal{A}_1, \mu_1, V_1\rangle$ and $\mathbb{M}_2 = \langle S_2, \mathcal{A}_2, \mu_2, V_2\rangle$. These models are $\mathcal{L}_{PPL}(\mathsf{AtProp})$-equivalent iff they are $\mathcal{L}^*_{PPL}(\mathsf{AtProp})$-equivalent.

Proof If $\mathbb{M}_1$ and $\mathbb{M}_2$ are $\mathcal{L}_{PPL}(\mathsf{AtProp})$-equivalent, then they are trivially also $\mathcal{L}^*_{PPL}(\mathsf{AtProp})$-equivalent, since $\mathcal{L}^*_{PPL}(\mathsf{AtProp}) \subseteq \mathcal{L}_{PPL}(\mathsf{AtProp})$. We now prove the other direction.

Consider an arbitrary propositional formula $\varphi \in \mathcal{L}_{BL}(\mathsf{AtProp})$, and suppose that $\mu_1([\![\varphi]\!]^{\mathbb{M}_1}) \neq \mu_2([\![\varphi]\!]^{\mathbb{M}_2})$. Without loss of generality, assume that $\mu_1([\![\varphi]\!]^{\mathbb{M}_1}) > \mu_2([\![\varphi]\!]^{\mathbb{M}_2})$ (the other case is completely analogous). Since $\mathbb{Q}$ is dense in $\mathbb{R}$, there exists a $k \in \mathbb{Q}$ such that $\mu_1([\![\varphi]\!]^{\mathbb{M}_1}) > k > \mu_2([\![\varphi]\!]^{\mathbb{M}_2})$. It now follows that $\mathbb{M}_1 \models P(\varphi) \geq k$, while $\mathbb{M}_2 \not\models P(\varphi) \geq k$, which contradicts the assumption that these models are $\mathcal{L}^*_{PPL}(\mathsf{AtProp})$-equivalent. We therefore conclude that $\mu_1([\![\varphi]\!]^{\mathbb{M}_1}) = \mu_2([\![\varphi]\!]^{\mathbb{M}_2})$. Since $\varphi \in \mathcal{L}_{BL}(\mathsf{AtProp})$ was chosen arbitrarily, this holds for *all* propositional formulas. Hence, for any atomic probability formula $a_1 P(\varphi_1) + \cdots + a_n P(\varphi_n) \geq b$, we have:

$$\begin{aligned}
&\mathbb{M}_1 \models a_1 P(\varphi_1) + \cdots + a_n P(\varphi_n) \geq b \\
&\Leftrightarrow a_1 \mu_1([\![\varphi_1]\!]^{\mathbb{M}_1}) + \cdots + a_n \mu_1([\![\varphi_n]\!]^{\mathbb{M}_1}) \geq b \\
&\Leftrightarrow a_1 \mu_2([\![\varphi_1]\!]^{\mathbb{M}_2}) + \cdots + a_n \mu_2([\![\varphi_n]\!]^{\mathbb{M}_2}) \geq b \\
&\Leftrightarrow \mathbb{M}_2 \models a_1 P(\varphi_1) + \cdots + a_n P(\varphi_n) \geq b.
\end{aligned}$$

Since $\mathbb{M}_1$ and $\mathbb{M}_2$ agree on all atomic probability formulas, they also agree on all Boolean combinations of such formulas, and are thus $\mathcal{L}_{PPL}(\mathsf{AtProp})$-equivalent. ⊣

⊣

Lemma 4.2
The class of probabilistic models $\{\mathbb{M} \in \mathcal{C}^{PPL} \mid \mathbb{M} = \langle S, \mathcal{A}, \mu, V\rangle, \mu(V(p)) \geq \mu(V(q))\}$ can be characterized by a formula in $\mathcal{L}_{PPL}(\mathsf{AtProp})$, but not by one in $\mathcal{L}^*_{PPL}(\mathsf{AtProp})$. ⊣

Proof

First, note that this class is trivially characterized by the $\mathcal{L}_{PPL}(\mathsf{AtProp})$-formula $P(p) \geq P(q)$. We will now show that this class cannot be characterized by any formula in $\mathcal{L}^*_{PPL}(\mathsf{AtProp})$. The formula $P(p) \geq P(q)$ is equivalent to the formula $P(p \wedge \neg q) - P(\neg p \wedge q) \geq 0$. Let

$$
\begin{array}{lll}
x_{00} & \text{be the probability value of} & \neg p \wedge \neg q, \\
x_{01} & \text{be the probability value of} & \neg p \wedge q, \\
x_{10} & \text{be the probability value of} & p \wedge \neg q, \\
x_{11} & \text{be the probability value of} & p \wedge q.
\end{array}
$$

Thus our formula is equivalent to asserting $x_{10} - x_{01} \geq 0$, subject to the constraint that the x_{ij}'s are non-negative and sum to 1. The projection A of the solution set onto the x_{10}-x_{01} plane is then the area enclosed by the equations:

$$
\begin{aligned}
x_{10} - x_{01} &\geq 0, \\
x_{10} + x_{01} &\leq 1, \\
x_{01} &\geq 0.
\end{aligned}
$$

The enclosed region A is depicted by the shaded region of the following:

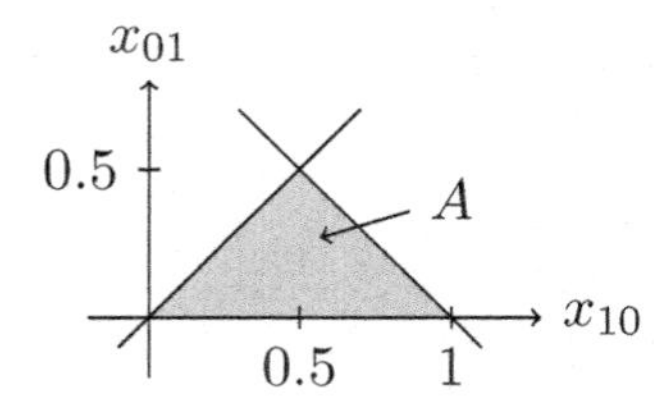

A formula f in $\mathcal{L}^*_{PPL}(\mathsf{AtProp})$ can be placed in its disjunctive normal form; each disjunct is a conjunction of atomic probability formulas or negations of atomic probability formulas. Let Q be the set of atomic propositions that appear in f together with p and q, and let A_Q be the set of formulas of the form $\bigwedge\{p' \mid p' = p \text{ or } p' = \neg p, \text{ with } p \in Q\}$. Then any probability formula $P(\varphi) \geq r$ appearing inside f is equivalent to some linear combination formula $\sum_{\chi \in A_Q} c_\chi P(\chi) \geq r$, where each c_χ is either 0 or 1. This conversion of any formula into its disjunctive normal form is similar to the analysis that is given in the standard completeness proof for probabilistic propositional logic. Each model assigns probability to formulas in A_Q, and x_{10} is the sum of probabilities of all formulas in A_Q that imply $p \land \neg q$, and similarly x_{01} the sum of those that imply $\neg p \land q$.

If $f \leftrightarrow P(p) \geq P(q)$ is valid, then $d \to P(p) \geq P(q)$ is valid for each disjunct d of the disjunctive normal form of f. Let d be such a disjunct. Let θ be a function mapping each value a that x_{10} can attain given the constraints in d to the supremum of the values x_{01} can attain when $x_{10} = a$. As each constraint given by a conjunct of d is equivalent to $\sum_{\chi \in A_Q} c_\chi P(\chi) \geq r$ or $\sum_{\chi \in A_Q} c_\chi P(\chi) < r$ for $c_\chi \in \{0, 1\}$ (significantly, where the coefficients are all non-negative), and as the probability constraints $P(\chi) \geq 0$ and $\sum_{\chi \in A_Q} P(\chi) \leq 1$ are of this form too, θ is non-increasing. Let c be the infimum of values x_{10} can obtain given the constraints in d. As θ is non-increasing, c is greater than or equal to the supremum of the values x_{01} can obtain.

We thus see that if $d \to P(p) \geq P(q)$ is valid, then there exists a c, such that $d \to P(p \land \neg q) \geq c \land P(\neg p \land q) \leq c$ is valid. Thus the models that satisfy d must be contained in regions that we depict as follows:

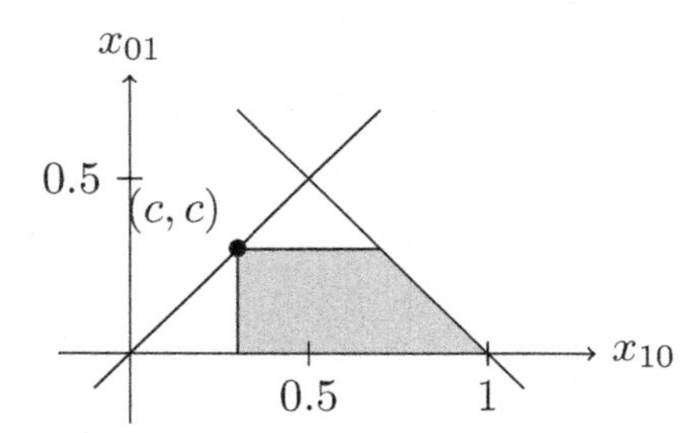

Clearly no finite set of regions subject to such constraints has a union equal to A. Thus no finite disjunction of conjunctions d, such that $d \to P(p) \geq P(q)$ can be equivalent to $P(p) \geq P(q)$. ⊣

4.2.4 Proof System

A proof system for probabilistic propositional logic is given in Figure 4.1. Beyond the propositional component, there is a probabilistic component, which is a straightforward translation into the language $\mathcal{L}_{PPL}(\mathsf{AtProp})$ of the well-known Kolmogorov axioms of probability, together with a rule stating that provably equivalent formulas have identical probabilities. This

Figure 4.1: Componentwise axiomatization of probabilistic propositional logic.

1. propositional component
   - all propositional tautologies and the modus ponens rule
2. probabilistic component
   - $P(\varphi) \geq 0$
   - $P(\top) = 1$
   - $P(\varphi \wedge \psi) + P(\varphi \wedge \neg\psi) = P(\varphi)$
   - if $\vdash \varphi \leftrightarrow \psi$ then $\vdash P(\varphi) = P(\psi)$
3. linear inequalities component
   - $\sum_{\ell=1}^{n} a_\ell P(\varphi_\ell) \geq b \leftrightarrow \sum_{\ell=1}^{n} a_\ell P(\varphi_\ell) + 0P(\varphi_{n+1}) \geq b$
   - $\sum_{\ell=1}^{n} a_\ell P(\varphi_\ell) \geq b \leftrightarrow \sum_{\ell=1}^{n} a_{p(\ell)} P(\varphi_{p(\ell)}) \geq b$ for any permutation p of $1, \ldots, n$
   - $\sum_{\ell=1}^{n} a_\ell P(\varphi_\ell) \geq b \wedge \sum_{\ell=1}^{n} a'_\ell P(\varphi_\ell) \geq b' \rightarrow \sum_{\ell=1}^{n} (a_\ell + a'_\ell) P(\varphi_\ell) \geq b + b'$
   - $\sum_{\ell=1}^{n} a_\ell P(\varphi_\ell) \geq b \leftrightarrow \sum_{\ell=1}^{n} d a_\ell P(\varphi_\ell) \geq db$ (for any $d > 0$)
   - $\sum_{\ell=1}^{n} a_\ell P(\varphi_\ell) \geq b \vee \sum_{\ell=1}^{n} a_\ell P(\varphi_\ell) \leq b$
   - $\sum_{\ell=1}^{n} a_\ell P(\varphi_\ell) \geq b \rightarrow \sum_{\ell=1}^{n} a_\ell P(\varphi_\ell) > b'$ (for any $b' < b$)

component thus ensures that the formal symbol $P(\cdot)$ behaves like a real probability function. Finally, the linear inequalities component is mainly a technical tool to ensure that the logic is strong enough to capture the behavior of linear inequalities of probabilities.

This logic is sound and complete:

Theorem 4.1
Probabilistic propositional logic, as axiomatized in Figure 4.1, is sound and weakly complete with respect to the class $\mathcal{C}^{PPL}$ of probabilistic models, and also with respect to the class $\mathcal{C}_d^{PPL}$ of discrete probabilistic models. ⊣

The notion of completeness used in this theorem is *weak* completeness ($\vdash \varphi$ iff $\models \varphi$), rather than *strong* completeness ($\Gamma \vdash \varphi$ iff $\Gamma \models \varphi$). These two notions do not coincide in probabilistic propositional logic, because this logic is not *compact*; for example, every finite subset of the set $\{P(p) >$

$0\} \cup \{P(p) \leq k \mid k > 0\}$ is satisfiable, but the entire set is not. (Similar remarks apply to the other logics that will be discussed later in this chapter.)

The proof of Theorem 4.1 involves establishing the existence of a satisfying model for a consistent probability formula f. To do this, it is shown that f is provably equivalent to a conjunction of atomic probability formulas or negations of atomic probability formulas. Hence, f is satisfiable iff the system of linear inequalities corresponding to this conjunction of (negated) atomic probability formulas, together with the equalities and inequalities given by the Kolmogorov axioms, has a solution. The satisfying model has a finite number of states, its σ-algebra is the powerset of its domain, and probabilities are assigned to singletons according to the solution of the linear system. This model is based on a probability space, and is thus a *non-discrete* probabilistic model. However, since all subsets of its domain are measurable, it can also be viewed as a *discrete* probabilistic model. It follows that the axiomatization in Figure 4.1 is complete with respect to both kinds of probabilistic models ($\mathcal{C}^{PPL}$ and $\mathcal{C}_d^{PPL}$).

4.2.5 Decidability and Complexity

Probabilistic propositional logic is decidable, and its satisfiability problem is NP-complete:

Theorem 4.2
Let $\mathcal{C}$ be either $\mathcal{C}^{PPL}$ or $\mathcal{C}_d^{PPL}$. The problem of deciding whether a probability formula is satisfiable in a model in $\mathcal{C}$ is NP-complete. $\dashv$

The proof is essentially the same for both classes of models. For the lower bound, we see that the satisfiability problem for probabilistic propositional logic is at least as hard as the Boolean satisfiability problem: a propositional formula φ is satisfiable if and only if the probability formula $P(\varphi) > 0$ is. Boolean satisfiability is well-known to be NP-complete. For the upper bound, one can non-deterministically select a 'small model', which justified by the small model theorem:

Theorem 4.3
Suppose f is a satisfiable probability formula, $|f|$ is its length (number of symbols), and $\|f\|$ is the length of the longest coefficient appearing in f (the length of the numerator plus the length of the denominator of that coefficient). Then f is satisfiable in a probabilistic model (i) which has most $|f|$ states, (ii) where every set of states is measurable, and (iii) where the probability assigned to each state is a rational number with size $O(|f|\,\|f\| + |f| \log(|f|))$. $\dashv$

The small model described in this theorem is a significant improvement upon the model constructed in the completeness proof, which has a state for every conjunction of literals of atomic propositions appearing in f. If there are k propositions in f, then there are 2^k such conjunctions, making the size of the models *exponential* in the length of the formula. In the small model, on the other hand, the number of states is *polynomial* in the length of the formula. Checking whether f is satisfiable in such a model (the model checking problem) can be done in polynomial time.

4.3 Probabilistic Epistemic Logic

In this section we introduce the static framework of probabilistic epistemic logic, which will be 'dynamified' in Sections 4.5 and 4.6. Subsection 4.3.1 discusses the models on which the logic is interpreted. Subsection 4.3.2 defines the formal language and its semantics, and Subsection 4.3.3 discusses its expressivity. Subsection 4.3.4 provides a complete axiomatization, and Subsection 4.3.5 discusses decidability and complexity properties.

4.3.1 Probabilistic Relational Models

In addition to the countable set AtProp of proposition letters, we consider a finite set I of agents, which we also assume as given throughout this chapter (and will hence often leave implicit).

Definition 4.3
A *probabilistic relational frame* is a tuple $\mathbb{F} = \langle W, R_i, \mathbb{P}_i \rangle_{i \in I}$, where W is a non-empty set of states, $R_i \subseteq W \times W$ is agent i's epistemic accessibility relation, and $\mathbb{P}_i$ assigns to each state $w \in W$ a probability space $\mathbb{P}_i(w) \stackrel{\text{def}}{=} \langle S_i(w), \mathcal{A}_i(w), \mu_i(w) \rangle$, with $S_i(w) \subseteq W$, $\mathcal{A}_i(w)$ a σ-algebra over $S_i(w)$, and $\mu_i(w) \colon \mathcal{A}_i(w) \to [0, 1]$ a probability measure.

Note that in principle, no conditions are imposed on the agents' epistemic accessibility relations. However, as is usually done in the literature on (probabilistic) epistemic logic, we will henceforth assume these relations to be *equivalence relations* (so that the corresponding knowledge operators satisfy the principles of the modal logic S5).

We now turn to the probabilistic component. This is highly general: it is based on probability spaces (instead of discrete probability structures), and can thus accommodate uniform probability distributions over infinite sets. Furthermore, the probability space $\mathbb{P}_i(w)$ depends on the agent i and state w, so which sets are measurable can vary from agent to agent, and, more importantly, from state to state.

For many applications, however, this high level of generality is not necessary, and various simplifying assumptions can be made. First of all, we can often assume that $\mathcal{A}_i(w) = \wp(S_i(w))$ (for every $w \in W$), i.e. that all subsets of $S_i(w)$ are measurable. As was explained in Subsection 4.2.1, this essentially means that we can view $\mathbb{P}_i(w)$ as a discrete probability structure $\langle S_i(w), \mu_i(w)\rangle$, with $\mu_i(w)\colon S_i(w) \to [0,1]$ and $\sum_{v\in S_i(w)} \mu_i(w)(v) = 1$, and put for any $X \subseteq S_i(w)$: $\mu_i(w)(X) = \sum_{x\in X} \mu_i(w)(x)$. Note that this simplification does not yet guarantee that *all* subsets of W are measurable; after all, if $v \in W - S_i(w)$, then any set $X \subseteq W$ such that $v \in X$ does not belong to $\wp(S_i(w))$, and hence does not get assigned a probability by $\mu_i(w)$. To eliminate such cases, we can make the additional simplifying assumption that $S_i(w) = W$ for all states $w \in W$. Note that the sample space now no longer depends on w; hence, we can view $\mathbb{P}_i(w)$ not just as the discrete probability structure $\langle W, \mu_i(w)\rangle$, but identify it with $\mu_i(w)$ altogether.

If a probabilistic relational frame satisfies these two simplifying assumptions, it will be called 'simplified':

Definition 4.4
A *simplified probabilistic relational frame* is a tuple $\mathbb{F} = \langle W, R_i, \mu_i\rangle_{i\in I}$, where W is a non-empty set of states, $R_i \subseteq W \times W$ is agent i's epistemic accessibility relation, and μ_i assigns to each state $w \in W$ a function $\mu_i(w)\colon W \to [0,1]$ which has countable support (i.e. the set $\{v \in W \mid \mu_i(w)(v) > 0\}$ is countable) and satisfies $\sum_{v\in W} \mu_i(w)(v) = 1$. ⊣

As usual, we obtain a model by adding a valuation to a frame. This construction applies to 'general' probabilistic relational frames and simplified probabilistic relational frames alike.

Definition 4.5
A *probabilistic relational model* is a tuple $\mathbb{M} = \langle \mathbb{F}, V\rangle$, where $\mathbb{F}$ is a probabilistic relational frame (with set of states W), and $V\colon \mathsf{AtProp} \to \wp(W)$ is a valuation. The model $\langle \mathbb{F}, V\rangle$ is said to be *simplified* iff the frame $\mathbb{F}$ on which it is based is simplified. The class of all probabilistic relational models will be denoted $\mathcal{C}^{PEL}$, and the class of all simplified probabilistic relational models will be denoted $\mathcal{C}_s^{PEL}$. ⊣

4.3.2 Language and Semantics

The language $\mathcal{L}_{PEL}(\mathsf{AtProp})$ of probabilistic epistemic logic is defined by means of the following grammar:

$$\varphi \ ::= \ p \mid \neg\varphi \mid (\varphi \wedge \varphi) \mid K_i\varphi \mid a_1P_i(\varphi) + \cdots + a_nP_i(\varphi) \geq b$$

where $p \in \mathsf{AtProp}, i \in I, 1 \leq n < \omega$, and $a_1, \ldots, a_n, b \in \mathbb{Q}$. We only allow rational numbers as values for $a_1, \ldots, a_n, b$ in order to keep the language countable. As usual, $K_i\varphi$ means that agent i knows that φ, or, more generally, that φ follows from agent i's information. Its dual is defined as $\hat{K}_i\varphi \stackrel{\text{def}}{=} \neg K_i \neg\varphi$, and means that φ is consistent with agent i's information.

Formulas of the form $a_1 P_i(\varphi_1) + \cdots + a_n P_i(\varphi_n) \geq b$ are called *i-probability formulas*. Note that mixed agent indices are not allowed; for example, $P_a(p) + P_b(q) \geq b$ is *not* a well-formed formula. Intuitively, $P_i(\varphi) \geq b$ means that agent i assigns probability at least b to φ. We adopt the same abbreviations as in the previous section.

It should be emphasized that because of its recursive definition, the language $\mathcal{L}_{PEL}(\mathsf{AtProp})$ can express the agents' higher-order information of any sort: higher-order knowledge (for example $K_a K_b \varphi$), but also higher-order probabilities (for example $P_a(P_b(\varphi) \geq 0.5) = 1$), and higher-order information that mixes knowledge and probabilities (for example, $K_a(P_b(\varphi) \geq 0.5)$ and $P_a(K_b\varphi) = 1$).

For simplicity and generality, the definition of a probabilistic relational model $\mathbb{M}$ in the previous subsection did not contain any restrictions to ensure that every formula corresponds to a measurable set in $\mathbb{M}$. We will discuss such restrictions after introducing the semantics.

To ensure a well-defined semantics, we extend each probability measure to a total function using inner or outer measures. Given a probability space $\langle W, \mathcal{A}, \mu \rangle$, the inner measure $\mu_* \colon \wp(W) \to [0,1]$ is defined by

$$\mu_*(X) \stackrel{\text{def}}{=} \sup\{\mu(Y) \mid Y \subseteq X, Y \in \mathcal{A}\}.$$

It is easy to see that $\mu_* \upharpoonright \mathcal{A} = \mu$, i.e. if $X \in \mathcal{A}$, then $\mu_*(X) = \mu(X)$. The outer measure μ^* can be defined as the dual of the inner measure ($\mu^*(X) \stackrel{\text{def}}{=} 1 - \mu_*(W - X)$ for all $X \subseteq W$), and it would be just as natural to use the outer measure in the semantics.

The formal semantics for $\mathcal{L}_{PEL}(\mathsf{AtProp})$ is defined as follows. Consider an arbitrary probabilistic relational model $\mathbb{M}$ (with set of states W) and a state $w \in W$. We will often abbreviate $[\![\varphi]\!]^{\mathbb{M}} \stackrel{\text{def}}{=} \{v \in W \mid \mathbb{M}, v \models \varphi\}$. Then:

$\mathbb{M}, w \models p$	iff	$w \in V(p)$,
$\mathbb{M}, w \models \neg\varphi$	iff	$\mathbb{M}, w \not\models \varphi$,
$\mathbb{M}, w \models \varphi \wedge \psi$	iff	$\mathbb{M}, w \models \varphi$ and $\mathbb{M}, w \models \psi$,
$\mathbb{M}, w \models K_i\varphi$	iff	for all $v \in W$: if $(w,v) \in R_i$ then $\mathbb{M}, v \models \varphi$,
$\mathbb{M}, w \models \sum_{\ell=1}^{n} a_\ell P_i(\varphi_\ell) \geq b$	iff	$\sum_{\ell=1}^{n} a_\ell (\mu_i(w))_*([\![\varphi_\ell]\!]^{\mathbb{M}} \cap S_i(w)) \geq b$.

To understand the clause for i-probability formulas better, note that the probability measure $\mu_i(w)$ is defined on the σ-algebra $\mathcal{A}_i(w) \subseteq \wp(S_i(w))$, and hence, its inner measure $(\mu_i(w))_*$ is defined on all of $\wp(S_i(w))$. Furthermore, the truth set $[\![\varphi_\ell]\!]^{\mathbb{M}}$ is intersected with $S_i(w)$ to ensure that the resulting set is a subset of $S_i(w)$ (i.e. belongs to the domain of $(\mu_i(w))_*$).

In *simplified* probabilistic relational models, there is no σ-algebra, and hence no need for inner measures. Furthermore, in such models it holds that $S_i(w) = W$, and hence there is no need to intersect $[\![\varphi_\ell]\!]^{\mathbb{M}}$ with $S_i(w)$. As a result, the semantic clause for i-probability formulas simplifies to the following:

$$\mathbb{M}, w \models \sum_{\ell=1}^{n} a_\ell P_i(\varphi_\ell) \geq b \qquad \text{iff} \qquad \sum_{\ell=1}^{n} a_\ell \mu_i(w)([\![\varphi_\ell]\!]^{\mathbb{M}}) \geq b.$$

Furthermore, in both cases we also define:

- $\mathbb{M} \models \varphi$ iff $\mathbb{M}, w \models \varphi$ for all $w \in W$,
- $\mathbb{F} \models \varphi$ iff $\langle \mathbb{F}, V \rangle \models \varphi$ for all valuations V on the frame $\mathbb{F}$,
- $\models \varphi$ iff $\mathbb{F} \models \varphi$ for all frames $\mathbb{F}$.

To illustrate the language and its semantics, we describe a small-scale scenario and show how it can be modeled naturally in the framework of probabilistic epistemic logic. (More elaborate examples will be discussed in Sections 4.5 and 4.6.)

Example 4.2
Consider the following scenario. An agent does not know whether p is the case, i.e. she cannot distinguish between p-states and $\neg p$-states. (In fact, p happens to be true.) Furthermore, the agent has no specific reason to think that one state is more probable than any other; therefore it is reasonable for her to assign equal probabilities to all states. This example can be formalized by the following (simplified) probabilistic relational model: $\mathbb{M} = \langle W, R, \mu, V \rangle, W = \{w, v\}, R = W \times W, \mu(w)(w) = \mu(w)(v) = \mu(v)(w) = \mu(v)(v) = 0.5$, and $V(p) = \{w\}$. (We work with only one agent in this example, so agent indices can be dropped.) One can easily check that this model is a faithful representation of the scenario described above; for example:

$$\mathbb{M}, w \models \neg Kp \land \neg K \neg p \land P(p) = 0.5 \land P(\neg p) = 0.5. \qquad \dashv$$

Note that this focuses on the agents' knowledge and probabilities at a single moment. In Sections 4.5 and 4.6, we will discuss how informational events (such as public announcements) can trigger changes in the agents' knowledge and probabilities, and return to this example to illustrate how such changes can naturally be modeled (see Example 4.3).

We will now define a restriction on probabilistic relational frames and models which ensures that $[\![\varphi]\!]^{\mathbb{M}} \cap S_i(w) \in \mathcal{A}_i(w)$ for every formula φ, agent i and state w. Hence, for probabilistic relational models satisfying this restriction, the semantic clause for i-probability formulas can be stated using the probability measure $\mu_i(w)$, rather than its inner measure $(\mu_i(w))_*$.

Definition 4.6
A probabilistic relational frame $\mathbb{F} = \langle W, R_i, \mathbb{P}_i \rangle_{i \in I}$ is said to satisfy *meas* if and only if there exists a σ-algebra $\mathcal{A} \subseteq \wp(W)$ such that the following conditions hold for each agent i:

- $\{A \cap S_i(w) \mid A \in \mathcal{A}\} \subseteq \mathcal{A}_i(w)$ for all states $w \in W$,
- $\{w \in W \mid R_i[w] \subseteq A\} \in \mathcal{A}$ for each $A \in \mathcal{A}$,[6]
- $\mathbb{P}_i$ is a measurable function[7] from $(W, \mathcal{A})$ to $(\mathsf{spaces}(W), \mathcal{B})$, where
  - $\mathsf{spaces}(W)$ is the set of all probability spaces $(S, \mathcal{C}, \nu)$ such that $S \subseteq W$ and $\{A \cap S \mid A \in \mathcal{A}\} \subseteq \mathcal{C}$,
  - $\mathcal{B}$ is the σ-algebra generated from the sets

$$\{(S, \mathcal{C}, \nu) \mid \sum_{k=1}^{n} a_k \nu(A_k \cap S) \geq r\}$$

  for each $n \geq 1$, $A_k \in \mathcal{A}$, and $a_k, r \in \mathbb{Q}$ $(1 \leq k \leq n)$.

A σ-algebra $\mathcal{A}$ satisfying these conditions is called a *general σ-algebra* of $\mathbb{F}$. A probabilistic relational model $\mathbb{M} = \langle \mathbb{F}, V \rangle$ is said to satisfy *meas* if and only if the frame $\mathbb{F}$ on which it is based satisfies *meas*, and the image of the valuation V is some general σ-algebra $\mathcal{A}$ of $\mathbb{F}$ (i.e. $V \colon \mathsf{AtProp} \to \mathcal{A}$). The class of all probabilistic relational models that satisfy *meas* will be denoted $\mathcal{C}_m^{PEL}$. ⊣

If $\mathcal{A}$ is a general σ-algebra for the underlying frame of a model $\mathbb{M} = \langle \mathbb{F}, V \rangle \in \mathcal{C}_m^{PEL}$, such that $\mathcal{A}$ contains the image of the valuation function (i.e. $V \colon \mathsf{AtProp} \to \mathcal{A}$), then one can prove (by induction on formula complexity) that $[\![\varphi]\!]^{\mathbb{M}} \in \mathcal{A}$ for all formulas $\varphi \in \mathcal{L}_{PEL}(\mathsf{AtProp})$. From this it follows immediately that $[\![\varphi]\!]^{\mathbb{M}} \cap S_i(w) \in \mathcal{A}_i(w)$ for every for every formula φ, agent i and state w, as desired.

We will now discuss some typical principles about the interaction between knowledge and probability, and see how they correspond to various properties of (simplified) probabilistic relational frames.

---

[6] As usual, we abbreviate $R_i[w] \overset{\text{def}}{=} \{v \in W \mid (w, v) \in R_i\}$. Note that the condition that $\{w \in W \mid R_i[w] \subseteq A\} \in \mathcal{A}$ for each $A \in \mathcal{A}$ implies that $\mathcal{A}$ is not only a σ-algebra, but also a modal algebra on W (recall Footnote 3).

[7] A function f from $(X, \mathcal{A})$ to $(Y, \mathcal{B})$ is said to be *measurable* iff it reflects measurable sets, i.e. $f^{-1}[B] \in \mathcal{A}$ for all $B \in \mathcal{B}$.

Definition 4.7
Let $\mathbb{F} = \langle W, R_i, \mathbb{P}_i \rangle_{i \in I}$ be a probabilistic relational frame, and $\mathbb{G} = \langle W, R_i, \mu_i \rangle_{i \in I}$ a simplified probabilistic relational frame; then we we define:

1. *i-state determined probability* (i-SDP):
   - $\mathbb{F}$ satisfies i-SDP iff $\forall w, v \in W$: if $(w, v) \in R_i$ then $\mathbb{P}_i(w) = \mathbb{P}_i(v)$,
   - $\mathbb{G}$ satisfies i-SDP iff $\forall w, v \in W$: if $(w, v) \in R_i$ then $\mu_i(w) = \mu_i(v)$,
2. *i-consistency*:
   - $\mathbb{F}$ is i-consistent iff $\forall w, v \in W$: if $(w, v) \notin R_i$ then $v \notin S_i(w)$,
   - $\mathbb{G}$ is i-consistent iff $\forall w, v \in W$: if $(w, v) \notin R_i$ then $\mu_i(w)(v) = 0$,
3. *i-prudence*:
   - $\mathbb{F}$ is i-prudent iff $\forall w, v \in W$: if $(w, v) \in R_i$ then $\{v\} \in \mathcal{A}_i(w)$ and $\mu_i(w)(\{v\}) > 0$,
   - $\mathbb{G}$ is i-prudent iff $\forall w, v \in W$: if $(w, v) \in R_i$ then $\mu_i(w)(v) > 0$,
4. *i-liveness*:
   - $\mathbb{F}$ is i-live iff $\forall w \in W$: $\{w\} \in \mathcal{A}_i(w)$ and $\mu_i(w)(\{w\}) > 0$,
   - $\mathbb{G}$ is i-live iff $\forall w \in W$: $\mu_i(w)(w) > 0$. $\dashv$

Lemma 4.3
Let $\mathbb{F}$ be a probabilistic relational frame.

1. $\mathbb{F}$ satisfies i-SDP $\Rightarrow \mathbb{F} \models (\neg)\varphi \to K_i(\neg)\varphi$ for all i-probability formulas φ,
2. $\mathbb{F}$ is i-consistent $\Rightarrow \mathbb{F} \models K_i p \to P_i(p) = 1$,
3. $\mathbb{F}$ is i-prudent $\Leftrightarrow \mathbb{F} \models \hat{K} p \to P_i(p) > 0$,
4. $\mathbb{F}$ is i-live $\Leftrightarrow \mathbb{F} \models p \to P_i(p) > 0$.

Furthermore, if $\mathbb{F}$ is simplified, then the $\Rightarrow$-arrow in the first two items can be strengthened to a $\Leftrightarrow$-arrow (this is not possible for probabilistic relational frames in general, since the formulas cannot characterize some fine-grained aspects of the sample spaces and σ-algebras). $\dashv$

The SDP condition asserts that the agents' probabilities are entirely determined by their epistemic information: if an agent cannot epistemically distinguish between two states, then she should have the same probability spaces (or just probability functions in the simplified setting) at those states. This property corresponds to an epistemic-probabilistic introspection principle, stating that agents know their own probabilistic setup (i.e. probability formulas and their negations).

Consistency asserts that states that an agent does not consider possible are left out of her sample space (i.e. $S_i(w) \subseteq R_i[w]$); in the simplified setting, this boils down to the agent assigning probability 0 to such states. This seems rational: if an agent knows that a certain state is not actual, then it would be a 'waste' to assign any non-zero probability to it. This property corresponds to the principle that knowledge implies certainty (probability 1). Allowing for the sample space to be strictly smaller than the set of possibilities may enable us to represent an agent being uncertain among different probability spaces.

Prudence asserts that agents assign non-zero probability to all states that are epistemically indistinguishable from the actual state. After all, it would be quite 'bold' for an agent to assign probability 0 to a state that, to the best of her knowledge, might turn out to be the actual state.[8] This property corresponds to the principle that epistemic possibility implies probabilistic possibility (non-zero probability), or, read contrapositively, that an agent can only assign probability 0 to propositions that she knows to be false ($P_i(p) = 0 \rightarrow K_i \neg p$). Yet another formulation is: an agent can only be certain of known propositions ($P_i(p) = 1 \rightarrow K_i p$). This last formula is exactly the converse of the formula corresponding to consistency, thus revealing the close connection between prudence and consistency.[9]

Liveness asserts that agents assign non-zero probability to the actual state. If one assumes that each state is indistinguishable from itself (i.e. that the epistemic indistinguishability relation R_i is reflexive), then liveness is a direct consequence of prudence. Furthermore, liveness corresponds to the

[8]However, there also exist counterexamples to this prudence principle. For example, consider tossing a fair coin an infinite number of times. A state in which the coin lands tails every single time is epistemically possible (we can perfectly imagine that this would happen), yet probabilistically impossible (it seems perfectly reasonable to assign probability 0 to it).

[9]It should be noted that if a simplified probabilistic relational model $\langle W, R_i, \mu_i, V\rangle_{i\in I}$ satisfies consistency as well as prudence, then for all states $w, v \in W$, it holds that $(w, v) \in R_i$ iff $\mu_i(w)(v) > 0$. This means that the relation R_i is definable in terms of the probability function μ_i, and can thus be dropped from the models altogether. The resulting structures are thus of the form $\langle W, \mu_i, V\rangle_{i\in I}$, and are essentially a type of *probabilistic transition systems*, which have been extensively studied in theoretical computer science.

principle that agents should assign non-zero probability to all true propositions ($p \rightarrow P_i(p) > 0$). If one assumes that knowledge is factive (which is exactly the principle corresponding to the reflexivity of R_i), then this principle follows immediately from the principle corresponding to prudence ($\hat{K}_i p \rightarrow P_i(p) > 0$).

Finally, note that the definitions of prudence and liveness for probabilistic relational frames require that certain individual states (or rather, certain singletons of states) be measurable, and hence partially reduce these frames to *simplified* frames. However, if the frames are assumed to satisfy *meas* (recall Definition 4.6), and frame validity only ranges over models in $\mathcal{C}_m^{PEL}$, i.e. for each frame $\mathbb{F}$ that satisfies *meas*,

$$\mathbb{F} \models_m \varphi \Leftrightarrow \mathbb{M} \models \varphi \text{ for each } \mathbb{M} \in \mathcal{C}_m^{PEL} \text{ extending } \mathbb{F},$$

then one can give alternative definitions of these conditions, which do *not* require individual states to be measurable: a frame $\mathbb{F}$ that satisfies *meas* is said to be *meas-i-prudent* iff

$$\forall w, v \in W\colon (w, v) \in R_i \Rightarrow \begin{cases} \exists A \in \mathcal{A}_i(w).(v \in A); \text{ and} \\ \forall A \in \mathcal{A}_i(w).(v \in A \text{ implies } \mu_i(w)(A) > 0) \end{cases}$$

and it is said to be *meas-i-live* iff

$$\forall w \in W\colon \begin{cases} \exists A \in \mathcal{A}_i(w).(w \in A); \text{ and} \\ \forall A \in \mathcal{A}_i(w).(\text{if } w \in A \text{ then } \mu_i(w)(A) > 0). \end{cases}$$

It is easy to see that if $\mathbb{F}$ is i-prudent, then it is also *meas*-i-prudent, but not vice versa. Similar remarks apply to i-liveness and *meas*-i-liveness. In other words, *meas*-i-prudence and *meas*-i-liveness are strictly weaker than i-prudence and i-liveness, respectively. However, one can prove that these weaker conditions correspond to exactly the same formulas as their strong versions:

Lemma 4.4
Let $\mathbb{F}$ be a probabilistic relational frame that satisfies *meas*.

1. $\mathbb{F}$ is *meas*-i-prudent $\Leftrightarrow \mathbb{F} \models_m \hat{K}p \rightarrow P_i(p) > 0$,

2. $\mathbb{F}$ is *meas*-i-live $\Leftrightarrow \mathbb{F} \models_m p \rightarrow P_i(p) > 0$. ⊣

4.3.3 Expressivity of Linear Combinations

Just like probabilistic propositional logic is more expressive with linear combinations than without (recall Lemma 4.2), it holds that probabilistic epistemic logic is more expressive with linear combinations than without. Let

$\mathcal{L}^*_{PEL}(\mathsf{AtProp})$ be the language that is obtained from $\mathcal{L}_{PEL}(\mathsf{AtProp})$ by only allowing atomic probability formulas of the form $P_i(\varphi) \geq r$ (linear combinations are not allowed). We then have the following:

Lemma 4.5
The class of (pointed) probabilistic relational models $\{\mathbb{M}, w \mid \mathbb{M} = \langle W, R_i, \mathbb{P}_i, V\rangle_{i \in I} \in \mathcal{C}^{PEL}_m, w \in W, \mu_i(w)(V(p)) \geq \mu_i(w)(V(q))\}$ can be characterized by a formula in $\mathcal{L}_{PEL}(\mathsf{AtProp})$, but not by one in $\mathcal{L}^*_{PEL}(\mathsf{AtProp})$. ⊣

The proof of Lemma 4.2 needs only minor modification in order to apply here. The primary difference is that given any formula $f \in \mathcal{L}^*_{PEL}(\mathsf{AtProp})$, we here consider all conjunctions of maximally consistent subsets of the set of subformulas of f, rather than the conjunction of literals A_Q used in the proof of Lemma 4.2. These conjunctions are essentially the same as those used in the standard completeness proof.

4.3.4 Proof System

Probabilistic epistemic logic can be axiomatized in a highly modular fashion. An overview is given in Figure 4.2. The propositional and epistemic components shouldn't need any further comments. The probabilistic and linear inequalities components are the same as in the axiomatization of probabilistic propositional logic (see Figure 4.1), except that all probability formulas now have agent indices.

Using standard techniques, the following theorem can be proved:

Theorem 4.4
Probabilistic epistemic logic, as axiomatized in Figure 4.2, is sound and weakly complete with respect to the class $\mathcal{C}^{PEL}_m$ of probabilistic relational models that satisfy *meas*, and also with respect to the class $\mathcal{C}^{PEL}_s$ of simplified probabilistic relational models.

The cases for $\mathcal{C}^{PEL}_m$ and $\mathcal{C}^{PEL}_s$ each follow from the same proof. A model is constructed to satisfy a consistent formula, such that for each agent i and state w, the set $S_i(w)$ consists of all states in the model, and $\mathcal{A}_i(w) = \wp(S_i(w))$. Probabilities are assigned to singletons according to a system of linear inequalities. Although the constructed model belongs to $\mathcal{C}^{PEL}_m$, its structure is essentially the same as a *simplified* probabilistic relational model. It follows that the axiomatization in Figure 4.2 is complete with respect to both kinds of probabilistic relational models ($\mathcal{C}^{PEL}_m$ and $\mathcal{C}^{PEL}_s$).

Figure 4.2: Componentwise axiomatization of probabilistic epistemic logic.

1. propositional component
   - all propositional tautologies and the modus ponens rule
2. epistemic component
   - the S5 axioms and rules for the K_i-operators
3. probabilistic component
   - $P_i(\varphi) \geq 0$
   - $P_i(\top) = 1$
   - $P_i(\varphi \wedge \psi) + P_i(\varphi \wedge \neg\psi) = P_i(\varphi)$
   - if $\vdash \varphi \leftrightarrow \psi$ then $\vdash P_i(\varphi) = P_i(\psi)$
4. linear inequalities component
   - $\sum_{\ell=1}^{n} a_\ell P_i(\varphi_\ell) \geq b \leftrightarrow \sum_{\ell=1}^{n} a_\ell P_i(\varphi_\ell) + 0P_i(\varphi_{n+1}) \geq b$
   - $\sum_{\ell=1}^{n} a_\ell P_i(\varphi_\ell) \geq b \leftrightarrow \sum_{\ell=1}^{n} a_{p(\ell)} P_i(\varphi_{p(\ell)}) \geq b$ for any permutation p of $1, \ldots, n$
   - $\sum_{\ell=1}^{n} a_\ell P_i(\varphi_\ell) \geq b \wedge \sum_{\ell=1}^{n} a'_\ell P_i(\varphi_\ell) \geq b' \rightarrow \sum_{\ell=1}^{n} (a_\ell + a'_\ell) P_i(\varphi_\ell) \geq b + b'$
   - $\sum_{\ell=1}^{n} a_\ell P_i(\varphi_\ell) \geq b \leftrightarrow \sum_{\ell=1}^{n} d a_\ell P_i(\varphi_\ell) \geq db$ (for any $d > 0$)
   - $\sum_{\ell=1}^{n} a_\ell P_i(\varphi_\ell) \geq b \vee \sum_{\ell=1}^{n} a_\ell P_i(\varphi_\ell) \leq b$
   - $\sum_{\ell=1}^{n} a_\ell P_i(\varphi_\ell) \geq b \rightarrow \sum_{\ell=1}^{n} a_\ell P_i(\varphi_\ell) > b'$ (for any $b' < b$)

4.3.5 Decidability and Complexity

Probabilistic epistemic logic is decidable, and complexity results are available for probabilistic epistemic logics under various constraints. We discuss only two classes here, each of which follow from the same proof.

Theorem 4.5
Let $\mathcal{C}$ be either $\mathcal{C}_m^{PEL}$ or $\mathcal{C}_s^{PEL}$. The satisfiability problem of probabilistic epistemic logic with respect to $\mathcal{C}$ is PSPACE-complete. ⊣

The lower bound follows from the fact that basic epistemic logic, which is already known to be PSPACE-hard, reduces to probabilistic epistemic logic (it is a fragment of probabilistic epistemic logic). The upper bound can be achieved by adapting a tableau construction for the complexity of modal logic to one where new branches are non-deterministically created at certain nodes for the purpose of non-deterministically assigning positive probabilities. By Theorem 4.3, the number of new branches and the size of the probability values are both polynomial in the size of the input formula. The tableau can be checked using polynomial space, and the PSPACE upper bound for the satisfiability problem follows from the well-known fact that PSPACE = NPSPACE.

4.4 Probabilistic Interpreted Systems

In this section we discuss probabilistic interpreted systems. Subsection 4.4.1 introduces interpreted systems, and shows how probabilities can be added to them. Subsection 4.4.2 defines the language that is interpreted on such systems. Finally, Subsection 4.4.3 provides some conceptual remarks about the relationship between (probabilistic) interpreted systems and (probabilistic) dynamic epistemic logic.

4.4.1 Adding Probabilities to Interpreted Systems

In Sections 4.2 and 4.3, the notion of 'state' or 'possible world' was taken as primitive: states are not assumed to have any internal structure. However, in many applications, the states can fruitfully be thought of as being structured. Consider, for example, a card game involving 3 players (1, 2 and 3) and 4 cards (A, B, C and D), where each player picks a card, and the fourth card (that was not picked by any player) remains face-down at the table. It makes sense to model this game using $4! = 24$ states of the form $(c_e, c_1, c_2, c_3) \in \{A, B, C, D\}^4$, where c_e represents the card that remains on the table, and c_i represents the card that player i has picked ($1 \leq i \leq 3$). One can then represent that player i knows her own card, but not those of the other players nor the one on the table, by requiring that she can epistemically distinguish between states (c_e, c_1, c_2, c_3) and (c'_e, c'_1, c'_2, c'_3) iff $c_i \neq c'_i$.

This perspective is often used to analyze multi-agent systems. Each agent i is supposed to be in some *internal (local) state* $s_i \in S_i$, which represents all the information that is accessible to her. Additionally, the *environment* is in some state $s_e \in S_e$, which represents all relevant information that is not in any of the agents' local states (in the example above, this was the card that remained face-down at the table). If there are n agents, a *global state* can then be taken to be an $(n+1)$-tuple $(s_e, s_1, \ldots, s_n)$, and the entire set $\mathcal{G}$ of global states can be seen as

$$\mathcal{G} = S_e \times S_1 \times \cdots \times S_n.$$

A global state represents all information in a system at a given point in time. However, the system changes over time. This is modeled using *runs*: we assume that time is discrete, and define a run as a function $r\colon \mathbb{N} \to \mathcal{G}$. Intuitively, $r(k) = g \in \mathcal{G}$ means that if the system evolves according to run r, then at time k it will be in the global state g. A pair $(r, k) \in \mathcal{G}^{\mathbb{N}} \times \mathbb{N}$ is often called a *point*. Furthermore, if $r(k) = (s_e, s_1, \dots, s_n)$, then we define $r_e(k) \stackrel{\text{def}}{=} s_e$ as the environment state and $r_i(k) \stackrel{\text{def}}{=} s_i$ as agent i's local state at the point (r, k)

In general, a system can evolve over time in various ways. Each of these ways is represented as a run, and therefore a system is defined as a non-empty set of runs. As usual, an interpreted system is obtained by adding a valuation.

Definition 4.8
An *interpreted system* is a tuple $\mathbb{I} = \langle \mathcal{R}, \pi \rangle$, where $\mathcal{R}$ is a non-empty set of runs, i.e. $\emptyset \neq \mathcal{R} \subseteq \mathcal{G}^{\mathbb{N}}$, and $\pi\colon \mathsf{AtProp} \to \wp(\mathcal{G})$ is a valuation. ⊣

We will now describe how to add probabilities to interpreted systems. The key idea is to add probabilities over the entire runs, and then use these to define probabilities over the individual points. First, however, we need to introduce a few auxiliary notions. We will use $[r, k]_i$ to denote the set of points that agent i considers epistemically possible at (i.e. cannot epistemically distinguish from) the point (r, k); formally:

$$[r, k]_i \stackrel{\text{def}}{=} \{(r', k') \in \mathcal{R} \times \mathbb{N} \mid r_i(k) = r'_i(k')\}.$$

Next, given a set of runs $\mathcal{S} \subseteq \mathcal{R}$ and a set of points $U \subseteq \mathcal{R} \times \mathbb{N}$, we use $\mathcal{S}(U)$ to denote the set of runs in $\mathcal{S}$ that go through some point in U, i.e.

$$\mathcal{S}(U) \stackrel{\text{def}}{=} \{r \in \mathcal{S} \mid \exists k \in \mathbb{N}\colon (r, k) \in U\},$$

and $U(\mathcal{S})$ to denote the set of points in U that lie on some run in $\mathcal{S}$, i.e.

$$U(\mathcal{S}) \stackrel{\text{def}}{=} \{(r, k) \in U \mid r \in \mathcal{S}\}.$$

Note that if we take $U = [r, k]_i$ (for some point $(r, k) \in \mathcal{R} \times \mathbb{N}$), then we find that $\mathcal{S}([r, k]_i)$ consists of the runs in $\mathcal{S}$ that agent i considers possible at (r, k), and $[r, k]_i(\mathcal{S})$ consists of the points (lying on runs) in $\mathcal{S}$ that agent i considers possible at (r, k). Formally:

$$\mathcal{S}([r, k]_i) = \{r' \in \mathcal{S} \mid \exists k' \in \mathbb{N}\colon r_i(k) = r'_i(k')\},$$
$$[r, k]_i(\mathcal{S}) = \{(r', k') \in \mathcal{S} \times \mathbb{N} \mid r_i(k) = r'_i(k')\}.$$

With these auxiliary abbreviations in place, we are now ready to formally introduce the notion of a probabilistic interpreted system.

Definition 4.9
A *probabilistic interpreted system* is a tuple $\mathbb{I} = \langle \mathcal{R}, \mathcal{A}_i, \mu_i, \pi \rangle_{i \in I}$, such that $\langle \mathcal{R}, \pi \rangle$ is an ('ordinary') interpreted system, and for all agents $i \in I$, $\langle \mathcal{R}, \mathcal{A}_i, \mu_i \rangle$ is a probability space, where $\mathcal{A}_i$ is a σ-algebra over $\mathcal{R}$, and $\mu_i \colon \mathcal{A}_i \to [0,1]$ is a probability measure. Furthermore, for all $r \in \mathcal{R}$, $k \in \mathbb{N}$ and $i \in I$, it is required that $\mathcal{R}([r,k]_i)$ is measurable and has non-zero probability, i.e. $\mathcal{R}([r,k]_i) \in \mathcal{A}_i$ and $\mu_i(\mathcal{R}([r,k]_i)) > 0$. ⊣

The condition that $\mu_i(\mathcal{R}([r,k]_i)) > 0$ will be needed when defining probabilities over points; this will be discussed in more detail in the next subsection.

4.4.2 Language and Semantics

We now describe how to turn a probabilistic interpreted system into a probabilistic relational model.

Definition 4.10
Let $\mathbb{I} = \langle \mathcal{R}, \mathcal{A}_i, \mu_i, \pi \rangle_{i \in I}$ be an arbitrary probabilistic interpreted system. Then the tuple $\mathsf{model}(\mathbb{I}) \stackrel{\text{def}}{=} \langle W, R_i, \mathbb{P}_i, V \rangle$ is defined as follows:

- $W = \mathcal{R} \times \mathbb{N}$,
- $R_i = \{((r,k),(r',k')) \in W \times W \mid (r',k') \in [r,k]_i\}$,
- $V(p) = \{(r,k) \in W \mid r(k) \in \pi(p)\}$,
- $\mathbb{P}_i$ assigns to each state $(r,k) \in W$ a probability space $\mathbb{P}_i(r,k) \stackrel{\text{def}}{=} \langle S_i(r,k), \mathcal{A}_i(r,k), \mu_i(r,k) \rangle$, with
  - $S_i(r,k) = [r,k]_i$,
  - $\mathcal{A}_i(r,k) = \{[r,k]_i(\mathcal{S}) \mid \mathcal{S} \in \mathcal{A}_i\}$,
  - $\mu_i(r,k) \colon \mathcal{A}_i(r,k) \to [0,1] \colon U \mapsto \mu_i(\mathcal{R}(U) \mid \mathcal{R}([r,k]_i))$. ⊣

Lemma 4.6
The tuple $\mathsf{model}(\mathbb{I})$ is a probabilistic relational model, in the sense of Definition 4.5. ⊣

The states of $\mathsf{model}(\mathbb{I})$ are thus the points of $\mathbb{I}$. Furthermore, Definition 4.10 implements exactly the perspective on knowledge described at the beginning of Subsection 4.4.1: two states of $\mathsf{model}(\mathbb{I})$ are epistemically indistinguishable for agent i iff her local information at both points is identical.

It is easy to see that the resulting epistemic indistinguishability relation is an equivalence relation (and hence, the logic governing the knowledge operators will be S5). The model's probabilities over points ($\mu_i(r,k)$) are defined in terms of the underlying system's probabilities over runs (μ_i). The sample space $S_i(r,k)$ at the point (r,k) consists of the points that agent i considers possible at (r,k). The definition of the σ-algebra $\mathcal{A}_i(r,k)$ states that a set of points $U \subseteq [r,k]_i$ is measurable iff it is of the form $[r,k]_i(\mathcal{S})$ for some $\mathcal{S} \in \mathcal{A}_i$, i.e. iff it consists of the points in $[r,k]_i$ lying on the runs in some measurable set of runs $\mathcal{S}$. Finally, the probability $\mu_i(r,k)(U)$ of some set of points U is defined to be the conditional probability of $\mathcal{R}(U)$, given $\mathcal{R}([r,k]_i)$, i.e. the probability of the set of runs going through $U \subseteq [r,k]_i$ conditioned on the set of runs going through all of $[r,k]_i$. Calculating this conditional probability requires dividing by $\mu_i(\mathcal{R}([r,k]_i))$, which explains the requirement that this probability be non-zero in Definition 4.9. We could have left out the condition that $\mathcal{R}([r,k]_i) \in \mathcal{A}_i$ from Definition 4.9, but then we would have to involve inner or outer measures in the definition of $\mathbb{P}_i(r,k)$.

Probabilistic interpreted systems can interpret the probabilistic temporal epistemic language $\mathcal{L}_{PTEL}(\mathsf{AtProp})$, which is obtained from the static probabilistic epistemic language $\mathcal{L}_{PEL}(\mathsf{AtProp})$ by adding temporal operators, such as $\bigcirc$ ('next point') and $\Box$ ('henceforth'). For the non-temporal part of $\mathcal{L}_{PTEL}$, we can simply make use of Lemma 4.6:

$$\mathbb{I},(r,k) \models \varphi \quad \text{iff} \quad \mathsf{model}(\mathbb{I}),(r,k) \models \varphi.$$

The temporal formulas $\bigcirc\varphi$ and $\Box\varphi$ state that φ will be true at the *next* point, resp. *all* later points (including the current one), of the current run; formally:

$$\begin{array}{lll} \mathbb{I},(r,k) \models \bigcirc p & \text{iff} & \mathbb{I},(r,k+1) \models \varphi, \\ \mathbb{I},(r,k) \models \Box\varphi & \text{iff} & \text{for all } k' \geq k\colon \mathbb{I},(r,k') \models \varphi. \end{array}$$

4.4.3 Temporal and Dynamic Epistemic Logic

Since probabilistic interpreted systems can model probabilities as well as time, it is natural to ask how the agents' probabilities change over time, i.e. when the system moves from point (r,k) to point $(r,k+1)$. Using Definition 4.10, we can view both $\mu_i(r,k)$ and $\mu_i(r,k+1)$ in terms of the probability function μ_i over runs. However, one might also wonder whether there is a direct relation between $\mu_i(r,k)$ and $\mu_i(r,k+1)$ (without a 'detour' over the runs). In Bayesian epistemology, the standard answer is that the agents' probabilities change via *Bayesian updating*. We will now show that under certain assumptions, the same holds for probabilistic interpreted systems.

For any run $r: \mathbb{N} \to \mathcal{G}$ and $k \in \mathbb{N}$, we will write $r \restriction k$ for the initial segment of r up to $r(k)$, i.e. $r \restriction k \stackrel{\text{def}}{=} \langle r(0), r(1), r(2), \ldots, r(k-1), r(k)\rangle \in \mathcal{G}^{k+1}$. Furthermore, for any sequence σ, we will write σ° for the sequence that is obtained by deleting successive identical elements from σ; for example, $\langle a, a, b, b, a, a, b, a\rangle^\circ = \langle a, b, a, b, a\rangle$. Finally, for any set of points $U \subseteq \mathcal{R} \times \mathbb{N}$ in a probabilistic interpreted system we will write $U - 1$ for the set of points that precede some point in U, i.e. $U - 1 \stackrel{\text{def}}{=} \{(r, k) \mid (r, k+1) \in U\}$. Using this terminology, one can show the following:

Theorem 4.6
Consider a probabilistic interpreted system $\langle \mathcal{R}, \mathcal{A}_i, \mu_i, \pi\rangle_{i \in I}$ and suppose that it satisfies perfect recall and synchronicity, i.e. for all $r \in \mathcal{R}$, $k \in \mathbb{N}$ and $i \in I$:

- if $r_i(k) = r'_i(k')$, then $(r \restriction k)^\circ = (r' \restriction k')^\circ$,
- if $r_i(k) = r'_i(k')$, then $k = k'$.

Then for all sets $U \in \mathcal{A}_i(r, k+1)$, it holds that $U - 1 \in \mathcal{A}_i(r, k)$ and

$$\mu_i(r, k+1)(U) = \mu_i(r, k)(U - 1 \mid [r, k+1]_i - 1). \qquad \dashv$$

This theorem states that after the system has moved from point (r, k) to point $(r, k+1)$, the probability that agent i assigns to some set U equals the probability that she assigned at (r, k) to the set of points preceding U, conditional on the set of points preceding those she cannot distinguish from $(r, k+1)$.

Since probabilistic interpreted systems can model not only the agents' epistemic attitudes (such as knowledge and probabilities), but also time, it leads to the question about how these attitudes should change over time (as new information becomes available). For example, Theorem 4.6 describes changes in probabilities in terms of Bayesian conditionalization. In this sense, probabilistic interpreted systems are somewhat analogous to dynamic extensions of probabilistic epistemic logic, which will be discussed in detail in Sections 4.5 and 4.6. After all, the latter can also model how the agents' knowledge and probabilities change in the face of new information; in particular, probabilistic public announcement logic also describes changes in probabilities in terms of Bayesian conditionalization.

The main difference between interpreted systems—and *temporal epistemic logic* in general—on the one hand and *dynamic epistemic logic* on the other is that these two approaches provide fundamentally different perspectives on time. Temporal epistemic logic provides a *global* perspective on time. A (probabilistic) interpreted system consists of multiple alternative histories (runs), that are infinitely long (each run is a function with domain $\mathbb{N}$). Dynamic epistemic logic, however, provides a much more *local* perspective on time. In Sections 4.5 and 4.6, we will see that a (probabilistic) relational model $\mathbb{M}$ is taken to represent a single moment in time, and the alternative histories starting in that moment are modeled as alternative (sequences of) model-changing operations to be performed on $\mathbb{M}$ (cf. Definitions 4.12 and 4.16).

4.5 Probabilistic Public Announcement Logic

In this section we discuss a first 'dynamification' of the probabilistic epistemic logic discussed in Section 4.3, by introducing public announcements into the logic. Subsection 4.5.1 discusses updated probabilistic relational models, and introduces a public announcement operator into the formal language to talk about these models. Subsection 4.5.2 provides a complete axiomatization, and Subsection 4.5.3 addresses complexity issues. Finally, Subsection 4.5.4 focuses on the role of higher-order information in public announcement dynamics.

4.5.1 Language and Semantics

Public announcements form one of the simplest types of epistemic dynamics. They concern the truthful and public announcement of some piece of information φ. That the announcement is *truthful* means that the announced information φ has to be true; that it is *public* means that all agents $i \in I$ learn about it simultaneously and commonly.

Public announcement logic represents these announcements as updates that change relational models, and introduces a dynamic public announcement operator into the formal language to describe these updated models. This strategy can straightforwardly be extended into the probabilistic realm.

Syntactically, we add a dynamic operator $[!\cdot]\cdot$ to the static language $\mathcal{L}_{PEL}(\mathsf{AtProp})$, thus obtaining the new language $\mathcal{L}^!_{PEL}(\mathsf{AtProp})$. The formula $[!\varphi]\psi$ means that after any truthful public announcement of φ, it will be the case that ψ. Its dual is defined as $\langle !\varphi\rangle\psi \overset{\text{def}}{=} \neg[!\varphi]\neg\psi$, and means that φ can truthfully and publicly be announced, and afterwards ψ will be the case. These formulas thus allow us to express 'now' (i.e. *before* any dynamics has taken place) what will be the case 'later' (*after* the dynamics has taken place). Given a (potentially simplified) probabilistic relational model $\mathbb{M}$ and state w, these formulas are interpreted as follows:

$$\begin{array}{lll} \mathbb{M}, w \models [!\varphi]\psi & \text{iff} & \text{if } \mathbb{M}, w \models \varphi \text{ then } \mathbb{M}|\varphi, w \models \psi, \\ \mathbb{M}, w \models \langle !\varphi\rangle\psi & \text{iff} & \mathbb{M}, w \models \varphi \text{ and } \mathbb{M}|\varphi, w \models \psi. \end{array}$$

Note that these clauses involve not only the model $\mathbb{M}$, but also the updated model $\mathbb{M}|\varphi$. The model $\mathbb{M}$ represents the agents' information *before* the public announcement of φ; the model $\mathbb{M}|\varphi$ represents their information *after* the public announcement of φ; hence the public announcement of φ *itself* is represented by the update mechanism $\mathbb{M} \mapsto \mathbb{M}|\varphi$. This mechanism is defined for simplified as well as non-simplified probabilistic relational models:

Definition 4.11
Consider a simplified probabilistic relational model $\mathbb{M} = \langle W, R_i, \mu_i, V\rangle_{i\in I}$, a state $w \in W$, and a formula $\varphi \in \mathcal{L}^!_{PEL}(\mathsf{AtProp})$ such that $\mathbb{M}, w \models \varphi$. Then the *updated simplified probabilistic relational model* $\mathbb{M}|\varphi \stackrel{\text{def}}{=} \langle W^\varphi, R^\varphi_i, \mu^\varphi_i, V^\varphi\rangle_{i\in I}$ is defined as follows:

- $W^\varphi \stackrel{\text{def}}{=} W$,
- $R^\varphi_i \stackrel{\text{def}}{=} R_i \cap (W \times [\![\varphi]\!]^{\mathbb{M}})$ (for every agent $i \in I$),
- $\mu^\varphi_i \colon W^\varphi \to (W^\varphi \to [0,1])$ is defined (for every agent $i \in I$) by

$$\mu^\varphi_i(v)(u) \stackrel{\text{def}}{=} \begin{cases} \frac{\mu_i(v)(\{u\}\cap[\![\varphi]\!]^{\mathbb{M}})}{\mu_i(v)([\![\varphi]\!]^{\mathbb{M}})} & \text{if } \mu_i(v)([\![\varphi]\!]^{\mathbb{M}}) > 0 \\ \mu_i(v)(u) & \text{if } \mu_i(v)([\![\varphi]\!]^{\mathbb{M}}) = 0, \end{cases}$$

- $V^\varphi \stackrel{\text{def}}{=} V$. ⊣

Definition 4.12
Consider a probabilistic relational model $\mathbb{M} = \langle W, R_i, \mathbb{P}_i, V\rangle_{i\in I}$, with $\mathbb{P}_i = (S_i, \mathcal{A}_i, \mu_i)$, a state $w \in W$ and a formula $\varphi \in \mathcal{L}^!_{PEL}(\mathsf{AtProp})$ such that $\mathbb{M}, w \models \varphi$. Then the *updated probabilistic relational model* $\mathbb{M}|\varphi = \langle W^\varphi, R^\varphi_i, \mathbb{P}^\varphi_i, V\rangle^\varphi_{i\in I}$ is defined as follows:

- $W^\varphi \stackrel{\text{def}}{=} W$,
- $R^\varphi_i \stackrel{\text{def}}{=} R_i \cap (W \times [\![\varphi]\!]^{\mathbb{M}})$ (for every agent $i \in I$),
- $\mathbb{P}^\varphi_i \stackrel{\text{def}}{=} (S^\varphi_i, \mathcal{A}^\varphi_i, \mu^\varphi_i)$ for each $i \in I$, such that for each $v \in W$
  - $S^\varphi_i(v) \stackrel{\text{def}}{=} \begin{cases} S_i(v) \cap [\![\varphi]\!]^{\mathbb{M}} & \text{if } (\mu_i(v))_*(S_i(v) \cap [\![\varphi]\!]^{\mathbb{M}}) > 0 \\ S_i(v) & \text{if } (\mu_i(v))_*(S_i(v) \cap [\![\varphi]\!]^{\mathbb{M}}) = 0, \end{cases}$
  - $\mathcal{A}^\varphi_i(v) \stackrel{\text{def}}{=} \begin{cases} \{A \cap [\![\varphi]\!]^{\mathbb{M}} \mid A \in \mathcal{A}_i(v)\} & \text{if } (\mu_i(v))_*(S_i(v) \cap [\![\varphi]\!]^{\mathbb{M}}) > 0 \\ \mathcal{A}_i(v) & \text{if } (\mu_i(v))_*(S_i(v) \cap [\![\varphi]\!]^{\mathbb{M}}) = 0, \end{cases}$

– $\mu_i^\varphi(v) : \mathcal{A}_i^\varphi(v) \to [0,1]$, such that for all $A \in \mathcal{A}_i^\varphi(v)$

$$\mu_i^\varphi(v)(A) \stackrel{\text{def}}{=} \begin{cases} \frac{(\mu_i(v))_*(A \cap [\![\varphi]\!]^{\mathbb{M}})}{(\mu_i(v))_*(S_i(v) \cap [\![\varphi]\!]^{\mathbb{M}})} & \text{if } (\mu_i(v))_*(S_i(v) \cap [\![\varphi]\!]^{\mathbb{M}}) > 0 \\ \mu_i(v)(A) & \text{if } (\mu_i(v))_*(S_i(v) \cap [\![\varphi]\!]^{\mathbb{M}}) = 0, \end{cases}$$

- $V^\varphi \stackrel{\text{def}}{=} V$. ⊣

The main effect of the public announcement of φ in a model $\mathbb{M}$ is that all links to $\neg\varphi$-states are deleted; hence these states are no longer accessible for any of the agents. This procedure is standard; we will therefore focus on the probabilistic components.

First of all, it should be noted that the case distinction in the definition of $\mu_i^\varphi(v)$ and $\mathbb{P}_i^\varphi(v)$ in Definitions 4.11 and 4.12 is made for strictly technical reasons, viz. to ensure that there are no 'dangerous' divisions by 0. In all examples and applications, we will be using the 'interesting' cases $\mu_i(v)([\![\varphi]\!]^{\mathbb{M}}) > 0$ and $(\mu_i(v))_*(S_i(v) \cap [\![\varphi]\!]^{\mathbb{M}}) > 0$. Still, for general theoretical reasons, *something* has to be said about the cases $\mu_i(v)([\![\varphi]\!]^{\mathbb{M}}) = 0$ and $(\mu_i(v))_*(S_i(v) \cap [\![\varphi]\!]^{\mathbb{M}}) = 0$. Leaving $\mu_i^\varphi(v)$ undefined in this case would lead to truth value gaps in the logic, and thus greatly increase the difficulty of finding a complete axiomatization. The approach taken here is to define $\mu_i^\varphi(v) := \mu_i(v)$ and $\mathbb{P}_i^\varphi(v) := \mathbb{P}_i(v)$ in these cases—so the public announcement of φ simply has *no effect* whatsoever on $\mu_i(v)$ or $\mathbb{P}_i(v)$. The intuitive idea behind this definition is that an agent i simply *ignores* new information if she previously assigned probability 0 to it. Technically speaking, this definition will yield a relatively simple axiomatization.

One can easily check that if $\mathbb{M} \in \mathcal{C}_s^{PEL}$, then $\mathbb{M}|\varphi \in \mathcal{C}_s^{PEL}$ as well. We focus on $\mu^\varphi(v)$ (for some arbitrary state $v \in W^\varphi$). If $\mu_i(v)([\![\varphi]\!]^{\mathbb{M}}) = 0$, then $\mu_i^\varphi(v)$ is $\mu_i(v)$, which is a probability function on $W = W^\varphi$. If $\mu_i(v)([\![\varphi]\!]^{\mathbb{M}}) > 0$, then for any $u \in W^\varphi$,

$$\mu_i^\varphi(v)(u) = \frac{\mu_i(v)(\{u\} \cap [\![\varphi]\!]^{\mathbb{M}})}{\mu_i(v)([\![\varphi]\!]^{\mathbb{M}})},$$

which is positive because $\mu_i(v)(\{u\} \cap [\![\varphi]\!]^{\mathbb{M}})$ and $\mu_i(v)([\![\varphi]\!]^{\mathbb{M}})$ are positive, and at most 1, because $\mu_i(v)(\{u\} \cap [\![\varphi]\!]^{\mathbb{M}}) \le \mu_i(v)([\![\varphi]\!]^{\mathbb{M}})$—and hence $\mu_i^\varphi(v)(u) \in [0,1]$. Furthermore,

$$\sum_{u \in W^\varphi} \mu_i^\varphi(v)(u) = \sum_{u \in W} \frac{\mu_i(v)(\{u\} \cap [\![\varphi]\!]^{\mathbb{M}})}{\mu_i(v)([\![\varphi]\!]^{\mathbb{M}})} = \sum_{\mathbb{M},u \models \varphi} \frac{\mu_i(v)(u)}{\mu_i(v)([\![\varphi]\!]^{\mathbb{M}})} = 1.$$

In the non-simplified setting, one can check that if $\mathbb{M} \in \mathcal{C}_m^{PEL}$, then $\mathbb{M}|\varphi \in \mathcal{C}_m^{PEL}$ as well. The argument for this is similar to the one above, and one can see that measurability is preserved by observing that any general σ-algebra for $\mathbb{M}$ (in the sense of Definition 4.6) is also a general σ-algebra for $\mathbb{M}|\varphi$.

It should be noted that the definition of $\mu_i^\varphi(v)$—in the interesting cases when $\mu_i(v)([\![\varphi]\!]^{\mathbb{M}}) > 0$ or $(\mu_i(v))_*(S_i(v) \cap [\![\varphi]\!]^{\mathbb{M}}) > 0$—can also be expressed in terms of conditional probabilities. Given any $A \in \mathcal{A}_i^\varphi(v)$, we have:

$$\mu_i^\varphi(v)(A) = \frac{\mu_i(v)(A \cap S_i(v) \cap [\![\varphi]\!]^{\mathbb{M}})}{\mu_i(v)(S_i(v) \cap [\![\varphi]\!]^{\mathbb{M}})} = \mu_i(v)(A \,|\, S_i(v) \cap [\![\varphi]\!]^{\mathbb{M}}),$$

and in the simplified setting, for any $X \subseteq W^\varphi$ we have:

$$\mu_i^\varphi(v)(X) = \frac{\mu_i(v)(X \cap [\![\varphi]\!]^{\mathbb{M}})}{\mu_i(v)([\![\varphi]\!]^{\mathbb{M}})} = \mu_i(v)(X \mid [\![\varphi]\!]^{\mathbb{M}}).$$

In other words, after the public announcement of a formula φ, the agents calculate their new, updated probabilities by means of *Bayesian conditionalization* on the information provided by the announced formula φ. This connection between public announcements and Bayesian conditionalization will be explored more thoroughly in Subsection 4.5.4.

To illustrate the naturalness and explanatory power of this framework, we finish this subsection by discussing two examples. Example 4.3 is a continuation of the small-scale scenario that was described in Subsection 4.3.2. Example 4.4 illustrates that while simplified probabilistic relational models may suffice to model many problems, there certainly also exist problems whose representation requires the full power of probabilistic relational models (with σ-algebras).

Example 4.3
Recall the scenario that was described in Example 4.2. We modeled this using a (simplified) probabilistic relational model $\mathbb{M}$, in such a way that

$$\mathbb{M}, w \models \neg Kp \wedge \neg K\neg p \wedge P(p) = 0.5 \wedge P(\neg p) = 0.5.$$

Now suppose that p is publicly announced (this is indeed possible, since p was assumed to be actually true). Applying Definition 4.11, we obtain the updated model $\mathbb{M}|p$, with $W^p = W, R = \{(w, w)\}$, and

$$\mu^p(w)([\![p]\!]^{\mathbb{M}|p}) = \mu^p(w)(w) = \frac{\mu(w)(\{w\} \cap [\![p]\!]^{\mathbb{M}})}{\mu(w)([\![p]\!]^{\mathbb{M}})} = \frac{\mu(w)(w)}{\mu(w)(w)} = 1.$$

Using this updated model $\mathbb{M}|p$, we find that

$$\mathbb{M}, w \models [!p]\big(Kp \wedge P(p) = 1 \wedge P(\neg p) = 0\big).$$

So after the public announcement of p, the agent has come to know that p is in fact the case. She has also adjusted her probabilities: she now assigns probability 1 to p being true, and probability 0 to p being false. These are the results that one would intuitively expect, so Definition 4.11 seems to yield an adequate representation of the epistemic and probabilistic effects of public announcements. ⊣

Example 4.4

The *cable guy paradox* describes a situation where a cable guy is coming to your home between 8 a.m. and 4 p.m., and you must be at home when he arrives. Unfortunately, you do not know when exactly he will come. Now, you place a bet with someone as to whether the cable guy will come during the time interval $(8, 12]$ or the time interval $(12, 16)$. The observation of the paradox is as follows. Until 8 a.m., you consider both intervals equally appealing, as the cable guy is as likely to come in the morning interval as in the afternoon interval. But regardless of when the cable guy actually comes, there will some period of time after 8 a.m. and before his arrival, and during this period, the probability you should rationally assign to the cable guy coming in the morning gradually decreases. For example, if the cable guy has not yet arrived at 10 a.m., then the probability you assign to him coming in the morning has dropped from 0.50 to 0.25.

We can model this situation using probabilistic public announcement logic in the following way. The pointed model $\mathbb{M}, w$ is defined as follows: its set of states is the interval $(8, 16)$, with $w \in (8, 16)$ being the time the cable guy actually comes. There is just one agent (and hence we will drop agent subscripts throughout this example), whose prior probability $\mu(w)$ is the uniform distribution over the interval $(8, 16)$. Such a distribution requires the involvement of σ-algebras: $\mathcal{A}(w)$ is the Borel σ-algebra on $S(w) = W = (8, 16)$. The epistemic accessibility relation is the universal relation on W. The set of propositions is $\{p\} \cup \{q_r\}_{r \in \mathbb{Q} \cap (8,16)}$. The proposition letter p says that the cable guy comes somewhere in the morning, and hence $V(p) = (8, 12]$. Similarly, for each rational number $r \in (8, 16)$, the proposition letter q_r says that it is currently later than time r, and hence $V(q_r) = (r, 16)$.

Note that $\mathbb{M}, w \models P(p) = 0.5 \wedge P(\neg p) = 0.5$. This reflects the fact that initially, you consider it equally likely that the cable guy will come somewhere in the morning and that he will come somewhere in the afternoon (since p means that the cable guy will come somewhere in the morning, $\neg p$ means that he will come somewhere in the afternoon).

The cable guy actually comes at time $w > 8$, and hence there is some time $r < 12$ such that $8 < r < w$. After waiting until time r, the situation is represented by the updated pointed model $\mathbb{M}|q_r, w$, where $W^{q_r} = W$, $R^{q_r} =$

Figure 4.3: Axiomatization of probabilistic public announcement logic.

1. static base logic
   - probabilistic epistemic logic, as axiomatized in Figure 4.2
2. necessitation for public announcement
   - if $\vdash \psi$ then $\vdash [!\varphi]\psi$
3. reduction axioms for public announcement

$$\begin{aligned}
[!\varphi]p &\leftrightarrow \varphi \to p \\
[!\varphi]\neg\psi &\leftrightarrow \varphi \to \neg[!\varphi]\psi \\
[!\varphi](\psi_1 \wedge \psi_2) &\leftrightarrow [!\varphi]\psi_1 \wedge [!\varphi]\psi_2 \\
[!\varphi]K_i\psi &\leftrightarrow \varphi \to K_i[!\varphi]\psi \\
[!\varphi]\textstyle\sum_\ell a_\ell P_i(\psi_\ell) \geq b &\leftrightarrow \varphi \to \\
&\quad \big(P_i(\varphi) = 0 \wedge \textstyle\sum_\ell a_\ell P_i(\langle !\varphi\rangle\psi_\ell) \geq b\big) \vee \\
&\quad \big(P_i(\varphi) > 0 \wedge \textstyle\sum_\ell a_\ell P_i(\langle !\varphi\rangle\psi_\ell) \geq bP_i(\varphi)\big)
\end{aligned}$$

$R \cap (W \times [\![q^r]\!]^{\mathbb{M}}) = (8,16) \times (r,16)$, and $\mu(w)^{q_r}$ is the uniform probability distribution over $S^{q_r}(w) = (r,16)$. Note that $\mathbb{M}|q_r, w \models P(p) < 0.5$, and hence

$$\mathbb{M}, w \models P(p) = 0.5 \wedge \langle !q_r\rangle P(p) < 0.5. \qquad \dashv$$

4.5.2 Proof System

Public announcement logic can be axiomatized by adding a set of *reduction axioms* to the static base logic. These axioms allow us to recursively rewrite formulas containing dynamic public announcement operators as formulas without such operators; hence the dynamic language is not more expressive than the static language. Alternatively, reduction axioms can be seen as 'predicting' what will be the case *after* the public announcement has taken place in terms of what is the case *before* the public announcement has taken place.

This strategy can be extended into the probabilistic realm. For the static base logic, we do not take some system of epistemic logic (usually S5), but rather the system of probabilistic epistemic logic described in Subsection 4.3.4 (Figure 4.2), and add the reduction axioms shown in Figure 4.3. The first four reduction axioms are familiar from classical (non-probabilistic) public announcement logic. Note that the reduction axiom for i-probability formulas makes, just like Definition 4.12, a case distinction based on whether the agent assigns probability 0 to the announced formula φ. The significance of this reduction axiom, and its connection with Bayesian conditionalization, will be further explored in Subsection 4.5.4. One can easily check that all of these reduction axioms are sound with respect to the class $\mathcal{C}_m^{PEL}$ as well as the class $\mathcal{C}_s^{PEL}$.

A standard rewriting argument suffices to prove the following theorem:

Theorem 4.7
Probabilistic public announcement logic, as axiomatized in Figure 4.3, is sound and weakly complete with respect to the class of simplified probabilistic relational frames $\mathcal{C}_s^{PEL}$, and also with respect to the class of probabilistic relational frames $\mathcal{C}_m^{PEL}$ that satisfy *meas*.

4.5.3 Decidability and Complexity

Probabilistic public announcement logic is decidable. As to its complexity properties, since the validity (as well as satisfiability) problem for probabilistic epistemic logic (which is a fragment of probabilistic public announcement logic) is PSPACE-complete, we conclude that probabilistic public announcement logic is PSPACE-hard. An upper bound can be achieved by translating probabilistic public announcement logic into probabilistic epistemic logic. All known translations are exponential, so we have an exponential time upper bound for the complexity of the validity problem of probabilistic public announcement logic.

4.5.4 Higher-Order Information in Public Announcements

In this subsection we will discuss the role of higher-order information in probabilistic public announcement logic. This will further clarify the connection, but also the distinction, between (dynamic versions of) probabilistic epistemic logic and probability theory proper.

In Subsection 4.5.2, we introduced a reduction axiom for i-probability formulas. This axiom allows us to derive the following principle as a special case:

$$(\varphi \wedge P(\varphi) > 0) \longrightarrow \big([!\varphi]P_i(\psi) \geq b \leftrightarrow P(\langle !\varphi\rangle\psi) \geq bP_i(\varphi)\big). \tag{4.1}$$

The antecedent states that φ is true (because of the truthfulness of public announcements) and that agent i assigns a strictly positive probability to it (so that we are in the 'interesting' case of the reduction axiom). To see the meaning of the consequent more clearly, note that $\vdash \langle !\varphi\rangle\psi \leftrightarrow (\varphi \wedge [!\varphi]\psi)$, and introduce the following abbreviation of conditional probability into the formal language:

$$P_i(\beta \mid \alpha) \geq b \stackrel{\text{def}}{=} P_i(\alpha \wedge \beta) \geq bP_i(\alpha).$$

Principle (4.1) can now be rewritten as follows:

$$(\varphi \wedge P(\varphi) > 0) \longrightarrow \big([!\varphi]P_i(\psi) \geq b \leftrightarrow P([!\varphi]\psi \mid \varphi) \geq b\big). \tag{4.2}$$

A similar version can be proved for $\leq$ instead of $\geq$; combining these two we get:

$$(\varphi \wedge P(\varphi) > 0) \longrightarrow \big([!\varphi]P_i(\psi) = b \leftrightarrow P([!\varphi]\psi \mid \varphi) = b\big). \tag{4.3}$$

The consequent thus states a connection between the agent's probability of ψ after the public announcement of φ, and her conditional probability of $[!\varphi]\psi$, given the truth of φ. In other words, after a public announcement of φ, the agent updates her probabilities by Bayesian conditionalization on φ. The subtlety of principle (4.3), however, is that the agent does not take the conditional probability (conditional on φ) of ψ *itself*, but rather of the *updated* formula $[!\varphi]\psi$.

The reason for this is that $[!\varphi]P_i(\psi) = b$ talks about the probability that the agent assigns to ψ after the public announcement of φ has *actually* happened. If we want to describe this probability as a conditional probability, we cannot simply make use of the conditional probability $P_i(\psi \mid \varphi)$, because this represents the probability that the agent *would* assign to ψ if a public announcement of φ *were* to happen—hypothetically, not actually! Borrowing a slogan from van Benthem: "The former takes place once arrived at one's vacation destination, the latter is like reading a travel folder and musing about tropical islands." Hence, if we want to describe the agent's probability of ψ after an actual public announcement of φ in terms of conditional probabilities, we need to represent the effects of the public announcement of φ on ψ explicitly, and thus take the conditional probability (conditional on φ) of $[!\varphi]\psi$, rather than ψ.

One might wonder about the relevance of this subtle distinction between actual and hypothetical public announcements. The point is that the public announcement of φ can have effects on the truth value of ψ. For large classes of formulas ψ, this will not occur: their truth value is not affected by the public announcement of φ. Formally, this means that $\vdash \varphi \rightarrow (\psi \leftrightarrow [!\varphi]\psi)$, and thus (the consequent of) principle (4.3) becomes:

$$[!\varphi]P_i(\psi) = b \leftrightarrow P_i(\psi \mid \varphi) = b$$

—thus wiping away all differences between the agent's probability of ψ after a public announcement of φ, and her conditional probability of ψ, given

φ. A typical class of such formulas (whose truth value is unaffected by the public announcement of φ) is formed by the Boolean combinations of proposition letters, i.e. those formulas which express *ontic* or *first-order information*. Since probability theory proper is usually only concerned with first-order information ('no nested probabilities'), the distinction between actual and hypothetical announcements—or in general, between actual and hypothetical learning of new information—thus vanishes completely, and Bayesian conditionalization can be used as a universal update rule to compute new probabilities after (actually) learning a new piece of information.

However, in probabilistic epistemic logic (and its dynamic versions, such as probabilistic PAL), higher-order information *is* taken into account, and hence the distinction between actual and hypothetical public announcements has to be taken seriously. Therefore, the consequent of principle (4.3) should really use the conditional probability $P_i([!\varphi]\psi \mid \varphi)$, rather than just $P_i(\psi \mid \varphi)$. [10]

Example 4.5
To illustrate this, consider again the model defined in Examples 4.2 and 4.3, and put $\varphi \stackrel{\text{def}}{=} p \wedge P(\neg p) = 0.5$. It is easy to show that

$$\mathbb{M}, w \models P(\varphi \mid \varphi) = 1 \wedge P([!\varphi]\varphi \mid \varphi) = 0 \wedge [!\varphi]P(\varphi) = 0.$$

Hence the probability assigned to φ after the public announcement is the conditional probability $P([!\varphi]\varphi \mid \varphi)$, rather than just $P(\varphi \mid \varphi)$. Note that this example indeed involves higher-order information, since we are talking about the probability of φ, which itself contains the probability statement $P(\neg p) = 0.5$ as a conjunct. Finally, this example also shows that learning a new piece of information φ (via public announcement) does *not* automatically lead to the agents being certain about (i.e. assigning probability 1 to) that formula. This is to be contrasted with probability theory, where a new piece of information φ is processed via Bayesian conditionalization, and thus always leads to certainty: $P(\varphi \mid \varphi) = 1$. The explanation is, once again, that probability theory is only concerned with first-order information, whereas the phenomena described above can only occur at the level of

[10] There exist alternative analyses that stay closer in spirit to probability theory proper. Their proponents argue that the public announcement of φ induces a shift in the interpretation of ψ (in our terminology: from ψ to $[!\varphi]\psi$, i.e. from $[\![\psi]\!]^{\mathbb{M}}$ to $[\![\psi]\!]^{\mathbb{M}|\varphi}$), and show that such meaning shifts can be modeled using Dempster-Shafer belief functions. Crucially, however, this proposal is able to deal with the case of ψ expressing *second-order* information (e.g. when it is of the form $P_i(p) = b$), but not with the case of higher-order information *in general* (e.g. when ψ is of the form $P_j(P_i(p) = b) = a$, or involves even more deeply nested probabilities).

higher-order information.[11,12] ⊣

4.6 Probabilistic Dynamic Epistemic Logic

In this section we will move from a probabilistic version of public announcement logic to a probabilistic version of 'full' dynamic epistemic logic. Subsection 4.6.1 introduces a probabilistic version of the *product update* mechanism that is behind dynamic epistemic logic. Subsection 4.6.2 introduces dynamic operators into the formal language to talk about these product updates, and discusses an example in detail. Subsection 4.6.3, finally, shows how to obtain a complete axiomatization in a fully standard (though non-trivial) fashion.

4.6.1 Probabilistic Product Update

Classical (non-probabilistic) dynamic epistemic logic models epistemic dynamics by means of a product update mechanism. The agents' *static* information (what is the current state?) is represented by a relational model $\mathbb{M}$, and their *dynamic* information (what type of event is currently taking place?) is represented by an update model $\mathbb{E}$. The agents' new information (after the dynamics has taken place) is represented by means of a product construction $\mathbb{M}\otimes\mathbb{E}$. We will now show how to define a probabilistic version of this construction.

Before stating the formal definitions, we show how they naturally arise as probabilistic generalizations of the classical (non-probabilistic) notions. The (simplified) probabilistic relational models introduced in Definition 4.5 represent the agents' static information, in both its epistemic and its probabilistic aspects. This static probabilistic information is called the *prior probability*. We can thus say that when w is the actual state, agent i considers it epistemically possible that v is the actual state ($(w, v) \in R_i$), and, more specifically, that she assigns probability b to v being the actual state ($\mu_i(w)(v) = b$).

[11]Similarly, the *success postulate* for belief expansion in the (traditional) AGM framework states that after expanding one's belief set with a new piece of information φ, the updated (expanded) belief set should always contain this new information. Here, too, the explanation is that AGM is only concerned with first-order information. (Note that we talk about the success postulate for belief *expansion*, rather than belief *revision*, because the former seems to be the best analogue of public announcement in the AGM framework.)

[12]The occurrence of higher-order information is a *necessary* condition for this phenomenon, but not a *sufficient* one: there exist formulas φ that involve higher-order information, but still $\models [!\varphi]P_i(\varphi) = 1$ (or epistemically: $\models [!\varphi]K_i\varphi$).

Update models are essentially like relational models: they represent the agents' information about events, rather than states. Since probabilistic relational models represent both epistemic and probabilistic information about *states*, by analogy probabilistic update models should represent both epistemic and probabilistic information about *events*. Hence, they should not only have epistemic accessibility relations R_i over their set of events E, but also probability functions $\mu_i \colon E \to (E \to [0,1])$. (Formal details will be given in Definition 4.13.) We can then say that when e is the actually occurring event, agent i considers it epistemically possible that f is the actually occurring event ($(e, f) \in R_i$), and, more specifically, that she assigns probability b to f being the actually occurring event ($\mu_i(e)(f) = b$). This dynamic probabilistic information is called the *observation probability*.

Finally, how probable it is that an event e will occur, might vary from state to state. We assume that this variation can be captured by means of a set Φ of (pairwise inconsistent) sentences in the object language (so that the probability that th event e will occur can only vary between states that satisfy *different* sentences of Φ). This will be formalized by adding to the probabilistic update models a set of preconditions Φ, and probability functions $\mathsf{pre} \colon \Phi \to (E \to [0,1])$. The meaning of $\mathsf{pre}(\varphi)(e) = b$ is that if φ holds, then event e occurs with probability b. These are called *occurrence probabilities*.[13]

We are now ready to formally introduce probabilistic update models:

Definition 4.13
A *probabilistic update model* is a tuple $\mathbb{E} = \langle E, R_i, \Phi, \mathsf{pre}, \mu_i \rangle_{i \in I}$, where E is non-empty finite set of events, $R_i \subseteq E \times E$ is agent i's epistemic accessibility relation, $\Phi \subseteq \mathcal{L}^{\otimes}_{PEL}(\mathsf{AtProp})$ is a finite set of pairwise inconsistent sentences called *preconditions*, $\mu_i \colon E \to (E \to [0,1])$ assigns to each event $e \in E$ a probability function $\mu_i(e)$ over E, and $\mathsf{pre} \colon \Phi \to (E \to [0,1])$ assigns to each precondition $\varphi \in \Phi$ a probability function $\mathsf{pre}(\varphi)$ over E.

All components of a probabilistic update model have already been commented upon. Note that we use the same symbols R_i and μ_i to indicate agent i's epistemic and probabilistic information in a (potentially simplified) probabilistic relational model $\mathbb{M}$ and in a probabilistic update model $\mathbb{E}$—from the context it will always be clear which of the two is meant. The language $\mathcal{L}^{\otimes}_{PEL}(\mathsf{AtProp})$ that the preconditions are taken from will be formally defined in the next subsection. (As is usual in this area, there is a non-vicious simultaneous recursion going on here.)

We now introduce occurrence probabilities for events at states:

[13]Occurrence probabilities are often assumed to be *objective frequencies*. This is reflected in the formal setup: the function pre is not agent-dependent.

Definition 4.14
Consider a (potentially simplified) probabilistic relational model $\mathbb{M}$, a state w, a probabilistic update model $\mathbb{E}$, and an event e. Then the *occurrence probability of e at w* is defined as

$$\mathsf{pre}(w)(e) = \begin{cases} \mathsf{pre}(\varphi)(e) & \text{if } \varphi \in \Phi \text{ and } \mathbb{M}, w \models \varphi \\ 0 & \text{if there is no } \varphi \in \Phi \text{ such that } \mathbb{M}, w \models \varphi. \end{cases} \quad \dashv$$

Since the preconditions are pairwise inconsistent, $\mathsf{pre}(w)(e)$ is always well-defined. The meaning of $\mathsf{pre}(w)(e) = b$ is that in state w, event e occurs with probability b. Note that if two states w and v satisfy the same precondition, then always $\mathsf{pre}(w)(e) = \mathsf{pre}(v)(e)$; in other words, the occurrence probabilities of an event e can only vary 'up to a precondition' (cf. supra).

The probabilistic product update mechanism can now be defined as follows. We start with the simplified probabilistic relational models:

Definition 4.15
Consider a simplified probabilistic relational model $\mathbb{M} = \langle W, R_i, \mu_i, V\rangle_{i\in I}$ and a probabilistic update model $\mathbb{E} = \langle E, R_i, \Phi, \mathsf{pre}, \mu_i\rangle_{i\in I}$. Then the *updated model* $\mathbb{M} \otimes \mathbb{E} \stackrel{\text{def}}{=} \langle W', R'_i, \mu'_i, V'\rangle_{i\in I}$ is defined as follows:

- $W' \stackrel{\text{def}}{=} \{(w, e) \mid w \in W, e \in E, \mathsf{pre}(w)(e) > 0\}$,
- $R'_i \stackrel{\text{def}}{=} \{((w, e), (w', e')) \in W' \times W' \mid (w, w') \in R_i \text{ and } (e, e') \in R_i\}$ (for every agent $i \in I$),
- $\mu'_i \colon W' \to (W' \to [0, 1])$ is defined (for every agent $i \in I$) by

  $$\mu'_i(w, e)(w', e') \stackrel{\text{def}}{=} \frac{\mu_i(w)(w') \cdot \mathsf{pre}(w')(e') \cdot \mu_i(e)(e')}{\sum_{\substack{w''\in W \\ e''\in E}} \mu_i(w)(w'') \cdot \mathsf{pre}(w'')(e'') \cdot \mu_i(e)(e'')}$$

  if the denominator is strictly positive, and $\mu'_i(w, e)(w', e') \stackrel{\text{def}}{=} 0$ otherwise,
- $V'(p) \stackrel{\text{def}}{=} \{(w, e) \in W' \mid w \in V(p)\}$ (for every $p \in \mathsf{AtProp}$). $\dashv$

We will only comment on the probabilistic component of this definition (all other components are fully classical). After the dynamics has taken place, agent i calculates at state (w, e) her new probability for (w', e') by taking the arithmetical product of (i) her *prior probability* for w' at w, (ii) the *occurrence probability* of e' in w', and (iii) her *observation probability* for e' at e, and then normalizing this product. The factors in this product are not *weighted* (or equivalently, they all have weight 1); there are also weighted versions of this update mechanism, one of which corresponds to the rule of *Jeffrey conditioning* from probability theory. Finally, note that $\mathbb{M} \otimes \mathbb{E}$ might fail to be a simplified probabilistic relational model: if the denominator in the definition of $\mu'_i(w,e)$ is 0, then $\mu'_i(w,e)$ assigns 0 to all states in W'. We will not care here about the interpretation of this feature, but only remark that technically speaking it is harmless and, perhaps most importantly, still allows for a reduction axiom for i-probability formulas (cf. Subsection 4.6.3).

Toward a definition of product update on non-simplified probabilistic relational models, first consider how probabilities are defined on sets $A' \subseteq W'$ in the simplified setting (in case the denominator is strictly positive):

$$\mu'_i(w,e)(A') = \frac{\sum_{(w',e')\in A'} \mu_i(w)(w') \cdot \mathsf{pre}(w')(e') \cdot \mu_i(e)(e')}{\sum_{\substack{w''\in W\\ e''\in E}} \mu_i(w)(w'') \cdot \mathsf{pre}(w'')(e'') \cdot \mu_i(e)(e'')}.$$

If the original model is not simplified, it is no longer guaranteed that the probabilities $\mu_i(w)(w')$ of individual states are defined. However, the finite set Φ provides a useful partition of the domain of $\mathbb{M}$, which we will discuss in a moment. The updated σ-algebra $\mathcal{A}'_i(w,e)$ will be generated by sets of the form $(A \times \{e'\}) \cap W'$, where $e' \in E$, $A \in \mathcal{A}_i(w)$, and W' is the domain of the updated model. For any set A' in the updated σ-algebra, and any $e' \in E$, let $A'_{e'} \overset{\text{def}}{=} \{x \in A' \mid \pi_2(x) = e'\}$.[14] Because E is finite, each element A' in the updated σ-algebra can be written as a finite union: $A' = \bigcup_{e'\in E} A'_{e'}$. Because the occurrence probabilities of each e' depend on $\varphi \in \Phi$, we divide A' even further as $A' = \bigcup_{e'\in E} \bigcup_{\varphi\in\Phi} (A'_{e'} \cap ([\![\varphi]\!]^{\mathbb{M}} \times \{e'\}))$. This partition motivates the definition of the updated probability measure in the following.

Definition 4.16
Consider a probabilistic relational model $\mathbb{M} = \langle W, R_i, \mathbb{P}_i, V\rangle_{i\in I}$ and a probabilistic update model $\mathbb{E} = \langle E, R_i, \Phi, \mathsf{pre}, \mu_i\rangle_{i\in I}$. Then the *updated model* $\mathbb{M} \otimes \mathbb{E} \overset{\text{def}}{=} \langle W', R'_i, \mu'_i, V'\rangle_{i\in I}$ is defined as follows:

- $W' \overset{\text{def}}{=} \{(w,e) \mid w \in W, e \in E, \mathsf{pre}(w)(e) > 0\}$,
- $R'_i \overset{\text{def}}{=} \{((w,e),(w',e')) \in W' \times W' \mid (w,w') \in R_i \text{ and } (e,e') \in R_i\}$ (for every agent $i \in I$),
- $\mathbb{P}'_i \overset{\text{def}}{=} (S'_i, \mathcal{A}'_i, \mu'_i)$ for each $i \in I$, such that for each $(w,e) \in W'$:
  - $S'_i(w,e) \overset{\text{def}}{=} \{(w',e') \in W' \mid w' \in S_i(w)\}$

[14] We use π_1 and π_2 for the projections of pairs onto the first and second coordinates respectively, i.e., $\pi_1(x,y) = x$ and $\pi_2(x,y) = y$. For any function $f : X \to Y$, and subset $A \subseteq X$, we write $f[A] \overset{\text{def}}{=} \{f(a) \mid a \in A\}$.

– $\mathcal{A}'_i(w,e)$ is the smallest σ-algebra containing

$$\{(A \times B) \cap W' \mid A \in \mathcal{A}_i(w), B \subseteq E\}$$

– $\mu'_i(w,e) : \mathcal{A}'_i(w,e) \to [0,1]$ is defined for each $A' \in \mathcal{A}'_i(w,e)$ by

$$\mu'_i(w,e)(A') \stackrel{\text{def}}{=} \frac{\sum_{\substack{\varphi\in\Phi \\ e'\in E}} (\mu_i(w))_*([\![\varphi]\!]^{\mathbb{M}} \cap \pi_1[A'_{e'}]) \cdot \mathsf{pre}(\varphi)(e') \cdot \mu_i(e)(e')}{\sum_{\substack{\varphi\in\Phi \\ e''\in E}} (\mu_i(w))_*([\![\varphi]\!]^{\mathbb{M}} \cap S_i(w)) \cdot \mathsf{pre}(\varphi)(e'') \cdot \mu_i(e)(e'')}$$

if the denominator is strictly positive, and $\mu'_i(w,e)(A) \stackrel{\text{def}}{=} 0$ otherwise,

- $V'(p) \stackrel{\text{def}}{=} \{(w,e) \in W' \mid w \in V(p)\}$ (for every $p \in \mathsf{AtProp}$). ⊣

The language $\mathcal{L}^{\otimes}_{PEL}(\mathsf{AtProp})$ is equally expressive as $\mathcal{L}_{PEL}(\mathsf{AtProp})$, as we will see from the reduction axioms. Hence, if $\mathbb{M} \in \mathcal{C}^{PEL}_s$, then every formula $\varphi \in \mathcal{L}^{\otimes}_{PEL}(\mathsf{AtProp})$ is measurable, i.e. $[\![\varphi]\!]^{\mathbb{M}} \cap S_i(w) \in \mathcal{A}_i(w)$ for every agent i and world w. In particular, every formula $\varphi \in \Phi$ is measurable. It is not hard to see that given any general σ-algebra $\mathcal{A}$ for $\mathbb{M}$, the set $\{(A \times \{e\}) \cap W' \mid A \in \mathcal{A}, e \in E\}$ generates (via finite unions) a general σ-algebra for $\mathbb{M} \otimes \mathbb{E}$.[15]

4.6.2 Language and Semantics

To talk about these updated models, we add dynamic operators $[\mathsf{E},\mathsf{e}]$ to the static language $\mathcal{L}_{PEL}(\mathsf{AtProp})$, thus obtaining the new language $\mathcal{L}^{\otimes}_{PEL}$ (AtProp). Here, E and e are formal names for the probabilistic update model $\mathbb{E} = \langle E, R_i, \Phi, \mathsf{pre}, \mu_i\rangle_{i\in I}$ and the event $e \in E$, respectively (recall our remark about the mutual recursion of the dynamic language and the updated models). The formula $[\mathsf{E},\mathsf{e}]\varphi$ means that after the event e has occurred (assuming that it *can* occur, i.e. that its occurrence probability is non-zero), it will be the case that φ. It has the following semantics:

$$\mathbb{M}, w \models [\mathsf{E},\mathsf{e}]\psi \quad \text{iff} \quad \text{if } \mathsf{pre}(w)(e) > 0, \text{ then } \mathbb{M}\otimes\mathbb{E}, (w,e) \models \psi.$$

We will now illustrate the expressive power of this framework by showing how it can be used to adequately model a rather intricate scenario.

Example 4.6
Upon entering the kitchen, you catch Mary, your three-year-old daughter, with a blushing face, and suspect she has taken a cookie out of the jar.

[15]There exist alternative approaches to showing that measurability of formulas is preserved via updating.

You know that Mary has a sweet tooth, and therefore consider the chance that she has actually taken a cookie to be quite high, say 75%. However, you also know that she rarely lies, and therefore decide to simply ask her whether she took a cookie: you know that if she did, she will admit this with probability 90%. (Of course, if she did *not* take a cookie, she will certainly not lie about it.) Mary shyly mumbles an answer—you did not quite hear what is was: you think you just heard her say 'yes' with probability 50%, and 'no' with probability 50%. What chance should you assign to Mary having taken a cookie?

Your initial information (before processing Mary's answer) can be represented using the following simplified probabilistic relational model: $\mathbb{M} = \langle W, R, \mu, V\rangle$, where $W = \{w, v\}, R = W \times W, \mu(w)(w) = \mu(v)(w) = 0.75, \mu(w)(v) = \mu(v)(v) = 0.25$, and $V(\mathsf{took}) = \{w\}$. (We work with only one agent in this example, so agent indices can be dropped.) Hence, initially you do not know whether Mary has taken a cookie, but you consider it quite likely:

$$\mathbb{M}, w \models \hat{K}\mathsf{took} \wedge \hat{K}\neg\mathsf{took} \wedge P(\mathsf{took}) = 0.75.$$

The event of Mary mumbling an answer can be represented using the following update model: $\mathbb{E} = \langle E, R, \Phi, \mathsf{pre}, \mu\rangle$, where $E = \{e, f\}$, $R = E \times E$, $\Phi = \{\mathsf{took}, \neg\mathsf{took}\}$, $\mathsf{pre}(\mathsf{took})(e) = 0.9$, $\mathsf{pre}(\mathsf{took})(f) = 0.1$, $\mathsf{pre}(\neg\mathsf{took})(e) = 0$, $\mathsf{pre}(\neg\mathsf{took})(f) = 1$, and $\mu(e)(e) = \mu(f)(e) = \mu(e)(f) = \mu(f)(f) = 0.5$. The events e and f represent Mary answering 'yes (I took a cookie)' and 'no (I did not take a cookie)', respectively.

We now construct the updated model $\mathbb{M}\otimes\mathbb{E}$. Since $\mathbb{M}, v \not\models \mathsf{took}$, it holds that $\mathsf{pre}(v)(e) = \mathsf{pre}(\neg\mathsf{took})(e) = 0$, and hence (v, e) does not belong to the updated model. It is easy to see that the other states (w, e), (w, f) and (v, f) do belong to the updated model. Furthermore, one can easily calculate that $\mu'(w, e)(w, e) = \frac{0.3375}{0.5} = 0.675$ and $\mu'(w, e)(w, f) = \frac{0.0375}{0.5} = 0.075$, so $\mu'(w, e)([\![\mathsf{took}]\!]^{\mathbb{M}\otimes\mathbb{E}}) = 0.675 + 0.075 = 0.75$, and thus

$$\mathbb{M}, w \models [\mathsf{E}, \mathsf{e}]P(\mathsf{took}) = 0.75.$$

Hence, even though Mary actually took a cookie (w is the actual state), and also actually confessed this (e is the event that actually occurred), you still assign probability 75% to her having taken a cookie—i.e. Mary's answer is entirely uninformative: it does not lead to any change in the probability of took.

Note that if Mary had answered reasonably clearly—let's formalize this as $\mu(e)(e) = 0.95$ and $\mu(e)(f) = 0.05$—, then the same update mechanism would have implied that

$$\mathbb{M}, w \models [\mathsf{E}, \mathsf{e}]P(\mathsf{took}) \geq 0.98.$$

⊣

Figure 4.4: Axiomatization of probabilistic dynamic epistemic logic.

1. static base logic
   - probabilistic epistemic logic, as axiomatized in Figure 4.2
2. necessitation for $[\mathsf{E},\mathsf{e}]$
   - if $\vdash \psi$ then $\vdash [\mathsf{E},\mathsf{e}]\psi$
3. reduction axioms

$$\begin{array}{rcl} [\mathsf{E},\mathsf{e}]p & \leftrightarrow & \mathsf{pre}_{\mathsf{E},\mathsf{e}} \to p \\ [\mathsf{E},\mathsf{e}]\neg\psi & \leftrightarrow & \mathsf{pre}_{\mathsf{E},\mathsf{e}} \to \neg[\mathsf{E},\mathsf{e}]\psi \\ [\mathsf{E},\mathsf{e}](\psi_1 \wedge \psi_2) & \leftrightarrow & [\mathsf{E},\mathsf{e}]\psi_1 \wedge [\mathsf{E},\mathsf{e}]\psi_2 \\ [\mathsf{E},\mathsf{e}]K_i\psi & \leftrightarrow & \mathsf{pre}_{\mathsf{E},\mathsf{e}} \to \bigwedge_{(e,f)\in R_i} K_i[\mathsf{E},\mathsf{f}]\psi \\ [\mathsf{E},\mathsf{e}]\sum_\ell a_\ell P_i(\psi_\ell) \geq b & \leftrightarrow & \mathsf{pre}_{\mathsf{E},\mathsf{e}} \to \\ & & \big(\sum_{\substack{\varphi\in\Phi \\ f\in E}} k_{i,e,\varphi,f} P_i(\varphi) = 0 \wedge 0 \geq b\big) \\ & & \vee\big(\sum_{\substack{\varphi\in\Phi \\ f\in E}} k_{i,e,\varphi,f} P_i(\varphi) > 0 \wedge \chi\big) \end{array}$$

   using the following definitions:
   - $\mathsf{pre}_{\mathsf{E},\mathsf{e}} \stackrel{\text{def}}{=} \bigvee_{\substack{\varphi\in\Phi \\ \mathsf{pre}(\varphi)(e)>0}} \varphi$
   - $k_{i,e,\varphi,f} \stackrel{\text{def}}{=} \mathsf{pre}(\varphi)(f)\cdot\mu_i(e)(f) \in \mathbb{R}$
   - $\chi \stackrel{\text{def}}{=} \sum_{\substack{\ell \\ \varphi\in\Phi \\ f\in E}} a_\ell k_{i,e,\varphi,f} P_i(\varphi \wedge \langle\mathsf{E},\mathsf{f}\rangle\psi_\ell) \geq \sum_{\substack{\varphi\in\Phi \\ f\in E}} b k_{i,e,\varphi,f} P_i(\varphi)$

In other words, if Mary had answered more clearly, then her answer *would* have been highly informative: it would have induced a significant change in your probabilities, leading you to become almost certain that she indeed took a cookie.

4.6.3 Proof System

A complete axiomatization for probabilistic dynamic epistemic logic can be found using the standard strategy, viz. by adding a set of reduction axioms to static probabilistic epistemic logic. Implementing this strategy, however, is not entirely trivial. The reduction axioms for non-probabilistic formulas are familiar from classical (non-probabilistic) dynamic epistemic logic, but the reduction axiom for i-probability formulas is more complicated.

First of all, this reduction axiom makes a case distinction on whether a certain sum of probabilities is strictly positive or not. We will show that this corresponds to the case distinction made in the definition of the updated probability functions (Definitions 4.15 and 4.16). For ease of exposition, let's first focus on simplified probabilistic relational models. In the definition of $\mu'_i(w,e)$, a case distinction is made on the value of the denominator of a fraction, i.e. on the value of the following expression:

$$\sum_{\substack{v\in W\\ f\in E}} \mu_i(w)(v)\cdot \mathsf{pre}(v)(f)\cdot \mu_i(e)(f). \tag{4.4}$$

But this expression can be rewritten as

$$\sum_{\substack{v\in W\\ f\in E\\ \varphi\in\Phi\\ \mathbb{M},v\models\varphi}} \mu_i(w)(v)\cdot \mathsf{pre}(\varphi)(f)\cdot \mu_i(e)(f).$$

Using the definition of $k_{i,e,\varphi,f}$ (cf. Figure 4.4), this can be rewritten as

$$\sum_{\substack{\varphi\in\Phi\\ f\in E}} \mu_i(w)([\![\varphi]\!]^{\mathbb{M}})\cdot k_{i,e,\varphi,f}.$$

Since E and Φ are finite, this sum is finite and corresponds to an expression in the formal language $\mathcal{L}^{\otimes}_{PEL}(\mathsf{AtProp})$, which we will abbreviate as σ:

$$\sigma \stackrel{\text{def}}{=} \sum_{\substack{\varphi\in\Phi\\ f\in E}} k_{i,e,\varphi,f}P_i(\varphi).$$

This expression can be turned into an i-probability formula by 'comparing' it with a rational number b; for example $\sigma \geq b$. Particularly important are the formulas $\sigma = 0$ and $\sigma > 0$: exactly these formulas are used to make the case distinction in the reduction axiom for i-probability formulas.[16]

Next, the reduction axiom for i-probability formulas provides a statement in each case of the case distinction: $0 \geq b$ in the case $\sigma = 0$, and χ (as defined in Figure 4.4) in the case $\sigma > 0$. We will only explain the meaning of χ in the 'interesting' case $\sigma > 0$. If $\mathbb{M}, w \models \sigma > 0$, then the value of (4.4) is strictly positive (cf. supra), and we can calculate:

[16]Note that E and Φ are components of the probabilistic update model $\mathbb{E}$ named by E; furthermore, the values $k_{i,e,\varphi,f}$ are fully determined by the model $\mathbb{E}$ and event e named by E and e, respectively (consider their definition in Figure 4.4). Hence any i-probability formula involving σ is fully determined by $\mathbb{E}, e$, and can be interpreted at any probabilistic relational model $\mathbb{M}$ and state w.

$$\begin{aligned}
\mu_i'(w,e)([\![\psi]\!]^{\mathbb{M}\otimes\mathbb{E}}) &= \textstyle\sum_{\mathbb{M}\otimes\mathbb{E},(w',e')\models\psi} \mu_i'(w,e)(w',e') \\
&= \sum_{\substack{w'\in W, e'\in E \\ \mathbb{M},w'\models\langle \mathsf{E},\mathsf{e}'\rangle\psi}} \frac{\mu_i(w)(w')\cdot\mathsf{pre}(w')(e')\cdot\mu_i(e)(e')}{\sum_{\substack{v\in W \\ f\in E}} \mu_i(w)(v)\cdot\mathsf{pre}(v)(f)\cdot\mu_i(e)(f)} \\
&= \frac{\sum_{\substack{\varphi\in\Phi \\ f\in E}} \mu_i(w)([\![\varphi\wedge\langle \mathsf{E},\mathsf{f}\rangle\psi]\!]^{\mathbb{M}})\cdot k_{i,e,\varphi,f}}{\sum_{\substack{\varphi\in\Phi \\ f\in E}} \mu_i(w)([\![\varphi]\!]^{\mathbb{M}})\cdot k_{i,e,\varphi,f}}.
\end{aligned}$$

Hence, in this case ($\sigma > 0$) we can express that $\mu_i'(w,e)([\![\psi]\!]^{\mathbb{M}\otimes\mathbb{E}}) \geq b$ in the formal language, by means of the following i-probability formula:

$$\sum_{\substack{\varphi\in\Phi \\ f\in E}} k_{i,e,\varphi,f} P_i(\varphi\wedge\langle \mathsf{E},\mathsf{f}\rangle\psi) \geq \sum_{\substack{\varphi\in\Phi \\ f\in E}} b k_{i,e,\varphi,f} P_i(\varphi).$$

Moving to linear combinations, we can express that $\sum_\ell a_\ell \mu_i'(w,e)([\![\psi_\ell]\!]^{\mathbb{M}\otimes\mathbb{E}}) \geq b$ in the formal language using an analogous i-probability formula, namely χ (cf. the definition of this formula in Figure 4.4).

A similar argument can be made for the non-simplified case. We thus obtain the following theorem:

Theorem 4.8
Probabilistic dynamic epistemic logic, as axiomatized in Figure 4.4, is sound and weakly complete with respect to the class $\mathcal{C}_m^{PEL}$ of probabilistic relational frames that satisfy *meas* as well as the class $\mathcal{C}_s^{PEL}$ of simplified probabilistic relational frames.

4.7 Further Developments and Applications

Probabilistic extensions of epistemic logic are a recent development, and there are various open questions and potential applications to be explored. In this section we will therefore briefly discuss a selection of such topics for further research.

A typical technical problem that needs further research is the issue of *surprising information.* In the update mechanisms described in this chapter, the agents' new probabilities are calculated by means of a fraction whose denominator might take on the value 0. The focus has been on the 'interesting' (non-0) cases, and the 0-case has been treated as mere 'noise': a technical artefact that cannot be handled convincingly by the system. However, sometimes such 0-cases *do* represent intuitive scenarios; for example, one might think of an agent being absolutely certain that a certain proposition φ is false ($P(\varphi) = 0$), while that proposition is actually true, and can thus be announced! In such cases, the system of probabilistic public announcement logic described in Section 4.5 predicts that the agent will simply *ignore* the announced information (rather than performing some sensible form of *belief revision*). More can, and should be said about such cases.

Several fruitful applications of probabilistic extensions of (dynamic) epistemic logic can be expected in the field of *game theory.* In recent years, epistemic logic has been widely applied to explore the epistemic foundations of game theory. However, given the importance of probability in game theory (for example, in the notion of mixed strategy), it is surprising that rather few of these logical analyses have a probabilistic component. Two examples of the powerful potential for probabilistic (dynamic) epistemic logic in game theory are (i) the analysis of epistemic characterization theorems for several solution concepts for normal form games and extensive games (such as Nash equilibrium, iterated strict dominance, and backward induction), and (ii) the analysis of the role and significance of communication and common knowledge in Aumann's agreeing to disagree theorem.

Another potential field of application is *cognitive science.* The usefulness of (epistemic) logic for cognitive science has been widely recognized. Of course, as in any other empirical discipline, one quickly finds out that real-life human cognition is rarely a matter of all-or-nothing, but often involves degrees (probabilities). Furthermore, a recent development in cognitive science is toward probabilistic (Bayesian) models of cognition. If epistemic logic is to remain a valuable tool here, it will thus have to be a thoroughly 'probabilized' version. For example, probabilistic dynamic epistemic logic has been used to model the cognitive phenomenon of surprise and its epistemic aspects.

4.8 Conclusion

In this chapter we have surveyed various ways of combining epistemic logic and probability theory. These systems provide a standard modal (possible-worlds) analysis of the agents' hard information, and supplement it with a fine-grained probabilistic analysis of their soft information. Higher-order information of any kind (knowledge about probabilities, probabilities about knowledge, etc.) can be represented explicitly, which leads to a non-trivial connection between public announcements and Bayesian conditionalization. Mathematically speaking, these systems draw upon notions from modal logic and measure theory; their meta-theoretical properties (sound and complete axiomatizations, complexity, etc.) have been studied extensively.

Throughout the chapter, we have shown how these systems can be used to analyze various scenarios in which both knowledge and probability play a role, and how they can be applied to questions in philosophy, computer science, game theory, cognitive science, and other disciplines that are concerned with information and information flow.

4.9 Notes

The origins of epistemic logic can be found in Hintikka's book (1962), which already dealt with higher-order knowledge. A well-known example of a probabilistic principle that involves higher-order information is Miller's principle; this principle has been widely discussed in Bayesian epistemology and philosophy of science; for example, see Halpern (1991), Lewis (1980), Meacham (2010), Miller (1966), and van Fraassen (1984).

Probabilistic propositional logic was introduced by Fagin, Halpern, and Megiddo (1990), who address the system's soundness and completeness, decidability and computational complexity. The satisfiability problem is NP complete (Fagin et al., 1990, Theorem 2.9). Its proof makes use of the NP-completeness of Boolean satisfiability, which was established by Cook (1971) and Levin (1984), a small model theorem (Theorem 4.3) which is from Fagin et al. (1990, Theorem 2.6), and the polynomial time complexity of the model checking problem for modal logic is from Marx (2007). In Footnote 3 we draw an analogy between σ-algebras and modal algebras; the notion of a modal algebra is defined by Blackburn, de Rijke, and Venema (2001, Definition 1.32). An alternative axiomatization of probabilistic propositional logic, which does not make use of linear combinations, can be found in a paper by Heifetz and Mongin (2001). For a system that captures countable additivity by means of an infinitary rule in a finitary logic, see Goldblatt (2010).

The first book on Probabilistic epistemic logic is by Fagin and Halpern (1994), who address the system's soundness and completeness, decidability and computational complexity. For complexity results, see (Fagin and Halpern, 1994, Theorem 4.5). Its proof make use of a tableau construction for modal logic complexity, which can be found in (Halpern and Moses, 1992, Section 6), and a PSPACE lower bound to the complexity of basic epistemic logic, for which see the paper by Halpern and Moses (1992, Theorem 6.6). Concerning the correspondence results for knowledge/probability interaction that are discussed in Section 4.3.2, further discussion and more examples can be found in the textbook by Halpern (2003); for the notion of frame correspondence (as used in Lemmas 4.3 and 4.4), see van Benthem (1983, 2001a). Probabilistic transition systems have been studied extensively in theoretical computer science; for example, see de Vink and Rutten (1999), Jonsson, Yi, and Larsen (2001), and Larsen and Skou (1991). The counterexample to prudence (as mentioned in Footnote 8) is due to Kooi (2003, p. 384).

More information about interpreted systems and their probabilistic extensions can be found in the textbooks by Fagin, Halpern, Moses, and Vardi (1995) and by Halpern (2003), respectively. For example, Halpern (2003, Section 6.4) provides more information about the properties of perfect recall and synchronicity, and discusses examples that show that they are necessary conditions for the Bayesian conditionalization equality in Theorem 4.6, which is from Halpern (2003, Corollary 6.4.4). More information about Bayesian epistemology and Bayesian conditionalization can be found in papers by Hájek and Hartmann (2010), and Hartmann and Sprenger (2010). Although temporal and dynamic epistemic logic provide significantly different perspectives on time, much can be said about their theoretical relationship; for example, see Sack (2007), van Benthem, Gerbrandy, Hoshi, and Pacuit (2009), and van Ditmarsch, van der Hoek, and Ruan (2014). See also Chapter 5 in this handbook on the relationship between knowledge and time.

Public announcement logic was proposed independently by Plaza (1989) and Gerbrandy and Groeneveld (1997); see van Ditmarsch, van der Hoek, and Kooi (2007) for a textbook presentation which provides an overview of reduction axioms, and Chapter 6 of this handbook for a recent overview. Its probabilistic extension was first described by Kooi (2003). For an alternative analysis of updating that stays closer in spirit to probability theory proper (as mentioned in Footnote 10), see Romeijn (2012). For more information about the AGM framework for belief revision, see Alchourrón, Gärdenfors, and Makinson (1985), Gärdenfors (1988), and Chapter 7 for a discussion of belief revision in dynamic epistemic logic.

Product update dynamic epistemic logic was first described by Baltag and Moss (2004), and Baltag, Moss, and Solecki (1998); its probabilistic extensions can be found in the paper by van Benthem (2003) (from which our quote on page 183 was taken) and by van Benthem, Gerbrandy, and Kooi (2009) using simplified probabilistic relational models, and in papers by Aceto, van der Hoek, Ingolfsdottir, and Sack (2011), and Sack (2009) using probabilistic relational models with σ-algebras. For example, van Benthem et al. (2009) introduce the distinction between prior, observation, and occurrence probabilities to this line of work, defines weighted versions of the probabilistic product update, and shows how one of these weighted versions corresponds to the rule of Jeffrey conditioning (Jeffrey, 1983). For an alternative proof that product update preserves measurability of formulas (as mentioned in Footnote 15), see Aceto et al. (2011, Theorem 1).

Section 4.7 discusses some open questions and applications; more suggestions can be found in the paper by van Benthem et al. (2009) and the book by van Benthem (2011, ch. 8). Recent proposals for dealing with probability-0 updates can be found in publications by Aucher (2003), Baltag and Smets (2008), and Rott and Girard (2014). Some examples of the use of epistemic logic to explore the epistemic foundations of game theory can be found in publications by Bonanno and Dégremont (2014) and van Benthem (2001b, 2007). Probabilistic epistemic logic has been used to analyze game-theoretical notions by de Bruin (2008a,b, 2010), and Demey (2014). The usefulness of (epistemic) logic for cognitive science is addressed by Isaac, Szymanik, and Verbrugge (2014), Pietarinen (2003), and van Benthem (2008). For the recent development in cognitive science toward probabilistic (Bayesian) models of cognition, see Oaksford and Chater (2008). Demey (forthcoming) and Lorini and Castelfranchi (2007) use probabilistic dynamic epistemic logic to analyze the cognitive and epistemic aspects of surprise.

Example 4.1 is loosely based on the work of Geiss and Geiss (2009, Example 1.2.5). The scenario of Mary and the cookies (Example 4.6) is based on an example originally due to Kooi (2007). Note that this example is based on the sense of *hearing*. Similar examples by van Benthem et al. (2009) and Demey and Kooi (2014) are based on the senses of *touch* and *sight*, respectively. This is no coincidence: because of the fallibility of sense perception, examples based on the senses can easily be used to illustrate the notion of observation probability, which is central in the field of probabilistic dynamic epistemic logic.[17] The cable guy paradox (Example 4.4) was first described by Hájek (2005).

Finally, it should be emphasized that in all the systems discussed in this chapter, probabilities are represented in the logic's object language. Other proposals provide a probabilistic semantics for an object language that is itself fully classical (i.e. that does not explicitly represent probabilities). A recent overview of the various ways of combining logic and probability can be found in the entry by Demey, Kooi, and Sack (2013).

Acknowledgements The first author holds a postdoctoral fellowship of the Research Foundation-Flanders (FWO). The research of the second author has been funded by the Netherlands Organisation for Scientific Research VIDI grant 639.072.904. We would like to thank Alexandru Baltag, Jan Heylen, Barteld Kooi, Margaux Smets and Sonja Smets for their valuable feedback.

[17] We are thus eagerly awaiting examples based on the senses of smell and taste.

References

Aceto, L., W. van der Hoek, A. Ingolfsdottir, and J. Sack (2011). Sigma algebras in probabilistic epistemic dynamics. In *Proceedings of the 13th Conference on Theoretical Aspects of Rationality and Knowledge*, TARK XIII, New York, NY, USA, pp. 191–199. ACM.

Alchourrón, C., P. Gärdenfors, and D. Makinson (1985). On the logic of theory change: Partial meet contraction and revision functions. *Journal of Symbolic Logic 50*, 510–530.

Aucher, G. (2003). A combined system for update logic and belief revision. Master's thesis, Institute for Logic, Language and Computation, Universiteit van Amsterdam.

Baltag, A. and L. S. Moss (2004). Logics for epistemic programs. *Synthese 139*, 1–60.

Baltag, A., L. S. Moss, and S. Solecki (1998). The logic of common knowledge, public announcements, and private suspicions. In I. Gilboa (Ed.), *Proceedings of the 7th Conference on Theoretical Aspects of Rationality and Knowledge (TARK '98)*, pp. 43–56. Morgan Kaufmann Publishers.

Baltag, A. and S. Smets (2008). Probabilistic dynamic belief revision. *Synthese 165*, 179–202.

van Benthem, J. (1983). *Modal Logic and Classical Logic.* Napoli: Bibliopolis.

van Benthem, J. (2001a). Correspondence theory. In D. M. Gabbay and F. Guenthner (Eds.), *Handbook of Philosophical Logic (second revised edition)*, Volume 3, pp. 325–408. Dordrecht: Kluwer.

van Benthem, J. (2001b). Games in dynamic epistemic logic. *Bulletin of Economic Research 53*, 219–248.

van Benthem, J. (2003). Conditional probability meets update logic. *Journal of Logic, Language and Information 12*, 409–421.

van Benthem, J. (2007). Rational dynamics and epistemic logic in games. *International Game Theory Review 9*, 13–45.

van Benthem, J. (2008). Logic and reasoning: Do the facts matter? *Studia Logica 88*, 67–84.

van Benthem, J. (2011). *Logical Dynamics of Information and Interaction.* Cambridge: Cambridge University Press.

van Benthem, J., J. Gerbrandy, T. Hoshi, and E. Pacuit (2009). Merging frameworks for interaction. *Journal of Philosophical Logic 38*, 491–526.

van Benthem, J., J. Gerbrandy, and B. P. Kooi (2009). Dynamic update with probabilities. *Studia Logica 93*, 67–96.

Blackburn, P., M. de Rijke, and Y. Venema (2001). *Modal Logic.* Cambridge: Cambridge University Press.

Bonanno, G. and C. Dégremont (2014). Logic and game theory. In A. Baltag and S. Smets (Eds.), *Outstanding Contributions: Johan F. A. K. van Benthem on Logical and Informational Dynamics*, pp. 421–449. Dordrecht: Springer.

de Bruin, B. (2008a). Common knowledge of payoff uncertainty in games. *Synthese 163*, 79–97.

de Bruin, B. (2008b). Common knowledge of rationality in extensive games. *Notre Dame Journal of Formal Logic 49*, 261–280.

de Bruin, B. (2010). *Explaining Games: The Epistemic Programme in Game Theory.* Dordrecht: Springer.

Cook, S. A. (1971). The complexity of theorem proving procedures. In *Proceedings of the Third Annual ACM Symposium on Theory of Computing*, pp. 151–158.

Demey, L. (2014). Agreeing to disagree in probabilistic dynamic epistemic logic. *Synthese 191*, 409–438.

Demey, L. (2015). The dynamics of surprise. To appear in *Logic et Analyse.*

Demey, L. and B. P. Kooi (2014). Logic and probabilistic update. In A. Baltag and S. Smets (Eds.), *Outstanding Contributions: Johan F. A. K. van Benthem on Logical and Informational Dynamics*, pp. 381–404. Dordrecht: Springer.

Demey, L., B. P. Kooi, and J. Sack (2013). Logic and probability. In E. N. Zalta (Ed.), *Stanford Encyclopedia of Philosophy.*

van Ditmarsch, H., W. van der Hoek, and B. Kooi (2007). *Dynamic Epistemic Logic.* Dordrecht: Springer.

van Ditmarsch, H., W. van der Hoek, and J. Ruan (2014). Connecting dynamic epistemic and temporal epistemic logics. *Logic Journal of the IGPL 21(3)*, 380–403.

Fagin, R. and J. Halpern (1994). Reasoning about knowledge and probability. *Journal of the ACM 41*, 340–367.

Fagin, R., J. Y. Halpern, and N. Megiddo (1990). A logic for reasoning about probabilities. *Information and Computation 87*, 78–128.

Fagin, R., J. Y. Halpern, Y. Moses, and M. Y. Vardi (1995). *Reasoning about Knowledge.* Cambridge, MA: MIT Press.

van Fraassen, B. (1984). Belief and the will. *Journal of Philosophy 81*, 235–256.

Gärdenfors, P. (1988). *Knowledge in Flux.* Cambridge, MA: MIT Press.

Geiss, C. and S. Geiss (2009). An introduction to probability theory. Course notes.

Gerbrandy, J. and W. Groeneveld (1997). Reasoning about information change. *Journal of Logic, Language and Information 6*, 147–169.

Goldblatt, R. (2010). Deduction systems for coalgebras over measurable spaces. *Journal of Logic and Computation 20*, 1069–1100.

Hájek, A. (2005). The cable guy paradox. *Analysis 65*, 112–119.

Hájek, A. and S. Hartmann (2010). Bayesian epistemology. In J. Dancy, E. Sosa, and M. Steup (Eds.), *A Companion to Epistemology*, pp. 93–106. Oxford: Blackwell.

Halpern, J. Y. (1991). The relationship between knowledge, belief, and certainty. *Annals of Mathematics and Artificial Intelligence 4*, 301–322 (errata in the same journal, 26:59–61, 1999).

Halpern, J. Y. (2003). *Reasoning about Uncertainty.* Cambridge, MA: MIT Press.

Halpern, J. Y. and Y. Moses (1992). A guide to completeness and complexity for modal logics of knowledge and belief. *Artificial Intelligence 54*(3), 319–379.

Hartmann, S. and J. Sprenger (2010). Bayesian epistemology. In S. Bernecker and D. Pritchard (Eds.), *Routledge Companion to Epistemology*, pp. 609–620. London: Routledge.

Heifetz, A. and P. Mongin (2001). Probability logic for type spaces. *Games and Economic Behavior 35*, 31–53.

Hintikka, J. (1962). *Knowledge and Belief. An Introduction to the Logic of the Two Notions.* Ithaca, NY: Cornell University Press.

Isaac, A., J. Szymanik, and R. Verbrugge (2014). Logic and complexity in cognitive science. In A. Baltag and S. Smets (Eds.), *Outstanding Contributions: Johan F. A. K. van Benthem on Logical and Informational Dynamics*, pp. 787–824. Dordrecht: Springer.

Jeffrey, R. (1983). *The Logic of Decision (2nd ed.).* Chicago, IL: University of Chicago Press.

Jonsson, B., W. Yi, and K. G. Larsen (2001). Probabilistic extensions of process algebras. In J. A. Bergstra, A. Ponse, and S. A. Smolka (Eds.), *Handbook of Process Algebra*, pp. 685–710. Amsterdam: Elsevier.

Kooi, B. P. (2003). Probabilistic dynamic epistemic logic. *Journal of Logic, Language and Information 12*, 381–408.

Kooi, B. P. (2007). Dynamic epistemic logic. Tutorial delivered at the Dynamic Logic workshop, Montréal.

Larsen, K. G. and A. Skou (1991). Bisimulation through probabilistic testing. *Information and Computation 94*, 1–28.

Levin, L. (1973 (translated into English by B. A. Trakhtenbrot: A survey of Russian approaches to perebor (brute-force searches) algorithms. *Annals of the History of Computing* 6:384–400, 1984)). Universal search problems (in Russian). *Problems of Information Transmission 9*, 265–266.

Lewis, D. (1980). A subjectivist's guide to objective chance. In R. C. Jeffrey (Ed.), *Studies in Inductive Logic and Probability. Volume 2*, pp. 263–293. Berkeley, CA: University of California Press.

Lorini, E. and C. Castelfranchi (2007). The cognitive structure of surprise: Looking for basic principles. *Topoi 26*, 133–149.

Marx, M. (2007). Complexity of modal logic. In P. Blackburn, J. van Benthem, and F. Wolter (Eds.), *Handbook of Modal Logic*, Volume 3 of *Studies in Logic and Practical Reasoning*, pp. 139–179. Elsevier.

Meacham, C. J. G. (2010). Two mistakes regarding the principal principle. *British Journal for the Philosophy of Science 61*, 407–431.

Miller, D. (1966). A paradox of information. *British Journal for the Philosophy of Science 17*, 59–61.

Oaksford, M. and N. Chater (2008). *The Probabilistic Mind: Prospects for Bayesian Cognitive Science.* Oxford: Oxford University Press.

Pietarinen, A.-V. (2003). What do epistemic logic and cognitive science have to do with each other? *Cognitive Systems Research 4*, 169–190.

Plaza, J. (1989). Logics of public communications. In M. L. Emrich, M. S. Pfeifer, M. Hadzikadic, and Z. W. Ras (Eds.), *Proceedings of the Fourth International Symposium on Methodologies for Intelligent Systems: Poster Session Program*, Oak Ridge, TN, pp. 201–216. Oak Ridge National Laboratory. Reprinted in: Synthese 158, 165–179 (2007).

Romeijn, J.-W. (2012). Conditioning and interpretation shifts. *Studia Logica 100*, 583–606.

Rott, H. and P. Girard (2014). Belief revision and logic. In A. Baltag and S. Smets (Eds.), *Outstanding Contributions: Johan F. A. K. van Benthem on Logical and Informational Dynamics.* Dordrecht: Springer.

Sack, J. (2007). *Adding Temporal Logic to Dynamic Epistemic Logic.* Ph. D. thesis, Indiana University, Bloomington, IN.

Sack, J. (2009). Extending probabilistic dynamic epistemic logic. *Synthese 169*(2), 241–257.

de Vink, E. P. and J. Rutten (1999). Bisimulation for probabilistic transition systems: a coalgebraic approach. *Theoretical Computer Science 221*, 271–293.

Part II

Dynamics of Informational Attitudes

Chapter 5

Knowledge and Time

Clare Dixon, Cláudia Nalon and Ram Ramanujam

Contents

Abstract In this chapter we discuss the dynamic aspects of knowledge, which can be characterised by a combination of temporal and epistemic logics. Time is restricted to a linear order of discrete moments and knowledge is given, as usual, as equivalence classes over moments in time. We present the language and axiomatisation for such a combination, and discuss complexity and expressiveness issues, considering both the independently axiomatisable combinations and interacting ones.

Chapter 5 of the *Handbook of Epistemic Logic*, H. van Ditmarsch, J.Y. Halpern, W. van der Hoek and B. Kooi (eds), College Publications, 2015, pp. 205–259.

We also present two different proof methods – resolution and tableaux – for reasoning within the temporal logics of knowledge. Levels of knowledge and the relation between knowledge and communication in distributed protocols are also discussed and we provide an automata theoretic characterisation of the knowledge of finite state agents. A brief survey on applications is given and we conclude by discussing related work.

5.1 Introduction

Epistemic logics are useful for describing situations where we are interested in reasoning about the knowledge of agents or groups of agents. However, on their own, they only describe *static* aspects of reasoning, that is, those aspects related to a specific situation. In order to formalise aspects of complex systems, for instance the specification of concurrent/distributed systems, it is also desirable to describe how the knowledge of a group of agents evolves over time. One way to achieve this is to extend the epistemic language by adding a temporal component. The *dynamic* aspects of reasoning are often described by a combination of the multi-modal logic of knowledge (or belief) with a temporal logic. We concentrate on knowledge in this chapter. The dynamics of belief is discussed in the next chapters.

5.1.1 A temporal logic of knowledge

Varieties of temporal logic differ in their underlying language (propositional versus first-order) and in their underlying model of time (branching versus linear time, and dense versus discrete). Here, we concentrate on Propositional Temporal Logic (PTL), a discrete, linear, temporal logic with finite past and infinite future. This particular logic can be seen as a multi-modal language with two modalities: one to represent the 'next' moment in time; the other, 'until', to represent conditions over a sequence of time moments. The temporal operators supplied in the language of PTL operate over a sequence of distinct 'moments' in time. Each 'moment' actually corresponds to a state (a set of propositional symbols) over a discrete linear structure, for example the natural numbers $\mathbb{N}$. Here, only future-time operators are used. It is possible to include past-time operators in the definition of PTL, but such operators add no extra expressive power.

Combinations of other temporal logics with epistemic logics have also been considered. In particular, combinations of the branching-time temporal logics CTL and CTL* with epistemic logic have been studied. The logics CTL and CTL* use the temporal operators described in this chapter and in addition allow the operators $\mathbb{A}$ and $\mathbb{E}$ that quantify over paths. Additionally different flows of time have been considered for example using the reals or the rationals, or allowing infinite past.

The Temporal Logic of Knowledge we present here, denoted by $\mathsf{KL}_{(n)}$, is obtained by a *fusion* of the temporal logic PTL and the multi-modal epistemic logic $\mathsf{S5}_{(n)}$. The language of $\mathsf{KL}_{(n)}$ is presented in Section 5.2. An important characteristic of modal fusions, such as the one we now consider, is that when there are no axioms containing operators from both languages, the component logics remain *independent*, and so a decision procedure for the resulting logic can be obtained by taking the union of the decision procedures for its components and making sure that enough information is generated and passed to each part. In the case of $\mathsf{KL}_{(n)}$, if enough care is taken, the combined procedure inherits the complexity of the most expressive component, that is, the complexity of the satisfiability problem does not increase.

Contexts where we are particularly interested in how the knowledge of an agent evolves over time is a typical example where *interactions* are required. For instance, whether an agent may gain knowledge as time goes by or whether an agent forgets or not about her past. For *interacting logics*, besides the characterisation of both logics (given by their axiomatic systems) further axioms, including operators of both logics, are needed in order to model a specific situation. We discuss complexity issues and axiomatisations for both $\mathsf{KL}_{(n)}$ and interacting logics in Sections 5.3 and 5.5.

We present two procedures for reasoning within $\mathsf{KL}_{(n)}$: tableaux and resolution calculi are given in Section 5.6. For the combined logics of knowledge and time, these proof methods might seem, at first, rather simple, as they consist of the same inference rules for the logics considered alone. Nevertheless, in the resolution method, its apparent simplicity comes from the separation of the epistemic and temporal dimensions (via the normal form) and from making sure that all relevant information is made available to these different dimensions (through the propositional language, which is shared by all logics, via simplification rules). Separation provides, in this case, a very elegant way to deal with the combined logics. The tableau method presented here combines analytical cut rules (i.e. the introduction of subformulae of the formula being tested for satisfiability) with the usual inference rules for epistemic and temporal logics. By doing so, unnecessary constructs are avoided, providing an efficient method for the combined logics.

5.1.2 Structural investigations

For any logic, the first technical questions we ask of the logic relate to proof systems and decision procedures. Answering these questions gives us a basic understanding of the patterns of reasoning embodied in the logic, and structure in models. Going further, the next set of questions are on structure in formulae and how they relate to the underlying notions for which the logic is designed. For instance, in temporal logic, analysing modal structure leads to a topological characterisation of temporal properties. Further investigation delineates the expressiveness of not only the logic but also of its structural fragments. Again, for temporal logic, we have Kamp's theorem which shows that PTL is as expressive as the monadic first order logic on infinite sequences, and the until-hierarchy. Expressiveness of the temporal logic of knowledge is briefly discussed in Section 5.4.

In the context of the Temporal Logic of Knowledge, a natural question on structure relates to the nesting of knowledge operators. Note that not only are $K_a K_b$ and $K_b K_a$ distinct when $a \neq b$ but these are also different from $K_a K_b K_a$ and $K_b K_a K_b$. Thus we have an *unbounded* set of operators but they are *not unrelated*. What precisely is the structure of such sets, and the temporal evolution of such sets? Clearly the operator $K_a K_b K_a$ can be considered to be at *one higher level* of knowledge than $K_a K_b$. The notion of common knowledge sits on top of all these levels. Do these levels form a hierarchy? These are interesting questions, and in Section 5.8 we discuss some characterisations of this kind.

Note that a temporal progression from $K_a p \wedge \neg K_b K_a p$ to $K_b K_a p$ can be seen as learning on the part of b or as an implicit *communication* from a to b. Indeed, this is an issue of central concern in the study of epistemic protocols: given an initial and final (desired) specification of information states given by knowledge formulas, can we find a sequence of communications that can take us from the former to the latter? When we can, is there an optimal such sequence? Investigations of this nature are important not only for applications (as in the study of distributed computing) but also yield considerable insight into the class of models of the Temporal Logic of Knowledge, and we sketch some in this in Sections 5.7 and 5.9.

While a precise delineation of expressiveness is as yet elusive for the Temporal Logic of Knowledge, it is clear that a study of *memory* structure is essential for understanding expressiveness. As an agent accumulates information over time, representation of knowledge involves structuring memory appropriately. In temporal reasoning, this is illustrated by automata constructions, and a natural question in the temporal epistemic context is a characterisation of the knowledge of *finite state* agents. We sketch attempts at an automata theory of knowledge and time in Section 5.10.

We discuss applications for time and knowledge in Section 5.11. Related work and final remarks are presented in Section 5.13.

5.2 Language

When talking about time, we need syntactic mechanisms to refer to facts that occur at specific moments. The language of $\mathsf{KL}_{(n)}$ extends the epistemic logic $\mathsf{S5}_{(n)}$ with two new operators. The 'next' operator, denoted by '$\bigcirc$', refers to the next moment in time. The 'until' operator, denoted by '$\mathcal{U}$', is more powerful: it allows us to talk about facts that hold until a specific condition is met. For instance, consider a very simple protocol for exchanging messages, where the messages are guaranteed to be delivered instantly or at the next moment in time. Thus, if a message is sent by agent 1, she knows that either agent 2 immediately knows that the message has been sent or at the next moment agent 2 knows that the message has been sent. This situation can be formalised, for instance, as $p_1 \to K_1(K_2p_1 \vee \bigcirc K_2p_1)$, where p_1 represents the fact that agent 1 sends a message. In a protocol where the messages are not guaranteed to be delivered, agent 1 can keep on sending the same message until she knows that agent 2 has received it, which can be formalised in our language as $(p_1 \mathcal{U} K_1r_2)$, where p_1 is as before and r_2 means that agent 2 has received the message.

5.2.1 Syntax

Basically, the syntax of $\mathsf{KL}_{(n)}$ comprises the set of modal operators as in $\mathsf{S5}_{(n)}$ together with the set of temporal operators of PTL.

Let $\mathscr{P} = \{p, q, p', q', p_1, q_1, \dots\}$ be the set of propositional symbols (or atoms). The set of agents is defined as $\mathscr{A} = \{1, \dots, n\}$. The language of $\mathsf{KL}_{(n)}$ is given by the following BNF:

$$\varphi ::= \mathbf{true} \mid p \mid \neg\varphi \mid \varphi \wedge \varphi \mid K_a\varphi \mid \bigcirc\varphi \mid \varphi \mathcal{U} \psi$$

where $a \in \mathscr{A}$. The abbreviations for classical and modal operators are defined as usual. For temporal operators, we have the following abbreviations:

description/name	*definiendum*	*definiens*
sometime	$\Diamond\varphi$	$\mathbf{true}\,\mathcal{U}\,\varphi$
always	$\Box\varphi$	$\neg\Diamond\neg\varphi$
unless	$\varphi \mathcal{W} \psi$	$(\varphi \mathcal{U} \psi) \vee \Box\varphi$

We denote by $\mathsf{WFF}_{\mathsf{KL}_{(n)}}$ the set of well-formed formulae of $\mathsf{KL}_{(n)}$. As in the classical case, a *literal* is either a proposition or its negation. Also, K_al or $\neg K_al$ (where $a \in \mathscr{A}$ and l is a literal) are *modal literals*, and an *eventuality* is a formula in the form $\Diamond\varphi$ (where φ is in $\mathsf{WFF}_{\mathsf{KL}_{(n)}}$).

5.2.2 Semantics

Semantics of $\mathsf{KL}_{(n)}$ interpret formulae over a set of temporal lines, where each temporal line corresponds to a discrete, linear model of time with finite past and infinite future and the accessibility relations $\sim_a$ are restricted to be equivalence relations. A *timeline* t is an infinitely long, linear, discrete sequence of states, indexed by the natural numbers. Let $TLines$ be the set of all timelines. A *point* q is a pair $q = (t, u)$, where $t \in TLines$ is a timeline and $u \in \mathbb{N}$ is a temporal index to t. Let $Points$ be the set of all points.

Recall that a *state* is defined as a set of propositional symbols. In the temporal case, a state is also associated with an index that gives its position in a model for PTL. Here, a *point* is defined as a particular position in the model, that is, it refers to a particular temporal index in a particular timeline. In the following, as no confusion should arise, we often use the term *point* to refer to the set of propositional symbols whose valuation (as defined below) is true at that particular position.

Definition 5.1
A *Kripke structure* (or a *model*) for $\mathsf{KL}_{(n)}$ is a tuple $M = (TL, \sim_1, \ldots, \sim_n, V)$ where:

- $TL \subseteq TLines$ is a set of timelines with a distinguished timeline t_0;
- $\sim_a$, for all $a \in \mathscr{A}$, is the accessibility relation over points, i.e., $\sim_a \subseteq Points \times Points$ where each $\sim_a$ is an equivalence relation; and
- V is a valuation, that is, a function $V : (Points \times \mathscr{P}) \rightarrow \{true, false\}$. $\dashv$

Intuitively, the distinguished timeline t_0 represents the first timeline in the model. As given below, the satisfiability and validity of a formula are defined with respect to the first point of the first timeline. Given two points, (t, m) and (t', m'), if $((t, m), (t', m')) \in \sim_a$, then we say that the points are *indistinguishable to agent* a (written $(t, m) \sim_a (t', m')$).

Definition 5.2
Truth of a formula in a model M, at a particular point (t, u), is given as follows:

- $(M, (t, u)) \models$ **true**;
- $(M, (t, u)) \models p$ if, and only if, $V(t, u)(p) = true$, where $p \in \mathscr{P}$;
- $(M, (t, u)) \models \neg\varphi$ if, and only if, $(M, (t, u)) \not\models \varphi$;

- $(M, (t, u)) \models (\varphi \wedge \psi)$ if, and only if, $(M, (t, u)) \models \varphi$ and $(M, (t, u)) \models \psi$;
- $(M, (t, u)) \models \bigcirc\varphi$ if, and only if, $(M, (t, u + 1)) \models \varphi$;
- $(M, (t, u)) \models \varphi \mathcal{U} \psi$ if, and only if, $\exists u', u' \in \mathbb{N}, u' \geq u, (M, (t, u')) \models \psi$ and $\forall u'', u'' \in \mathbb{N}$, if $u \leq u'' < u'$, then $(M, (t, u'')) \models \varphi$;
- $(M, (t, u)) \models K_a\varphi$ if, and only if, for all (t', u') such that $(t, u) \sim_a (t', u')$, $(M, (t', u')) \models \varphi$, for $a \in \mathscr{A}$. ⊣

Let $M = (TL, \sim_1, \ldots, \sim_n, V)$ be a Kripke structure. A formula φ is *satisfiable* if, and only if, $(M, (t_0, 0)) \models \varphi$, for some M. A formula φ is *valid* (written $\models \varphi$), if $(M, (t_0, 0)) \models \varphi$, for every M. The Kripke structure given here is similar to that of *interpreted systems*, where the local states of agents and the local state of the environment are explicitly given.

The following example is an instance of the well-known muddy children puzzle. The common knowledge operator C_A, where $A \subseteq \mathscr{A}$, can be defined (see Chapter 1) as:

$$(M, (t, s)) \models C_A\varphi \quad \text{iff} \quad \text{for all } (t', s') \text{ such that } (t, s) \sim_{C_A} (t', s'), \ (M, (t', s')) \models \varphi$$

where φ is a formula and $\sim_{C_A}$ is given by $(\bigcup_{a \in A} \sim_a)^*$, the reflexive transitive closure of the equivalence relations for all agents in the group A.

Example 5.1
Two children are told by their mother that they could play outside the house, but that they should not get themselves muddy. While playing, some of the children get mud on their forehead. Each child can see if the others are muddy, but they cannot see their own foreheads. Obviously, if a child sees that another is muddy, she will keep quiet about it. Then the father comes and says that at least one of them has a muddy forehead. In our example, both children have a muddy forehead in reality. As most children, the ones in our example are completely honest and perfect, logically omniscient reasoners. The father asks repeatedly if the children know if they have a muddy forehead. It can be shown that, after the father asks for the first time, both children know they have a muddy forehead. The initial situation, which is true at the initial state, can be represented by the following formula:

$$1. \quad m_1 \wedge m_2$$

where m_a, $a = 1, 2$, represents the fact that child a is muddy. The next formulae (2-8) hold at every state of the model. Formulae 2 to 5 say that if a child gets muddy, she stays muddy; also, that if she is not muddy, she stays clean:

$$\begin{array}{ll} 2. & m_1 \leftrightarrow \Box m_1 \\ 3. & m_2 \leftrightarrow \Box m_2 \\ 4. & \neg m_1 \leftrightarrow \Box \neg m_1 \\ 5. & \neg m_2 \leftrightarrow \Box \neg m_2 \end{array}$$

Formulae 1, 2, and 3 imply that the initial situation will not change. The next formulae represent the knowledge each child has about the others, that is, if a child has a muddy forehead the other children know about it. For instance, formula 6 says that if the second child has a muddy forehead, then the first child knows that the second child has a muddy forehead (and vice versa).

$$\begin{array}{ll} 6. & m_2 \leftrightarrow K_1 m_2 \\ 7. & m_1 \leftrightarrow K_2 m_1 \end{array}$$

We note that before the father tells the children that at least one of them has a muddy forehead, the children do not know if their own forehead is muddy. Figure 5.1 shows how the knowledge of the children evolve over time. Circles represent states of a model we try to build, arrows represent the flow of time, and the remaining edges are labelled by the agents. Note that the figure does *not* show a model, as not all states have temporal successors and they cannot be part of a model. Also, note that edges denoting the accessibility relations $\sim_1$ or $\sim_2$ are symmetric and all states have reflexive edges which are not shown in the figure. At the initial time, $(t_0, 0)$, it is easy to see that although both children know that at least one of them has a muddy forehead, this fact is not common knowledge.

If, however, we introduce the following formula

$$8. \quad \bigcirc(m_1 \vee m_2)$$

which holds at all states and represents the statement given by the father that at least one child had a muddy forehead, then common knowledge of this fact is attained. In Figure 5.1, at state $(t_0, 1)$, we can check that the first (resp. second) child knows that the second (resp. first) child is muddy. However, the children do not know if their own forehead is muddy. After the father asks for the first time and both children say that they do not know about their own situation $(\neg K_1 m_1 \wedge \neg K_2 m_2)$, this also becomes common knowledge. Note that if $C_{\mathscr{A}}\varphi$ holds, then $K_a\varphi$ also holds, for all $a \in \mathscr{A}$. Thus, from $C_{\mathscr{A}}(\neg K_1 m_1 \wedge \neg K_2 m_2)$, we have that both $K_1 \neg K_2 m_2$ and $K_2 \neg K_1 m_1$ hold. The first child reasons that if the second child does not know she is muddy, it is because she can see that there is some mud on

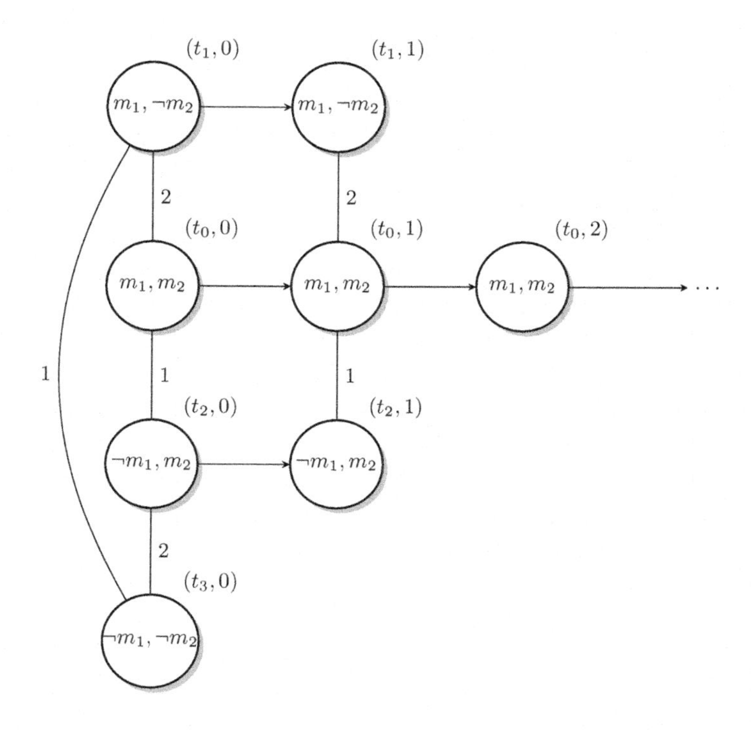

Figure 5.1: Trying to construct a model for the muddy children instance given in Example 5.1.

the first child's forehead. The second child reasons in a similar way. Both children can then infer that they are both muddy, as represented by state $(t_0, 2)$. ⊣

5.3 Complexity

The satisfiability problem for $\mathsf{KL}_{(n)}$ is PSPACE-complete, as both components, $\mathsf{S5}_{(n)}$ (where $n \geq 2$) and PTL, are PSPACE-complete. If $n = 1$, as the satisfiability problem for $\mathsf{S5}_{(1)}$ is NP-complete, the problem for the combined logic inherits the complexity of the temporal component. We note that these results apply to the combination given here, which corresponds to the fusion of the particular logics we have chosen to describe the epistemic and the dynamic components. If the underlying epistemic language allows for group operators, for instance, the complexity of the satisfiability problem might increase. The satisfiability problem is EXPTIME-complete, if we add the common knowledge operator to the language of $\mathsf{KL}_{(n)}$, for instance.

If instead of a linear temporal logic, we had chosen a branching-time logic, as CTL, the complexity would again increase: the satisfiability problem for the branching-time logic of knowledge is EXPTIME-complete.

Fusions of logics have quite good behaviour in the sense that some nice properties from the independent logics, as, for instance, *finite model property* and *decidability*, transfer to the combined logic. Lower bound results also transfer as, obviously, deciding the satisfiability of a formula in the combined logic is as hard as deciding this in the component logics. Upper bound results, however, do not transfer immediately. For instance, $\mathsf{S5}_{(n)}$ is PSPACE-complete for $n > 1$ and is the fusion of several $\mathsf{S5}_{(1)}$, which is NP-complete. It is an open question under which conditions the complexity results transfer to a fusion of logics. See Section 5.13 for references.

If the combined logic also allows for *interactions*, then the complexity of the resulting logic usually increases. Interactions are often described by axioms which contain operators of both component logics. Several properties expressed by interacting axioms between knowledge and time have been discussed in the literature. *No learning* says that an agent will not increase her knowledge over time. *Perfect recall* (also known as *no forgetting*) says that an agent keeps memory of her knowledge. *Synchrony* says that agents have access to a global clock, which means that they know the time. The agents are said to have a *unique initial state* if the knowledge they have at the beginning of time is the same, independent of the timeline.

Properties of these systems are given in Table 5.1, where $\sim_a$ is the epistemic accessibility relation of agent a; and s and t are timelines. Examples of models that have those properties are shown in Figures 5.2–5.5. In these figures each dot represents a point. Timelines are shown as horizontal lines with the time increasing from left to right. The grey shaded areas represent the epistemic accessibility relation for an agent a. Any two points (s, m) and (t, n) within some grey region satisfy $(s, m) \sim_a (t, n)$. The shape of the grey regions illustrates the conditions on the epistemic accessibility relation. For example in Figure 5.3 all the grey regions relate to some moment in time and do not cover several moments in time. This illustrates the condition for synchrony stating that two points related by the epistemic accessibility relation must be at the same moment in time.

Interacting systems of knowledge and time allow us to consider particular aspects of how knowledge evolves over time. These properties can be combined, for example, perfect recall with synchrony means that agents can distinguish between the same timelines as time goes on. In particular, logical characterisations of distributed systems are often based on the fact that parts of a protocol are taken in turns by the agents, which requires some mechanism for synchronisation, and that those agents do not forget what happened in previous turns. In this case, when synchrony and perfect

Property (abbreviation)	*Semantic Characterisation*
Unique Initial State (*uis*)	$(s, 0) \sim_a (t, 0)$, for all $a \in \mathscr{A}$.
Synchrony (*sync*)	for all points (s, m) and (t, n), if $(s, m) \sim_a (t, n)$ then $m = n$.
Perfect Recall (*pr*)	for all points (s, m) and (t, n), if $m > 0$ and $(s, m) \sim_a (t, n)$, either $(s, m-1) \sim_a (t, n)$ or there exists $l < n$ such that $(s, m-1) \sim_a (t, l)$ and for all k, $l < k \leq n$, then $(s, m) \sim_a (t, k)$.
No Learning (*nl*)	for all points (s, m) and (t, n), if $(s, m) \sim_a (t, n)$, either $(s, m+1) \sim_a (t, n)$ or there exists $l > n$ such that $(s, m+1) \sim_a (t, l)$ and for all k, $l > k \geq n$, then $(s, m) \sim_a (t, k)$.

Table 5.1: Properties of Interacting Systems.

recall is assumed, the knowledge of the agents does not decrease over time, that is, they might distinguish between more situations as time goes on. In blindfold games, the opposite is expected: agents can distinguish between less situations as time goes on. For these games, synchrony and no learning are assumed.

Example 5.2
The muddy children puzzle, shown in Example 5.1, can be viewed as a system with *synchrony* and *perfect recall*. In the general case, where there are n children with muddy foreheads, it can be shown that, after the father asks $n-1$ times, the children that have muddy foreheads will say they know about being muddy. The children answer simultaneously, meaning that their knowledge of time is implicit in the characterisation of the puzzle. The reason why they achieve the right answer in the $n-1$-th round is that they can remember everything that happened in the previous rounds. Synchrony and perfect recall is characterised by the axiom $K_a \bigcirc \varphi \to \bigcirc K_a \varphi$, where $a \in \mathscr{A}$ and φ is a formula. Note that if two points are in the same equivalence class for an agent a, then they share the same time index, that is, if $(t, m) \sim_a (t', m')$, then $m = m'$. Also, in systems with synchrony and perfect recall, if $(t, m) \sim_a (t', m)$, then their predecessors are also in the same equivalence class, i.e. $(t, m-1) \sim_a (t', m-1)$, for all $m > 0$. $\dashv$

In the following, we use the abbreviations given in Table 5.1 as superscripts in the name of the logic to talk about a temporal logic of knowledge

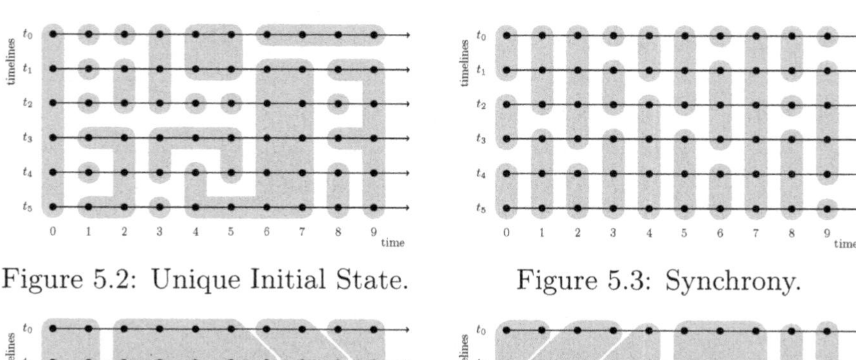

Figure 5.2: Unique Initial State.

Figure 5.3: Synchrony.

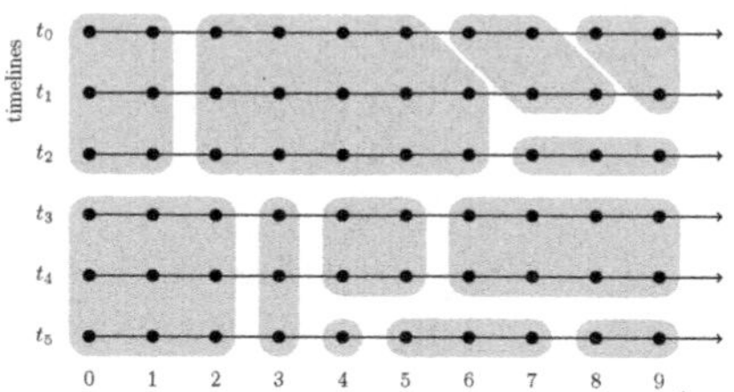

Figure 5.4: Perfect Recall.

Figure 5.5: No Learning.

where a particular property (or their combination) holds. For instance, $\mathsf{KL}^{nl,uis}_{(n)}$ is the temporal logic of knowledge with no learning and a unique initial state. Also, we use calligraphic letters to denote the class of models of a particular logic. For instance $\mathcal{KL}^{nl,uis}_{(n)}$ is the class of models for $\mathsf{KL}^{nl,uis}_{(n)}$.

Theorem 5.1 (Satisfiability)
The complexity of the satisfiability problem for the class of models $\mathcal{X}$ is

1. PSPACE-complete if $\mathcal{X} \in \{\mathcal{KL}^{sync}_{(n)}, \mathcal{KL}^{sync,uis}_{(n)}, \mathcal{KL}^{uis}_{(n)}\}$ and $n \geq 1$.

2. EXPSPACE-complete if

   (a) $\mathcal{X} \in \{\mathcal{KL}^{nl,sync,uis}_{(n)}, \mathcal{KL}^{nl,pr,sync,uis}_{(n)}\}$ and $n \geq 1$;

   (b) $\mathcal{X} \in \{\mathcal{KL}^{nl}_{(1)}, \mathcal{KL}^{nl,pr}_{(1)}, \mathcal{KL}^{nl,pr,sync}_{(1)}, \mathcal{KL}^{nl,sync}_{(1)}, \mathcal{KL}^{nl,pr,uis}_{(1)}, \mathcal{KL}^{nl,uis}_{(1)}\}$.

3. 2EXPTIME-complete if $\mathcal{X} \in \{\mathcal{KL}^{pr}_{(1)}, \mathcal{KL}^{pr,sync}_{(1)}, \mathcal{KL}^{pr,uis}_{(1)}, \mathcal{KL}^{pr,sync,uis}_{(1)}\}$.

4. NON-ELEMENTARY TIME-complete if
   $\mathcal{X} \in \{\mathcal{KL}^{pr}_{(n)}, \mathcal{KL}^{pr,sync}_{(n)}, \mathcal{KL}^{pr,uis}_{(n)}, \mathcal{KL}^{pr,sync,uis}_{(n)}\}$ and $n > 1$.

5. NON-ELEMENTARY SPACE-complete if
   $\mathcal{X} \in \{\mathcal{KL}^{nl}_{(n)}, \mathcal{KL}^{nl,pr}_{(n)}, \mathcal{KL}^{nl,pr,sync}_{(n)}, \mathcal{KL}^{nl,sync}_{(n)}\}$ and $n > 1$.

6. RE-complete if $\mathcal{X} \in \{\mathcal{KL}^{nl,uis}_{(n)}\}$ and $n > 1$.

7. Σ_1^1-complete if $\mathcal{X} \in \{\mathcal{KL}_{(n)}^{nl,pr,uis}\}$ and $n > 1$. ⊣

Note that systems with RE or Σ_1^1 complexity are undecidable. It can be shown, for instance, that the language of $\mathsf{KL}_{(n)}^{nl,pr,uis}$, $n > 1$, is expressive enough to encode the halting problem. The language of $\mathsf{KL}_{(n)}^{nl,uis}$, $n > 1$, can be used to define a formula, for every deterministic Turing machine A, that encodes an infinite computation of A.

While the satisfiability problem for the logics of knowledge and time is hard, as evidenced above, often we are interested in the following model checking problem. Suppose that we are given a finite model M and a formula φ, does $M \models \varphi$? The system we are interested in modelling is presented as M, and we wish to check the truth of φ at the initial state of the system.

The model checking problem for linear time temporal logic is already PSPACE-complete, so it is not surprising that the problem continues to be PSPACE-hard for the temporal logic of knowledge as well. A matching upper bound can also be given, in a manner similar to the tableaux construction to be given in Section 5.6.

Model checking for the subclasses presented is not as clear, since the conditions to be imposed on system transitions can be complex. While some results can be found in the literature, a systematic unifying study is still awaited. Model checking for epistemic and temporal logic is further discussed in Chapter 8 of this volume.

5.4 Expressiveness

The foregoing discussion shows that reasoning about knowledge and time entails significant complexity. A natural question then is to delineate the expressiveness of these logics. Unfortunately, we have few satisfactory answers to this question as yet.

5.4.1 First order fragments

Propositional modal logics are typically fragments of first order logics. A classic theorem of Johan van Benthem characterises propositional modal logic as the bisimulation invariant fragment of monadic first order logic over one binary relation. Over discrete linear orders, Kamp's theorem shows that propositional temporal logic has exactly the same expressiveness as first order logic, and this allows a characterisation of temporal logic's models as the star-free fragment of regular languages.

The expressiveness of fragments of temporal logic is also interesting. It can be shown that the next modality cannot be eliminated, that the until modality is strictly more expressive than the eventuality modality. Levels of nesting the until modality give rise to a hierarchy of sublogics within temporal logic, and these languages admit algebraic characterisations as well.

Compared to the richness of this theory of temporal logics, the theory of epistemic temporal logics is much less developed. The semantics presented in Section 5.2 is clearly first order. Consider the following monadic first order logic with $n+1$ binary relations. Fix a countable supply of monadic predicate symbols, $P_0, P_1, \ldots$ and of variables $x_0, x_1, \ldots$. The syntax of the logic is then given by the following scheme:

$$P_j(x) \mid x = y \mid x < y \mid x \sim_a y, a \in \mathscr{A} \mid \neg\alpha \mid \alpha \vee \beta \mid \exists x.\alpha$$

The semantics of the logic over models $M = (TL, \sim_1, \ldots, \sim_n, V, \pi)$ should be clear, where V maps each P_j to a subset of M-points and π each variable to an M-point. We define $(t, u) < (t', u')$ when $t = t'$ and $u < u'$. It is then easy to see that there is a uniform translation of every formula in the temporal logic of knowledge φ to a formula $\tau(\varphi)(x)$ with one free variable x such that $(M, (t, u)) \models \varphi$ iff $(M, \pi') \models \tau(\varphi)$, where π' is the same as π except that $\pi'(x) = (t, u)$.

Thus we can interpret the temporal logic of knowledge in an easily defined first order logic of order and equivalence, but it is clearly less expressive than the latter. This follows from the fact that satisfiability of the former is decidable whereas the latter is an undecidable logic. (To see this, let $n = 2$ and code up the tiling problem in the logic.) So we have a fragment, but which fragment is it?

While there are many interesting results on first order logics over linear orders, over equivalence relations and on linear orders equipped with equivalence relations, the expressiveness of first order logics over collections of linear orders equipped with equivalence relations on points across them seems to be much less studied.

Moreover expressiveness of structural fragments of the temporal logic of knowledge is also unclear. Can we show that the until modality cannot be expressed using the eventuality and equivalence modalities? Can one show that a logic with $k+1$ equivalence modalities is strictly more expressive than one with k equivalence modalities, for $k > 2$? And so on. Interestingly, over discrete linear orders, the temporal logic with past and future modalities is no more expressive than its future fragment. In the presence of epistemic modalities, it is no longer clear whether the past can be eliminated.

We merely remark here that the deeper exploration of expressiveness results may give rise to newer epistemic and temporal modalities, with interesting properties and new applications.

5.4.2 Quantified extensions

It is easy to define a quantified extension of temporal logic with unary predicates, with models as discrete linear orders. The extended logic is undecidable and simple variants are not finitely axiomatisable. Therefore, sublogics have been proposed: one of them works with quantification over propositions and another restricts quantification so that only one free variable is ever in the scope of an eventuality (the so-called *monodic* fragment). A very nice theory of such quantified temporal logics has been developed in recent times.

Knowledge extensions of these quantified temporal logics, along the same lines, of quantification over propositions and of monodic quantification, have fortunately been successful. While the details are tricky, it is reassuring that the overall research programme carries over, albeit at the price of considerable complexity (which is only to be expected). However, delineating the expressiveness of these logics is not easy. A warning sign exists in this regard: Kit Fine showed that the basic modal logic of propositional quantification is as expressive as second order arithmetic, so constraining the expressiveness of these logics is interesting and challenging. One solution is that of *local* quantification but there is much room for exploration in this terrain.

5.5 Axiomatisation

The axiomatisation of $\mathsf{KL}_{(n)}$, denoted by $\mathbf{AX}_{\mathsf{KL}_{(n)}}$, is given by the union of axiomatic systems for $\mathsf{S5}_{(n)}$ and PTL:

1	all classical tautologies
K	$\vdash K_a(\varphi \rightarrow \psi) \rightarrow (K_a\varphi \rightarrow K_a\psi)$
T	$\vdash K_a\varphi \rightarrow \varphi$
D	$\vdash \neg K_a \neg \mathbf{true}$
4	$\vdash K_a\varphi \rightarrow K_aK_a\varphi$
5	$\vdash \neg K_a\varphi \rightarrow K_a\neg K_a\varphi$
$\mathbf{K}_\Box$	$\vdash \Box(\varphi \rightarrow \psi) \rightarrow (\Box\varphi \rightarrow \Box\psi)$
$\mathbf{K}_\bigcirc$	$\vdash \bigcirc(\varphi \rightarrow \psi) \rightarrow (\bigcirc\varphi \rightarrow \bigcirc\psi)$
Fun	$\vdash \bigcirc\neg\varphi \leftrightarrow \neg\bigcirc\varphi$
Mix	$\vdash \Box\varphi \rightarrow (\varphi \wedge \bigcirc\Box\varphi)$
Ind	$\vdash \Box(\varphi \rightarrow \bigcirc\varphi) \rightarrow (\varphi \rightarrow \Box\varphi)$
$\mathcal{U}_1$	$\vdash (\varphi\mathcal{U}\psi) \rightarrow \Diamond\psi$
$\mathcal{U}_2$	$\vdash (\varphi\mathcal{U}\psi) \leftrightarrow (\psi \vee (\varphi \wedge \bigcirc(\varphi\mathcal{U}\psi)))$.

The inference rules are *modus ponens* (from φ and $\varphi \rightarrow \psi$ infer ψ), *necessitation* (from φ infer $K_a\varphi$, for all a in $\mathscr{A}$), and *generalisation* (from φ infer $\Box\varphi$).

The axiomatic system for knowledge was presented in Chapter 1 and we will skip the discussion here, but note that there are alternative axiomatic systems for $\mathsf{S5}_{(n)}$. It can be shown, for instance, that the axiomatic system comprising classical tautologies, axioms **K**, **T**, and **5**, together with *modus ponens* and *necessitation* is a complete and sound axiomatisation for the multi-modal logics of knowledge. The rule *generalisation* is just an instance of *necessitation* applied to the temporal component of the language. The axioms $\mathbf{K}_\Box$ and $\mathbf{K}_\bigcirc$ correspond to the distribution axiom for the temporal part. **Fun** characterises the interpretation of the 'next' operator by a total function. The axioms **Mix** and **Ind** correspond to the interpretation of 'always' by the reflexive transitive closure of the interpretation of $\bigcirc$. The axiom **Ind**, as its name suggests, expresses the induction principle: if something is always true, then it is true now and it is always true in the future. The axiom $\mathcal{U}_1$ says that if $\varphi \mathcal{U} \psi$ occurs, then ψ will eventually occur. The axiom $\mathcal{U}_2$ shows the separation in terms of present formulae and future formulae in the interpretation of 'until'.

Example 5.3
We show a proof for an instance of $\mathbf{4}_\Box$, given by $\Box\varphi \to \Box\Box\varphi$, which illustrates the use of temporal induction.

1.	$\Box\varphi \to (\varphi \wedge \bigcirc\Box\varphi)$	**Mix**
2.	$\Box\varphi \to \bigcirc\Box\varphi$	Classical Reasoning, 1
3.	$\Box(\Box\varphi \to \bigcirc\Box\varphi)$	generalisation, 2
4.	$\Box(\Box\varphi \to \bigcirc\Box\varphi) \to (\Box\varphi \to \Box\Box\varphi)$	**ind**
5.	$\Box\varphi \to \Box\Box\varphi$	modus ponens, 3, 4

$\dashv$

Most of the interacting logics given in Section 5.3 also have sound and complete axiomatisations. The exceptions are $\mathsf{KL}_{(n)}^{nl,uis}$ and $\mathsf{KL}_{(n)}^{nl,pr,uis}$, where $n > 1$. Before giving the results concerning the possible axiomatisations, we introduce the following schemata:

$$\begin{array}{ll}
\mathbf{KT2}: & \vdash K_a \bigcirc \varphi \to \bigcirc K_a \varphi \\
\mathbf{KT3}: & \vdash K_a\varphi_1 \wedge \bigcirc(K_a\varphi_2 \wedge \neg K_a\varphi_3) \to \neg K_a \neg (K_a\varphi_1 \mathcal{U} (K_a\varphi_2 \mathcal{U} \neg\varphi_3)) \\
\mathbf{KT4}: & \vdash (K_a\varphi_1 \mathcal{U} K_a\varphi_2) \to K_a(K_a\varphi_1 \mathcal{U} K_a\varphi_2) \\
\mathbf{KT5}: & \vdash \bigcirc K_a\varphi \to K_a \bigcirc \varphi \\
\mathbf{KT6}: & \vdash K_a\varphi \leftrightarrow K_1\varphi
\end{array}$$

The following theorem presents the sound and complete axiomatisations for the combined logics we have seen in this chapter. Note that there are no extra axioms for $\mathsf{KL}_{(n)}^{sync}$, $\mathsf{KL}_{(n)}^{uis}$, and $\mathsf{KL}_{(n)}^{sync,uis}$, as the set of formulae valid in these systems is the same as the set of formulae valid in $\mathsf{KL}_{(n)}$. Also, when the system is synchronous and has a unique initial state together with no learning, the formula $K_a\varphi \rightarrow K_b\varphi$, for all agents $a, b \in \mathscr{A}$, is valid, which means that not only agents consider the same set of initial states, but they consider the same set of states possible at all times. This results in reduction to the single-agent case, as shown by axiom **KT6**. If **AX** is an axiomatic system and $\mathcal{C}$ is a class of models, we say that **AX** characterises $\mathcal{C}$, if **AX** is a complete and sound axiomatisation with respect to each $\mathcal{C}$.

Theorem 5.2

- $\mathbf{AX}_{\mathsf{KL}_{(n)}}$ characterises $\mathcal{KL}_{(n)}$, $\mathcal{KL}_{(n)}^{sync}$, $\mathcal{KL}_{(n)}^{sync,uis}$, $\mathcal{KL}_{(n)}^{uis}$.
- $\mathbf{AX}_{\mathsf{KL}_{(n)}}$ + **KT2** characterises $\mathcal{KL}_{(n)}^{pr,sync}$, $\mathcal{KL}_{(n)}^{pr,sync,uis}$.
- $\mathbf{AX}_{\mathsf{KL}_{(n)}}$ + **KT3** characterises $\mathcal{KL}_{(n)}^{pr}$, $\mathcal{KL}_{(n)}^{pr,uis}$.
- $\mathbf{AX}_{\mathsf{KL}_{(n)}}$ + **KT4** characterises $\mathcal{KL}_{(n)}^{nl}$.
- $\mathbf{AX}_{\mathsf{KL}_{(n)}}$ + **KT5** characterises $\mathcal{KL}_{(n)}^{nl,sync}$.
- $\mathbf{AX}_{\mathsf{KL}_{(n)}}$ + **KT2** + **KT5** characterises $\mathcal{KL}_{(n)}^{nl,pr,sync}$.
- $\mathbf{AX}_{\mathsf{KL}_{(n)}}$ + **KT2** + **KT5** + **KT6** characterises $\mathcal{KL}_{(n)}^{nl,sync,uis}$, $\mathcal{KL}_{(n)}^{nl,pr,sync,uis}$.
- $\mathbf{AX}_{\mathsf{KL}_{(n)}}$ + **KT3** + **KT4** characterises $\mathcal{KL}_{(n)}^{nl,pr}$, $\mathcal{KL}_{(1)}^{nl,pr,uis}$, $\mathcal{KL}_{(1)}^{nl,uis}$. $\dashv$

5.6 Reasoning

In general, given a set of formulae Γ and a formula φ, the reasoning problem is concerned with answering the following question: is every model of Γ also a model for φ (i.e. $\Gamma \models \varphi$)? As discussed in Chapter 1, this question can be related to the *derivation* of φ from Γ ($\Gamma \vdash \varphi$): if an inference system is sound and strongly complete with respect to a class of models, then we have that $\Gamma \models \varphi$ if and only if $\Gamma \vdash \varphi$; therefore, finding a proof for φ from Γ also provides an answer to the satisfiability problem (when Γ is consistent). Axiomatic systems, such as the ones we have seen in Section 5.5, are formal tools to answer such a question. However, even when they are complete and sound with respect to a class of models, they are not easily implementable, nor provide efficient means for automatically finding proofs.

Efficient computational tools for reasoning usually involve specialised procedures for either *model checking* or *theorem proving*. Given a model and a formula, model checking consists in answering the question whether such formula is satisfied in the given model. Broadly speaking, theorem proving tools answer the question whether a given formula can be *derived*, by applications of inference rules, from a set of formulae. When this set of formulae is empty, a formula for which there is a proof is called a *theorem*. *Theorem proving tools* are often associated with fully automated methods, which implement either forward or backward search for a proof of the validity of a given formula. Popular theorem proving methods include natural deduction, axiomatic Hilbert's systems, Gentzen sequent systems, tableaux systems, and resolution. Here, we concentrate on resolution and tableaux for the combined logics of time and knowledge. Model checking procedures for $\mathsf{KL}_{(n)}$ are discussed in Chapter 8.

Resolution and tableaux are both refutational procedures and work in a backward way. A backward search starts with the formula we want to prove and the method is applied until an axiom is found. Recall that given a set of formulae $\Gamma = \{\gamma_0, \ldots, \gamma_n\}$, $n \in \mathbb{N}$, and a formula φ, we have that $\gamma_0 \wedge \ldots \wedge \gamma_n \rightarrow \varphi$ is valid if and only if its negation $\gamma_0 \wedge \ldots \wedge \gamma_n \wedge \neg\varphi$ is unsatisfiable. Thus, for refutational methods that work in a backward way, the input is the negation of a formula we want to prove and, instead of an axiom, the method is applied until either a contradiction is found or no inferences rules can be further applied. If a contradiction is found, the negated formula is unsatisfiable and the original formula is valid. Satisfiability can be tested by applying those methods to the original formula.

Resolution methods fall into two classes: clausal and non-clausal. In the clausal resolution method, the negation of the formula is firstly translated into a normal form before the resolution rules can be applied. As formulae are in a specific form, usually there are few inference rules in clausal resolution methods. For propositional logic, for instance, there is just one inference rule and this is applied to the set of clauses (disjunctions of literals). The purpose of the non-clausal resolution method is to extend the resolution proof to formulae in arbitrary (not just clausal) form. Nevertheless, the number of inference rules to be applied often increases and makes the non-clausal method more difficult to implement. Here, we concentrate on clausal resolution methods.

In a tableaux system, the negation of the original formula is replaced by a set (or sets) of its subformulae, generated by applications of a set of inference rules. The method is applied until all the formulae in the sets are atomic formulae. If all generated sets are unsatisfiable, that is, they contain a formula and its negation, then the original formula is a theorem.

The theorem proving methods for $\mathsf{KL}_{(n)}$ we present here essentially combine the proof methods for the component logics, making sure that enough information is passed for each component of the calculus via the propositional language that the epistemic and temporal languages share.

5.6.1 Resolution

The resolution-based proof method for $\mathsf{KL}_{(n)}$ combines the inference rules for temporal and multi-modal knowledge logics when considered alone. A formula is first translated into a normal form, called *Separated Normal Form for Logics of Knowledge* (SNF_K). A nullary connective, **start**, which represents the beginning of time, is introduced. Formally, $(M, (t, u)) \models$ **start** if, and only if, $t = t_0$ and $u = 0$, where M is a model and (t, u) is a point. Formulae are represented by a conjunction of clauses, $\bigwedge_i A_i$, where A_i is a clause in one of the following forms, l, l_a, k_a are literals, and m_{a_j} are literals or modal literals in the form $K_a l$ or $\neg K_a l$:

$$\textbf{Initial clause:} \quad \textbf{start} \rightarrow \bigvee_{b=1}^{r} l_b$$

$$\textbf{Sometime clause:} \quad \bigwedge_{a=1}^{g} k_a \rightarrow \Diamond\, l$$

$$\textbf{Step clause:} \quad \bigwedge_{a=1}^{g} k_a \rightarrow \bigcirc \bigvee_{b=1}^{r} l_b$$

$$K_a\textbf{-clause:} \quad \textbf{true} \rightarrow \bigvee_{b=1}^{r} m_{a_b}$$

$$\textbf{Literal clause:} \quad \textbf{true} \rightarrow \bigvee_{b=1}^{r} l_b$$

The transformation of any given formula into its separated normal form is satisfiability preserving and depends on three main operations: the renaming of complex subformulae; the removal of temporal operators; and classical style rewrite operations. The resulting set of clauses is linear on the size of the original formula. The transformation ensures that clauses hold everywhere in a model. Once a formula has been transformed into SNF_K, the resolution method can be applied. The method consists of two main procedures: the first performs initial, modal and step resolution; the second performs temporal resolution. Each procedure is performed until a contradiction (either **true** $\rightarrow$ **false** or **start** $\rightarrow$ **false**) is generated or no new clauses can be generated. In the following l is a literal; m_a are literals or modal literals in the form of either $K_a l$ or $\neg K_a l$; D, D' are disjunctions of literals; M, M' are disjunction of literals or modal literals; and C, C' are conjunctions of literals.

Initial Resolution is applied to clauses that hold at the beginning of time:

$$[\text{IRES1}] \quad \frac{\begin{array}{rcl}\mathbf{true} & \rightarrow & (D \lor l) \\ \mathbf{start} & \rightarrow & (D' \lor \neg l)\end{array}}{\mathbf{start} \rightarrow (D \lor D')} \qquad [\text{IRES2}] \quad \frac{\begin{array}{rcl}\mathbf{start} & \rightarrow & (D \lor l) \\ \mathbf{start} & \rightarrow & (D' \lor \neg l)\end{array}}{\mathbf{start} \rightarrow (D \lor D')}$$

Modal Resolution is applied between clauses referring to the same agent (i.e. two K_a-clauses; a literal and a K_a-clause; or two literal clauses):

$$[\text{MRES1}] \quad \frac{\begin{array}{l}\mathbf{true} \rightarrow (M \lor m_a) \\ \mathbf{true} \rightarrow (M' \lor \neg m_a)\end{array}}{\mathbf{true} \rightarrow (M \lor M')} \qquad [\text{MRES2}] \quad \frac{\begin{array}{l}\mathbf{true} \rightarrow (M \lor K_a l) \\ \mathbf{true} \rightarrow (M' \lor K_a \neg l)\end{array}}{\mathbf{true} \rightarrow (M \lor M')}$$

$$[\text{MRES3}] \quad \frac{\begin{array}{l}\mathbf{true} \rightarrow (M \lor K_a l) \\ \mathbf{true} \rightarrow (M' \lor \neg l)\end{array}}{\mathbf{true} \rightarrow (M \lor M')} \qquad [\text{MRES4}] \quad \frac{\begin{array}{l}\mathbf{true} \rightarrow (M \lor \neg K_a l) \\ \mathbf{true} \rightarrow (M' \lor l)\end{array}}{\mathbf{true} \rightarrow (M \lor mod_a(M'))}$$

$$[\text{MRES5}] \quad \frac{\mathbf{true} \rightarrow (M \lor K_a l)}{\mathbf{true} \rightarrow (M \lor l)}$$

where $mod_a(A \lor B) = mod_a(A) \lor mod_a(B)$, $mod_a(K_a l) = K_a l$, $mod_a(\neg K_a l)$ $= \neg K_a l$, and $mod_a(l) = \neg K_a \neg l$.

MRES1 corresponds to classical resolution. MRES2 is justified by the axiom **D**. The rules MRES3 and MRES5 are justified by the axiom **T**. Note that MRES2 and MRES3 are not needed for completeness, as they can both be simulated by applications of MRES5 and MRES1. As mentioned before, clauses are true at all states, therefore from a literal clause $\mathbf{true} \rightarrow l_1 \lor l_2 \lor \ldots \lor l_n$, for some $n \in \mathbb{N}$, we can infer $K_a(\mathbf{true} \rightarrow l_1 \lor l_2 \lor \ldots \lor l_n)$ and, from that, $\mathbf{true} \rightarrow \neg K_a \neg l_1 \lor \neg K_a \neg l_2 \lor \ldots \lor K_a l_n$. MRES4 is, then, applied between the clauses containing the complementary modal literals $\neg K_a l$ from the first premise and $K_a l$ from the transformation of the second premise. The function mod_a performs this transformation and it is justified by **K** (for distributing the knowledge operator over M'), **4**, and **5** (for modal simplification) to generate the clausal form of $\neg K_a \neg M'$. For instance, if MRES4 is applied between $\mathbf{true} \rightarrow t_1 \lor \neg K_1 m_1$ and $\mathbf{true} \rightarrow m_1 \lor m_2 \lor m_3$, then the resolvent is $\mathbf{true} \rightarrow t_1 \lor mod_1(m_2 \lor m_3)$, that is, $\mathbf{true} \rightarrow t_1 \lor \neg K_1 \neg m_2 \lor \neg K_1 \neg m_3$.

Step Resolution is applied to clauses that hold at the same moment in time:

$$[\text{SRES1}] \quad \frac{\begin{array}{rcl}C & \rightarrow & \bigcirc(D \lor l) \\ C' & \rightarrow & \bigcirc(D' \lor \neg l)\end{array}}{(C \land C') \rightarrow \bigcirc(D \lor D')} \qquad [\text{SRES2}] \quad \frac{\begin{array}{rcl}\mathbf{true} & \rightarrow & (D \lor l) \\ C & \rightarrow & \bigcirc(D' \lor \neg l)\end{array}}{C \rightarrow \bigcirc(D \lor D')}$$

together with the following simplification rule:

$$[\text{SIMP1}] \quad \frac{C \;\rightarrow\; \bigcirc\, \mathbf{false}}{\mathbf{true} \;\rightarrow\; \neg C}$$

Temporal Resolution is applied between an eventuality $\Diamond l$ and a *set of clauses* which forces l always to be false. In detail, the temporal resolution rule is (where A_j is a conjunction of literals, B_j is a disjunction of literals, and C and l are as above):

$$[\text{TRES}] \quad \frac{\begin{array}{c} A_0 \rightarrow \bigcirc B_0 \\ \vdots \\ A_n \rightarrow \bigcirc B_n \\ C \rightarrow \Diamond l \end{array}}{C \rightarrow \left(\bigwedge_{i=0}^{n} (\neg A_i) \right) \mathcal{W} l} \qquad \begin{array}{l} \text{where:} \\ \\ \forall i, 0 \leq i \leq n, \vdash B_i \rightarrow \neg l \\ \forall i, 0 \leq i \leq n, \vdash B_i \rightarrow \bigvee_{j=0}^{n} A_j \end{array}$$

The set of clauses that satisfy the side conditions are together known as *a loop in* $\neg l$. We note that each $A_j \rightarrow \bigcirc B_j$ are step clauses in *merged* SNF_K, that is, they correspond to a conjunction of step and literal clauses in SNF_K. A translation of the resolvent into the normal form is given by the following clauses (where t is a new proposition): $\mathbf{true} \rightarrow (\neg C \vee \neg A_i \vee l)$, $t \rightarrow \bigcirc(\neg A_i \vee l)$, $\mathbf{true} \rightarrow (\neg C \vee t \vee l)$, and $t \rightarrow \bigcirc(t \vee l)$. Clauses are kept in their simplest form by performing classical style simplification, that is, literals in conjunctions and/or disjunctions are always pairwise different; constants **true** and **false** are removed from conjunctions and disjunctions with more than one conjunct/disjunct, respectively; conjunctions (resp. disjunctions) with either complementary literals or **false** (resp. **true**) are simplified to **false** (resp. **true**). Classical subsumption is also applied, that is, if $\models C \rightarrow C'$, for clauses C, C', and the set of clauses contains both C and C', then the latter is removed from the set of clauses. Also, valid formulae can be removed during simplification as they cannot contribute to the generation of a contradiction. This method is sound, complete, and terminating.

Example 5.4

We consider the instance of the muddy children puzzle given in Example 5.1. Firstly, we show the clausal representation of that problem. Then, we show a proof that the children know they both have a muddy forehead after the father asks them for the first time. As we have two muddy children in our example, the next three clauses define three moments in time: t_0, before the father says that there is at least one child muddy; t_1, when the father says so; and t_2, when the father asks the children whether they know that they have a muddy forehead:

$$\begin{array}{lrcl} 1. & \textbf{start} & \rightarrow & t_0 \\ 2. & t_0 & \rightarrow & \bigcirc t_1 \\ 3. & t_1 & \rightarrow & \bigcirc t_2 \end{array}$$

The initial situation is given by the next two clauses, where both children have a muddy forehead:

$$\begin{array}{lrcl} 4. & \textbf{start} & \rightarrow & m_1 \\ 5. & \textbf{start} & \rightarrow & m_2 \end{array}$$

The initial situation does not change, i.e. if a child is muddy, then she will remain muddy.

$$\begin{array}{lrcl} 6. & m_1 & \rightarrow & \bigcirc m_1 \\ 7. & \neg m_1 & \rightarrow & \bigcirc \neg m_1 \\ 8. & m_2 & \rightarrow & \bigcirc m_2 \\ 9. & \neg m_2 & \rightarrow & \bigcirc \neg m_2 \end{array}$$

The next clause represents the statement given by the father that at least one child has a muddy forehead:

$$\begin{array}{lrcl} 10. & \textbf{true} & \rightarrow & \bigcirc (m_1 \vee m_2) \end{array}$$

If a child has a muddy forehead the other children know about it.

$$\begin{array}{lrcl} 11. & \textbf{true} & \rightarrow & \neg m_2 \vee K_1 m_2 \\ 12. & \textbf{true} & \rightarrow & m_1 \vee K_2 \neg m_1 \\ 13. & \textbf{true} & \rightarrow & m_2 \vee K_1 \neg m_2 \\ 14. & \textbf{true} & \rightarrow & \neg m_1 \vee K_2 m_1 \end{array}$$

When the children answer they do not know about their muddy foreheads, their ignorance becomes common knowledge. The clausal forms of $t_2 \rightarrow K_1 \neg K_2 m_2$ and $t_2 \rightarrow K_2 \neg K_1 m_1$ are given by the following clauses:

$$\begin{array}{lrcl} 15. & \textbf{true} & \rightarrow & \neg t_2 \vee K_1 x \\ 16. & \textbf{true} & \rightarrow & \neg x \vee \neg K_2 m_2 \\ 17. & \textbf{true} & \rightarrow & \neg t_2 \vee K_2 y \\ 18. & \textbf{true} & \rightarrow & \neg y \vee \neg K_1 m_1 \end{array}$$

The refutation follows:

19.	$\neg m_2$	$\rightarrow$	$\bigcirc m_1$	[10, 9, SRES1]
20.	$\neg m_1 \wedge \neg m_2$	$\rightarrow$	$\bigcirc$**false**	[19, 7, SRES1]
21.	**true**	$\rightarrow$	$m_1 \vee m_2$	[20, SIMP1]
22.	**true**	$\rightarrow$	$\neg x \vee \neg K_2 \neg m_1$	[21, 16, MRES4]
23.	**true**	$\rightarrow$	$\neg x \vee m_1$	[22, 12, MRES1]
24.	**true**	$\rightarrow$	$\neg y \vee \neg K_1 x$	[23, 18, MRES4]
25.	**true**	$\rightarrow$	$\neg t_2 \vee \neg y$	[24, 15, MRES1]
26.	**true**	$\rightarrow$	$\neg t_2$	[25, 17, MRES3]
27.	t_1	$\rightarrow$	$\bigcirc$**false**	[26, 3, SRES2]
28.	**true**	$\rightarrow$	$\neg t_1$	[27, SIMP1]
29.	t_0	$\rightarrow$	$\bigcirc$**false**	[28, 2, SRES2]
30.	**true**	$\rightarrow$	$\neg t_0$	[29, SIMP1]
31.	**start**	$\rightarrow$	**false**	[30, 1, IRES1]

$\dashv$

5.6.2 Tableau

We describe a tableau-based proof method for the Temporal Logics of Knowledge, $\mathsf{KL}_{(n)}$. Such methods involve the systematic construction of a structure, for a formula φ, from which a model for φ can be obtained. Any parts of the structure from which a model cannot be built are deleted. To show a formula φ valid the tableau algorithm is applied to $\neg\varphi$. If a model cannot be extracted from the resulting structure, then $\neg\varphi$ is unsatisfiable and φ valid.

The decision procedure for $\mathsf{KL}_{(n)}$ represents a generalisation of both the tableau method for the underlying temporal logic and for the modal systems associated with the epistemic modalities. The tableau construction is carried out by interleaving the unwinding of the temporal and epistemic relations. Once the structure corresponding to both dimensions has been built, inconsistent states are removed and a simple procedure checks for satisfiability of eventualities.

We say that an *atom* is either a literal or a formula of the form $K_a\varphi$ or $\neg K_a\varphi$. If φ is of the form $\bigcirc\psi$, then φ is a *next-time formula.* If $\Delta \subseteq \mathsf{WFF}_{\mathsf{KL}_{(n)}}$, then $next(\Delta) \stackrel{\text{def}}{=} \{\varphi \mid \bigcirc\varphi \in \Delta\}$.

We assume formulae are in simplified form and with negations pushed through formulae until they precede atoms. This simplifies the presentation and avoids the need for additional rules to deal with negations. For instance, $\neg\neg\varphi$ is identified with φ. Transformation rules for classical connectives are assumed (e.g. $\neg(\varphi \vee \psi) \equiv (\neg\varphi \wedge \neg\psi)$) as well as the following rules for temporal formulae: $\neg\Box\varphi \equiv \Diamond\neg\varphi$, $\neg\Diamond\varphi \equiv \Box\neg\varphi$, $\neg\bigcirc\varphi \equiv \bigcirc\neg\varphi$,

$\neg(\varphi \mathcal{W} \psi) \equiv \neg\psi \mathcal{U} (\neg\varphi \wedge \neg\psi)$, and $\neg(\varphi \mathcal{U} \psi) \equiv \neg\psi \mathcal{W} (\neg\varphi \wedge \neg\psi)$. In particular, when adding a formula $\neg\psi$ to a set of formulae or performing membership tests, we use a formula equivalent to $\neg\psi$ with the negation pushed through until it precedes an atom.

As with other tableau-based decision procedures, the procedure relies upon *alpha* and *beta* equivalences, shown in Figure 5.6. We omit the rules for classical connectives, as these are standard.

α	α_1	α_2
$\Box\varphi$	φ	$\bigcirc\Box\varphi$
$K_a\varphi$	φ	$K_a\varphi$

β	β_1	β_2
$\Diamond\varphi$	φ	$\neg\varphi \wedge \bigcirc\Diamond\varphi$
$\varphi \mathcal{U} \psi$	ψ	$\neg\psi \wedge \varphi \wedge \bigcirc(\varphi \mathcal{U} \psi)$
$\varphi \mathcal{W} \psi$	ψ	$\neg\psi \wedge \varphi \wedge \bigcirc(\varphi \mathcal{W} \psi)$

Figure 5.6: Alpha and Beta Equivalences

The alpha and beta rules applied to temporal operators separate the formula in its present and future parts. The alpha rule for the epistemic operator corresponds to reflexivity.

A *propositional tableau* is a set of formulae where no further alpha and beta rules may be applied. If $\varphi \in \mathsf{WFF}_{\mathsf{KL}_{(n)}}$, let $sub(\varphi)$ be the set of all subformulae of φ. A set of formulae Δ is said to be *subformula complete* iff for every $K_a\varphi \in \Delta$ and for all $K_a\psi \in sub(\varphi)$ either $K_a\psi \in \Delta$ or $\neg K_a\psi \in \Delta$. A *propositional subformula complete tableau* or *PC-tableau* is a set of formulae Δ that is both a propositional tableau and subformula complete.

The internal consistency of states during tableau generation is established by checking whether they are *proper*. If $\Delta \subseteq \mathsf{WFF}_{\mathsf{KL}_{(n)}}$ then Δ is *proper* (notation $proper(\Delta)$) if and only if (1) **false** $\notin \Delta$; and (2) if $\varphi \in \Delta$, then $\neg\varphi \notin \Delta$. Obviously, improper sets are unsatisfiable.

We now describe how to construct the set of *PC-tableaux* for a set of formulae Δ. In order to ensure termination, rules are applied only if they modify the structure, that is, only once to any formula in Δ.

Constructing the set of PC-tableaux. Given a set of formulae Δ, the set of PC-tableaux can be constructed by applying the following rules (1) and (2) to $\mathcal{F} = \{\Delta\}$ until they cannot be applied any further.

1. *Forming a propositional tableau.* For all Δ in $\mathcal{F}$, do:

   (a) If $\varphi \in \Delta$ is an α formula let $\mathcal{F} = \mathcal{F} \setminus \{\Delta\} \cup \{\Delta \cup \{\alpha_1, \alpha_2\}\}$

   (b) If $\varphi \in \Delta$ is an β formula let $\mathcal{F} = \mathcal{F} \setminus \{\Delta\} \cup \{\Delta \cup \{\beta_1\}\} \cup \{\Delta \cup \{\beta_2\}\}$

2. *Forming a subformula complete tableau.*

   For any $\Delta \in \mathcal{F}$ such that $proper(\Delta)$, if $K_a\varphi \in \Delta$ and $K_a\psi \in sub(\varphi)$ such that neither $K_a\psi \in \Delta$ nor $\neg K_a\psi \in \Delta$ then let $\mathcal{F} = \mathcal{F} \setminus \{\Delta\} \cup \{\Delta \cup \{K_a\psi\}\} \cup \{\Delta \cup \{\neg K_a\psi\}\}$.

3. *Deleting improper sets.*

   Delete any $\Delta \in \mathcal{F}$ such that $\neg proper(\Delta)$.

Step 2 corresponds to the use of the *analytic cut* rule. Thus formulae are added to the sets being constructed which are subformulae of existing formulae. The model-like structures which are generated by the tableau algorithm is defined below, where $State = \{s, s', \ldots\}$ is the set of all states.

Definition 5.3
A *structure*, H, is a tuple $H = (S, \eta, \sim_1, \ldots, \sim_n, L)$, where:

- $S \subseteq State$ is a set of states;
- $\eta \subseteq S \times S$ is a binary *next-time* relation on S;
- $\sim_a \subseteq S \times S$ represents an accessibility relation over S for agent $a \in \mathscr{A}$;
- $L : S \to \mathcal{P}(\mathsf{WFF}_{\mathsf{KL}_{(n)}})$ labels each state with a set of $\mathsf{WFF}_{\mathsf{KL}_{(n)}}$ formulae. ⊣

Definition 5.4
If $\varphi \in \mathsf{WFF}_{\mathsf{KL}_{(n)}}$ is of the form $\chi \mathcal{U} \psi$ or $\Diamond\psi$ then φ is said to have *eventuality* ψ. If $(S, \eta, \sim_1, \ldots, \sim_n, L)$ is a structure, $s \in S$ is a state, η^* is the reflexive transitive closure of η, and $\varphi \in \mathsf{WFF}_{\mathsf{KL}_{(n)}}$, then φ is said to be *resolvable* in $(S, \eta, \sim_1, \ldots, \sim_n, L)$ from s, (notation $resolvable(\varphi, s, (S, \eta, \sim_1, \ldots, \sim_n, L))$), iff if φ has eventuality ψ, then $\exists s' \in S$ such that $(s, s') \in \eta^*$ and $\psi \in L(s')$. ⊣

The tableau algorithm first expands the structure and then contracts it. We try to construct a structure from which a model may possibly be extracted, and then delete states in this structure that are labelled with formulae such as $\Diamond p$ or $\neg K_a p$ which are not satisfied in the structure. Expansion uses the formulae in the labels of each state to build η and $\sim_a$ successors.

Temporal successors (η-successors) to a state s are constructed by taking the set of formulae in the label of s whose main connective is $\bigcirc$, removing the outermost $\bigcirc$ operator, labelling a new state s' by this set if one does not already exist and adding (s, s') to η.

The $\sim_a$ successors are slightly more complicated. If the label for a state s contains the formula $\neg K_a\psi$ then we must build an $\sim_a$ successor s' containing $\neg\psi$. If s also contains $K_a\chi$ formulae then the label of s' must also contain χ and $K_a\chi$, as all $\sim_a$ successors of s must contain χ, to satisfy the semantics of the epistemic modality, and $K_a\chi$, to satisfy axiom **4**. Further, for all formulae $\neg K_a\chi$ in s, in order to satisfy axiom **5** the label of s' must also contain $\neg K_a\chi$.

Note that if s contains no formulae of the form $\neg K_a\chi$ but does contain $K_a\chi$ formulae we must build an $\sim_a$ successor containing χ and $K_a\chi$, as above, to satisfy the seriality conditions imposed by axiom **D**. The **T** axiom is incorporated by an extra alpha rule.

Having built a structure, states must be deleted that can never be part of a model. Considering the temporal dimension, states with unresolvable eventualities are deleted. Also, states containing next-time formulae without η successors are deleted. For the epistemic dimension we must ensure that any state labelled by a formula of the form $\neg K_a\chi$ has an $\sim_a$ successor containing $\neg\chi$, and also that every modal state containing $K_a\chi$ has at least one $\sim_a$ successor and all its $\sim_a$ successors contain χ.

The Tableau Algorithm Given the $\mathsf{KL}_{(n)}$ formula φ_0 to be shown unsatisfiable, perform the following steps.

1. *Initialisation.*

   Set $S = \eta = \sim_1 = \cdots = \sim_n = L = \emptyset$.

   Construct $\mathcal{F}$, the set of PC-tableaux for $\{\varphi_0\}$. For each $\Delta_i \in \mathcal{F}$ create a new state s_i and let $L(s_i) = \Delta_i$ and $S = S \cup \{s_i\}$. For each $\Delta_i \in \mathcal{F}$ repeat steps (2)–(3) below, until none apply and then apply step 4.

2. *Creating $\sim_a$ successors.*

   For any state s labelled by formulae $L(s)$, where $L(s)$ is a PC-tableau, for each formula of the form $\neg K_a\psi \in L(s)$ create a set of formulae

   $$\begin{aligned} \Delta \ = \ & \{\neg\psi\} \cup \{\chi \mid K_a\chi \in L(s)\} \cup \\ & \{K_a\chi \mid K_a\chi \in L(s)\} \cup \{\neg K_a\chi \mid \neg K_a\chi \in L(s)\}. \end{aligned}$$

   If s contains no such formulae but there exists $K_a\psi \in L(s)$ then construct the set of formulae

   $$\Delta = \{\chi \mid K_a\chi \in L(s)\} \cup \{K_a\chi \mid K_a\chi \in L(s)\}.$$

   For each Δ above construct $\mathcal{F}$, the set of PC-tableaux for Δ, and for each member $\Delta' \in \mathcal{F}$ if $\exists s'' \in S$ such that $\Delta' = L(s'')$ then add (s, s'')

to $\sim_a$, otherwise add a new state s' to S, labelled by $L(s') = \Delta'$, and add (s, s') to $\sim_a$.

3. *Creating η successors.*

   For any state s labelled by formulae $L(s)$, where $L(s)$ is a PC-tableau, if $\bigcirc\psi \in L(s)$ create the set of formulae $\Delta = next(L(s))$. For each Δ construct $\mathcal{F}$ the set of PC-tableaux for Δ, and for each member $\Delta' \in \mathcal{F}$ if $\exists s'' \in S$ such that $\Delta' = L(s'')$ then add (s, s'') to η, otherwise add a state s' to the set of states, labelled by $L(s') = \Delta'$ and add (s, s') to η.

4. *Contraction.*

   Continue deleting any state s where

   (a) $\exists\psi \in L(s)$ such that $\neg resolvable(\psi, s, (S, \eta, \sim_1, \ldots, \sim_n, L))$; or

   (b) $\exists\psi \in L(s)$ such that ψ is of the form $\bigcirc\chi$ and $\nexists s' \in S$ such that $(s, s') \in \eta$; or

   (c) $\exists\psi \in L(s)$ such that ψ is of the form $\neg K_a\chi$ and $\nexists s' \in S$ such that $(s, s') \in\sim_a$ and $\neg\chi \in L(s')$; or

   (d) $\exists\psi \in L(s)$ such that ψ is of the form $K_a\chi$ and $\nexists s' \in S$ such that $(s, s') \in\sim_a$ and $\chi \in L(s')$

   until no further deletions are possible.

Note that we cannot interleave expansion steps with deletion steps. For example to determine whether an eventuality is resolvable, the structure must be fully expanded, otherwise states may be wrongly deleted.

If $\varphi_0 \in \mathsf{WFF}_{\mathsf{KL}_{(n)}}$, then we say the tableau algorithm is *successful* iff the structure returned contains a state s such that $\varphi_0 \in L(s)$. A formula φ_0 is $\mathsf{KL}_{(n)}$ satisfiable iff the tableau algorithm performed on φ_0 is successful. The tableau procedure for $\mathsf{KL}_{(n)}$ is sound, complete, and terminating.

Example 5.5
We show that (1) $K_a \Box p \rightarrow \Box p$ is a valid formula in $\mathsf{KL}_{(1)}$. By negating (1), we obtain $K_a \Box p \wedge \Diamond\neg p$. Let $\Delta = \{K_a \Box p \wedge \Diamond\neg p\}$. Then $\mathcal{F}_0$, the set of PC-tableaux for Δ is

$$\mathcal{F}_0 = \{\Delta_0 = \{K_a \Box p \wedge \Diamond\neg p, K_a \Box p, \Diamond\neg p, \Box p, p, \bigcirc \Box p, \bigcirc\Diamond\neg p\}\}$$

Let s_0 be the state such that $L(s_0) = \Delta_0$. We start now constructing the $\sim_a$ relation. Δ_0 does not contain formulae of the form $\neg K_a\varphi$, but it contains $K_a \Box p$. The set of PC-tableaux $\mathcal{F}_1$ for $\Delta' = \{K_a \Box p, \Box p\}$ is

$$\mathcal{F}_1 = \{\Delta_1 = \{K_a \Box p, \Box p, p, \bigcirc \Box p\}\}$$

Let s_1 be the state such that $L(s_1) = \Delta_1$. We include in $\sim_a$ the edge (s_0, s_1). Now, there is only one epistemic formula in Δ_1, which is again $K_a \Box p$. The set of PC-tableaux for $\Delta'' = \{K_a \Box p, \Box p\}$ is exactly as $\mathcal{F}_1$. Therefore, instead of adding new states to the tableau, we add the edge (s_1, s_1) to $\sim_a$. In order to build the temporal relation, note that $next(\Delta_0) = \{\Box p, \Diamond\neg p\}$. Building the PC-tableau $\mathcal{F}_2$ for $\{\Box p, \Diamond\neg p\}$ results in $\{\Delta_2 = \{\Box p, p, \bigcirc \Box p, \bigcirc\Diamond\neg p\}\}$. Let s_2 be the state such that $L(s_2) = \Delta_2$. We include edge (s_0, s_2) in η. Now, $next(\Delta_1) = \{\Box p\}$. Building the PC-tableaux $\mathcal{F}_3$ for $\{\Box p\}$ results in $\{\Delta_3 = \{\Box p, p, \bigcirc \Box p\}\}$. As $next(\Delta_2) = next(\Delta_1)$, instead of adding a new state, we only include the edge (s_2, s_2) in η. Let s_3 be the state such that $L(s_3) = \Delta_3$ and include (s_1, s_3) in η. As $next(\Delta_3) = next(\Delta_1)$, instead of adding a new state, we only include the edge (s_3, s_3) in η. This finishes the expansion of the tableau, as shown in Figure 5.7. Contraction consists in removing from the tableau states which

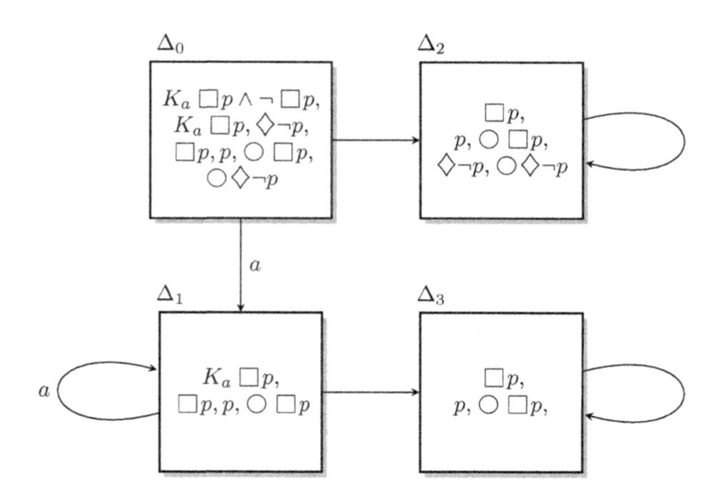

Figure 5.7: Tableau for $K_a \Box p \wedge \Diamond\neg p$.

contain formulae which are not satisfied in the resulting structure. Note that $\Diamond\neg p$ is a formula in Δ_0, but there is no η-successor of Δ_0 that contains $\neg p$. As Δ_0 is removed from the tableau, no remaining states contain the original formula, i.e. $K_a \Box p \wedge \Diamond\neg p$. As the tableau is not successful, this formula is unsatisfiable and, therefore, the original formula $K_a \Box p \rightarrow \Box p$ is valid. $\dashv$

The tableau construction given here also provides a sound and complete method for checking the satisfiability of formulae in the combined logics of time and belief (the fusion of $\mathsf{S4}_{(n)}$ and PTL). The only difference in the construction is that the alpha rule associated with the epistemic modalities is not applied, as models in $\mathsf{S4}_{(n)}$ do not need to satisfy the axiom **T**. If the tableau procedure is applied to $\mathsf{KL}_{(n)}$, then some improvements can be adopted. In step 2 of the tableau algorithm, when there is no formulae of the form $\neg K_a\psi \in L(s)$, we build a set of formulae containing χ and $K_a\chi$, for $K_a\chi \in L(s)$. This is not needed for $\mathsf{KL}_{(n)}$, as reflexivity is captured by the alpha rule. Note also that if this refinement is adopted, we must also delete step (4d) of the tableau algorithm.

5.7 Knowledge and communication

The temporal logic of knowledge and time offers interesting ways of viewing how and why an agent's knowledge changes over time. In the one-agent case, this could be discovery of some factual information about the world. In the multi-agent case, such change could be the result of some communication. As we saw in the example of muddy children, communication in the presence of all the children causes immediate common knowledge. On the other hand, if the father had whispered a statement to one of the children, that child would have acquired knowledge but the other children would only have known that she learnt something but not what she learnt. Considering that common knowledge is an infinite conjunction of mutual knowledge assertions, we may well wonder how the announcement achieved it instantaneously, and this leads us to study the relationship between knowledge and communication.

Several critical issues arise in reasoning about knowledge in the context of communicating agents:

- Common knowledge, defined by an infinite conjunction, is closely linked to coordination and consensus; on the other hand, it's hard to achieve. Hence one might ask, if we look at weaker *levels of knowledge* given by finite conjunctions, what can we say about them?

- Communication in these systems is purposive: a sender sends a message so that the recipient's knowledge is updated in a particular way. Thus, the semantics of communication is in epistemic terms: rather than discussing knowledge change due to communication, we could simply define the meaning of communications by epistemic change.

- Distributed protocols are analysed in terms of communication histories. When agents have perfect recall, they have complete memory of past events. However, in the context of agents realisable in distributed systems, it is more natural to consider *limited recall*, based on memory limitations of agents. A natural restriction in this regard is that of *finite state* agents synthesisable as automata.

There are further issues, such as the *strategic* nature of communication (again epistemic in nature), that of information revealing and hiding in communications, and so on.

We will look at some of these issues in the sections below. Before we do that, it is worth taking a closer look at communication histories mentioned above.

5.7.1 The history based model

Models were presented in Section 5.2.2 as collections of *timelines*, which are countably infinite sequences of world *states*. We did not explicitly specify the set of states: in general, we could consider these to be defined by a set of *event types*, and a state being the condition of the world after the occurrence of such an event. In this sense, the set of timelines can be viewed as a subset of E^ω, where E is a fixed (at most countable) set of events, and we will speak of a model M over E.

We worked with points which are pairs of the form (t, u), where t is a timeline and u is a temporal instant that acts as an index to t. In the context of distributed systems, a point (t, u) can be seen as a historical record of system states up till that point: the sequence $h = (t, 1), (t, 2), \ldots, (t, u) \in E^*$. Let $h \preceq h'$ denote that h is a finite prefix of h'. For a set of timelines TL, let $\mathcal{P}(TL)$ denote the set $\{h \mid h \preceq t \text{ for some } t \in TL\}$ containing all finite prefixes of timelines in TL. We often speak of $\mathcal{P}$ as a protocol. We write $h \cdot t'$ to denote the concatenation of the finite history h with the rest of its timeline: in this case, the timeline $t \in TL$ of the form $h \cdot t'$. In particular, when $e \in E$, and $t = h \cdot e \cdot t'$, the occurrence of e after the finite history h is a matter of interest.

The models we studied were equipped with equivalence relations over points. Clearly $\sim_a \subseteq (E^* \times E^*)$ for every agent a. For $h \in E^*$, let $[h]_{\sim_a}$ denote the equivalence class of h under $\sim_a$. These relations offer us a notion of local histories of agents, or information available to agents at the time point when a global history has occurred. For $h \in \mathcal{P}(\mathcal{T})$, let $\lambda_a(h) = [h]_{\sim_a}$. Let L_a denote the range of λ_a, referred to as the set of *local* histories of agent a.

Why should we consider histories at all? This is merely to explicate a sense of temporal continuity experienced by an agent. This is an underlying assumption of temporal reasoning; however, in Hintikka-style reasoning about knowledge, points across multiple timelines obscure this aspect, and local histories are a way of recovering such temporal continuity.

Consider a history h and its extension $h \cdot e$ by an event e. If $h \sim_a h; e$, we interpret this as an event in which agent a does not participate. Thus we can speak of events observable by an agent and events external to it. This becomes important when we wish to consider *how* an agent's knowledge may change. Let $lh = [h]_{\sim_a}$, where $h \in E^*$ and $e \in E$. Can we speak of the extension of the local history lh by e? Let $lh; e$ denote $[h \cdot e]_{\sim_a}$: but this is not well-defined as a function. In the models we studied in Section 5.2, we could well have the following situation: let e be the same u^{th} element (state or event) of two timelines t and t'; we could well have $(t, v) \sim_a (t', v)$ for all $v < u$ and yet $(t, u) \not\sim_a (t', u)$. History-based models serve to highlight such instances so that natural restrictions may be imposed as required by applications.

5.8 Levels of knowledge

We remarked that common knowledge is required for coordination in distributed systems. For many algorithms, level two or level three knowledge (2 knows that 1 knows that 2 knows) may be sufficient. In Example 5.4, for instance, only two levels of knowledge was required to reason about the muddy children puzzle. Indeed characterising the level of knowledge *required* for a particular application is an interesting and challenging computational problem.

Note that even in this simple example, the specific level of knowledge referred to is not a number but a collection of *strings*: ("2 knows that 1 knows that 2 knows" is different from "1 knows that 2 knows that 1 knows"). In general, what sets of strings are we speaking of?

Consider the proposition p to mean that agent 1 has bought a house. At (t, u), it may well be the case that $K_1(p \wedge \neg K_2 p)$ holds (perhaps because agent 2 is in another city). Hence 1 sends a message to 2 informing her that p holds. When this message reaches at (t, v), $K_2 K_1 p$ holds but $K_1 K_2 K_1 p$ may still not hold (since the communication may not have reached), so 2 sends an acknowledgement to 1. When this reaches, $K_1 K_2 K_1 p$ becomes true but $K_2 K_1 K_2 K_1 p$ is not true yet, and so on.

This suggests that at any state of knowledge there are infinitely many such "knowledge strings" that might be relevant. But these strings have further structure as well: in any state in which $K_2 K_3 K_1 p$ holds, so also does $K_2 K_1 p$. Thus every state describes some "level" of knowledge for each formula. What can we say structurally about these levels?

If $u = a_1 a_2 \dots a_m$ let $K_u \varphi$ abbreviate $K_{a_1} K_{a_2} \dots K_{a_m} \varphi$. Consider the ordering $u \leq v$ if u is a subsequence of v, given by the embeddability ordering, that is, if u is of length k, then there exist indexes $a_1, a_2, \ldots a_k$ such that the word $v_{a_1} v_{a_2} \ldots v_{a_k} = u$: $132 \leq 1243112$ but $132 \not\leq 1241123$.

Below, fix a model $M = (TL, \sim_1, \ldots, \sim_n, V)$ and let h, h' etc to range over histories in M.[1] The following proposition easily follows from reflexivity and transitivity of the knowledge operators. (See axioms **T** and **4** in Section 5.5).

Proposition 5.1
Let $u, v \in \mathscr{A}^*$, $a \in \mathscr{A}$ and φ be a formula. Then $\models K_u K_a K_v \varphi \equiv K_u K_a K_a K_v \varphi$. $\dashv$

This suggests that we need not consider repetition of elements in sequences that characterise levels of knowledge. Let Γ be the set of all strings in $\mathscr{A}^*$ without repetition.

Proposition 5.2
For all $u, v \in \mathscr{A}^*$, for all h, if $u \leq v$ then for every formula φ, we have: if $(M, h) \models K_v \varphi$ then $(M, h) \models K_u \varphi$. $\dashv$

Definition 5.5
The **level** of φ at h in M, denoted $L(\varphi, h) = \{u \in \Gamma \mid (M, h) \models K_u \varphi\}$. $\dashv$

The previous proposition asserts that levels are downward closed sets with respect to embeddability. The complement of $L(\varphi, h)$ (in Γ) is an upward closed set.

At this point we observe that, by Higman's Lemma, embeddability is not merely a partial ordering but a *well-partial ordering*: it has no infinite decreasing chains and no infinite anti-chains. Moreover, every upward closed set of a well-partial order can be represented by a *finite* set of minimal elements. This has a pleasant consequence.

Corollary 1
There are only countably many levels of knowledge and all of them are *regular* subsets of $\mathscr{A}^*$. $\dashv$

The latter observation stems from the fact that given the minimal elements, a finite state automaton can easily check embeddability of a minimal element in a given string.

Call a formula φ *persistent* if whenever $(M, h) \models \varphi$ and $h \preceq h'$, we also have $(M, h') \models \varphi$. Clearly when φ is persistent, so is $K_a \varphi$ for any a. Given $U, V \subseteq \Gamma$, let $U * V$ denote the downward closure of $\{uv \mid u \in U, v \in V\}$.

[1]The notion $(M, h) \models \varphi$ is defined exactly as that of $(M, (t, u)) \models \varphi$ in Section 5.2.2.

We now address the question of knowledge change. In terms of levels, how does $L(\varphi, h')$ relate to $L(\varphi, h)$ when $h \preceq h'$? Is there some way we can assert that all communications in between can make only a "finite difference"? Indeed, we can show that for persistent formulae φ, when $h \preceq h'$, there exists a finite $U \subseteq \Gamma$ such that $L(\varphi, h) \subseteq L(\varphi, h') \subseteq U * L(\varphi, h)$.

What is interesting is that downward closed sets in well-partial orders can be seen as general characterisations of levels of knowledge, as the following 'completeness' theorem asserts.

Theorem 5.3
Every downward closed $L \subseteq \Gamma$ is a level $L(\varphi, h)$ for a suitable formula φ and a history h in a model M. ⊣

The purport of this theorem is that we can, in principle, *synthesise* communications that can take agents from a level of knowledge to another, a rather surprising assertion.

5.9 Knowledge based semantics

In general, a communication can be thought of as a transformation on knowledge states, not only of those who are involved in that communication, but also those who observe the act of communication. Indeed, this idea can form the basis of assigning semantics to messages in a general fashion. The temporal logic of knowledge is just the right framework for description of such semantics, and we discuss this application in this section.

If a lion could talk, we could not understand him.
Ludwig Wittgenstein

Not knowing lion language, could we still assign meanings to lions' communications? If a lion growls, surely we would run. In some sense, such epistemically induced behavioural change can be seen as semantics of a kind. For another example, we do not share baby language with babies (in the sense that we share English or Portuguese), but can reliably assign meanings to certain sounds and act on them, to play, to comfort, to feed, etc. More importantly, babies learn to produce the same sounds to achieve the intended effect.

In general, a communication is an intensional act, intended to produce certain epistemic change in its recipient. The meaning may well depend on who says it, when, and on the state of the recipient. Moreover, if the participants have a prior arrangement, the meaning may be entirely different from what is inferred by observers of that communication (as in the case of encrypted communications).

We now present an attempt at epistemic semantics of events. The 'protocol' dependence of meaning is made explicit below.

Definition 5.6
Fix E, a model $M = (TL, \sim_1, \ldots, \sim_n, V)$ over E, $m \in E$, $a \in \mathscr{A}$ and L_a, the set of local histories of a.

$$Sem_M(a, m) \stackrel{\text{def}}{=} \{(\varphi, \psi) \mid \forall [h; m]_{\sim_a} \in L_a : (h \models \varphi \rightarrow h; m \models K_a \psi)\}. \quad \dashv$$

Intuitively, the pair of formulae (φ, ψ) is in the meaning of m if whenever event m takes place, then if φ was *true* before, a always *knows* ψ after m has taken place.

It is easy to see that the definition can be generalised to $Sem_M(a, u)$ where $u \in E^*$ is a finite sequence of events. We further see that:

- $Sem_M(a, u)$ is closed under conjunction and weakening: whenever $(\varphi, \psi_1) \in Sem_M(a, u)$ and $(\varphi, \psi_2) \in Sem_M(a, u)$ then $(\varphi, \psi_1 \wedge \psi_2) \in Sem_M(a, u)$ and for all ψ', $(\varphi, \psi_1 \vee \psi')$ is also in $Sem_M(a, u)$.

- Since knowledge of ψ implies the truth of ψ, if $(\varphi, \psi_1) \in Sem_M(a, u)$ and $(\psi_1, \psi_2) \in Sem_M(a, u')$, then $(\varphi, \psi_2) \in Sem_M(a, u; u')$.

This suggests that meanings of communication sequences can be studied in a Hoare-like logic of knowledge and time.

The use of temporal modalities like $\Diamond \alpha$ ensures that what is communicated by messages in the sense of $Sem_M(a, m)$ are not only invariant properties but also eventuality properties, promises of what will occur in future. These are essential in situations where the purpose of communication is in itself to *set up a protocol* to be followed subsequently. This is important, because we need not only consider the receiver of a message interpreting it according to an agreed protocol, but the message itself may be part of setting up the protocol between the parties in a bootstrapping manner.

The meaning of m is given in terms of properties expressible in the logical language, which is a meta-language. This approach, of giving semantics for messages (themselves semantic entities, in the sense that they are elements of models and are not syntactic objects in a specified language) using syntactic means (formulae in a logical language) may seem curious. But this has definite and interesting implications, and allows for us to consider communication between parties who do not share a language or even signals that do not come from any apparently recognisable language! The use of a logical language here emphasises that the discussion of what a communication means to those involved in the communication is limited only by the level of reasoning carried out at the meta-level.

Formulae as messages

It is interesting to consider systems where the messages sent are themselves formulae of the logic considered. That is, we consider that some formulae of the temporal logic of knowledge are themselves sent as strings and the sending or receipt of such a formula φ are themselves events.

What is interesting in this situation is that receipt of a communication can be seen as theory updating: the knowledge of the recipient is the new logical theory obtained by 'adding' the received formula to what was known earlier. But this requires notions of honesty and trust, and the updates are complex.

To a great extent, this is the programme carried out by work on Dynamic Epistemic Logic, which studies the dynamics of information change (see Chapter 6). In its simplest version, Public Announcement Logic considers the effect of truthful broadcasts of epistemic formulae, and in the most general version, 'action models' are specified by pre-conditions and post-conditions stated in epistemic logic. Such a description of 'logical dynamics' is, of course, closely related to temporal reasoning about knowledge, and this relationship can be established formally.

5.10 Automata theory of knowledge

In the models we have been studying, timelines represent temporal evolution of *global system states*, and knowledge is defined via equivalence relations on points, which are global states. Local states and local histories are derived notions in such a presentation,

A natural question that arises is whether we can have a *compositional* presentation of knowledge: in this picture, properties are local and associated with agents. We only describe agents, and the mechanisms by which they acquire information about local properties relating to other agents. Global properties (and thus epistemic indistinguishability relations) are compositionally derived from local properties. In terms of systems, we obtain global behaviour by taking *products* of local behaviours.

Another question relates to how much memory is required for knowledge, especially interesting in the context of temporal reasoning about knowledge. Every agent observes the system evolution, depending on her limited visibility, and records her observations, and learns from communications. However, such observations and learning are clearly limited by the agent's memory. If agents had bounded memory, such as in the case of finite state agents, how does it affect their knowledge?

We suggest that automata theory provides a natural formalism for such local/product presentations. This is the topic we take up in this section;

here, we consider information exchange between agents only in the *synchronous* mode, i.e. with bounded delay or instantaneous communication.

5.10.1 Epistemic automata

In the automaton model we discuss here, each agent is finite state, and the local state of an agent is annotated with what the agent knows about the rest of the system at that state. Clearly, this depends on what the agent can observe about the system, and since such observations have to be held in the agent's memory, they must be finite as well. Each agent's knowledge (at any state) would then be tuples of observables (which can be seen as abstract projections of states).

Thus the description of epistemic agents constrains global system behaviour: the possible evolutions of the system must unfold in such a manner that agents do observe (and thus know) in their local histories what the annotations assert. Therefore, the global system is not a simple product of automata but an epistemic product that satisfies natural conditions: when an event occurs, the state (and knowledge) of those not participating in it remains unchanged, and among those who do participate, information exchange is perfect and maximal. We can see this as causality information being transferred between agents during a communication event.

5.10.2 Finite state automata

Before we present epistemic automata some preliminaries setting up the notation are in order. Let Σ be a finite and nonempty alphabet. Σ^* is the set of all finite strings over Σ. We use $x, y, z, \ldots$ for the letters of Σ and $u, v, w, \ldots$ for strings. The empty string is denoted as ϵ. Concatenation of strings u and v is denoted either by $u \cdot v$ or just uv. Any subset $L \subseteq \Sigma^*$ is called a language over Σ.

A *transition system* (TS) over Σ is a tuple $M = (Q, \longrightarrow, I)$, where Q is a *finite* set of states, $\longrightarrow \subseteq (Q \times \Sigma \times Q)$ is the transition relation and $I \subseteq Q$ is the set of initial states. We use letters p, q etc. for the states of a TS. When $(q, x, q') \in \longrightarrow$, we write it as $q \xrightarrow{x} q'$.

The one step transition on letters is extended to transitions on strings over Σ^* and the extended transition relation is denoted by $\Longrightarrow : Q \times \Sigma^* \times Q$. When $p \stackrel{u}{\Longrightarrow} q$, we say that there is a *run of M* from p to q on u.

We call a TS *deterministic* (DTS) if $\longrightarrow$ is a function on $(Q \times \Sigma)$ and I is a singleton, namely $\{q^0\}$. This entails that whenever $p \xrightarrow{x} q$ and $p \xrightarrow{x} q'$ we have $q = q'$. Notice that if M is a DTS, the extended transition relation is a function, and given any $p \in Q$ and $u \in \Sigma^*$, there is a unique q such that $p \stackrel{u}{\Longrightarrow} q$. In this case, we denote by $(u)_M$ the state q such that $q^0 \stackrel{u}{\Longrightarrow} q$.

A *finite state automaton* (FA) is a tuple $A = (M, F)$, where M is is a TS and $F \subseteq Q$, is the set of its final states. When $q \overset{u}{\Longrightarrow} q_k$ for some $q \in I$ and $q_k \in F$, we say that the string u is accepted by A. The set of all strings accepted by A (also called the language of A) is denoted as $L(A)$ or $L(M, F)$.

Definition 5.7
Given Σ, the class of languages accepted by FA's over Σ is defined as:

$$\mathscr{L}FA_\Sigma \overset{\text{def}}{=} \{L \subseteq \Sigma^* \mid \text{ there is an FA } (M, F) \text{ such that } L = L(M, F)\}.$$

We call this class of languages **recognisable** and denote it as Rec_Σ. ⊣

An FA (M, F) is called *deterministic* (DFA) if M is a DTS. We denote the class of languages accepted by DFA over Σ as $\mathscr{L}DFA_\Sigma$. It is easy to show that $\mathscr{L}DFA_\Sigma = \mathscr{L}FA_\Sigma = Rec_\Sigma$. Rec_Σ has many nice properties: it is closed under Boolean operations, concatenation and finite iteration (the Kleene star "*").

5.10.3 Distributed alphabets

Since we will be considering agents as finite state automata, we will associate agents' views with strings over finite alphabets that automata take as inputs. For systems of autonomous agents, we have distributed alphabets.

We model the distribution of actions among the agents (also referred to as processes) by a *distributed alphabet* $\widetilde{\Sigma}$. It is a tuple $(\Sigma_1, \cdots, \Sigma_n)$, where each Σ_a is a finite nonempty set of actions and is called an *agent alphabet.* These alphabets are not required to be disjoint. In fact, when $x \in \Sigma_a \cap \Sigma_b, a \neq b$, we think of it as a potential synchronisation action between a and b.

Given a distributed alphabet $\widetilde{\Sigma}$, we call the set $\Sigma \overset{\text{def}}{=} \Sigma_1 \cup \ldots \cup \Sigma_n$ as the alphabet of the system since the overall behaviour of the system is expressed by strings from Σ^*. For any action in Σ, we have the notion of agents participating in this action. Let $loc : \Sigma \to 2^{\{1,\ldots,n\}}$ be defined by $loc(x) \overset{\text{def}}{=} \{a \mid x \in \Sigma_a\}$. So $loc(x)$ (called "locations of x") gives the set of agents that participate (or, synchronise) in the action x. By definition, for all $x \in \Sigma$, $loc(x) \neq \emptyset$. The notion of locations (or, participating processes) is easily extended to strings over Σ^*. If $u \in \Sigma^*$, $alph(u) \overset{\text{def}}{=} \{x \in \Sigma \mid x \text{ occurs in } u\}$. Then, $loc(u) \overset{\text{def}}{=} \bigcup_{x \in alph(u)} loc(x)$.

From the definition of locations, we can derive an irreflexive and symmetric relation $\mathcal{I}$ on Σ. We call this an *independence* relation. The independence relation is defined to be $\mathcal{I} \overset{\text{def}}{=} \{(x, y) \in \Sigma \times \Sigma \mid loc(x) \cap loc(y) = \emptyset\}$.

Informally, $(x, y) \in \mathcal{I}$ if the participants in the actions x and y are disjoint. Since processes are spatially distributed, it entails that the participants of y remain unaffected by the execution of x and vice versa. Hence in the system, independent actions may proceed asynchronously. Extending the same idea to strings over Σ^*, we say two strings u and v are independent, i.e., $(u, v) \in \mathcal{I}$ iff $loc(u) \cap loc(v) = \emptyset$. Two independent strings can be thought of as independent execution sequences of two disjoint sets of processes.

Lastly, we say that the distributed alphabet $\widetilde{\Sigma}$ is *non-trivial* iff the independence relation on Σ is non-empty. For example, the distributed alphabet $(\Sigma_1 = \{x, y\}, \Sigma_2 = \{y, z\})$ is non-trivial because $(x, z) \in \mathcal{I}$. On the other hand, the distributed alphabet $\widetilde{\Sigma} = (\Sigma_1 = \{x, y\}, \Sigma_2 = \{y, z\}, \Sigma_3 = \{z, x\})$ is trivial because $loc(x) \cap loc(y) = \{1\} \neq \emptyset$ and so on for every pair of letters and hence $\mathcal{I}$ is empty. Since trivial alphabets constrain systems so that agents cannot proceed asynchronously, knowledge properties are essentially the same as in the case of one reasoning agent. Hence we consider only systems over non-trivial distributed alphabets.

5.10.4 Epistemic transition systems

In the definition of epistemic automata attempted below, we seek an alternative to global indistinguishability relations; this can be achieved by enriching local states with structure.

If Q_a is the states of agent a, $q \in Q_a$ can be thought of as a tuple $(q[1], \ldots, q[n])$, where $q[b]$ is the information agent a has *about* agent b. But then the question arises: what about what a knows about what b knows about c? Should $q[b]$ be another tuple? But this is infinite regress.

Automata theory offers a standard technique for this problem: use an *alphabet of observations*, which we call an epistemic alphabet.

Fix a distributed alphabet $\widetilde{\Sigma}=(\Sigma_1, \cdots, \Sigma_n)$. Additionally consider an **epistemic alphabet**

$$\widetilde{\mathcal{C}} = ((\mathcal{C}_1, \preceq_1), \cdots, (\mathcal{C}_n, \preceq_n)).$$

where each $\mathcal{C}_a$ is a nonempty set and for all $a \neq b$, $\mathcal{C}_a \cap \mathcal{C}_b = \{\bot\}$. $\preceq_a$ is a (reflexive and transitive) pre-ordering on $\mathcal{C}_a$ with the least element $\bot$. The element $\bot$ stands for trivial knowledge, or null information. We call $\mathcal{C} = \mathcal{C}_1 \cup \ldots \cup \mathcal{C}_n$ the *information set.* Since we work with finite state systems, it suffices to consider finite $\mathcal{C}_a$. An alternate presentation of $\widetilde{\mathcal{C}}$ is given by the information map:

$$\Phi \stackrel{\text{def}}{=} \{\chi : loc \mapsto \mathcal{C} \mid \forall a \in loc, \chi(a) \in \mathcal{C}_a\}.$$

It may be convenient to view elements of $\mathcal{C}$ as formulae of some logic of knowledge and implication as the ordering. In general, the modelling above corresponds to observations of each agent's state being recorded in a separate vocabulary.

Note that the entire hierarchy of a-propositions (what is true about a, what a knows about b, what a knows about what b knows about c etc) is collapsed into $(\mathcal{C}_a, \preceq_a)$.

Definition 5.8
Let $a \in \{1, 2, \ldots, n\}$. An **epistemic agent** (EA) over $(\Sigma_a, \mathcal{C}_a)$ is a tuple (M_a, f_a) where $M_a = (Q_a, \longrightarrow_a, q_a^0)$ is a TS over Σ_a and $f_a : Q_a \to \Phi$ is called an **information map**. ⊣

At a state $p \in Q_a$, if $f_a(p) = \chi$ then $\chi(b)$ is what M_a knows about M_b at p; strictly speaking, $\chi(b)$ is irrelevant in the theory presented here when $b = a$, but it is notationally simpler to have a uniform map.

Definition 5.9
An **epistemic transition system** (ETS) over the alphabet $(\widetilde{\Sigma}, \mathcal{C})$ is given by a tuple

$$\widetilde{M} = (M_1, \cdots, M_n, \langle f_1, f_2, \cdots, f_n \rangle, F),$$

where for each $a \in loc, (M_a = (Q_a, \longrightarrow_a, q_a^0), f_a)$ is an EA over $(\Sigma_a, \mathcal{C}_a)$, and $F \subseteq (Q_1 \times \ldots \times Q_n)$. ⊣

The global behaviour of $\widetilde{M}$ is given below as that of the product automaton $\widehat{M}$ associated with the system. The product automaton is not over the whole global state space but over global states where information at local states is mutually compatible.

Definition 5.10
Let $\widetilde{M} = (M_1, \cdots, M_n, \langle f_1, \cdots, f_n \rangle, F)$ be an ETS and let $Q = (Q_1 \times \ldots \times Q_n)$. A global state $(p_1, p_2, \cdots, p_n) \in Q$ is called **coherent** iff

$$\text{for all } a, b \in loc: \ f_a(p_a)(b) \preceq_b f_b(p_b)(b).$$

Denote the set of all coherent states of the ETS as $\widehat{Q}$. From now on, we consider only ETS's whose initial state $(q_1^0, \cdots, q_n^0)$ is coherent.

Definition 5.11
We definite the epistemic product automaton of an ETS $\widetilde{M}$ as

$$\widehat{M} = (\widehat{Q}, \longrightarrow, (q_1^0, \cdots, q_n^0), \widehat{F}) \text{ where}$$

1. $\widehat{Q}$ is the set of all coherent global states,

2. $(q_1^0, \cdots, q_n^0)$ is the initial state,
3. $\widehat{F} \subseteq \widehat{Q} \cap F$ is the set of final states,
4. $\longrightarrow \subseteq (\widehat{Q} \times \Sigma \times \widehat{Q})$ is the transition relation defined as: $(p_1, \cdots, p_n) \xrightarrow{x} (q_1, \cdots, q_n)$ iff
   (a) for all $a \notin loc(x), p_a = q_a$.
   (b) for all $a \in loc(x), p_a \xrightarrow{x}_a q_a$.
   (c) for all $a, b \in loc(x), f_a(q_a) = f_b(q_b)$. ⊣

The last condition (4c) above represents *perfect exchange*, a *synchrony* condition: immediately after a communication, agents participating in the action have complete information on the observables of all other participants, as well as a consensus on the observables of all non-participating agents. Now this is obviously a strong condition, and motivated by the characterisation we seek below.

Example 5.6
We illustrate epistemic products by an example. Consider the two processor mutual exclusion problem, whereby two processors sharing a resource must ensure that only one uses it at any point of time. Each process can be in one of two states: w_i (for waiting) or c_i (for being 'in the critical section'), where $i \in \{1, 2\}$. Processor i may be permitted to use the resource, or not, and can enter the critical section only when it knows that it is permitted and the other process is not permitted. This is achieved by using an epistemic alphabet $\{p_i, np_i, \perp\}$ for process i, as given in Figure 5.8, with obvious meaning. When process i waits, it knows it is not permitted, but nothing about the other process. The automata for processes are shown in Figures 5.9 and 5.10. A synchronisation can reveal the state of the process, thus learning that the other is not permitted and gaining permission for itself. The coherence condition on product states rules out the state (c_1, c_2) from the product shown in Figure 5.11.

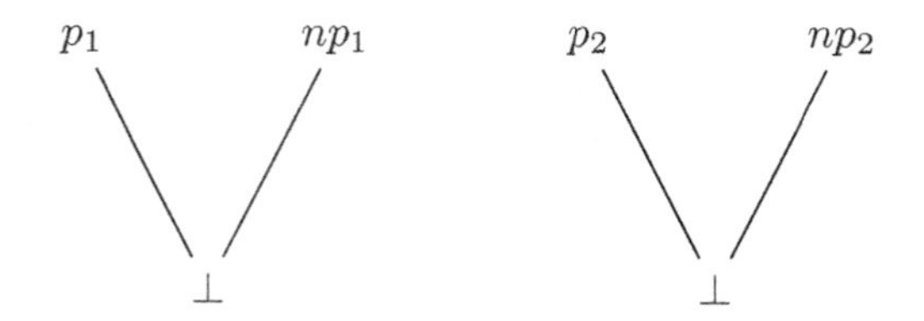

Figure 5.8: Alphabets for processes 1 and 2.

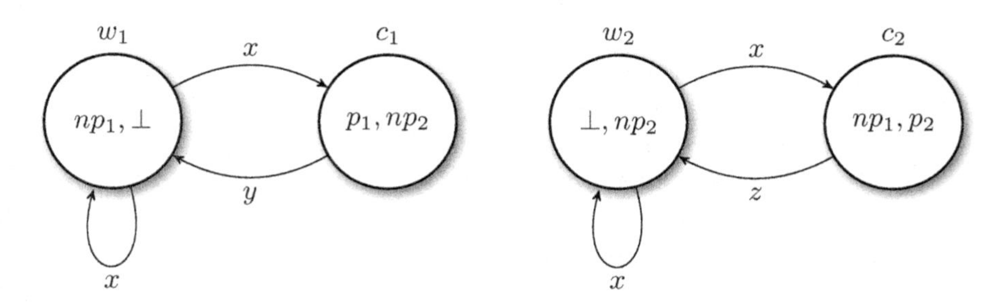

Figure 5.9: Process 1.

Figure 5.10: Process 2.

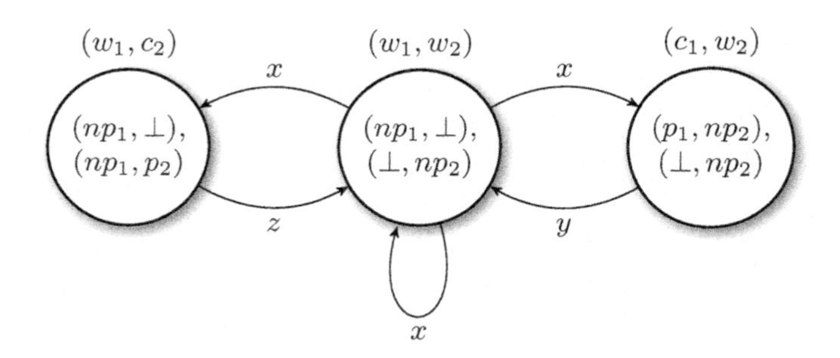

⊣

Figure 5.11: The epistemic product automaton for processes 1 and 2.

Note the explicit mechanism by which agents acquire information from others. In fact, this is an important motivation for studying these automata.

The class of languages over $\widetilde{\Sigma}$ accepted by ETS's is denoted as $ETS_{\widetilde{\Sigma}}$. Formally,

$$ETS_{\widetilde{\Sigma}} = \{L \subseteq \Sigma^* \mid \exists \widetilde{\mathcal{C}} \text{ and an ETS } \widetilde{M} \text{ over } (\widetilde{\Sigma}, \mathcal{C}) \text{ such that } L = L(\widehat{M}, F)\}.$$

From the definition of the epistemic product above, the following theorem can be proved.

Theorem 5.4
Given an ETS $\widetilde{M}$, checking whether $L(\widehat{M}, F) = \emptyset$ is decidable in time linear in the size of the epistemic product. ⊣

5.10.5 Behaviours

What kind of languages do we see inside $ETS_{\widetilde{\Sigma}}$? To answer this question, it is useful to consider them as being closed under a natural equivalence relation.

Given a distributed alphabet $\widetilde{\Sigma}$, let $\lceil$ be the component projection map: $\lceil: (\Sigma^* \times loc) \to \Sigma^*$ defined as:

$$u\lceil a = \begin{cases} \epsilon & \text{if } u = \epsilon, \\ v\lceil a & \text{if } u = vx \text{ and } a \notin loc(x), \text{ and} \\ (v\lceil a) \cdot x & \text{if } u = v \cdot x \text{ and } a \in loc(x). \end{cases}$$

We wish to study systems in which an agent, at a history h considers another h' to be possible because her information cannot distinguish between the two. This suggests that the behaviour of systems should be closed under the equivalence relation, where strings u and v are equated when they are indistinguishable by *any* agent in the system. The resulting languages are called *Regular consistent languages* (RCL) below, and these are essentially the same as *Regular trace languages* (RTL) of Mazurkiewicz.

Definition 5.12
Define the relation $\sim$ on Σ^* as: for all $u, v \in \Sigma^*$, $u \sim v \stackrel{\text{def}}{=}$ for all $a \in loc$, $u\lceil a = v\lceil a$. It is easy to see that $\sim$ is an equivalence. The equivalence classes of $\sim$ are called *traces.* $\dashv$

Note that the closure of a regular language under $\sim$ need not be regular. For example, if the distributed alphabet is $(\{x\}, \{y\})$ and $L = (xy)^*$ then the closure of L under $\sim$ is the language containing strings with equal number of x's and y's, which is not regular any more.

Definition 5.13
A language $L \subseteq \Sigma^*$ is said to be a *regular consistent language* iff $L \in Reg_\Sigma$ and is closed under $\sim$. $\dashv$

Let $\mathscr{L}RCL_{\widetilde{\Sigma}}$ denote the class of all regular consistent languages over $\widetilde{\Sigma}$. With a little bit of work, one can show that this class is closed under Boolean operations as well as a special kind of concatenation and its iteration.

Since the definition of epistemic product does include the asynchrony condition (4a), it can be easily proved that that $ETS_{\widetilde{\Sigma}} \subseteq RCL_{\widetilde{\Sigma}}$. Interestingly, we can show that the perfect exchange condition above implies that epistemic products exactly characterise regular consistent languages.

Theorem 5.5
$ETS_{\widetilde{\Sigma}} = RCL_{\widetilde{\Sigma}}$. $\dashv$

An important corollary of the theorem above is that the class $ETS_{\widetilde{\Sigma}}$ is closed under all the Boolean operations, and under homomorphisms, so that a Kleene-type theorem can be proved showing that epistemic automata can be equivalently described by a class of 'knowledge regular expressions'.

Another interesting application of these automata is that we can employ these automata to decide the satisfiability of a temporal logic of knowledge:

the models need to be restricted to be temporal evolutions of these systems, though the syntax and semantics of modalities remains the same. We can then construct an automaton A_φ for every formula φ such that the models of φ can be placed in bijection with the language of A_φ. Such a connection is important and interesting for the following reason: the formula only describes how agents' knowledge changes with time whereas the automaton explicitly specifies the mechanism by which agents acquire knowledge. Thus one can see the automata as finite state implementations of knowledge specifications in temporal logic.

The foregoing discussion is intended not to argue for the specific class of epistemic automata discussed here, but to show the relevance of automata theory to reasoning about knowledge and time, and its relation to concurrency theory. We remark that notions like epistemic homomorphisms may lead to an algebraic theory of knowledge composition.

5.11 Applications

The modal theory of knowledge and time has many applications, and these have been studied extensively by computer scientists. The major applications are in the theory of distributed systems, artificial intelligence, information security and game theory. Many of these latter applications are discussed in detail in this volume, in terms of information change. We mention here some applications where temporal reasoning about agents' knowledge is essential.

With the advent of the Internet on the one hand, and multicore architectures on the other, concurrent and distributed computation has emerged as a leading paradigm. One major area of concern with distributed programs is that they are hard to analyse and ensuring correctness is difficult. A principal reason for this is that a process in a distributed system has only a limited view of the global system and attempts to learn more about global states from communications. Ensuring that processes have the requisite knowledge when they need to act, is a design challenge.

Consider the following example that serves to illustrate the connection between epistemic reasoning and distributed algorithms. Consider 4 processes, that must agree whether to set one bit b (or not). For instance, this could be a closed valve that should be opened (or not). None of them can act autonomously. Each process i has its private value b_i, reflecting its opinion. Let $f : \{0,1\}^4 \rightarrow \{0,1\}$ be a decision function that is known to all the processes and is to be applied to aggregate opinions into a decision. We can set up this entire scenario in our model and show that there is consensus on the value b if and only if it is common knowledge among the 4 processes that b is the decided value.

A natural solution is for each process to broadcast its private value. Once every process has the same 4-bit vector, they all apply f and arrive at the consensus value. But such a solution depends on the assumption that communications and processes are reliable: that every communication is received, and by all intended recipients simultaneously, and that this fact is common knowledge among processes; that no process crashes in the middle of all this, that every process sends only correct values, and so on.

At first glance it is unclear that there is any need to spell out such assumptions in detail. In fact, relaxing these assumptions can make the difference between consensus being achievable or not. For instance, when communication is asynchronous (meaning that message delivery may be delayed unboundedly), even if only one process may crash during the algorithm execution, consensus is not achievable, according to the celebrated theorem of Fischer, Lynch and Paterson. Thanks to our remark above, we can arrive at the theorem by an epistemic route: we can show that in any run of a system that displays unbounded message delays, it can never be common knowledge that a message has been delivered.

On the other hand, when communications are instantaneous (or when message delays are uniformly bounded), even if some of the processes are unreliable, consensus can indeed be achieved. The unreliable processes could even be malicious or *Byzantine*: they could mislead other reliable processes about each other's values. If there are at most t Byzantine processes, consensus can be achieved with $t + 1$ rounds of communication. That $t + 1$ rounds are necessary is a theorem of Lynch and Fischer; once again this proof can be cast entirely in epistemic terms, and proved in the logic of knowledge and time.

Indeed, we need not use epistemic temporal logics only for the analysis of distributed algorithms and proving lower bounds. We could go further and use epistemic reasoning to *design* distributed algorithms as well. The structure of a distributed program consists of rules that specify when an agent may send which communication, and what action an agent should take on receipt of a communication. Clearly the former depends on the knowledge of the sender and the latter impacts on the knowledge of the recipient. These can be considered *knowledge based programs*, as suggested by Halpern, Fagin, Moses and Vardi. Such a design may be a significant improvement over intuitive design. For instance, in the case of Byzantine consensus, we can give an optimal knowledge based design that achieves consensus with $r + 1$ rounds where r is the number of processes that are *actually* Byzantine during the execution, as opposed to 'classical' algorithms that use $t+1$ rounds where t is the assumed bound on number of Byzantine processes (which might be significantly higher than r).

In the theory of distributed systems, agreement problems are not the only ones that are naturally modelled as knowledge. A fundamental problem is that of **mutual exclusion**, whereby processes sharing a resource must ensure that only one process uses the resource at a time. In the absence of a central coordinating / enforcing agency, the actions of processes are determined epistemically: a process taking control must know that no other process has access. More, when we wish to ensure some form of *fairness*, that every process in need of the resource eventually gets access, the epistemic analysis is tricky. Most distributed algorithms are based on such considerations of knowledge and eventuality.

There are also applications of immense practical benefit. The Internet-standard routing protocol for networks, called the Routing Information Protocol (RIP) uses *routers* that forward messages to neighbour routers. The idea is to forward each message along the path that will incur the lowest cost. Initially the network has distributed knowledge of the minimum cost path between any two nodes, and the objective of the protocol is to make this knowledge explicit for all routers. This knowledge transformation is naturally specified in the temporal logic of knowledge.

Another important aspect of distributed or multi-agent systems is that of **secrecy**, and more generally, security (see Chapter 12). Facts need to be shared among some agents, but also to be kept hidden from other agents. This requires public communication to reveal information to some but not to others, clearly a transformation on knowledge states of agents. Such formalisation can be carried out in the temporal logic of knowledge, but there are many tricky issues to be sorted out.

In the area of artificial intelligence, an important theme is that of general problem solving. One paradigm for this is distributed search of a problem's solution space. Wooldridge has suggested that an alternative is to use distributed problem solving agents who each have a knowledge base, and communicate. He proves that any problem that has a solution can be solved by agents that send their complete state in every message, and that distributed knowledge of such a system is non-diminishing. There is an extensive variety of applications of the temporal logic of knowledge to multi-agent systems.

5.12 In conclusion

Propositional temporal logic and propositional multi-agent epistemic logic are both formalisms that have been extensively studied from a theoretical perspective and have found many applications, especially in computer science and formal system verification. Combining them in a temporal logic of knowledge is natural, and gives rise to an interesting class of models, axiomatised and reasoned about in natural inference systems. However, the complexity of decision procedures is high in general.

The temporal logic of knowledge has a natural connection with the study of multi-agent systems and distributed algorithms, whereby agents that operate autonomously must communicate to learn about system states. This can be seen as temporal evolution of knowledge states, and this idea leads to a fine analysis of such algorithms. It also opens up epistemic semantics for message passing.

While the range of applications for this logic expands rapidly, many questions related to its mathematical foundations, especially relating to underlying computational models (or automata) and to expressiveness, need to be answered yet. Our knowledge about logics of knowledge and time can be expected to increase considerably in the near future.

5.13 Notes

Temporal logic was originally developed to represent tense in natural language (Prior, 1967). It has been used in the formal specification and verification of concurrent and distributed systems (Pnueli, 1977; Manna and Pnueli, 1995), program specification (Manna and Pnueli, 1992), temporal databases (Tansel, 1993), knowledge representation (Artale and Franconi, 1999), executable temporal logics (Barringer, Fisher, Gabbay, Owens, and Reynolds, 1996), and natural language (Steedman, 1997).

The axiomatisation of PTL, the discrete, linear, temporal logic with finite past and infinite future which is the temporal component of the combined logic presented in this chapter is given by Gabbay, Pnueli, Shelah, and Stavi (1980). The language is an extension of classical logic with the next and until operators. The until operator was proposed in the seminal work of Kamp (1968), who also proved that the temporal logic with the until and since (the past counterpart of until) operators is as expressive as first-order logic over words. Gabbay et al. (1980) show that any temporal formula can be rewritten as their past, present, and future component and that the future fragment is expressively complete. Topological characterisation of temporal properties is by Baier and Kwiatkowska (2000) and the until hierarchy is investigated by Etessami and Wilke (1996).

Computation tree logic, CTL, was introduced by Emerson and Clarke (1982). In the underlying model of time, as its name suggests, every state may have many successors in a tree-like structure. Formulae in CTL express properties over *paths*, where the path quantifier operators $\mathbb{A}$ (for all paths) and $\mathbb{E}$ (exists a path) can only occur by immediately preceding a temporal operator. PTL and CTL are expressively incomparable (Clarke and Draghicescu, 1989; Kupferman and Vardi, 1998), that is, there are formulae in PTL that cannot be expressed in CTL, and vice versa.

Formalisations including both time and knowledge have appeared in (Sato, 1977; Lehmann, 1984; Fagin, Halpern, and Vardi, 1991; Parikh and Ramanujam, 1985; Ladner and Reif, 1986). In a series of papers by Halpern and Vardi (1986, 1988b,a, 1989) and by Halpern, van der Meyden, and Vardi (2004), a uniform treatment of earlier approaches is given; also axiomatisations and issues of complexity for combined logics of linear time and knowledge, including interacting systems, are discussed. Axiomatisations and complexity results for epistemic branching-time logics are by van der Meyden and Wong (2003). The perfect recall and synchrony axiom was introduced by Lehmann (1984). The perfect recall axiom was introduced by van der Meyden (1994). No learning was first discussed in the context of blindfold games by Ladner and Reif (1986). Complete axiomatisations for temporal logics of knowledge with perfect recall over different flows of time, including arbitrary linear orders; the integers; the rationals; the reals; and for uniform flows of time (where the linear order is common knowledge to all agents), can be found in (Mikulas, Reynolds, and French, 2009). Results about axiomatisation and decidability of the propositional logics of knowledge with perfect recall under concrete semantics, i.e. where propositions are evaluated according to the set of observations of a given agent at a given world, is by Dima (2011).

Combinations of logics and their properties are discussed by Gabbay, Kurucz, Wolter, and Zakharyaschev (2003) and references therein. Properties that are inherited by fusions are given in (Kracht and Wolter, 1991; Spaan, 1993; Fine and Schurz, 1996).

Model checking knowledge and time has been studied by way of translation into temporal logic by van der Hoek and Wooldridge (2002); and Engelhardt, Gammie, and van der Meyden (2007) give a PSPACE algorithm.

The invariance theorem, which asserts that modal logic is the bisimulation invariant fragment of monadic first order logic (with one binary relation), was proved by van Benthem (1983). Fine (1970) studied the modal logic of propositional quantification, and Kaminski and Tiomkin (1996) showed that it is as expressive as full second order logic.

Propositional quantification for temporal logics was initially studied by Kesten and Pnueli (1995), and the logic of local propositions was proposed by Engelhardt, van der Meyden, and Moses (1998). For monodic fragments, see Hodkinson, Kontchakov, Kurucz, Wolter, and Zakharyaschev (2003).

The proof methods for temporal logics of knowledge combine inference rules for each of component logic and, of course, the inference rules for the classical language they share. An introduction to the tableau method for classical logic can be found in Smullyan (1968b). Resolution was first proposed by Robinson (1965).

The development of proof methods for temporal logic have followed three main approaches: tableaux (Wolper, 1983; Gough, 1984; Wolper, 1985), automata (Sistla, Vardi, and Wolper, 1987), and resolution (Cavalli and del Cerro, 1984; Abadi and Manna, 1985; Venkatesh, 1986; Fisher, 1991; Fisher, Dixon, and Peim, 2001). Each of these methods is a refutational method.

Tableaux for epistemic logics can be found in (Fitting, 1983; Goré, 1992; Halpern and Moses, 1992). The analytic cut rule for classical and modal logics and its properties are discussed in (Smullyan, 1968a; Fitting, 1983; Goré, 1999).

The resolution-based proof method for $\mathsf{KL}_{(n)}$ presented in this chapter is by Dixon and Fisher (2000b). The transformation rules and satisfiability preserving results are by Dixon, Fisher, and Wooldridge (1998), and by Dixon and Fisher (2000b). Algorithms for finding loops, used in the application of the temporal resolution inference rule, are by Dixon (1998). Termination, soundness and completeness results are by Dixon and Fisher (2000b). Extensions of this resolution method to deal with synchronous systems and either perfect recall or no learning can be found in (Dixon and Fisher, 2000a; Nalon, Dixon, and Fisher, 2004).

Tableaux for $\mathsf{KL}_{(n)}$ are by Wooldridge, Dixon, and Fisher (1998). Clausal tableaux methods for $\mathsf{KL}_{(1)}^{nl,sync}$ and $\mathsf{KL}_{(1)}^{pr,sync}$ are by Dixon, Nalon, and Fisher (2004).

History-based models for knowledge were introduced by Parikh and Ramanujam (1985). The relation between such models and interpreted systems is by Pacuit (2007). The formal relationship between Dynamic Epistemic Logics and temporal logics of knowledge is explicated by van Benthem, Gerbrandy, Hoshi, and Pacuit (2009).

Chandy and Misra (1986) first studied how chains of communications can be used to cause changes in agents' knowledge. Levels of knowledge were formally studied by Parikh (1986), where the regularity property is proved. Parikh and Ramanujam (2003) studied knowledge based semantics of messages, and Pacuit and Parikh (2004) developed the model further. Mohalik and Ramanujam (2010) proposed epistemic automata and proved the theorem characterising regular consistent languages.

Fagin, Halpern, Moses, and Vardi (1995) discuss knowledge-based programs and the applications of the temporal logic of knowledge to distributed algorithms, Halpern and Moses (1990) proved the formal relationship between common knowledge and achieving consensus. Dwork and Moses (1990) derived an *optimal* knowledge-based algorithm for byzantine agreement. The impossibility theorem of Fischer, Lynch and Paterson, the lower bound for Byzantine agreement by Lynch and Fischer, and other theorems on distributed agreement can be found in (Lynch, 1996).

For applications of the temporal logic of knowledge to information security, see Halpern and van der Meyden (2001), van Ditmarsch (2003), Ramanujam and Suresh (2005), and Baskar, Ramanujam, and Suresh (2007). For applications to problem solving in artificial intelligence, see Wooldridge and Jennings (1999), and Dixon (2006).

The quote from Wittgenstein appears in page 223 of (Wittgenstein, 2001).

Acknowledgements The authors thank Bożena Woźna-Szcześniak for her comments on the previous version of this chapter. Cláudia Nalon would like to thank the Department of Computer Science at the University of Liverpool for hosting her during the time she worked on this paper. She was partially funded by the National Council for the Improvement of Higher Education (CAPES Foundation, BEX 8712/11-5). Ram Ramanujam would like to thank Rohit Parikh for discussions on the chapter's contents.

References

Abadi, M. and Z. Manna (1985). Nonclausal Temporal Deduction. *Lecture Notes in Computer Science 193*, 1–15.

Artale, A. and E. Franconi (1999). Introducing Temporal Description Logics. In C. Dixon and M. Fisher (Eds.), *Proceedings of the 6th International Workshop on Temporal Representation and Reasoning (TIME-99)*, Orlando, Florida. IEEE Computer Society Press.

Baier, C. and M. Kwiatkowska (2000). On topological hierarchies of temporal properties. *Fundam. Inf. 41*(3), 259–294.

Barringer, H., M. Fisher, D. Gabbay, R. Owens, and M. Reynolds (1996, May). *The Imperative Future: Principles of Executable Temporal Logic.* Research Studies Press.

Baskar, A., R. Ramanujam, and S. P. Suresh (2007). Knowledge-based modelling of voting protocols. In D. Samet (Ed.), *TARK*, pp. 62–71.

van Benthem, J. (1983). *Modal Logic and Classical Logic.* Bibliopolis.

van Benthem, J., J. Gerbrandy, T. Hoshi, and E. Pacuit (2009). Merging frameworks for interaction. *J. Philosophical Logic 38*(5), 491–526.

Cavalli, A. and L. F. del Cerro (1984). A Decision Method for Linear Temporal Logic. In R.E.Shostak (Ed.), *Proceedings of the 7th International Conference on Automated Deduction*, Volume 170 of *Lecture Notes in Computer Science*, pp. 113–127. Springer-Verlag.

Chandy, K. M. and J. Misra (1986). How processes learn. *Distributed Computing 1*(1), 40–52.

Clarke, E. M. and I. A. Draghicescu (1989, June). Expressibility results for linear-time and branching-time logics. In *Linear Time, Branching Time and Partial Order in Logics and Models for Concurrency.*, Berlin - Heidelberg - New York, pp. 428–437. Springer.

Dima, C. (2011). Non-axiomatizability for linear temporal logic of knowledge with concrete observability. *Journal of Logic and Computation 21*(6), 939–958.

van Ditmarsch, H. P. (2003). The russian cards problem. *Studia Logica 75*(1), 31–62.

Dixon, C. (1998). Temporal Resolution using a Breadth-First Search Algorithm. *Annals of Mathematics and Artificial Intelligence 22*, 87–115.

Dixon, C. (2006). Using Temporal Logics of Knowledge for Specification and Verification–a Case Study. *Journal of Applied Logic 4*(1), 50–78.

Dixon, C. and M. Fisher (2000a). Clausal Resolution for Logics of Time and Knowledge with Synchrony and Perfect Recall. In *Proceedings of ICTL 2000*, Leipzig, Germany.

Dixon, C. and M. Fisher (2000b, July). Resolution-Based Proof for Multi-Modal Temporal Logics of Knowledge. In S. Goodwin and A. Trudel (Eds.), *Proceedings of the Seventh International Workshop on Temporal Representation and reasoning (TIME'00)*, Cape Breton, Nova Scotia, Canada, pp. 69–78. IEEE Computer Society Press.

Dixon, C., M. Fisher, and M. Wooldridge (1998). Resolution for Temporal Logics of Knowledge. *Journal of Logic and Computation 8*(3), 345–372.

Dixon, C., C. Nalon, and M. Fisher (2004). Tableaux for logics of time and knowledge with interactions relating to synchrony. *Journal of Applied Non-Classical Logics 14*(4/2004), 397–445.

Dwork, C. and Y. Moses (1990). Knowledge and common knowledge in a byzantine environment: Crash failures. *Inf. Comput. 88*(2), 156–186.

Emerson, E. A. and E. M. Clarke (1982). Using branching time temporal logic to synthesize synchronization skeletons. *Science of Computer Programming 2*(3), 241 - 266.

Engelhardt, K., P. Gammie, and R. van der Meyden (2007). Model checking knowledge and linear time: Pspace cases. In S. N. Artëmov and A. Nerode (Eds.), *LFCS*, Volume 4514 of *Lecture Notes in Computer Science*, pp. 195–211. Springer.

Engelhardt, K., R. van der Meyden, and Y. Moses (1998). Knowledge and the logic of local propositions. In I. Gilboa (Ed.), *TARK*, pp. 29–41. Morgan Kaufmann.

Etessami, K. and T. Wilke (1996). An until hierarchy for temporal logic. In *In 11th Annual IEEE Symposium on Logic in Computer Science*, pp. 108–117. IEEE Computer Society Press.

Fagin, R., J. Y. Halpern, Y. Moses, and M. Y. Vardi (1995). *Reasoning About Knowledge.* MIT Press.

Fagin, R., J. Y. Halpern, and M. Y. Vardi (1991). A model-theoretic analysis of knowledge. *Journal of the ACM 38*(2), 382–428.

Fine, K. (1970). Propositional quantifiers in modal logic. *Theoria 36*, 336–346.

Fine, K. and G. Schurz (1996). Transfer theorems for stratified modal logics. In J. Copeland (Ed.), *Logic and Reality: Essays in Pure and Applied Logic. In Memory of Arthur Prior*, pp. 169–213. Oxford University Press.

Fisher, M. (1991, August). A Resolution Method for Temporal Logic. In *Proceedings of the Twelfth International Joint Conference on Artificial Intelligence (IJCAI)*, Sydney, Australia, pp. 99–104. Morgan Kaufman.

Fisher, M., C. Dixon, and M. Peim (2001, January). Clausal Temporal Resolution. *ACM Transactions on Computational Logic 2*(1), 12–56.

Fitting, M. (1983). *Proof Methods for Modal and Intuitionistic Logics.* Synthese Library. Springer.

Gabbay, D., A. Pnueli, S. Shelah, and J. Stavi (1980, January). The Temporal Analysis of Fairness. In *Proceedings of the 7th ACM Symposium on the Principles of Programming Languages*, Las Vegas, Nevada, pp. 163–173.

Gabbay, D. M., A. Kurucz, F. Wolter, and M. Zakharyaschev (2003). *Many-Dimensional Modal Logics: Theory and Applications.* Amsterdam: Elsevier.

Goré, R. (1992, June). *Cut-Free Sequent Systems for Propositional Normal Modal Logics.* Ph. D. thesis, Computer Laboratory, University of Cambridge, U.K. (Technical report no. 257).

Goré, R. (1999). Tableau methods for modal and temporal logics. In D. Gabbay, R. Hähnle, and J. Posegga (Eds.), *Handbook of Tableau Methods.* Springer.

Gough, G. D. (1984, October). Decision Procedures for Temporal Logic. Master's thesis, Manchester University. Also University of Manchester, Department of Computer Science, Technical Report UMCS-89-10-1.

Halpern, J. Y. and R. van der Meyden (2001). A logic for SDSI's linked local name spaces. *Journal of Computer Security 9*(1/2), 105–142.

Halpern, J. Y., R. van der Meyden, and M. Y. Vardi (2004, March). Complete axiomatizations for reasoning about knowledge and time. *SIAM J. Comput. 33*(3), 674–703.

Halpern, J. Y. and Y. Moses (1990). Knowledge and common knowledge in a distributed environment. *J. ACM 37*(3), 549–587.

Halpern, J. Y. and Y. Moses (1992, April). A guide to completeness and complexity for modal logics of knowledge and belief. *Artif. Intell. 54*(3), 319–379.

Halpern, J. Y. and M. Y. Vardi (1986, May). The Complexity of Reasoning about Knowledge and Time: Extended Abstract. In *Proceedings of the Eighteenth Annual ACM Symposium on Theory of Computing*, Berkeley, California, pp. 304–315.

Halpern, J. Y. and M. Y. Vardi (1988a, May). Reasoning about Knowledge and Time in Asynchronous Systems. In *Proceedings of the Twentieth Annual ACM Symposium on Theory of Computing*, Chicago, Illinois, pp. 53–65.

Halpern, J. Y. and M. Y. Vardi (1988b, April). The Complexity of Reasoning about Knowledge and Time: Synchronous Systems. Technical Report RJ 6097, IBM Almaden Research Center, San Jose, California.

Halpern, J. Y. and M. Y. Vardi (1989). The Complexity of Reasoning about Knowledge and Time. I Lower Bounds. *Journal of Computer and System Sciences 38*, 195–237.

Hodkinson, I. M., R. Kontchakov, A. Kurucz, F. Wolter, and M. Zakharyaschev (2003). On the computational complexity of decidable fragments of first-order linear temporal logics. In *TIME*, pp. 91–98. IEEE Computer Society.

van der Hoek, W. and M. Wooldridge (2002). Model checking knowledge and time. In D. Bosnacki and S. Leue (Eds.), *SPIN*, Volume 2318 of *Lecture Notes in Computer Science*, pp. 95–111. Springer.

Kaminski, M. and M. L. Tiomkin (1996). The expressive power of second-order propositional modal logic. *Notre Dame Journal of Formal Logic 37*(1), 35–43.

Kamp, H. (1968). *On tense logic and the theory of order*. Ph. D. thesis, UCLA.

Kesten, Y. and A. Pnueli (1995). A complete proof system for qptl. In *LICS*, pp. 2–12. IEEE Computer Society.

Kracht, M. and F. Wolter (1991, December). Properties of independently axiomatizable bimodal logics. *The Journal of Symbolic Logic 56*(4), 1469–1485.

Kupferman, O. and M. Y. Vardi (1998). Freedom, weakness, and determinism: From linear-time to branching-time. In *LICS*, pp. 81–92. IEEE Computer Society.

Ladner, R. E. and J. H. Reif (1986). The Logic of Distributed Protocols (Preliminary Report). In J. Y. Halpern (Ed.), *Theoretical Aspects of Reasoning about Knowledge: Proceedings of the First Conference*, Los Altos, California, pp. 207–222. Morgan Kaufmann Publishers, Inc.

Lehmann, D. (1984). Knowledge, common knowledge and related puzzles (extended summary). In *Proceedings of the third annual ACM symposium on Principles of distributed computing*, PODC '84, New York, NY, USA, pp. 62–67. ACM.

Lynch, N. A. (1996). *Distributed Algorithms*. Morgan Kaufmann.

Manna, Z. and A. Pnueli (1992). *The Temporal Logic of Reactive and Concurrent Systems: Specification*. New York: Springer-Verlag.

Manna, Z. and A. Pnueli (1995). *Temporal Verification of Reactive Systems: Safety*. New York: Springer-Verlag.

van der Meyden, R. (1994). Axioms for Knowledge and Time in Distributed Systems with Perfect Recall. In *Logic in Computer Science*, pp. 448–457.

van der Meyden, R. and K. Wong (2003). Complete axiomatizations for reasoning about knowledge and branching time. *Studia Logica 75*, 93–123.

Mikulas, S., M. Reynolds, and T. French (2009). Axiomatizations for temporal epistemic logic with perfect recall over linear time. In *Proceedings of the 2009 16th International Symposium on Temporal Representation and Reasoning*, TIME '09, Washington, DC, USA, pp. 81–87. IEEE Computer Society.

Mohalik, S. and R. Ramanujam (2010). Automata for epistemic temporal logic with synchronous communication. *Journal of Logic, Language and Information 19*(4), 451–484.

Nalon, C., C. Dixon, and M. Fisher (2004, September 9–11). Resolution for synchrony and no learning. In R. A. Schmidt, I. Pratt-Hartmann, M. Reynolds, and H. Wansing (Eds.), *AiML-2004: Advances in Modal Logic*, Volume 04-09-01 of *UMCS*, Manchester, UK, pp. 303–317. University of Manchester.

Pacuit, E. (2007). Some comments on history based structures. *J. Applied Logic 5*(4), 613–624.

Pacuit, E. and R. Parikh (2004). The logic of communication graphs. In J. A. Leite, A. Omicini, P. Torroni, and P. Yolum (Eds.), *DALT*, Volume 3476 of *Lecture Notes in Computer Science*, pp. 256–269. Springer.

Parikh, R. (1986). Levels of knowledge in distributed computing. In *LICS*, pp. 314–321. IEEE Computer Society.

Parikh, R. and R. Ramanujam (1985). Distributed processes and the logic of knowledge. In R. Parikh (Ed.), *Logic of Programs*, Volume 193 of *Lecture Notes in Computer Science*, pp. 256–268. Springer.

Parikh, R. and R. Ramanujam (2003). A knowledge based semantics of messages. *Journal of Logic, Language and Information 12*(4), 453–467.

Pnueli, A. (1977, November). The Temporal Logic of Programs. In *Proceedings of the 18th Symposium on the Foundations of Computer Science*, Providence.

Prior, A. (1967). *Past, Present and Future.* Oxford University Press.

Ramanujam, R. and S. P. Suresh (2005). Deciding knowledge properties of security protocols. In R. van der Meyden (Ed.), *TARK*, pp. 219–235. National University of Singapore.

Robinson, J. A. (1965, January). A Machine–Oriented Logic Based on the Resolution Principle. *ACM Journal 12*(1), 23–41.

Sato, M. (1977). A study of Kripke-style methods for some modal logics by Gentzen's sequential method. *Publications Research Institute for Mathematical Sciences, Kyoto University 13*(2), 381–468.

Sistla, A. P., M. Vardi, and P. Wolper (1987). The Complementation Problem for Büchi Automata with Applications to Temporal Logic. *Theoretical Computer Science 49*, 217–237.

Smullyan, R. M. (1968a, December). Analytic Cut. *Journal of Symbolic Logic 33*(4), 560–564.

Smullyan, R. M. (1968b). *First-Order Logic*, Volume 43 of *Ergebnisse der Mathematik und ihrer Grenzgebiete.* Springer-Verlag, New York.

Spaan, E. (1993). *Complexity of Modal Logics.* Ph. D. thesis, University of Amsterdam.

Steedman, M. (1997). Temporality. In *Handbook of Logic and Language*, North Holland, pp. 895–935. Elsevier.

Tansel, A. (1993). *Temporal Databases: Theory, Design, and Implementation.* Benjamin/Cummings.

Venkatesh, G. (1986). A Decision Method for Temporal Logic based on Resolution. *Lectures Notes in Computer Science 206*, 272–289.

Wittgenstein, L. (1953 / 2001). *Philosophical Investigations.* Blackwell Publishing.

Wolper, P. (1983). Temporal Logic Can Be More Expressive. *Information and Control 56*, 72–99.

Wolper, P. (1985, June-Sept). The Tableau Method for Temporal Logic: An overview. *Logique et Analyse 110–111*, 119–136.

Wooldridge, M., C. Dixon, and M. Fisher (1998). A tableau-based proof method for temporal logics of knowledge and belief. *Journal of Applied Non-Classical Logics 8*(3), 225–258.

Wooldridge, M. and N. R. Jennings (1999). The cooperative problem-solving process. *J. Log. Comput. 9*(4), 563–592.

Chapter 6

Dynamic Epistemic Logic

Lawrence S. Moss

Contents

Abstract

Dynamic Epistemic Logic (**DEL**) extends the logical system of Chapter 1 by adding operators corresponding to *epistemic actions* of various sorts. The most basic epistemic action is a public announcement of a given sentence to all agents. We add an announcement operator to the syntax of epistemic logic. The semantics works in a fairly natural way, but even here there is a subtlety: what should we do about announcements in a given state of sentence which are false there? Once we settle on the semantics, we begin the traditional work of providing a logical system for the valid sentences, providing a proof theory, looking at the expressive power, etc. The system we study is called **PAL**, and we discuss much of what is known about it. It turns out that **PAL** without common knowledge is reducible to standard epistemic logic: the announcement operators may be

Chapter 6 of the *Handbook of Epistemic Logic*, H. van Ditmarsch, J.Y. Halpern, W. van der Hoek and B. Kooi (eds), College Publications, 2015, pp. 261–312.

translated away. However, this changes when we have common knowledge operators.

The second part of our chapter is devoted to more general epistemic actions, such as private announcements of various sorts. It is more complicated to propose a logical system with syntax, semantics, and proof theory. It is also more interesting, since the subject comes alive with extensions that go beyond public announcements. At the same time, much more is known about **PAL** than about the more general logics in the **DEL** family.

This chapter is intended for the reader familiar with epistemic logic who wants to learn enough about the subject to then read the main sources. In this tutorial introduction, we have tried to emphasize pictures, ideas, and motivations, and to de-emphasize technical details.

A students' proverb says 'No one knows everything but true wisdom is to know whom to ask'. This emphasizes that knowledge can consist not only of ground facts but also can involve statements about someone else's knowledge. Reasoning about knowledge is characteristic especially for the situations where information is exchanged.

6.1 Introduction

The quote above is the very opening of one of the primary sources for the topic of this chapter, Dynamic Epistemic Logic (**DEL**). As the quote indicates, **DEL** is a set of extensions to the basic forms of epistemic logic which we saw in the Introduction (Chapter 1). The overall aim is to introduce *epistemic actions* of various sorts along with models, and then to formulate logical languages which explicitly incorporate those actions. The emphasis on actions is what puts the D in **DEL**, and the overall way that we think about actions in **DEL** is strongly influenced by propositional dynamic logic. So in this chapter we are going to introduce logics semantically (rather than proof-theoretically), and the guiding principle is that we should attempt to provide logical languages suitable for the understanding of "situations where information is exchanged." And then once we do this, we may seek results of the kind that we would pursue in all semantically-introduced logics: complete axioms systems, decidability and expressivity results, etc.

As we understand it, **DEL** is a field driven by examples and focused on various features of information interchange, especially by people. These features are shared by epistemic logic (**EL**) more generally, of course. And as in other areas of **EL**, the notion of *knowledge* that is used is always *provisional.* One frequently makes assumptions that are too strong to model flesh-and-blood people, such as logical omniscience; we do this knowing that this assumption is an over-simplification, but that the simple models which we use are illuminating despite it. For this reason, and for others as well, we take **DEL** to be a *family* of logics, not a single one, because once we start to model and use the notion of an epistemic action, we naturally formulate a basic system and then consider many extensions of it.

6.1.1 Scenarios

One of the charming features of **DEL** is that the subject may be introduced using various *epistemic scenarios.* These are short stories featuring one or more agents – usually there are two or three agents – and usually with some change in the knowledge states of the agents. The most basic of these features indifference of two agents about one proposition.

Scenario 1 (Two agents enter a room)
There are two agents, called here A and B. Their actual names are *Amina (A)* (female) and *Bao (B)* (male) in this story. (Of course, nothing hinges on this, and they might as well be the more traditional *Ann* and *Bob.*) A and B together enter a room. At the other side lies a remote-control mechanical coin flipper. One of the agents presses a button, and a coin spins through the air, landing in a small box on a table in front. The box closes. The two people are much too far to see the coin. In reality, the coin shows heads.

We are going to present this scenario in some detail, not because it is terribly interesting or important, but rather because it is simple enough for us to do a thorough job. Our later scenarios are going to be treated in a more abbreviated manner. Let us first make a chart of our intuitions in English and their symbolizations in **EL**. We take At to consist of atomic sentences H and T. We also use lower-case letters a and b in the knowledge modalities K_a and K_b corresponding to A and B.

The coin shows heads.	H
The coin doesn't show tails.	$\neg\mathsf{T}$
A doesn't know that the coin shows heads.	$\neg K_a\mathsf{H}$
B doesn't know that the coin shows heads.	$\neg K_b\mathsf{H}$
A knows that she doesn't know that the coins shows heads.	$K_a\neg K_a\mathsf{H}$
A knows that B doesn't know that the coins shows heads.	$K_a\neg K_b\mathsf{H}$

Here is how we draw a picture of this model:

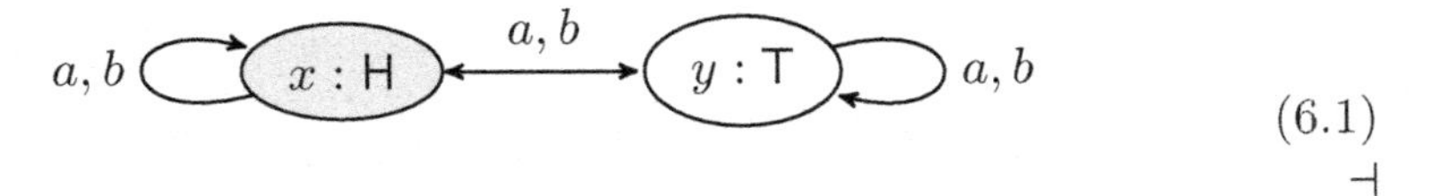

(6.1)

⊣

Note that the model has a real world, and it is x, the shaded world. Now the *justification* for the model is that it makes our intuitions come out true. More precisely, the translations of our intuitions are formally true at the real world. This is easy to check, and we assume that the reader has enough experience with **EL** to understand this point.

Scenario 2 (The coin revealed to show heads)
Now A and B walk up to the box. One opens the box and puts the coin on the table for both to see. It's H, as we know. The model now is a single point.

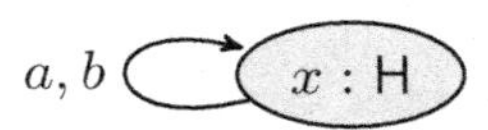

We can write intuitions in English, formalize them, and then check that the model satisfies the formal sentences.

English	formalization
A knows that the coin shows heads	$K_a\mathsf{H}$
B does not know that the coin shows tails	$\neg K_b\mathsf{T}$
A knows that B knows that the coin shows heads	$K_a K_b\mathsf{H}$

Incidentally, there is an alternative representation of this same scenario.

One can again check that the "real world" of this model satisfies the exact same sentences as the "real world" of the one-point model. (Even better: there is a bisimulation between the models which relates the real worlds, and so we have a very general reason to call the models equivalent.)

The opening of the box is our first example of an epistemic action. We shall model this as a *public announcement.*

To be sure, the very first scenario contained an epistemic action, too: before they entered the room, Amina and Bao might not have known what they were in for. And so *raising the issue* of the coin is itself an interesting epistemic action.

Scenario 3 (Amina looks)
After the first scenario "Two people enter a room", A opens the box herself. The coin is lying heads up, as we know. B observes A open the box but does not see the coin. And A also does not disclose whether it is heads or tails.

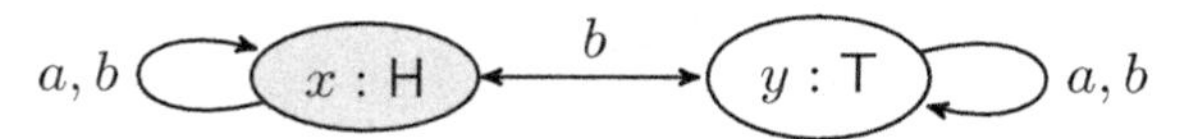

No matter which alternative holds, B would consider both as possible, and A would be certain which was the case. Again, we make a chart of our intuitions:

English	formalization
A knows that the coin shows heads	$K_a\mathsf{H}$
B does not know that the coin shows heads	$\neg K_b\mathsf{H}$
B does not know that A knows that the coin shows heads	$\neg K_b K_a\mathsf{H}$
B does not know that A doesn't know that the coin doesn't show heads	$\neg K_b\neg K_a\neg\mathsf{H} \equiv \neg K_b L_a\mathsf{H}$
B knows that either A knows it's heads, or that A knows it's tails	$K_b(K_a\mathsf{H} \vee K_a\mathsf{T})$
A knows that B doesn't know that the coin shows heads	$K_a\neg K_b\mathsf{H}$

One can check these in the real world of the model, and this lends credence to the model. In the other direction, the fact that we can make models like this adds credence to the enterprise of **EL**.

This kind of semi-private, semi-public looking is another epistemic action. We eventually will abstract this kind of action from this particular scenario. That is, we'll develop a formal definition of what it means to apply this action to a given model and thereby obtain a new model. ⊣

Scenario 4 (Cheating)
After the opening scenario, A secretly opens the box herself. B does not observe A open the box, and indeed A is certain that B did not suspect that anything happened after they sat down.

The main point is that after A looks, B does not think that the actual state of the world is even possible! So in addition to the way the world actually is, *or rather as part of it*, we need to consider other worlds to represent the possibilities that B must entertain.

Here is our proposal for a model:

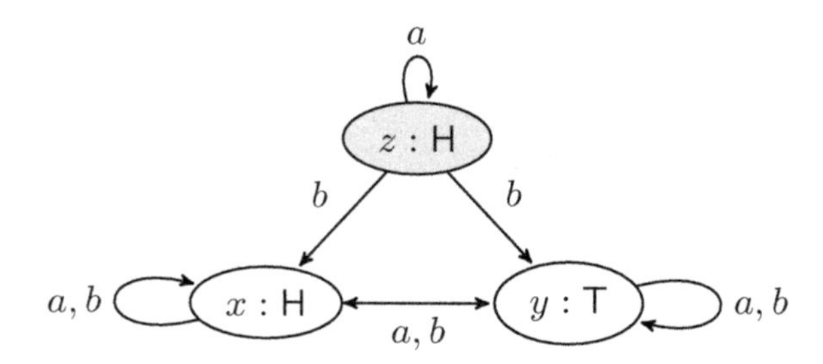

It is by no means obvious that this model is "correct". Notice that Bao's accessibility relation is not an equivalence relation: it is not even reflexive. It makes sense to change our intuitions to refer to his *belief* rather than his knowledge. That is, we use the same logical symbol K_b, but we understand this to be "knowledge" (with scare quotes), or belief. At this point, we justify the model by making intuitions and checking them in the real world.

English	formalization
A knows that the coin shows heads	$K_a\mathsf{H}$
B does not know that the coin shows heads	$\neg K_b\mathsf{H}$
B "knows" that A does not know that the coin shows heads	$K_b\neg K_a\mathsf{H}$
A knows that B doesn't know that the coin shows heads	$K_a\neg K_b\mathsf{H}$

The model indeed matches our intuitions about the sentences above, and we are not aware of *any* sentence that seems intuitively correct but whose translation fails in the model. The only exceptions to this are sentences whose meaning challenges the rendering of English words by the logical symbols (for example, rendering natural language conditionals by the material conditional $\rightarrow$), or ones which trade on the problem of logical omniscience. Still, even with a model that seems to be what we want, the question invariably comes up concerning the worlds x and y. Why do we need them to model cheating by A?

Scenario 5 (Double Cheating)
After A sneaks a peak, B does the same thing. Neither even think it is

possible that the other one looked.

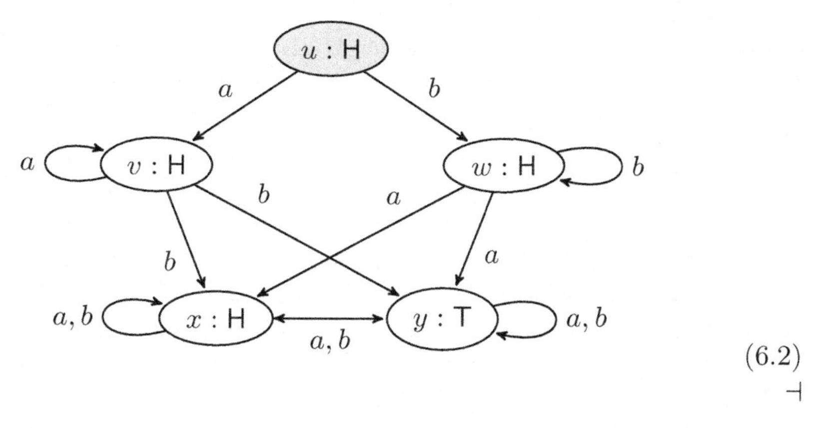

(6.2)

⊣

This can be *checked* just as we checked the previous models: by formalizing intuitions and checking them at the "real world" u. But it's a challenge to come up with the model in the first place. We will see a systematic way of getting this model, and many similar ones.

6.1.2 General notes on this chapter

Dynamic epistemic logic attempts to model and study changes to epistemic models, following what we saw in the scenarios of Section 6.1.1. The idea is to formulate logical languages which have the nice features of epistemic logic, such as decidability, and at the same time are expressive enough to formulate *epistemic actions* in a convenient way.

The simplest kind of epistemic action is a public announcement. We saw an example of this in Scenario 2, when two agents were informed together about the state of a hidden coin. There are a few features of our modeling which make this especially simple. The main one is that our description of the action is that the agents see the coin with their own eyes, together. Thus they both can "update their state" by deleting the world the coin lies tails up. This deletion applies to the other agent as well, and this is important. If the communication came over a loudspeaker and was a little garbled, for example, then each agent might not be 100% sure that it got the message clearly. Thus, a representation might need to keep "ghost states" where the coin lies tails up, if only to represent one agent's unsure state regarding the other's state. For this reason, a "hard" announcement is the simplest kind. We study these in Section 6.2, and the theme of the section is that even though the idea behind public announcements is very easy, the resulting logical system turns out to lead to a rich mathematical theory. For example, there are several ways to axiomatize its valid sentences, and these ways correspond to interesting directions that are taken in other aspects of the theory. Section 6.2 is intended to mention most of what is known about Public Announcement Logic (**PAL**). A few of the additions to **PAL** that have received some attention in the literature are found in Section 6.3. However, there are many more ways to extend **PAL** than I could cover in a reasonable way, and so Section 6.3 is partly a set of pointers.

The scenarios of Section 6.1.1 were also intended to show more complicated kinds of epistemic actions. These are the topic of Section 6.4. The goal of the work is to study epistemic actions in general, and to propose very expressive languages. In a very rough way, the difference between the study of public announcements and the much more general field is analogous to the difference between biography and sociology: they both have their place, and those places are different. Accordingly, my goal in Section 6.4 is different from the goal in earlier parts of the chapter. I am much more concerned with intuitions and basic ideas, and less with the technical details.

Intended readers This chapter is mainly for people new to the area, and secondarily for people who are familiar with some of it and who want an introduction to new topics. I assume that readers are familiar with the Introduction to this Handbook, and for the most part I try to adhere to the notations and terminology set out there. I tried to include more examples and pictures than papers usually feature, mostly because I find them to be an attractive feature of the subject. In fact, much of the lighter material in this chapter derives from course notes that I have given over the years, teaching **DEL** as the last topic in a class on modal logic for undergraduates, including many who have not studied logic. Getting back to the style of the chapter, many results are merely stated, a few have proof sketches, and when the proofs are short, I gave them in detail.

6.2 Public Announcements and PAL

The most basic epistemic action is a *public announcement.* By this we mean an action of informing all of the agents of some sentence φ (stated in some logic – more on that below), and having all of the agents simultaneously accept φ. The way that this is modeled is by taking a given model M and throwing out all of the worlds where φ is false. This means that in any world w, all of the agents will only consider as possible the worlds where φ is true. Moreover, the publicity of the announcement means that each agent will know that she herself and all of the the others have done the same things.

Two remarks Before we turn to the modeling of this notion of an announcement, we should elaborate a bit and discuss the more delicate matter of whether announcements need to be true. Of course, in real life, an announcement is not taken to be absolutely true; usually one typically keeps in mind the fallibility of both the speaker and hearer. And then the modeling would involve *belief revision.* In this chapter, we are going to ignore all of these considerations, because it simplifies the theory. As a result, much more is known about the formal properties of logical systems for representing true public announcements than about any other type of epistemic action. For more on modeling that uses belief revision, see Chapter 6 of this handbook, "Belief Revision in Dynamic Epistemic Logic."

We also should comment that in a group setting, making a public announcement of some sentence φ is different than privately announcing φ to all of the agents individually. To see this, let us return to the scenarios at the beginning of our chapter. We started with Scenario 1, representing two agents in a situation where there is common knowledge of ignorance about a single proposition. Announcing to each privately that the coin lies heads up is tantamount to having each of them "cheat" and look on their own, and so a reasonable representation would be that of Scenario 5. However, a public announcement of this same fact should result in the model of Scenario 2.

What we described above is an informal account of public announcements, and at this point we can give a formal story. If φ is a sentence in some language $\mathscr{L}$, and $M = (S, R, V)$ a multi-agent model[1] in which we can evaluate $\mathscr{L}$-sentences at worlds, then we write M_φ for the *relativization* of M to the worlds where φ holds. In more detail, M_φ would be a model, call it $N = (S, R, V)$ given as follows:

$$\begin{array}{lcl} N & = & \{w \in M : w \models \varphi \text{ in } M\} \\ w\ R_a w' & \text{iff} & w\ R_a w' \text{ in } M \\ V(s)(p) & = & V(s)(p) \text{ in } M \end{array}$$

Example 6.1

Examples of relativization are provided by the *Muddy Children*[2] scenarios. A scenario with three children is shown in Figure 6.1. We have three children who played in the mud. Thus each may be muddy or not, and so there are $2^3 = 8$ possibilities. To model this, we take atomic sentences D_A, D_B, and D_C. (D stands for "dirty"; using M would be confusing since we use this for "model.") The eight possible worlds are named $w_0, \ldots, w_7$, as shown in the bottom of Figure 6.1. We assume that all three children are muddy. One

[1] We are adopting the notation from Definition 3 in the Introduction.

[2] Although this example is over-used, we sense that no survey article on this subject can avoid it.

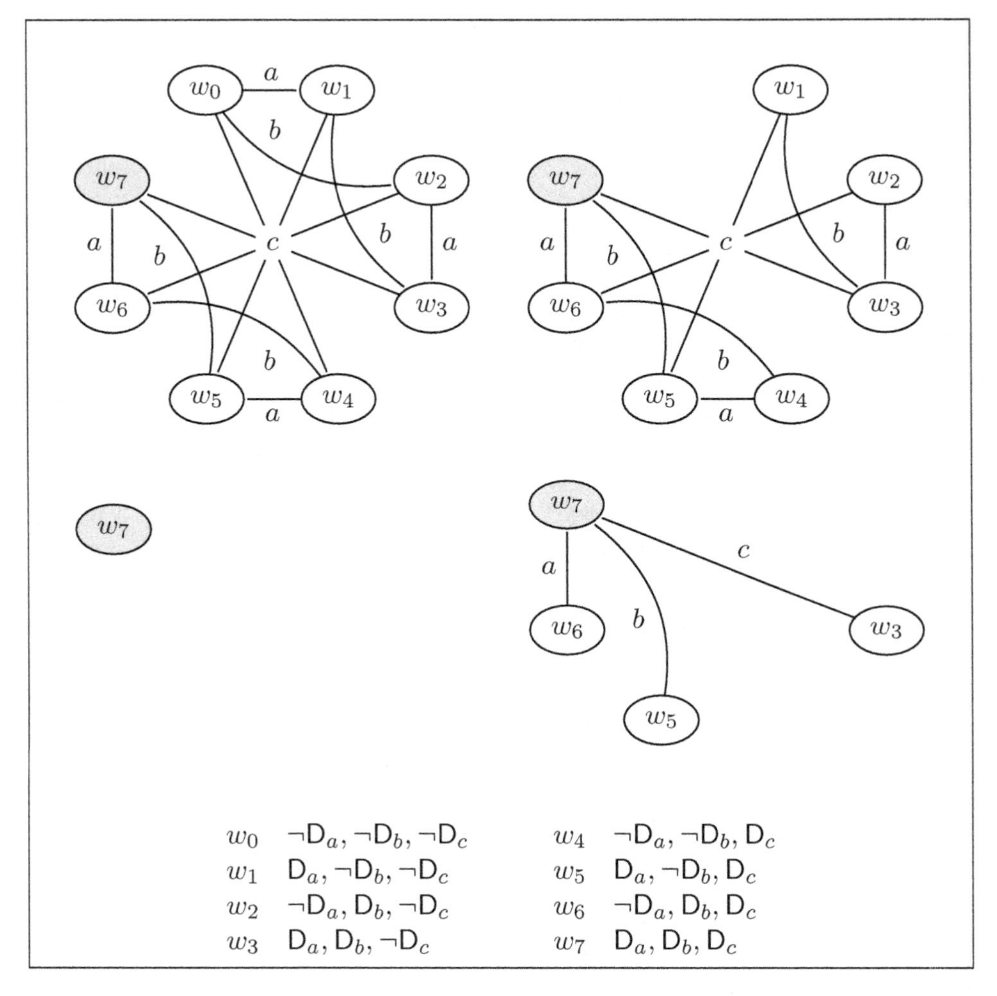

Figure 6.1: Three muddy children, a, b, and c and their successive situations represented as four models. The atomic sentences true in the various worlds are shown at the bottom. The accessibility relations are shown, with the understanding that each world is reflexive for all agents, and the relations for c go across the center of each model.

of the children's parents announces to them that at least one is muddy, and then asks all of them whether they know their state. They all simultaneously say no. On the third questioning, they all know that they are indeed muddy.

We shall discuss various aspects of this scenario in this chapter. First, we want to draw pictures of the various situations that are described, and also to connect things to the matter of public announcements that we are introducing. The figure shows four models; these are the connected graphs. We discuss the models in turn.

The first model, M^0, has eight worlds. The children are called a, b, and c, and the three atomic sentences are D_a, D_b, and D_c; these say that a, b, and c (respectively) are dirty. Thus, world w_7 is the real world. The arrows for the three children are shown, and for c the idea is that the arrows go across the figure. In all the figures, we have left off the reflexive arrows on all the worlds for all the agents.

The second model, M^1, is the relativization of M^0 to the sentence $\varphi_{\text{at least one}}$ saying that at least one child is muddy. Formally, $\varphi_{\text{at least one}}$ is

$$\varphi \quad \equiv \quad \mathsf{D}_a \vee \mathsf{D}_a \vee \mathsf{D}_c \tag{6.3}$$

Then $M^1 = (M^0)_{\varphi_{\text{at least one}}}$. This model is shown in the upper-right corner of the figure.

Further, let $\varphi_{\text{nobody knows}}$ be the sentence that says that nobody knows their state:

$$\neg K_a\, \mathsf{D}_a \wedge \neg K_a\, \neg\mathsf{D}_a \quad \wedge \quad \neg K_b\, \mathsf{D}_b \wedge \neg K_b \neg\mathsf{D}_b \quad \wedge \quad \neg K_c\, \mathsf{D}_c \wedge \neg K_c \neg\mathsf{D}_c \tag{6.4}$$

That is, nobody knows whether or not they are muddy. The story continues with successive public announcements of just this assertion. And turning to our formalism, we have models $M^2 = (M^1)_{\varphi_{\text{nobody knows}}}$, $M^3 = (M^2)_{\varphi_{\text{nobody knows}}}$, and $M^4 = (M^3)_{\varphi_{\text{nobody knows}}}$.

The implicit claim here is that these formal models are good representations of the situations in the story. ⊣

6.2.1 Public Announcement Logic PAL

We now take a decisive step and add the action of public announcement to epistemic logic. This is precisely what makes the logic "dynamic", and so gives the D in **DEL**.

The language **PAL** (for *Public Announcement Logic*) is now given as follows:

$$\varphi ::= p \mid \neg\varphi \mid \varphi \wedge \varphi \mid K_a\varphi \mid [!\varphi]\varphi \tag{6.5}$$

The semantics is *not* given by fixing a model and then (setting that model in the background) defining the relation $w \models \varphi$ by recursion on the language.

The problem is that the semantics of sentences $[!\varphi]\psi$ involves *changing the model* by relativization. So in effect we are defining a relation $M, w \models \varphi$, where w is a world in M, and we are defining the update operation simultaneously for all *pointed models*. (A pointed model is a model together with a specified world in it.) Again, a key feature of the definition of the semantics is that it involves a change in models and hence the model is a "first-class participant" in the semantics, not a parameter which can be suppressed.

Returning to the definition of the semantics, the clauses for all the features of **EL** are what we know from that logic. The extra clause for sentences $[!\varphi]\psi$ reads as follows:

$$M, w \models [!\varphi]\psi \quad \text{iff} \quad M, w \not\models \varphi, \text{ or else } M_\varphi, w \models \varphi. \tag{6.6}$$

We have already seen that we *change the model* when evaluating whether a sentence $[!\varphi]\psi$ is true or not in a given world of a given model. The next point to comment on is that if φ is false, then we take $[!\varphi]\psi$ to be *true*. This is a parallel with taking a $K\varphi$ sentence to be true in a world w of a given Kripke model when w has no successors in the model; for that matter, it is a parallel to the standard semantics of the (material) conditional in propositional logic.

The dual $\langle!\varphi\rangle\psi$ of $[!\varphi]\psi$ It is conventional to introduce the dual operators $\langle!\varphi\rangle$. We define $\langle!\varphi\rangle\psi$ to abbreviate $\neg[!\varphi]\neg\psi$. (This is a parallel to taking $\Diamond\varphi$ as an abbreviation of $\neg K\neg\varphi$.) Then we have the following characterization:

$$M, w \models \langle!\varphi\rangle\psi \quad \text{iff} \quad M, w \models \varphi, \text{ and } M_\varphi, w \models \varphi. \tag{6.7}$$

For most modeling purposes, it will be more natural to use the "diamond" forms.

Example 6.2

We can now go back to our original M^0 for the muddy children and check some intuitively satisfying facts. In the statements below, recall $\varphi_{\text{at least one}}$ from (6.3) and $\varphi_{\text{nobody knows}}$ from (6.4).

1. $M^0, w_1 \models [!\varphi_{\text{at least one}}]K_a\, \mathsf{D}_a$.
2. $M^0, w_1 \not\models [!\varphi_{\text{at least one}}]K_b\, \mathsf{D}_b$.
3. $M^0, w_0 \models [!\varphi_{\text{at least one}}]\psi$ for all ψ.
4. $M^0, w_7 \models \langle!\varphi_{\text{at least one}}\rangle\varphi_{\text{nobody knows}}$.

5. $M^0, w_7 \models \langle !\varphi_{\text{at least one}}\rangle\langle !\varphi_{\text{nobody knows}}\rangle\varphi_{\text{nobody knows}}$.

6. $M^0, w_7 \models \langle !\varphi_{\text{at least one}}\rangle\langle !\varphi_{\text{nobody knows}}\rangle\langle !\varphi_{\text{nobody knows}}\rangle K_a \mathsf{D}_a$.

The last of these tells the whole story. In w_7 of the original model, it is possible to truthfully announce that at least one child is muddy, then announce that nobody knows their state, then announce this again, and at this point, a (for example) knows that she is muddy. With only one announcement of ignorance, the children do not know their state, as (5) states. ⊣

Common knowledge operators We formulated the syntax of **PAL** in (6.5) without the common knowledge operators. This is mostly to match the literature.

6.2.2 Announcing a true fact might falsify it

This section contains an important reflection on the semantics of **PAL**: sentences of the form $[!\varphi]\varphi$ are *not* in general valid. For example, let φ be $\neg K\neg p$. Then $[!\neg K\neg p]\neg K\neg p$ is not valid. The simplest counter-model is

We do not take the self-loops on the nodes x and y. The sentence $Lp \equiv \neg K\neg p$ is true at the real world x but not at y, and so the relativization of the model to x is the model with one world and no arrow. In this model, the world does not satisfy Lp.

More interestingly, the muddy children scenarios also provide examples where a sentence is true, but the act of announcing that fact renders the sentence false.

Here are some definitions and results related to these notions. A sentence φ is *successful* if $[!\varphi]\varphi$ is valid. Also, φ is *preserved (under submodels)* if whenever $M, w \models \varphi$ and N is a submodel of M which contains w, then also $N, w \models \varphi$.

For example, every sentence φ which is *universal* is preserved under submodels. (These are the sentences that can be built from atomic sentences and their negations using the boolean operations $\neg$, $\wedge$, $\vee$, etc., and also the knowledge operators K_a and the common knowledge operators C_G.) Also, every sentence preserved under submodels is successful.

Here is what is known on this topic:

Proposition 6.1
The following are equivalent:

1. φ is successful.

2. $[\varphi]K_a\varphi$ is valid for some agent a.

3. $[\varphi]C\varphi$ is valid.

Moreover, every sentence built in the following way is successful:

$$p \mid \neg p \mid \varphi \wedge \psi \mid \varphi \vee \psi \mid K_a\varphi \mid C\varphi \mid [\neg\varphi]\psi \mid$$

$\dashv$

Moore's paradox: a related sentence that has no models Here is a sentence which would be called *self-refuting*, *unsuccessful*, or even *failing*: $p \wedge \neg Kp$. This is related to *Moore's paradox*: I cannot know the sentence "p is true and I don't know it." In the setting of public announcement logic, the point is that

$$\langle !(p \wedge \neg Kp)\rangle(p \wedge \neg Kp)$$

has no models. That is, the sentence $[!(p \wedge \neg Kp)]\neg(p \wedge \neg Kp)$ is valid.

It might be worth going through the semantic argument for this, since we shall later be concerned with formal derivations of this sentence. Let M be a model and w be a world in it. Assume that $M, w \models p \wedge \neg Kp$. Let N be the updated model $M_{p \wedge \neg Kp}$. We show that $N, w \models \neg(p \wedge \neg Kp)$. For suppose that $N, w \models p \wedge \neg Kp$. Let v be a world in N such that $w \; R \; v$ in N and $N, v \models \neg p$. However, since $v \in N$, $v \models p \wedge \neg Kp$ in M. In particular, $v \models p$ in M. And as the truth of atomic sentences is the same in M and N, we have a contradiction.

Digression: the knowledge flip-flop It might be interesting to see how complex the results of successive truthful public announcements can be, especially when those announcements are about what players do and do not know.

Suppose we have four players: Amina, Bao, Chandra, and Dianne. They have a deck with two indistinguishable $\spadesuit$ cards, one $\diamondsuit$ and one $\clubsuit$. The cards are dealt, and the players look at their own cards (and no others).

We assume that the following are common knowledge: the distribution of cards in the deck, the fact that each player knows which card was dealt to them, and that they do not initially know any other player's card.

We take a modal language with atomic sentences

$$A\spadesuit, A\diamondsuit, A\clubsuit, B\spadesuit, B\diamondsuit, B\clubsuit, C\spadesuit, C\diamondsuit, C\clubsuit, D\spadesuit, D\diamondsuit, D\clubsuit.$$

We have agents a, b, c, and d, and so we have knowledge sentences involving those agents. We shall be concerned with the sentence φ, given as

$$\neg K_b\, A\spadesuit \;\wedge\; \neg K_b\, A\diamondsuit \;\wedge\; \neg K_b\, A\clubsuit.$$

It says that Bao doesn't know Amina's card.

The deal is represented by the world w shown below:

$$\begin{array}{rcl} w & = & (A\spadesuit, B\spadesuit, C\diamondsuit, D\clubsuit) \\ w' & = & (A\clubsuit, B\spadesuit, C\spadesuit, D\diamondsuit) \end{array}$$

For later purposes, we also wish to consider the second world shown above, w'.

Notice that $w \models K_c\,\varphi$. It is easy to check this formally in the model, and the informal reason is that since there are three card values and Bao sees only his own, then he can't know anyone else's card. So φ is indeed common knowledge at this point. For the same reason, $w' \models K_c\,\varphi$.

Amina declares, "I do not have $\diamondsuit$." At this point, $w \models \neg K_c\,\varphi$. The reason is that Chandra considers it possible that Bao has $\clubsuit$. If this were the case, then Amina's announcement would tip Bao off to the fact that Amina has $\spadesuit$. And again, we also have $w' \models \neg K_c\,\varphi$.

Following Amina's declaration, Dianne announces, "I do not have $\spadesuit$." We claim that in the resulting model, $w \models K_c\,\varphi$. For now Chandra knows that Dianne must have $\clubsuit$, and Chandra also knows that Amina and Bao have the two $\spadesuit$ cards. In particular, Bao has $\spadesuit$. And at this point, Bao does not know Amina's card, since he (Bao) thinks that the world could be $w' = (A\clubsuit, B\spadesuit, C\spadesuit, D\diamondsuit)$.

To summarize things in our notation, let M be the original model. Then in w,

$$\begin{array}{l} K_c\varphi \\ \langle !\neg A\diamondsuit\rangle \neg K_c\varphi \\ \langle !\neg A\diamondsuit\rangle\langle !\neg D\spadesuit\rangle K_c\varphi \end{array}$$

So Chandra's knowledge of φ has "flip-flopped": Chandra knew φ at the outset, then Amina's announcement destroyed that knowledge, and finally Dianne's announcement restored it.

6.2.3 Sound and complete logical systems

A logic is born when it can used to say interesting things, and its childhood is an examination of its salient properties. In the case of public announcement logic (**PAL**), the infancy begins with completeness. We make the evident semantic definition:

$$\Gamma \models \varphi \quad \text{iff} \quad \text{for all models } M \text{ and all worlds } s \text{ of } M, \text{ we have } M, s \models \varphi$$

Here Γ is a set of sentences of **PAL**, and φ a sentence. We aim to have a proof system that matches this, in the sense that $\Gamma \vdash \varphi$ implies $\Gamma \models \varphi$ (this is called the *soundness* of the proof system), and conversely (this is called *completeness*). This section presents *three* logical systems which are sound and complete for **PAL**. All of the logical systems are built on top of the system K.

Figure 6.2 presents one set of axioms for **PAL**. It defines a proof relation $\Gamma \vdash \varphi$ in the usual way. Let us check that the system is sound in the weaker sense that all axioms are true at every world in every model.

The Basic Axioms are sound for very general reasons having nothing to do with public announcements: they are sound principles of propositional logic and modal logic, and so they are sound for **PAL** as well. For the same reason, the rule of inference which we call $[!\varphi]$-necessitation is also sound.

For the soundness of the other axioms in Figure 6.2, it is more natural to discuss the duals[3]:

$$\begin{array}{l} \langle !\varphi\rangle p \leftrightarrow (\varphi \wedge p) \\ \langle !\varphi\rangle \neg\psi \leftrightarrow (\varphi \wedge \neg\langle !\varphi\rangle\psi) \\ \langle !\varphi\rangle L_a\psi \leftrightarrow (\varphi \wedge L_a(\varphi \wedge \langle !\varphi\rangle\psi)) \\ \langle !\varphi\rangle\langle !\psi\rangle\chi \leftrightarrow \langle\langle !\varphi\rangle\psi\rangle\chi \end{array} \tag{6.8}$$

The Atomic Permanence axiom is special in that the p in it really must be an atomic proposition, not an arbitrary sentence. To see this, we must examine the soundness a bit. Fix a model M and some world w in it. First, assume that $w \models \langle !\varphi\rangle p$. Then by our semantics, $w \models \varphi$, and also in the model M_φ, $w \models p$. Here is the key point: the definition of M_φ did not change the valuation of atomic sentences. So $w \models p$ in M_φ iff $w \models p$ in M. Thus $w \models \varphi \wedge p$. The converse of this is basically the same argument.

The second axiom above $\langle !\varphi\rangle\neg\psi \leftrightarrow (\varphi \wedge \neg\langle !\varphi\rangle\psi)$ amounts to the fact that the result of a true announcement is uniquely defined: it is a single state of a single model. The rest of the verification is easy.

Let us check the last axiom in (6.8). Assume first that $M, s \models \langle !\varphi\rangle L_a\psi$. Then $M, s \models \varphi$ by the semantics of announcements. Let t be any world in M_φ such that $s\ R_a\ t$ in M_φ and $M_\varphi, t \models \psi$. We use t to show that $M, s \models L_a(\varphi \wedge \langle !\varphi\rangle\psi)$. Since t is a world in M_φ, it is a world in M and $M, t \models \varphi$. And since $M, t \models \varphi$ and $M_\varphi, t \models \psi$, we have $M, t \models \varphi \wedge \langle !\varphi\rangle\psi$. Thus indeed $M, s \models L_a(\varphi \wedge \langle !\varphi\rangle\psi)$. This is half of the last axiom in (6.8). The other direction of the biconditional is similar.

We also encourage the reader to check the soundness of last axiom $\langle !\varphi\rangle\langle !\psi\rangle\chi \leftrightarrow \langle\langle !\varphi\rangle\psi\rangle\chi$.

[3]We are writing L for the dual of K rather than M (as in the rest of this handbook), since M is used for "model" here.

Basic Axioms

($[!\chi]$-normality)	$[!\chi](\varphi \to \psi) \to ([!\chi]\varphi \to [!\chi]\psi)$

Announcement Axioms

(Atomic Permanence)	$[!\varphi]p \leftrightarrow (\varphi \to p)$
(Partial Functionality)	$[!\varphi]\neg\psi \leftrightarrow (\varphi \to \neg[!\varphi]\psi)$
(Action-Knowledge)	$[!\varphi]K_a\psi \leftrightarrow (\varphi \to K_a(\varphi \to [!\varphi]\psi))$
(Composition)	$[!\varphi][!\psi]\chi \leftrightarrow [!(\varphi \wedge [!\varphi]\psi)]\chi$

Figure 6.2: The first logical system for **PAL**

Completeness by translation and reduction We hinted above that there are many complete logical systems for **PAL**. We are presenting several of them partly to illustrate different techniques. The first technique is by *translation back to (multi-agent) epistemic logic* L, the logical language containing knowledge operators but not announcements. That is, we have the following result:

Theorem 6.1
For every sentence φ of **PAL** there is a sentence $t(\varphi)$ of L such that $\models \varphi \leftrightarrow t(\varphi)$.

To get a complete logical system for **PAL**, all we need is a logical system which includes a complete logic for epistemic logic and which is strong enough so that for all φ in **PAL**, $\vdash \varphi \leftrightarrow t(\varphi)$. We check this concerning *weak completeness* (no hypotheses Γ); the stronger assertion is argued similarly. For suppose that $\models \varphi$. Then in our system we have $\vdash \varphi \leftrightarrow t(\varphi)$. Moreover, by soundness we have $\models t(\varphi)$, and so by completeness of epistemic logic, we have $\vdash t(\varphi)$. Using all of this and propositional reasoning, we have $\vdash \varphi$.

The idea behind Theorem 6.1 is that the logical system in Figure 6.2 should be used to *rewrite* a sentence. One needs several steps, but eventually all of the announcement operators will be eliminated. Because more than one step is needed, Theorem 6.1 is not exactly obvious. In fact, the main point is to show that the rewriting given by the logic eventually stops. To get a hint at what is going, consider $[!p][!q][!r]s$, where p, q, r and s are all atomic sentences. Using (Composition) and (Atomic Permanence), we have the following equivalent sentences:

$$\begin{array}{l} [p \wedge [p]q][r]s \\ [p \wedge [p]q \wedge [p \wedge [p]q]r]s \\ (p \wedge [p]q \wedge [p \wedge [p]q]r) \to s \end{array} \tag{6.9}$$

Basic Axioms

($[!\chi]$-normality)	$[!\chi](\varphi \to \psi) \to ([!\chi]\varphi \to [!\chi]\psi)$

Announcement Axioms

(Atomic Permanence)	$[!\varphi]p \leftrightarrow (\varphi \to p)$
(Partial Functionality)	$[!\varphi]\neg\psi \leftrightarrow (\varphi \to \neg[!\varphi]\psi)$
(Action-Knowledge)	$[!\varphi]K_a\psi \leftrightarrow (\varphi \to K_a(\varphi \to [!\varphi]\psi))$

Rules of Inference

(Announcement Necessitation)	from ψ, infer $[!\varphi]\psi$

Figure 6.3: The second logical system for **PAL**

Then we use (Atomic Permanence) to replace $[p]q$ with $p \to q$. And by the same reasoning, we see that $[p \wedge [p]q]r$ is equivalent to $(p \wedge (p \to q)) \to r$. Thus using propositional reasoning, we see that the last line of (6.9), is equivalent to $(p \wedge q \wedge r) \to s$.

Another point concerning the biconditionals in the Announcement Axioms is that the right-hand sides are longer than the left. This, too, is an issue for the proof. Turning to the details, we first define a *complexity function* which assigns numbers to formulas:

$$\begin{array}{lcl} c(p) & = & 1 \\ c(\neg\varphi) & = & 1 + c(\varphi) \\ c(\varphi \wedge \psi) & = & 1 + \max(c(\varphi), c(\psi)) \\ c(K_a\varphi) & = & 1 + c(\varphi) \\ c([!\varphi]\psi) & = & (4 + c(\varphi))c(\psi) \end{array}$$

It is important to see that $c(\varphi)$ is defined by recursion on φ, and that the definition is not problematic in any way. An easy induction also shows that the complexity of any sentence is strictly greater than that of any proper subsentence.

Theorem 6.2
For every $\varphi \in$ **PAL** there is an equivalent formula $t(\varphi) \in \mathsf{L}$.

Proof We define t by recursion on $c(\varphi)$.

$$\begin{array}{lcllcl} t(p) & = & p & t([!\varphi]p) & = & t(\varphi) \to p \\ t(\neg\varphi) & = & \neg t(\varphi) & t([!\varphi]\neg\psi) & = & t(\varphi \to \neg[!\varphi]\psi) \\ t(\varphi \wedge \psi) & = & t(\varphi) \wedge t(\psi) & t([!\varphi](\psi \wedge \chi)) & = & t([!\varphi]\psi \wedge [!\varphi]\chi) \\ t(K_a\varphi) & = & K_a t(\varphi) & t([!\varphi]K_a\psi) & = & t(\varphi \to K_a[!\varphi]\psi) \\ & & & t([!\varphi][!\psi]\chi) & = & t([!(\varphi \wedge [!\varphi]\psi)]\chi) \end{array} \quad (6.10)$$

Announcement Axioms

(Atomic Permanence)	$[!\varphi]p \leftrightarrow (\varphi \to p)$
(Partial Functionality)	$[!\varphi]\neg\psi \leftrightarrow (\varphi \to \neg[!\varphi]\psi)$
(Conjunction)	$[!\varphi](\psi \wedge \chi) \leftrightarrow [!\varphi]\psi \wedge [!\varphi]\chi$
(Action-Knowledge)	$[!\varphi]K_a\psi \leftrightarrow (\varphi \to K_a(\varphi \to [!\varphi]\psi))$

Rules of Inference

(Replacement)	from $\varphi \leftrightarrow \psi$, infer $\chi(\varphi/p) \leftrightarrow \chi(\psi/p)$

Figure 6.4: The third logical system for **PAL**

To check that this works, let us verify for each equation that the complexity of the argument of t on the right hand side is less than the complexity of the argument of t on the left.

For the last of these,

$$\begin{array}{rcl} c([!\varphi][!\psi]\chi) & = & (4+c(\varphi))(4+c(\psi))c(\chi) \\ & = & 16c(\chi)+4c(\varphi)c(\chi)+4c(\psi)c(\chi)+c(\varphi)c(\psi)c(\chi) \\ & > & 4c(\chi)+4c(\psi)c(\chi)+c(\varphi)c(\psi)c(\chi) \\ & = & [4+(4+c(\varphi)c(\psi)]c(\chi) \\ & = & (4+c([!\varphi]\psi))c(\chi) \\ & = & c([!(\varphi \wedge [!\varphi]\psi)]\chi) \end{array}$$

The other steps are similar. ⊣

This proves Theorem 6.1.

A second system We present a slightly different logical system in Figure 6.3. The difference here is that we have traded in the Composition Axiom for the Announcement Necessitation rule. This rule is very natural, since the announcement operators $[!\varphi]$ are "box-like." Using the system in Figure 6.3, we again prove Theorem 6.2, but the details are rather different. To see why, let us again discuss $[!p][!q][!r]s$. This time, we are able to work "inside-out". Instead of the equivalences of (6.9), we have

$$\begin{array}{l} [!p][!q](r \to s) \\ [!p]([!q]r \to [!q]s) \\ [!p](q \to (r \to s)) \\ p \to (q \to (r \to s)) \end{array}$$

In fact, to show that this second system gives rise to a *terminating rewriting system* calls on work from term rewriting theory.

Basic Axioms	
($[!\chi]$-normality)	$[!\chi](\varphi \to \psi) \to ([!\chi]\varphi \to [!\chi]\psi)$
Announcement Axioms	
(Atomic Permanence)	$p \to [!\varphi]p$
(Atomic Permanence 2)	$\neg p \to \neg[!\varphi]p$
(Partial Functionality)	$\langle !\varphi\rangle\psi \leftrightarrow (\varphi \wedge [!\varphi]\psi)$
	$L_a\langle !\varphi\rangle\psi \to [!\varphi]L_a\psi$
	$\langle !\varphi\rangle L_a\psi \to L_a\langle !\varphi\rangle\psi$
Rules of Inference	
(Announcement Necessitation)	from ψ, infer $[!\varphi_a]\psi$

Figure 6.5: The fourth logical system for **PAL**.

A third system: biconditionals only A third logical system is shown in Figure 6.4. It differs from the other axiomatizations in that all of the axioms shown are biconditionals, and there is a rule of *Replacement* (of equivalents).

A fourth system: completeness by examining the canonical model We finally present a different completeness proof for **PAL**. In addition to giving an interesting result on its own, the topic allows us to discuss some other work, a characterization of the *substitution core* of **PAL**. The logic of this section is shown in Figure 6.5.

An *extended model* of **PAL** is a Kripke model which, in addition to having accessibility relations R_a for agents a also has accessibility relations R_φ for sentences φ of **PAL**. On an extended model M, we have two semantic relations $M, s \models \varphi$ and $M, s \models^* \varphi$. In the first, we ignore the new accessibility relations R_φ and just use the semantics we already have. In the second, we use those new relations and write

$$M, s \models^* [\varphi]\psi \qquad \text{iff} \quad \text{for all } t \text{ such that } s\ R_\varphi\ t,\ M, t \models^* \psi$$

We can think of extended models as Kripke models with formulas as new "agents"; as such, it is immediate what the definition of bisimulation should be for extended models, and it is clear that all sentences of **PAL** are preserved by bisimulations.

An extended model is *normal* if it meets the following conditions:

functionality Each R_φ is the graph of a partial function. Moreover, s has an outgoing edge under R_φ iff $s \models^* \varphi$.

atomic invariance If $s\ R_\varphi\ t$, then for atomic sentences p, $s \models^* p$ iff $s \models^* p$.

zig If $s\ R_a\ s'$, $s\ R_\varphi\ t$, and $s'\ R_\varphi\ t'$, then $t\ R_a\ t'$.

zag If $s\ R_\varphi\ t$ and $t\ R_a\ t'$, then there is some s' such that $s\ R_a\ s'$ and $s'\ R_\varphi\ t'$.

Lemma 6.1
Let M be a normal extended model. Then for every sentence φ of **PAL**,

$$s \models \varphi \quad \text{iff} \quad s \models^* \varphi$$

⊣

The lemma is proved by induction on φ, and the proof makes a key use of the notion of bisimulation.

Recall from Chapter 1 that given any multi-modal logic $\mathcal{L}$, the *canonical model* is the Kripke model whose points are the maximal consistent sets in $\mathcal{L}$, and with $s\ R_a\ t$ iff for all $K_a\varphi \in s$, $\varphi \in t$. This applies in the setting of extended models, since they are just Kripke models with a more liberal notion of "agent". Concretely, we would have $s\ R_\varphi t$ iff whenever $[!\varphi]\psi \in s$, then $\psi \in t$. The main point of the canonical model of **PAN** is to prove the Truth Lemma. In our setting, this would say:

$$s \models \varphi \text{ in the canonical model of } \mathbf{PAN} \quad \text{iff} \quad \varphi \in s$$

The point of taking the axioms and rules of inference for **PAN** as we listed them in Figure 6.5 is to obtain the following result:

Lemma 6.2
The canonical model C of **PAN** is a normal extended model. ⊣

Theorem 6.3
PAN is strongly complete for **PAL**. ⊣

Proof Suppose that Γ is a set of sentences in **PAL**, and $\Gamma \models \varphi$. Then we claim that $\Gamma \vdash \varphi$ in the logic **PAN**. If not, $\Gamma \cup \{\neg\varphi\}$ is consistent in the logic, and hence has a maximal consistent superset, say s. Consider the canonical model, C. By the Truth Lemma, $C, s \models^* \psi$ for all $\psi \in \Gamma \cup \{\neg\varphi\}$. But by Lemma 6.1, we may return to the original semantics of the logic. That is, $C, s \models \psi$ for $\psi \in \Gamma \cup \{\neg\varphi\}$. But then $\Gamma \cup \{\neg\varphi\}$ has a model, contradicting our original assumption that $\Gamma \models \varphi$. ⊣

Further results on axiomatization of PAL Let **PA** (for *public announcement*, not *Peano Arithmetic*), be the system in Figure 6.4, but without the (Replacement) rule. As always, we work on top of multi-modal K.

Theorem 6.4
The following systems are complete:

1. **PA+** (Announcement Necessitation) + $((\psi \to [\psi]\varphi) \to [\psi]\varphi)$.

The following systems are not complete:

1. **PA+** (Announcement Necessitation)

2. **PA+** $([!\varphi](\psi \to \chi) \to ([!\varphi](\psi) \to [!\varphi](\chi)))$. ⊣

6.2.4 Succinctness and complexity

A number of results in **DEL** pertain to the succinctness of various logics. The main point of the exercise is to show that even if a larger language is equivalent to a smaller one, the larger one might enable one to say things in a more succinct way. We outline the method and present most of the details. We deal with two agents, a and b. (We use the letter c to denote a or b.) Consider the following sentences:

$$\begin{array}{lllllll} \varphi_0 & = & \mathsf{true} & \quad & \psi_0 & = & \mathsf{true} \\ \varphi_{i+1} & = & \langle\langle\varphi_i\rangle L_a \mathsf{true}\rangle L_b \mathsf{true} & \quad & \psi_{i+1} & = & \psi_i \wedge L_a\psi \wedge L_b(\psi_i \wedge L_a\psi_i) \end{array}$$

It is easy to check that φ_i and ψ_i are equivalent for all i. Note that the size of φ_i is linear in i, whereas the size of ψ_i is exponential. Indeed, define the *path set* $P(\alpha)$ of an epistemic logic formula α by

$$\begin{array}{lll} P(\mathsf{true}) & = & \{\epsilon\} \\ P(p) & = & \{\epsilon\} \\ P(\neg\alpha) & = & P(\alpha) \\ P(\alpha \wedge \beta) & = & P(\alpha) \cup P(\beta) \\ P(K_c\alpha) & = & \{\epsilon\} \cup \{cw : w \in P(\alpha)\} \end{array}$$

(Here and below, ϵ is the empty world.) Then observe that $P(\psi_i) \supseteq \{a, b\}^i$. Also, the number of symbols in α is at least the size of $P(\alpha)$.

Theorem 6.5
No sentence α of L with fewer than 2^i symbols is equivalent to ψ_i.

Proof Suppose towards a contradiction that χ were equivalent to ψ_i, and yet $P(\chi)$ is not a superset of $\{a, b\}^i$. Fix some word w^* of length $\leq i$ not in $P(\chi)$.

We turn the set $P(\psi_i)$ into a model by $w\ R_a\ wa$ for $w, wa \in P(\psi_i)$, and similarly for b. We also set $V(p) = \emptyset$ for all atomic sentences p. Next, we form a second model M by taking the submodel of all words $w \in P(\psi_i)$ such that w^* is not a prefix of w. It is easy to check that $\epsilon \models \psi_i$ in $P(\psi_i)$, but $\epsilon \not\models \psi_i$ in M. We obtain our contradiction by showing that $P(\psi_i), \epsilon \models \psi_i$ iff $M, \epsilon \models \psi_i$. This follows from the following claim, taking w to be ϵ and α to be χ.

Claim If $w \in M$ and α is a subformula of χ, and if $\{ws : s \in P(\alpha)\} \subseteq P(\chi)$, then $w \models \alpha$ in $P(\psi_i)$ iff $w \models \alpha$ in M.

This claim is proved by induction on α. The only interesting step is when we prove that $w \models K_c\alpha$ in $P(\psi_i)$ on the basis of the induction hypothesis and the assumptions that $K_c\alpha$ is a subformula of χ, that $w \models K_c\alpha$ in $P(\psi_i)$, and that $\{ws : s \in P(K_c\alpha)\} \subseteq P(\chi)$. Since $wc \in \{ws : s \in P(K_c\alpha)\} \subseteq P(\chi)$, $wc \neq w^*$. Moreover, w^* is not a prefix of w because $w \in M$. Thus w^* is not a prefix of wc, and so $wc \in M$. The induction hypothesis applies to wc and α, since $\{wcs : s \in P(\alpha)\} = \{ws : s \in P(K_c\alpha)\}$. Using the induction hypothesis and $wc \models \alpha$ in $P(\psi)$, we have $wc \models \alpha$ in M. Thus we have $w \models K_c\alpha$ in M.

This concludes the proof of Theorem 6.5. ⊣

Complexity The thrust of Theorem 6.5 is that although the addition of announcement operators to the basic language of modal operators does not add expressive power, it adds succinctness: intuitively, a formula of a given size in **PAL** can say what would take an exponentially larger formula in L to say.

We also mention two complexity results. These are for satisfiability on $\mathcal{S}5$ models.

1. single-agent **PAL** satisfiability is in NP.

2. multi-agent **PAL** is in PSPACE.

In both cases, known results also show the matching hardness results, and so we see that single-agent **PAL** satisfiability on S5 models is NP complete. In the multi-agent setting, the complexity is PSPACE-complete.

6.2.5 The substitution core

The atomic sentences in logical systems like **PAL** do not function the same way that they do in logical systems like epistemic logic **EL**. In **EL** as in most logical systems, an atomic sentence p can function as an *arbitrary* sentence, at least for purposes of formal reasoning. What is proved for a given sentence is provable when p is replaced by an arbitrary sentence. Not so for **PAL**. For example, $[p]p$ is valid, but $[p \wedge \neg Kp](p \wedge \neg Kp)$ is not valid. As a result **PAL** is in this sense an unusual logical system. The root of the problem is the (Atomic Permanence) laws: $[!\varphi]p \to (\varphi \to p)$. This holds only when p is atomic. For other p, there is no general reason to believe that we have a validity.

There are two axiomatizions of the substitution core of **PAL**. One uses axioms that are close to what we have seen, except that we may use atomic proposition letters p, q, and r instead of the variable letters φ, ψ, and χ that we have seen. We have the following reduction axioms:

$$\begin{array}{l}
[!p](\neg q) \leftrightarrow (p \to \neg[!p]q) \\
[!p](q \wedge r) \leftrightarrow ([!p]q \wedge [!p]r) \\
[!p]K_a q \leftrightarrow (p \to K_a[!p]q) \\
[!p][!q]r \leftrightarrow [!(p \wedge [!p]q)]r
\end{array}$$

We also have some additional axioms: $[!\mathsf{true}]p \to p$, and $[!p]\neg\mathsf{true} \leftrightarrow \neg p$. For the rules,

$$\frac{\varphi^\sigma}{\sigma} \qquad \frac{[p]\varphi \leftrightarrow [p]\psi}{\varphi \leftrightarrow \psi} \qquad \frac{[\varphi]p \leftrightarrow [\psi]p}{\varphi \leftrightarrow \psi}$$

This defines a logical system on top of the system **K** of epistemic logic.

Theorem 6.6
Assume that the set of agents is infinite. Then for all sentences φ, φ is in the substitution core iff φ is provable in the logical system described above. $\dashv$

6.3 Additions to PAL

Section 6.2 presented public announcements and then considered the logical system **PAL** obtained from epistemic logic by adding announcement operators. As with all applied logics, as soon as some phenomenon is understood, moments later people propose extensions. This section presents some of the extensions to **PAL**. In a sense, the bigger extension is to more general types of epistemic actions, and we consider those in Section 6.4. But that would be an orthogonal extension; it would be possible to extend both ways.

Basic Axioms

(C-normality)	$C(\varphi \to \psi) \to (C\varphi \to C\psi)$

Rules of Inference

(C-necessitation)	from $\vdash \varphi$, infer $\vdash C\varphi$
(Announcement Rule)	from $\vdash \chi \to [!\varphi]\psi$ and $\vdash (\chi \wedge \varphi) \to K_a\chi$ infer $\vdash \chi \to [!\varphi]C\psi$

Figure 6.6: Additions to the logic of **PAL** for common knowledge. We also add the (Everyone) and (Mix) axioms from the logic of common knowledge in Chapter 1.

6.3.1 Common knowledge and relativized common knowledge

Figure 6.6 shows sound principles for the logic obtained by adding common knowledge operators to **PAL**. For simplicity, we only deal here with common knowledge to the entire group of all agents; it is straightforward to extend the results to the more general setting of common knowledge among a proper subset of the agents. The logic is complete, but we are not going to show this. Before going on, we check that this logic is not translatable into any of the smaller logics which we have seen.

In Theorem 6.7, we write $\hat{C}$ for the dual of the common knowledge operator. The semantics is given in a concrete form by saying that $w \models \hat{C}\varphi$ if there is a path of finite length (and possibly of length 0) from w following the union of the accessibility relations of all agents, ending in a state x, such that $x \models \varphi$.

Theorem 6.7
The sentence $\langle !p\rangle\hat{C}q$ is not expressible in L_C, even by a set of sentences.

Proof In this proof, we assume that there is only one agent, and so we drop agents from the notation.

We show the weaker assertion. Define a *rank function* on sentences of L_C as follows: $|p| = 0$ for p atomic, $\mathrm{rk}(\neg\varphi) = \mathrm{rk}(\varphi)$, $\mathrm{rk}(\varphi \wedge \psi) = \max(\mathrm{rk}(\varphi), \mathrm{rk}(\psi))$, $\mathrm{rk}(K\varphi) = 1 + \mathrm{rk}(\varphi)$, and $\mathrm{rk}(C\varphi) = 1 + \mathrm{rk}(\varphi)s$.

For fixed n, we show that $\langle !p\rangle\hat{C}q$ is not expressible by any single sentence of L_C of rank n. We assume familiarity with the technique of Ehrenfeucht-Fraïssé games. There are two players, *I* and *II*. We consider games of n plays, and at each round, we have worlds in each of the structures. *I* picks one of the designated worlds, and either moves along the accessibility relation, or along its transitive closure. *II* must respond by a similar move on the opposite model. The standard results in this setting is that two pointed structures satisfy the same sentences of L_C rank at most n, just in case the second player (player *II*) has a winning strategy in the n-round game.

Let A_n be the cycle

$$a_0 \to a_1 \to \cdots \to a_{n+2} \to a_{n+3} \to \cdots a_{2n+4} = a_0.$$

We set p true everywhere except a_{n+2} and q true only at a_0.

Announcing p means that we delete a_{n+2}. So $(A_n)_p$ splits into two disjoint pieces. this means that in $(A_n)_p$, a_1 does not satisfy $\hat{C}q$. But a_{n+3} does satisfy it.

We show that a_1 and a_{n+3} agree on all sentences of L_C of rank $\leq n$. For this, we show that *II* has a winning strategy in the n-round game on between (A_n, a_{n+1}) and (A_n, a_{n+3}). Then we appeal to the connection between games and expressivity that we mentioned above. *II*'s strategy is as follows: if *I* ever makes a $\hat{C}$ move, *II* should make a move on the other side to the exact same point. (Recall that A_n is a cycle.) Thereafter, *II* should mimic *I*'s moves exactly. Since the play will end with the same point in the two structures, *II* wins. But if *I* never makes a $\hat{C}$ move, the play will consist of n $\Diamond$-moves. *II* should simply make the same moves in the appropriate structures. Since a_{n+2} is $n+1$ steps from a_1, and $a_0 = a_{2n+4}$ is $n+1$ steps from a_{n+3}, *II* will win the play in this case. ⊣

The same sentence $\langle !p\rangle\hat{C}q$ is not expressible by a set of sentences using epistemic logic and common knowledge, *even when the accessibility relations are equivalence relations.*

With this result in place, we return to the logical system of Figure 6.6. It is easy to see that the C-normality axioms are sound, as is the C-necessitation rule. The main one to discuss is the Announcement Rule. Let χ be as in the hypothesis of the rule. We check that $\models \chi \to [!\varphi]C\psi$. Let (M, w) be such that $w \models \chi$. To see that $w \models [!\varphi]C\chi$, we assume that $w \models \varphi$ in M. We check by induction on n that if v is reachable in M_φ from w by a path of length at most n, then in M_φ, $v \models \psi$.

We continue with examples that illustrate the use of the logic in Figure 6.6.

Example 6.3
Perhaps the easiest example is a fact which we noted in Section 6.2.2: if φ is successful, ($\models [!\varphi]\varphi$), then also $\models [!\varphi]C\varphi$. We use Announcement Rule with $\chi = \mathsf{true}$ and $\psi = \varphi$. ⊣

Example 6.4
We check that for all φ,

$$\vdash \varphi \rightarrow [!\mathsf{true}]\varphi. \tag{6.11}$$

⊣

Intuitively, a public announcement of a triviality changes nobody's epistemic state. Our fact is proved by induction on φ. We get the result for atomic sentences easily, and the only induction steps worth discussing are those for sentences $K_a\varphi$ and for $C\varphi$.

Assuming that $\vdash \varphi \rightarrow [!\mathsf{true}]\varphi$, we use propositional modal reasoning to get $\vdash K_a\varphi \rightarrow K_a[!\mathsf{true}]\varphi$. Even more, we get

$$\vdash K_a\varphi \rightarrow (\mathsf{true} \rightarrow K_a[!\mathsf{true}]\varphi)).$$

By the Action-Knowledge Axiom,

$$\vdash K_a\varphi \rightarrow [!\mathsf{true}]K_a\varphi$$

Finally, we use the same assumption $\vdash \varphi \rightarrow [!\mathsf{true}]\varphi$ to see that $\vdash C\varphi \rightarrow [!\mathsf{true}]C\varphi$. We use the Announcement Rule, with $\chi = C\varphi$. By Mix and the induction hypothesis, $\vdash C\varphi \rightarrow [!\mathsf{true}]\varphi$. And $\vdash (C\varphi \wedge \mathsf{true}) \rightarrow K_aC\varphi$, by Mix again.

Example 6.5
Generalizing Example 6.4

$$\vdash C\varphi \rightarrow ([!\varphi]\psi \leftrightarrow \psi).$$

Intuitively, if φ is common knowledge, then there is no point in announcing it: an announcement of φ changes nothing. ⊣

Example 6.6
We next have an application of the Announcement Rule to our coin scenarios from Section 6.1.1. Once again, we work with the atomic sentences H and T for heads and tails, and with the set $\{A, B\}$ of agents. We show

$$\vdash C(\mathsf{H} \leftrightarrow \neg\mathsf{T}) \rightarrow [!\mathsf{H}]C\neg\mathsf{T}.$$

That is, on the assumption that it is common knowledge that heads and tails are mutually exclusive, then as a result of a public announcement of heads it will be common knowledge that the state is not tails.

To use the Announcement Rule, we take χ to be $C(\mathsf{H} \leftrightarrow \neg\mathsf{T})$. We we must show that

1. $\vdash C(\mathsf{H} \leftrightarrow \neg\mathsf{T}) \rightarrow [!\mathsf{H}]\neg\mathsf{T}$.

2. For all agents A, $\vdash (C(\mathsf{H} \leftrightarrow \neg\mathsf{T}) \wedge \mathsf{H}) \rightarrow K_A C(\mathsf{H} \leftrightarrow \neg\mathsf{T})$

From these assumptions, we may infer $\vdash C(\mathsf{H} \leftrightarrow \neg\mathsf{T}) \rightarrow [!\mathsf{H}]C\neg\mathsf{T}$.

For the first statement,

(a)	$\mathsf{T} \leftrightarrow [!\mathsf{H}]\mathsf{T}$	Atomic Permanence
(b)	$(\mathsf{H} \leftrightarrow \neg\mathsf{T}) \rightarrow (\mathsf{H} \rightarrow \neg[!\mathsf{H}]\mathsf{T})$	(a), Prop. reasoning
(c)	$[!\mathsf{H}]\neg\mathsf{T} \leftrightarrow (\mathsf{H} \rightarrow \neg[!\mathsf{H}]\mathsf{T})$	Functionality
(d)	$C(\mathsf{H} \leftrightarrow \neg\mathsf{T}) \rightarrow (\mathsf{H} \leftrightarrow \neg\mathsf{T})$	Mix
(e)	$C(\mathsf{H} \leftrightarrow \neg\mathsf{T}) \rightarrow [!\mathsf{H}]\neg\mathsf{T}$	(d), (b), (c), Prop. reasoning

And the second statement is an easy consequence of the Mix Axiom. ⊣

Relativized common knowledge There is an important generalization of common knowledge called *relativized common knowledge.* We present the main definition and results. As with our treatment of common knowledge, we simplify the presentation by only working with common knowledge to the group of all agents rather than to a proper subset G. Instead of adding the one-place operator $C(\varphi)$, we instead add a *two-place* connective $C(\varphi, \psi)$. The intuition is that this means that if φ is announced publicly, then it would be common knowledge that ψ was true.

The formal semantics is given as follows (we write $w = w_0 \rightarrow w_1 \rightarrow \cdots \rightarrow w_n$ for a path $w_0, w_1, \ldots, w_n$ starting at $w = w_0$):

$$w \models C(\varphi, \psi) \quad \text{iff} \quad \text{for every path } w = w_0 \rightarrow w_1 \rightarrow \cdots \rightarrow w_n \text{ if } w_i \models \varphi \text{ for all } i, \text{ then } w_n \models \psi$$

6.3.2 Iterated announcements

It is possible to add *announcement iteration* to **PAL**. One reason for doing so would be to model sequences of instructions which are repeated, such as the one in the muddy children scenario. In that setting, the children might be told to repeatedly announce whether they know their state, and the repetition would go on until someone does indeed know their state. We present the simplest formalization of ideas along these lines in this section, and a more general formalization will come later.

Let us add to **PAL** the ability to form sentences by announcement iteration: if φ and ψ are sentences, then so is $[!\varphi]^*\psi$. The idea is that $[!\varphi]^*\psi$ should be semantically equivalent to the infinite conjunction

$$\psi \ \wedge \ [!\varphi]\psi \ \wedge \ [!\varphi][!\varphi]\psi \ \wedge \ \cdots \ \wedge \ [!\varphi]^n\psi \ \wedge \ \cdots$$

However useful this would be, the unfortunate fact is that the resulting logical system is not so nicely behaved. It lacks the finite model property, and even small fragments of it are undecidable. Here are the details on this:

Proposition 6.2
$[!L\mathsf{true}]^* LK\mathsf{false}$ is satisfiable, but not in any finite model.

Proof A state s in a model $\mathbf{S}$ is called an *end state* if s has no successors; that is, if $s \models K\mathsf{false}$. For each model $\mathbf{S}$, let $\mathbf{S}'$ be the same model, except with the end states removed. $\mathbf{S}'$ is isomorphic to what we have written earlier as $\mathbf{S}_{L\mathsf{true}}$, the relativization of $\mathbf{S}$ to the states which are not end states. So $s \models [!L\mathsf{true}]^* LK\mathsf{false}$ holds of s just in case the following hold for all n:

1. $s \in \mathbf{S}^{(n)}$, the n-fold application of the operation of dropping end states from $\mathbf{S}$.

2. s has some child t which is an end state (and hence t would not belong to $\mathbf{S}^{(n+1)}$).

It is clear that any model of $[!L\mathsf{true}]^* LK\mathsf{false}$ must be infinite, since the sets $S^{(n+1)} \setminus S^{(n)}$ are pairwise disjoint and nonempty.

There are well-known models of $[!L\mathsf{true}]^* LK\mathsf{false}$. One would be the set of decreasing sequences of natural numbers, with $s \to t$ iff t is a one-point extension of s. The end states are the sequences that end in 0. For each n, $\mathbf{S}^{(n)}$ is the submodel consisting of the sequences which end in n. ⊣

This result has several dramatic consequences for this work. First, it means that the logics of iterated public announcements cannot be translated into any logic with the finite model property.

We also have the following extension of Proposition 6.2:

Theorem 6.8
Let $\mathbf{PAL}^*$ be the language of iterated public announcements.

1. $\{\varphi \in \mathbf{PAL}^* : \varphi \text{ is satisfiable}\}$ is Σ^1_1-complete. As a result, there is no computably presented logical system for the valid sentences in $\mathbf{PAL}^*$.

2. $\{\varphi \in \mathbf{PAL}^* : \varphi \text{ has a finite model}\}$ is undecidable. ⊣

The proof of Theorem 6.8 is to long to present here. It involves encoding *tiling* problems into $\mathbf{PAL}^*$. The proof also shows that the negative results above even go through for small fragments of $\mathbf{PAL}^*$, such as the fragment where the only iterated announcement is $L\mathsf{true}$ (but using common knowledge), or the fragment without common knowledge but instead allowing iterations of the announcement of two sentences in the language.

6.3.3 Arbitrary announcement logic

We close this section on **PAL** by discussing an extension involving quantification over arbitrary announcements, in a slightly limited sense. The syntax adds to **PAL** the syntactic construct $\Box\varphi$, with the idea that this means "no matter what formula is announced, φ will hold". We would like to say """no matter what formula in the entire world is announced ...", but this would result in a circular definition that does not seem to admit a straightening. (At least I am not aware of any work that gives a non-circular definition accomplishing the same thing.) And in order to make the topic tractable, one usually limits attention to the sentences in pure epistemic logic L, with no common knowledge operators. The semantics is that

$$M, s \models K\psi \quad \text{iff} \quad \text{for all } \psi \in \mathsf{L},\ [!\psi]\varphi$$

Of course, we have a dual $\Diamond$ operator with the obvious semantics:

$$M, s \models K\psi \quad \text{iff} \quad \text{for some } \psi \in \mathsf{L},\ [!\psi]\varphi$$

Let us check that $\mathcal{L}_{apal}$ is more expressive than ordinary modal logic. Consider the sentence

$$\varphi \quad \equiv \quad \Diamond(K_a p \wedge \neg K_b K_a p)$$

On the (removable) assumption that we have infinitely many atomic sentences, we check that φ is not equivalent to any modal sentence. For suppose that φ is equivalent to the modal sentence φ', and that q is an atomic sentence not in φ'. Consider the following two models:

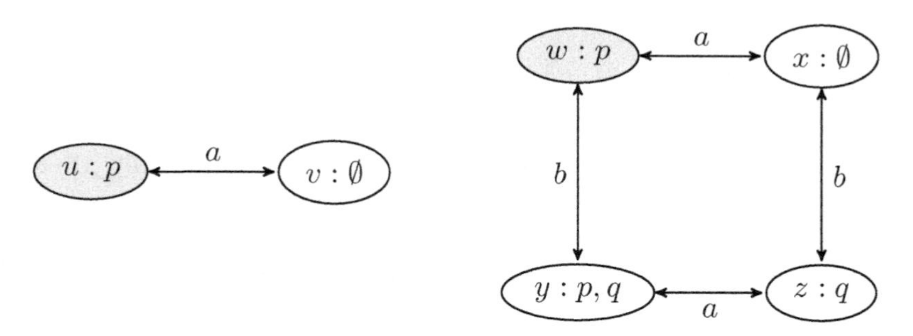

We call these M_1 (on the left) and M_2. We did not indicate it in the diagrams, but we intend that all atomic sentences other than p and q are true in all worlds. We also did not indicate self-loops on all nodes for both agents. Announcing $p \vee q$ in the real world w of M_2 drops x, and so a knows that p holds, but b does not know this. Thus $M_2, w \models \varphi$. On the other hand, $u \not\models \varphi$ because all announcements either result in M_1 or a singleton;

this implies easily that $M_1, u \models \neg\varphi$. However, u, w, and y satisfy all the same sentences without q, as do v, x, and z.

Here are some additional facts,

1. With only one agent, $\mathcal{L}_{apal}$ is translatable into L (on $\mathcal{S}5$ models).
2. There is a sound and complete logical system for the valid sentences of $\mathcal{L}_{apal}$, and so the set of validities is computably enumerable.
3. The set of validities of $\mathcal{L}_{apal}$ is undecidable.

The axiomatization is interesting. The completeness results goes via a canonical model construction, and the undecidability uses tiling.

6.4 More General Epistemic Actions

The main part of this chapter has been about **PAL** because this is the most well-understood dynamic epistemic logic. We now make our way towards to the ultimate generalization of this, dynamic epistemic logic using a much wider class of *epistemic actions*. The leading idea is that the *public* in *public announcement* means that all agents are certain about the action; indeed it is common knowledge. So we could think of the action itself as a kind of Kripke model: one world with a loop for all agents. (Of course, we need some "content", to distinguish a public announcement of one sentence from a public announcement of another.) By generalizing this Kripke model to an arbitrary one (with the appropriate "contents"), we arrive at what we call an *action model*. But we need an account of how an action model combines *acts on* multi-agent Kripke models. All this will come in this section, in an abbreviated presentation.

6.4.1 Private announcements and more general epistemic actions

Suppose that we have two agents, a and b, and we want to privately announce a sentence φ to a, while keeping b completely in the dark. This is an example of what we mean by an *epistemic action*. Indeed, a completely private announcement is perhaps the simplest form of epistemic action beyond a public announcement.

A concrete example of such a private announcement may be found in Scenario 4 earlier in the chapter. We want to think A looking privately, as a private announcement to A of H. (That is, there is no distinction here between A looking herself in a private way and some person informing A in a private way.) We have already exhibited a model of what the states of A

and B should be after this private announcement. But we have not shown the structure of the action itself. This is given below:

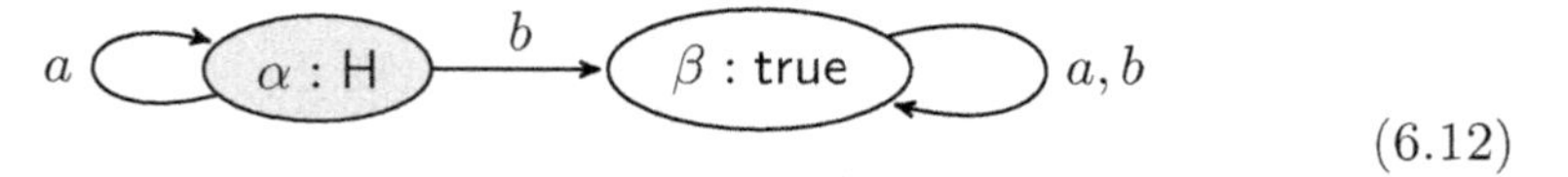

(6.12)

Again, this is our representation of a *private announcement of heads to a*. The idea is that α is the private announcement itself, and β is the action of a public announcement of true. (Here and below, we take true to be some fixed sentence which is valid on the kinds of semantics that we are considering. Since our logic is classical, we might as well take true to be $p \vee \neg p$. An announcement of true is tantamount to an announcement of "nothing," a *null announcement*, because a public announcement of a valid sentence does not change an agent's epistemic state.) The "real action" is α, as indicated by the shading. (We say this even though β represents one of the agents being "in the dark.") Moreover, a is "sure" that the action α happens; this sureness is due to the fact that $\alpha \xrightarrow{a} \alpha$, but $\neg(\alpha \xrightarrow{a} \beta)$. The fact that β represents the public announcement of true is reflected in the fact that the submodel induced by β is the one-point model, just as we mentioned in the first paragraph of Section 6.4. And the fact that B is in the dark about what is going on is reflected by the fact that from the real action α, B has an arrow only to β.

It is possible to vary this by allowing B to "suspect" that A received was informed of H. This would be rendered as

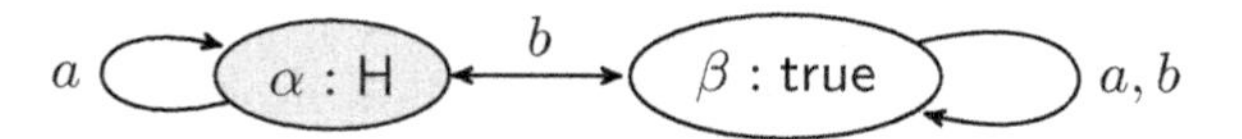

The only difference is that $\beta \xrightarrow{b} \alpha$. By changing the "real action" (the "real world" of the model), we show an epistemic action where nothing in fact happens, but B suspect that A was informed of H:

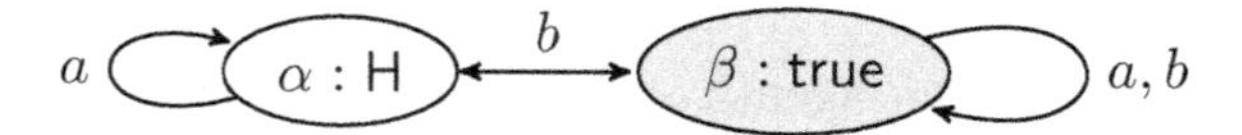

We also can model a "fair game" announcement. Suppose that someone whispers *something* into A's ear in the presence of B; all the while A and B are "publicly aware" of what is going on. B comes to learn that A learned H or learned T, but not which. This gives us a model quite like one we have already seen:

a, b — α : H ←b→ β : T — a, b

(6.13)

(Recall that T means "tails", while true means "some fixed tautology".)

In all of the action models above, A is certain of what was happening. This is reflected by the presence of one and only one arrow for A from the real action, a loop. It is possible to be yet more subtle. Suppose we want to model the following scenario: A receives H but believes (wrongly) that B thinks nothing has happened; for his part, B believes that either nothing happened or that A learned H. Here is a representation of this epistemic action:

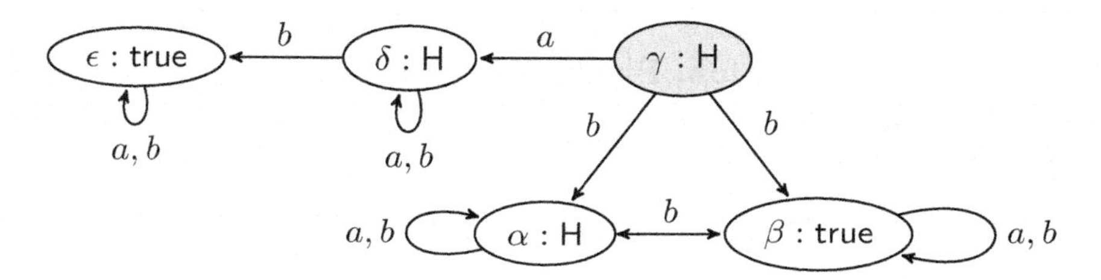

One noteworthy feature of this last model is that the real action, γ, has no loops for either agent. That is, both agents believe that some other action is taking place.

Although we have presented

6.4.2 The update product

At this point, we have given some illustrations of action models. Quite out of order, we now give the definition. As with all of our work, there is a silent dependence on As usual, we assume background parameters in the form of a set of agents $\mathscr{A}$.

Definition 6.1 (Action model)
Let $\mathcal{L}$ be a logical language. An *action model over* $\mathscr{L}$ is a structure $\mathsf{U} = \langle \mathsf{S}, \mathsf{R}, \mathsf{pre} \rangle$ such that S is a domain of *action points*, which we typically denote using Greek letters; for each $a \in \mathscr{A}$, R_a is an accessibility relation on S, and $\mathsf{pre} : \mathsf{S} \to \mathcal{L}$ is a *precondition function* that assigns a *precondition* $\mathsf{pre}(\sigma) \in \mathcal{L}$ to each $\sigma \in \mathsf{S}$. An *epistemic action* is a pointed action model (U, σ), with $\sigma \in \mathsf{S}$. ⊣

The previous section contains a few examples. Here is the simplest one.

Example 6.7
For every sentence φ in $\mathscr{L}$, The *public announcement action model* consists of only one action point, accessible to all agents, and having φ as its precondition. We call this action model *Pub* φ, and denote by $!\varphi$ the (action

corresponding to the) unique point of this model. Here is a picture:

(6.14)

⊣

At this point, we have a definition of an action model, and the decisive next step is to define the operation of combining a Kripke model M with an action model U. This operation gives us a new model. We call this the *execution* of U on M, and we write it as $M \otimes \mathsf{U}$.

Definition 6.2 (Execution, Product Update)
Let M be a Kripke model, and U be an action model. We define $M \otimes \mathsf{U}$ to be the model whose set of states is:

$$M \otimes \mathsf{U} \quad = \quad \{(w, \alpha) \in M \times \mathsf{S} \ : M, w \models \text{PRE}(\alpha)\}.$$

The accessibility relation for each agent is the product of the relations in M and U:

$$(w, \alpha) \xrightarrow{a} (v, \beta) \qquad \text{iff} \quad w \xrightarrow{a} v \text{ in } M, \text{ and } \alpha \xrightarrow{a} \beta \text{ in } \mathsf{U}$$

Finally, the valuation is determined by the model.

$$V(w, \alpha) \quad = \quad V(s)$$

When we are dealing with a pointed model (M, m) and a pointed action model (U, α), with the property that $(m, \alpha) \in M \otimes \mathsf{U}$, then we obtain a pointed model $(M \otimes \mathsf{U}, (m, \alpha))$. ⊣

The symbol $\otimes$ is used to suggest "tensor products", but the connection is scant. We think of $M \otimes \mathsf{U}$ as a "restricted product", since its set of worlds is a restriction (i.e., a subset) of the product set $M \times \mathsf{S}$. The idea is that for a pair $(w, \alpha) \in M \times \mathsf{S}$ to be a world in the restricted product, w must satisfy the precondition of the action α. The rest of the structure is pretty much what one would expect. Using the product arrows for the accessibility relation reflects the intuition that an agent's indifference about the states is independent of their indifference about the epistemic action. More to the point: if there were settings in which this intuition seemed doubtful, then one would want to reconsider this aspect of the definition of the update product. But with no compelling examples in evidence, we fall back on the mathematically simplest definition. Concerning the valuation, the intuition here is that executing an "epistemic action" is not supposed to change the atomic facts, it's supposed to change the knowledge states of the agents. For example, telling the Muddy Children to wipe off their foreheads would not be an epistemic action in our sense; telling them to each look in a mirror privately would be such an action. Having said this, we should mention that there are interesting extensions of the logics that we define which do allow for actions which change the world.

We refer to the operation

$$M, \mathsf{U} \quad \mapsto \quad M \otimes \mathsf{U}$$

taking models and action models to models as the *update product*. In addition to the worlds, we have specified the accessibility relations and the valuation. The intuition is that indistinguishable actions performed on indistinguishable input-states yield indistinguishable output-states: whenever the real state is s, agent a thinks it is possible that the state might be s', and if when action σ is happening, agent a thinks it is possible that action σ' might be happening, then after this, when the real state is (s, σ), agent a thinks it is possible that the state might be (s', σ').

Example 6.8
Perhaps the simplest family of examples concerns the recovery of public announcements in our earlier sense via update product using the model in (6.14). For this, let M be any model, let φ be a sentence, and let U be the action model from (6.14), where we use our given sentence φ as the precondition of the one action. Then $M \otimes \mathsf{U}$ is isomorphic to what we earlier called M_φ, the *relativization* of M to φ (see page 269). That is, $M \otimes \mathsf{U}$ is not literally equal to M_φ, since the former model has pairs $(m, !)$ for those $m \in M$ satisfying φ in M. But modulo this identification, the models are the same.

Example 6.9
Let G be an arbitrary non-empty set of agents that we'll call the *insiders*. We wish to model a completely private announcement of some sentence φ to these insiders, done in a way that the outsiders to not even suspect. The announcement is meant public for the insiders. We model this epistemic action using the action model U shown below:

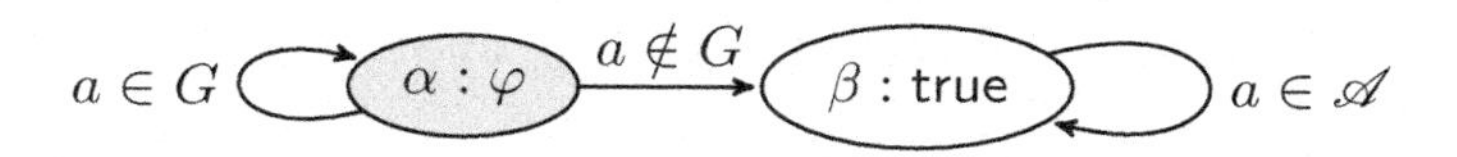

This generalizes examples we saw before where $G = \{A\}$. Now let M be an arbitrary Kripke model. Let us say what $M \otimes \mathsf{U}$ looks like informally. It would consist of two parts: (a) M_φ, but with the arrows only for the insiders; and (b) the original model M. The arrows from the first part to the second are those for the outsiders only. More specifically, if $x \xrightarrow{a} y$ is an arrow in M, and x happens to belong to M_φ, then in $M \otimes \mathsf{U}$ we have an arrow from x in M_φ to y in the copy of M.

The last example discussed the update product of a given action model with an *arbitrary* model M. We should emphasize that this generality is an important feature of the approach. The idea is to obtain a mathematical model of an epistemic action that may be applied to *any* model. This has some real value when it comes to writing down complicated models.

Example 6.10
Recall the "double-cheating" model which we saw in Scenario 5. The model is rather hard to come up with from first principles. But the update product machinery allows us to factor the model of interest as a product. We can think of this as the original coin model, followed by a private announcement to A of H, followed by a private announcement to B of the same thing. That is, the model we want is

$$(M \otimes \mathsf{U}_1) \otimes \mathsf{U}_2,$$

where M is from Scenario 1, U_1 is from (6.12), and U_2 is the same as U_1, except with all a's and b's interchanged.

6.4.3 Slower version of the update product

For those challenged by the lack of details in Examples 6.8–6.10, we present a fuller discussion of the technical matters.

The starting point is the notion of the *product of sets*. Given two sets A and B, we form the *product set* $A \times B$ by taking all ordered pairs (a, b), where $a \in A$ and $b \in B$. For example, $\{1,2,3\} \times \{1,2\} = \{(1,1),(1,2),(2,1),(2,2),(3,1),(3,2)\}$.

We upgrade this product operation to next define the *product of graphs*. Given two graphs $G = (G, \rightarrow_1)$ and $H = (H, \rightarrow_2)$, we form a new graph $G \times H$ by taking the product set $G \times H$ as the set of nodes, and the edge relation in $G \times H$ is given by

$$(g, h) \rightarrow (g', h') \quad \text{iff} \quad g \rightarrow_1 g' \text{ and } h \rightarrow_2 h'$$

The new graph $G \times H$ is called the *product of G and H*.

Here is an example:

We get a disconnected graph:

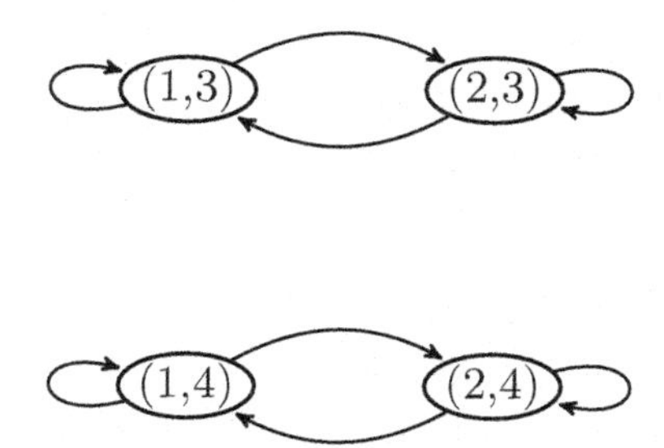

A multi-agent graph has agents on the arrows. To form the product of multi-agent graphs, we work agent-by-agent. For example,

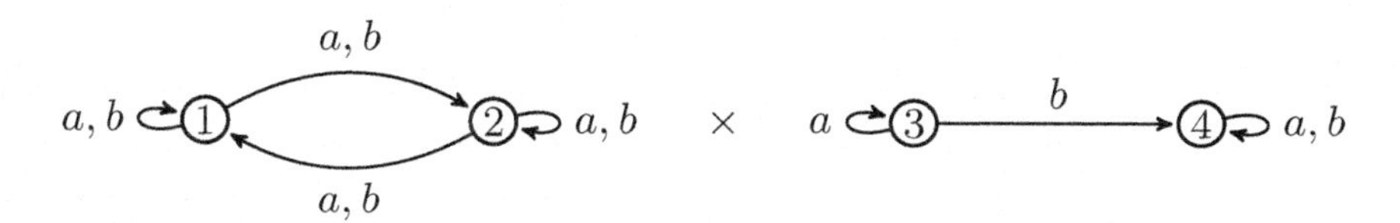

is

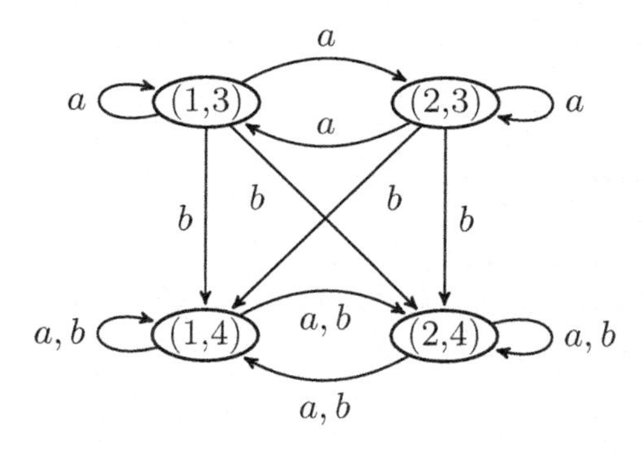

We almost have everything we need to define the action product of a multi-agent *model* and an *action model.* Recall that an action model comes with atomic propositions on the nodes, and an action model comes with a pre-condition function.

Let's illustrate the computation by an example:

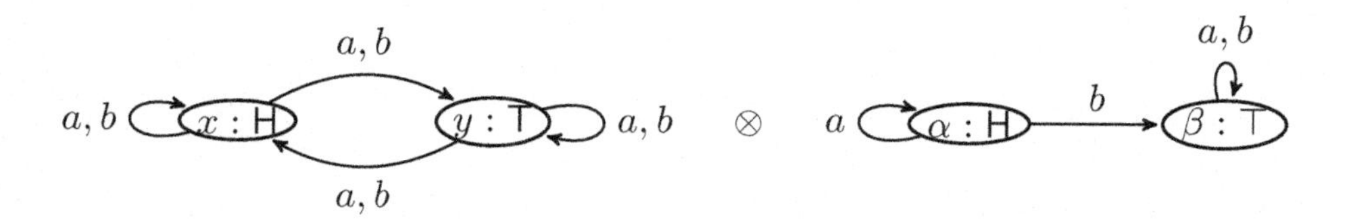

The first step is to compute the set of nodes in $W \otimes \Sigma$. We make a chart:

node z	action σ	$\mathsf{pre}(\sigma)$	$z \models \mathsf{pre}(\sigma)$?
x	α	H	✓
x	β	T	✓
y	α	H	×
y	β	T	✓

Thus our node set for $W \otimes \Sigma$ is $\{(x, \alpha), (x, \beta), (y, \beta)\}$. We restrict the product operation that we have already seen, and we also determine the valuation in the resulting model by "inheritance from the first component." We get

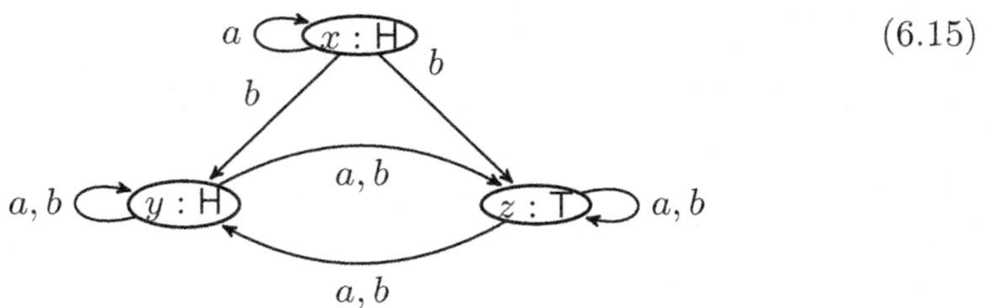

(6.15)

Technically, the nodes are (x, α), (x, β), and (y, β). To save space, we have renamed them to x, y, and z.

Finally, let us make good on what we said in Example 6.10. We derive the double-cheating model that we exhibited in (6.2) on page 267. We would take the action product of the model shown above with the action model shown below

The node set will be $\{(x, \alpha), (x, \beta), (y, \alpha), (y, \beta), (z, \alpha)\}$. We omit the rest of the calculation. The reader may check that the update product is indeed the model shown in (6.2).

Non-deterministic actions In many applications, it is natural to think of *non-deterministic actions*. These are like the actions which we have seen, but they might not have a unique "real action". Instead, there might be more than one real action. We have already mentioned the example of telling the Muddy Children to look in a mirror. This action is not the same as a private announcement of "clean", and it is also not the same as a private announcement of "dirty." Nevertheless, it is the kind of thing we mean by an epistemic action. We would capture such an action by an epistemic action with a *set* S_0 of possible actions. The update product is defined in the same way. The additional point which non-deterministic actions bring is that there is a natural relation between a model M and the product $M \otimes \mathsf{U}$: relate each world $w \in M$ to all worlds $(w, \alpha) \in M \otimes \mathsf{U}$ such that $\alpha \in S_0$.

6.4.4 DEL based on epistemic actions

We have seen action models and their pointed variants, the epistemic actions. We formulate a logical system using epistemic actions as modal operators. Before we do this, we fix (as always) a set $\mathscr{A}$ and (for the time being) a set AtSen of atomic sentences. (It will be important later in this section to note that AtSen may be any collection of atomic sentences; the only thing we require is that these sentences should be interpretable on the kinds of models which we have seen throughout this chapter.) Let $\mathscr{L}$ be epistemic logic formulated over $\mathscr{A}$ and AtSen.

We wish to consider all of the finite action models, where we take our preconditions to be sentences in $\mathscr{L}$. (As a fine point, note that there is a *proper class* of such action models, but we can easily find a *set* $\mathcal{S}$ of finite action models with the property that every action model is isomorphic to one in our set. For example, $\mathcal{S}$ could be the finite action models whose simple actions are natural numbers.)

We are going to consider epistemic actions over $\mathcal{S}$. These are pointed action models (U, α) with $\alpha \in \mathsf{U}$. We usually elide the action model and denote these by their second component α. And we write

$$\alpha \xrightarrow{a} \beta \tag{6.16}$$

to mean that there is some action model U which we are eliding such that both α and β are actions in it, and $\alpha \xrightarrow{a} \beta$ in U.

Now we define the language $\mathscr{L}(\alpha)$ in the following way:

$$\varphi ::= p \mid \neg\varphi \mid \varphi \wedge \varphi \mid K_a\varphi \mid C_A\varphi \mid [\alpha]\varphi \tag{6.17}$$

Compare this with (6.5). We have added the common knowledge operators for all groups A of agents, and we also have epistemic action modalities $[\alpha]\varphi$ for all epistemic actions φ.

The semantics of this language is given in the obvious way for all aspects of the language except the action modalities. And here is how they work. Let M be a Kripke model, and let $w \in M$. Then we have

$$M, w \models [\alpha]\varphi \quad \text{iff} \quad M, w \models \textsc{Pre}(\alpha) \text{ implies } M \otimes \mathsf{U}, (w, \alpha) \models \varphi. \tag{6.18}$$

Again, compare with the semantics of **PAL** in (6.6). Here is an explanation of the notation in (6.6): α on the left is an abbreviation for (U, α); that is, we elide the underlying action model. And (w, α) is one of the worlds in the model $M \otimes \mathsf{U}$.

It should be clear that practically all of the formal sentences in this paper up until now may be stated in this language $\mathscr{L}(\alpha)$. (We leave it to the diligent reader to find the few exceptions; some will be mentioned below.)

Basic Axioms

([α]-normality) $\vdash [\alpha](\varphi \to \psi) \to ([\alpha]\varphi \to [\alpha]\psi)$

Additional Axioms

(Atomic Permanence) $\vdash [\alpha]p \leftrightarrow (\textsc{Pre}(\alpha) \to p)$

(Partial Functionality) $\vdash [\alpha]\neg\chi \leftrightarrow (\textsc{Pre}(\alpha) \to \neg[\alpha]\chi)$

(Action-Knowledge) $\vdash [\alpha]K_a\varphi \leftrightarrow (\textsc{Pre}(\alpha) \to \bigwedge_{\alpha \xrightarrow{a} \beta} K_a[\beta]\varphi)$

Modal Rules

([α]-necessitation) From $\vdash \varphi$, infer $\vdash [\alpha]\varphi$

Action Rule

Let ψ be sentence, and let A be a group of agents. Consider sentences χ_β for all β such that $\alpha \xrightarrow{A} \beta$ (including α itself). Assume that:

1. $\vdash \chi_\beta \to [\beta]\psi$.
2. If $A \in A$ and $\beta \xrightarrow{a} \gamma$, then $\vdash \chi_\beta \wedge \textsc{Pre}(\beta) \to K_a\chi_\gamma$.

From these assumptions, infer $\vdash \chi_\alpha \to [\alpha]C_A\psi$.

Figure 6.7: The logical system for $\mathscr{L}(\alpha)$. We have omitted the parts of the system inherited from multi-agent modal epistemic logic.

Logic We now turn to the logical system for the validities in our language **DEL**. The logical system for valid sentences for **DEL** may be found in Figure 6.7. Perhaps the most interesting axioms are the Action-Knowledge Axioms; see Example 6.11 below for illustrations. If we set aside the common knowledge modalities, then the system in Figure 6.7 is a generalization of the second logical system for **PAL** in Figure 6.3. The completeness in this setting would again be by re-writing, showing that all of the epistemic modalities may be eliminated. However, the parallel result for the system in Figure 6.7 is more involved.

We next turn to the Action Rule, the rule of the system which allows us to infer common knowledge facts. The statement of the rule in Figure 6.7 uses some notation. If G is a group of agents and α and β are actions, then we write $\alpha \xrightarrow{A} \beta$ to mean that there is a sequence of actions $\gamma_1, \ldots, \gamma_k$ so that $\gamma_1 = \alpha$, $\gamma_k = \beta$, and for $1 \leq i < k$, $\gamma_i \xrightarrow{a} \gamma_{i+1}$ for some $a \in G$. In other words, there is a path from α to β in the "transition system of all actions" defined by (6.16), where the labels on the path are taken from the group G. Returning to the statement of the Action Rule, the sentences χ_β are a bit like "loop invariants" in logics of programs. We hope that an example will give the reader the essential features of this rule.

Example 6.11
Here are some examples of the Action Rule. Let $\mathscr{A}$ be a two element set $\{a, b\}$. We take $A = \mathscr{A}$. Consider the action model from (6.13), repeated below:

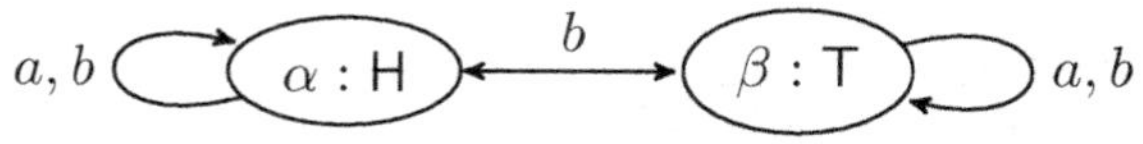

$$a, b \circlearrowleft (\alpha : \mathsf{H}) \xleftrightarrow{b} (\beta : \mathsf{T}) \circlearrowright a, b \tag{6.19}$$

⊣

We have two pointed versions of this model, one where the real action is α (this is what we saw in (6.13), and the other where it is β. We continue our practice of referring to these as α and β. So α (and β) represent actions where A comes to learn H (T), and B is aware that either α or β is happening but cannot tell which.

The intuition behind executing α is that any situation where α could be executed, A comes to learn H, and similarly for β with A learning T; and moreover that if *either* α or β was executed, then it would be common knowledge that either A learned H or that A learned T. We could easily give a semantic argument for this. The point of the Action Rule is to carry out the reasoning syntactically, inside a proof system. Here is how this works. We claim that $\vdash [\alpha]C(K_a\mathsf{H} \vee K_a\mathsf{T})$. Let ψ abbreviate $K_a\mathsf{H} \vee K_a\mathsf{T}$. To use the Action Rule to show that $\vdash [\alpha]C(K_a\psi)$, we must come up with sentences χ_α and χ_β with such that all of the following sentences are provable in our system:

$$\begin{array}{ll} \chi_\alpha \to [\alpha]\psi & (\chi_\alpha \wedge \mathsf{H}) \to K_b\chi_\beta \\ \chi_\beta \to [\beta]\psi & (\chi_\beta \wedge \mathsf{T}) \to K_a\chi_\beta \\ (\chi_\alpha \wedge \mathsf{H}) \to K_a\chi_\alpha & (\chi_\beta \wedge \mathsf{T}) \to K_b\chi_\beta \\ (\chi_\alpha \wedge \mathsf{H}) \to K_b\chi_\alpha & (\chi_\beta \wedge \mathsf{T}) \to K_b\chi_\alpha \end{array}$$

We take χ_α and χ_β to both be a propositional tautology, say true. Then all of our requirements are very easy to check, save for the first two. And for those, it is sufficient to check that in our system

$$\vdash [\alpha]K_a\mathsf{H} \qquad \text{and} \qquad \vdash [\beta]K_a\mathsf{T}.$$

Both of these are easy to check using other axioms of the system. For example, let us check the second assertion. One of the Action-Knowledge

Axioms is

$$[\beta]K_a\mathsf{T} \leftrightarrow (\mathsf{T} \rightarrow K_a[\beta]\mathsf{T}).$$

Here we have used the structure of the model in (6.19): we have $\beta \xrightarrow{a} \beta$ but not $\beta \xrightarrow{a} \alpha$. And so we need only show $\vdash \mathsf{T} \rightarrow K_a[\beta]\mathsf{T}$. This is due to the Atomic Permanence Axiom, the fact that $\text{PRE}(\beta) = \mathsf{T}$, and the Necessitation Rule for K_a. In this way, we can indeed check that $\vdash [\alpha]C(K_a\psi)$, and with similar work we check $\vdash [\beta]C(K_a\psi)$.

At first glance, one might also want to show, for example, that executing α results in common knowledge that B does *not* know H. However, a few moments of thought will show that this is not quite right: if we have a pointed model where B knows H to begin with, then executing α will not erase this knowledge. So we need an additional hypothesis. This is all for the good, since the Action Rule calls makes use of additional hypotheses in the form of the sentences χ_α and χ_β. For this particular assertion, we take them both to be

$$C(\neg K_b\mathsf{H} \wedge \neg K_b\mathsf{T} \wedge (\mathsf{H} \rightarrow \neg\mathsf{T})).$$

Then the Mix Axiom easily implies facts such as $\vdash \chi_\alpha \rightarrow K_a\chi_\alpha$. And once again, to use the Action Rule we need only verify that $\vdash \chi_\alpha \rightarrow [\alpha]\neg K_b\mathsf{H}$ and $\vdash \chi_\beta \rightarrow [\beta]\neg K_b\mathsf{H}$. We omit the details on this.

What has been done, and what has not been done We started our construction of a logic **DEL** with epistemic actions in careful way, by fixing a language $\mathscr{L}$ and then considering all of the epistemic action models over that language. (That is, we only considered action models whose precondition function took values in a language $\mathscr{L}$ that we considered at the outset.) And then from this we constructed a new language, gave its semantics, and then exhibited a logical system for it.

In a sense, we have not done everything that one would ideally want to do. We might like to go further and construct a logical language which the property that every sentence φ can itself occur as the precondition of an action in an action model, and every action model can be associated with an action in the language. The examples in Section 6.4.2 were deceptively simple in the sense that when we had concrete sentences as preconditions, those sentences were very simple indeed. We did not see "announcements of announcements" or anything of the type.

When we allow preconditions to be sentences in the *same* language we are defining, we must define the action models inductively in tandem with the language itself. For this reason (and others), the task of finding a natural general syntax for epistemic actions is not an easy problem.

Interlude: private announcements add expressive power Whenever we introduce a new piece of syntax, it is an obvious question to justify the addition on the grounds of expressive power. The first place where this becomes interesting is in *the logic of private announcements to one of two agents* This would be the logic formulated like **PAL** but with an additional action $Pri_a\ p$ of announcing the atomic sentence p to agent A. (We are taking the set of agents be a two-element set $\{A, B\}$ here.) As we know, without common knowledge operators, every sentence in this language is equivalent to a purely modal sentence; that is, the epistemic actions may be eliminated. The point is that allowing the common knowledge operators changes the situation.

Theorem 6.9
The sentence

$$[Pri_a\ p]C_a K_b\ p$$

is not expressible in the logic of public announcements and common knowledge, even allowing common knowledge operators C_a, C_b, and $C_{a,b}$.

The proof adapts Ehrenfeucht-Fraíssé games in the style of finite model theory. The overall conclusion is that a logical system with private announcements can say more than a logical system which only had public announcements, even in the presence of common knowledge operators.

Composition of actions There is a natural notion of *composition of epistemic actions.* Let U and V be action models. Then we obtain a new action model $\mathsf{U}\cdot\mathsf{V}$ in the following way: the set of action points in $\mathsf{U}\cdot\mathsf{V}$ is the Cartesian product $U \times V$; the arrows are the product arrows: $(\alpha, \beta) \xrightarrow{a} (\alpha', \beta')$ iff $\alpha \xrightarrow{a} \alpha'$ in U and $\beta \xrightarrow{a} \beta'$ in V; and the precondition function works as

$$\textsc{Pre}((\alpha, \beta)) \quad = \quad \langle\alpha\rangle\textsc{Pre}(\beta).$$

where $\langle\alpha\rangle\textsc{Pre}(\beta)$ abbreviates $\neg[(\mathsf{U}, \alpha)]\neg\textsc{Pre}_{\mathsf{V}}(\beta)$. This defines a composition of action models. We compose epistemic actions (= pointed action models) in the obvious way, by taking the real action in the composition to be the pair of the real actions.

Example 6.12
Let (U, α) be the epistemic action in (6.12), where A "cheated" by looking at the state of a concealed coin. Let (U', α') be the epistemic action where B cheats the same way:

Then the composition $(\mathsf{U}, \alpha) \cdot (\mathsf{V}, \alpha')$ is shown below:

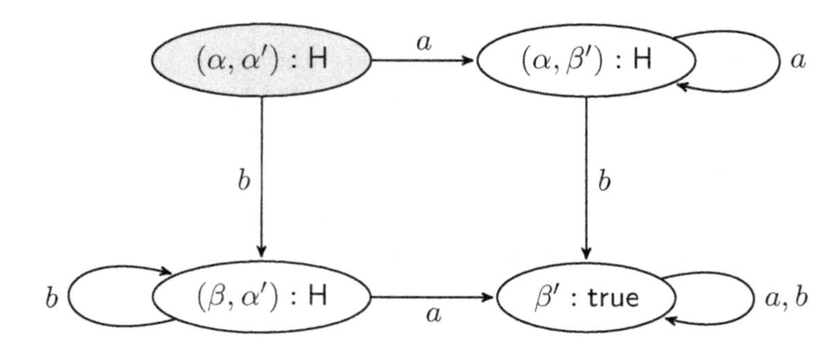

We have simplified the precondition function by using equivalences in our logic. Then

$$M \otimes ((\mathsf{U}, \alpha) \cdot (\mathsf{V}, \alpha')) \;\cong\; N,$$

where M is the model shown in (6.1) in Scenario 1, N is the "double cheating" model in Scenario 6.2, and $\cong$ means "isomorphic." One point that this makes is that the work we did in Section 6.4.3 could have been done in a different way: first compose the actions, and then apply the composition to the model. That is,

$$M \otimes ((\mathsf{U}, \alpha) \cdot (\mathsf{V}, \alpha')) \;\cong\; (M \otimes (\mathsf{U}, \alpha)) \otimes (\mathsf{V}, \alpha').$$

⊣

Now that we have seen composition of actions, the next step is to add this to the language of **DEL**. This can be done in two ways, as a "first-class" operation, or a "second-class" operation. At this point in our development, the second-class way is more natural. One would expand the syntax of the language to include not only sentences of the form $[\alpha]\varphi$ but also ones of the form $[\alpha_1 \cdot \alpha_2 \cdot \cdots \cdot \alpha_n]\varphi$. Then the syntax would work as we have seen it, and in the proof theory we would need a new axiom:

$$[\alpha \cdot \beta]\varphi \;\leftrightarrow\; [\alpha][\beta]\varphi.$$

The soundness of this *Composition Axiom* is suggested by what we saw in Example 6.12.

Program-based formulations of DEL We formulated **DEL** in stages, starting with the notion of an epistemic action over a basic language $\mathscr{L}$, then we alluded to the construction of languages which are "closed under epistemic actions", and finally we added composition in a second-class way.

Our discussion of **DEL** would not be complete if we did not at least mention the "first-class" way to add composition, and much more besides. This is the move from a language of sentences to a language of sentences and *programs*. So the logic here is quite reminiscent of *propositional dynamic logic* **PDL**. The syntax looks as follows:

sentences φ	true	p_i	$\neg\varphi$	$\varphi \wedge \psi$	$K_a\varphi$	$C_A\varphi$	$[\pi]\varphi$
programs π	*skip*	*crash*	$\sigma\psi_1, \ldots, \psi_n$	$\pi \cup \rho$	$\pi \; ; \; \rho$	π^*	

Here the program constructs include the standard programs *skip* (do nothing) and *crash*, and also non-deterministic choice $\pi \cup \rho$, composition $\pi \; ; \; \rho$; and iteration π^*. The construct $\sigma\psi_1, \ldots, \psi_n$ is basically an abbreviation for an action model; one lists the preconditions of the action points in some pre-assigned order.

The point of this language is that one can write *epistemic programs* of a rather general form and then study them.

6.5 Conclusions

As we close this handbook chapter on **DEL** and related matters, we now have a few comments on the area as a whole. As its presence in this handbook indicates, we should view the development **DEL** as a "chapter" in the development of epistemic logic (**EL**) as a whole. **DEL** is concerned with a special set of issues and ideas: group-level epistemic phenomena, logical languages adapting dynamic logic into epistemic settings, common knowledge understood via unbounded iterations of knowledge, and the update product. Each of those is worthy of extended discussion on its own, and of course this handbook contains quite a bit on these points. For the most part, **DEL** has seen a lot of technical progress largely because it set aside many of the conceptually difficult background issues of **EL**. To make the case for this, I wish to mention what I call the **DEL** Thesis. First, a related thesis:

The EL Thesis Let S be a real-world scenario involving people and some set of well-understood sentences about S. Then there is a finite set T of natural language sentences which explains S in a precise enough manner so that all possible statements about knowledge, higher-order knowledge, and common knowledge have determined truth values.

Furthermore, corresponding to S is a mathematical object, a finite multi-agent Kripke model $\hat{S}$, a world $s_0 \in \hat{S}$, and a set $\hat{T}$ of sentences in the formal language of multi-agent epistemic logic $\mathcal{L}$ such that for each natural language sentence A about the same atomic facts and the same participants in S, the following are equivalent:

1. A is intuitively true about S.

2. $s_0 \models \hat{A}$ in $\hat{S}$.

Although this thesis does not seem to be discussed much, it does seem to be a reasonable starting point. It seems fair to say that most presentations of **EL** either assume it, or act "as if" it were valid. Once one does either of these things, then the subject takes off. To some extent, the results that flow from the **EL** Thesis lend the subject some support. However, the technical successes of a subject do not necessarily cover for philosophical difficulties. In the case of **EL**, two of the primary difficulties are *logical omniscience* and the problem of *justifying knowledge.* To make technical progress in **EL**, one usually sets these problems (and perhaps others) aside.

We can propose a parallel thesis for **DEL**:

The DEL Thesis Let A be an epistemic action in the real world. Then A can be modeled as an epistemic action in our sense. Moreover, the notion of applying A to a real-world setting is captured by the update product.

Just as some of the technical results in **EL** lend some support to the basic semantics (Kripke models), the same is true for **DEL**. In more detail, the semantics of **PAL** via restriction works for all the examples that anyone has come up with. (The only objections that I can imagine are objections to **EL** more generally, such as the difficulties I mentioned above.) Similarly, the modeling of epistemic actions like private announcements and others which we have seen succeeds in the same way: in every example, our intuitions are matched by the formal details. Further, even though it seems hard to say what an "epistemic action" is in detail, it also seems fair to say that nobody has proposed an epistemic action which did not look and act like what we study in **DEL**.

There are two main lines of work in **DEL**. One is concerned with stronger and stronger languages, starting from **PAL** and adding other epistemic actions and going all the way up to program-based versions of **DEL**. This was the content of the final part of our chapter. The first parts of our chapter represent most of the actual work in the subject, where one studies **PAL** and its interesting special features. What the two lines have in common is some relation to *reduction principles* relating announcements (or more general types of actions) and knowledge. The growing body of work on these principles motivates the different axiomatizations of **PAL**, the different syntaxes of **DEL**, and the main principles in the proof systems. And taken together, this set of ideas is the lasting contribution of **DEL** to **EL**.

6.6 Notes

The quote at the opening is taken from Plaza (1989).

Proposition 6.1 is due to van Ditmarsch and Kooi (2003). Concerning modal logic itself, we might mention a related result: Andréka, Németi, and van Benthem (1998) showed that a sentence is preserved under submodels iff it is equivalent to a universal sentence. The matter is open when one adds the common knowledge operators.

Term rewriting theory is implicit in the work of Section 6.2.3 on translations. For connections of term rewriting theory to the work in this chapter, see the manuscript by Baltag, Moss, and Solecki (unpublished). Their work is more general than would be needed for the particular logical systems.

The third logical system in Section 6.2.3 is basically the system in the first paper on the subject, by Plaza (1989). The fourth system comes from Wang and Cao (2013). The name **PAN** derives from Wang and Cao (2013).

Theorem 6.4 is also due to Wang and Cao (2013). This paper in addition proves a number of other fundamental results concerning the axiomatizations of **PAL**. We have followed their notation in our presentation.

The succinctness result in Theorem 6.5 is due to Lutz (2006). The complexity results which we mentioned for satisfiability in single-agent **PAL** and multi-agent **PAL** are also from Lutz (2006). As Lutz (2006) puts it, "the question arizes whether a penalty has to be paid for this succinctness in terms of computational complexity: is reasoning in **PAL** more expensive than reasoning in epistemic logic? Interestingly, this is not the case."

The work on the substitution core of **PAL** in Section 6.2.5 comes from Holliday, Hoshi, and Icard (2012).

Theorem 6.7 is due to Baltag et al. (unpublished), as is Proposition 6.2.

The notion of relativized common knowledge comes from van Benthem, van Eijck, and Kooi (2006).

Theorem 6.8 comes from Miller and Moss (2005).

The work on arbitrary announcement logic in Section 6.3.3 comes primarily from Balbiani, Baltag, van Ditmarsch, Herzig, Hoshi, and de Lima (2007), and one of the facts at the end of the section comes from French and van Ditmarsch (2008).

The action model framework for **DEL** has been developed by Baltag, Solecki, and Moss, and has appeared in various forms (Baltag, Moss, and Solecki, 1998, 1999; Baltag, 2002; Baltag and Moss, 2004). The signature-based languages are introduced by Baltag and Moss (2004). The completeness of this system is worked out in the as-yet-unpublished paper by Baltag, Moss, and Solecki (2003a).

More extensive discussion of the conceptual matters concerning the update product may be found in the paper by Baltag and Moss (2004), one of the primary sources of this material, and in the handbook article by Baltag, van Ditmarsch, and Moss (2006).

Van Benthem, van Eijck and Kooi (2006) use a syntax which is richer than that of Baltag et al. (2003a) and of Baltag and Moss (2004). Compared to what we have presented, the main point is that the syntax allows for reduction axioms for common knowledge, and thus it avoids special induction rules for combinations of common knowledge and epistemic action modalities. It is not known whether the system is actually stronger than the logical systems by Baltag et al. (2003a) and Baltag and Moss (2004).

Completeness and decidability results for **DEL** may be found in the paper by Baltag et al. (1998) and the book by van Ditmarsch, van der Hoek, and Kooi (2007).

We also wish to mention other approaches to the syntax, semantics, and proof theory of **DEL**. The textbook by van Ditmarsch et al. (2007) is the only book-length treatment of these topics in any form.

One paper which allows actions to change the world is by van Benthem et al. (2006).

We mentioned that it is difficult to even formulate the syntax of action model logics in a correct manner. A number of different languages have been proposed, see e.g., work by Gerbrandy (1999), Baltag (2002) Baltag et al. (2003a), Baltag and Moss (2004), van Benthem et al. (2006), van Ditmarsch (2000, 2002), and by van Ditmarsch, van der Hoek, and Kooi (2003).

Some of the earliest work in this direction includes the papers by Gerbrandy (1999) and by Gerbrandy and Groeneveld (1997). Gerbrandy's action language was interpreted on non-wellfounded set theoretical structures; these correspond to bisimilarity classes of pointed Kripke models. Van Ditmarsch (2000; 2002) proposed another logical language. His semantics is restricted to $S5$ model transformations. Van Ditmarsch et al. (2003) proposed *concurrent epistemic actions.*

In other directions, there is some interest in connections of the action model framework of **DEL** with *algebra* (cf. Baltag, Coecke, and Sadrzadeh (2005) and Baltag, Cooke, and Sadrzadeh (2007)), and even more with *coalgebra* a (cf. Cîrstea and Sadrzadeh (2007)).

Theorem 6.9 comes from Baltag et al. (2003a).

The **DEL** Thesis comes from Moss (2010). The **EL** Thesis is new here.

Acknowledgements This work was partially supported by a grant from the Simons Foundation (#245591 to Lawrence Moss). I am grateful to Wiebe van der Hoek and Hans van Ditmarsch for their great patience during the time that I was (not) writing this chapter. I thank Hans van Ditmarsch and Yanjing Wang for conversations which improved the chapter considerably.

References

Andréka, H., I. Németi, and J. van Benthem (1998). Modal languages and bounded fragments of predicate logic. *Journal of Philosophical Logic 27*, 217–274.

Balbiani, P., A. Baltag, H. van Ditmarsch, A. Herzig, T. Hoshi, and T. de Lima (2007). What can we achieve by arbitrary announcements? - a dynamic take on fitch's knowability. In D. Samet (Ed.), *Proceedings of TARK 2007*, pp. 42–51. Presses Universitaires de Louvain.

Baltag, A. (2002). A logic for suspicious players: epistemic actions and belief updates in games. *Bulletin Of Economic Research 54*(1), 1–46.

Baltag, A., B. Coecke, and M. Sadrzadeh (2005). Algebra and sequent calculus for epistemic actions. *Electronic Notes in Theoretical Computer Science 126*, 27–52.

Baltag, A., B. Cooke, and M. Sadrzadeh (2007). Epistemic actions as resources. *Journal of Logic and Computation 17*, 555–585. Also in LiCS 2004 Proceedings of Logics for Resources, Programs, Processes (LRPP).

Baltag, A. and L. Moss (2004). Logics for epistemic programs. *Synthese 139*, 165–224. Knowledge, Rationality & Action 1–60.

Baltag, A., L. Moss, and S. Solecki (1998). The logic of common knowledge, public announcements, and private suspicions. In I. Gilboa (Ed.), *Proceedings of the 7th Conference on Theoretical Aspects of Rationality and Knowledge (TARK 98)*, pp. 43–56.

Baltag, A., L. Moss, and S. Solecki (1999). The logic of public announcements, common knowledge, and private suspicions. Technical report, Centrum voor Wiskunde en Informatica, Amsterdam. CWI Report SEN-R9922. This is a longer version of Baltag et al. (1998), and the ultimate version of this paper is Baltag et al. (2003a).

Baltag, A., L. Moss, and S. Solecki (2003a). The logic of public announcements, common knowledge and private suspicions. Manuscript, originally presented at TARK 98.

Baltag, A., L. Moss, and S. Solecki (2003b). Logics for epistemic actions: Completeness, decidability, expressivity. unpublished ms., Indiana University.

Baltag, A., H. van Ditmarsch, and L. Moss (2006). Epistemic logic and information update. In J. van Benthem and P. Adriaans (Eds.), *Handbook on the Philosophy of Information*, Amsterdam. Elsevier. In progress.

van Benthem, J., J. van Eijck, and B. Kooi (2006). Logics of communication and change. *Information and Computation 204*(11), 1620–1662.

Cîrstea, C. and M. Sadrzadeh (2007). Coalgebraic epistemic update without change of model. In T. Mossakowski (Ed.), *Proceedings of 2nd Conference on Algebra and Coalgebra in Computer Science*, needed. needed.

van Ditmarsch, H. P. (2000). *Knowledge Games.* Ph. D. thesis, University of Groningen. ILLC Dissertation Series DS-2000-06.

van Ditmarsch, H. P. (2002). Descriptions of game actions. *Journal of Logic, Language and Information 11*, 349–365.

van Ditmarsch, H. P. and B. P. Kooi (2003). Unsuccessful updates. In E. Álvarez, R. Bosch, and L. Villamil (Eds.), *Proceedings of the 12th International Congress of Logic, Methodology, and Philosophy of Science (LMPS)*, pp. 139–140. Oviedo University Press.

van Ditmarsch, H. P., W. van der Hoek, and B. Kooi (2007). *Dynamic Epistemic Logic*, Volume 337 of *Synthese Library.* Springer.

van Ditmarsch, H. P., W. van der Hoek, and B. P. Kooi (2003). Concurrent dynamic epistemic logic. In V. Hendricks, K. Jørgensen, and S. Pedersen (Eds.), *Knowledge Contributors*, Dordrecht, pp. 45–82. Kluwer Academic Publishers. Synthese Library Volume 322.

French, T. and H. van Ditmarsch (2008). Undecidability for arbitrary public announcement logic. In C. Areces and R. Goldblatt (Eds.), *AiML-2008: Advances in Modal Logic*, pp. 23–42. College Publications.

Gerbrandy, J. (1999). *Bisimulations on Planet Kripke.* Ph. D. thesis, University of Amsterdam. ILLC Dissertation Series DS-1999-01.

Gerbrandy, J. and W. Groeneveld (1997). Reasoning about information change. *Journal of Logic, Language, and Information 6*, 147–169.

Holliday, W. H., T. Hoshi, and I. Icard, Th F. (2012). A uniform logic of information dynamics. In *Proceedings of AiML'2012 (Advances in Modal Logic)*, Volume 9, pp. 348–367. College Publications.

Lutz, C. (2006). Complexity and succinctness of public announcement logic. In P. Stone and G. Weiss (Eds.), *Proceedings of the Fifth International Joint Conference on Autonomous Agents and Multiagent Systems (AAMAS'06)*, pp. 137–144. Association for Computing Machinery (ACM).

Miller, J. and L. Moss (2005). The undecidability of iterated modal relativization. *Studia Logica 79(3)*, 373–407.

Moss, L. S. (2010). Interview on epistemic logic. In *Epistemic Logic, 5 Questions.* Automatic Press /VIP.

Plaza, J. (1989). Logics of public communications. In M. Emrich, M. Pfeifer, M. Hadzikadic, and Z. Ras (Eds.), *Proceedings of the 4th International Symposium on Methodologies for Intelligent Systems*, pp. 201–216.

Wang, Y. and Q. Cao (2013). On axiomatizations of public announcement logic. *Synthese 190*(1), 103–134. DOI 10.1007/s11229-012-0233-5.

Chapter 7

Dynamic Logics of Belief Change

Johan van Benthem and Sonja Smets

Contents

Abstract

This chapter gives an overview of current dynamic logics that describe belief update and revision, both for single agents and in multi-agent settings. We employ a mixture of ideas from AGM belief revision theory and dynamic-epistemic logics of information-driven agency. After describing the basic background, we review logics of various kinds of beliefs based on plausibility

Chapter 7 of the *Handbook of Epistemic Logic*, H. van Ditmarsch, J.Y. Halpern, W. van der Hoek and B. Kooi (eds), College Publications, 2015, pp. 313–393.

models, and then go on to various sorts of belief change engendered by changes in current models through hard and soft information. We present matching complete logics with dynamic-epistemic recursion axioms, and develop a very general perspective on belief change by the use of event models and priority update. The chapter continues with three topics that naturally complement the setting of single steps of belief change: connections with probabilistic approaches to belief change, long-term temporal process structure including links with formal learning theory, and multi-agent scenarios of information flow and belief revision in games and social networks. We end with a discussion of alternative approaches, further directions, and windows to the broader literature, while links with relevant philosophical traditions are discussed throughout.

Human cognition and action involve a delicate art of living dangerously. Beliefs are crucial to the way we plan and choose our actions, even though our beliefs can be very wrong and refuted by new information. What keeps the benefits and dangers in harmony is our ability to revise beliefs as the need arises. In this chapter, we will look at the logical structure of belief revision, and belief change generally. But before we can do this, we need background of two kinds: (a) the pioneering AGM approach in terms of postulates governing belief revision which showed that this process has clear formal structures regulating its behavior, and (b) the basics of dynamic-epistemic logics of information flow which showed that change of attitudes for agent and the events triggering such changes are themselves susceptible to exact logical analysis. This is what we will provide in the first two sections of this chapter. With this material in place, Section 7.3 will then start our main topic, the logical treatment of belief revision.

7.1 Basics of belief revision

7.1.1 The AGM account of belief revision

What happens when an agent is confronted with a new fact φ that goes against her prior beliefs? If she is to accept the new fact φ and maintain a consistent set of beliefs, she will have to give up some of her prior beliefs. But which of her old beliefs should she give up? More generally, what policy should she follow to revise her beliefs? As we will see in this chapter, several answers to this question are possible. The standard answer in the literature says that our agent should accept the new fact and at the same time maintain as many as possible of her old beliefs without arriving at a contradiction. Making this more precise has been the driving force behind Belief Revision Theory. Standard Belief Revision Theory, also called *AGM theory* (after the pioneering authors Alchourrón, Gärdenfors and Makinson) has provided us with a series of "rationality conditions", that are meant to precisely govern the way in which a rational agent should revise her beliefs.

AGM theory The AGM theory of belief revision is built up from three basic ingredients: 1) the notion of a *theory* (or "belief set") T, which is a logically closed set of sentences $\{\psi, \gamma ...\}$ belonging to a given language $\mathcal{L}$; 2) the *input of new information*, i.e., a syntactic formula φ; and 3) a *revision operator* $*$ which is a map associating a theory $T * \varphi$ to each pair (T, φ) consisting of a theory T and an input sentence φ. The construct $T * \varphi$ is taken to represent the agent's new theory after learning φ. Hence $T * \varphi$ is the agent's new set of beliefs, given that the initial set of beliefs is T and that the agent has learnt that φ.

Expansion The AGM authors impose a number of postulates or rationality conditions on the revision operation $*$. To state these postulates, we first need an auxiliary *belief expansion operator* $+$, that is often considered an unproblematic form of basic update. Belief expansion is intended to model the simpler case in which the new incoming information φ does not contradict the agent's prior beliefs. The *expansion* $T + \varphi$ of T with φ is defined as the closure under logical consequence of the set $T \cup \{\varphi\}$. AGM provides a list of 6 postulates that exactly regulate the expansion operator, but instead of listing them here we will concentrate on belief revision. However, later on, we will see that even expansion can be delicate when complex epistemic assertions are added.

Revision Now, belief revision goes beyond belief expansion in its intricacies. It is regulated by the following famous AGM Belief Revision Postulates:

(1)	Closure	$T * \varphi$ is a belief set
(2)	Success	$\varphi \in T * \varphi$
(3)	Inclusion	$T * \varphi \subseteq T + \varphi$
(4)	Preservation	If $\neg\varphi \notin T$, then $T + \varphi \subseteq T * \varphi$
(5)	Vacuity	$T * \varphi$ is inconsistent iff $\vdash \neg\varphi$
(6)	Extensionality	If $\vdash \varphi \leftrightarrow \psi$, then $T * \varphi = T * \psi$
(7)	Subexpansion	$T * (\varphi \wedge \psi) \subseteq (T * \varphi) + \psi$
(8)	Superexpansion	If $\neg\psi \notin T * \varphi$, then $T * (\varphi \wedge \psi) \supseteq (T * \varphi) + \psi$.

These postulates look attractive, though there is more to them than meets the eye. For instance, while the success postulate looks obvious, in our later dynamic-epistemic logics, it is the most controversial one in this list. In a logical system allowing complex epistemic formulas, the truth value of the target formula can change in a revision step, and the Success Postulate would recommend incorporating a falsehood φ into the agent's theory T. One important case in which this can occur is when an introspective agent revises her beliefs on the basis of new information that refers to beliefs or higher-order beliefs (i.e., beliefs about beliefs). Because the AGM setting does not incorporate "theories about theories", i.e., it ignores an agent's higher-order beliefs, this problem is side-stepped. All the beliefs covered by AGM are so-called factual beliefs about ontic facts that do not refer to the epistemic state of the agent. However any logic for belief change that does allow explicit belief-operators in the language, will have to pay attention to success conditions for complex updates.

A final striking aspect of the Success Postulate is the heavy emphasis placed on the last incoming proposition φ, which can abruptly override long accumulated earlier experience against φ. This theme, too, will return later when we discuss connections with formal theories of inductive learning.

Contraction A third basic operation considered in AGM is that of *belief contraction* $T - \varphi$, where one removes a given assertion φ from a belief set T, while removing enough other beliefs to make φ underivable. This is harder than expansion, since one has to make sure that there is no other way within the new theory to derive the target formula after all. And while there is no unique way to construct a contracted theory, AGM prescribes the following formal postulates:

(1)	Closure	$T - \varphi$ is a belief set
(2)	Contraction	$(T - \varphi) \subseteq T$
(3)	Minimal Action	If $\varphi \notin T$, then $T - \varphi = T$
(4)	Success	If $\nvdash \varphi$ then $\varphi \notin (T - \varphi)$
(5)	Recovery	If $\varphi \in T$, then $T \subseteq (T - \varphi) + \varphi$
(6)	Extensionality	If $\vdash \varphi \leftrightarrow \psi$, then $T - \varphi = T - \psi$
(7)	Min-conjunction	$T - \varphi \cap T - \psi \subseteq T - (\varphi \wedge \psi)$
(8)	Max-conjunction	If $\varphi \notin T - (\varphi \wedge \psi)$, then $T - (\varphi \wedge \psi) \subseteq T - \varphi$

Again, these postulates have invited discussion, with Postulate 5 being the most controversial one. The Recovery Postulate is motivated by the intuitive principle of minimal change, which prescribes that a contraction should remove as little as possible from a given theory T.

The three basic operations on theories introduced here are connected in various ways. A famous intuition is the Levi-identity

$$T * \varphi = (T - \neg\varphi) + \varphi$$

saying that a revision can be obtained as a contraction followed by an expansion.

An important result in this area is a theorem by Gärdenfors which shows that if the contraction operation satisfies postulates (1-4) and (6), while the expansion operator satisfies its usual postulates, then the revision operation defined by the Levi-identity will satisfy the revision postulates (1-6). Moreover, if the contraction operation satisfies the seventh postulate, then so does revision, and likewise for the eight postulate.

7.1.2 Conditionals and the Ramsey Test

Another important connection runs between belief revision theory and the logic of conditionals. The *Ramsey Test* is a key ingredient in any study of this link. In 1929, F.P. Ramsey wrote:

> "If two people are arguing 'If A, will B?' and are both in doubt as to A, they are adding A hypothetically to their stock of knowledge and arguing on that basis about B; so that in a sense 'If A, B' and 'If $A, \overline{B}$' are contradictories."

Clearly, this evaluation procedure for conditional sentences $A > B$ uses the notion of belief revision. Gärdenfors formalised the connection with the

Ramsey Test as the following statement:

$$A > B \in T \text{ iff } B \in T * A$$

which should hold for all theories T and sentences A, B. In a famous impossibility result, he then showed that the existence of such Ramsey conditionals is essentially incompatible with the AGM postulates for the belief revision operator $*$. The standard way out of Gärdenfors' impossibility result is to weaken the axioms of $*$, or else to drop the Ramsey test.

Most discussions in this line are cast in purely syntactic terms, and in a setting of propositional logic. However, in section 7.3 we will discuss a semantic perspective which saves much of the intuitions underlying the Ramsey Test.

This is in fact a convenient point for turning to the modal logic paradigm in studying belief revision. Like we saw with belief expansion, it may help to first introduce a simpler scenario. The second part of this introductory section shows how modal logics can describe information change and its updates[1] in what agents know. The techniques found in this realm will then be refined and extended in our later treatment of belief revision.

7.2 Modal logics of belief revision

Starting in the 1980s, several authors have been struck by analogies between AGM revision theory and modal logic over suitable universes. Belief and related notions like knowledge could obviously be treated as standard modalities, while the dynamic aspect of belief change suggested the use of ideas from Propositional Dynamic Logic of programs or actions to deal with update, contraction, and revision. There is some interest in seeing how long things took to crystallise into the format used in this chapter, and hence we briefly mention a few of these proposals before introducing our final approach.

Propositional dynamic logic over information models Propositional Dynamic Logic (PDL) is a modal logic that has both static propositions φ and programs or actions π. It provides dynamic operators $[\pi]\varphi$ that one can use to reason about what will be true after an action takes place. One special operator of PDL is the "test of a proposition φ" (denoted as $\varphi?$): it

[1] Our use of the term "update" in this chapter differs from a common terminology of "belief update" in AI, due to Katsuno and Mendelzon. The latter notion of update refers to belief change in a factually changing world, while we will mainly (though not exclusively) consider epistemic and doxastic changes but no changes of the basic ontic facts. This is a matter of convenience though, not of principle.

takes a proposition φ into a program that tests if the current state satisfies φ. Using this machinery over tree-like models of successive information states ordered by inclusion, in 1989, van Benthem introduced dynamic operators that mirror the operations of AGM in a modal framework. One is the addition of φ (also called "update", denoted as $+\varphi$), interpreted as moving from any state to a minimal extension satisfying φ. Other operators included "downdates" $-\varphi$ moving back to the first preceding state in the ordering where φ is not true. Revision was defined via the Levi-identity. In a modification of this approach by de Rijke in the 1990s, these dynamic operators were taken to work on universes of theories.

Dynamic doxastic logic over abstract belief worlds These developments inspired Segerberg to develop the logical system of Dynamic Doxastic Logic (DDL), which operates at a higher abstraction level for its models. DDL combines a PDL dynamics for belief change with a static logic with modalities for knowledge K and belief B. The main syntactic construct in DDL is the use of the dynamic modal operator $[*\varphi]\psi$ which reads "ψ holds after revision with φ", where $*\varphi$ denotes a relation (often a function) that moves from the current world of the model to a new one. Here φ and ψ were originally taken to be factual formulas only, but in later versions of DDL they can also contain epistemic or doxastic operators. This powerful language can express constructs such as $[*\varphi]B\psi$ stating that after revision with φ the agent believes ψ. In what follows, we will take a more concrete modal approach to DDL's abstract world, or state, changes involved in revision – but a comparison will be given in Section 7.9.1.

Degrees of belief and quantitative update rules In this chapter, we will mainly focus on qualitative logics for belief change. But historically, the next step were quantitative systems for belief revision in the style of Dynamic Epistemic Logic, where the operations change current models instead of theories or single worlds. Such systems were proposed a decade ago, using labelled operators to express degrees of belief for an agent. In 2003, van Ditmarsch and Labuschagne gave a semantics in which each agent has associated accessibility relations corresponding to labeled preferences, and a syntax that can express degrees of belief. Revision of beliefs with new incoming information was modeled using a binary relation between information states for knowledge and degrees of belief. A more powerful system by Aucher in 2003 had degrees of belief interpreted in Spohn ranking models, and a sophisticated numerical "product update rule" in the style of Baltag, Moss and Solecki (BMS, see Section 7.5.2 below) showing how ranks of worlds change under a wide variety of incoming new information.

Belief expansion via public announcement logic An early qualitative approach, due to van Ditmarsch, van der Hoek and Kooi in 2005, relates AGM belief expansion to the basic operation of public announcement in Dynamic Epistemic Logic. The idea is to work with standard relational modal models M for belief (in particular, these need not have a reflexive accessibility relation, since beliefs can be wrong), and then view the action of getting new information φ as a public announcement that takes M to its submodel consisting only of its φ-worlds. Thus, an act of belief revision is modeled by a transformation of some current epistemic or doxastic model. The system had some built-in limitations, and important changes were made later by van Benthem and Baltag & Smets to the models and update mechanism to achieve a general theory – but it was on the methodological track that we will follow now for the rest of this chapter.

Public announcement logic To demonstrate the methodology of Dynamic Epistemic Logic to be used in this chapter, we explain the basics of Public Announcement Logic (PAL). The language of PAL is built up as follows:

$$\varphi ::= p \mid \neg\varphi \mid \varphi \wedge \varphi \mid K_i\varphi \mid [!\varphi]\varphi$$

Here we read the K_i-modality as the knowledge of agent i and we read the dynamic construct $[!\varphi]\psi$ as "ψ holds after the public announcement of φ". We think of announcements $!\varphi$ as public events where indubitable hard information that φ is the case becomes available to all agents simultaneously, whether by communication, observation, or yet other means. In what follows, we define and study the corresponding transformations of models, providing a constructive account of how information changes under this kind of update.

Semantics for PAL We start with standard modal models $M = (W, R_i, V)$ where W is a non-empty set of possible worlds. For each agent i, we have an epistemic accessibility relation R_i, while V is a valuation which assigns sets of possible worlds to each atomic sentence p. The satisfaction relation can be introduced as usual in modal logic, making the clauses of the non-dynamic fragment exactly the standard ones for the multi-agent epistemic logic $S5$. For the case of knowledge (only an auxiliary initial interest in this chapter), we take R_i to be an equivalence relation, so that the underlying base logic is a multi-agent version of the modal logic $S5$. We now concentrate on the dynamics.

The clause for the dynamic modality goes as follows:

$$(M, w) \models [!\varphi]\psi \text{ iff } (M, w) \models \varphi \text{ implies } (M|\varphi, w) \models \psi$$

where $M|\varphi = (W', R'_i, V')$ is obtained by relativising the model M with φ as follows (here, as usual, $[\![\varphi]\!]_M$ denotes the set of worlds in M where φ is true):

$$W' = [\![\varphi]\!]_M$$

$$R'_i = R_i \cap ([\![\varphi]\!]_M \times [\![\varphi]\!]_M)$$

$$V'(p) = V(p) \cap [\![\varphi]\!]_M$$

Example 7.1 (Public announcement by world elimination) Figure 7.1 illustrates the update effect of a public announcement in the state w of model M such that in model $M|p$ only the p-worlds survive:

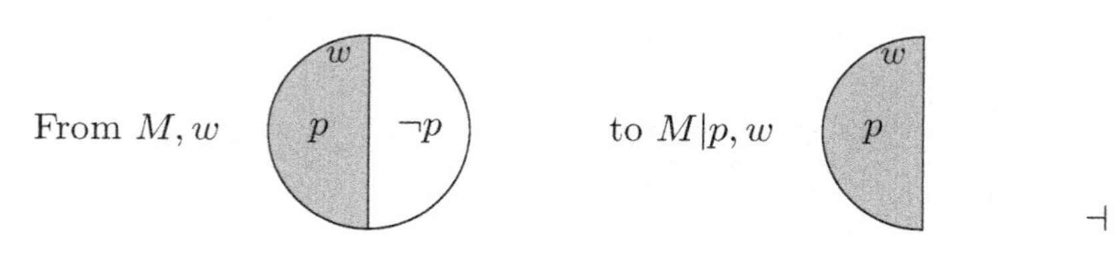

⊣

Figure 7.1: A public announcement of p.

Proof system for PAL We start with the proof system for $S5$, including all its standard axioms and rules (including replacement of provable equivalents), and in addition we have the following recursion axioms:

$$\begin{array}{rcl} [!\varphi]p & \leftrightarrow & (\varphi \rightarrow p) \\ [!\varphi]\neg\psi & \leftrightarrow & (\varphi \rightarrow \neg[!\varphi]\psi) \\ [!\varphi](\psi \wedge \gamma) & \leftrightarrow & ([!\varphi]\psi \wedge [!\varphi]\gamma) \\ [!\varphi]K_i\psi & \leftrightarrow & (\varphi \rightarrow K_i[!\varphi]\psi) \end{array}$$

Completeness The axiomatic system for PAL is sound and complete. One can easily show this by using the recursion axioms to translate every sentence that contains a dynamic modality to an equivalent one without it, and then using the completeness of the static base logic.

Public announcements and belief expansion The effect of a public announcement $!\varphi$ on a given model is of a very specific type: all non-φ worlds are deleted. One can easily see that when models contract under truthful announcements, the factual knowledge of the agent expands. As we stated already, these restrictions were lifted in the work of van Benthem in 2007 and by Baltag & Smets in 2008, who deal with arbitrary updates of both plain and conditional beliefs, actions of revision and contraction, as well as other triggers for all these than public announcements. The latter systems will be the heart of this chapter, but before we go there, we also need to discuss the optimal choice of an underlying base logic in more detail than we have done so far.

Summary Modal logic approaches to belief revision bring together three traditions: 1) modal logics for static notions of knowledge and belief, 2) the AGM theory of belief revision, and 3) the modal approach to actions of Propositional Dynamic Logic. Merging these ideas opens the door to a study of knowledge update and belief change in a standard modal framework, without having to invent non-standard formalisms, allowing for smooth insertion into the body of knowledge concerning agency that has been accumulated in the modal paradigm.

7.3 Static base logics

7.3.1 Static logic of knowledge and belief

In line with the literature in philosophical logic, we want to put knowledge and belief side by side in a study of belief change. But how to do this, requires some thinking about the best models. In particular, Hintikka's original models with a binary doxastic accessibility relation that drive such well-known systems as "KD45" are not what we need – as we shall see in a moment.

Reinterpreting PAL One easy route tries to reinterpret the dynamic-epistemic logic PAL that we have presented in the previous section. Instead of knowledge, we now read the earlier K_i-operators as beliefs, placing no constraints on the accessibility relations: using just pointed arrows. One test for such an approach is that it must be possible for beliefs to be wrong:

Example 7.2 (A mistaken belief)
Consider the model in Figure 7.2 with two worlds that are epistemically accessible to each other via dashed arrows, but where the pointed arrow is the only belief relation. Here, in the actual world to the left, marked in black, the proposition p is true, but the agent mistakenly believes that $\neg p$.

With this view of doxastic modalities, PAL works exactly as before. But as pointed out by van Benthem around 2005, there is a problem, i.e., that of overly drastic belief change.

Consider an announcement $!p$ of the true fact p. The PAL result is the one-world model where p holds, with the inherited empty doxastic accessibility relation.

But on the universal quantifier reading of belief, this means the following: the agent believes that p, but also that $\neg p$, in fact $B\bot$ is true at such an end-point. Clearly, this is not what we want: agents who have their beliefs contradicted would now be shattered and they would start believing anything. ⊣

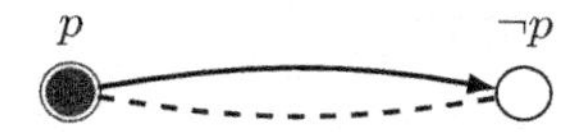

Figure 7.2: A mistaken belief.

While some ad-hoc patches persist in the literature, a better way is to change the semantics to allow for more intelligent responses, by using more structured models for conditional beliefs, as we shall show now.

7.3.2 Plausibility models

A richer view of belief follows the intuition that an agent believes the things that are true, not in all her accessible worlds, but only in those that are "best" or most relevant to her. Static models for this setting are easily defined.

Definition 7.1 (Epistemic plausibility models)
Epistemic plausibility models are structures $M = (W, \{\sim_i\}_{i\in I}, \{\leq_i\}_{i\in I}, V)$, where W is a set of states or possible worlds. The epistemic accessibility $\{\sim_i\}_{i\in I}$ relations for each agent $i \in I$ are equivalence relations. The family of plausibility relations $\{\leq_i\}_{i\in I}$ consists of binary comparison relations for agents and is read as follows, $x \leq_i y$ if agent i considers y at least as plausible as x. The plausibility relations are assumed to satisfy two conditions: (1) $\leq_i$-comparable states are $\sim_i$-indistinguishable (i.e. $s \leq_i t$ implies $s \sim_i t$) and (2) the restriction of each plausibility relation $\leq_i$ to each $\sim_i$-equivalence class is a well-preorder.[2] ⊣

Notation We use the notation $Max_{\leq_i}P$ for the set of $\leq_i$-maximal elements of a given set $P \subseteq W$ and use this to denote the "most plausible

[2]Here, a "preorder" is a reflexive and transitive binary relation. A "well-preorder" over W is a preorder guaranteeing that every non-empty subset of W has maximal (most plausible) elements. Note also that the conditions (1) and (2) are useful in an epistemic-doxastic context but can be relaxed in various ways to yield a more general setting. We will return to this point below.

elements of P under the given relation". There are two other relations that are useful to name: the strict plausibility relation $s <_i t$ iff $s \leq_i t$ but $t \not\leq_i s$, and the equiplausibility relation $s \cong_i t$ iff both $s \leq_i t$ and $t \leq_i s$. We denote by $s(i)$ the $\sim_i$-equivalence class of s, i.e. $s(i) := \{t \in W : s \sim_i t\}$

Simplifying the setting The definition of epistemic plausibility models contains superfluous information. The epistemic relation $\sim_i$ can actually be recovered from the plausibility relation $\leq_i$ via the following rule:

$$s \sim_i t \text{ iff either } s \leq_i t \text{ or } t \leq_i s$$

provided that the relation $\leq_i$ is connected.[3] This makes two states epistemically indistinguishable for an agent i iff they are *comparable* with respect to $\leq_i$. Accordingly, the most economic setting are the following simplified semantic structures that we will use in the remainder of this chapter.

Definition 7.2 (Plausibility models)
A *plausibility model* $M = (W, \{\leq_i\}_{i \in I}, V)$ consists of a set of possible worlds W, a family of locally well-preordered relations $\leq_i$ and a standard valuation V.[4] ⊣

Fact 7.1
There is a bijective correspondence between Epistemic Plausibility Models and the above Plausibility Models. ⊣

Baltag & Smets show in 2006 how every plausibility model can be canonically mapped into an epistemic plausibility model and vice versa.[5]

Generalisation to arbitrary preorders The above simplification to plausibility models only works because all epistemically accessible worlds are comparable by plausibility and all plausibility comparisons are restricted to epistemically accessible worlds. These conditions are restrictive, though they can be justified in the context of belief revision.[6] However, many

[3] A relation is "connected" if either $s \leq_i t$ or $t \leq_i s$ holds, for all $s, t \in W$.

[4] Here a *locally well-preordered* relation $\leq_i$ is a preorder whose restriction to each corresponding comparability class $\sim_i$ is well-preordered. In case the set W is *finite*, a locally well-preordered relation becomes a *locally connected preorder*: a preorder whose restriction to any comparability class is connected.

[5] In one direction this can be shown by defining the epistemic relation as $(\sim_i := \leq_i \cup \geq_i)$. In the other direction, all that is needed is a map that "forgets" the epistemic structure.

[6] One advantage is that connected pre-orders ensure the AGM rule of "rational monotonicity", capturing the intuitive idea that an agent who believes two propositions p and q will still hold a belief in q after revision with $\neg p$.

authors have also used plausibility models with non-connected orderings, allowing for genuinely incomparable (as opposed to indifferent) situations. With a few modifications, what we present in this chapter also applies to this more general setting that reaches beyond belief revision theory.

In this chapter we will use plausibility models as our main vehicle.

Languages and logics

One can interpret many logical operators in plausibility models. In particular, knowledge can be interpreted as usual. However, there is no need to stick to just the knowledge and belief operators handed down by the tradition. First of all, plain belief $B_i\varphi$ is a modality interpreted as follows:

Definition 7.3 (Belief as truth in the most plausible worlds)
In plausibility models, we interpret *belief* by putting $M, s \models B_i\varphi$ iff $M, t \models \varphi$ for all worlds t in $s(i)$ that are in $Max_{\leq_i}[\![\varphi]\!]$. ⊣

But information flow and action also involve richer conditional beliefs $B_i^{\psi}\varphi$, with the intuitive reading that, conditional on ψ, the agent believes that φ. Roughly speaking, conditional beliefs "pre-encode" the beliefs that agents would have if they were to learn certain things expressed by the restricting proposition.

Definition 7.4 (Interpreting conditional beliefs)
In plausibility models, $M, s \models B_i^{\psi}\varphi$ iff $M, t \models \varphi$ for all worlds t that are maximal for the order $x \leq_i y$ in the set $\{u \mid M, u \models \psi\} \cap s(i)$. ⊣

Plain belief $B_i\varphi$ can now be recovered as the special case $B_i^T\varphi$ with a trivially true condition T. It can be shown that conditional belief is not definable in terms of plain belief, so we have obtained a genuine language extension.

Example 7.3 (Conditional beliefs depicted)
Figure 7.3 shows a plausibility model containing both the plausibility relations and the indistinguishability relations (represented via the dashed arrows) for a given agent. To keep the picture simple, we draw neither reflexive arrows nor arrows that can be obtained via transitivity. In every world, the agent believes p and q_3, i.e. $B_i(p \wedge q_3)$. Note that while this agent currently holds a true belief in p, her belief in p is rather fragile because it can easily be given up when new information is received. Indeed, conditional on the information $\neg q_3$, the agent would believe that $\neg p$ was the case, i.e. $B_i^{\neg q_3}\neg p$. ⊣

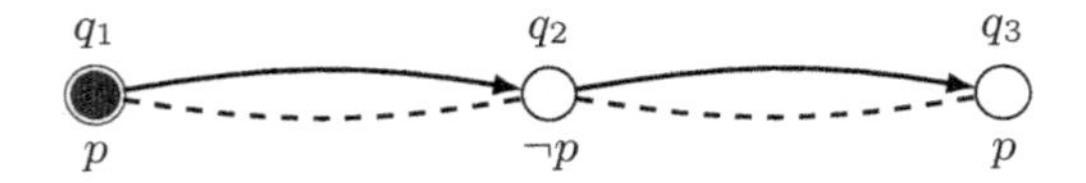

Figure 7.3: Illustrating conditional beliefs.

Epistemic-doxastic introspection We already noted that in plausibility models, epistemic accessibility is an equivalence relation, and plausibility a pre-order over the equivalence classes, the same as viewed from any world inside that class. This reflects the fact that in such models, all agents know their beliefs. Formally, it is easy to see that the following axiom is valid:

$$B_i\varphi \to K_iB_i\varphi \qquad \text{Epistemic-Doxastic Introspection}$$

This assumption is of course debatable, since not all doxastic agents may have this type of introspective capacity. Abandoning this assumption via ternary orderings will be explored briefly in Section 7.3.2.

As an illustration of the semantic framework introduced here, we return to an earlier issue about belief revision.

The Ramsey Test revisited Recall the Ramsey Test and Gärdenfors' impossibility result from Section 7.1. In 2010, Baltag & Smets gave a semantic re-analysis that uses modalities for knowledge, belief and conditional belief. In this setting, the main question becomes: "Are there any truth conditions for the conditional $A > B$ that are compatible with the Ramsey Test – given the usual modal semantics for belief, and some reasonable semantics for belief revision?". It can be shown that, if the Ramsey test is to hold for *all* theories (including those representing future belief sets, after possible revisions) and if some reasonable rationality conditions are assumed (such as full introspection of beliefs and of dispositions to believe, plus unrevisability of beliefs that are known to be true), then the answer is "no". The reason is this: the Ramsey Test treats conditional beliefs about beliefs in the same way as hypothetical beliefs about facts. The test would succeed only if, when making a hypothesis, agents revise their beliefs about their own belief revision in the same way as they revise their factual beliefs. But the latter requirement is inconsistent with the restrictions posed by introspection. Introspective agents know their own hypothetical beliefs, and so cannot accept hypotheses that go against their knowledge. Thus, beliefs about one's own belief revision policy cannot be revised.

But this is not the end of the story, and in a sense, the logics to be presented later on show a "Possibility Result". A dynamic revision of the sort pursued in the coming sections, that represents agents' revised beliefs about the situation after the revision, can consistently satisfy a version of the Ramsey Test.

World-based plausibility models

Plausibility models as defined here have uniform plausibility relations that do not vary across epistemically accessible worlds. This reflects an intuition that agents know their own plain and conditional beliefs. However, it is possible, at a small technical cost, to generalise our treatment to ternary world-dependent plausibility relations, and such relations are indeed common in current logics for epistemology. Stated equivalently, the assumption of epistemic-doxastic introspection can be changed in models that keep track of the different beliefs of the agent in every possible world.

Definition 7.5 (World-based plausibility models)
World-based plausibility models are structures of the form $M = (W, \{\sim_i\}_{i\in I}, \{\leq_i^s\}_{i\in I}, V)$ where the relations $\sim_i$ stand for epistemic accessibility as before, but the $(\leq_i^s)$ are ternary comparison relations for agents that read as: $x \leq_i^s y$ if, in world s, agent i considers y at least as plausible as x. ⊣

Models like this occur in conditional logic, logics of preference, and numerical graded models for beliefs. One can again impose natural conditions on ternary plausibility relations, such as reflexivity, or transitivity. Adding connectedness yields the well-known nested spheres from the semantics of conditional logic, with pictures of a line of equiplausibility clusters, or of concentric circles. But there are also settings that need a fourth option of incomparability: $\neg s \leq t \wedge \neg t \leq s$. This happens, for instance, when comparing worlds with conflicting criteria for preference. And sometimes also, non-connected pre-orders or partially ordered graphs are a mathematically more elegant approach.

Essentially the same truth conditions as before for plain and conditional beliefs work on these more general models, but it is easy to find models now where agents are not epistemically introspective about their own beliefs.

One thing to note in the logic of world-based plausibility models is how their treatment of conditional beliefs is very close to conditional logic:

Digression on conditionals Recall that conditional beliefs pre-encode beliefs we would have if we were to learn new things. A formal analogy is a well-known truth condition from conditional logic. A conditional $\varphi \Rightarrow \psi$ says that ψ is true in the closest worlds where φ is true, along some comparison order on worlds.[7] Thus, results from conditional logic apply to

[7] On infinite models, this clause needs some modifications, but we omit details here.

doxastic logic. For instance, on reflexive transitive plausibility models, we have a completeness theorem whose version for conditional logic is due to Burgess in 1981 and Veltman in 1985:

Theorem 7.1
The logic of conditional belief $B^\psi\varphi$ on world-based plausibility models is axiomatised by the laws of propositional logic plus obvious transcriptions of the following principles of conditional logic:

(a) $\varphi \Rightarrow \varphi$
(b) $\varphi \Rightarrow \psi$ implies $\varphi \Rightarrow \psi \vee \chi$
(c) $\varphi \Rightarrow \psi, \varphi \Rightarrow \chi$ imply $\varphi \Rightarrow \psi \wedge \chi$,
(d) $\varphi \Rightarrow \psi, \chi \Rightarrow \psi$ imply $(\varphi \vee \chi) \Rightarrow \psi$,
(e) $\varphi \Rightarrow \psi, \varphi \Rightarrow \chi$ imply $(\varphi \wedge \psi) \Rightarrow \chi$. ⊣

Richer attitudes: safe and strong belief

We now return to uniform plausibility models. In epistemic reality, agents have a rich repertoire of attitudes concerning information beyond just knowledge and belief, such as being certain, being convinced, assuming, etcetera. Among all options in this plethora, of special interest to us are notions whose definition has a dynamic intuition behind it. The following notion makes particular sense, intermediate between knowledge and belief. It is a modality of true belief which is stable under receiving new true information:

Definition 7.6 (Safe belief)
The modality of *safe belief* $\Box_i\varphi$ is defined as follows: $M, s \models \Box_i\varphi$ iff for all worlds t in the epistemic range of s with $t \geq_i s, M, t \models \varphi$. Thus, φ is true in all epistemically accessible worlds that are at least as plausible as the current one. ⊣

The modality $\Box_i\varphi$ is stable under hard information updates, at least for factual assertions φ that do not change their truth value as the model changes. In fact, it is just the universal base modality $[\leq_i]\varphi$ for the plausibility ordering. In what follows, we will make safe belief part of the static doxastic language, treating it as a pilot for a richer theory of attitudes in the background. Pictorially, one can think of this as illustrated in the following example:

Example 7.4 (Three degrees of doxastic strength)
Consider the model in Figure 7.4, with the actual world in the middle.

$K_i\varphi$ describes what the agent knows in an absolute sense: φ must be true in all worlds in the epistemic range, less or more plausible than the current one. In Figure 7.4, the agent knows q and all tautologies in this way. $\Box_i p$ describes her safe beliefs in further investigation: p is true in all worlds of the model from the middle towards the right. Thus, we have a safe belief $\Box_i p$ at the black node and at all more plausible worlds which is not knowledge: $\neg K_i p$. Finally, $B_i\varphi$ describes the most fragile thing: her beliefs as true in all worlds in the current rightmost position. In the model of Figure 7.4, the agent holds a true belief in r, i.e. $B_i r$, which is not a safe belief, and a fortiori, not knowledge. ⊣

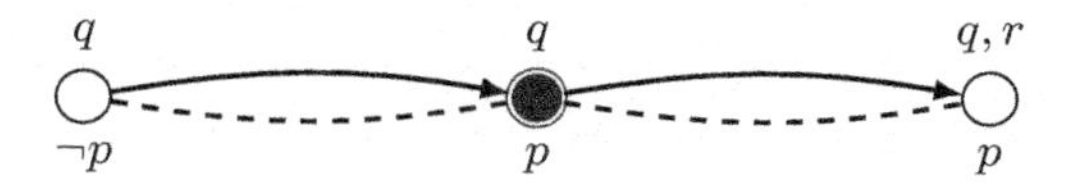

Figure 7.4: Illustrating safe beliefs.

In addition to its intuitive merits, safe belief simplifies the logic when used as a technical device, since it can define other notions of interest:

Fact 7.2
The following holds on finite connected epistemic plausibility models:

(a) Safe belief can define its own conditional variant,

(b) With a knowledge modality, safe belief can define conditional belief. ⊣

Indeed, conditional belief $B_i^{\varphi}\psi$ is equivalent to the modal statement

$$\widetilde{K_i}\varphi \rightarrow \widetilde{K_i}(\varphi \wedge \Box_i(\varphi \rightarrow \psi))$$

where $\widetilde{K_i}\varphi := \neg K_i \neg\varphi$ is the diamond modality for K_i. Alternative versions use modalities for the strict plausibility relation. This definition needs reformulation on non-connected plausibility models, where the following formula works:

$$K_i((\psi \wedge \varphi) \rightarrow \widetilde{K_i}(\psi \wedge \varphi \wedge \Box_i(\psi \rightarrow \varphi)))$$

Failure of negative introspection Safe belief also has less obvious features. For instance, since its accessibility relation is not Euclidean, it fails to satisfy Negative Introspection. The reason is that safe belief mixes purely epistemic information with procedural information about what may happen later on in the current process of inquiry. In the above example of Figure 7.4 it is easy to see that in the left most node $\neg\Box_i p$ holds while $\Box_i\neg\Box_i p$ does not.

Strong belief Safe belief is not the only natural new doxastic notion of interest in plausibility models. Another important doxastic attitude is this:

Definition 7.7 (Strong belief)
The proposition φ is a *strong belief* at a state w in a given model M iff φ *is epistemically possible* and also, all epistemically possible φ-states at w are strictly more plausible than all epistemically possible non-φ states. ⊣

Example 7.5 (Strong, safe, and plain belief)
Consider the model in Figure 7.5, with the actual world in the middle. In this model, the agent holds a strong belief in p but not in q. She holds a safe belief at the actual world in p but not in q, and a mere belief in $\neg q \wedge p$.⊣

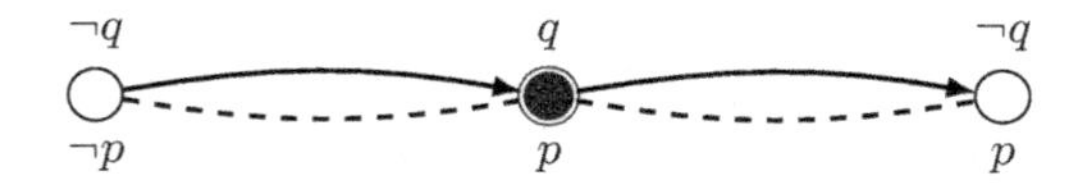

Figure 7.5: Illustrating strong beliefs.

Summary The above notions of plain, conditional, safe and strong belief can be found in various places in the logical, philosophical and economical literature. Yet there are many further epistemic and doxastic notions in our cognitive repertoire as encoded in natural language, and more notions can be interpreted on plausibility models than what we have shown.

But even with what we have shown, this section will have established its point that plausibility models are a natural way of representing a wide range of interesting doxastic notions. Hence they form a good basis for the dynamics of belief change, the main topic of this chapter, to which we now turn.

7.4 Belief revision by model transformations

Knowledge and varieties of belief express attitudes of an agent in its current epistemic-doxastic state. But belief revision is about changes in such states, and in this section, we turn to modeling this crucial feature, first on the analogy of the earlier Public Announcement Logic, and after that, with a new mechanism of changing plausibility orderings.

7.4.1 From knowledge to belief, hard and soft information

The original dynamic-epistemic logics such as PAL deal with knowledge, information and truth. Now, however, we also look at beliefs that agents form beyond the hard information that they possess. The result is a *tandem* of "jumping ahead" and "self-correction": agents believe more then they know, but this creative ability has to be kept in check by a capacity for self-correction, or stated more positively: for learning by trial and error.

Hard versus soft information With this new setting comes a richer dynamics of information. A public announcement $!\varphi$ of a fact φ was an event of *hard information* that changes irrevocably what an agent knows. Such events of hard information may also change beliefs. But in line with the greater softness and flexibility of belief over knowledge, there are also events that convey *soft information*, affecting agents' beliefs in less radical ways than world elimination.

The earlier plausibility models are well-suited for modeling the dynamics of both hard and soft information through suitable update operations.

7.4.2 Belief change under hard information

Our first dynamic logic of belief revision puts together the logic PAL of public announcement, a paradigmatic event producing hard information, with our static models for conditional belief, following the methodology explained above.

A complete axiomatic system For a start, we locate the key recursion axiom that governs belief change under hard information:

Fact 7.3
The following formula is valid for beliefs after hard information:

$$[!\varphi]B_i\psi \leftrightarrow (\varphi \to B_i^{\varphi}([!\varphi]\psi))$$

This is still somewhat like the PAL recursion axiom for knowledge. But the conditional belief in the consequent does not reduce to any obvious conditional plain belief of the form $B(\varphi \to ...)$. Therefore, we also need to state which conditional beliefs are formed after new information.[8]

Theorem 7.2
The logic of conditional belief under public announcements is axiomatised completely by (a) any complete static logic for the model class chosen, (b) the PAL recursion axioms for atomic facts and Boolean operations, (c) the following new recursion axiom for conditional beliefs:

$$[!\alpha]B_i^{\psi}\varphi \leftrightarrow (\alpha \to B_i^{\alpha\wedge[!\alpha]\psi}[!\alpha]\varphi).$$

[8] Classical belief revision theory says only how new plain beliefs are formed. The resulting "Iteration Problem" for consecutive belief changes cannot arise in a logic that covers revising conditional beliefs.

To get a joint version with knowledge, we just combine with the PAL axioms. ⊣

The Ramsey Test once more The above recursion axioms distinguish between formulas ψ before the update and after: $[!\varphi]\psi$. In this context, we can revisit our earlier discussion on why the existence of Ramsey conditionals of the form $A > B$ is incompatible with the AGM postulates.[9] Recall that the Ramsey Test says: "A conditional proposition $A > B$ is true, if, after adding A to your current stock of beliefs, the minimal consistent revision implies B." In our logic, this is ambiguous, as the consequent B *need no longer say the same thing after the revision.* As was already noted earlier, in a truly dynamic setting the truth-value of epistemic and doxastic sentences can change. Even so, the above axioms become Ramsey-like for the special case of factual propositions ψ without modal operators, that do not change their truth value under announcement:

$$[!\varphi]B_i\psi \leftrightarrow (\varphi \rightarrow B_i^{\varphi}\psi)$$
$$[!\varphi]B_i^{\alpha}\psi \leftrightarrow (\alpha \rightarrow B_i^{\varphi\wedge\alpha}\psi)$$

The reader may worry how this can be the case, given the Gärdenfors Impossibility Result. The nice thing of our logic approach, however, is that every law we formulate is sound. In other words, Ramsey-like principles do hold, provided we interpret the modalities involved in the way we have indicated here.[10]

Belief change under hard update links to an important theme in agency: variety of attitudes. We resume an earlier theme from the end of Section 7.3.

7.4.3 From dynamics to statics: safe and strong belief

Scenarios with hard information, though simple, contain some tricky cases:

Example 7.6 (Misleading with the truth)
Consider again the model in Figure 7.3 where the agent believed that p, which was indeed true in the actual world to the far right. This is a correct belief for a wrong reason: the most plausible world is not the actual world. For convenience, we assumed that each world verifies a unique proposition letter q_i.

---

[9] As before, we use the notation $>$ to denote a binary conditional operator.

[10] In line with this, a weaker, but arguably the correct, version of the Ramsey test offers a way out of the semantic impossibility result. We must restrict the validity of the Ramsey test only to "theories" T that correspond to actual belief sets (in some possible world s) about the current world (s itself), excluding the application of the test to already revised theories.

Now giving the true information that we are not in the final world (i.e., the announcement $!\neg q_3$) updates the model to the one shown in Figure 7.6, in which the agent believes mistakenly that $\neg p$.[11] ⊣

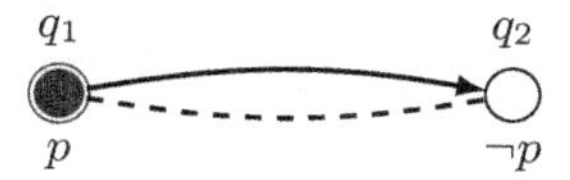

Figure 7.6: Updated belief model.

Example 7.6, though tricky, is governed precisely by the complete logic of belief change under hard information. This logic is the one whose principles were stated before, with the following simple recursion law added for safe belief:

$$[!\varphi]\Box_i\psi \leftrightarrow (\varphi \rightarrow \Box_i(\varphi \rightarrow [!\varphi]\psi))$$

In fact, this axiom for safe belief under hard information implies the earlier more complex-looking one for conditional belief, by unpacking the modal definition of Section 7.3.2 and applying the relevant recursion laws. This shows once more how safe belief can simplify the total logic of belief and belief change.[12]

7.4.4 Belief change under soft information: radical upgrade

Soft information and plausibility change We have seen how a hard attitude like knowledge or a soft attitude like belief changes under hard information. But an agent can also take incoming signals in a softer manner, without throwing away options forever. Then public announcement is too strong:

Example 7.7 (No way back)
Consider the earlier model in Figure 7.6 where the agent believed that $\neg p$, though p was in fact the case. Publicly announcing p removes the $\neg p$-world, making any later belief revision reinstating $\neg p$ impossible. ⊣

We need a mechanism that just makes incoming information P more plausible, without removing the $\neg P$ worlds. There are many sources for this. Here is one from the work of Veltman in the 1990s on default rules in natural language.

[11] Cases like this occur independently in philosophy, computer science and game theory.

[12] A recursion axiom can also be found for strong belief, though we need to introduce a stronger conditional variant there. Moreover, safe and strong belief also yield natural recursion axioms under the more general dynamic operations to be discussed in the following sections.

Example 7.8 (Default reasoning)
A default rule $A \Rightarrow B$ of the form "A's are normally B's" does not say that all A-worlds must be B-worlds. Accepting it just makes counter-examples to the rule (the $A \wedge \neg B$-worlds) less plausible. This "soft information" does not eliminate worlds, it changes their order. More precisely, a triggering event that makes us believe that φ need only rearrange worlds making the most plausible ones φ: by "promotion" or "upgrade" rather than by elimination of worlds. ⊣

Thus, in our models $M = (W, \{\sim\}_i, \{\leq_i\}_i, V)$, we must now change the plausibility relations $\leq_i$, rather than the world domain W or the epistemic accessibilities $\sim_i$. Indeed, rules for plausibility change have been considered in earlier semantic "Grove models" for AGM-style belief revision as different policies that agents can adopt toward new information. We now show how the above dynamic logics deal with them in a uniform manner.

Radical revision One very strong, but widely used revision policy effects an upheaval in favor of some new proposition φ:

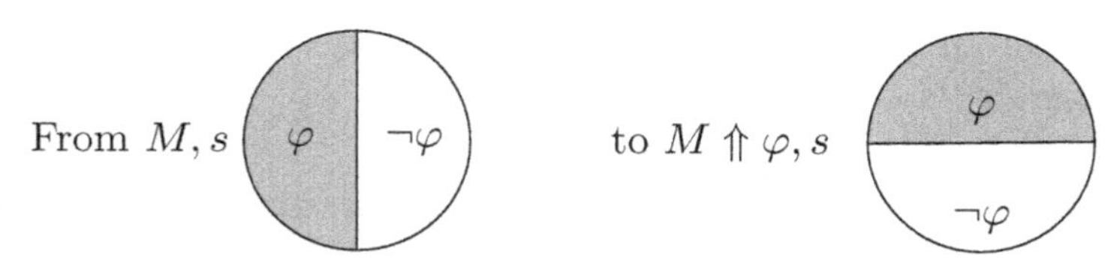

Figure 7.7: A radical revision step: all φ-worlds are moved to the top.

Definition 7.8 (Radical upgrade)
A *radical* (also called "lexicographic") *upgrade* $\Uparrow\varphi$ changes the current order $\leq_i$ between worlds in the given model M, s to a new model $M\Uparrow\varphi, s$ as follows: all φ-worlds in the current model become better than all $\neg\varphi$-worlds, while, within those two zones, the old plausibility order remains. ⊣

Language and logic With this definition in place, our earlier methodology applies. Like we did with public announcement, we introduce a corresponding "upgrade modality" in our dynamic doxastic language:

$$M, s \models [\Uparrow\varphi]\psi \text{ iff } M\Uparrow\varphi, s \models \psi$$

The resulting system can be axiomatised completely, in the same style as for our dynamic logics with hard information change. Here is a complete account of how agents' beliefs change under soft information, in terms of the key recursion axiom for changes in conditional belief under radical revision:

Theorem 7.3
The dynamic logic of lexicographic upgrade is axiomatised completely by

(a) any complete axiom system for conditional belief on the static models, plus

(b) the following recursion axioms[13]:

$$
\begin{array}{lll}
[\Uparrow\varphi]q & \leftrightarrow & q \quad \text{for all atomic proposition letters } q \\
[\Uparrow\varphi]\neg\psi & \leftrightarrow & \neg[\Uparrow\varphi]\psi \\
[\Uparrow\varphi](\psi \wedge \alpha) & \leftrightarrow & [\Uparrow\varphi]\psi \wedge [\Uparrow\varphi]\alpha \\
[\Uparrow\varphi]K_i\varphi & \leftrightarrow & K_i[\Uparrow\varphi]\varphi \\
[\Uparrow\varphi]B_i^{\alpha}\psi & \leftrightarrow & (\widetilde{K}_i(\varphi \wedge [\Uparrow\varphi]\alpha) \wedge B_i^{\varphi\wedge[\Uparrow\varphi]\alpha}[\Uparrow\varphi]\psi) \vee \\
 & & (\neg\widetilde{K}_i(\varphi \wedge [\Uparrow\varphi]\alpha) \wedge B_i^{[\Uparrow\varphi]\alpha}[\Uparrow\varphi]\psi)
\end{array}
$$

For plain beliefs $B_i\varphi$ with $\alpha =$ 'True', things simplify to:

$$[\Uparrow\varphi]B_i\psi \leftrightarrow (\widetilde{K}_i\varphi \wedge B_i^{\varphi}[\Uparrow\varphi]\psi) \vee (\neg\widetilde{K}_i\varphi \wedge B_i[\Uparrow\varphi]\psi)$$

And here is the simplified recursion axiom for factual propositions:

$$[\Uparrow\varphi]B_i^{\alpha}\psi \leftrightarrow ((\widetilde{K}_i(\varphi \wedge \alpha) \wedge B_i^{\varphi\wedge\alpha}\psi) \vee (\neg\widetilde{K}_i(\varphi \wedge \alpha) \wedge B_i^{\alpha}\psi))$$

Things get easier again with our earlier safe belief:

Fact 7.4
The following recursion axiom is valid for safe belief under radical revision:

$$[\Uparrow\varphi]\Box_i\psi \leftrightarrow (\varphi \wedge \Box_i(\varphi \rightarrow [\Uparrow\varphi]\psi)) \vee (\neg\varphi \wedge \Box_i(\neg\varphi \rightarrow [\Uparrow\varphi]\psi) \wedge K_i(\varphi \rightarrow [\Uparrow\varphi]\psi))$$

Static pre-encoding We have shown now how complete modal logics exist for belief revision mechanisms, without any need for special-purpose formalisms. But our design in terms of recursion laws has a striking side effect, through its potential for successive reduction of dynamic modalities. As was the case with knowledge in PAL, our analysis of belief change says that any statement about epistemic-doxastic effects of hard or soft information is encoded in the initial model: the epistemic present contains the epistemic future. While this looks appealing, the latter feature may be too strong in the end.

In Section 7.7 of this chapter, we will look at extended models for belief revision encoding global informational procedures ("protocols"), that preserve the spirit of our recursion laws, but without a reduction to the static base language.

[13]Here, as before, $\widetilde{K_i}$ is the dual existential epistemic modality $\neg K_i \neg$.

7.4.5 Conservative upgrade and other revision policies

The preceding logic was a proof-of-concept. Radical revision is just one way of taking soft information. Here is a well-known less radical policy for believing a new proposition. The following model transformation puts not all φ-worlds on top, but just the most plausible φ-worlds.

Definition 7.9 (Conservative plausibility change)
A *conservative upgrade* $\uparrow\varphi$ replaces the current order $\leq_i$ in a model M by the following: the best φ-worlds come on top, but apart from that, the old plausibility order of the model remains.[14] ⊣

The complete dynamic logic of conservative upgrade can be axiomatised in the same style as radical upgrade, this time with the following valid key recursion axiom for conditional belief:

$$[\uparrow\varphi]B_i^{\alpha}\psi \leftrightarrow (B_i^{\varphi}\neg[\uparrow\varphi]\alpha \wedge B_i^{[\uparrow\varphi]\alpha}[\uparrow\varphi]\psi) \vee (\neg B_i^{\varphi}\neg[\uparrow\varphi]\alpha \wedge B_i^{\varphi\wedge[\uparrow\varphi]\alpha}[\uparrow\varphi]\psi)$$

Clearly, this is a rather formidable principle, but then, there is no hiding the fact that belief revision is a delicate process, while it should also be kept in mind that recursion axioms like this can often be derived from general principles.

Variety of revision policies Many further changes in plausibility order can happen in response to an incoming signal. This reflects the many belief revision policies in the literature.[15] The same variety occurs in other realms of ordering change, such as the preference changes induced by the many deontic speech acts in natural language such as hard commands or soft suggestions. In fact, no uniform choice of revision policy is enforced by the logic: our language of belief change can also describe different sorts of revising behavior together, as in mixed formulas $[\Uparrow\varphi][\uparrow\psi]\alpha$. Does this dissolve logic of belief revision into a jungle of options? In Section 7.5, we will look at this situation in greater generality.

Summary This Section has extended the dynamic approach for updating knowledge to revising beliefs. The result is one merged theory of information update and belief revision, using standard modal techniques instead of ad-hoc formalisms.

---

[14]Technically, $\uparrow\varphi$ is a special case of radical revision: $\Uparrow(best(\varphi))$.

[15]Maybe "policy" is the wrong term, as it suggests a persistent habit over time. But our events describe local responses to particular inputs. Moreover, speech act theory has a nice distinction between information per se (what is said) and the *uptake*, how a recipient reacts. In that sense, the softness of our scenarios is in the response, rather than in the signal itself.

7.5 General formats for belief revision

As we have seen, acts of belief revision can be modeled as transforming current epistemic-doxastic models, and there is a large variety of such transformations.

We have also seen how, given a definition of model change, one can write a matching recursion axiom, and then a complete dynamic logic. But how far does this go? In this section, we discuss two general formats that have been proposed to keep the base logic of belief change simple and clean.

7.5.1 Relation transformers as PDL programs.

One general method uses programs in propositional dynamic logic to define the new relations via standard program constructs including the test of a proposition φ (denoted as $?\varphi$), the arbitrary choice of two programs $\alpha \cup \beta$, and sequential program composition $\alpha; \beta$. Many examples fit in this format:

Fact 7.5
Radical upgrade $\Uparrow P$ is definable as the following program in propositional dynamic logic, with 'T' the universal relation between all worlds:

$$\Uparrow P(R) := (?P; T; ?\neg P) \cup (?P; R; ?P) \cup (?\neg P; R; ?\neg P)$$

Van Benthem & Liu then introduced the following general notion in 2007:

Definition 7.10 (PDL-format for relation transformers)
A definition for a new relation on models is in *PDL-format* if it can be stated in terms of the old relation R, union, composition, and tests. ⊣

A further example is a weaker act that has been studied in the logic of preference change, as well as that of "relevant alternatives" in formal epistemology.

Example 7.9 (Suggestions as order transformations)
A *suggestion* merely takes out R-pairs with '$\neg P$ over P'. This transformation is definable as the PDL program

$$\sharp P(R) = (?P; R) \cup (R; ?\neg P)$$

This generalises our earlier procedure with recursion axioms considerably:

Theorem 7.4
For each relation change defined in PDL-format, there is a complete set of recursion axioms that can be derived via an effective procedure. ⊣

Example 7.10
Instead of a proof, here are two examples of computing modalities for the new relation after the model change, using the well-known recursive program axioms of PDL. Note how the second calculation uses the existential epistemic modality $<>$ for the occurrence of the universal relation:

$$\text{(a)} < \sharp P(R) >< R > \varphi \leftrightarrow < (?P; R) \cup (R; ?\neg P) > \varphi$$
$$\leftrightarrow < (?P; R) > \varphi \vee < (R; ?\neg P) > \varphi$$
$$\leftrightarrow <?P >< R > \varphi \vee < R ><?\neg P > \varphi \leftrightarrow (P \wedge < R > \varphi) \vee < R > (\neg P \wedge \varphi).$$

$$\text{(b)} <\Uparrow P(R) > \varphi \leftrightarrow < (?P; T; ?\neg P) \cup (?P; R; ?P) \cup (?\neg P; R; ?\neg P) > \varphi \leftrightarrow$$
$$< (?P; T; ?\neg P) > \varphi \vee < (?P; R; ?P) > \varphi \vee < (?\neg P; R; ?\neg P) > \varphi \leftrightarrow$$
$$<?P >< T ><?\neg P > \varphi \vee <?P >< R ><?P > \varphi \vee <?\neg P >< R ><$$
$$?\neg P > \varphi \leftrightarrow (P \wedge E(\neg P \wedge \varphi)) \vee (P \wedge < R > (P \wedge \varphi)) \vee (\neg P \wedge < R >$$
$$(\neg P \wedge \varphi)).$$ [16]

The final formula arrived at easily transforms into the axiom that was stated earlier for safe belief after radical upgrade $\Uparrow P$. $\dashv$

In this style of analysis, logic of belief revision becomes a form of propositional dynamic logic, and PDL then serves as the "mother logic" of belief revision. Much more complex PDL mechanisms for model change have been proposed recently, for which we refer to our Section 7.10 on further literature.

However, some forms of belief revision seem to require methods going beyond the PDL frarmework, and in order to explain the resulting general logic, we need to take a closer look at the heart of current Dynamic Epistemic Logic.

7.5.2 Product update in general dynamic-epistemic logic

Dynamic Epistemic Logic (DEL) goes far beyond public announcement logic, whose earliest version is from 1989. In the 1990s, more complex informational scenarios were investigated by Groeneveld, Gerbrandy, and van Ditmarsch, involving mixtures of public and private information. The crucial mechanism in use today is that of *product update* introduced by Baltag, Moss and Solecki, where current models need not shrink in size: they can even *grow* under update, reflecting increased complexities of an informational setting. Currently, the term DEL is used to denote a collection of logical systems that deal with complex multi-agent scenarios in which individual agents or groups of agents update their knowledge and beliefs when new information comes in a variety of public or private events.

[16] Here the operator E stands for the existential modality in some world.

In their original setting, these logics were designed to model only cases in which the newly received information is consistent with the agent's prior doxastic or epistemic state. But in recent years, it has become clear that ideas from Belief Revision Theory fit naturally with DEL, allowing us to model a richer set of scenarios in which agents can be confronted with surprising information that may contradict their prior beliefs.

In line with our treatment so far in this chapter, DEL interprets all epistemic actions or events as "model transformers", i.e. ways to transform a given input model into an output model, where the actual transformation captures the effect of the epistemic action that took place. A powerful mechanism of this sort is the above product update that covers highly sophisticated scenarios. In this subsection, we will present its epistemic version, where it transforms a current epistemic model (usually, though not necessarily, satisfying the constraints of the modal logic $S4$ or $S5$) using a further "event model" that collects all relevant events insofar as the agents' observational abilities are concerned:

Definition 7.11 (Event models)
An *event model* over a given language $\mathcal{L}$ is a structure $\Sigma = (E, R_i, PRE)$ such that E is the set of relevant actions (or events) in our domain of investigation. For each agent $i \in I$ we have an accessibility relation R_i on E, and instead of a valuation we now have a precondition map $PRE : E \to \mathcal{L}$ which assigns the precondition $PRE(e)$ to each $e \in E$. ⊣

Here is what these structures represent. In public announcement logic, the event taking place was totally clear to every agent. In scenarios with private information, agents may not be sure exactly what it is they are observing, and an event model captures this epistemic horizon. The precondition map encodes the information that observed events can convey, while the accessibility relations encode, as in epistemic models, the extent to which the agents can observe what is the event actually taking place. A concrete example will follow shortly, after we have stated the update mechanism.

Product update Consider a situation in which we start from a given epistemic model $M = (W, R_i, V)$ which captures a full description of the epistemic states of all agents $i \in I$. Let the possible world s in M be our point of evaluation (alternatively, the "actual world" where the agents are) in which a given epistemic event e happens, coming from an event model $\Sigma = (E, R_i, PRE)$. To see how the epistemic event (Σ, e) affects the state (M, s), we first observe that this event can only take place when the precondition holds, i.e. $M, s \models PRE(e)$. The result of performing this event is then a new epistemic state (s, e) belonging to the direct product of the model M with the model Σ. We denote the effect of the product update by $(M \otimes \Sigma) = (W', R'_i, V')$ and define it as:

$$W' = \{(s,e) | s \in W, e \in E, \text{ and } M, s \models PRE(e)\}$$

$$R'_i = \{((s,e),(t,f)) \mid R_i(s,t) \text{ and } R_i(e,f)\}$$

$$(s,e) \in V'(p) \text{ iff } s \in V(p)$$

Here the stipulation for the valuation says that base facts do not change under information update.[17] The stipulation for the new accessibility is perhaps best understood negatively: the only ways in which an agent can distinguish situations after the update is, either the situations were already distinguishable before, or they were indistinguishable, but the signal observed makes a difference: the agent has learnt from the observation.

To see that this rule reflects our intuitions about information change, we look at a concrete example in the style of Baltag and Moss:

Example 7.11 (Private announcements)
Imagine that a coin lies on the table in such a position that two agents Alice (a) and Bob (b) cannot see the upper face of the coin. We assume that it is common knowledge that a and b are both uncertain about whether the upper face is Heads or Tails. This scenario is represented in the $S4$ model of Figure 7.8, where the pointed arrows represent the agent's epistemic uncertainty.

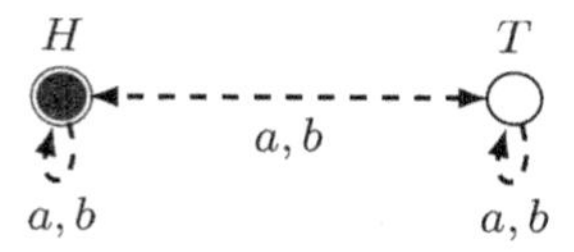

Figure 7.8: Initial coin scenario.

Assume now that the following action takes place: Bob takes a peek at the coin, while Alice doesn't notice this. We assume that Alice thinks that nothing happened and that Bob knows that Alice didn't see him take a peek at the coin. This action is a typical example of a "fully private announcement" in DEL. Bob learns the upper face of the coin, while the outsider Alice believes nothing is happening. In Figure 7.9, we depict the epistemic event model for this situation.

We take two 'actions' into account: the actual action e in which Bob takes a peek and sees Heads up; and the action τ where 'nothing is really happening', which has $\top$ as precondition. The accessibility relations drawn in the picture show that τ is the only action that agent a considers possible.

[17] There are also versions of DEL that model real factual changes by "postconditions".

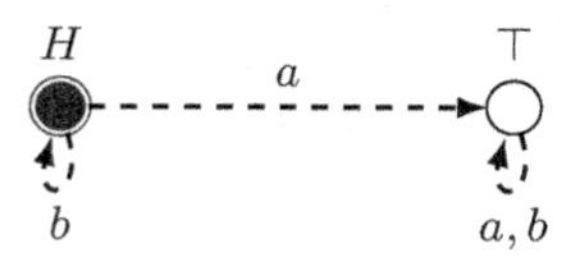

Figure 7.9: Epistemic event model.

The product update of the previous two models yields the result depicted in Figure 7.10. In this model, the real state of affairs is represented on top, where Heads is true. In this world b knows the upper fact of the coin is Heads while a thinks that nobody knows the upper face of the coin.

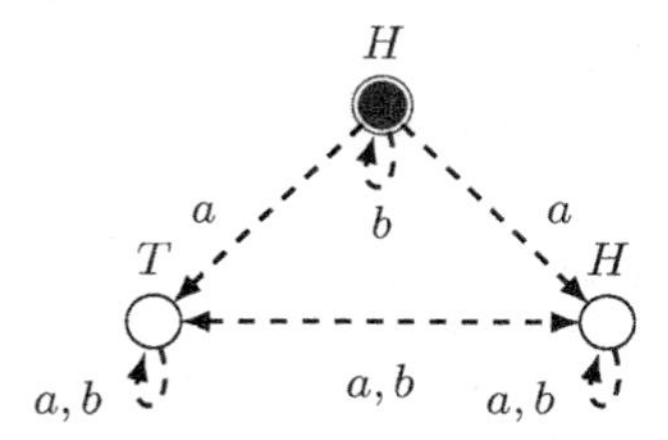

Figure 7.10: The product update result depicted.

Scenarios like this may look a bit contrived, but this is a mistaken impression. Differences in information between agents are precisely what drives communication, and when we get to belief revision later on, this is often a private process, not undergone by all agents in the same manner. ⊣

Recursion axioms and complete logic The proof theory for DEL can be seen as a generalisation of the setting for public announcement logic. Like we saw with PAL, recursion axioms can be provided to prove the completeness of the system.[18] As in the completeness proof for PAL, we can provide a reduction of every dynamic operator in the logic to obtain an equivalent formula in the static epistemic base language. For the Boolean connectives, the recursion laws are straightforward, so let us focus on the K_i operator. The basic Action-Knowledge Axiom is the following equivalence:

$$[\Sigma, e]K_i\varphi \leftrightarrow (PRE(e) \rightarrow \bigwedge_{f \in \Sigma, eR_i f} K_i[\Sigma, f]\varphi)$$

[18]Things are more delicate with group modalities such as common knowledge. The first completeness proofs for PAL and DEL with common knowledge were intricate because of the lack of a recursion axiom for a common knowledge operator. These matters have been solved since in terms of suitable PDL style extensions of the epistemic base language. For more discussion, see Section 7.9, as well as the references in Section 7.10.

Here the events f with eR_if in Σ form the "appearance" of e to agent i in the current scenario represented by Σ. As our example of private announcement showed, not every event appears in the same way to each agent.

The PAL reduction axiom for $[!\varphi]K_i\psi$ is a special case of the general Action-Knowledge Axiom where the event model consists of one action with a precondition φ seen by all the agents. But the new axiom covers many further scenarios. The product update mechanism of DEL has been successfully applied to many different scenarios involving epistemic uncertainty, private and public communication, public learning-theoretic events, and game solution procedures. We will see a few of these later on in this chapter.

However, to handle more sophisticated scenarios involving doxastic uncertainty and belief revision, we need to extend product update to deal with plausibility order change of the kind introduced in the preceding Section 7.4.

7.5.3 Priority update

Event models as triggers of plausibility change Given the private nature of many belief changes, but also, the social settings that induce these (we often change our beliefs under pressure from others), it makes sense to use the general DEL methodology of interacting different agents. Here is a powerful idea from the work of Baltag & Smets, where the reader may now want to view the events in event models as "signals" for information strength:

Definition 7.12 (Plausibility event models)
Plausibility event models are event models as in the preceding section, now expanded with a plausibility relation over their epistemic equivalence classes. ⊣

This setting covers the earlier concrete examples from Section 7.4.

Example 7.12 (Radical upgrade)
A radical upgrade $\Uparrow\varphi$ can be implemented in a plausibility event model. We do not throw away worlds, so we use two 'signals' $!\varphi$ and $!\neg\varphi$ with obvious preconditions φ, $\neg\varphi$ that will produce an isomorphic copy of the input model. But we now say that signal $!\varphi$ is more plausible than signal $!\neg\varphi$, relocating the revision policy in the nature of the input. With a suitable update rule, to be defined below, this will proceed the output model depicted in Figure 7.11. ⊣

Different event models will represent a great variety of update rules. But we still need to state the update mechanism itself more precisely, since the required treatment of plausibility order is not quite the same as that for

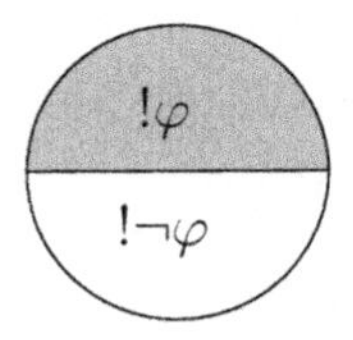

Figure 7.11: Implementing radical upgrade.

epistemic accessibility in the preceding subsection. The following proposal has been called 'One Rule To Rule Them All', i.e., one new rule that replaces earlier separate update rules for plausibility. It places the emphasis on the last event observed, but is conservative with respect to everything else:

Definition 7.13 (Priority update)
Consider an epistemic plausibility model (M, s) and a plausibility event model (Σ, e). The *priority product model* $(M \times \Sigma)$, (s, e) is defined entirely as its earlier epistemic version, with the following additional rule for the plausibility relation $\leq_i$, which also refers to its strict version $<_i$:

$$(s, e) \leq_i (t, f) \text{ iff } (s \leq_i t \wedge e \leq_i f) \vee e <_i f$$

Thus, if the new incoming information induces a strong preference between signals, that takes precedence: otherwise, we go by the old plausibility order. The emphasis on the last observation or signal received is like in belief revision theory, where receiving just one signal $*\varphi$ leads the agent to believe that φ, even if all of her life, she had been receiving evidence against φ.[19]

Theorem 7.5
The dynamic logic of priority update is axiomatisable completely. ⊣

As before, it suffices to state the recursion axioms reflecting the above rule. We display just one case here, and to make things simple, we do so for the notion of safe belief, written in an existential format:

$$< \Sigma, e > \Diamond_i \varphi \leftrightarrow$$
$$PRE(e) \wedge (\bigvee_{e \leq f \text{ in } \Sigma} \Diamond_i < \Sigma, f > \varphi \vee (\bigvee_{e < f \text{ in } \Sigma} \tilde{K}_i < \Sigma, f > \varphi))$$

where $\tilde{K}_i$ is again the existential epistemic modality.

[19]Priority Update is also in line with "Jeffrey Update" in probability theory that imposes a new probability for some specified proposition, while adjusting all other probabilities proportionally.

Example 7.13 (The coin scenario revisited)
Consider again the coin example of the preceding subsection, which showed how DEL handles public and private knowledge updates of agents. The given scenario might lead to problems if Alice finds out that Bob took a peak at the coin. So let us now also introduce beliefs. Imagine again that there is a coin on the table in such a position that the two agents Alice (a) and Bob (b) cannot see the upper face of the coin. We now assume that it is common knowledge that a and b believe that the upper face is Heads (see Figure 7.12).[20]

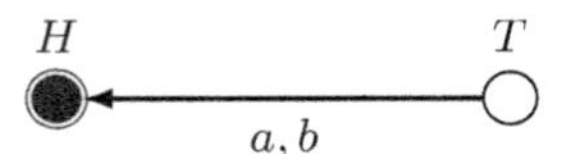

Figure 7.12: Initial coin scenario revisited.

Assume now that the following action takes place: Bob takes a peek at the coin, while Alice does not notice. We assume that Alice believes that nothing happened. Then Bob learns what is the upper face of the coin, while the outsider Alice believes nothing has happened. In Figure 7.13, we represent the plausibility event model for this situation. We take three 'actions' into account: the actual action e in which Bob takes a peek and sees Heads up; the alternative action f in which Bob sees Tails up; and the action τ in which nothing is really happening. The plausibility marking in the event model shows that τ is the most plausible action for Alice, while, if she had to choose between Bob taking a peek and seeing Heads or seeing Tails, she would give preference to him seeing Heads.

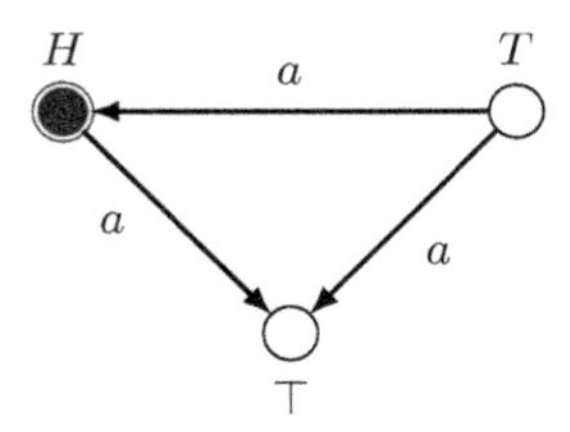

Figure 7.13: Plausibility event model for a private peep.

Now taking the Priority Update of the preceding two models yields the result depicted in Figure 7.14. In this model with four worlds, the most plausible world for Alice is the lower left one where Heads is true. In this

[20] To make our drawings more transparent, reflexive plausibility arrows have been omitted as well as the epistemic uncertainty relations for agents, which can be computed on the basis of the plausibility relations as we have shown earlier.

model, Alice believes indeed that everyone believes that Heads is the case and nobody knows the upper face of the coin. The real world however is the top left world, in which Bob does know the upper face of the coin to be Heads.

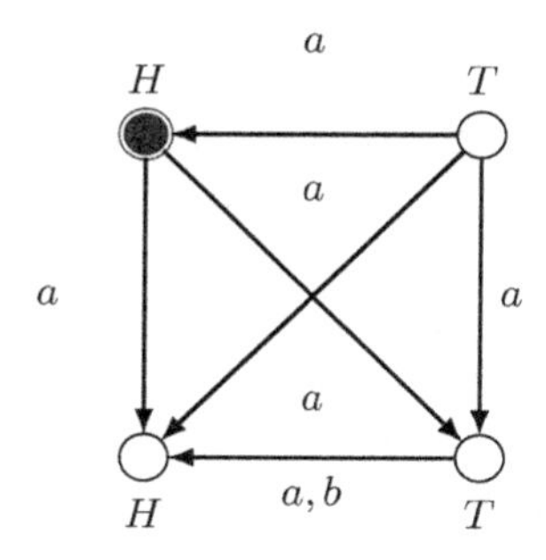

⊣

Figure 7.14: The computed result of priority update.

The general advantage of this approach is that, instead of having to introduce many different policies for processing an incoming signal, each with its own logic, we can now put the policy in the input Σ. Accordingly, the one logic for Priority Update that we gave before is the unique dynamic logic of belief revision. Admittedly, this introduces some artificial features. The new event models are rather abstract - and, to describe even simple policies like conservative upgrade, the language of event models must be extended to event preconditions of the form 'best φ'. But the benefit is clear, too. Infinitely many policies can be encoded in the choice of plausibility event models, while belief change works with just one update rule, and the common objection that belief revision theory is non-logical and messy for its proliferation of policies evaporates.[21]

7.6 Belief revision and probability dynamics

The main technical alternative to using logic in studies of belief revision have been probabilistic methods. The interface of logic and probability is a vast and flourishing topic, well beyond the confines of this chapter. Even so, we provide a brief presentation of a way in which probabilistic perspectives merge well with the dynamic logical methodology of this chapter. This

[21]More can still be said on general plausibility update rules, since it is not obvious how the stated priority format covers all PDL program changes discussed earlier. Also, one might question the emphasis on the last signal received, rather than engaging in more cautious processes of learning. These issues will return in Sections 7.6, 7.7, and 7.10.

section presents the basics of the system of Probabilistic Epistemic Dynamic Logic, be it in a somewhat terse manner. For details, the reader may consult the literature references in Section 7.10, while Section 7.8 will contain a concrete application to the phenomenon of informational cascades.

Language and models We can think of subjective probabilities as a numerical measure for degrees of belief, though more objective interpretations as observed frequencies make sense as well in our setting of informational inquiry. The following language, due to the work of Halpern and Fagin, allows for a wide range of expressions that mix knowledge and probability.

Definition 7.14 (Epistemic-probabilistic language)
The syntax of *epistemic-probabilistic logic* has the full syntax of the standard epistemic modal language of Section 7.3 plus the probabilistic construct

$$\alpha_1 \cdot P_i(\varphi) + \ldots + \alpha_n \cdot P_i(\varphi) \geq \beta$$

where $\alpha_1, \ldots, \alpha_n, \beta$ stand for arbitrary rational numbers. ⊣

The reason for having these inequalities is technical, having to do with treating conditional probabilities, and a smooth completeness proof for the base logic. Standard probability assertions like $P(\varphi) \leq \alpha$ are an obvious special case. Using the new construct we can also add useful abbreviations to our language that express notions often encountered in discussing probabilistic beliefs:

$$P_i(\varphi) > P_i(\psi)(\text{ also written } [\varphi : \psi]_i > 1) \;:=\; \neg(P_i(\psi) - P_i(\varphi) \geq 0),$$

It is easy then to also define $P_i(\varphi) = P_i(\psi)$ and similar comparative notions.

Next, here are the semantic structures that this language refers to.

Definition 7.15 (Probabilistic epistemic models)
A *probabilistic epistemic model* $\mathcal{M}$ is a tuple $(W, I, (\sim_i)_{i\in I}, (P_i)_{i\in I}, \Psi, [\![\bullet]\!])$ consisting of a set W of worlds; a set I of agents; and for each agent i, an equivalence relation $\sim_i \subseteq W \times W$ representing i's epistemic indistinguishability as in the above. Also, for each agent i, $P_i : W \to [0, 1]$ is a map that induces a probability measure on each $\sim_i$-equivalence class.[22] Finally, Ψ is a given set of atomic propositions and $[\![\bullet]\!] : \Psi \to \mathcal{P}(W)$ a standard valuation map, assigning to each $p \in \Psi$ some set of worlds $[\![p]\!] \subseteq W$. ⊣

On these models, a standard truth definition can be given for the epistemic-probabilistic language. We merely note that formulas of the form

[22] That is, we have $\sum\{P_i(s') : s' \sim_i s\} = 1$ for each $i \in I$ and each $s \in S$.

$P_i(\psi) \leq k$ are to be interpreted conditionally on agent i's knowledge at s, i.e., as

$$P_i(\llbracket\psi\rrbracket \cap \{s' \in W : s' \sim_i s\}) \leq k$$

The setting also yields other useful notions. In particular, the *relative likelihood* (or "odds") of state s against state t according to agent i, $[s : t]_i$ is

$$[s : t]_i := \frac{P_i(s)}{P_i(t)}$$

For concrete illustrations of these models and their transformations to follow, we refer to the references in Section 7.10, and the Urn Scenario in Section 7.8.

Dynamic updates In line with the main theme of this chapter, our next concern is how models like this change under the influence of new information. Again, the simplest case would be public announcements, or soft upgrades – but for many serious applications, the complete generality is needed of dynamic-epistemic product update in the sense of Section 7.5.

As before, the update rule requires forming products with event models that capture the relevant informational scenario as it appears toy the agents. This time, their impressions also have a probabilistic character.

Definition 7.16 (Probabilistic event models)
A *probabilistic event model* $\mathcal{E}$ is a structure $(E, I, (\sim_i)_{i\in I}, (P_i)_{i\in I}, \Phi, pre)$ consisting of a set of possible events E; a set of agents I, equivalence relations $\sim_i \subseteq E \times E$ representing agent i's epistemic indistinguishability between events; and probability assignments P_i for each agent i and each $\sim_i$-information cell. Moreover, Φ is a finite set of mutually inconsistent propositions[23] called *preconditions*. Here *pre* assigns a probability distribution $pre(\bullet|\varphi)$ over E for every proposition $\varphi \in \Phi$.[24] ⊣

A word of explanation is needed here. Event models crucially contain two probabilities: one for the certainty or quality of the *observation* made by the agent, and another for the probability of *occurrence* of an event given a precondition. The latter sort of information, often overlooked at first glance, is crucial to analysing famous scenarios of probabilistic reasoning such as Monty Hall or Sleeping Beauty, since it represents the agents' experience of, or belief about, the process they are in (cf. Section 7.8 for more on this theme of procedure).

[23]Preconditions usually come from the above static probabilistic-epistemic language.
[24]Alternatively, P_i is the probabilistic odds $[e : e']_i$ for any events e, e' and agent i.

Now we formulate an update rule that weighs all factors involved: prior probability of worlds, observation probabilities, and occurrence probabilities:

Definition 7.17 (Probabilistic product update)
Given a probabilistic epistemic model $\mathcal{M} = (W, I, (\sim_i)_{i\in I}, (P_i)_{i\in I}, \Psi, [\![\bullet]\!])$ and a probabilistic event model $\mathcal{E} = (E, I, (\sim_i)_{i\in I}, (P_i)_{i\in I}, \Phi, pre)$, the *probabilistic product model* $\mathcal{M} \otimes \mathcal{E} = (W', I, (\sim'_i)_{i\in I}, (P'_i)_{i\in I}, \Psi', [\![\bullet]\!]')$, is given by:

$$W' = \{(s, e) \in W \times E \mid pre(e \mid s) \neq 0\},$$

$$\Psi' = \Psi,$$

$$[\![p]\!]' = \{(s, e) \in W' : s \in [\![p]\!]\},$$

$$(s, e) \sim'_i (t, f) \text{ iff } \; s \sim_i t \text{ and } e \sim_i f,$$

$$P'_i(s, e) = \frac{P_i(s) \cdot P_i(e) \cdot pre(e \mid s)}{\sum\{P_i(t) \cdot P_i(f) \cdot pre(f \mid t) : s \sim_i t, e \sim_i f\}},$$

where we use the notation

$$pre(e \mid s) := \sum\{pre(e \mid \varphi) : \varphi \in \Phi \text{ such that } s \in [\![\varphi]\!]_{\mathcal{M}}\}$$

Here $pre(e \mid s)$ is either $= pre(e|\varphi_s)$ where φ_s is the unique precondition in Φ such that φ_s is true at s, or $pre(e \mid s) = 0$ if no such precondition φ_s exists.[25] ⊣

This combination of probabilistic logic and dynamic update sheds new light on many issues in probabilistic reasoning, such as the status of Bayes' Theorem. Moreover, its use of occurrence probabilities allows for careful modeling of probabilistic scenarios in areas such as learning theory (cf. Section 7.7), where we may have probabilistic information about the nature of the process giving us a stream of evidence about the world. denominator However, the above update mechanism does not solve all interface problems. For instance, like our earlier update mechanisms in Sections 7.4 and 7.5, the last signal received, represented by the event model, gets a huge weight – and alternatives have been considered, closer to standard systems of "inductive logic", that weigh the three probabilities involved differently.

---

[25] Note some subtleties apply to the definition of Probabilistic Product Update, which in addition has to build in the fact that $P'_i(s, e)$ excludes the deonomator from being 0.

Dynamic Probabilistic Logic As with our earlier systems of dynamic epistemic logic, there is a complete static and dynamic logic for the system defined here, including a complete set of recursion axioms for the new probability inequalities that hold after the application of an event model. We omit technical formulations, but the point is that our earlier logical methodology fully applies to the present more complex setting.

But there are many further logical issues about connections with our earlier sections. One is that we could add qualitative plausibility-based belief as well, giving us both quantitative and qualitative versions of beliefs. In recent years, such combinations have attracted increasing interest. Qualitative logical notions are often considered competitors to quantitative probabilistic ones, and indeed there are interesting issues as to whether the above dynamic update mechanisms can be stated entirely in terms of plausibility order.[26] But perhaps the more interesting issue for this chapter would be whether qualitative notions emerge naturally in cognition as companions to underlying probabilistic ones, a line of thought to which several recent authors have given new impetus.

Summary We have shown how probabilistic views of belief can be merged naturally with the main dynamic logic approach in this chapter. This combination is useful for applications to areas that heavily depend on probabilistic methods, such as game theory or the social sciences, but it also raises interesting broader conceptual issues that are far from resolved.

7.7 Time, iterated update, and learning

Belief revision theory has mainly focused on single steps of changing beliefs, and the same is true for the events of information change in the dynamic-epistemic logics that we have used. Of course, we can iterate such steps to form longer sequences, but so far, we have not looked at the global temporal dimension per se. And yet, single steps of belief revision may lack direction, like leaves in the wind. Many serious scenarios where belief revision plays a role are global processes of inquiry or learning that have an intrinsic temporal structure, sometimes even consisting of infinite histories.

With this temporal setting come constraints on how these histories can unfold, not necessarily captured by the local preconditions of events that we have used so far. These constraints on the procedure are often called

[26] Numbers play different roles in the above rules: as strengths of beliefs, as weights for "gluing" beliefs, and others. Thus, qualitative versions of the update rule may have to involve different mechanisms, such as "order merge" for the three probabilities above made qualitative.

temporal protocols. For instance, information systems may demand that only true information is passed, or that each request is answered eventually. And civilised conversation has rules like "do not repeat yourself", or "let others speak as well".

Restricting the legitimate sequences of announcements is not just an extra, it affects our earlier dynamic logics.

Example 7.14 (Contracting consecutive assertions: admissible, or no PAL can suppress longer sequences of announcements into one. A well-known PAL-validity states that two consecutive announcements $!\varphi, !\psi$ have the same effect as the single announcement

$$!(\varphi \wedge [!\varphi]\psi)$$

However, the new assertion may well be more complex than what is admissible by current rules for conversation or investigation, and hence this law may fail in protocol-based models. ⊣

In this section, we will introduce some temporal logics of knowledge that form a natural extension to the dynamic epistemic logics used so far. After that we show how dynamic epistemic logics lie embedded here, including their protocol versions. Then we show how the same is true for logics of belief, and finally, bringing together all ideas developed so far, we provide some recent connections with formal learning theory, a natural continuation of belief revision theory.

This is a vast area, and we only provide some windows, referring the reader to the further literature referenced in Section 7.10.

7.7.1 Epistemic temporal logic

Branching temporal models are a Grand Stage view of agency, as depicted in Figure 7.15, with histories as complete runs of some information-driven process that can be described by languages with epistemic and temporal operators.

Temporal logics for epistemic and doxastic agents come in different flavors, and we will only briefly discuss one of them, interpreted over a modal universe of finite histories and indices of evaluation.

Models and language Start with two sets I of agents and E of events. A *history* is a finite sequence of events, and E^* is the set of all histories. Here he is history h followed by event e, the unique history after e has happened in h. We write $h \leq h'$ if h is a prefix of h', and $h \leq_e h'$ if $h' = he$.

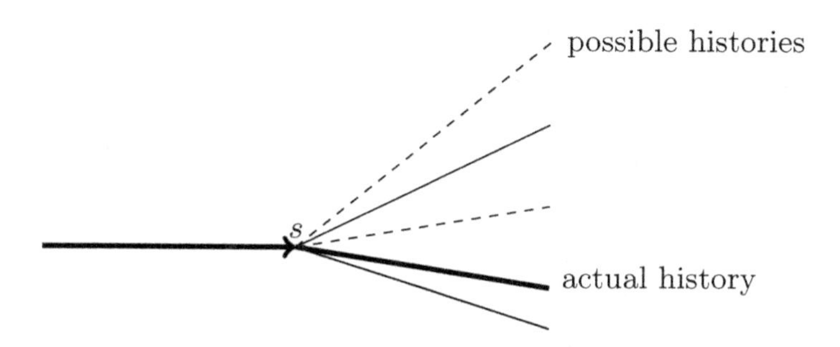

Figure 7.15: A branching temporal tree of histories.

Definition 7.18 (Epistemic-temporal ETL models)
A *protocol* is a set of histories $H \subseteq E^*$ closed under prefixes. An *ETL model* is a tuple $(E, H, \{\sim_i\}_{i \in I}, V)$ with a protocol H, accessibility relations $\{\sim_i\}_{i \in I}$ plus a valuation map V sending proposition letters to sets of histories in H. ⊣

An ETL model describes how knowledge evolves over time in some informational process. The relations $\sim_i$ represent uncertainty of agents about the current history, due to their limited powers of observation or memory. $h \sim_i h'$ means that from agent i's point of view, history h' looks the same as history h.

An epistemic temporal language LETL for these structures is generated from a set of atomic propositions At by the following syntax:

$$\varphi ::= p \mid \neg\varphi \mid \varphi \wedge \psi \mid K_i\varphi \mid < e > \varphi$$

where $i \in I, e \in E$, and $p \in At$. Booleans, and dual modalities $\tilde{K}_i$, $[e]$ are defined as usual. Let $M = (E, H, \{\sim_i\}_{i \in I}, V)$ be an ETL model. The truth of a formula φ at a history $h \in H$, denoted as $M, h \models \varphi$, is defined inductively as usual. We display the two key clauses, for knowledge and events:

(a) $M, h \models K_i\varphi$ iff for each $h' \in H$, if $h \sim_i h'$, then $M, h' \models \varphi$

(b) $M, h \models < e > \varphi$ iff there exists $h' = he \in H$ with $M, h' \models \varphi$.

This language can express many properties of agents and their long-term behavior over time. It has a base logic that we will not formulate here, though we will make a few comments on it later.

Agent properties In particular, the language LETL can express interesting properties of agents or informational processes through constraints on ETL models. These often come as epistemic-temporal axioms matched by modal frame correspondences. Here is a typical example.

Fact 7.6
The axiom $K_i[e]\varphi \to [e]K_i\varphi$ corresponds to Perfect Recall:

$$\text{if } he \sim_i k, \text{ then there is a history } h' \text{ with } k = h'e \text{ and } h \sim_i h'$$

This says that agents' current uncertainties can only come from previous uncertainties: a constraint on ETL models that expresses a strong form of so-called *Perfect Recall.*[27] Note that the axiom as stated presupposes perfect observation of the current event e: therefore, it does not hold in general DEL, where uncertainty can also be created by the current observation, when some event f is indistinguishable from e for the agent.

In a similar fashion, the axiom $[e]K_i\varphi \to K_i[e]\varphi$ corresponds to *No Miracles*: for all ke with $h \sim_i k$, we also have $he \sim_i ke$. This says essentially that learning only takes place for agents by observing events resolving current uncertainties.

Digression: epistemic temporal logics and complexity Epistemic-temporal logics have been studied extensively, and we cannot survey their theory here. However, one feature deserves mention. In this chapter, we will not pay attention to *computational complexity* of our logics, except for noting the following. There is a delicate balance between expressive power and computational complexity of combined logics of knowledge and time which also extends to belief. A pioneering investigation of these phenomena was made in 1989 by Halpern & Vardi. Table 7.1 lists a few observations from their work showing where dangerous thresholds occur for the complexity of validity.

In Table 7.1 complexities run from decidable through axiomatisable (RE) to Π^1_1-complete, which is the complexity of truth for universal second-order statements in arithmetic. What we see here is that complexity of the logic depends on two factors: expressive power of the language (in particular, social forms of group knowledge matter), and so do special assumptions about the type of agents involved. In particular, the property of Perfect Recall, which seems a harmless form of good behavior, increases the complexity of the logic.[28]

---

[27]Perfect Recall implies synchronicity: uncertainties $h \sim_i k$ only occur between h, k at the same tree level. Weaker forms of perfect memory in games also allow uncertainty links that cross between tree levels.

[28]Technically, Perfect Recall makes epistemic accessibility and future moves in time

	K, P, F	K, C_G, $<e>$	$K, C_G, <e>$, $PAST_e$	K, C_G, F
ETL	decidable	decidable	decidable	RE
ETL + PR	RE	RE	RE	Π_1^1-complete
ETL + NM	RE	RE	RE	Π_1^1-complete

Table 7.1: Complexity of epistemic temporal logics. ETL denotes the class of all ETL models, PR denotes Perfect Recall and NM is No Miracles.

7.7.2 Protocols in dynamic epistemic logic

Before we connect DEL and ETL in technical detail, let us see how the crucial notion of protocol in temporal logic natural enters the realm of DEL, and its extensions to belief revision.

Definition 7.19 (DEL protocols)
Let Σ be the class of all pointed event models. A *DEL protocol* is a set $P \subseteq \Sigma^*$ closed under taking initial segments. Let M be any epistemic model. A *state-dependent DEL protocol* is a map P sending worlds in M to DEL protocols. If the protocol assigned is the same for all worlds, it is called "uniform". ⊣

DEL protocols induce TL models in a simple manner. Here is an illustration.

Example 7.15 (An ETL model generated by a uniform PAL protocol)
We use a public announcement protocol for graphical simplicity. Consider the epistemic model M in Figure 7.16 with four worlds and two agents 1 and 2.

The model comes with a protocol

$$P = \{<!p>, <!p, !q>, <!p, !r>\}$$

of available sequences of announcements or observations. The ETL forest model depicted in Figure 7.17 is then the obvious "update evolution" of M under the available announcements, with histories restricted by the event sequences in P. Note how some worlds drop out, while others 'multiply'. ⊣

behave like a *grid* of type $IN \times IN$, with cells satisfying a confluence property. Logics of such grids are known to have very high complexity since they can encode so-called "tiling problems" of a complex geometrical nature.

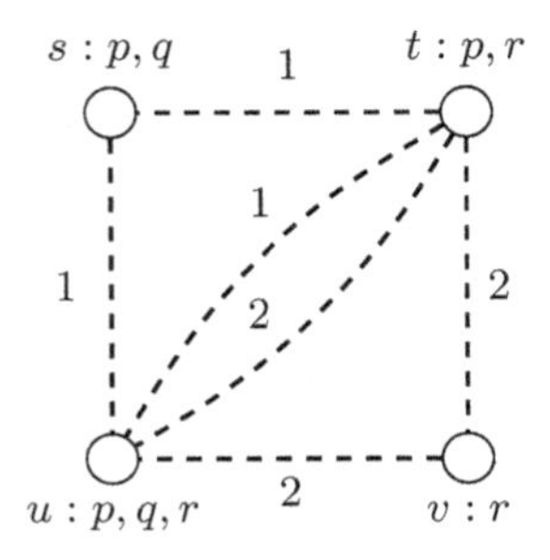

Figure 7.16: Initial epistemic model for a uniform PAL protocol.

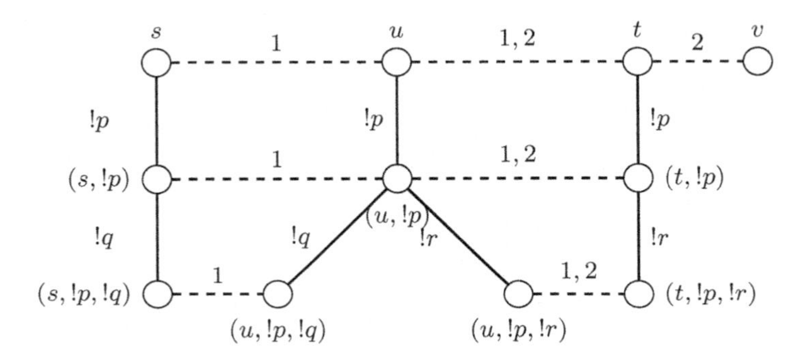

Figure 7.17: Generated ETL forest model.

The logic of PAL protocols Adding protocols changes the laws of dynamic epistemic logic. Again our example concerns public announcement.

Example 7.16 (Failures of PAL validities)
PAL had a valid axiom $<!\varphi> q \leftrightarrow \varphi \wedge q$. As a special case, this implied

$$<!\varphi> T \leftrightarrow \varphi$$

From left to right, this is valid with any protocol: an announcement $!\varphi$ can only be executed when φ holds. But the direction from right to left is no longer valid: φ may be true at the current world, but the protocol need not allow a public announcement of this fact at this stage. Similar observations can be made for the PAL recursion law for knowledge. ⊣

Thus, assertions $<!\varphi> T$ come to express genuine procedural information about the informative process agents are in, and hence, they no longer reduce to basic epistemic statements. However, there is still a decidable and complete logic TPAL for PAL protocol models, be it, that we need to modify the recursion axioms. The two key cases are as follows:

$$<!\varphi> q \leftrightarrow <!\varphi> T \wedge q \quad \text{for atomic facts } q$$

$$<!\varphi> K_i\psi \leftrightarrow <!\varphi> T \wedge K_i(<!\varphi> T \rightarrow <!\varphi> \psi)$$

Similar modifications yield protocol versions of DEL with product update.

7.7.3 Representing DEL inside temporal logic.

Now we state the more general upshot of the preceding observations. The Grand Stage is also a natural habitat for the local dynamics of DEL. Epistemic temporal trees can be created through constructive unfolding of an initial epistemic model M by successive product updates, and one can determine precisely which trees arise in this way. We only give a bare outline.

A basic representation theorem Consider a scenario of "update evolution": some initial epistemic model M is given, and it gets transformed by the gradual application of event models $\Sigma = \Sigma_1, \Sigma_2, \cdots$ to form a sequence

$$M_0 = M, M_1 = M_0 \times \Sigma_1, M_2 = M_1 \times \Sigma_2, \cdots$$

where stages are horizontal, while worlds may extend downward via one or more event successors. Through successive product update, worlds in the resulting "induced ETL forest model" *Forest*(M, Σ) are finite sequences starting with one world in the initial epistemic model M followed by a finite sequence of events, inheriting their standard epistemic accessibility relations.

Induced ETL forest models have three properties making them stand out, two of which generalise the above special agent properties. In what follows, quantified variables $h, h', k, \cdots$ range only over histories present in M:

(a) If $he \sim k$, then for some f, both $k = h'f$ and $h \sim h'$ (Perfect Recall)

(b) If $h \sim k$, and $h'e \sim k'f$, then $he \sim kf$ (Uniform No Miracles)

(c) The domain of any event e is definable in the epistemic base language.

Condition (c) of "Definable Executability" ensures admissible preconditions.

Theorem 7.6
For ETL models $\mathcal{H}$, the following two conditions are equivalent:[29]

(a) $\mathcal{H}$ is isomorphic to some DEL-induced model Forest(M, Σ)

(b) $\mathcal{H}$ satisfies Perfect Recall, Uniform No Miracles, and Definable Executability. ⊣

[29]The theorem still holds when we replace Definable Executability by closure of event domains under all purely epistemic bisimulations of the ETL-model $\mathcal{H}$.

DEL as an ETL-logic Now we can place DEL and understand its behavior. Its language is the $K, C_G, < e >$ slot in the earlier table, on models satisfying Perfect Recall and No Miracles. Thus, there is grid structure, but the expressive resources of DEL do not exploit it to the full, using only one-step future operators $<!P>$ or $< \Sigma, e >$. Adding unbounded future yields the same complexity as for ETL. Miller & Moss showed in 2005 that the logic of public announcement with common knowledge and Kleene iteration of assertions is Π^1_1-complete.

7.7.4 Beliefs over time

We have only discussed temporal perspectives on knowledge so far, but with a few suitable extensions, everything that we have said also applies to belief and belief revision. We show some technicalities here, and then continue with applications in later subsections.

Epistemic-doxastic temporal models So-called "DETL models" are branching event trees as before, with nodes in epistemic equivalence classes now also ordered by plausibility relations for agents. These models interpret belief modalities at finite histories, in the same style as in the earlier plausibility models.

The epistemic-doxastic language of these models is the natural temporal extension of the logics of belief discussed in earlier sections. It can express natural doxastic properties of agents (some are stated below), and it fits well with the temporal analysis of the AGM postulates that have been made by Bonanno (see his chapter on temporal belief revision in this Handbook for many further themes), as well as Dégrémont & Gierasimczuk.

As with knowledge, we can ask which DETL models arise as traces of the update scenarios that we have studied before. For this purpose, one can take sequences of plausibility event models applied to some initial epistemic-doxastic model, and compute their update evolution with the earlier product Rule for epistemic accessibilities and the Priority Rule of Section 7.5 for plausibility. An analogue to the earlier representation theorem then arises, in terms of two basic properties of plausibility between histories:

Fact 7.7
The histories h, h', j, j' arising from iterated Priority Update satisfy the following two principles for any events e, f:

(a) if $je \leq j'f$, then $he \geq h'f$ implies $h \geq h'$ ("Plausibility Revelation")

(b) if $je \leq j'f$, then $h \leq h'$ implies $he \leq h'f$ ("Plausibility Propagation") $\dashv$

Theorem 7.7
A DETL model can be represented as the update evolution of an epistemic-doxastic model under a sequence of epistemic-plausibility updates iff it satisfies the structural conditions of the epistemic DEL-ETL representation theorem[30], plus Plausibility Revelation and Plausibility Propagation. ⊣

7.7.5 Iterated belief upgrades and limit behavior

ETL models and DETL models are very abstract and general. Much more concrete temporal scenarios arise with iterated dynamic epistemic steps of knowledge update or belief revision. Scenarios of this kind are the well-known Puzzle of the *Muddy Children*, often cited in the dynamic epistemic literature, disagreement scenarios in the epistemic foundations of game theory, or the game solution procedures of Section 7.8 below. However, in this section, we will only consider a scenario from work by Baltag and Smets on propagating beliefs in groups of agents that makes sense in social networks.

We will be concerned with iterated truthful belief revision, namely, iterations of different upgrades with true assertions that lead to sequences or streams of models. When we do this with PAL style updates of the form $!\varphi$, at least on finite models, "stabilisation" occurs in a unique model since the sequence of submodels produced is monotonically non-increasing.[31] With iterated belief changes, however, the plausibility order has a third option of "cycling", resulting in endless oscillation, as we will show a bit later on. Therefore, our main question will be whether and when an iterated belief revision process induced by truthful upgrades converges to a fixed point or not.

We start with some basic definitions, and then state a basic result on limits of belief change from the work of Baltag and Smets.

Definition 7.20 (Joint upgrades)
Over the plausibility models of Section 7.4, we use the term *joint upgrade* for the effect that three different model transformers can have. We denote them in general by $\dagger\varphi$, where $\dagger \in \{!, \Uparrow, \uparrow\}$. ⊣

Now we define some technical types of behavior. A joint upgrade $\dagger\varphi$ is *redundant* on a pointed model M for a group of agents G if $\dagger\varphi(M)$ is bisimilar with M (written as $\dagger\varphi(M) \simeq_G M$.)[32] This means that, as far as

[30]Here one now needs invariance of event domains for epistemic-doxastic bisimulations.

[31]In fact, at the limit stage, only two options can occur. The sequence stabilises in a non-empty model where φ has become common knowledge, or the sequence breaks off at the first stage where φ has become false at the actual world (as happens with the Muddy Children).

[32]Here we mean bisimilarity in the usual sense of modal logic, with respect to all accessibility relations for all agents.

group G is concerned, $\dagger\varphi$ does not change anything essential when applied to model M): all the group G's mutual beliefs, conditional beliefs, strong beliefs, mutual knowledge, and common knowledge stay the same after the upgrade. By contrast, an upgrade $\dagger\varphi$ is *informative* on M for group G if it is not redundant with respect to G.[33] Finally, we say that a model M is a *fixed point* of $\dagger\varphi$ if $M \simeq \dagger\varphi(M)$, i.e. if $\dagger\varphi$ is redundant on M with respect to the group of all agents.

At this point, we can capture stabilisation in the limit.

Logical characterisations Redundancy and Fixed Points can be characterised in the following logical terms, using doxastic notions that were introduced in Section 7.4.

1. $!\varphi$ is redundant with respect to a group G iff φ is common knowledge in the group G; i.e., $M \simeq_G !\varphi(M)$ iff $M \models C_G\varphi$.[34]

2. $\Uparrow\varphi$ is redundant with respect to a group G iff it is common knowledge in the group G that φ is strongly believed by all G-agents. That is, $M \simeq_G \Uparrow\varphi(M)$ iff $M \models C_G(ESb_G\varphi)$.

3. $\uparrow\varphi$ is redundant with respect to a group G iff it is common knowledge in the group G that φ is believed by all G-agents. That is, $M \simeq_G \uparrow \varphi(M)$ iff $M \models C_G(EB_G\varphi)$.

Now we set up the machinery for iterations of upgrades starting from some initial model. The following auxiliary definitions lead up to our main result.

Upgrade streams An *upgrade stream* $\dagger\vec{\varphi} = (\dagger\varphi_n)_{n\in N}$ is an infinite sequence of joint upgrades $\dagger\varphi_n$ of the same type $\dagger \in \{!, \Uparrow, \uparrow\}$. Any upgrade stream $\dagger\vec{\varphi}$ induces a function mapping every pointed model M into an infinite sequence $\dagger\vec{\varphi}(M) = (M_n)_{n\in N}$ of pointed models, defined inductively by:

$$M_0 = M, \quad \text{and} \quad M_{n+1} = \dagger\varphi_n(M_n).$$

The upgrade stream $\dagger\vec{\varphi}$ is *truthful* if every $\dagger\varphi_n$ is truthful with respect to M_n, i.e., $M_n \models \varphi_n$. Next, a *repeated truthful upgrade* is a truthful upgrade stream of the form $(\dagger\varphi_n)_{n\in N}$, where $\varphi_n \in \{\varphi, \neg\varphi\}$ for some proposition φ. In other words, it consists in repeatedly learning the answer to the same question φ?

[33] As a special case, an upgrade $\dagger\varphi$ is redundant with respect to (or informative to) an agent i if it is redundant with respect to (or informative to) the singleton group $\{i\}$.

[34] Here are two special cases. $!\varphi$ is redundant with respect to an agent i iff i knows φ. Also, *M is a fixed point of $!\varphi$ iff $M \models C\varphi$*. Similar special cases apply for the next two clauses in the text.

Stabilisation A stream $\dagger\vec{\varphi}$ *stabilises* a pointed model M if there exists some $n \in N$ with $M_n \simeq M_m$ for all $m > n$. A repeated truthful upgrade stabilises M if it reaches a fixed point of either $\dagger\varphi$ or of $\dagger(\neg\varphi)$. Next, we say that $\dagger\vec{\varphi}$ *stabilises all simple beliefs* (i.e., non-conditional ones) on M if the process of belief-changing induced by $\dagger\vec{\varphi}$ on M reaches a fixed point; i.e., if there exists some $n \in N$ such that $M_n \models B_i\varphi$ iff $M_m \models B_i\varphi$, for all agents i, all $m > n$, and all doxastic propositions φ. Similarly, $\dagger\vec{\varphi}$ *stabilises all conditional beliefs* on a model M if the process of conditional-belief-changing induced by $\dagger\vec{\varphi}$ on M reaches a fixed point as before, but now with respect to conditional belief $B_i^{\psi}\varphi$ for all doxastic propositions φ, ψ.[35] Finally, $\dagger\vec{\varphi}$ *stabilises all knowledge* on the model M if the knowledge-changing process induced by $\dagger\vec{\varphi}$ on M reaches a fixed point, in an obvious sense modifying the preceding two notions.

At last, we can state some precise technical results.

Lemma 7.1
The following two assertions are equivalent:

- An upgrade stream $\dagger\vec{\varphi}$ stabilises a pointed model M,
- $\dagger\vec{\varphi}$ stabilises all conditional beliefs on M. $\dashv$

Theorem 7.8
Every truthful radical upgrade stream $(\Uparrow\varphi_n)_{n\in N}$ stabilises all simple, non-conditional beliefs – even if it does not stabilise the model. $\dashv$

This result has a number of interesting consequences. For instance, every iterated truthful radical upgrade definable by a formula in doxastic-epistemic logic (i.e., in the language of simple belief and knowledge operators, without conditional beliefs) stabilises every model with respect to which it is correct, and thus stabilises all conditional beliefs. The analogue of this result is not true for conservative upgrade, where updates can keep oscillating – so limit behavior depends in subtle manners on the sort of belief revision involved.

We have shown how limit behavior of belief revision steps fits in the scope of logic, provided we place our dynamic epistemic logics in a broader temporal setting. In particular, the specific protocol of only allowing upgrades for specified assertions turned out to have a lot of interesting properties.

[35] A similar definition can be formulated for stabilisation of strong beliefs.

7.7.6 From belief revision to formal learning theory

Continuing with the theme of long-term behavior, we conclude this section by pointing out links with *formal learning theory.* In this theory, learning methods are studied for identifying an as yet unknown world on the basis of evidence streams that it generates, where identification sometimes takes place by some finite stage, but often "only in the limit", in a sense to be defined below.

Belief revision and learning theory The DEL framework is well equipped to describe local learning of facts, but for the long-term identification of a total, possibly infinite, history of evidence, temporal notions need to enter as well, as happened in the iterated upgrade scenarios of the preceding subsection. But this combination indeed matches basic features of formal learning theory. Following a suggestion by Kelly that learning theory is a natural extension of belief revision, separating bad policies from optimal ones, Baltag, Smets & Gierasimczuk have found a number of precise links, where the equivalent of "learning methods" are upgrade methods forming beliefs about the actual history on the basis of growing finite sets of data, with a correct stable belief achieved on a history as the doxastic counterpart of the learning theoretic concept of "identifiability in the limit". In what follows, we briefly present a few highlights.

Revision policies for learning Already in Section 7.1, we have seen that the AGM postulates are conservative in the sense that an agent who adopts this method will keep holding on to her old beliefs as long as possible while incorporating the new incoming information. While this looks attractive as a learning method, conservative upgrade as defined in Section 7.4 is often unable to shake a current suboptimal belief out of its complacency, and other more radical methods turn out to work better in limit identification, as we shall see.

Modeling learning methods via belief revision We start from an *epistemic space* (W, Φ) consisting of a set W of epistemic possibilities (or "possible worlds"), together with a family of observable properties $\Phi \subseteq \mathcal{P}(W)$. We work with streams of successive observations, and denote an infinite such stream as

$$\epsilon = (\epsilon_1, \epsilon_2 \ldots) \in \Phi^\omega$$

A stream of observations is said to be *sound and complete* with respect to a given world $s \in W$ if the set $\{\epsilon_n : n \in N\}$ of all properties that are observed in the stream coincides with the set $\{\varphi \in \Phi : s \in \varphi\}$ of all observable properties φ that are true in s. Now comes the crucial notion driving the analysis.

Definition 7.21 (Learning methods)
A *learning method* L for an agent is a map that associates to any epistemic space (W, Φ) and any finite sequence of observations $\sigma = (\sigma_0, \ldots, \sigma_n)$, some *hypothesis* as a result, where hypotheses are just subsets of W. A world $s \in W$ is said to be *learnable* by the method L if, for every observation stream ϵ that is sound and complete for s, there is a finite stage N such that $L(W, \Phi; \epsilon_0, \ldots, \epsilon_n) = \{s\}$ for all $n \geq N$. The epistemic space (W, Φ) itself is learnable by method L if all the worlds in W are learnable by the same method. ⊣

Now we introduce belief structure via plausibility orderings. Starting from an epistemic space (W, Φ), we define a *plausibility space* $(W, \Phi, \leq)$ by equipping it with a total preorder $\leq$ on W. A *belief-revision method* will then be a function R that associates to $(W, \Phi, \leq)$ and any observation sequence $\sigma = (\sigma_0, \ldots, \sigma_n)$, a new plausibility space

$$R(W, \Phi, \leq; \sigma) := (W^\sigma, \Phi^\sigma; \leq^\sigma),$$

with $W^\sigma \subseteq W$ and $\Phi^\sigma = \{P \cap W^\sigma : P \in \Phi\}$. Such a belief-revision method R, together with a prior plausibility ordering over W, generates in a canonical way a learning method L via the stipulation

$$L(W, \Phi, \sigma) := Min\, R(W, \Phi, \leq_W, \sigma)$$

where $Min(W', \leq')$ is the set of all the least elements of W' with respect to $\leq'$ if such least elements exist, or $\emptyset$ otherwise.

Now we say that an epistemic space (W, Φ) is *learnable by a belief-revision method* R if there exists some prior plausibility assignment on W such that (W, Φ) is learnable by the associated learning method $L(W, \Phi, \leq_W)$.

Learning methods can differ in terms of strength, ranging from weak methods to those that are *universal* in the sense of being able to learn any epistemic state that is learnable at all. Here is a major result of the belief-based analysis.

Theorem 7.9
Conditioning and radical upgrade are universal AGM-like iterated belief revision methods. Conservative upgrade is not. ⊣

But there is more to this style of analysis, and we mention two further issues.

Well-founded prior. The preceding theorem is based on arbitrary initial epistemic plausibility spaces, and freedom in the choice of the prior turns out very important. It can be shown that there are learning problems where only a non-wellfounded prior plausibility ordering allows the revision methods of iterated conditioning and lexicographic revision to be universal.[36]

Errors in observations. As it is standard in Learning Theory, the easiest scenario is learning under truthful observations. Much harder to analyse are learning scenarios that allow for errors in observations. In such a setting, one needs a "fairness condition" for learning, which says that that errors occur only finitely often and are always eventually corrected. Under such conditions, there still exist belief-revision methods that are universal, but now only one remains, that allows for radical shifts in hypotheses.

Fact 7.8
Iterated radical upgrade is a universal learning method for fair evidence streams. Conditioning and minimal revision are not universal in this setting. $\dashv$

Logic of learning theory Learning Theory supports the logical languages of earlier sections, which can distinguish a wide range of learning goals. For instance, the formula $FK\varphi$ or modalised variants thereof expresses for suitable φ that there comes a stage where the agent will know that φ.[37]

But closer to the preceding analysis are weaker belief-based success principles for learning such as the following:

$F(B\psi \rightarrow \psi)$ says that my beliefs will eventually be true,

while $F(B\psi \rightarrow K\psi)$ says that they will turn into knowledge.

Using suitably extended languages, logical definitions have been given for most standard notions of learnability, ranging from identification in the limit to stronger forms of "finite identifiability" and "conclusive learning". Thus, epistemic-doxastic-temporal logic comes to express the basics of learning theory.

Summary We have seen how dynamic logics of belief change receive a natural continuation in iterative scenarios and temporal logics of inquiry

[36] In the class of standard well-founded plausibility spaces, which seemed rather natural in earlier sections, no AGM-like belief-revision method is universal.

[37] Stronger learning notions would be expressed by epistemic-temporal formulas like $FGK\varphi$ or $F(GK\varphi \vee GK\neg\varphi)$.

over time, leading to a better view of the role of belief revision in a general account of learning beings.

7.8 Belief, games, and social agency

Belief revision may seem an agent-internal process, with unobservable changes of mind taking place in utmost privacy. But in reality, belief revision has many important social aspects. For instance, while many people think of triggers for belief revision as some surprising fact that a person observes – very often, the trigger is something said by someone else, in a multi-agent setting of communication. Human beliefs are deeply influenced by social settings, and they enter into mutual expectations that drive behavior in scenarios such as games. Moreover, in addition to individual agents interacting, at various places in this chapter, we have even gone further, and also mentioned groups as collective actors that can have properties such as common or distributed knowledge or belief.

The logics that we have introduced in the preceding sections are well up to this extension, even though the technicalities of adding common knowledge or belief are not our main concern here.[38] Instead of developing this theory here (for which we provide literature references in Section 7.10), we discuss two samples of belief revision in basic social settings, viz. games and social networks. Some additional themes will be mentioned in Section 7.9 on further directions.

7.8.1 Iterated beliefs in games

An important area for interactive social agency are games. Games support the logics of this chapter in various ways, from modal logics of action to logics of knowledge and belief for the agents playing them. There is a fast-growing literature on these interfaces that lies far outside the scope of this chapter (see Section 7.10 for some references), but a concrete illustration is within reach.

Beliefs play a major role in various phases of game play, from prior deliberation to actual play, and from there to post-game analysis. In this subsection, we consider the game solution method of *Backward Induction*, a procedure for creating expectations about how a game will proceed. We will construe it as an iteration scenario like in our preceding section, first in terms of public announcements and knowledge, and in our final analysis,

[38]These technicalities occasionally involve a move to much more powerful static formalisms such as "epistemic PDL" – and also, some specific open questions remain, such as finding an optimal formalism with recursion laws for reasoning with common belief under Priority Update.

in terms of forming beliefs. The presentation to follow is from earlier work by van Benthem.

Backward Induction and announcing rationality Our first dynamic analysis of Backward Induction is as a process of silent deliberation by players whose minds proceed in harmony. The steps are announcements $!\varphi$, now interpreted as mental reminders to players that some relevant proposition φ is true. Here the driving assertion φ is *node rationality ("rat")*, defined as follows. At a turn for player i, call a move a dominated by a sibling move b (available at the same node) if every history through a ends worse, in terms of i's preference, than every history through b.

Now the key proposition *rat* says:

"Coming to the current node, no one ever chose a strictly dominated move"

Announcing this is informative, and it will in general make a current game tree smaller by eliminating nodes. But then we get a dynamics as with the earlier-mentioned Muddy Children, where repeated true assertions of ignorance eventually solve the puzzle. For, in the smaller game tree, new nodes become dominated, and so announcing *rat* again (saying that it still holds after this round of deliberation) now makes sense.

As we saw in Section 7.8, this process of iterated announcement reaches a limit, a smallest subgame where no node is dominated any more.

Example 7.17 (Solving games through iterated assertions of rational Consider the game depicted in Figure 7.18, with three turns, four branches, and pay-offs for players A, E marked in that order:

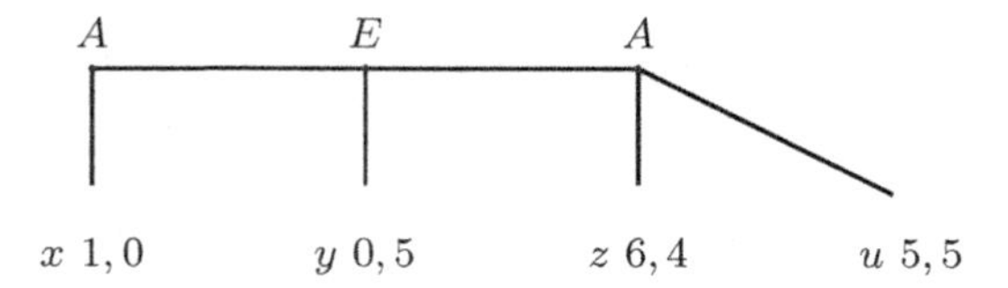

Figure 7.18: An extensive form game.

Stage 0 of the procedure rules out point u: the only point where *rat* fails, Stage 1 rules out z and the node above it: the new points where *rat* fails, and Stage 2 rules out y and the node above it. Each stage deletes part of the game tree so that in the remaining game, *rat* holds throughout. ⊣

The actual Backward Induction path for extensive games is obtained by repeated announcement of the assertion *rat* to its limit:

Theorem 7.10
In any game tree M, $(!rat, M)\sharp$ is the actual subtree computed by BI. ⊣

A more sensitive scenario: iterated plausibility change The preceding analysis of Backward Induction was in terms of knowledge. However, many foundational studies in game theory view rationality as choosing a best action given what one *believes* about the current and future behavior of the players. This suggests a refinement in terms of our soft updates of Section 7.4 that did not eliminate worlds, but rearrange their plausibility ordering. Now recall our observation that Backward Induction creates expectations for players. The information produced by the algorithm is then in the plausibility relations that it creates inductively for players among end nodes in the game, i.e., complete histories:

Solving a game via radical upgrades. Consider the preceding game once more. This time, we start with all endpoints of the game tree incomparable qua plausibility. Next, at each stage, we compare sibling nodes, using the following notion. A move x for player i *dominates its sibling* y *in beliefs* if the most plausible end nodes reachable after x along any path in the whole game tree are all better for the active player than all the most plausible end nodes reachable in the game after y. We now use the following driver for iterated upgrades:

Rationality* (rat^*): No player plays a move that is dominated in beliefs.

Then we can use an ordering change that is like a radical upgrade $\Uparrow rat^*$:

If x dominates y in beliefs, we make all end nodes from x more plausible than those reachable from y, keeping the old order inside these zones.

This changes the plausibility order, and hence the dominance pattern, so that the doxastic assertion rat^* can change truth value, and iteration can start. Figure 7.19 depicts the stages for this procedure in the preceding game example, where the letters x, y, z, u stand for the end nodes or histories of the game:

Theorem 7.11
On finite game trees, the Backward Induction strategy is encoded in the stable plausibility order for end nodes created in the limit by iterated radical upgrade with rationality-in-belief. ⊣

Notice that, at the end of this procedure, the players as a group have acquired *common belief in rationality*, a fundamental notion in the epistemic foundations of game theory. However, in line with the dynamic focus of this

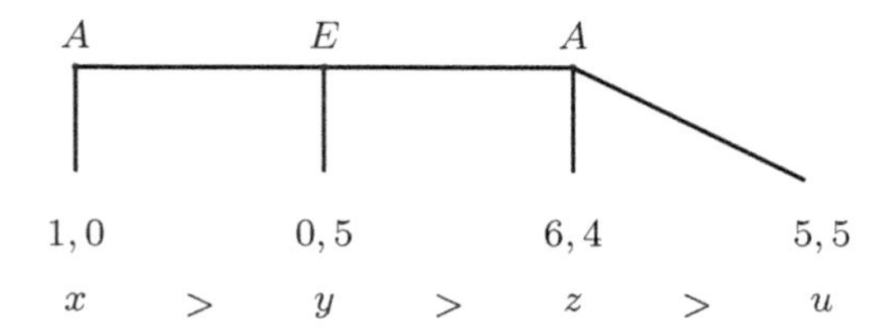

Figure 7.19: Creating plausibility order on histories of a game.

chapter, it is not so much this end result as the procedure itself that deserves attention. Rationality is not a state of grace, but a style of doing things.

7.8.2 Belief change in social networks

A further source of examples are recent studies of belief change in scenarios from social science, rather than economic game theory. Phenomena such as group polarisation, informational cascades, pluralistic ignorance, peer pressure, or epistemic bandwagoning all center around the epistemic and doxastic processes of agents in a social network. These mechanisms are more complex then those for individual agents. In a social network, the formation of opinions does not only depend, as it did in our earlier sections, on the individual agent's own prior epistemic-doxastic state, the new information it faces, and its individual belief revision mechanism. It is also crucially affected by the opinions and belief changes of other agents in the network.

The logical tools for belief change in this chapter can be adapted to incorporate the qualitative features of a social network, such as its topological configuration or its social hierarchy. We briefly present two instance of logical models for belief revision in this richer setting.

A first question that has been solved for social networks is whether informational cascades are due to irrational moves or mistakes in the reasoning steps of the agents or whether they are actually unavoidable by "rational" means.

7.8.3 Informational cascades

Informational cascades can occur when agents find themselves in a sequential network, obeying strict rules about what and how they can communicate with each other, and end up following the opinion of the preceding agents in the sequence, ignoring their own private evidence.

Example 7.18 (Choosing a restaurant)
A standard example is the choice between two restaurants A and B. You have prior private evidence that restaurant A is better than B, but you still

end up choosing for B based on the fact that it has many more customers. You interpret the other customers' choice for restaurant B as conveying the information that they somehow know better. It is however very well possible that all others made their decision in exactly the same way as you. ⊣

Several models for informational cascades have been developed from 1992 onwards, after the term was first introduced by Bikhchandani, Hirshleifer, and Welch, with analyses usually in terms of Bayesian reasoning. Recent work by Baltag, Christoff, Hansen and Smets has used a combined probabilistic and qualitative Dynamic-Epistemic Logic. It then turns out that what might seem an irrational form of influence, manipulation or irregularity in the process of opinion formation, is actually the result of fully rational inference process. Moreover, the modeling makes explicit the agents' higher-order reasoning about their own as well as other agents' knowledge and beliefs.

Example 7.19 (An urn guessing scenario)
Consider the following information cascade based on a standard urn example used in the literature. Consider a sequence of agents $(i_1, i_2, ..., i_n)$ lined up in front of a room. The room contains one of two non-transparent urns U_W and U_B. It is common knowledge among the agents that nobody knows which urn is actually placed in the room. It is also common knowledge that Urn U_W contains two white balls and one black ball, and urn U_B contains one white ball and two black balls.

Now the agents enter the room one by one, and each agent draws a ball from the urn, looks at it, puts it back, and leaves the room. After leaving the room she publicly communicates her guess: urn U_W or U_B, to all the other agents. We assume that each agent knows the guesses of the people preceding her in the sequence before entering the room herself. It is crucial here (and realistic in many social settings) that while the agents communicate their guess about U_W or U_B, they do not communicate their private evidence, namely, the color of the ball they observed in the room. The standard Bayesian analysis of this example shows that if U_B is the real urn in the room and the first two agents i_1 and i_2 draw white balls (which happens with probability $\frac{1}{9}$), then a cascade leads everyone to the wrong guess U_A.

We can model this cascade in the Probabilistic Epistemic Dynamic Logic of Section 7.6. The reader should refer back to the notions introduced there. ⊣

Urn scenario, probabilistic epistemic model Consider the preceding urn example, with U_B the real urn in the room and the two first agents drawing white balls. The probabilistic epistemic model $\mathcal{M}_0$ of Figure 7.20

has equal odds for the initial situation, encoded in two worlds making it equally probable that U_W or U_B are true, and all agents know this. The actual state of affairs s_B satisfies the proposition U_B, while the other possible world s_W satisfies U_W. The relative likelihood of s_B versus s_W is $[1 : 1]_i$ for all agents i.

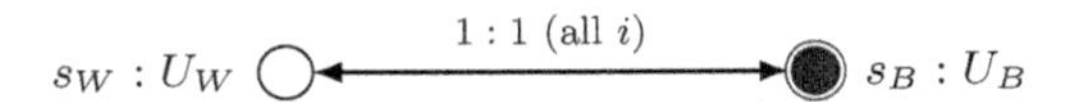

Figure 7.20: Probabilistic epistemic base model.

Urn Scenario, probabilistic event model To model the first observation of a ball by agent i_1, we use the probabilistic event model $\mathcal{E}_1$ depicted in Figure 7.21. At the moment of i_1's observation of a white ball, all other agents consider one of the following two events possible: either i_1 observes a white ball (event w_1) or she observes a black ball (event b_1). Only agent i_1 knows which event (w_1 or b_1) is actually taking place. For the event w_1 to happen, the preconditions are $pre(U_W) = \frac{2}{3}$ and $pre(U_B) = \frac{1}{3}$. All agents except i_1 consider both events equally likely and assign them the odds $1 : 1$. All this information in the successive visits by the agents is present pictorially in Figure 21.

$1 : 1$ (all $a \neq i_1$)

$b_1, pre(U_W) = 1/3, pre(U_B) = 2/3$

$w_1, pre(U_w) = 2/3, pre(U_B) = 1/3$

Figure 7.21: Probabilistic event model for the agent episodes.

To model the effect of the first observation, we combine the probabilistic epistemic model with the probabilistic event model using PDEL product update.

Urn scenario, the updates The product update of the initial epistemic probability model $\mathcal{M}_0$ with the event model $\mathcal{E}_1$ is the epistemic probability model $\mathcal{M}_0 \otimes \mathcal{E}_1$ illustrated in Figure 7.22, consisting of 4 possible worlds. In the world representing the actual situation, U_B is true and the first ball which has been observed was a white one w_1. The model makes it clear pictorially that agent i_1 *knows* which ball she observed: there are no a_1-arrows between the two top states in the model and the bottom states. The other agents (different from i_1) consider all four states possible and for them it is common knowledge that, if agent 1 observed a white ball (w_1), then she considers U_W to be twice as likely as U_B, and in case she would

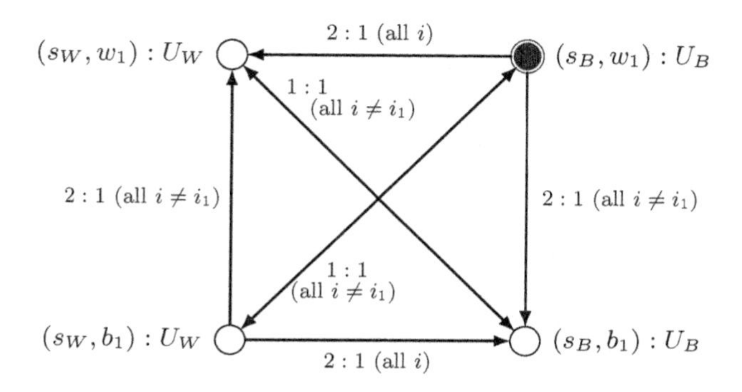

Figure 7.22: The result of probabilistic product update.

have observed a black ball b_1, then she would consider U_B twice as likely as U_W. In particular, the new model indicates that all the other agents cannot exclude any epistemic possibility as long as i_1 has not yet announced her guess publicly.

Next, the public announcement of a_1's guess is modeled as a further PAL update resulting in a third model $\mathcal{M}_1$ in which two states (those not compatible with the guess of the agent) are deleted.

The event of the second agent entering the room, drawing a ball from the urn, and announcing its guess publicly will induce further changes of the model $\mathcal{M}_1$, analogous to those in the first step, resulting in $\mathcal{M}_2$.

Things become more interesting when agent a_3 enters the room and observes a black ball. The event model $\mathcal{E}_3$ of this action will trigger an update of $\mathcal{M}_2$ to the model $\mathcal{M}_2 \otimes \mathcal{E}_3$. But in this model, a_3 still considers U_W more probable than U_B, irrespective of the result of her private observation. Hence, if i_3 then announces her guess of U_W, it deletes no more worlds: the model $\mathcal{M}_3$ after the announcement stays the same as before. From now on an informational cascade will unfold, and all further agents in the sequence announce the guess U_W.

Much more can be said about this example, but it will be clear how the machinery of this chapter can analyse agents' beliefs in social phenomena such as cascades. Moreover, it yields interesting insights such as the following. Agents who are logical omniscient and perfect reasoners, but only announce their private guesses, simply do not have the tools which can always prevent an informational cascade from happening.[39]

[39]Baltag, Christoff, Hansen and Smets also present a second formalisation in which probabilistic reasoning is replaced by a "counting" heuristics in terms of pieces of evidence.

7.8.4 Influence in social networks

To show the reach of the methods in our chapter, we also consider a second basic phenomenon in the social realm, that of agents in social networks.

Different types of epistemic and doxastic attitudes arise in, and maintain, social settings. Seligman, Liu & Girard have recently proposed a logical framework for investigating how agents' beliefs, or knowledge, are formed and changed under the influence of the beliefs of other agents who belong to the same social community. The system of *"Facebook Logic"* designed for this has a social as well as an epistemic dimension. Both these dimensions are needed to specify basic social relations between agents, such as friendship, agents' epistemic attitudes, and their entanglement. We describe a few basics here.

Epistemic social networks We start with situations where agents can be friends, though they need not know exactly who are their friends. An *epistemic social network model* $M =< W, I, \sim_i, \asymp_w, V >$ consists of a set of states W, a set of agents I, and an epistemic relation $\sim_i$ for each agent $i \in I$. Each state $w \in W$ comes equipped with a binary irreflexive and symmetric friendship relation $\asymp_w$ over the set of agents, and, as usual, V is a valuation map assigning subsets of $W \times A$ to propositional variables.

The *epistemic friendship language* is a multimodal formalism given by:

$$\varphi ::= p \mid n \mid \neg\varphi \mid \varphi \wedge \varphi \mid K\varphi \mid F\varphi \mid A\varphi$$

where K is the standard epistemic knowledge modality, F is a friendship modality which is read as "for all my friends", and A is an auxiliary universal modality which quantifies over all agents.[40] These operators are interpreted in epistemic social network models as follows:

$$\begin{array}{lll} M, w, i \models p & \text{iff} & (w, i) \in V(p), \quad \text{for } p \in Prop \\ M, w, i \models K\varphi & \text{iff} & M, v, i \models \varphi \text{ for every } v \sim_i w \\ M, w, i \models F\varphi & \text{iff} & M, w, j \models \varphi \text{ for every } j \asymp_w i \\ M, w, i \models A\varphi & \text{iff} & M, w, j \models \varphi \text{ for every } j \in I \end{array}$$

This simple language is quite expressive. Combinations of operators such as KFp or $FKFKp$ make complex statements about what members of a community know about their friends' knowledge. To boost power of defining social situations even further, Facebook Logic makes use of a technical device from "Hybrid Logic", the indexical "downarrow pointer" $\downarrow n$

[40] Obvious existential modalities exist, too, such as the dual $< F >$ of the modality F, defined as $\neg F \neg$.

which introduces a name n to refer to the current agent.[41] Using various defined notions in this hybrid setting, Seligman, Girard and Liu et al. can define interesting anaphoric expressions, such as $\downarrow x < F > K@_n < F > x$ which says that "I have a friend who knows that n is friends with me". While all this seems geared toward knowledge, analogous methods also apply to agents' *beliefs* in social networks.

Network dynamics This social epistemic setting supports various kinds of dynamics, such as agents learning facts that change their knowledge or beliefs. The relevant events here can be of the kinds we have discussed in our earlier hard and soft update scenarios for belief.

But the setting of Facebook Logic also includes interesting new dynamic phenomena, such as changes in agents' beliefs under social influence. For instance, a typical 'dynamic network rule' would be that

An agent comes to believe a proposition p iff all her friends believe p.

This induces changes in what is believed in the network, and hence iterations can start. As we saw in Section 7.8 on belief change over time, these iterations can either stabilise in a state where all agents have a acquired a permanent belief or a permanent disbelief in p, that could be viewed as the resulting *community opinion.* But updates can also start oscillating, as often happens in dynamical systems for population behavior, modeling cycles in 'public opinion'.

Of course, other qualitative network rules are possible, as well as more sophisticated quantitative ways of measuring the dynamics of influence.

Finally, going beyond that, there can also be changes in the network itself, like when friendship relations start changing by adding or deleting links.

Network logics All the preceding kinds of updates can be described in the same model transformation style as before. For instance, the new dynamic network rules induce dynamic updates that change the current truth value of propositions through a sort of "predicate re-assignments", or "substitutions", of the form

$$p := \varphi(p)$$

Accordingly, dynamic logics with modalities for local and global network evolution in terms of agents' knowledge and beliefs can be found, resembling

[41] Semantically, this involves an additional 'assignment function' g with clauses (a) $M, w, i \models n$ iff $g(n) = i$, for $n \in ANom$, (b) $M, w, i \models\downarrow n\varphi$ iff $M[\overset{i}{n}], w, a \models \varphi$, where $M[\overset{i}{n}]$ is the result of changing the model M so that n names agent i.

the epistemic-doxastic-temporal types that we have studied before.

However, there are also interesting new technical developments here. Seligman, Girard and Liu define a very general class of dynamic operators and actions on the above epistemic social models, leading to a system of "General Dynamic Dynamic Logic" that generalises our earlier Priority Update while still remaining reducible to PDL. Other relevant topics in this setting includes the work of Zhen and Seligman on peer pressure which investigates the effect of social relations on the logical dynamics of preference change. A few further recent references on dynamic network logics will be found in Section 7.10.

7.9 Further directions

There are many further streams in the literature on belief revision than what we have covered here, both in terms of topics and approaches. In this section, we will give a brief panorama of further directions.

7.9.1 Postulational and constructive approaches

In addition to dynamic-epistemic logics for belief change based on explicit model transformations, there is another tradition, that of Segerberg's Dynamic Doxastic Logic (DDL). It is closer to AGM by leaving the nature of the revision steps abstract, while incorporating intuitive *postulates* on belief revision as axioms in a modal logic. These are both valid styles of approach, and we briefly explore differences as well as ways in which the two complement each other.

Dynamic-doxastic logic has abstract modal operators describing update transitions in abstract universes, relational or functional.[42] The format of interpretation is as follows. Let M be a model, $[\![P]\!]$ the set of worlds in M satisfying P, and $M * [\![P]\!]$ some new updated model. Then we set

$$M, s \models [*P]\varphi \text{ iff } M * [\![P]\!], s \models \varphi$$

The minimal modal logic K is valid in this semantic framework, and further axioms constrain relation changes for special revision policies, leading to special modal logics extending K.

Instead of compiling a detailed list of differences with our dynamic-epistemic logics, we focus on an interesting bridge between the two approaches, proposed by van Benthem around 2007. It works in terms of

[42]Concrete DDL models can be pictured graphically as Lewis spheres for conditional logic or neighborhood models.

standard "frame correspondence" for modal axioms. Given some modal axioms on an abstract update operation, to which extent will these axioms enforce the precise constructive recipes employed in dynamic-epistemic logic?

Correspondence on update universes To make a modal correspondence analysis work, we need a suitably general semantic framework behind the concrete models of earlier sections, in the spirit of DDL. Consider any family $\mathbf{M}$ of pointed epistemic models (M, s), viewed as an "update universe". Possible changes are given as a family of update relations RP $(M, s)(N, t)$ relating pointed models, where the index set P is a subset of M: intuitively, the proposition triggering the update. One can think of the R as recording the action of some update operation $\heartsuit$ occurring in the syntax of our language that depends on the proposition P. Here different operations from earlier sections can have different effects: from hard updates $!\varphi$ to soft updates $\Uparrow\varphi$.

For each formula φ, let $[\![\varphi]\!]$ be the set of worlds in M satisfying φ. We set

$$M, s \models < \heartsuit\varphi > \psi \text{ iff there exists a model } (N, t) \text{ in } \mathbf{M}$$

where $R[\![\varphi]\!](M, s)(N, t)$ and $(N, t) \models \psi$

Now we can analyse given dynamic modal principles precisely, and we will give two examples of such a correspondence analysis.[43]

Correspondence for eliminative update One obvious choice for constraints on update operations are the earlier recursion axioms that gave us the heart of the dynamics of knowledge and belief change. Here is the outcome of a correspondence analysis for public announcement logic PAL.[44] By going through the content of all recursion axioms, with a special role for the knowledge clause, we can see that the PAL axioms essentially enforce world elimination as its interpretation of the update for $!\varphi$. Technically, however, there is some "slack", in terms of a well-known modal structure-preserving map:[45]

---

[43]More precisely, we are interpreting our language in a three-index format $\mathbf{M}, M, s$, and for the accessibility relations R in this update universe $\mathbf{M}$, we have that $(M, s)R(M, t)$ iff Rst in M, without any jumps out of the model M.

[44]In what follows, we must sidestep one particularity of PAL and related dynamic-epistemic logics: its use of non-schematic axioms such as $<!\varphi> q \leftrightarrow (\varphi \wedge q)$ that is not closed under substitutions of arbitrary complex formulas for the atomic proposition q. This can be solved by going to special cases such as $<!\varphi> T \leftrightarrow \varphi$. We refer to the literature for details.

[45]The recursion axiom for negation imposes functionality of the update relations. The left and right directions of the knowledge recursion axiom enforce precisely the two central

Theorem 7.12
An update universe satisfies the substitution-closed principles of PAL iff its transition relations F_P are partial p-morphisms defined on the sets P. ⊣

To force the p-morphisms to become embeddings as submodels, a few further twists are needed, analysing some further features of the treatment of propositional atoms ("context-dependent", or not), as well as a further modal recursion axiom for a global "existential modality". We omit these details here.

Discussion: refinements Update universes are reminiscent of the *protocol* version of PAL considered in Section 7.8, where available transitions can be restricted by the model. Our correspondence perspective also works for modified recursion axioms on protocol models, where update functions on domains may now become partial. Another generalisation of our semantics in Sections 7.4 and 7.5 suggested by the above is a possible context-dependent interpretation of proposition letters, not as sets of worlds, but as sets of pairs (M, w) where M is a model and w a world in M. In that case, the logic will become the substitution-closed schematic core of PAL – and similar observations holds for DEL in general.

Correspondence for belief change The same style of analysis applies to update principles for plausibility orderings. We merely state the outcome here:

Theorem 7.13
The recursion axioms of the dynamic logic of radical upgrade and conditional belief hold for an update operation on a universe of pointed plausibility models iff that operation is in fact radical upgrade. ⊣

Here is what this says. AGM-style postulates on changes in beliefs alone do not fix the relational transformation: we need to constrain the changes in conditional beliefs, since the new plausibility order encodes all of these. But there is an easier road as well, in line with observations in Sections 7.3 and 7.4:

Theorem 7.14
Radical upgrade is the only update operation validating the given recursion axioms for atoms, the Boolean operations, and safe belief. ⊣

One could also do correspondence analysis directly on the AGM postulates, but this would take us too far afield here. Likewise, a correspondence

relation-preserving clauses for a p-morphism.

analysis of update postulates can also be performed for dynamic-epistemic logic and recursion axioms for product update as defined in Section 7.5. One reason why these techniques work is the "Sahlqvist form" of many recursion axioms for belief revision, making them amenable to standard modal techniques.

7.9.2 Belief revision versus nonstandard consequence

Van Benthem pointed out in 2008 that update of beliefs under hard or soft information is an alternative to *nonstandard notions of consequence.* Classical consequence says that all models of premises P are models for the conclusion C. McCarthy in 1980 famously pointed out how problem solving goes beyond this. A "circumscriptive consequence" from P to C says that C is true in all the *minimal* models for P, where minimality refers to a relevant comparison order $\leq$ for models, Circumscription supports non-monotonic consequence relations that show resemblances to our earlier conditional logics of Section 7.3.

This is reminiscent of plausibility models for belief. Starting from initial information we must reach a goal as new information comes in. Nonmonotonic logics leave such events implicit in the background, while dynamic-epistemic logics provide an alternative in terms of the beliefs that problem solvers have as they go through their process of observation and inference.[46]

Thus, there is an interesting duality between explicit treatments of belief change in logic like the ones in this chapter, and approaching the same issues via new consequence relations.[47] The precise interplay between these two views of belief revision is far from being understood.

7.9.3 Belief revision and evidence

Belief revision has been studied on relatively coarse semantic models in this chapter. However, it is quite feasible, as has been done in the AGM tradition and also in dynamic-doxastic logic, to work with more fine-grained models for *evidence.* Staying close to modal logic, one obvious candidate are "neighborhood models" where each world is connected to a family of sets of worlds, its neighborhoods, that may be considered as standing for pieces of evidence that we have concerning the location of the actual world. Such pieces of evidence may be consistent, but they can also contradict each

[46]The latter approach even suggests new notions of circumscription, for instance, versions that take the information in the premises as soft rather than hard.

[47]This triangle of perspectives has also been noted in the classical AGM survey of Gärdenfors & Rott around 1995. Also relevant here is the modal-style analysis of update operations by Ryan & Schobbens around the same time.

other, leading to a much more realistic view of settings in which we need to form our beliefs.

In particular, van Benthem and Pacuit have recently proposed evidence models for belief that interpret notions such as having evidence for φ:

$$M, s \models \odot \varphi \text{ iff there is a set } X \text{ in } N \text{ with } M, t \models \varphi \text{ for all } t \in X$$

as well as cautious beliefs based on what is true in the intersections of maximal families of mutually consistent pieces of evidence.

The resulting static logics are simple, but not cheap, and combine standard modalities with generalised monotone evidence modalities. Such models lend themselves well to a generalised dynamic-epistemic treatment, in terms of actions of adding, removing, modifying, and combining evidence. For instance, the single update operation for PAL events $!\varphi$ will now decompose into several actions: adding evidence for φ, and removing evidence for $\neg\varphi$. We refer to the literature for the resulting recursion axioms. One interesting feature of this setting is that analysing belief contraction, traditionally one of the more difficult AGM operations, becomes quite perspicuous.

Fine-grained models like this have been proposed in formal epistemology recently, since they combine a finer representation of evidence and belief with the continued availability of the semantic techniques of this chapter.

7.9.4 Combining information and preference

In this chapter, we have shown how logical dynamics deals with agents' knowledge and beliefs and informational events that change these. But agency also involves a second major system, not of information but of "evaluation", as recorded in agents' *preferences* between worlds or actions. It is preference that guides our actions, rather than just possession of hard or soft information.

Preference logics can be studied with the modal and dynamic techniques that we have discussed here. In fact, one can reinterpret our plausibility models as models with worlds, or just any kind of objects, carrying a binary "betterness ordering" for each agent. These structures occur in decision theory, where worlds are outcomes of actions, and game theory, where worlds are histories, with preferences for different players.

When added to our earlier logics, a simple modality of betterness can define equilibrium notions for game solution, as well as normative deontic notions in general agency. Moreover, our earlier relation changes now can do duty as preference changes, for instance, in response to a suggestion or a command from some agent with sufficient authority. Finally, preference, knowledge, and belief often occur *entangled*, in notions such as "ceteris paribus preference", obligations based on current information, or qualitative versions of expected utility. The above dynamic-epistemic techniques will still work for such combined systems.

7.9.5 Group belief and merge

Our forays into multi-agent scenarios in Section 7.8 have only scratched the surface of social aspects of belief change. In addition to individual belief changing acts, a social setting suggests many further dynamic operations at the level of groups that can have information and act collectively. One basic notion that has received attention in the post-AGM literature on belief revision is "belief merge" of the individual agents forming a group. Related notions occur in the area of Judgment Aggregation, while the mentioned entanglement with preference also suggests strong analogies with Social Choice Theory (see below).

There are quite a few logical challenges here since the logical structure of collectives is quite intricate, witness the semantics of plurals and collective expressions in natural language, which is by no means a simple extension of the logic of individuals and their properties. Indeed, groups bring in new features, since they are usually much more than a flat set of individual agents. There can be hierarchical structure of informational status (trustworthiness), or of preferential status (authority and power), and this structure can crucially affect how collective behavior arises, and how beliefs of individual members change.

There are promising formal models for this sort of structure that are congenial to the dynamic-epistemic approach in this chapter. Andréka, Ryan & Schobbens noted around 2002 that merge operations typically need a structured view of groups as graphs with a dominance order. The resulting "prioritised relation merge" can be defined as follows. Given an ordered priority graph $G = (G, <)$ of indices for individual relations that may have multiple occurrences in the graph, the *merged group priority relation* is:

$x \leq_G y$ iff for all indices $i \in G$, either $x \leq_i y$, or
there is some $j > i$ in G with $x <_i y$.

This is slightly reminiscent of the 'priority update' in our Section 7.5.3, and this graph setting has turned out to apply in many areas, including the above topics, but also inquiry and learning in the presence of a structured "agenda" of investigation, or preference in the presence of an ordered graph of "criteria" for judging betterness. As has been shown by Girard & Liu around 2008, priority graphs lend themselves well to algebraic and logical treatment, especially, when we focus on two major dynamic operations of merging priority graphs: "parallel composition" with no links between the two graphs, and "sequential composition" where all nodes in one graph dominate all nodes in the other.

Ideas from this circle also apply to our earlier treatment of belief change. In 2009, van Benthem analysed belief change as a process of social choice, merging signals from different sources (past experience, current observation, etc.), and made a connection with some natural intuitions from Social Choice Theory:

Theorem 7.15
A preference aggregation function is a Priority Update iff it satisfies Permutation Invariance, Locality, Abstentions, Closed Agenda, and Overruling. ⊣

We do not provide details here, but the point of results like this is that they capture policies for belief revision or belief merge in terms of postulates from the economics and social science literature. Many further relevant connections, relating AGM postulates to choice principles in the foundations of decision theory and economics, were found earlier on in the work of Rott.

7.9.6 Belief revision and social choice

Dynamic epistemic logic fits well with social choice theory, where group actors form collective preferences from the preferences of individual members. Merging social choice with the main concerns of this chapter can provide two things: informational structure, and more finely-grained procedure. For the first, think of Arrow's Theorem, and the horror of a dictator whose opinions are the social outcome. Even if it is common "de dicto" knowledge that there is a dictator, this does no harm if there is no person whom we know "de re" to be the dictator. Not even the dictator herself may know she is one. To see the real issues of democracy, we need social choice plus epistemic logic.

Also, rules for voting represent just a fragment of richer practices of communication and debate. One can study how groups arrive at choices by deliberating, and ways in which agents then experience preference changes. This is reminiscent of two dynamic processes that we considered in Section 7.8: deliberation, and belief adaptation in social groups. Again, the preceding combined approach seems called for here, especially when we think of groups that also have means of communication, giving them designated information channels. Two concrete sources of relevant scenarios for this broader study might be Argumentation Theory, with its studies of rules for fair debate, and the Law.

7.10 Notes

In this section we provide some major references for the material presented in this chapter, offering also some pointers to the literature for further study.

Section 7.1: Basics of belief revision The AGM theory of belief revision goes back to a classic paper by Alchourrón, Gärdenfors, and Makinson (1985). This approach has given rise to an extended series of papers and books on the topic, of which we mention (Gärdenfors, 1988), and (Gärdenfors and Rott, 1995). The AGM postulates for revision, contraction and expansion have been the subject of much philosophical discussion. One basic example is the Recovery Postulate, which, as motivated by the principle of minimal change, prescribes that a contraction should remove as little as possible from a given theory T. Extended discussions of this principle can be found in (Hansson, 1991; Fuhrmann, 1991; Levi, 1991; Niederée, 1991; Hansson, 1997). A concise up to date overview of standard belief revision theory is found in the chapter on Theory Replacement in (Kyburg and Teng, 2001), while Rott (2001) provides an in depth study of standard belief revision theory in the context of nonmonotonic reasoning.

Belief revision theory has important links with the logic of conditionals, via the "Ramsey Test". The original source is a short note by Ramsey (1990), while basic modern results are provided by Gärdenfors (1986, 1988). The mentioned semantic modal perspective on Gärdenfors' Impossibility Result comes from Baltag and Smets (2010). A different concept of belief update as involving world change, mentioned in one of the footnotes, is studied in (Katsuno and Mendelzon, 1992).

Section 7.2: Modal logics of belief revision The original modal logic approach to static notions of knowledge and belief is ascribed to Hintikka (1962). Major sources for the Propositional Dynamic Logic of actions are (Harel, 1984; Harel, Kozen, and Tiuryn, 2000), while Segerberg (1995, 1998, 1991) provides classical references for Dynamic Doxastic Logic. In the context of PDL, we refer to van Benthem (1989) for dynamic operators that mirror the AGM operations, and to (Rijke, de, 1994) for extensions of this approach.

Early quantitative systems for belief revision in the style of Dynamic Epistemic Logic were proposed by Aucher (2003) and by van Ditmarsch and Labuschagne (2003). We also mentioned the "ranking models" of Spohn (1988). The BMS notion of "product update" refers to work by Baltag, Moss, and Solecki (1998). Further details and references on Public Announcement Logic (PAL), can be found in chapter 6 on DEL in this volume as well as in work by van Ditmarsch, van der Hoek, and Kooi (2007), and by van Benthem (2011). Landmark papers on PAL and its extensions include (Plaza, 1989; Gerbrandy and Groeneveld, 1997), and the dissertation (Gerbrandy, 1998). An early PAL-style approach to AGM belief expansion is found in van (van Ditmarsch, van der Hoek, and Kooi, 2005). The much more general approaches on which this chapter is based are by van Benthem (2007a) and Baltag and Smets (2008).

Section 7.3: Static base logics Basic modal logics for belief such as "KD45" and many others can be found in the textbook by Meyer and van der Hoek (1995).

The material on connected plausibility models is based on work by Baltag and Smets (2008), while Baltag and Smets (2006b) developed the correspondence between plausibility frames and epistemic plausibility frames. Plausibility models allowing non-connected orderings are used extensively by van Benthem (2007a), and by van Eijck and Sietsma (2008). World-based plausibility models with ternary relations are discussed by van Benthem (2007a), by Baltag and Smets (2006b), and by Board (2002). However, no doubt the classic predecessor to all of this is the first semantic modeling for AGM theory given by Grove (1988). For similar models in conditional logic for AI an philosophy, see Shoham (1988) and Spohn (1988). For background in classical conditional logic, cf. (Lewis, 1973) for completeness theorems on connected orderings, and contributions by Burgess (1981), and by Veltman (1985) on preorders. An elegant setting based on partially ordered graphs is found in (Andreka, Ryan, and Schobbens, 2002).

Safe belief on plausibility models has been discovered independently in many areas, as diverse as AI (Boutilier (1994), Halpern (1997)), multi-agent systems (Shoham and Leyton-Brown (2008)), and philosophy (Stalnaker (2006)). Safe belief is related to defeasible knowledge in formal epistemology (Baltag and Smets, 2006b, 2008), and to modal preference logics (Liu, 2008; Girard, 2008) – with some related work in (Lenzen, 1980). Our definition of strong belief refers to that of Baltag and Smets (2008, 2013), but predecessors are found in work by Battigalli and Siniscalchi (2002) and Stalnaker (1998) in economic game theory.

Section 7.4: Belief revision by model transformations This section is based on proposals by van Benthem (2007a) and by Baltag and Smets (2008) on complete dynamic-epistemic logics with recursion axioms for belief revision under hard and soft announcements. Earlier sources for soft announcements are the default logic of Veltman (1996), and the minimal conservative revision of Boutillier (1993). Different rules for plausibility change in models of belief revision can be found in (Grove, 1988) and Rott

(Rott, 2006), while radical and conservative upgrade in dynamic-epistemic style are highlighted by van Benthem (2007a), and by Baltag and Smets (2008).

Section 7.5: General formats for belief revision The PDL-format for relation transformers is due to van Benthem and Liu (2007). Much more complex mechanisms in this line have been studied by van Eijck and his collaborators (van Eijck and Wang (2008); van Eijck and Sietsma (2008); van Eijck (2008)), extending the methods of van Benthem, van Eijck, and Kooi (2006) to deal with a wide range of belief changes defined by PDL programs, and by Girard, Seligman, and Liu (2012), who merge the idea of PDL transformers with product models for DEL to describe information flow in complex social settings.

The classical source for the general DEL framework with product update and recursion axioms is (Baltag et al., 1998), while more examples and extensions can be found in (Baltag and Moss, 2004; Baltag and Smets, 2008, 2006b,a,c, 2007; van Benthem, 2007a). Larger treatises on the subject are offered by van Ditmarsch, van der Hoek, and Kooi (2007) and by van Benthem (2011). Recursion axioms for "relativised common knowledge" are found in (van Benthem et al., 2006). See also Chapter 6 in this volume on DEL for more material.

Our discussion of event models as triggers of soft information change is based on the Priority Update of Baltag and Smets (2006b) (with a precursor in (Aucher, 2003), and more broadly, in (van Benthem, 2002)). The link to "Jeffrey Update" refers to Jeffrey (1983). Note that the coin examples in this section are due to Baltag and Smets (2008).

Section 7.6: Belief revision and probability dynamics The probabilistic dynamic epistemic logic of this section, including probabilistic event models and a probabilistic product update mechanism, was developed by van Benthem, Gerbrandy, and Kooi (2006). A different approach by Baltag and Smets (2007) uses probabilistic models in line with the Popper-Rényi theory of conditional probabilities. A masterful overview of static logics to reason about uncertainty is by Halpern (2003). New developments include a logic for reasoning about multi-agent epistemic probability models, by van Eijck and Schwarzentruber (2014).

Section 7.7: Time, iterated update, and learning Hodkinson and Reynolds (2006) and van Benthem and Pacuit (2006) provide surveys of the basics of epistemic temporal logics (see also Chapter5 in this volume). Agents properties like perfect recall are discussed extensively by Halpern and Vardi (1989), including a finite automata perspective that is also discussed in a dynamic-epistemic setting by Liu (2008).

For epistemic temporal logics and complexity, the classic source is provided by Halpern and Vardi (1989). Connections to further work in computational logic are given by Parikh and Ramanujam (2003), and foundations of games are surveyed by van Benthem and Pacuit (2006) in a dynamic-epistemic format. Miller and Moss (2005) prove the high complexity of PAL with a PDL-style iteration operator.

Protocol models for DEL and their connections with epistemic-temporal logics are found in van (van Benthem and Liu, 2004), and in a more general setting in (van Benthem, Hoshi, Gerbrandy, and Pacuit, 2009), including a formalisation of state dependent protocols. Hoshi (2009) axiomatises the laws of special protocol logics and makes connections with procedural information in epistemology.

The extensions to belief refer to work by van Benthem and Dégrémont (2008) and by Dégrémont (2010). For related work by Bonanno and others, see Chapter 5 of this volume.

Our treatment of iterated belief upgrades and limit behavior and its main stabilisation results is based on work by Baltag and Smets (2009a,b), with van Benthem (2002) as an early predecessor. More information on bisimilarity between pointed Kripke models can be found in the textbook by Blackburn, de Rijke, and Venema (2001).

For formal learning theory, we refer to work by Kelly (1998a,b), by Kelly, Schulte, and Hendricks (1995) and by Hendricks (2003, 2001). Various links with dynamic-epistemic logic have been explored by Gierasimczuk (2010), by Baltag and Smets (2011), by Dégrémont and Gierasimczuk (2011, 2009), and by Gierasimczuk and de Jongh (2013). The major results on the learning power of different belief revision methods were taken from (Baltag, Gierasimczuk, and Smets, 2011). A brief overview of the use of temporal-epistemic logic in a learning theoretic context has been sketched in (Gierasimczuk, Hendricks, and de Jongh, 2014).

Finally, various connections with propositional dynamic logics can be found in the "knowledge programs" of Fagin, Halpern, Moses, and Vardi (1995), in the "epistemic-temporal PDL" of van Benthem and Pacuit (2007), and in the use of PDL-definable protocols by Wang, Sietsma, and van Eijck (2011).

Section 7.8: Belief, games, and social agency Logical tools and methods for solution concepts in game theory have been studied widely, see the surveys by van der Hoek and Pauly (2006), by Bonanno (1991, 1992), and by Battigalli, Brandenburger, Friedenberg, and Siniscalchi (2014). An explanation of the epistemic significance of the Muddy Children Puzzle is found in (Fagin et al., 1995).

The main sources for our treatment of Backward Induction are from van Benthem (2007b, 2011), while (van Benthem, 2014a) is an extensive exploration of the realm of logic in games. For the specific results on limits of iterated radical upgrades, see also work by Baltag, Smets, and Zvesper (2009), and by van Benthem and Gheerbrant (2010).

The logical analysis of informational cascades work presented here is from Baltag, Christoff, Ulrik Hansen, and Smets (2013). A new probabilistic logic of communication and change in cascades is proposed in Achimescu, Baltag, and Sack (2014). Facebook Logic and its recent developments can be found in a series of papers by Seligman, Liu, and Girard (2011); Liu, Seligman, and Girard (2014); Seligman, Liu, and Girard (2013a,b). Sano and Tojo (2013) apply ideas from the facebook logic setting in the context of DEL, while other merges are found in (Christoff and Ulrik Hansen, 2013). A study of the social effects on preference dynamics is presented by Liang and Seligman (2011).

Section 7.9: Further directions Classic sources for Dynamic Doxastic Logic, as mentioned earlier, are provided by Segerberg (1995, 1991, 1999). See also work by Segerberg and Leitgeb (2007) for an extensive discussion of the DDL research program. The work reported on connections with DEL via frame correspondence is from van Benthem (2014b). Another systematic approach relating modal logics and belief revision is by Ryan and Schobbens (1997).

For circumscriptive consequence, two main sources are by McCarthy (1980) and by Shoham (1988). Belief revision versus nonstandard consequence as an approach to belief is discussed by van Benthem (2011), but the interface of AGM Theory and nonmonotonic logic was already discussed extensively by Gärdenfors and Rott (1995), and by Rott (2001).

Neighborhood models for modal logic go back to Segerberg (1971). Evidence dynamics on neighborhood models was developed by van Benthem and Pacuit (2011). Further developments including links with justification logic are given by van Benthem, Fernández-Duque, and Pacuit (2014), while applications to formal epistemology are found in (Fiutek, 2013). For neighborhood models in Dynamic Doxastic Logic, see (Segerberg, 1995; Girard, 2008).

Preference change in dynamic-epistemic style has been studied in by van Benthem and Liu (2004), while the monograph of Liu (2011) is an extensive study with references to other streams in preference logic and deontic logic. For richer constraint-based views of preference in terms of ordered criteria, see also (Rott, 2001). An elegant technical framework are the priority graphs of Andreka et al. (2002). For entangled dynamic-epistemic systems that combine preference, knowledge, and belief see (Girard, 2008; Liu, 2011).

Belief merge has been studied using AGM techniques by Maynard-Reid and Shoham (1998), while List and Pettit (2004) on "judgment aggregation" is relevant too. In a DEL setting, a relevant system is the E-PDL of van van Benthem et al. (2006). The analysis of Priority Update as a form of social choice is from van Benthem (2007c).

Acknowledgements The authors thank Thomas Bolander, Barteld Kooi, Wiebe van der Hoek and Shane Steinert-Threlkeld for valuable feedback during the writing of this chapter. The contribution of Sonja Smets was funded in part by the VIDI grant 639.072.904 of the Netherlands Organisation for Scientific Research and by the European Research Council under the 7th Framework Programme FP7/2007-2013/ERC Grant no. 283963.

References

Achimescu, A., A. Baltag, and J. Sack (2014). The probabilistic logic of communication and change. Presented at LOFT 2014, Bergen.

Alchourrón, C. E., P. Gärdenfors, and D. Makinson (1985). On the logic of theory change: partial meet contraction and revision functions. *Journal of Symbolic Logic 50*, 510–530.

Andreka, H., M. Ryan, and P. Y. Schobbens (2002). Operators and laws for combining preference relations. *Journal of Logic and Computation 12*, 13–53.

Aucher, G. (2003). A combined system for update logic and belief revision. Master's thesis, Institute for Logic, Language and Computation, Universiteit van Amsterdam.

Baltag, A., Z. Christoff, J. Ulrik Hansen, and S. Smets (2013). Logical models of informational cascades. *Studies in Logic Volume, College Publications. 47*, 405–432.

Baltag, A., N. Gierasimczuk, and S. Smets (2011). Belief revision as a truth-tracking process. In *Proceedings of the 11th conference on Theoretical aspects of rationality and knowledge*, pp. 187–190. ACM.

Baltag, A. and L. S. Moss (2004). Logics for epistemic programs. *Synthese 139*(2), 165–224.

Baltag, A., L. S. Moss, and S. Solecki (1998). The logic of public announcements, common knowledge, and private suspicions. In *Proceedings of the 7th Conference on Theoretical Aspects of Rationality and Knowledge*, TARK '98, San Francisco, CA, USA, pp. 43–56. Morgan Kaufmann Publishers Inc.

Baltag, A. and S. Smets (2006a). Conditional doxastic models: a qualitative approach to dynamic belief revision. *Electronic Notes in Theoretical Computer Science 165*, 5–21.

Baltag, A. and S. Smets (2006b). Dynamic belief revision over multi-agent plausibility models. In W. van der Hoek and M. Wooldridge (Eds.), *Proceedings of LOFT'06*, pp. 11–24. Liverpool: University of Liverpool.

Baltag, A. and S. Smets (2006c). The logic of conditional doxastic actions: a theory of dynamic multi-agent belief revision.

Baltag, A. and S. Smets (2007). Probabilistic dynamic belief revision. In J. van Benthem, S. Ju, and F. Veltman (Eds.), *Proceedings of LORI'07*. College Publications, London.

Baltag, A. and S. Smets (2008). A qualitative theory of dynamic interactive belief revision. In G. Bonanno, W. van der Hoek, and M. Wooldridge (Eds.), *Logic and the Foundations of Game and Decision Theory (LOFT 7)*, Texts in Logic and Games, pp. 9–58. Amsterdam University Press.

Baltag, A. and S. Smets (2009a). Group belief dynamics under iterated revision: Fixed-points and cycles of joint upgrades. In *Proceedings of the 12th Conference on Theoretical Aspects of Rationality and Knowledge (TARK'09)*, pp. 41–50. ACM.

Baltag, A. and S. Smets (2009b). Learning by questions and answers: From belief- revision cycles to doxastic fixed points. *LNAI Lecture Notes in Computer Science 5514*, 124–139.

Baltag, A. and S. Smets (2010). A semantic view on ramsey's test. In X. Arrazola and M. Ponte (Eds.), *LogKCA-10, Proceedings of the Second ILCLI International Workshop on Logic and Philosophy of Knowledge, Communication and Action*, pp. 119–134. Univ. of the Basque Country Press.

Baltag, A. and S. Smets (2011). Keep changing your beliefs and aiming for the truth. *Erkenntnis 75(2)*, 255–270.

Baltag, A. and S. Smets (2013). Protocols for belief merge: Reaching agreement via communication. *Logic Journal of IGPL 21*(3), 468–487.

Baltag, A., S. Smets, and J. Zvesper (2009). Keep 'hoping' for rationality: a solution to the backward induction paradox. *Synthese 169*, 301–333.

Battigalli, P., A. Brandenburger, A. Friedenberg, and M. Siniscalchi (2014). *Epistemic Game Theory: Reasoning about Strategic Uncertainty.* Manuscript under construction.

Battigalli, P. and M. Siniscalchi (2002). Strong belief and forward induction reasoning. *Journal of Econonomic Theory 105*, 356–391.

van Benthem, J. (1989). Semantic parallels in natural language and computation. In H. D. Ebbinghaus (Ed.), *Logic Colloquium Granada 1987*, pp. 331–375. Amsterdam North Holland.

van Benthem, J. (2002). One is a lonely number: on the logic of communication. Technical report, ILLC, University of Amsterdam. Report PP-2002-27 (material presented at the Logic Colloquium 2002).

van Benthem, J. (2007a). Dynamic logic for belief revision. *Journal of Applied Non-classical Logics 17*(2), 129–155.

van Benthem, J. (2007b). Rational dynamics and epistemic logic in games. *International Game Theory Review 9*, 13–45.

van Benthem, J. (2007c). The social choice behind belief revision (lecture presentation). In *Workshop Dynamic Logic Montreal.*

van Benthem, J. (2011). *Logical Dynamics of Information and Interaction.* Cambridge University Press.

van Benthem, J. (2014a). *Logic in Games.* MIT Press.

van Benthem, J. (2014b). Two logical faces of belief revision. In R. Trypuz (Ed.), *Outstanding Contributions to logic: Krister Segerberg on Logic of Actions*, pp. 281–300. Springer.

van Benthem, J. and C. Dégrémont (2008). Multi-agent belief dynamics: bridges between dynamic doxastic and doxastic temporal logics. In G. Bonanno, B. Löwe, and W. van der Hoek (Eds.), *Logic and the Foundations of Game and Decision Theory (LOFT 8)*, Number 6006 in LNAI, pp. 151 – 173. Springer.

van Benthem, J., D. Fernández-Duque, and E. Pacuit (2014). Evidence and plausibility in neighborhood structures. *Annals of Pure and Applied Logic 165*, 106–133.

van Benthem, J., J. Gerbrandy, and B. Kooi (2006). Dynamic update with probabilities. In *Proceedings LOFT'06, Liverpool.* University of Liverpool.

van Benthem, J. and A. Gheerbrant (2010). Game solution, epistemic dynamics and fixed-point logics. *Fundamenta Informaticae 1*(4), 19–41.

van Benthem, J., T. Hoshi, J. Gerbrandy, and E. Pacuit (2009). Merging frameworks for interaction. *Journal of Philosophical logic 38*(5), 491–526.

van Benthem, J. and F. Liu (2004). Dynamic logic of preference upgrade. Technical report, University of Amsterdam. ILLC Research Report PP-2005-29.

van Benthem, J. and F. Liu (2007). Dynamic logic of preference upgrade. *Journal of Applied Non-classical Logics 17*(2), 157–182.

van Benthem, J., J. van Eijck, and B. Kooi (2006). Logics of communication and change. *Information and Computation 204*(11), 1620–1662.

van Benthem, J. and E. Pacuit (2006). The tree of knowledge in action: Towards a common perspective. In *Advances in Modal Logic (AiML).*

van Benthem, J. and E. Pacuit (2007). Modelling protocols in temporal models. ILLC, University of Amsterdam.

van Benthem, J. and E. Pacuit (2011). Dynamic logics of evidence-based beliefs. *Studia Logica 99(1)*, 61–92.

Blackburn, P., M. de Rijke, and Y. Venema (2001). *Modal Logic.* Cambridge University Press.

Board, O. (2002). Dynamic interactive epistemology. *Games and Economic Behaviour 49*, 49–80.

Bonanno, G. (1991). The logic of rational play in games of perfect information. *Economics and Philosophy 7(1)*, 37–65.

Bonanno, G. (1992). Rational beliefs in extensive games. In *Theory and Decision vol. 33.* Kluwer Academic Publishers.

Boutilier, C. (1994). Conditional logics of normality: A modal approach. *Artificial Intelligence 68*, 87–154.

Boutillier, C. (1993). Revision sequences and nested conditionals. In *IJCAI'93: Proceedings of the 13th International Joint conference on AI*, pp. 519–525. Chambery, France.

Burgess, J. (1981). Quick completeness proofs for some logics of conditionals. *Notre Dame Journal of Formal Logic 22:1*, 1–84.

Christoff, Z. and J. Ulrik Hansen (2013). *LORI Proceedings in Lecture Notes in Computer Science*, Chapter A Two-Tiered Formalization of Social Influence, pp. 68–81. Springer.

Dégrémont, C. (2010). *The Temporal Mind: Observations on Belief Change in Temporal Systems, Dissertation.* ILLC University of Amsterdam.

Dégrémont, C. and N. Gierasimczuk (2009). Can doxastic agents learn? on the temporal structure of learning. In X. He, J. Horty, and E. Pacuit (Eds.), *Logic, Rationality, and Interaction, Second International Workshop, LORI 2009, Chongqing, China, October 8–11, 2009, Proceedings*, Volume 5834 of *LNCS*. Springer.

Dégrémont, C. and N. Gierasimczuk (2011). Finite identification from the viewpoint of epistemic update. *Information and Computation 209*(3), 383–396.

van Ditmarsch, H. and W. A. Labuschagne (2003). A multimodal language for revising defeasible beliefs. In E. Alvarez, R. Bosch, and L. Villamil (Eds.), *Proceedings of the 12th International Congress of Logic, Methodology, and Philosophy of Science*, pp. 140–141. Oviedo university Press.

van Ditmarsch, H., W. van der Hoek, and B. Kooi (2007). *Dynamic Epistemic Logic*. Synthese Library. Springer.

van Ditmarsch, H., W. van der Hoek, and B. Kooi (2005). Public announcements and belief expansion. *Advances in Modal Logic 5*, 335–346.

van Eijck, J. (2008). Yet more modal logics of preference change and belief revision. In K. R. Apt and R. van Rooij (Eds.), *New Perspectives on Games and Interaction*. Amsterdam University Press.

van Eijck, J. and F. Schwarzentruber (2014). Epistemic probability logic simplified. In R. Goré, B. Kooi, and A. Kurucz (Eds.), *Advances in Modal Logic*, pp. 158–177. College Publications.

van Eijck, J. and F. Sietsma (2008). Multi-agent belief revision with linked plausibilities. In G. Bonanno, B. Löwe, and W. van der Hoek (Eds.), *Logic and the Foundations of Game and Decision Theory (LOFT 8)*, Number 6006 in LNAI, pp. 174–189. Springer.

van Eijck, J. and Y. Wang (2008). Propositional dynamic logic as a logic of belief revision. In W. Hodges and R. de Queiroz (Eds.), *Logic, Language, Information and Computation*, Volume 5110 of *LNAI*, pp. 136–148. Springer.

Fagin, R., J. Y. Halpern, Y. Moses, and M. Y. Vardi (1995). *Reasoning about knowledge*, Volume 4. MIT press Cambridge, MA.

Fiutek, V. (2013). *Playing with Knowledge and Belief, PhD Dissertation.* ILLC, University of Amsterdam.

Fuhrmann, A. (1991). Theory contraction through base contraction. *Journal of Philosophical Logic 20*, 175–203.

Gärdenfors, P. (1986). Belief revisions and the ramsey test for conditionals. *Philosophical Review 95*, 813–9.

Gärdenfors, P. (1988). *Knowledge in Flux: Modelling the Dynamics of Epistemic States.* MIT Press, Cambridge MA.

Gärdenfors, P. and H. Rott (1995). Belief revision. In D. M. Gabbay, C. J. Hogger, and J. A. Robinson (Eds.), *Handbook of Logic in Artificial Intelligence and Logic Programming*, pp. 35–132. Oxford University Press.

Gerbrandy, J. D. (1998). *Bisimulations on Planet Kripke.* Ph. D. thesis, University of Amsterdam. ILLC Dissertation Series DS-1999-01.

Gerbrandy, J. D. and W. Groeneveld (1997). Reasoning about information change. *Journal of Logic, Language, and Information 6*, 147–169.

Gierasimczuk, N. (2010). *PhD Thesis: Knowing one's limits. logical analysis of inductive inference. Knowing one's limits. logical analysis of inductive inference.* Universiteit van Amsterdam.

Gierasimczuk, N. and D. de Jongh (2013). On the complexity of conclusive update. *The Computer Journal 56*(3), 365–377.

Gierasimczuk, N., V. Hendricks, and D. de Jongh (2014). Logic and learning. In A. Baltag and S. Smets (Eds.), *Johan F. A. K. van Benthem on Logical and Informational Dynamics.* Springer.

Girard, P. (2008). *PhD Thesis: Modal Logic for Belief and Preference Change.* Stanford University and ILLC Amsterdam.

Girard, P., J. Seligman, and F. Liu (2012). General dynamic dynamic logic. In Th. Bolander, T. Braüner, S. Ghilardi, and L. S. Moss (Eds.), *Advances in Modal Logics*, pp. 239–260. College Publications.

Grove, A. J. (1988). Two modellings for theory change. *Journal of Philosophical Logic 17*, 157–170.

Halpern, J. and M. Vardi (1989). The complexity of reasoning about knowledge and time. i. lower bounds. *Journal of Computer and System Sciences 38*(1), 195 – 237.

Halpern, J. Y. (1997). Defining relative likelihood in partially-ordered preferential structure. *Journal of Artificial Intelligence Research 7*, 1–24.

Halpern, J. Y. (2003). *Reasoning about uncertainty.* MIT Press.

Hansson, S. O. (1991). Belief contraction without recovery. *Studia Logica 50*, 251–260.

Hansson, S. O. (1997). On having bad contractions, or: no room for recovery. *Journal of Applied Non-Classical Logics 7:1-2*, 241–266.

Harel, D. (1984). Dynamic logic. In D. Gabbay and F. Guenthner (Eds.), *Handbook of Philosophical Logic*, Volume II, Dordrecht, pp. 497–604. Kluwer Academic Publishers.

Harel, D., D. Kozen, and J. Tiuryn (2000). *Dynamic Logic (Foundations of Computing).* MIT Press.

Hendricks, V. F. (2001). *The Convergence of Scientific Knowledge: A View from The Limit.* Kluwer Academic Publishers.

Hendricks, V. F. (2003). Active agents. *Journal of Logic, Language and Information 12(4)*, 469–495.

Hintikka, J. (1962). *Knowledge and Belief. An Introduction to the Logic of the Two Notions.* Ithaca, NY: Cornell University Press.

Hodkinson, I. and M. Reynolds (2006). Temporal logic. In P. Blackburn, J. van Benthem, and F. Wolter (Eds.), *Handbook of Modal Logic*, pp. 655–720. Amsterdam Elsevier.

van der Hoek, W. and M. Pauly (2006). Modal logic for games and information. *Handbook of modal logic 3*, 1077–1148.

Hoshi, T. (2009). *Epistemic Dynamics and Protocol Information.* Ph. D. thesis, Stanford University.

Jeffrey, R. (1983). *The Logic of Decision (2nd ed.).* Chicago, IL: University of Chicago Press.

Katsuno, H. and A. Mendelzon (1992). On the difference between updating a knowledge base and revising it. In *Cambridge Tracts in Theoretical Computer Science*, pp. 183–203.

Kelly, K., V. Schulte, and V. Hendricks (1995). Reliable belief revision. In *Proceedings of the 10th International Congress of Logic, Methodology, and Philosophy of Science*, pp. 383–398. Kluwer Academic Publishers.

Kelly, K. T. (1998a). Iterated belief revision, reliability, and inductive amnesia. *Erkenntnis 50*(1), 11–58.

Kelly, K. T. (1998b). The learning power of belief revision. In *TARK'98: Proceedings of the 7th conference on Theoretical aspects of rationality and knowledge*, pp. 111–124. Morgan Kaufmann Publishers Inc.

Kyburg, H. E. and C. M. Teng (2001). *Uncertain Inference.* Cambrdige University Press.

Lenzen, W. (1980). *Glauben, Wissen, und Wahrscheinlichkeit.* Springer Verlag, Library of Exact Philosophy.

Levi, I. (1991). *The Fixation of Belief and Its Undoing.* Cambridge University Press.

Lewis, D. K. (1973). *Counterfactuals.* Oxford: Blackwell Publishing.

Liang, Z. and J. Seligman (2011). A logical model of the dynamics of peer pressure. *Electronic Notes in Theoretical Computer Science 278*, 275–288.

List, C. and Ph. Pettit (2004). Aggregating sets of judgements. two impossibility results compared. *Synthese 140*, 207–235.

Liu, F. (2008). *PhD Thesis: Changing for the better: Preference Dynamics and Preference Diversity.* University of Amsterdam.

Liu, F. (2011). *Reasoning about Preference Dynamics.* Synthese Library vol. 354, Springer.

Liu, F., J. Seligman, and P. Girard (2014). Logical dynamics of belief change in the community. *Synthese 191*(11), 2403–2431.

Maynard-Reid, II, P. and Y. Shoham (1998). From belief revision to belief fusion. In *Proceedings of LOFT98, Torino.*

McCarthy, J. (1980). Circumscription - a form of non-monotonic reasoning. *Artificial Intelligence 13*, 27–39.

Meyer, J.-J.Ch. and W. van der Hoek (1995). *Epistemic Logic for AI and Computer Science*, Volume 41 of *Cambridge Tracts in Theoretical Computer Science.* Cambridge University Press.

Miller, J. S. and L. S. Moss (2005). The undecidability of iterated modal relativization. *Studia Logica 79*(3), 373–407.

Niederée, R. (1991). Multiple contraction: A further case against gaerdenfors' principle of recovery. In A. Fuhrmann and M. Morreau (Eds.), *The Logic of Theory Change*, pp. 322–334. Springer-Verlag.

Parikh, R. and R. Ramanujam (2003). A knowledge based semantics of messages. *Journal of Logic, Language, and Information 12*, 453–467.

Plaza, J. (1989). Logics of public communications. In M. L. Emrich, M. S. Pfeifer, M. Hadzikadic, and Z. W. Ras (Eds.), *Proceedings of the Fourth International Symposium on Methodologies for Intelligent Systems: Poster Session Program*, Oak Ridge, TN, pp. 201–216. Oak Ridge National Laboratory. Reprinted in: Synthese 158, 165–179 (2007).

Ramsey, F. (1929 (1990)). General propositions and causality. In H. A. Mellor (Ed.), *F. Ramsey, Philosophical Papers*. Cambridge University Press.

Rijke, de, M. (1994). Meeting some neighbours. In J. van Eijck and A. Visser (Eds.), *Logic and information flow*, Cambridge, Massachusetts, pp. 170–195. The MIT Press.

Rott, H. (2001). *Change, Choice and Inference*. Oxford University Press.

Rott, H. (2006). Shifting priorities: Simple representations for 27 iterated theory change operators. In H. Lagerlund, S. Lindström, and R. Sliwinski (Eds.), *Modality Matters: Twenty-Five Essays in Honour of Krister Segerberg*. Uppsala Philosophical Studies.

Ryan, M. and P. Y. Schobbens (1997). Counterfactuals and updates as inverse modalities. *Journal of Logic, Language and Information 6*, 123–146.

Sano, K. and S. Tojo (2013). Dynamic epistemic logic for channel-based agent communication. In K. Lodaya (Ed.), *LNAI Proceedings of ICLA 2013*, pp. 109–120. Springer-Verlag Berlin Heidelberg.

Segerberg, K. (1971). *An Essay in Classical Modal Logic*. Uppsala,Filosofiska Föreningen Och Filosofiska Institutionen Vid Uppsala Universitet.

Segerberg, K. (1991). The basic dynamic doxastic logic of AGM. In *The Goldblatt Variations*, Volume 1, pp. 76–107. Uppsala Prints and Preprints in Philosophy.

Segerberg, K. (1995). Belief revision from the point of view of doxastic logic. *Bulletin of the IGPL 3*, 535–553.

Segerberg, K. (1998). Irrevocable belief revision in Dynamic Doxastic Logic. *Notre Dame Journal of Formal Logic 39*(3), 287–306.

Segerberg, K. (1999). Two traditions in the logic of belief: bringing them together. In H. J. Ohlbach and U. Reyle (Eds.), *Logic, Language, and Reasoning*, Dordrecht, pp. 135–147. Kluwer Academic Publishers.

Segerberg, K. and H. Leitgeb (2007). Dynamic doxastic logic - why, how and where to? *Synthese 155*(2), 167–190.

Seligman, J., F. Liu, and P. Girard (2011). Logic in the community. *Lecture Notes in Computer Science 6521*, 178–188.

Seligman, J., F. Liu, and P. Girard (2013a). Facebook and the epistemic logic of friendship. In *Proceedings of the 14th TARK Conference*, pp. 229–238.

Seligman, J., F. Liu, and P. Girard (2013b). Knowledge, friendship and social announcement. In *Logic Across the University: Foundations and Applications, Proceedings of the Tsinghua Logic Conference*. College Publications.

Shoham, Y. (1988). *Reasoning About Change: Time and Change from the standpoint of Artificial Intelligence.* MIT.

Shoham, Y. and K. Leyton-Brown (2008). *Multiagent Systems: Algorithmic, Game Theoretical and Logical Foundations.* Cambridge University Press.

Spohn, W. (1988). Ordinal conditional functions: a dynamic theory of epistemic states. In W. L. Harper and B. Skyrms (Eds.), *Causation in Decision, Belief Change, and Statistics*, Volume II, pp. 105–134.

Stalnaker, R. (1998). Belief revision in games: forward and backward induction. *Mathematical Social Sciences 36*, 31–56.

Stalnaker, R. (2006). On logics of knowledge and belief. *Philosophical Studies 128*, 169–199.

Veltman, F. (1985). *Logics for Conditionals (dissertation).* Philosophical Institute, University of Amsterdam.

Veltman, F. (1996). Defaults in update semantics. *Journal of Philosophical Logic 25*, 221–226.

Wang, Y., F. Sietsma, and J. van Eijck (2011). Logic of information flow on communication channels. In A. Omicini, S. Sardina, and W. Vasconcelos (Eds.), *Declarative Agents Languages and Technologies VIII*, pp. 130–147. Springer.

Part III

Applications

Chapter 8

Model Checking Temporal Epistemic Logic

Alessio Lomuscio and Wojciech Penczek

Contents

Abstract We survey some of our work on model checking systems against temporal-epistemic specifications, i.e., systems specified in Temporal Epistemic Logic (TEL). We discuss both OBDD-based and SAT-based approaches for verifying systems against TEL specifications. The presentation is grounded on various notions of interpreted systems, including one for modelling real-time systems. We also present a partial-order reduction approach to reduce the models and discuss several alternative and advanced techniques.

Chapter 8 of the *Handbook of Epistemic Logic*, H. van Ditmarsch, J.Y. Halpern, W. van der Hoek and B. Kooi (eds), College Publications, 2015, pp. 397–441.

8.1 Introduction

The study of epistemic logics, or logics for the representation of knowledge, has a long and successful tradition in Logic, Computer Science, Economics, and Philosophy. Its main motivational thrust is the observation that the knowledge of the actors (or *agents*) in a system is fundamental not only to the study of the information they have at their disposal, but also to the analysis of their rational actions and, consequently, the overall behaviour of the system. It is often remarked that the first systematic attempts to develop modal formalisms for knowledge date back to the sixties and seventies and in particular to the works of Hintikka.

At the time considerable attention was given to the adequacy of specific principles, expressed as axioms of modal logic, representing certain properties of knowledge in a rational setting. The standard framework that emerged consisted of the propositional normal modal logic $S5_n$ (see Chapter 1) built on top of the propositional calculus by considering the axioms $K : K_i(p \rightarrow q) \rightarrow K_ip \rightarrow K_iq$, $T : K_ip \rightarrow p$, $4 : K_ip \rightarrow K_iK_ip$, $5 : \neg K_ip \rightarrow K_i\neg K_ip$, together with the rules of necessitation Nec : From φ infer $K_i\varphi$, where K_i denotes the operator for knowledge of agent i, and Modus Ponens. Since then several other formalisms have been introduced accounting for weaker notions of knowledge as well as subtly different informational attitudes such as belief, explicit knowledge, and others.

While in the sixties soundness and completeness of these formalisms were shown, the standard semantics considered was that of plain Kripke models. These are models of the form $M = (W, \{R_i\}_{i \in A}, V)$, where W is a set of "possible worlds", $R_i \subseteq W \times W$ is a binary relation between worlds expressing epistemic indistinguishability between them, A is the set of agents, and $V : W \rightarrow 2^{PV}$ is an interpretation function for a set of propositional variables PV. Much of the theory of modal logic was developed in this setting up to relatively recent times. However, in the eighties and nineties attention was given to finer grained semantics that explicitly referred to particular states of computation in a system. In terms of epistemic logic, the challenge was to develop a semantics that accounted both for the low-level models of (a-)synchronous actions and protocols, and that at the same time was amenable to simple yet intuitive notions of knowledge. The key formalism put forward at the time satisfying these considerations was the one which became popular with the name of "interpreted system". Originally developed independently by, among others, Parikh and Ramanujam, Halpern and Moses, and Rosenschein, and later popularised in the seminal book "Reasoning about Knowledge" by Fagin, Halpern, Moses, and Vardi, interpreted systems offer a natural yet powerful formalism to represent the temporal evolution of a system as well as the evolution of knowledge of the agents in a run. The development of this model, succinctly described in the next section, triggered an acceleration in the study of logics for knowledge in the context of Computer Science leading to several results including axiomatisations with respect to several different classes of models of agents (synchronous, asynchronous, perfect recall, no learning, etc.) as well as applications of these to standard problems such as coordinated attack, communication, security, and others.

In this setting logic was most often seen as a formal reasoning tool. Attention was given to the exploration of metaproperties of the various formalisms including their completeness, decidability, and computational complexity. Attempts were made to verify systems automatically by exploring the relation $\Gamma \vdash_L \varphi$, where φ is a specification for the system, L is a logic suitably axiomatised representing the system under analysis, and Γ is a set of formulae expressing the initial conditions. However, partly due to the inherent complexity of some of the epistemic formalisms, verification of concrete systems via theorem proving for epistemic logics did not attract significant attention.

At the same time (the early nineties) the area of verification by model checking began acquiring considerable importance with a stream of results being produced for a variety of temporal logics. It was prominently suggested by Halpern and Vardi that model checking, not theorem proving, may provide the suitable verification technique for epistemic concepts. However, it was not before the very end of the nineties that model checking techniques were applied to the verification of multi-agent systems via temporal-epistemic formalisms. To our knowledge the first contribution in the area dates back to 1999 when van der Meyden and Shilov explored the complexity of model checking perfect recall semantics against epistemic specifications. Attention then switched to the use of ad-hoc local propositions for translating the verification of temporal-epistemic into plain temporal logic. Developments of bounded model checking algorithms and BDD-based labelling procedures followed.

The aim of this chapter is to summarise some of the results obtained by the authors in this area. The emphasis is mostly placed on BDD and SAT-based techniques as they constitute the underlying building blocks of many further refinements that followed. The area has grown considerably in recent years and this chapter cannot provide a complete survey of the area. Some additional approaches are discussed in Section 8.7, but others, inevitably, are omitted. In particular, here we only consider approaches where knowledge is treated as a modality interpreted on sets of global states in possible executions and not as a simple predicate as in other approaches.

The rest of the chapter is organised as follows. In Section 8.2 we present syntax and semantics of CTLK, the branching-time combination between knowledge and time we use throughout the paper, as well as the special class of the interleaved interpreted systems and two known scenarios from the literature. In Section 8.3 we introduce and discuss an OBDD-based approach to the verification of temporal-epistemic logic. In Section 8.4 an alternative yet complementary approach based on bounded and unbounded model checking is discussed. In Section 8.5 extensions to real-time are summarised briefly. In Section 8.6 we present an abstraction technique based on partial orders that is shown to be correct against CTLK. We discuss several alternative and related approaches in Section 8.7.

8.2 Syntax and Semantics

Model checking approaches differ, among other characteristics, by the specification languages they support. In the following we introduce a branching-time version of temporal-epistemic logic.

8.2.1 Syntax

Given a set of agents $A = \{1, \ldots, n\}$ and a set of propositional variables PV, we define the language $\mathcal{L}$ of CTLK as the fusion between the branching time logic CTL and the epistemic logic $S5_n$ for n modalities of knowledge $K_i, i \in A$ and group epistemic modalities E_Γ, D_Γ, and C_Γ ($\Gamma \subseteq A$):

$$\varphi ::= p \mid \neg\varphi \mid \varphi \wedge \varphi \mid K_i\varphi \mid E_\Gamma\varphi \mid D_\Gamma\varphi \mid C_\Gamma\varphi \mid AX\varphi \mid AG\varphi \mid A(\varphi U \varphi)$$

where $p \in PV$.

In addition to the standard Boolean connectives the syntax above defines two fragments: an epistemic and a temporal one. The epistemic part includes formulas of the form $K_i\varphi$ representing "agent i knows that φ", $E_\Gamma\varphi$ standing for "everyone in group Γ knows that φ", $D_\Gamma\varphi$ representing "it is distributed knowledge in group Γ that φ is true″, C_Γ formalising "it is common knowledge in group Γ that φ". The temporal fragment defines formulas of the form $AX\varphi$ meaning "in all possible paths, φ holds at next step"; $AG\varphi$ standing for "in all possible paths φ is always true"; and $A(\varphi U \psi)$ representing "in all possible paths at some point ψ holds true and before then φ is true along the path".

Whenever $\Gamma = A$ we omit the subscript from the group modalities E, D, and C. As customary we also use "diamond modalities", i.e., modalities dual to the ones defined. In particular, for the temporal part we use $\mathrm{EF}\varphi = \neg\mathrm{AG}\neg\varphi$, $\mathrm{EX}\varphi = \neg\mathrm{AX}\neg\varphi$, representing "there exists a path where at some point φ is true" and "there exists a path in which at the next step φ is true" respectively. We will also use the $\mathrm{E}(\varphi \mathrm{U} \psi)$ with obvious meaning.

For the epistemic fragment we use overlines to indicate the dual epistemic operators; in particular we use $\overline{\mathrm{K}}_i\varphi$ as a shortcut for $\neg K_i\neg\varphi$, meaning "agent i considers it possible that φ" and similarly for $\overline{\mathbf{E}_{\mathbf{K}\Gamma}}$, $\overline{\mathbf{C}_{\mathbf{B}\Gamma}}$, and $\overline{\mathrm{C}}_\Gamma$.

Formulas including both temporal and epistemic modalities can represent expressive specifications including the evolution of private and group knowledge over time, knowledge about a changing environment as well as knowledge about other agents' knowledge.

8.2.2 Interpreted systems semantics

In what follows the syntax of the specification language above is interpreted on the multi-agent semantics of interpreted systems. Interpreted systems are a fine-grained semantics often employed to represent the temporal evolution and knowledge in multi-agent systems. Although initially developed for linear time, given the applications of this paper we here present it in its branching time version.

Assume a set of *possible local states* L_i for each agent i in a set $A = \{1, \dots, n\}$ and a set L_e of possible local states for the environment e. The set of *possible global states* $G \subseteq L_1 \times \dots \times L_n \times L_e$ is the set of all possible tuples $g = (l_1, \dots, l_n, l_e)$ representing instantaneous snapshots of the system as a whole. The formalism stipulates that each agent i performs one of the enabled actions in a given state according to a *protocol function* $P_i : L_i \to 2^{Act_i}$. P_i maps local states to sets of possible local actions for agent i within its repertoire of local actions Act_i, which includes a special action ϵ known as the *null-action*. Similarly, the environment e is assumed to perform actions following its protocol $P_e : L_e \to 2^{Act_e}$. A *joint action* $a = (act_1, \dots, act_n, act_e)$ is a tuple of actions performed jointly by all agents and the environment in accordance with their respective protocols. The joint actions form part of the domain of the transition function $T : G \times Act_1 \times \dots \times Act_n \times Act_e \to G$ which gives the evolution of a system from an initial global state $g^0 \in G$. A *path* $\pi = (g_0, g_1, \dots)$ is a maximal sequence of global states such that $(g_k, g_{k+1}) \in T$ for each $k \geq 0$ (if π is finite then the range of k is restricted accordingly). For a path $\pi = (g_0, g_1, \dots)$, we take $\pi(k) = g_k$. By $\Pi(g)$ we denote the set of all the paths starting at $g \in G$.

The model above can be enriched in several ways by expressing explicitly observation functions for the agents in the system or by taking more concrete definitions for the sets of local states thereby modelling specific classes of systems (perfect recall, no learning, etc.). We do not discuss these options here.

To interpret the formulas of the language $\mathcal{L}$ for convenience we define models simply as tuples $M = (G, g^0, T, \sim_1, \dots, \sim_n, V)$, where G is the set of the global states reachable from the initial global state g^0 via T; $\sim_i \subseteq G \times G$

is an epistemic relation for agent i defined by $g \sim_i g'$ iff $l_i(g) = l_i(g')$, where $l_i : G \rightarrow L_i$ returns the local state of agent i given a global state; and $V : G \times PV \rightarrow \{true, false\}$ is an interpretation for the propositional variables PV in the language.

The intuition behind the definition of the relations $\sim_i$ above is that the global states whose local components are the same for agent i are not distinguishable for the agent in question. This definition is standard in epistemic logic.

Let M be a model, $g = (l_1, \ldots, l_n)$ a global state, and φ, ψ formulas in $\mathcal{L}$, the satisfaction relation $\models$ is inductively defined as follows:

- $(M, g) \models p$ iff $V(g, p) = true$,
- $(M, g) \models \mathrm{K}_i\varphi$ iff for all $g' \in G$ if $g \sim_i g'$, then $(M, g') \models \varphi$,
- $(M, g) \models D_\Gamma\varphi$ iff for all $i \in \Gamma$ and $g' \in G$ if $g \sim_i g'$, then $(M, g') \models \varphi$,
- $(M, g) \models E_\Gamma\varphi$ iff $(M, g) \models \bigwedge_{i\in\Gamma} K_i\varphi$,
- $(M, g) \models C_\Gamma\varphi$ iff for all $k \geq 0$ we have $(M, g) \models E^k_\Gamma\varphi$,
- $(M, g) \models AX\varphi$ iff for all $\pi \in \Pi(g)$ we have $(M, \pi(1)) \models \varphi$,
- $(M, g) \models AG\varphi$ iff for all $\pi \in \Pi(g)$ and for all $k \geq 0$ we have $(M, \pi(k)) \models \varphi$,
- $(M, g) \models A(\varphi U \psi)$ iff for all $\pi \in \Pi(g)$ there exists a $k \geq 0$ such that $(M, \pi(k)) \models \psi$ and for all $0 \leq j < k$ we have $(M, \pi(j)) \models \varphi$.

The definitions for the Boolean connectives and the other inherited modalities are given as usual and are not repeated here. $E^k\varphi$ is to be understood as a shortcut for k occurrences of the E modality followed by φ, i.e., $E^0\varphi = \varphi$; $E^1\varphi = E\varphi$; $E^{k+1}\varphi = EE^k\varphi$.

8.2.3 Interleaved interpreted systems

Interleaved interpreted systems are a restriction of interpreted systems, where all the joint actions are of a special form. More precisely, we assume that if more than one agent is active at a given global state, i.e., executes a non null-action, then all the active agents perform the same (shared) action in that round. Formally, let $Act = \bigcup_{i=1}^n Act_i \cup Act_e$ and for each action $a \in Act$ by $Agent(a)$ we mean the set $\{j \in A \cup \{e\} \mid a \in Act_j\}$, i.e., the set of all agents and environment having a in their repertoires. A tuple $(a_1, \ldots, a_n, a_e)$ is a joint action iff there exists a non-null action $a \in Act \setminus \{\epsilon_1, \ldots, \epsilon_n, \epsilon_e\}$ such that $a_j = a$ for all $j \in Agent(a)$, and $a_j = \epsilon_j$

for all $j \in \{1, .., n, e\} \setminus Agent(a)$. By $g \xrightarrow{a} g'$ we denote that there is a transition from g to g' by means of the joint action a.

Similarly to blocking synchronisation in automata, the condition above insists on all agents performing the same non-null action in a global transition; additionally, note that if an agent has the action being performed in its repertoire, it must be performed, for the global transition to be allowed.

8.2.4 Temporal-epistemic specifications

The formalism of interpreted systems has been used successfully to model a variety of scenarios including communication protocols (e.g., the bit transmission problem, message passing systems), coordination protocols (e.g., the attacking generals setting), and cache coherence protocols. We briefly present only two scenarios here and refer the reader to the specialised literature for more details. Our key consideration here is that temporal-epistemic languages seem particularly attractive to express natural and precise specifications involving the information states of the agents in the system.

The dining cryptographers protocol (DCP) for anonymous broadcast is a well-known anonymity protocol in the security literature. We report it in its original wording given by Chaum.

> Three cryptographers are sitting down to dinner at their favorite three-star restaurant. Their waiter informs them that arrangements have been made with the maitre d'hotel for the bill to be paid anonymously. One of the cryptographers might be paying for dinner, or it might have been NSA (U.S. National Security Agency). The three cryptographers respect each other's right to make an anonymous payment, but they wonder if NSA is paying. They resolve their uncertainty fairly by carrying out the following protocol:
>
> Each cryptographer flips an unbiased coin behind his menu, between him and the cryptographer on his right, so that only the two of them can see the outcome. Each cryptographer then states aloud whether the two coins he can see – the one he flipped and the one his left-hand neighbor flipped – fell on the same side or on different sides. If one of the cryptographers is the payer, he states the opposite of what he sees. An odd number of differences uttered at the table indicates that a cryptographer is paying; an even number indicates that NSA is paying (assuming that dinner was paid for only once). Yet if a cryptographer is paying, neither of the other two learns anything from the utterances about which cryptographer it is ((Chaum, 1988, p. 65)).

Temporal-epistemic logic can be used to express the specification of the protocol, which, in turn, can be modelled as an interpreted system. For each agent i in the set of cryptographers A we can consider a local state consisting of the triple (l_i^1, l_i^2, l_i^3), representing, respectively, whether or not the coins observed are the same, whether agent i paid for the bill, and whether the announcements have an even or odd parity. A local state for the environment can be taken as a tuple $(l_e^1, l_e^2, l_e^3, l_e^4)$ where l_e^1, l_e^2, l_e^3 represent the coin tosses for each agent and l_e^4 represents whether or not the agent in question paid for the bill. Actions and protocols for the agents and the environment can easily be given following Chaum's narrative above and so can the transition function.

The following specifications can be considered for the protocol.

$$AG(\bigwedge_{i \in A}(odd \wedge \neg paid_i) \rightarrow AX(K_i(\bigvee_{j \neq i} paid_j) \bigwedge_{k \neq i} \neg K_i paid_k)) \quad (8.1)$$

$$AG(\bigwedge_{i \in A}(even \rightarrow AX(C_A(\bigwedge_{j \in A} \neg paid_j) \quad (8.2)$$

Specification 8.1 states that when an agent $i \in A$ observes an odd parity in a situation when he did not cover the bill, then in all next states (i.e., when the announcements have been made) he will know that one of the others paid for dinner but without knowing who it was. Specification 8.2 states that when an even parity has been observed, then the cryptographers acquire common knowledge of the fact that none of them paid the bill. Both specifications represent the precise and intuitive requirements of the protocol in question; both formulas can be shown to hold on the protocol.

The DCP is amenable to be scaled to represent any number of cryptographers. The corresponding number of states grows exponentially in the number of cryptographers considered and therefore may only be verified by automated techniques, such as model checking.

The train-gate-controller problem (TGC) is a deadlock protocol in which a number of trains share a tunnel regulated by a traffic signal. While each train runs on its own track, the tunnel can only accommodate one train. A single controller operates a system of traffic lights at both ends of the tunnel, thereby regulating access to the tunnel.

The scenario can be modelled as interpreted system by associating the possible local states *away* from the tunnel, *wait*, and *train-in-tunnel* to each train and the possible local states *green* and *red* to the controller. The local actions for the train consist of *enter the tunnel*, *leave the tunnel*, and *wait* with the obvious effects implemented by the transition function. Assuming the trains work correctly, a train may attempt to enter the tunnel when it is in the *wait* state and the controller signals the tunnel is free of trains. Protocols can be given for the agents to perform actions following the description above and the transition function can similarly be defined accordingly.

Specifications that can be checked on the TGC include the following.

$$AG(train_1\_in\_tunnel \rightarrow K_1 \neg train_2\_in\_tunnel) \tag{8.3}$$

$$AG(\neg train_1\_in\_tunnel \rightarrow (\neg K_1(train_2\_in\_tunnel) \wedge \neg K_1 \neg train_2\_in\_tunnel)) \tag{8.4}$$

Property 8.3 states that when train 1 is in the tunnel, it knows that train 2 is not in the tunnel. Property 8.4 states that when train 1 is in the tunnel, it does not know whether or not train 2 is in the tunnel. Both specifications can be shown to hold on the TGC model. Variants of the scenario where trains may develop faults have been studied and other specifications have been analysed.

Like DCP the TGC is scalable to any number of trains and may be verified by means of the techniques discussed below.

8.3 OBDD-based Symbolic Model Checking

Given a system S and a specification P to be checked, the model checking approach involves representing S as a logical model M_S, the specification P as a logic formula φ_P, and investigating whether $M_S \models \varphi_P$. In the traditional approach the model M_S is finite and represents all the possible evolutions of the system S; the specification φ_P is a temporal logic formula expressing some property to be checked on the system, e.g., liveness, safety, etc. When the formula φ_P is given in LTL or CTL, checking φ_P on an explicitly given M_S is a tractable problem. It is, however, impractical to present M_S explicitly; instead, M_S is normally given implicitly by means of a program in a dedicated modelling language. This is convenient for the engineer, but the number of states in the resulting model grows exponentially with the number of variables used in the program describing M_S, causing what is commonly referred to as the *state explosion problem.*

Since state-spaces of real systems can be too large to be checked, much of the model checking literature deals with methodologies to limit the impact of the state explosion problem. The most prominent techniques include partial order reduction, symmetry reduction, ordered-binary decision diagrams (OBDDs), bounded and unbounded model checking, and various forms of abstraction. By using partial-order reduction techniques the model M_S

is pruned and provably redundant states are eliminated or collapsed with others depending on the formula to be checked, thereby reducing the size of the state space to be considered. Symmetry reduction techniques are used to reduce the state space of distributed systems composed of many similar processes which can be suitably abstracted. OBDDs are a compact and canonical representation for Boolean formulas, and traditionally offer the underpinnings for the mainstream symbolic approaches to model checking. Bounded and unbounded model checking exploit recent advances in the efficiency of checking satisfiability for appropriate Boolean formulas representing the model and the specification. Abstraction techniques are used to generate smaller models, typically simulations or bisimulations, that can alternatively be checked against the same specification. Predicate abstraction is based on the identification of key predicates, often generated automatically via calls to SMT checkers, which can be used to construct a smaller model for the verification of the formula in question; crucially it is used in verification of infinite-state systems. Several tools have been developed for model checking systems against temporal specifications. These include SPIN, which provides an on-the-fly automata-based approach combined with partial-order reduction for LTL, SMV and NuSMV supporting OBDD-based model checking and bounded model checking for LTL and CTL, POEM supporting partial-order semantic reduction. Several other tools exist for other varieties of temporal logic, e.g., real-time logics and probabilistic temporal logic.

The tools mentioned above are nowadays very sophisticated and support expressive input languages; however they are limited to temporal logics only. In the rest of the chapter we summarise work by the authors towards techniques and tools supporting specifications given in temporal-epistemic logic.

8.3.1 State space representation and labelling

At the heart of the OBDDs approach is the symbolic representation of sets and functions paired with the observation that to assess whether $(M, g) \models \varphi$ it is sufficient to evaluate whether $g \in SAT(\varphi)$ where $SAT(\varphi)$ is the set of states in the model M satisfying φ. To introduce the main ideas of the approach we proceed in three stages: first, we observe we can encode sets as Boolean formulas; second, we show how OBDDs offer a compact representation to Boolean functions; third we give algorithms for the calculation of $SAT(\varphi)$.

To begin, observe that given a set G of size $|G|$ it is straightforward to associate uniquely a vector of Boolean variables $(w_1, \ldots, w_m)$ to any element $g \in G$ where $m = \lceil log_2 |G| \rceil$ (a tuple of m places can represent

2^m different elements). Any subset $S \subseteq G$ can be represented by using a characteristic function $f_S : (g_1, \ldots, g_m) \to \{0, 1\}$, expressing whether or not the element as encoded is in S. Note that functions and relations can also be encoded as Boolean functions; for instance to encode that two states are related by some relation we can consider a vector of Boolean functions comprising of two copies of the representation of the state to which we add a further Boolean variable expressing whether or not the states are related. Vectors designed in this way represent conjunctions of Boolean atoms or their negation and, as such, denote a Boolean formula.

Given this, Boolean formulas can be used to represent a given interpreted system as follows.

- Sets of local states, global states, actions, and initial global states can be encoded as Boolean formulas for the respective sets.

- Protocols for each agent, the local evolution function for each agent, and the valuation for the atoms can be expressed as Boolean formulas for the respective functions.

- Following this, the global temporal relation and the n epistemic relations for the agents can also be suitably represented as Boolean formulas for the respective relations. The Boolean formula encoding the temporal relation needs to reflect the fact that joint actions are composed of enabled local actions: $f_T(g, g') = \bigvee_{a \in JointAct}(g, a, g') \in T \bigwedge_{i \in A} a_i \in P_i(l_i(g))$, where $a = (a_1, \ldots, a_n)$ is a joint action for the system and all individual action components a_i are enabled by the local protocols at the corresponding local state $l_i(g)$ in g. The epistemic relations for the agents can be represented simply by imposing equality on the corresponding local state component.

- The set of reachable global states can be represented by a Boolean formula by calculating the fix-point of the operator $\tau(Q) = (I(g) \vee \exists g'(T(g, a, g') \wedge Q(g'))$.

Boolean functions are a convenient representation to perform certain logical operations on them (e.g., $\wedge, \vee$); however, calculating their satisfiability and validity can be expensive. Truth tables themselves do not offer any advantage in this respect: for instance checking satisfiability on them may involve checking 2^n rows of the table where n is the number of atoms present. OBDDs constitute a symbolic representation for Boolean functions. Observe that every Boolean function we can be associated to a binary decision tree (BDT), in which each level represents a different atom appearing in the Boolean function. Taking a different path along the tree corresponds to

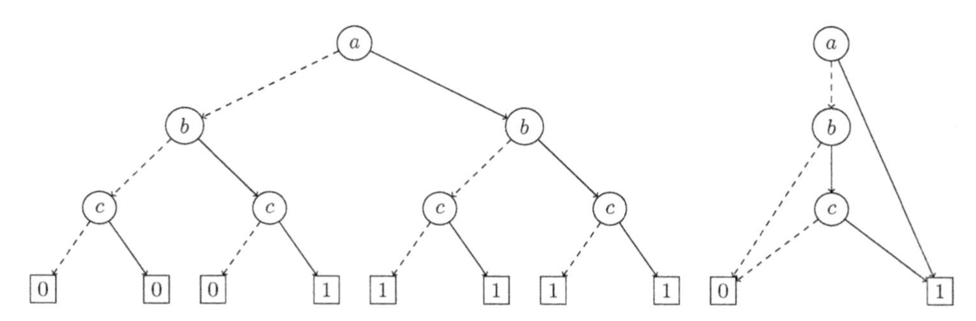

Figure 8.1: A BDT for the Boolean function $a \vee (b \wedge c)$ (left) and its corresponding BDD (right). The dotted lines correspond to assigning the value *false* to the atom whose name the edge leaves from. Conversely the solid lines represent assignments to *true*.

selecting a particular combination of values for the atoms (see Figure 8.1), thereby determining the truth value of the formula.

BDTs are not efficient representations of their corresponding Boolean function. However, a series of operations can be performed on them to reduce them to binary decision diagrams (BDDs). A BDD is a directed acyclic graph with an initial node, and in which each node (representing a Boolean atom) has two edges (corresponding to decision points "true" and "false") originating from it and with the final leaves labelled either as "true" (marked with 1) or "false" (marked with 0) (see Figure 8.1). The order in which operations are applied on the initial BDT affects the resulting BDD and comparing BDDs is also an expensive operation. However, if the ordering of the variables in the BDT is fixed, the resulting reduced BDD is unique (or canonical). This leads to an alternative technique to comparing Boolean functions: compute their canonical BDDs; if they are the same they represent the same Boolean function; if not they represent different functions. The canonical BDDs in which the variables are ordered obtained as above are normally referred to as reduced OBDDs and constitute one of the leading data structures in symbolic model checking.

The reason for this is that operations on Boolean functions and specific set operations such as existential pre-images can be efficiently executed directly on the corresponding OBDDs.

We now present the algorithms for the calculation of the set of states $SAT(\varphi)$ satisfying a formula φ in $\mathcal{L}$. In the OBDD approach all sets of states below are computed symbolically on the corresponding OBDDs.

```
SAT(φ) {
  φ is an atomic formula: return {g | V(g, φ) = true};
  φ is ¬φ1: return S \ SAT(φ1);
  φ is φ1 ∧ φ2: return SAT(φ1) ∩ SAT(φ2);
  φ is EXφ1: return SAT_EX(φ1);
  φ is E(φ1 U φ2): return SAT_EU(φ1, φ2);
  φ is EFφ1: return SAT_AF(φ1);
  φ is K_i φ1: return SAT_K(φ1, i);
  φ is Eφ: return SAT_E(φ);
  φ is Cφ: return SAT_C(φ);
}
```

In the algorithm above, the auxiliary procedures SAT_{EX}, SAT_{EU}, and SAT_{AF} follow the standard algorithms used in temporal logic[1]. For instance the set of global states satisfying $EX\varphi$ is computed as follows (in what follows G is the set of reachable states).

```
SAT_EX(φ) {
  X = SAT(φ);
  Y = {g ∈ G | ∃g' ∈ X and T(g, a, g')}
  return Y;
}
```

Note that the calculation of EX involves computing the pre-image of T. The set of states satisfying the epistemic modalities are defined as follow (note that below we use $\sim_\Gamma^E = \bigcup_{i\in\Gamma} \sim_i$ and $\sim_\Gamma^D = \bigcap_{i\in\Gamma} \sim_i$).

```
SAT_K(φ, i) {
  X = SAT(¬φ);
  Y = {g ∈ S | ∃g' ∈ X and ~_i (g, g')}
  return ¬Y ∩ G;
}
```

```
SAT_E(φ, Γ) {
  X = SAT(¬φ);
  Y = {g ∈ G | ~^E_Γ (g, g') and g' ∈ X}
  return ¬Y ∩ G;
}
```

```
SAT_D(φ, Γ) {
  X = SAT(¬φ);
  Y = {g ∈ G | ~^D_Γ (g, g') and g' ∈ X}
  return ¬Y ∩ G;
}
```

[1] For efficiency reasons the CTL modalities implemented are typically EX, EU, and AF.

```
SAT_C(φ,Γ) {
  Y = SAT(¬φ);
  X = G;
  while ( X ≠ Y ) {
    X = Y;
    Y = {g ∈ G |~^E_Γ (g,g') and g' ∈ X}
  return ¬Y ∩ G;
}
```

The algorithm for $K_i\varphi$ is similar in spirit to the CTL algorithm for computing $AX\varphi$: essentially we compute the pre-image under the epistemic relation of the set of formulas not satisfying φ and negate the result. $E_\Gamma\varphi$ (respectively $D_\Gamma\varphi$ is computed similarly but on $\sim^\Gamma_E$ ($\sim^\Gamma_D$, respectively). For C we use a fix-point construction; note that fix-point constructions already appear in the algorithm to compute the satisfiability of the CTL until operator. All sets operations above are implemented on the corresponding OBDDs thereby producing the OBDD for $SAT(\varphi)$. We can now recast the model checking query $(M, g^0) \models \varphi$ into $g^0 \in SAT(\varphi)$ where g^0 and $SAT(\varphi)$ are suitably encoded as OBDDs.

8.3.2 MCMAS

MCMAS is an open-source toolkit that implements the OBDD-based procedures described above. The model checker takes as input a program describing the evolutions of a multi-agent system and a set of specifications to be checked and returns as output whether or not the specifications are satisfied and witnesses or counterexamples for them. The program is given in ISPL (Interpreted Systems Programming Language), a modelling language suited to represent interpreted systems. An ISPL program consists of a sequence of declarations for the agents in the system, an evaluation for the atomic propositions, and a set of specifications in CTLK (other specification languages including ATL are also supported) to be checked.

In line with interpreted systems an agent is modelled by describing the variables that define it (bounded integers, Boolean, and enumeration types), the set of local actions, protocols, and the local evolution function. This is given in terms of a set of rules governing the value of the target local states when global actions are performed in given sets of local states. A very simple ISPL agent is described in Figure 8.2.

Upon invocation the tool parses the input, builds the OBDD for transition relation and the OBDD for the set of reachable states. This is then used in the calculation of the OBDD for the sets of states satisfying the formula to be verified. By comparing whether the initial state belongs to this set the output is displayed. MCMAS can be used within Eclipse where

```
1    Agent Sender
2      Vars:
3        bit : {b0, b1};
4        ack : boolean;
5      end Vars
6      Actions = {sb0, sb1, null};
7      Protocol:
8        bit=b0 and ack=false : {sb0};
9        bit=b1 and ack=false : {sb1};
10       ack=true : {null};
11     end Protocol
12     Evolution:
13       (ack=true) if (ack=false) and
14          ( ( (Receiver.Action=sendack) and
15              (Environment.Action=sendSR) )
16           or
17          ( (Receiver.Action=sendack) and
18            (Environment.Action=sendR) )
19          );
20     end Evolution
21   end Agent
```

Figure 8.2: A simple ISPL Agent representing the sender agent in the bit transmission problem.

various functionalities, including counterexample and witness generation, are supported.

Through MCMAS several scenarios from areas like web-services, cache-coherence protocols, diagnosis, and security protocols, have been verified. In line with other BDD-based checkers, the size of the model that can be usefully verified depends on the specific example but is normally around 10^{12} reachable global states. The corresponding number of possible global states can be greater than 10^{30}.

8.4 SAT-based Symbolic Model Checking

SAT-based model checking is the most recent symbolic approach for modal logic. It has been motivated by a dramatic increase in efficiency of SAT-solvers, i.e., algorithms solving the satisfiability problem for propositional formulas. The main idea of SAT-based methods consists in translating the model checking problem for a temporal-epistemic logic to the problem of satisfiability of a formula in propositional logic. This formula is typically obtained by combining an encoding of the model and of the temporal-epistemic property. In principle, the approaches to SAT-based symbolic verification can be viewed as bounded (BMC) or unbounded (UMC). BMC applies to an existential fragment of a logic (here ECTLK) on a part of the model, whereas UMC is for an unrestricted logic (here CTL_pK) on the whole model.

8.4.1 Bounded Model Checking

BMC is based on the observation that some properties of a system can be checked over a part of its model only. In the simplest case of reachability analysis, this approach consists in an iterative encoding of a finite symbolic path as a propositional formula. The satisfiability of the resulting propositional formula is then checked using an external SAT-solver. We present here the main definitions of BMC for ECTLK and later discuss extensions to more expressive logics. We refer the reader to the literature cited at the end of this chapter.

To explain how the model checking problem for an ECTLK formula is encoded as a propositional formula, we first define k-models, bounded semantics over k-models, and then propositional encodings of k-paths in the k-model and propositional encodings of the formulas. In order to define a bounded semantics for ECTLK we start with defining k-models. Let $M = (G, g^0, T, \sim_1, \ldots, \sim_n, V)$ be a model and $k \in \mathbb{N}_+$. The $k-$model for M is defined as a structure $M_k = (G, g^0, P_k, \sim_1, \ldots, \sim_n, \mathcal{V})$, where P_k is the set of all the k-paths of M over G, where a k-path is the prefix of length k of a path.

We need to identify k-paths that represent infinite paths so that satisfaction of EG formulas in the bounded semantics implies their satisfaction in the unbounded one. To this aim define the function $loop : P_k \to 2^{\mathbb{N}}$ as: $loop(\pi) = \{l \mid 0 \leq l \leq k \; and \; (\pi(k), \; \pi(l)) \in T\}$, which returns the set of indices l of π for which there is a transition from $\pi(k)$ to $\pi(l)$.

Let M_k be a $k-$model and φ, ψ be ECTLK formulas. $(M_k, g) \models \alpha$ denotes that φ is true at the state g of M_k. The bounded semantics is summarised as follows. $(M_k, g) \models \mathrm{EX}\varphi$ has the same meaning as for unbounded models, for $k > 0$. $(M_k, g) \models \mathrm{E}(\varphi \mathrm{U} \psi)$ has the same meaning as for unbounded models. $(M_k, g) \models \mathrm{EG}\varphi$ states that there is a k-path π, which starts at g, all its states satisfy φ and π is a loop, which means that one of the states of π is a T-successor of the last state of π. $loop(\pi)$ returns the indexes of such states of π. For the epistemic modalities $(\overline{K}_i\varphi, \overline{E}_\Gamma\varphi, \overline{D}_\Gamma\varphi, \overline{C}_\Gamma\varphi)$ the bounded semantics is the same as unbounded, but insisting on reachability of the state satisfying φ on a k-path starting in g^0.

Model checking over models can be reduced to model checking over k-models, called BMC. The main idea of BMC for ECTLK is that checking φ over M_k is replaced by checking the satisfiability of the propositional formula $[M, \varphi]_k := [M^{\varphi, g^0}]_k \wedge [\varphi]_{M_k}$. The formula $[M^{\varphi, g^0}]_k$ represents the k-model under consideration, whereas $[\varphi]_{M_k}$ - a number of constraints that must be satisfied on M_k for φ to be satisfied. Checking satisfiability of an ECTLK formula can be done by means of a SAT-solver. Typically, we start with $k := 1$, test satisfiability for the translation, and increase k by one until either $[M^{\varphi,g^0}]_k \wedge [\varphi]_{M_k}$ becomes satisfiable, or k reaches the maximal depth of M, which is bounded by $|G|$. It can be shown that if $[M^{\varphi,g^0}]_k \wedge [\varphi]_{M_k}$ is satisfiable for some k, then $(M, g^0) \models \varphi$, where M is the full model. If $[M^{\varphi,g^0}]_k \wedge [\varphi]_{M_k}$ is not satisfiable for some k, then we cannot infer that $(M, g^0) \not\models \varphi$ unless k reaches the size of the model M.

Translation to SAT

We provide here some details of the translation. The states and the transitions of the system under consideration are encoded similarly as for BDDs in Section 8.3. Let $w = (w[1], \ldots, w[m])$ be sequence of propositions (called a *global state variable*) for encoding global states. A sequence $w_{0,j}, \ldots, w_{k,j}$ of global state variables is called a symbolic k-path j. Since a model for a branching time formula is a tree (a set of paths), we need to use a set of symbolic k-paths to encode it. The number of them depends on the value of k and the formula φ, and it is computed using the function f_k. This function determines the number of k-paths sufficient for checking an ECTLK formula. Intuitively, each nesting of an epistemic or temporal formula in φ increases the value of $f_k(\varphi)$ by 1, whereas the subformulas EU, EG, and $\overline{\mathrm{C}}_\Gamma$ add more $k-$paths.

The propositional formula $[M^{\varphi,g^0}]_k$, representing the k-paths in the k-model, is defined as follows:

$$[M^{\varphi,g^0}]_k := I_{g^0}(w_{0,0}) \wedge \bigwedge_{j=1}^{f_k(\varphi)} \bigwedge_{i=0}^{k-1} T(w_{i,j}, w_{i+1,j}),$$

where $w_{0,0}$ and $w_{i,j}$ for $0 \leq i \leq k$ and $1 \leq j \leq f_k(\varphi)$ are global state variables, and $T(w_{i,j}, w_{i+1,j})$ is a formula encoding the transition relation T.

An intuition behind this encoding is as follows. The vector $w_{0,0}$ encodes the initial state g^0 and for each symbolic k-path, numbered $1, \ldots, f_k(\varphi)$, each pair of the consecutive vectors on this path encodes pairs of states that are in the transition relation T. The formula $T(w, v)$ is typically a logical disjunction of the encodings of the transitions corresponding to all the actions of the model M. This way, one symbolic k-path encodes all the (concrete) k-paths.

The next step of the algorithm consists in translating an ECTLK formula φ into a propositional formula. Let w, v be global state variables. We make use of the following propositional formulas in the encoding:

- $p(w)$ encodes a proposition p of ECTLK over w.

- $H(w, v)$ represents the logical equivalence between the global state encodings u and v (i.e., encodes that u and v represent the same global states).
- $HK_i(w, v)$ represents the logical equivalence between the i-local state encodings u and v (i.e., encodes that u and v share the i-local states).
- $L_{k,j}(l)$ encodes a backward loop connecting the k-th state to the l-th state in the symbolic $k-$path j, for $0 \leq l \leq k$.

The translation of each ECTLK formula is directly based on its bounded semantics. The translation of φ at the state $w_{m,n}$ into the propositional formula $[\varphi]_k^{[m,n]}$ is as follows, where we give the translation of some selected formulas only. Let w be $w_{m,n}$ and let v denote $w_{0,i}$.

$$
\begin{array}{lll}
[\mathrm{EX}\alpha]_k^{[m,n]} & := & \bigvee_{i=1}^{f_k(\varphi)} \left(H(\mathsf{w},\mathsf{v}) \wedge [\alpha]_k^{[1,i]}\right) \\
[\mathrm{EG}\alpha]_k^{[m,n]} & := & \bigvee_{i=1}^{f_k(\varphi)} \left(H(\mathsf{w},\mathsf{v}) \wedge (\bigvee_{l=0}^{k} L_{k,i}(l)) \wedge \bigwedge_{j=0}^{k} [\alpha]_k^{[j,i]}\right) \\
[\mathrm{E}(\alpha \mathrm{U} \beta)]_k^{[m,n]} & := & \bigvee_{i=1}^{f_k(\varphi)} \left(H(\mathsf{w},\mathsf{v}) \wedge \bigvee_{j=0}^{k} \left([\beta]_k^{[j,i]} \wedge \bigwedge_{t=0}^{j-1} [\alpha]_k^{[t,i]}\right)\right) \\
[\overline{\mathrm{K}}_l\alpha]_k^{[m,n]} & := & \bigvee_{i=1}^{f_k(\varphi)} \left(I_{g^0}(\mathsf{v}) \wedge \bigvee_{j=0}^{k} \left([\alpha]_k^{[j,i]} \wedge HK_l(\mathsf{w}, w_{j,i})\right)\right)
\end{array}
$$

Intuitively, $[\mathrm{EG}\alpha]^{[m,n]_k}$ is translated to all the $f_k(\varphi)$-symbolic k-paths ($\mathrm{EG}\alpha$ is considered as a subformula of φ) that start at the states encoded by w, satisfy α, and are loops. $[\overline{\mathrm{K}}_l\alpha]_k^{[m,n]}$ is translated to all the $f_k(\varphi)$-symbolic k-paths such that each symbolic k-path starts at the initial state g^0, one of its states satisfies α and shares the l-th state with these encoded by w. Given the translations above, verification of φ over M_k reduces to checking the satisfiability of the propositional formula $[M^{\varphi,g^0}]_k \wedge [\varphi]_{M_k}$, where $[\varphi]_{M_k} = [\varphi]_k^{[0,0]}$.

Example 8.1
Below we show a part of the model encoding of the TGC protocol for two trains and the controller. Each train can be in 3 different local states, so we need 2 bits for representing its local states: $away_i$ by $(0,0)$, $wait_i$ by $(0,1)$, and $train_i\_in\_tunnel$ by $(1,0)$, where $i \in \{1,2\}$. The controller can be in 2 local states, so the binary representation of these states requires only one bit as follows: $green$ by (0) and red by (1). Thus, a global state, which is composed of a local state for train 1, a local state for the controller, and a local state for train 2, requires 5 bits to be represented, e.g., the initial state $g^0 = (away_1, green, away_2)$ is represented by $(0,0,0,0,0)$. In order to encode global states in the propositional logic we use global state variables being vectors composed of 5 propositional variables. Let

$w = (w[1], ..., w[5])$, $v = (v[1], ..., v[5])$ be two global state variables, and $A = \{1, 2, 3\}$, $D_1 = \{1, 2\}$, $D_2 = \{3\}$, and $D_3 = \{4, 5\}$. The initial state encoding over w is as follows: $I_{g^0}(w) := \bigwedge_{i \in D_1 \cup D_2 \cup D_3} \neg w[i]$.

In order to encode the global transition relation T in propositional logic we need first to encode sets of the global states sharing one local state. To this aim we number all the local states according to their position from 1 to 8 in the sequence $(away_1, wait_1, train_1\_in\_tunnel, green, red, away_2, wait_2, train_2\_in\_tunnel)$.

The formula $p_i(w)$ for $i \in \{1, \ldots, 8\}$ encodes all the global states, containing the local state numbered i, where: $p_1(w) := \neg w[1] \wedge \neg w[2]$ for $away_1$, $p_2(w) := \neg w[1] \wedge w[2]$ for $wait_1$, $p_3(w) := w[1] \wedge \neg w[2]$ for $train_1\_in\_tunnel$, $p_4(w) := w[3]$ for $green$, $p_5(w) := \neg w[3]$ for red, $p_6(w) := \neg w[4] \wedge \neg w[5]$ for $away_2$, $p_7(w) := \neg w[4] \wedge w[5]$ for $wait_2$, $p_8(w) := w[4] \wedge \neg w[5]$ for $train_2\_in\_tunnel$.

Let $agent(t)$ be the set of the numbers of agents whose local states are changed by executing the global transition t. For $t \in T$ let $B_t := \bigcup_{j \in A \setminus agent(t)} D_j$, be the set of the numbers of the bits of a global state that are not changed by t, $pre(t)$ be the set of the numbers of local states from which t has to be executed and $post(t)$ be the set of the numbers of local states reached after executing t.

It is easy to check that for the transition $t = enter_1\_in\_tunnel$, we have $pre(t) = \{2, 4\}$ and $post(t) = \{3, 5\}$, for the transition $t = leave_1\_in\_tunnel$, $pre(t) = \{3, 5\}$ and $post(t) = \{1, 4\}$, and for the transition $t = wait_1$, $pre(t) = \{1\}$ and $post(t) = \{2\}$.

The encoding $T(w, v)$ of the global transition relation T of TGC is as follows:

$$T(w, v) := \bigvee_{t \in T} \Big(\bigwedge_{i \in pre(t)} p_i(w) \wedge \bigwedge_{i \in post(t)} p_i(v) \wedge \bigwedge_{i \in B_t} (w[i] \Leftrightarrow v[i]) \Big).$$

Several improvements have been suggested to the above encoding of ECTLK such that the length of the formula $[\varphi]_{M_k}$ is reduced. They are listed in the final section. These approaches show an improved performance over the original encoding for some subclasses of ECTLK composed mainly of long and deeply nested formulas.

8.4.2 Unbounded Model Checking

UMC was originally introduced by McMillan for verification of CTL as an alternative to BMC and approaches based on BDDs. Then, UMC was extended to $\mathrm{CTL_pK}$ as well as to other more expressive logics.

We begin by extending the syntax and semantics of CTLK to $\mathrm{CTL_pK}$ by adding the past operators AY and AH. The operators including *Since* are omitted. A backward path $\pi = (g_0, g_1, \ldots)$ is a maximal sequence of global states such that $(g_{k+1}, g_k) \in T$ for each $k \geq 0$ (if π is finite, then k needs to be restricted accordingly). Let $\overline{\Pi}(g)$ denote the set of all the backward paths starting at $g \in G$.

- $(M, g) \models \mathrm{AY}\varphi$ iff for all $\pi \in \overline{\Pi}(g)$ we have $(M, \pi(1)) \models \varphi$,
- $(M, g) \models \mathrm{AH}\varphi$ iff for all $\pi \in \overline{\Pi}(g)$ and for all $k \geq 0$ we have $(M, \pi(k)) \models \varphi$.

Intuitively, $\mathrm{AY}\varphi$ specifies that for all the predecessor states φ holds, whereas $\mathrm{AH}\varphi$ expresses that for all the states in the past φ holds.

Unlike BMC, UMC is capable of handling the whole language of the logic. Our aim is to translate $\mathrm{CTL_pK}$ formulas into propositional formulas in the conjunctive normal form, accepted as an input by SAT-solvers.

Specifically, for a given $\mathrm{CTL_pK}$ formula φ, a corresponding propositional formula $[\varphi](w)$ is computed, where w is a global state variable (i.e., a vector of propositional variables for representing global states) encoding these states of the model where φ holds. The translation is not operating directly on temporal-epistemic formulas. Instead, to calculate propositional formulas either the Quantified Boolean Formula (QBF) or the fix-point characterisation of $\mathrm{CTL_pK}$ formulas (see Section 8.3) is used. More specifically, three basic algorithms are exploited. The first one, implemented by the procedure *forall*, defined by McMillan, is used for translating the formulas $\mathrm{Z}\alpha$ such that $\mathrm{Z} \in \{\mathrm{AX}, \mathrm{AY}, \mathrm{K}_i, \mathrm{D}_\Gamma, \mathrm{E}_\Gamma\}$. This procedure eliminates the universal quantifiers from a QBF formula characterising a $\mathrm{CTL_pK}$ formula, and returns the result in the conjunctive normal form. The second algorithm, implemented by the procedure gfp_O is applied to formulas $\mathrm{Z}\alpha$ such that $\mathrm{Z} \in \{\mathrm{AG}, \mathrm{AH}, \mathcal{C}_\Gamma\}$. This procedure computes the greatest fix-point, in the standard way, using Boolean representations of sets rather than sets themselves. For formulas of the form $\mathrm{A}(\varphi \mathrm{U} \psi)$ the third procedure, called lfp_{AU}, computing the least fix-point (in a similar way), is used. In so doing, given a formula φ, a propositional formula $[\varphi](w)$ is obtained such that φ is valid in the model M iff the propositional formula $[\varphi](w) \wedge I_{g^0}(w)$ is satisfiable.

In the following section we show how to represent the $\mathrm{CTL_pK}$ formulas in QBF and then translate them to propositional formulas in CNF.

From a fragment of QBF to CNF

The *Quantified Boolean Formulas* (QBF) are an extension of propositional logic by means of quantifiers ranging over propositions. The BNF syntax of a QBF formula is given by:

$$\alpha ::= p \mid \neg\alpha \mid \alpha \wedge \alpha \mid \exists p.\alpha \mid \forall p.\alpha.$$

The semantics of the quantifiers is defined as follows:

- $\exists p.\alpha$ iff $\alpha(p \leftarrow \mathbf{true}) \vee \alpha(p \leftarrow \mathbf{false})$,
- $\forall p.\alpha$ iff $\alpha(p \leftarrow \mathbf{true}) \wedge \alpha(p \leftarrow \mathbf{false})$,

where $\alpha \in$ QBF, $p \in PV$ and $\alpha(p \leftarrow q)$ denotes substitution with the variable q of every occurrence of the variable p in the formula α.

For example, the formula $[\mathrm{AX}\alpha](w)$ is equivalent to $\forall v.(T(w,v) \Rightarrow [\alpha](v))$ in QBF. Similar equivalences are obtained for the formulas $\mathrm{AY}\alpha$, $\mathrm{K}_i\alpha$, $\mathrm{D}_\Gamma\alpha$, and $\mathrm{E}_\Gamma\alpha$ by replacing $T(w,v)$ with suitable encodings of the relations T^{-1}, $\sim_i$, $\sim_\Gamma^D$, and $\sim_\Gamma^E$.

For defining a translation from a fragment of QBF (resulting from the translation of $\mathrm{CTL_pK}$) to propositional logic, one needs to know how to compute a CNF formula which is equivalent to a given propositional formula φ. While the standard algorithm toCNF, which transforms a propositional formula to one in CNF, preserving satisfiability only, is of linear complexity, a translation to an equivalent propositional formula in CNF is NP-complete. For such a translation, one can use the algorithm *equCNF* - a version of the algorithm toCNF, known as a *cube reduction*. The algorithm *equCNF* is a slight modification of the DPLL algorithm checking satisfiability of a CNF formula, but it can be presented in a general way, abstracting away from its specific realisation.

Assume that φ is an input formula. Initially, the algorithm *equCNF* builds a satisfying assignment for the formula toCNF$(\varphi) \wedge \neg l_\varphi$ (l_φ is a literal used in toCNF(φ)), i.e., the assignment which falsifies φ. If one is found, instead of terminating, the algorithm constructs a new clause that is in conflict with the current assignment (i.e., it rules out the satisfying assignment). Each time a satisfying assignment is obtained, a blocking clause is generated by the algorithm `blocking_clause` and added to the working set of clauses. This clause rules out a set of cases where φ is false. Thus, on termination, when there is no satisfying assignment for the current set of clauses, the conjunction of the blocking clauses generated precisely characterises φ.

A blocking clause could in principle be generated using the conflict-based learning procedure. If we require a blocking clause to contain only input variables, i.e., literals used in φ, then one could either use an (alternative) implication graph of McMillan, in which all the roots are input literals or a method introduced by Szreter, which consists in searching a directed acyclic graph representing the formula.

Now, our aim is to compute a propositional formula equivalent to a given QBF formula $\forall p_1 \ldots \forall p_n.\varphi$. The algorithm constructs a formula ψ equivalent to φ and eliminates from ψ the quantified variables on-the-fly, which is correct as ψ is in CNF. The algorithm differs from *equCNF* in one step only, where the procedure `blocking_clause` generates a blocking clause and deprives it of the quantified propositional variables. On termination, the resulting formula is a conjunction of the blocking clauses without the quantified propositions and precisely characterises $\forall p_1 \ldots \forall p_n.\varphi$.

8.4.3 VerICS

VerICS is a verification tool, developed at ICS PAS, that implements the SAT-based BMC and UMC procedures described above as well as in Section 8.5. It offers three complementary methods of model checking: SAT-based BMC, SAT-based UMC, and an on-the-fly verification while constructing abstract models of systems. A network of communicating (timed) automata (together with a valuation function) is the basic VerICS's formalism for modelling a system to be verified. Timed automata are used to specify real time systems, whereas timed or untimed automata are applied to model MAS. VerICS translates a network of automata and a temporal-epistemic formula into a propositional formula in CNF and invokes a SAT-solver in order to check for its satisfiability. VerICS has been implemented in C++; its internal functionalities are available via a interface written in Java. In line with other SAT-based model checkers, the size of the state space, which can be efficiently verified depends on the specific example and ranges between 10^6 and 10^{50}, which corresponds to encoding and checking $k-$models with k ranging from 10 to 40 approximately.

8.5 Extensions to Real-Time Epistemic Logic

In this section we briefly discuss some extensions to real-time to the ECTLK framework analysed so far. The timed temporal-epistemic logic TECTLK was introduced to deal with situation where time is best assumed to be dense and hence modelled by real numbers. The underlying semantics uses networks of *timed automata* to specify the behaviour of the agents. These automata extend standard finite state automata by a set of clocks $\mathcal{X}$ (to measure the flow of time) and time constrains built over $\mathcal{X}$ that can be used for defining guards on the transitions as well invariants on their locations. When moving from a state to another, a timed automaton can either execute action transitions constrained by guards and invariants, or time transitions constrained by invariants only. Crucial for automated verification of timed automata is the definition of an equivalence relation $\equiv \subseteq \mathbb{R}^{|\mathcal{X}|} \times \mathbb{R}^{|\mathcal{X}|}$ on clocks valuations, which identifies two valuations v and v' in which either all the clocks exceed some value c_{max}[2], or two clocks x and y with the same integer part in v and v' and either their fractional parts are equal to 0, or are ordered in the same way, i.e., $fractional(v(x)) \leq fractional(v(y))$ iff $fractional(v'(x)) \leq fractional(v'(y))$. The equivalence classes of $\equiv$ are called *zones*. Since $\equiv$ is of finite index, there is only finitely many zones for each timed automaton.

In addition to the standard epistemic operators, the language of TECTLK contains the temporal operators EG and EU combined with time intervals I on reals in order to specify when precisely formulas are supposed to hold. Note that TECTLK does not include the next step operator EX as this operator is meaningless on dense time models. The formal syntax of TECTLK in BNF is as follows:

$$\varphi ::= p \mid \neg p \mid \varphi \wedge \varphi \mid \varphi \vee \varphi \mid \overline{\mathrm{K}}_i\varphi \mid \overline{\mathbf{E}_{\mathbf{K}}}_\Gamma\varphi \mid \overline{\mathbf{C}_{\mathbf{B}}}_\Gamma\varphi \mid \overline{\mathrm{C}}_\Gamma\varphi \mid \mathrm{EG}_I\varphi \mid \mathrm{E}(\varphi \mathrm{U}_I \varphi)$$

with $p \in PV$. A (real-time interpreted) model for TECTLK over a timed automaton is defined as a tuple $M = (Q, s^0, T, \sim_1, \ldots, \sim_n, V)$, where Q is the subset of $G \times \mathbb{R}^{|\mathcal{X}|}$ such that G is the set of locations of the timed automaton, all the states in Q are reachable from $s^0 = (g^0, v^0)$ with g^0 being the initial location of the timed automaton and v^0 the valuation in which all the clocks are equal to 0; T is defined by the action and timed transitions of the timed automaton, $\sim_i \subseteq Q \times Q$ is an epistemic relation for agent i defined by $(g, v) \sim_i (g', v)$ iff $g \sim_i g'$ and $v \equiv v'$; and $V : Q \times PV \rightarrow \{true, false\}$ is a valuation function for PV. Intuitively, in the above model two states are in the epistemic relation for agent i if their locations are in this relation according to the standard definition in Section 8.2 and their clocks valuations belong to the same zone.

In what follows, we give the semantics of $\mathrm{E}(\varphi \mathrm{U}_I \psi)$ and $\mathrm{EG}_I\varphi$ of TECTLK and discuss how BMC is applied to this logic. Differently from the paths of temporal-epistemic models, the paths in real-time models consist of action transitions interleaved with timed transitions. The time distance to a state s from the initial one at a given path can be computed by adding the times of all the timed transitions that have occurred up to this state. Following this intuition the semantics is formulated as follows:

- $(M, s) \models \mathrm{E}(\varphi \mathrm{U}_I \psi)$ iff there is a path in M starting at s which contains a state where ψ holds, reached from s within the time distance of I, and φ holds at all the earlier states,

- $(M, s) \models \mathrm{EG}_I\varphi$ iff there is a path in M starting at s such that φ holds at all the states within the time distance of I.

[2]This constant is computed from a timed automaton and a formula to be verified.

The idea of BMC for $(M, s^0) \models \varphi$, where φ is TECTLK formula, is based on two translations and on the application of BMC for ECTLK. An infinite real-time model M is translated to a finite epistemic model M_d and each formula φ of TECTLK is translated to the formula $cr(\varphi)$ of the logic ECTLK$_y$, which is a slight modification of ECTLK. The above two translations guarantee that $(M, s^0) \models \varphi$ iff $(M_d, s^0) \models cr(\varphi)$.

Assume we are given a timed automaton A and a TECTLK formula φ. We begin by translating the real-time model M (for A) to M_d. First, the automaton A is extended with one special clock y, an action a_y, and the set of transitions E_y labelled with a_y going from each location to itself and resetting the clock y. These transitions are used to start the paths over which sub-formulas of φ are checked. Then, the finite model M_d for the extended timed automaton is built. The model $M_d = (Q_d, q^0, T_d, \sim_1^d, \ldots, \sim_n^d, \mathcal{V}_d)$, where Q_d is a suitably selected (via discretisation) finite subset of Q, the relations $T_d, \sim_i^d$ are suitably defined restrictions of the corresponding relations in M, and $\mathcal{V}_d = \mathcal{V}|Q_d$.

The above translation cr of the temporal modalities is non-trivial only. Applying cr to $\mathrm{E}(\alpha \mathrm{U}_I \beta)$ we get the formula $\mathrm{EX}_y\mathrm{E}(cr(\alpha)\mathrm{U}cr((\beta) \wedge p))$, where the operator EX_y is interpreted over the transitions corresponding to the action a_y, and p is a propositional formula characterising zones. A similar translation applies to $\mathrm{EG}_I\alpha$.

After the above two translations have been defined, the model checking of a TECTLK formula φ over M is reduced to model checking of $cr(\varphi)$ over M_d, for which BMC can be used as presented in Section 8.4.1.

8.5.1 Example

To exemplify the expressive power of TECTLK we specify a correctness property for an extension of the *Railroad Crossing System* (RCS), a well-known example in the literature of real-time verification. Below, we give its short description.

The system consists of three agents: Train, Gate, and Controller running in parallel and synchronising through the events: *approach, exit, lower* and *raise*. When a train approaches the crossing, Train sends the signal **approach** to Controller and enters the crossing between 300 and 500 milliseconds (ms) from this event. When Train leaves the crossing, it sends the signal **exit** to Controller. Controller sends the signal **lower** to Gate exactly 100ms after the signal **approach** is received, and sends the signal **raise** signal within 100ms after **exit**. Gate performs the transition **down** within 100ms of receiving the request **lower**, and responds to **raise** by moving **up** between 100ms and 200ms.

Consider the following correctness property: there exists a behaviour of

RCS such that agent Train considers possible a situation in which it sends the signal **approach** but agent Gate does not send the signal **down** within 50 ms. This property can be formalised by the following TECTLK formula:

$$\varphi = \mathrm{EF}_{[0,\infty]}\overline{\mathrm{K}}_{Train}(\mathbf{approach} \wedge \mathrm{EF}_{[0,50]}(\neg\mathbf{down})).$$

By using BMC techniques we can verify the above property for RCS.

8.6 Partial Order Reductions

Several approaches are available to alleviate the difficulty of verifying large state spaces. Partial order reductions have extensively been used in the verification of reactive systems specified against $\mathrm{LTL}_{-\mathrm{X}}$ and $\mathrm{CTL}_{-\mathrm{X}}$ formulas. So far, the only approach which has been implemented and proved efficient for MAS was defined over interleaved interpreted systems. We follow this approach in our presentation.

The main idea behind the partial order reductions is the observation that two paths that differ only in the ordering of independent actions will satisfy the same temporal properties, provided the next step operator is not used. Therefore, rather than dealing with the full model, one can generate a reduced one, which does not contain all the interleavings of the independent actions and still satisfies the same properties. In order to describe partial order reductions used for generating reduced models, we need the following relations and definitions.

Let $i \in A$, $g, g' \in G$, and $J \subseteq A$. The relation $\sim_J = \bigcap_{j \in J} \sim_j$ corresponds to the indistinguishably relation for the epistemic modality of distributed knowledge in group J, whereas the relation $I = \{(a,b) \in Act \times Act \mid Agent(a) \cap Agent(b) = \emptyset\}$ is referred to as the *independence relation* in partial order approaches. Notice that $\sim_\emptyset = G \times G$ while $\sim_{\mathrm{A}} = id_G$. Two actions $a, a' \in Act$ are *dependent* if $(a, a') \notin I$. An action $a \in Act$ is *invisible* in a model M if whenever $g \xrightarrow{a} g'$ for any two states $g, g' \in G$ we have that $V(g) = V(g')$. Given two models $M = (G, g, T, \sim_1, .., \sim_n, V)$ and $M = (G', g', T', \sim'_1, .., \sim'_n, V')$. If $G' \subseteq G$, $g' = g$ and $V' = V|G'$, then we write $M' \subseteq M$ and say that M' is a submodel of M, or that M' is a *reduced* model of M.

As mentioned above, the idea of verification by model checking with partial order reductions is to define an algorithm which generates reduced models which preserve the satisfaction of a class of formulae. In general, the algorithm can be defined for several classes of formulae, which are subsets of CTL*K. In what follows we present an algorithm for CTLK_{-X}, i.e., CTLK without the next step operator. Observe that the formula $EX(executed_a \wedge EXexecuted_b)$, where the proposition $executed_x$ denotes that action $x \in$

$\{a, b\}$ is executed, can distinguish between two paths in which the ordering of the actions a and b is different. This explains why the next step operator is not used in this context.

For $J \subseteq A$, we write CTLK^J_{-X} for the restriction of the logic CTLK_{-X} such that for each subformula $K_i\varphi$ we have $i \in J$. One can define a notion of equivalence on models, called J-stuttering bisimulation, which guarantees the preservation of the formulas in CTLK^J_{-X}. The algorithm presented explores the given model M and returns a reduced one, which is J-stuttering bisimilar to M. Traditionally, in partial order reduction the exploration is carried out by depth-first-search (DFS). In this context DFS is used to compute successor states that will make up the reduced model by exploring systematically the possible computation tree and selecting only some of the possible paths generated. In the following a stack represents the path $\pi = g_0 a_0 g_1 a_1 \cdots g_n$ currently being visited. For the top element of the stack g_n the following three operations are computed in a loop:

1. The set $en(g_n) \subseteq Act$ of enabled actions (not including the ϵ action) is identified and a subset $E(g_n) \subseteq en(g_n)$ of possible actions is heuristically selected (see below).

2. For any action $a \in E(g_n)$ compute the successor state g' such that $g_n \xrightarrow{a} g'$, and add g' to the stack thereby generating the path $\pi' = g_0 a_0 g_1 a_1 \cdots g_n a g'$. Recursively proceed to explore the submodel originating at g' in the same way by means of the present algorithm beginning at step 1.

3. Remove g_n from the stack.

The algorithm begins with a stack comprising of the initial state and terminates when the stack is empty. The model generated by the algorithm is a submodel of the original one. The choice of $E(q)$ is constrained by the class of formulae of CTLK^J_{-X} that must be preserved. Below we give the conditions defining a heuristics for the selection of $E(g)$ (such that $E(g) \neq en(g)$) while visiting state g in the algorithm above.

C1 No action $a \in Act \setminus E(g)$ that is dependent on an action in $E(g)$ can be executed before an action in $E(g)$ is executed.

C2 For every cycle in the constructed state graph there is at least one node g in the cycle for which $E(g) = en(g)$, i.e., for which all the successors of g are expanded.

C3 All actions in $E(g)$ are invisible.

C4 $E(g)$ is a singleton set.

CJ For each action $a \in E(g)$, $Agent(a) \cap J = \emptyset$, i.e., no action in $E(g)$ changes local states of the agents in J.

The conditions **C1** – **C3** are used for preserving $\text{LTL}_{-\text{X}}$, **C4** for preserving $\text{CTL}_{-\text{X}}$, whereas **CJ** is aimed at preserving the truth value of subformulae of the form $K_i\varphi$ for $i \in J$.

8.6.1 Evaluation

In order to evaluate partial order reductions generated by the implementation of the above method, a prototype tool, based on MCMAS, was implemented. Experimental results have been obtained for three well-knows systems in the literature on MAS: The Train, Gate, and Controller (TGC),the Dining Cryptographers Protocol (DCP) introduced earlier, and the Write-Once cache coherence protocol (WO).

Starting with TGC, the property checked expresses that whenever the train 1 is in the tunnel, it knows that no other train is in the tunnel at the same time:

$$AG(\text{in\_tunnel}_1 \rightarrow K_{\text{train}_1} \bigwedge_{i=2}^{n} \neg\text{in\_tunnel}_i), \tag{8.5}$$

where n is the number of trains in the system, and the proposition in_tunnel$_i$ holds in the states where the train i is in the tunnel. The results showed that the size of the reduced state space $R(n)$ generated by the algorithm is a function of the number of trains n, for $1 \leq n \leq 10$, which turns out to be exponentially smaller when compared to the size of the full state space $F(n)$ below:

- $F(n) = c_n \times 2^{n+1}$, for some $c_n > 1$,
- $R(n) = 3 + 4(n-1)$.

Regarding the DCP scenario and the Write-Once cache coherence protocol, it was found that the algorithm for CTLK^J_{-X} brings negligible benefits for the tested properties. However, in both the cases, substantial reductions are obtained if the properties are expressed in LTLK^J_{-X}. This can be explained by the fact that in order to preserve LTLK^J_{-X} the set $E(g)$ does not need to satisfy the condition **C4**, thereby leading to smaller reduced models

8.7 Notes

Background. Treatments of epistemic logic (see Chapter 1) in terms of the modal system $S5_n$ are normally attributed to the work of Hintikka (1962). In these approaches the semantics is given in terms of Kripke models as introduced by Kripke (1959). The formalism of interpreted systems, introduced in Section 8.2 as a semantics for epistemic logic was put forward independently by Parikh and Ramanujam (1985), Rosenschein (1985), and by Halpern and Moses (1990), and later popularised by Fagin, Halpern, Moses, and Vardi (1995). Interleaved interpreted systems were studied by Lomuscio, Penczek, and Qu (2010) in the context of partial order reductions. They are inspired by standard synchronisation patterns used in network of automata. Other variants of interpreted systems have been discussed in the literature: see the work of Jamroga and Ågotnes (2007), and of Kouvaros and Lomuscio (2013).

Over the years, several scenarios have been modelled as interpreted systems and specified by means of temporal-epistemic properties. Fagin, Halpern, Moses, and Vardi (1995) present an in-depth analysis of a number of scenarios, including communication protocols, against epistemic specifications. The dining cryptographers problem was originally introduced by Chaum (1988) and first analysed in a temporal-epistemic setting by van der Meyden and Su (2004). The formulation presented in Section 8.2 was first discussed by Raimondi and Lomuscio (2007). A reformulation of the protocol including non trustworthy cryptographers was done by Kacprzak, Lomuscio, Niewiadomski, Penczek, Raimondi, and Szreter (2006). The original wording reported in Section 8.2 was first cited by van der Meyden and Su (2004). The train-gate-controller scenario was presented in the context of alternating temporal logic by Alur, Henzinger, Mang, Qadeer, Rajamani, and Tasiran (1998) and recast in terms of epistemic properties by van der Hoek and Wooldridge (2002b). The description adopted here was encoded as an interpreted system in a paper by Jones and Lomuscio (2010). The Write-Once cache coherence protocol described by Baukus and van der Meyden (2004) and by Archibald and Baer (1986), used as a benchmark in Section 8.6, was encoded as an interleaved interpreted system by Lomuscio et al. (2010).

Research on verification of epistemic specifications by model checking was spurred by Halpern and Vardi (1991). At the time theorem proving was the leading verification technology and approaches to verifying epistemic specifications relied on various proof-theoretical techniques for the epistemic logic $S5_n$ enriched by operators for group knowledge. A case in point is the well-known BAN approach put forward by Burrows, Abadi, and Needham (1990) whereby authentication properties could be shown, it was argued, through automatic proof procedures based on axiomatisations inspired by epistemic logic. Halpern and Vardi (1991) presented the rationale for verifying systems against epistemic properties by model theoretic approaches instead. While in the key example discussed in the paper (the muddy children problem) the temporal evolution is captured through updates of static epistemic models, various remarks on the possible theoretical advantage of model checking procedures over theorem proving are given and used to motivate research into model-theoretical approaches.

Halpern and Vardi (1991) initially spurred considerable research in the area of updates for static epistemic models, like the work by Baltag, Moss, and Solecki (1998), Lomuscio and Ryan (1999), and by Gerbrandy (1999). As most of the updates first studied concerned announcements and information sharing, this in turn lead to considerable research on logics for public announcements, as first suggested by Plaza (1989). As far as we are aware, it was not before the early 2000s that logics of knowledge and time were used as specification languages on which model checking procedures could be applied to. Ten years on from then, while Halpern and Vardi (1991) only expressed the view that "In summary, we do not expect the model checking approach to supplement the theorem proving approach." (p. 19), it appears that model checking (see the volume by Clarke, Grumberg, and Peled (1999)) has now become the de-facto technique for the verification of systems against temporal-epistemic specifications.

This development was profoundly influenced by at least two decades of successful research in model checking for purely temporal specifications where several techniques have been put forward with the aim of mitigating the state-explosion problem. The most prominent methodologies include partial order reductions as proposed by Valmari (1990), Godefroid (1991), and by Peled (1993), symmetry reductions as studied by Clarke, Filkorn, and Jha (1993), Emerson and Jutla (1993), and by Emerson and Sistla (1995), ordered-binary decision diagrams as described by Burch, Clarke, McMillan, Dill, and Hwang (1990) and by McMillan (1993), bounded and unbounded model checking as suggested by Biere, Cimatti, Clarke, and Zhu (1999) and by McMillan (2002), and various forms of abstraction as, among others, discussed by Dams, Gerth, Dohmen, Herrmann, Kelb, and Pargmann (1994) and by Ball, Podelski, and Rajamani (2001). A number of tools have been released implementing these techniques; some of these have reached a high level of maturity and are used in industrial settings. These include SPIN, see the work of Holzmann (1997), which adopts an on-the-fly automata-based approach combined with partial-order reductions for LTL specifications and NuSMV, for which we refer to work by McMillan (1993) and Cimatti, Clarke, Giunchiglia, and Roveri (1999), which used OBDDs and bounded model checking for symbolic model checking against

LTL and CTL specifications. Other implementations are available for either the same or different classes of temporal specifications including real-time logics, probabilistic logic, etc. The work reported in this paper is influenced by the successful methodologies employed against temporal specifications.

To our knowledge the first treatment of model checking against epistemic specifications, in the sense discussed in this chapter, was undertaken by van der Meyden and Shilov (1999a) where the problem is formalised and complexity results for perfect recall semantics are given. In what follows we briefly summarise other notable approaches to the verification of systems against temporal-epistemic specification and relate them to the material presented earlier. Due to the recent growth in the area the discussion is incomplete.

Reduction-based approaches. One of the first approaches to model checking temporal-epistemic logic that we are aware of was put forward by van der Hoek and Wooldridge (2002a, 2003), who proposed a reduction from temporal-epistemic specifications to plain temporal specifications. The approach uses local propositions to identify and fully characterise the local states of the agents. A feature of this approach is that no automatic procedure for the automatic synthesis of local propositions from the model is given. As stated by the authors, the approach is inspired by Engelhardt, van der Meyden, and Moses (1998); in their paper a logic for local proposition is developed, and the basic principles through which epistemic formulas can be rewritten via quantification of local propositions are put forward. It should be noted that, differently from van der Hoek and Wooldridge (2002a), the technique of Engelhardt et al. (1998) focuses on representational issues in static epistemic logic. In this context it is shown that negative results such as non-axiomatisability and undecidability follow in some settings because of reductions to second-order logic even when assuming a weak semantics.

More recently, temporal-epistemic logic on discrete time was reduced to a special case of action-restricted CTL (ARCTL for short) by Lomuscio, Pecheur, and Raimondi (2007b). ARCTL as proposed by Pecheur and Raimondi (2007) is an extension of CTL whereby path quantifiers are labelled with actions. The work of Lomuscio, Pecheur, and Raimondi (2007a) concerns a reduction of CTLK to ARCTL and a compiler into NuSMV implementing the translation. In this work the relations corresponding to the epistemic modalities are effectively recast as special actions in the corresponding action-based transition system and special labels are used for the temporal relations. Experimental results on the dining cryptographers showed that the approach was as efficient as the other toolkits available at the time.

OBDD-based approaches. The MCMAS toolkit developed by Raimondi and Lomuscio (2007), Lomuscio and Raimondi (2006b), and by Lomuscio, Qu, and Raimondi (2009) described in Section 8.3 implements interpreted systems semantics and supports the verification of systems against not only temporal-epistemic specifications but also properties based on deontic concepts as done by Raimondi and Lomuscio (2004a) and by Woźna, Lomuscio, and Penczek (2005a), explicit knowledge as discussed by Lomuscio, Raimondi, and Woźna (2007) and ATL cooperation primitives as put forward by Lomuscio and Raimondi (2006c). It is implemented in C++ and relies on the latest BDD package of Somenzi (2005). An in-depth description of the tool is given by Raimondi (2006). More details on OBDDs and related techniques can be found in papers by Bryant (1986) and by Huth and Ryan (2000).

Van der Meyden and colleagues were the first to propose and implement the use of OBDD-based model checking in a temporal-epistemic setting, see the paper by Gammie and van der Meyden (2004). These were preceded by theoretical studies on the computational complexity of the model checking problem by, e.g., van der Meyden and Shilov (1999b); further results on this can be found in work by Lomuscio and Raimondi (2006a), by Engelhardt, Gammie, and van der Meyden (2007), and by Huang and van der Meyden (2010). Like MCMAS, MCK has undergone several versions since its original version. At the time of writing it supports CTL*K specifications and a variant of probabilistic knowledge, i.e., epistemic modalities defined on probabilistic systems as by Huang, Luo, and Meyden (2011). MCK is released in binary form and is implemented in Haskell by using using Long's BDD library. A difference with respect to MCMAS and VerICS is that MCK includes specialised implementations for different evolution semantics including perfect recall and clock semantics. While these can also be encoded on specific examples with MCMAS and VerICS it is possible that specialised semantics can provide the user with efficiency gains; no comparison has been made in this respect so far.

A further BDD-based checker, MCTK, has recently been released as described by Luo (2009). MCTK follows the local propositions approach discussed above to reduce the verification of epistemic specifications to temporal formulas only. MCTK uses NuSMV as the underlying temporal checker. Differently from the approach of van der Hoek and Wooldridge (2002a), a technique for the automatic calculation of local propositions is here given by Su (2004). Experimental results against other checkers are not available at the time of writing.

SAT-based approaches. The SAT-based BMC approach presented in Section 8.4.1 predates the OBDD-based approaches here described. While techniques using OBDDs became prevalent in the ten years up to 2010, there has recently been an increased activity on SAT-based methods. The VerICS tool of Kacprzak, Nabialek, Niewiadomski, Penczek, Pólrola, Szreter,

Woźna, and Zbrzezny (2008) and by Penczek and Półrola (2006), described in Section 8.4, implements timed automata in the sense of Alur and Dill (1994) and interleaved interpreted systems semantics as proposed in Lomuscio, Penczek, and Qu (2010) to support the SAT-based verification of systems against not only temporal-epistemic specifications but also properties based on deontic concepts, see the work of Woźna, Lomuscio, and Penczek (2005a), and of Woźna, Lomuscio, and Penczek (2005b), and real time systems as described by Lomuscio, Woźna, and Penczek (2007).

SAT-based BMC was originally introduced for the verification of LTL specifications by Biere et al. (1999) and by Biere, Cimatti, Clarke, Strichman, and Zhu (2003) as an alternative to approaches based on OBDDs. Then, BMC was defined for ECTL - the existential fragment of CTL, first by Penczek, Woźna, and Zbrzezny (2002) and then refined by Zbrzezny (2008) such that a specific symbolic k-path is allocated to each subformula of the tested formula starting with a modality. Moreover, reduced Boolean circuits as described by Abdulla, Bjesse, and Eén (2000) are used in the encoding of Zbrzezny (2008). A reduced Boolean circuit represents subformulas of the encoding by fresh propositions such that each two identical subformulas correspond to the same proposition. BMC for ECTL was extended to ECTLK by Penczek and Lomuscio (2003) and further to ECTLKD by Woźna, Lomuscio, and Penczek (2005a). The solution of Zbrzezny (2008) for ECTL was extended by Huang, Luo, and van der Meyden (2010) to ECTLK, but without using reduced Boolean circuits.

The approach based on UMC, discussed in Section 8.4.2, was originally introduced by McMillan (2002) for the verification of CTL as an alternative to BMC and the approaches based on BDDs. Then, UMC was extended to $\mathrm{CTL_pK}$ by Kacprzak, Lomuscio, and Penczek (2004). The reader is referred to (Kacprzak, Lomuscio, and Penczek, 2003) for more details on UMC, especially on computing the fix-points over propositional representations of sets. The standard algorithm toCNF transforming a propositional formula to one in CNF, preserving satisfiability only, was presented by McMillan (2002) and by Penczek and Półrola (2006). The translation *equCNF* to an equivalent propositional formula in CNF was given by Penczek and Półrola (2006). We refer the reader to work by Chauhan, Clarke, and Kroening (2003) and by Ganai, Gupta, and Ashar (2004), where alternative solutions can be found. Blocking clauses used in the algorithm *equCNF* can be computed using the methods discussed by McMillan (2002) and by Szreter (2006, 2005).

In addition to the BMC approaches for extensions of CTLK discussed in this chapter, BMC approaches for LTLK (Męski, Penczek, Szreter, Woźna-Szcześniak, and Zbrzezny, 2014) have been put forward, like for instance by Penczek, Woźna-Szcześniak, and Zbrzezny (2012), using a translation to

SAT and also by Meşki, Penczek, and Szreter (2012), using operations on BDDs. Moreover, the latest release of the MCK checker reported above, now supports BMC through SAT. Experimental results tend to show that OBDDs and SAT-based BMC methods are complementary to one another with BMC working better for reachability and checking small epistemic specifications on very large models, and OBDDs outperforming BMC with complex specifications and models with reachable state spaces in the region of 10^6 to 10^{10}.

Abstraction. While OBDDs and SAT-based approaches can be very effective for representing large state spaces, the state-space grows exponentially with the number of variables in the system. To alleviate this difficulty various forms of abstraction have been put forward to verify systems against temporal-epistemic specifications. The first abstraction technique that we are aware of in this context was developed by Enea and Dima (2007), where a number of abstractions for Kripke models with epistemic relations are defined. The methodology is defined for a specification language based on branching-time temporal (both past and future) and epistemic modalities, interpreted on a Kleene's three valued semantics. This enables the authors to give weak-preservation and error-preservation results for temporal-epistemic specifications with respect to the three-valued semantics given. Somewhat related to this is the approach by Dechesne, Orzan, and Wang (2008) where a notion of refinement is developed in the context of public announcement logic, thereby enabling the authors to give abstractions for which a preservation theorem can be shown. In this case, however, two-valued interpretations are used.

While Enea and Dima (2007) and Dechesne, Orzan, and Wang (2008) developed their work for Kripke models, other techniques have adopted a modular, agent-based view and used interpreted systems as the underlying semantics for Kripke models. The first approach following this line appears to be that of Cohen, Dam, Lomuscio, and Russo (2009),where an existential abstraction technique is developed and a preservation theorem shown. The approach involves taking a quotient of an interpreted system by defining abstract local states, actions, protocols and the transition relation on the abstract model. This enables the authors to show that if a specification in the universal fragment of CTLK holds in the abstract model, it also holds in the concrete one. The approach was later extended by Lomuscio, Qu, and Russo (2010), where a data abstraction methodology for interpreted systems specified against CTLK formulas was put forward and implemented in conjunction with MCMAS. More recently Al-Bataineh and van der Meyden (2011), presented abstraction results applicable to the verification of the dining cryptographers scenario and applied them to knowledge-based programs.

Symmetry reduction is a well-established form of abstraction whereby symmetry considerations are exploited to produce abstract models preserving a given specification. Cohen, Dam, Lomuscio, and Qu (2009b) use a counterpart semantics to interpret epistemic modalities on abstract models by means of symmetry considerations. Experimental results presented by Cohen, Dam, Lomuscio, and Qu (2009b) show a linear reduction in the memory requirements for BDD-based verification. While they relied on manual identification of symmetries, Cohen, Dam, Lomuscio, and Qu (2009a) presented an automatic technique for the reduction and application to data symmetry as well. The technique was implemented on an ad-hoc, extended version of ISPL system descriptions; the benchmarks reported showed an exponential reduction in the time and memory footprint in some scenarios amenable to symmetry reduction.

Partial order reductions have extensively been used for the verification of reactive systems specified against $\mathrm{LTL}_{-\mathrm{X}}$ by e.g., Peled (1994) and against $\mathrm{CTL}_{-\mathrm{X}}$ formulas by Gerth, Kuiper, Peled, and Penczek (1999) and by Penczek, Szreter, Gerth, and Kuiper (2000). Lomuscio, Penczek, and Qu (2009) present theoretical results in the context of interpreted systems and temporal-epistemic logic. In Section 8.6 we summarised the work of Lomuscio, Penczek, and Qu (2010), as this is the only approach we are aware of which has been implemented and shown to be efficient. Traditionally, in partial order reductions the exploration is carried out by depth-first-search (DFS), as done by e.g., Gerth, Kuiper, Peled, and Penczek (1999), or by double-depth-first-search, as done by e.g., Courcoubetis, Vardi, Wolper, and Yannakakis (1992). The conditions $C1-C3$, used in the algorithm, are inspired by Peled (1993) and by Gerth, Kuiper, Peled, and Penczek (1999).

Optimised algorithms. Some results in the literature have focused on novel, optimised algorithms for the verification of temporal-epistemic specifications. Sometimes these algorithms are distributed or parallel. For example, Kwiatkowska, Lomuscio, and Qu (2010) present parallel versions of the labelling algorithms for the automatic verification of temporal-epistemic properties. The results point to a significant speed-up in the labelling of formulas although the performance is strongly dependent on the number of cores available and the type of specification to be checked. The work by Jones and Lomuscio (2010), discussed above in the context of the combination between BMC and OBDDs, also includes a distributed algorithm for bounded satisfaction based on the notion of seed states for state-space partitioning. In a different context Cohen and Lomuscio (2010) put forward an algorithm for the non-elementary speed-up of model checking synchronous systems with perfect recall. An improved encoding for the BMC problem via SAT, which was shown to generate a polynomially smaller number of propositions in the encodings thereby allowing faster verification times, was presented by Huang, Luo, and Meyden (2010). An approach to synthesising groups of agents satisfying an epistemic specification on a given system was explored by Jones, Knapik, Lomuscio, and Penczek (2012).

Extensions to other agent-based specifications. Model checking approaches have also been investigated for specifications richer than the temporal-epistemic logics discussed here. Raimondi and Lomuscio (2004b) presented an OBDD-based approach to the verification of deontic interpreted systems (Lomuscio and Sergot, 2003); the BMC case was analysed by Woźna, Lomuscio, and Penczek (2005a). Deontic interpreted systems are a formalism for the representation of correct functioning behaviour of the agents in a system. Local states are partitioned into correct and incorrect ones, and a further agent-index modality representing "at all the correct states for agent i" is introduced. The modality is interpreted by considering the global states in which the agent in question is operating correctly. By means of this formalism, one can analyse scenarios where some agents may display faulty behaviour. For example, the properties of a variant of the dining cryptographers scenario where some cryptographers are intruders saying the opposite of what they should was verified through this formalism by Kacprzak, Lomuscio, Niewiadomski, Penczek, Raimondi, and Szreter (2006).

Extensions to epistemic logic to include explicit knowledge have also been discussed and implemented, e.g., by Lomuscio and Woźna (2006) and by Lomuscio, Raimondi, and Woźna (2007). Both VerICS and MCMAS support these features.

Model checking of systems against alternating-temporal logic (ATL) specifications has been pursued by Alur, Henzinger, and Kupferman (2002). ATL extends CTL by adding strategies in the semantics and explicit constructs for representing what groups of agents can enforce. Its model checking problem is considerably harder under partial observability. MOCHA, see (Alur, de Alfaro, Henzinger, Krishnan, Mang, Qadeer, Rajamani, and Tasiran, 2000) is an explicit state model checker supporting ATL modalities. Even if strategies and knowledge can interact in subtle ways as argued by Jamroga and van der Hoek (2004)[3], progress has been made both in the definition of combinations between ATL with knowledge and their verification. Specifically, Lomuscio and Raimondi (2006c) put forward a symbolic, OBDD-based model checking algorithm for the verification of ATLK specifications and discussed experimental results. An alternative approach using MOCHA is discussed by Wooldridge, Agotnes, Dunne, and van der Hoek (2007) in combination with the local propositions construction referenced above.

[3] For a detailed discussion on this interaction, see Chapter 11.

Epistemic concepts have been used in a broader context to reason about multi-agent systems modelled by other attitudes (such as norms, beliefs, desires, goals, or intentions). These properties are often treated simply as propositions in a temporal language and not as first-citizens like the modalities discussed above. Given this, they are technically very different and not discussed here.

The conclusion we can draw from the results above is that temporal-epistemic logic specifications can now be verified effectively with appropriate symbolic model checking techniques.

Acknowledgements Much of the work described in this chapter is based on joint research with Magdalena Kacprzak, Hongyang Qu, Franco Raimondi, Maciej Szreter, Boèna Woźna Szcześniak, and Andrzej Zbrzezny.

We would like to thank Catalin Dima, Masoud Koleini and Ji Ruan for valuable feedback on a preliminary version of this chapter.

References

Abdulla, P. A., P. Bjesse, and N. Eén (2000). Symbolic reachability analysis based on SAT-solvers. In *Proceedings of the 6th Int. Conf. on Tools and Algorithms for the Construction and Analysis of Systems (TACAS'00)*, Volume 1785 of *LNCS*, pp. 411–425. Springer-Verlag.

Al-Bataineh, O. I. and R. van der Meyden (2011). Abstraction for epistemic model checking of dining cryptographers-based protocols. In *Proceedings of the 13th Conference on Theoretical Aspects of Rationality and Knowledge (TARK-2011)*, pp. 247–256.

Alur, R., L. de Alfaro, T. Henzinger, S. Krishnan, F. Mang, S. Qadeer, S. Rajamani, and S. Tasiran (2000). MOCHA user manual. Technical report, University of California at Berkeley. http://www-cad.eecs.berkeley.edu/~mocha/doc/c-doc/c-manual.ps.gz.

Alur, R. and D. Dill (1994). A theory of timed automata. *Theoretical Computer Science 126(2)*, 183–235.

Alur, R., T. Henzinger, and O. Kupferman (2002). Alternating-time temporal logic. *Journal of the ACM 49(5)*, 672–713.

Alur, R., T. Henzinger, F. Mang, S. Qadeer, S. Rajamani, and S. Tasiran (1998). MOCHA: Modularity in model checking. In *Proceedings of the 10th International Conference on Computer Aided Verification (CAV'98)*, Volume 1427 of *LNCS*, pp. 521–525. Springer-Verlag.

Archibald, J. and J.-L. Baer (1986). Cache coherence protocols: Evaluation using a multiprocessor simulation model. *ACM Transactions on Computer Systems 4*, 273–298.

Ball, T., A. Podelski, and S. K. Rajamani (2001). Boolean and cartesian abstraction for model checking c programs. In *Tools and Algorithms for the Construction and Analysis of Systems, 7th International Conference (TACAS 2001)*, pp. 268–283.

Baltag, A., L. S. Moss, and S. Solecki (1998). The logic of public announcement, common knowledge, and private suspicions. In I. Gilboa (Ed.), *Proceedings of the 7th Conference on Theoretical Aspects of Rationality and Knowledge (TARK-98)*, San Francisco, pp. 125–132. Morgan Kaufmann.

Baukus, K. and R. van der Meyden (2004). A knowledge based analysis of cache coherence. In Proceedings of 6th International Conference on Formal Engineering Methods (ICFEM'04), Volume 3308 of *LNCS*, pp. 99–114. Spriger-Verlag.

Biere, A., A. Cimatti, E. Clarke, O. Strichman, and Y. Zhu (2003). Bounded model checking. In *Highly Dependable Software*, Volume 58 of *Advances in Computers*. Academic Press. Pre-print.

Biere, A., A. Cimatti, E. Clarke, and Y. Zhu (1999). Symbolic model checking without BDDs. In *Proceedings of the 5th Int. Conf. on Tools and Algorithms for the Construction and Analysis of Systems (TACAS'99)*, Volume 1579 of *LNCS*, pp. 193–207. Springer-Verlag.

Bryant, R. (1986). Graph-based algorithms for boolean function manipulation. *IEEE Transaction on Computers 35(8)*, 677–691.

Burch, J. R., E. Clarke, K. L. McMillan, D. L. Dill, and L. J. Hwang (1990). Symbolic model checking: 10^{20} states and beyond. *Information and Computation 98(2)*, 142–170.

Burrows, M., M. Abadi, and R. Needham (1990). A logic of authentication. *ACM Transactions on Computer Systems 8*(1), 18–36.

Chauhan, P., E. Clarke, and D. Kroening (2003, July). Using SAT-based image computation for reachability analysis. Technical Report CMU-CS-03-151, Carnegie Mellon University.

Chaum, D. (1988). The dining cryptographers problem: Unconditional sender and recipient untraceability. *Journal of Cryptology 1(1)*, 65–75.

Cimatti, A., E. M. Clarke, F. Giunchiglia, and M. Roveri (1999). NUSMV: A new symbolic model verifier. In *Proceedings of the 11th International Conference on Computer Aided Verification (CAV'99)*, Volume 1633 of *LNCS*, pp. 495–499. Springer.

Clarke, E., T. Filkorn, and S. Jha (1993). Exploiting symmetry in temporal logic model checking. In *Proceedings of the 5th Int. Conf. on Computer Aided Verification (CAV'93)*, Volume 697 of *LNCS*, pp. 450–462. Springer-Verlag.

Clarke, E., O. Grumberg, and D. Peled (1999). *Model Checking.* MIT Press.

Cohen, M., M. Dam, A. Lomuscio, and H. Qu (2009a). A data symmetry reduction technique for temporal-epistemic logic. In Z. Liu and A. P. Ravn (Eds.), *Proceedings of the 7th International Symposium on Automated Technology for Verification and Analysis (ATVA 09)*, Volume 5799 of *Lecture Notes in Computer Science*, Macao, China, pp. 69–83. Springer.

Cohen, M., M. Dam, A. Lomuscio, and H. Qu (2009b). A symmetry reduction technique for model checking temporal-epistemic logic. In C. Boutilier (Ed.), *Proceedings of the 21st International Joint Conference on Artificial Intelligence (IJCAI)*, Pasadena, USA, pp. 721–726.

Cohen, M., M. Dam, A. Lomuscio, and F. Russo (2009). Abstraction in model checking multi-agent systems. In *Proceedings of the 8th International Conference on Autonomous Agents and Multiagent Systems (AAMAS09)*, Budapest, Hungary, pp. 945–952. IFAAMAS Press.

Cohen, M. and A. Lomuscio (2010). Non-elementary speedup for model checking synchronous perfect recall. In *Proceedings of the 19th European Conference on Artificial Intelligence (ECAI10)*, Lisbon,Portugal, pp. 1077–1078. IOS Press.

Courcoubetis, C., M. Vardi, P. Wolper, and M. Yannakakis (1992). Memory-efficient algorithms for the verification of temporal properties. *Formal Methods in System Design 1(2/3)*, 275–288.

Dams, D., R. Gerth, G. Dohmen, R. Herrmann, P. Kelb, and H. Pargmann (1994). Model checking using adaptive state and data abstraction. In *Proceedings of the 6th Int. Conf. on Computer Aided Verification (CAV'94)*, Volume 818 of *LNCS*, pp. 455–467. Springer-Verlag.

Dechesne, F., S. Orzan, and Y. Wang (2008). Refinement of kripke models for dynamics. In J. S. Fitzgerald, A. E. Haxthausen, and H. Yenigün (Eds.), *ICTAC*, Volume 5160 of *Lecture Notes in Computer Science*, pp. 111–125. Springer.

Emerson, E. A. and C. S. Jutla (1993). Symmetry and model checking. In *Proceedings of the 5th Int. Conf. on Computer Aided Verification (CAV'93)*, Volume 697 of *LNCS*, pp. 463–478. Springer-Verlag.

Emerson, E. A. and A. P. Sistla (1995). Symmetry and model checking. *Formal Methods in System Design 9*, 105–131.

Enea, C. and C. Dima (2007). Abstractions of multi-agent systems. In H.-D. Burkhard, G. Lindemann, R. Verbrugge, and L. Z. Varga (Eds.), *CEEMAS*, Volume 4696 of *Lecture Notes in Computer Science*, pp. 11–21. Springer.

Engelhardt, K., P. Gammie, and R. van der Meyden (2007). Model checking knowledge and linear time: PSPACE cases. In *LFCS*, pp. 195–211.

Engelhardt, K., R. van der Meyden, and Y. Moses (1998). Knowledge and the logic of local propositions. In *Proceedings of the 7th International Conference on Theoretical Aspects of Rationality and Knowledge (TARK'98)*, pp. 29–41.

Fagin, R., J. Y. Halpern, Y. Moses, and M. Vardi (1995). *Reasoning about Knowledge.* Cambridge: MIT Press.

Gammie, P. and R. van der Meyden (2004). MCK: Model checking the logic of knowledge. In *Proceedings of the 16th Int. Conf. on Computer Aided Verification (CAV'04)*, Volume 3114 of *LNCS*, pp. 479–483. Springer-Verlag.

Ganai, M., A. Gupta, and P. Ashar (2004). Efficient SAT-based unbounded symbolic model checking using circuit cofactoring. In *Proceedings of the Int. Conf. on Computer-Aided Design (ICCAD'04)*, pp. 510–517.

Gerbrandy, J. (1999). *Bisimulations on planet Kripke.* Ph. D. thesis, Institute for Logic, Language and Computation, University of Amsterdam.

Gerth, R., R. Kuiper, D. Peled, and W. Penczek (1999). A partial order approach to branching time logic model checking. *Information and Computation 150*, 132–152.

Godefroid, P. (1991). Using partial orders to improve automatic verification methods. In *Proceedings of the 2nd Int. Conf. on Computer Aided Verification (CAV'90)*, Volume 3 of *ACM/AMS DIMACS Series*, pp. 321–340.

Halpern, J. and Y. Moses (1990). Knowledge and common knowledge in a distributed environment. *Journal of the ACM 37*(3), 549–587. A preliminary version appeared in *Proc. 3rd ACM Symposium on Principles of Distributed Computing*, 1984.

Halpern, J. and M. Vardi (1991). *Model checking vs. theorem proving: a manifesto*, pp. 151–176. Artificial Intelligence and Mathematical Theory of Computation. Academic Press, Inc.

Hintikka, J. (1962). *Knowledge and Belief, An Introduction to the Logic of the Two Notions.* Ithaca (NY) and London: Cornell University Press.

van der Hoek, W. and M. Wooldridge (2002a). Model checking knowledge and time. In *Proceedings of the 9th Int. SPIN Workshop (SPIN'02)*, Volume 2318 of *LNCS*, pp. 95–111. Springer-Verlag.

van der Hoek, W. and M. Wooldridge (2002b). Tractable multiagent planning for epistemic goals. In M. Gini, T. Ishida, C. Castelfranchi, and W. L. Johnson (Eds.), *Proceedings of the 1st Int. Conf. on Autonomous Agents and Multi-Agent Systems (AAMAS'02)*, Volume III, pp. 1167–1174. ACM Press.

van der Hoek, W. and M. Wooldridge (2003, September). Model checking cooperation, knowledge, and time - a case study. *Research In Economics 57*(3), 235–265.

Holzmann, G. J. (1997). The model checker SPIN. *IEEE transaction on software engineering 23*(5), 279–295.

Huang, X., C. Luo, and R. v. Meyden (2010). Improved bounded model checking for a fair branching-time temporal epistemic logic. In W. van der Hoek, G. A. Kaminka, Y. Lespérance, M. Luck, and S. Sen (Eds.), *Proceedings of the 9th International Conference on Autonomous Agents and Multiagent Systems (AAMAS 2010)*, pp. 1403–1404. IFAAMAS.

Huang, X., C. Luo, and R. v. Meyden (2011). Symbolic model checking of probabilistic knowledge. In K. R. Apt (Ed.), *Proceedings of the 13th Conference on Theoretical Aspects of Rationality and Knowledge (TARK-2011)*, pp. 177–186. ACM.

Huang, X., C. Luo, and R. van der Meyden (2010). Improved bounded model checking for a fair branching-time temporal epistemic logic. In *Proceedings of the 6th Int. Workshop on Model Checking and Artificial Intelligence (MoChArt'10)*, pp. 95–111.

Huang, X. and R. van der Meyden (2010). The complexity of epistemic model checking: Clock semantics and branching time. In *Proceedings of the 19th European Conference on Artificial Intelligence (ECAI 2010)*, pp. 549–554.

Huth, M. R. A. and M. D. Ryan (2000). *Logic in Computer Science: Modelling and Reasoning about Systems.* Cambridge, England: Cambridge University Press.

Jamroga, W. and T. Ågotnes (2007). Modular interpreted systems. In *Proceedings of the 6th International Conference on Autonomous Agents and Multi-Agent systems (AAMAS 2007)*, pp. 131–138.

Jamroga, W. and W. van der Hoek (2004). Agents that know how to play. *Fundamenta Informaticae 63(2-3)*, 185–219.

Jones, A. V., M. Knapik, A. Lomuscio, and W. Penczek (2012). Group synthesis for parametric temporal-epistemic logic. In *Proceedings of the 11th International Conference on Autonomous Agents and Multi-Agent systems (AAMAS'12)*, Valencia, Spain, pp. 1107–1114. IFAAMAS Press.

Jones, A. V. and A. Lomuscio (2010). Distributed BDD-based BMC for the verification of Multi-Agent Systems. In *Proceedings of the 9th International Conference on Autonomous Agents and Multi-Agent systems (AAMAS 2010)*, pp. 675–682. IFAAMAS Press.

Kacprzak, M., A. Lomuscio, A. Niewiadomski, W. Penczek, F. Raimondi, and M. Szreter (2006). Comparing BDD and SAT based techniques for model checking Chaum's dining cryptographers protocol. *Fundamenta Informaticae 63*(2,3), 221–240.

Kacprzak, M., A. Lomuscio, and W. Penczek (2003, December). Unbounded model checking for knowledge and time. Technical Report 966, ICS PAS, Ordona 21, 01-237 Warsaw.

Kacprzak, M., A. Lomuscio, and W. Penczek (2004). From bounded to unbounded model checking for temporal epistemic logic. *Fundamenta Informaticae 63(2-3)*, 221–240.

Kacprzak, M., W. Nabialek, A. Niewiadomski, W. Penczek, A. Pólrola, M. Szreter, B. Woźna, and A. Zbrzezny (2008). VerICS 2007 - a model checker for knowledge and real-time. *Fundamenta Informaticae 85*(1-4), 313–328.

Kouvaros, P. and A. Lomuscio (2013). A cutoff technique for the verification of parameterised interpreted systems with parameterised environments. In *Proceedings of the 23rd International Joint Conference on Artificail Intelligence (IJCAI'13)*, pp. 2013–2019. AAAI Press.

Kripke, S. (1959). Semantic analysis of modal logic (abstract). *Journal of Symbolic Logic 24*, 323–324.

Kwiatkowska, M., A. Lomuscio, and H. Qu (2010). Parallel model checking for temporal epistemic logic. In *Proceedings of the 19th European Conference on Artificial Intelligence (ECAI'10)*, Lisbon, Portugalu, pp. 543–548. IOS Press.

Lomuscio, A., C. Pecheur, and F. Raimondi (2007a, January). Automatic verification of knowledge and time with NuSMV. In *Proceedings of the Twentieth International Joint Conference on Artificial Intelligence*, Hyderabad, India, pp. 1384–1389. AAAI.

Lomuscio, A., C. Pecheur, and F. Raimondi (2007b, January). Verification of knowledge and time with nusmv. In *Proceedings of the 20th International Joint Conference on Artificial Intelligence*, Hyderabad, India, pp. 1384–1389. AAAI.

Lomuscio, A., W. Penczek, and H. Qu (2009). Towards partial order reduction for model checking temporal epistemic logic. In *Proceedings of the 7th International Workshop on Model Checking and Artificial Intelligence (MoChArt 2009)*, Volume 5348 of *LNAI*, pp. 106–121. Springer-Verlag.

Lomuscio, A., W. Penczek, and H. Qu (2010). Partial order reductions for model checking temporal-epistemic logics over interleaved multi-agent systems. *Fundamenta Informaticae 101*(1-2), 71–90.

Lomuscio, A., H. Qu, and F. Raimondi (2009). Mcmas: A model checker for the verification of multi-agent systems. In A. Bouajjani and O. Maler (Eds.), *Proceedings of the 21th International Conference on Computer Aided Verification (CAV 2009)*, Volume 5643 of *Lecture Notes in Computer Science*, pp. 682–688. Springer.

Lomuscio, A., H. Qu, and F. Russo (2010). Automatic data-abstraction in model checking multi-agent systems. In R. van der Meyden and J.-G. Smaus (Eds.), *MoChArt*, Volume 6572 of *Lecture Notes in Computer Science*, pp. 52–68. Springer.

Lomuscio, A. and F. Raimondi (2006a). The complexity of model checking concurrent programs against CTLK specifications. In *Proceedings of the 5th International Joint Conference on Autonomous Agents and Multiagent Systems (AAMAS'06)*, Hakodake, Japan, pp. 548–550. ACM Press.

Lomuscio, A. and F. Raimondi (2006b). MCMAS: A model checker for multi-agent systems. In H. Hermanns and J. Palsberg (Eds.), *Proceedings of the 12th International Conference on Tools and Algorithms for Construction and Analysis of Systems (TACAS 2006)*, Volume 3920, pp. 450–454. Springer Verlag.

Lomuscio, A. and F. Raimondi (2006c). Model checking knowledge, strategies, and games in multi-agent systems. In *Proceedings of the 5th International Joint Conference on Autonomous Agents and Multiagent Systems (AAMAS'06)*, Hakodake, Japan, pp. 161–168. ACM Press.

Lomuscio, A., F. Raimondi, and B. Woźna (2007). Verification of the Tesla protocol in MCMAS-X. *Fundamenta Informaticae 79*(3–4), 473–486.

Lomuscio, A. and M. Ryan (1999). An algorithmic approach to knowledge evolution. *Artificial Intelligence for Engineering Design, Analysis and Manufacturing (AIEDAM) 13*(2), 119–132.

Lomuscio, A. and M. Sergot (2003). Deontic interpreted systems. *Studia Logica 75*(1), 63–92.

Lomuscio, A. and B. Woźna (2006). A complete and decidable security-specialised logic and its application to the tesla protocol. In P. Stone and G. Weiss (Eds.), *Proceedings of the 5th International Joint Conference on Autonomous Agents and Multiagent Systems (AAMAS'06)*, Hakodake, Japan, pp. 145–152. ACM Press.

Lomuscio, A., B. Woźna, and W. Penczek (2007). Bounded model checking for knowledge over teal time. *Artificial Intelligence 171*(16-17), 1011–1038.

Luo, H. (2009). `https://sites.google.com/site/cnxyluo/MCTK/`.

McMillan, K. (1993). *Symbolic model checking: An approach to the state explosion problem.* Kluwer Academic Publishers.

McMillan, K. L. (2002). Applying SAT methods in unbounded symbolic model checking. In *Proceedings of the 14th Int. Conf. on Computer Aided Verification (CAV'02)*, Volume 2404 of *LNCS*, pp. 250–264. Springer-Verlag.

Meşki, A., W. Penczek, and M. Szreter (2012). Bounded model checking for linear time temporal-epistemic logic. In A. V. Jones (Ed.), *ICCSW*, Volume 28 of *OASICS*, pp. 88–94. Schloss Dagstuhl - Leibniz-Zentrum fuer Informatik, Germany.

Meşki, A., W. Penczek, M. Szreter, B. Woźna-Szcześniak, and A. Zbrzezny (2014). BDD-versus SAT-based bounded model checking for the existential fragment of linear temporal logic with knowledge: algorithms and their performance. *Autonomous Agents and Multi-Agent Systems 28*(4), 558–604.

van der Meyden, R. and H. Shilov (1999a). Model checking knowledge and time in systems with perfect recall. In *Proceedings of Proceedings of FST&TCS*, Volume 1738 of *Lecture Notes in Computer Science*, Hyderabad, India, pp. 432–445.

van der Meyden, R. and N. V. Shilov (1999b). Model checking knowledge and time in systems with perfect recall. In *Proceedings of the 19th Conf. on Foundations of Software Technology and Theoretical Computer Science (FSTTCS'99)*, Volume 1738 of *LNCS*, pp. 432–445. Springer-Verlag.

van der Meyden, R. and K. Su (2004). Symbolic model checking the knowledge of the dining cryptographers. In *Proceedings of the 17th IEEE Computer Security Foundations Workshop (CSFW-17)*, pp. 280–291. IEEE Computer Society.

Parikh, R. and R. Ramanujam (1985). Distributed processes and the logic of knowledge. In *Logic of Programs*, pp. 256–268.

Pecheur, C. and F. Raimondi (2007). Symbolic model checking of logics with actions. In *Proceedings of MoChArt 2006*, Volume 4428 of *Lecture Notes in Computer Science*, pp. 113–128. Springer.

Peled, D. (1993). All from one, one for all: On model checking using representatives. In *Proceedings of the 5th Int. Conf. on Computer Aided Verification (CAV'93)*, Volume 697 of *LNCS*, pp. 409–423. Springer-Verlag.

Peled, D. (1994). Combining partial order reductions with on-the-fly model-checking. In *Proceedings of the 6th Int. Conf. on Computer Aided Verification (CAV'94)*, Volume 818 of *LNCS*, pp. 377–390. Springer-Verlag.

Penczek, W., B. Woźna-Szcześniak, and A. Zbrzezny (2012). Towards SAT-based BMC for LTLK over interleaved interpreted systems. *Fundamenta Informaticae 119*(3-4), 373–392.

Penczek, W. and A. Lomuscio (2003). Verifying epistemic properties of multi-agent systems via bounded model checking. *Fundamenta Informaticae 55(2)*, 167–185.

Penczek, W. and A. Półrola (2006). *Advances in Verification of Time Petri Nets and Timed Automata: A Temporal Logic Approach*, Volume 20 of *Studies in Computational Intelligence.* Springer-Verlag.

Penczek, W., M. Szreter, R. Gerth, and R. Kuiper (2000). Improving partial order reductions for universal branching time properties. *Fundamenta Informaticae 43*, 245–267.

Penczek, W., B. Woźna, and A. Zbrzezny (2002). Bounded model checking for the universal fragment of CTL. *Fundamenta Informaticae 51(1-2)*, 135–156.

Plaza, J. (1989). Logics of public communications. In *Proceedings of the 4th International Symposium on Methodologies for Intelligent Systems (ISMIS 1989)*, pp. 201–216.

Raimondi, F. (2006). *Model Checking Multi-Agent Systems.* Ph. D. thesis, University of London.

Raimondi, F. and A. Lomuscio (2004a). Automatic verification of deontic interpreted systems by model checking via OBDDs. In *Proceedings of the Sixteenth European Conference on Artificial Intelligence (ECAI04)*, pp. 53–57. IOS PRESS.

Raimondi, F. and A. Lomuscio (2004b). Symbolic model checking of deontic interpreted systems via OBDDs. In *Proceedings of the 7th International Workshop on Deontic Logic in Computer Science (DEON 2004)*, Volume 3065 of *LNCS*, pp. 228–242. Springer Verlag.

Raimondi, F. and A. Lomuscio (2007). Automatic verification of multi-agent systems by model checking via OBDDs. *Journal of Applied Logic 5*, 235 – 251.

Rosenschein, S. J. (1985). Formal theories of AI in knowledge and robotics. *New generation computing 3*, 345–357.

Somenzi, F. (2005). CUDD: CU decision diagram package - release 2.4.0. http://vlsi.colorado.edu/ fabio/CUDD/cuddIntro.html.

Su, K. (2004). Model checking temporal logics of knowledge in distributed systems. In D. L. McGuinness and G. Ferguson (Eds.), *Proceedings of the 19th National Conference on Artificial Intelligence, 16th Conference on Innovative Applications of Artificial Intelligence*, pp. 98–103. AAAI Press / The MIT Press.

Szreter, M. (2005). Selective search in bounded model checking of reachability properties. In *Proceedings of the 3rd Int. Symp. on Automated Technology for Verification and Analysis (ATVA'05)*, Volume 3707 of *LNCS*, pp. 159–173. Springer-Verlag.

Szreter, M. (2006). Generalized blocking clauses in unbounded model checking. In *Proceedings of the 3rd Int. Workshop on Constraints in Formal Verification (CFV'05)*.

Valmari, A. (1990). A stubborn attack on state explosion. In *Proceedings of the 2nd Int. Conf. on Computer Aided Verification (CAV'90)*, Volume 531 of *LNCS*, pp. 156–165. Springer-Verlag.

Wooldridge, M., T. Agotnes, P. E. Dunne, and W. van der Hoek (2007). Logic for automated mechanism design - a progress report. In *Proceedings of the Twenty-Second Conference on Artificial Intelligence (AAAI-07)*.

Woźna, B., A. Lomuscio, and W. Penczek (2005a). Bounded model checking for deontic interpreted systems. In *Proceedings of the 2nd Int. Workshop on Logic and Communication in Multi-Agent Systems (LCMAS'04)*, Volume 126 of *ENTCS*, pp. 93–114. Elsevier.

Woźna, B., A. Lomuscio, and W. Penczek (2005b). Bounded model checking for deontic interpreted systems. *Electronic Notes inTheoretical Computer Science 126*, 93–114.

Zbrzezny, A. (2008). Improving the translation from ECTL to SAT. *Fundamenta Informaticae 85*(1-4), 513–531.

Chapter 9

Epistemic Foundations of Game Theory

Giacomo Bonanno

Contents

Abstract This chapter provides an introduction to the so-called epistemic foundation program in game theory, whose aim is to characterize, for any game, the behavior of rational and intelligent players who know the structure of the game and the preferences of their opponents and who recognize each other's rationality and reasoning abilities. The analysis is carried out both semantically and syntactically, with a focus on the implications of common belief of rationality in strategic-form games and in dynamic games with perfect information.

Chapter 9 of the *Handbook of Epistemic Logic*, H. van Ditmarsch, J.Y. Halpern, W. van der Hoek and B. Kooi (eds), College Publications, 2015, pp. 443–487.

9.1 Introduction

Game theory provides a formal language for the representation of interactive situations, that is, situations where several "entities" - called players - take actions that affect each other. The nature of the players varies depending on the context in which the game theoretic language is invoked: in evolutionary biology players are non-thinking living organisms; in computer science players are artificial agents; in behavioral game theory players are "ordinary" human beings, etc. Traditionally, however, game theory has focused on interaction among intelligent, sophisticated and rational individuals.

The focus of this chapter is a relatively recent development in game theory, namely the so-called *epistemic foundation program*. The aim of this program is to characterize, for any game, the behavior of rational and intelligent players who know the structure of the game and the preferences of their opponents and who recognize each other's rationality and reasoning abilities. The two fundamental questions addressed in this literature are: (1) Under what circumstances can a player be said to be rational? and (2) What does 'mutual recognition' of rationality mean? Since the two main ingredients of the notion of rationality are beliefs and choice and the natural interpretation of 'mutual recognition' of rationality is in terms of common belief, it is clear that the tools of epistemic logic are the appropriate tools for this program.

It is useful to distinguish three related notions that have emerged in the analysis of games. The first notion is that of a solution concept, which is a map that associates with every game a set of strategy profiles that constitute a prediction of how the game will be played. Examples of solution concepts are Nash equilibrium, correlated equilibrium, perfect equilibrium, etc. The second notion is that of an algorithm that computes, for every game, a set of strategy profiles. The algorithm is often presented as an attempt to capture the steps in the reasoning process of the players. An example is the iterated deletion of dominated strategies. The third notion is that of an explicit epistemic hypothesis that describes the players' state of mind. An example is the hypothesis of common belief of rationality. Epistemic game theory is concerned with the third notion and seeks to provide an understanding of existing solution concepts in terms of explicit epistemic conditions, as well as a framework within which new solution concepts can be generated.

The chapter is organized as follows. In Sections 9.2 and 9.3 we begin with the semantic approach to rationality in simultaneous games with ordinal payoff. In Sections 9.4 and 9.5 we turn to the syntactic approach and explore the difference between common belief and common knowledge of rationality. In Section 9.6 we briefly discuss probabilistic beliefs and cardinal preferences. In Sections 9.7, 9.8 and 9.9 we turn to a semantic analysis of rationality in dynamic games with perfect information, based on dispositional belief revision (or subjective counterfactuals). Section 9.10 lists the most important contributions in the literature for the topics discussed in this chapter and gives references for additional solution concepts that could not be covered in this chapter because of space constraints.

9.2 Epistemic Models of Strategic-Form Games

Traditionally, game-theoretic analysis has been based on the assumption that the game under consideration is common knowledge among the players. Thus not only is it commonly known who the players are, what choices they have available and what the possible outcomes are, but also how each player ranks those outcomes. While it is certainly reasonable to postulate that a player knows his own preferences over the possible outcomes, it is much more demanding to assume that a player knows the preferences of his opponents. If those preferences are expressed as ordinal rankings of the outcomes, this assumption is less troublesome than in the case where preferences also incorporate attitudes to risk (that is, the payoff functions that represent those preferences are Bernoulli, or von Neumann Morgenstern, utility functions: see Section 9.6). We will thus begin by considering the case where preferences are expressed by *ordinal* rankings.

We first consider games where each player chooses in ignorance of the choices of the other players (as is the case, for example, in simultaneous games).

Definition 9.1
A *finite strategic-form game with ordinal payoffs* is a quintuple

$G = \left\langle \mathsf{Ag}, \{S_i\}_{i \in \mathsf{Ag}}, O, z, \{\succsim_i\}_{i \in \mathsf{Ag}} \right\rangle$

where

$\mathsf{Ag} = \{1, 2, \ldots, n\}$ is a finite set of *players*,

S_i is a finite set of *strategies* (or choices) of player $i \in \mathsf{Ag}$,

O is a finite set of *outcomes*,

$z : S \to O$ (where $S = S_1 \times ... \times S_n$) is a function that associates with every strategy profile $s = (s_1, ..., s_n) \in S$ an outcome $z(s) \in O$,

$\succsim_i$ is player i's *ranking* of O, that is, a binary relation on O which is complete (for all $o, o' \in O$, either $o \succsim_i o'$ or $o' \succsim_i o$) and transitive (for all $o, o', o'' \in O$, if $o \succsim_i o'$ and $o' \succsim_i o''$ then $o \succsim_i o''$). The interpretation of $o \succsim_i o'$ is that player i considers outcome o to be at least as good as outcome o'. The corresponding strict ordering, denoted by $\succ_i$, is defined by: $o \succ_i o'$ if and only if $o \succsim_i o'$ and not $o' \succsim_i o$. The interpretation of $o \succ_i o'$ is that player i strictly prefers outcome o to outcome o'.

Remark 9.1

Games are often represented in *reduced form*, which is obtained by replacing the triple $\langle O, z, \{\precsim_i\}_{i\in\mathsf{Ag}}\rangle$ with a set of *payoff functions* $\{\pi_i\}_{i\in\mathsf{Ag}}$ where $\pi_i : S \to \mathbb{R}$ is any real-valued function that satisfies the property that, $\forall s, s' \in S$, $\pi_i(s) \geq \pi_i(s')$ if and only if $z(s) \succsim_i z(s')$. In the following we will adopt this more succinct representation of strategic-form games. It is important to note, however, that (with the exception of Section 9.6) the payoff functions are taken to be purely ordinal and one could replace π_i with any other function obtained by composing π_i with an arbitrary strictly increasing function on the reals. ⊣

Part a of Figure 9.1 shows a two-player strategic-form game where the sets of strategies are $S_1 = \{A, B, C, D\}$ and $S_2 = \{e, f, g, h\}$. The game is represented as a table where the rows are labeled with the possible strategies of Player 1 and the columns with the possible strategies of Player 2. Each cell in the table corresponds to a strategy-profile, that is, an element of $S = S_1 \times S_2$; inside each cell the first number is the payoff of Player 1 and the second number is the payoff of Player 2; thus, for example, $\pi_1(A, e) = 6$ and $\pi_2(A, e) = 3$.

A strategic-form game provides only a partial description of an interactive situation, since it does not specify what choices the players make, nor what beliefs they have about their opponents' choices. A specification of these missing elements is obtained by introducing the notion of an epistemic model of a game, which represents a possible context in which the game is played. The players' beliefs are represented by means of a $\mathcal{KD}45$ Kripke frame $\langle W, \{R_i\}_{i\in\mathsf{Ag}}\rangle$, where W is a set of *states* (or possible worlds) and, for every player i, R_i is a binary relation on W which is serial ($\forall w \in W,\ R_i(w) \neq \varnothing$, where $R_i(w)$ denotes the set $\{w' \in W : wR_iw'\}$), transitive (if $w' \in R_i(w)$ then $R_i(w') \subseteq R_i(w)$) and euclidean (if $w' \in R_i(w)$ then $R_i(w) \subseteq R_i(w')$).[1] Given a state w, $R_i(w)$ is interpreted as the set of states that are doxastically accessible to player i at w, that is, the states that she considers possible according to her beliefs. The player, at a state w, is said to believe a formula φ if and only if φ is true at every state that she considers possible at w. Seriality of the accessibility relation R_i guarantees that the player's beliefs are consistent (it is not the case that she believes φ and also $\neg\varphi$), while transitivity corresponds to positive introspection (if the player believes φ then she believes that she believes φ) and Euclideaness corresponds to negative introspection (if the player does not believe φ then

[1] In the game-theoretic literature, it is more common to view R_i as a function that associates with every state $w \in W$ a set of states $R_i(w) \subseteq W$ and to call such a function a *possibility correspondence* or information correspondence. Of course, the two views (binary relation and possibility correspondence) are equivalent.

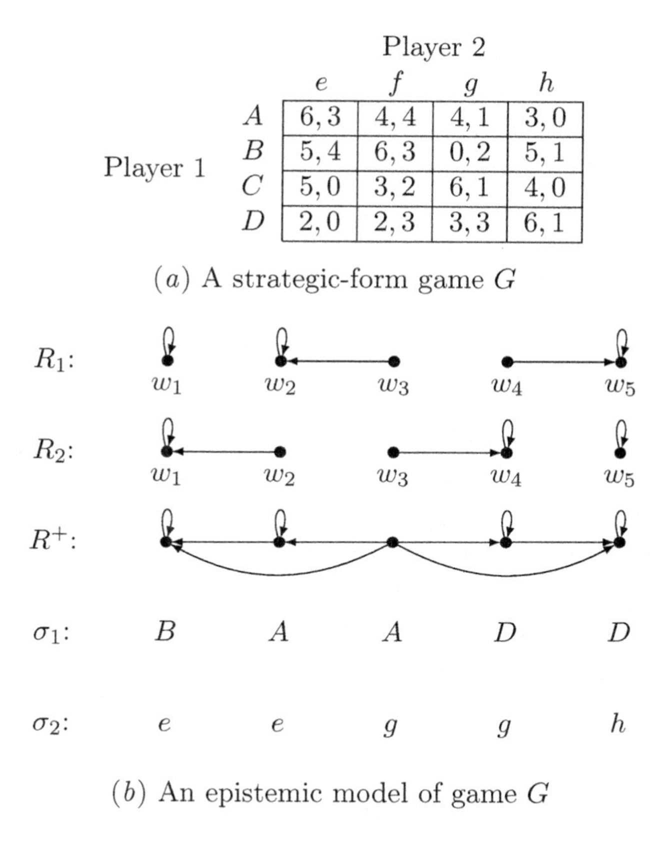

Player 1 \ Player 2	e	f	g	h
A	6,3	4,4	4,1	3,0
B	5,4	6,3	0,2	5,1
C	5,0	3,2	6,1	4,0
D	2,0	2,3	3,3	6,1

(a) A strategic-form game G

(b) An epistemic model of game G

Figure 9.1: A strategic-form game and an epistemic model of it

she believes that she does not believe φ). Note that erroneous beliefs are not ruled out: it is possible that a player believes φ even though φ is actually false.[2]

Definition 9.2
Given a strategic-form game with ordinal payoffs $G = \left\langle \mathsf{Ag}, \{S_i, \pi_i\}_{i \in \mathsf{Ag}} \right\rangle$ an *epistemic model of* G is a tuple $\langle W, \{R_i\}_{i\in\mathsf{Ag}}, \{\sigma_i\}_{i\in\mathsf{Ag}}\rangle$ where $\langle W, \{R_i\}_{i\in\mathsf{Ag}}\rangle$ is a $\mathcal{KD}45$ Kripke frame and, for every player $i \in \mathsf{Ag}$, $\sigma_i : W \to S_i$ is a function that satisfies the following property: if $w' \in R_i(w)$ then $\sigma_i(w') = \sigma_i(w)$. $\dashv$

The interpretation of $\sigma_i(w) = s_i \in S_i$ is that, at state w, player i chooses

[2]Erroneous beliefs are ruled out if one imposes the restriction that R_i be reflexive ($w \in R_i(w), \forall w \in W$). If reflexivity is added to the above assumptions, then R_i gives rise to a partition of W and in such a case it is common to use the term 'knowledge' rather than 'belief'. In the game-theoretic literature, partitional structures tend to be more common than $\mathcal{KD}45$ frames.

strategy s_i and the requirement that if $w' \in R_i(w)$ then $\sigma_i(w') = \sigma_i(w)$ expresses the assumption that a player is always certain about what choice he himself makes. On the other hand, a player may be uncertain about the choices of the other players.

Remark 9.2
In an epistemic model of a game the function $\sigma : W \to S$ defined by $\sigma(w) = (\sigma_i(w))_{i \in \mathsf{Ag}}$ associates with every state a strategy profile. Given a state w and a player i, we will often denote $\sigma(w)$ by $(\sigma_i(w), \sigma_{-i}(w))$, where $\sigma_{-i}(w) \in S_{-i} = S_1 \times ... \times S_{i-1} \times S_{i+1} \times ... \times S_n$. Thus $\sigma_{-i}(w)$ is the strategy profile of the players other than i at state w. ⊣

Part b of Figure 9.1 shows an epistemic model for the game of Part a. The relations R_i ($i = 1, 2$) are represented by arrows: for player i there is an arrow from state w to state w' if and only if $w' \in R_i(w)$. The relation R^+, which is discussed below, is the transitive closure of $R_1 \cup R_2$. [3]

In the game-theoretic literature individual beliefs and common belief are typically represented by means of semantic operators on events. Given a $\mathcal{KD}45$ Kripke frame $\langle W, \{R_i\}_{i \in \mathsf{Ag}} \rangle$, an *event* is any subset of W and one can associate with the doxastic accessibility relation R_i of player i a *semantic belief operator* $\mathbb{B}_i : 2^W \to 2^W$ and a *semantic common belief operator* $\mathbb{CB} : 2^W \to 2^W$ as follows:

$$\begin{aligned} \mathbb{B}_i E &= \{w \in W : R_i(w) \subseteq E\}, \quad \text{and} \\ \mathbb{CB} E &= \{w \in W : R^+(w) \subseteq E\} \end{aligned} \tag{9.1}$$

where R^+ is the transitive closure of $\bigcup_{i \in \mathsf{Ag}} R_i$.[4,5] $\mathbb{B}_i E$ is interpreted as the

[3] Thus in Figure 9.1 we have that

$$\begin{aligned} R_1 &= \{(w_1, w_1), (w_2, w_2), (w_3, w_2), (w_4, w_5), (w_5, w_5)\}, \\ R_2 &= \{(w_1, w_1), (w_2, w_1), (w_3, w_4), (w_4, w_4), (w_5, w_5)\}, \text{and} \\ R^+ &= \{(w_1, w_1), (w_2, w_1), (w_2, w_2), (w_3, w_1), (w_3, w_2), \\ &\quad (w_3, w_4), (w_3, w_5), (w_4, w_4), (w_4, w_5), (w_5, w_5)\}. \end{aligned}$$

Hence, for example, in terms of our notation, $R_1(w_3) = \{w_2\}$, $R_2(w_3) = \{w_4\}$ and $R^+(w_3) = \{w_1, w_2, w_4, w_5\}$.

[4] In the game-theoretic literature the transitive closure of the union of the accessibility relations is called the 'finest common coarsening'.

[5] The intuitive and prevalent definition of common belief is as follows. Let $\mathbb{B}_{all} E = \bigcap_{i \in \mathsf{Ag}} \mathbb{B}_i E$ denote the event that everybody believes E. Then the event that E is commonly believed is defined as the infinite intersection $\mathbb{CB} E = \mathbb{B}_{all} E \cap \mathbb{B}_{all} \mathbb{B}_{all} E \cap \mathbb{B}_{all} \mathbb{B}_{all} \mathbb{B}_{all} E \cap \ldots$, that is, the event that everybody believes E and everybody believes that everybody believes E and everybody believes that everybody believes that everybody believes E, and so on. Let us call this the infinitary definition of common belief. It can be shown that, for every state w and every event E, $w \in \mathbb{CB} E$ according to the infinitary definition of $\mathbb{CB}$ if and only if $R^+(w) \subseteq E$.

event that (that is, the set of states at which) player i believes event E and $\mathbb{CB}E$ as the event that E is commonly believed.[6]

The analysis of the consequences of common belief of rationality in strategic-form games was first developed in the game-theoretic literature from a semantic point of view. We will review the semantic approach in the next section and turn to the syntactic approach in Section 9.4.

9.3 Semantic Analysis of Common Belief of Rationality

A player's choice is considered to be rational if it is "optimal", given the player's beliefs about the choices of the other players. When beliefs are expressed probabilistically and payoffs are taken to be von Neumann-Morgenstern payoffs, a choice is optimal if it maximizes the player's expected payoff. We shall discuss the notion of expected payoff maximization in Section 9.6. In this section we will focus on the non-probabilistic beliefs represented by the qualitative Kripke frames introduced in Definition 9.2.

Within the context of an epistemic model of a game, a rather weak notion of rationality is the following.

Definition 9.3
Fix a strategic-form game G and an epistemic model of G. At state w player i's strategy $s_i = \sigma_i(w)$ is *rational* if it is not the case that there is another strategy $s_i' \in S_i$ of player i which yields a higher payoff than s_i against *all* the strategy profiles of the other players that player i considers possible, that is, if

$$\{s_i' \in S_i : \pi_i(s_i', \sigma_{-i}(w')) > \pi_i(\sigma_i(w), \sigma_{-i}(w')),\ \forall w' \in R_i(w)\} = \varnothing$$

[recall that, by Definition 9.2, the function $\sigma_i(\cdot)$ is constant on the set $R_i(w)$]. Equivalently, $s_i = \sigma_i(w)$ is rational at state w if, for every $s_i' \in S_i$, there exists a $w' \in R_i(w)$ such that $\sigma_i(w)$ is at least as good as s_i' against the strategy profile $\sigma_{-i}(w')$ of the other players, that is, $\pi_i(\sigma_i(w), \sigma_{-i}(w')) \geq \pi_i(s_i', \sigma_{-i}(w'))$. $\dashv$

Given an epistemic model of a strategic-form game G, using Definition 9.3 one can compute the event that player i's choice is rational. Denote that event by RAT_i. Let $RAT = \bigcap_{i \in \mathsf{Ag}} RAT_i$. Then RAT is the event that (the

[6] The operator $\mathbb{B}_i$ satisfies the following properties: $\forall E \subseteq W$, (i) Consistency: if $E \neq \varnothing$ then $\mathbb{B}_i E \neq \varnothing$, (because of seriality of R_i), (ii) Positive Introspection: $\mathbb{B}_i E \subseteq \mathbb{B}_i \mathbb{B}_i E$ (because of transitivity of R_i), (iii) Negative Introspection: $\neg\mathbb{B}_i E \subseteq \mathbb{B}_i \neg\mathbb{B}_i E$ (because of Euclideaness of R_i, where $\neg F$ denotes the complement of event F). Among the properties of the common belief operator $\mathbb{CB}$ we highlight one that we will use later, which is a consequence of transitivity of R^+: $\mathbb{CB}E \subseteq \mathbb{CB}\,\mathbb{CB}E$.

set of states at which) the choice of every player is rational. One can then also compute the event $\mathbb{CB}RAT$, that is, the event that it is common belief among the players that every player's choice is rational. For example, in the epistemic model of Part *b* of Figure 9.1, $RAT_1 = \{w_2, w_3, w_4, w_5\}$ and $RAT_2 = \{w_1, w_2, w_3, w_4\}$, so that $RAT = \{w_2, w_3, w_4\}$. Hence $\mathbb{B}_1 RAT = \{w_2, w_3\}$, $\mathbb{B}_2 RAT = \{w_3, w_4\}$ and $\mathbb{CB}RAT = \varnothing$. Thus at state w_3 each player makes a rational choice and believes that also the other player makes a rational choice, but it is not common belief that both players are making rational choices (indeed we have that $\mathbb{B}_1\mathbb{B}_2 RAT = \mathbb{B}_2\mathbb{B}_1 RAT = \varnothing$, that is, neither player believes that the other player believes that both players are choosing rationally).

Remark 9.3
It follows from Definition 9.2 (in particular, from the requirement that a player always knows what choice he is making) that, for every player i, $\mathbb{B}_i RAT_i = RAT_i$, that is, the set of states where player i makes a rational choice coincides with the set of state where she believes that her own choice is rational.

The central question in the literature on the epistemic foundations of game theory is: What strategy profiles are compatible with common belief of rationality? The question can be restated as follows.

Problem 9.4
Given a strategic-form game G, determine the subset $\tilde{S}$ of the set of strategy profiles S that satisfies the following properties:

(A) given an arbitrary epistemic model of G, if w is a state at which there is common belief of rationality, then the strategy profile chosen at w belongs to $\tilde{S}$: if $w \in \mathbb{CB}RAT$ then $\sigma(w) \in \tilde{S}$, and

(B) for every $s \in \tilde{S}$, there exists an epistemic model of G and a state w such that $\sigma(w) = s$ and $w \in \mathbb{CB}RAT$. $\dashv$

A set $\tilde{S}$ of strategy profiles that satisfies the two properties of Problem 9.4 is said to *characterize* the notion of common belief of rationality in game G.

In order to obtain an answer to Problem 9.4 we introduce the notion of strictly dominated strategy and an algorithm known as the Iterated Deletion of Strictly Dominated Strategies.

Definition 9.4
Given a strategic-form game with ordinal payoffs $G = \left\langle \mathsf{Ag}, \{S_i, \pi_i\}_{i \in \mathsf{Ag}} \right\rangle$ we say that strategy $s_i \in S_i$ of player i is *strictly dominated in* G if there is another strategy $t_i \in S_i$ of player i such that – no matter what strategies the other players choose – player i prefers the outcome associated with t_i

to the outcome associated with s_i, that is, if, for all $s_{-i} \in S_{-i}$, $\pi_i(t_i, s_{-i}) > \pi_i(s_i, s_{-i})$. ⊣

For example, in the game of Figure 9.1*a*, for Player 2 strategy h is strictly dominated (by g).

Let $G = \left\langle \mathsf{Ag}, \{S_i, \pi_i\}_{i \in \mathsf{Ag}} \right\rangle$ and $G' = \left\langle \mathsf{Ag}, \{S'_i, \pi'_i\}_{i \in \mathsf{Ag}} \right\rangle$ be two games. We say that G' is a *subgame* of G if for every player i, $S'_i \subseteq S_i$ (so that $S' \subseteq S$) and π'_i is the restriction of π_i to S' (that is, for every $s' \in S'$, $\pi'_i(s') = \pi_i(s')$).

Definition 9.5
The Iterated Deletion of Strictly Dominated Strategies (IDSDS) is the following procedure. Given a game $G = \left\langle \mathsf{Ag}, \{S_i, \pi_i\}_{i \in \mathsf{Ag}} \right\rangle$ let $\langle G^0, G^1, \ldots, G^m, \ldots \rangle$ be the sequence of subgames of G defined recursively as follows. For all $i \in \mathsf{Ag}$,

1. Let $S_i^0 = S_i$ and let $D_i^0 \subseteq S_i^0$ be the set of strategies of player i that are strictly dominated in $G^0 = G$;

2. For $m \geq 1$, let $S_i^m = S_i^{m-1} \backslash D_i^{m-1}$ and let G^m be the subgame of G with strategy sets S_i^m. Let $D_i^m \subseteq S_i^m$ be the set of strategies of player i that are strictly dominated in G^m.

Let $S_i^\infty = \bigcap_{m \in \mathbb{N}} S_i^m$ (where $\mathbb{N}$ denotes the set of non-negative integers) and let G^∞ be the subgame of G with strategy sets S_i^∞. Let $S^\infty = S_1^\infty \times \ldots \times S_n^\infty$.[7] ⊣

Figure 9.2 shows the application of the IDSDS procedure to the game of Figure 9.1*a*. In the initial game strategy h of Player 2 is strictly dominated by g; deleting h we obtain game G^1 where $S_1^1 = \{A, B, C, D\}$ and $S_2^1 = \{e, f, g\}$. In G^1 strategy D of Player 1 is strictly dominated by C; deleting D we obtain game G^2 where $S_1^2 = \{A, B, C\}$ and $S_2^2 = \{e, f, g\}$. In G^2 strategy g of Player 2 is strictly dominated by f; deleting g we obtain game G^3 where $S_1^3 = \{A, B, C\}$ and $S_2^3 = \{e, f\}$. In G^3 strategy C of Player 1 is strictly dominated by A; deleting C we obtain game G^4 where $S_1^4 = \{A, B\}$ and $S_2^4 = \{e, f\}$. In G^4 there are no strictly dominated strategies and, therefore, the procedure stops, so that $G^\infty = G^4$; thus $S_1^\infty = \{A, B\}$ and $S_2^\infty = \{e, f\}$.

The following proposition states that the answer to Problem 9.4 is provided by the output of the IDSDS procedure.

[7] Note that, since the strategy sets are finite, there exists an integer r such that $G^\infty = G^r = G^{r+k}$ for every $k \in \mathbb{N}$.

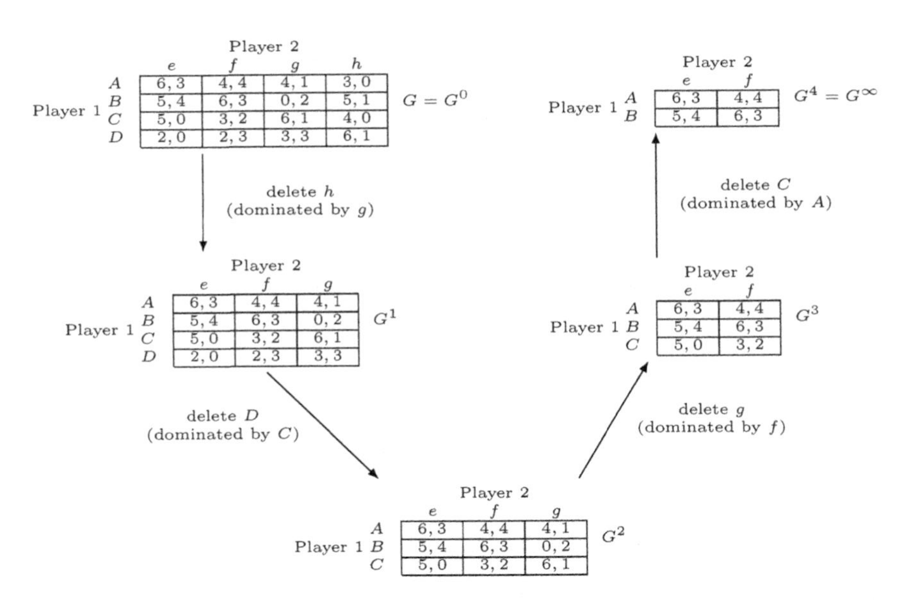

Figure 9.2: Application of the IDSDS procedure to the game of Figure 9.1a

Proposition 9.1
Fix a strategic-form game with ordinal payoffs $G = \left\langle \mathsf{Ag}, \{S_i, \pi_i\}_{i\in\mathsf{Ag}} \right\rangle$ and let $S^\infty \subseteq S$ be the set of strategy profiles obtained by applying the IDSDS algorithm. Then:

(A) given an arbitrary epistemic model of G, if w is a state at which there is common belief of rationality, then the strategy profile chosen at w belongs to S^∞: if $w \in \mathbb{CB}RAT$ then $\sigma(w) \in S^\infty$, and

(B) for every $s \in S^\infty$, there exists an epistemic model of G and a state w such that $\sigma(w) = s$ and $w \in \mathbb{CB}RAT$. ⊣

Proof (A) Fix a game G, an epistemic model of it and a state w_0 and suppose that $w_0 \in \mathbb{CB}RAT$. We want to show that $\sigma(w_0) \in S^\infty$.

First we prove by induction that

$$\forall w \in R^+(w_0), \forall i \in \mathsf{Ag}, \forall m \geq 0,\ \sigma_i(w) \notin D_i^m \tag{9.2}$$

(recall that R^+ is the transitive closure of $\bigcup_{i\in\mathsf{Ag}} R_i$ and D_i^m is the set of strategies of player i that are strictly dominated in game G^m: see Definition 9.5).

1. Base step ($m = 0$). Fix an arbitrary $w \in R^+(w_0)$ and an arbitrary player i. If $\sigma_i(w) \in D_i^0$, then there is a strategy $\hat{s}_i \in S_i$ such that, for all $s_{-i} \in S_{-i}$, $\pi_i(\sigma_i(w), s_{-i}) < \pi_i(\hat{s}_i, s_{-i})$; thus, in particular, for all $w' \in R_i(w)$, $\pi_i(\sigma_i(w), \sigma_{-i}(w')) < \pi_i(\hat{s}_i, \sigma_{-i}(w'))$. Hence, by Definition 9.3, $w \notin$

RAT_i so that, since $RAT \subseteq RAT_i$, $w \notin RAT$, contradicting - since $w \in R^+(w_0)$ - the hypothesis that $w_0 \in \mathbb{CB}RAT$.

2. Inductive step: assume that (9.2) holds for all $k \leq m$; we want to show that it holds for $k = m+1$. Suppose that $\forall w \in R^+(w_0), \forall i \in \mathsf{Ag}, \forall k \leq m$, $\sigma_i(w) \notin D_i^k$. Then (see Definition 9.5)

$$\forall w \in R^+(w_0), \sigma(w) \in S^{m+1}. \tag{9.3}$$

Fix an arbitrary $w \in R^+(w_0)$ and an arbitrary player i and suppose that $\sigma_i(w) \in D_i^{m+1}$. Then, by definition of D_i^{m+1} (see Definition 9.5) there is a strategy $\hat{s}_i \in S_i$ such that, for all $s_{-i} \in S_{-i}^{m+1}$, $\pi_i(\sigma_i(w), s_{-i}) < \pi_i(\hat{s}_i, s_{-i})$. By transitivity of R^+, since $w \in R^+(w_0)$, $R^+(w) \subseteq R^+(w_0)$. Thus, by (9.3) and the fact that $R_i(w) \subseteq R^+(w)$, we have that $\pi_i(\sigma_i(w), \sigma_{-i}(w')) < \pi_i(\hat{s}_i, \sigma_{-i}(w'))$ for all $w' \in R_i(w)$, so that, by Definition 9.3, $w \notin RAT_i$, contradicting the hypothesis that $w_0 \in \mathbb{CB}RAT$.

Thus (9.2) holds and therefore, by Definition 9.5,

$$\forall w \in R^+(w_0), \forall i \in \mathsf{Ag},\ \sigma_i(w) \in S_i^\infty. \tag{9.4}$$

The proof is not yet complete, since it may be the case that $w_0 \notin R^+(w_0)$. Fix an arbitrary player i and an arbitrary $w \in R_i(w_0)$ (recall the assumption that R_i is serial). By definition of epistemic model (see Definition 9.2) $\sigma_i(w_0) = \sigma_i(w)$. By (9.4) $\sigma_i(w) \in S_i^\infty$. Thus $\sigma_i(w_0) \in S_i^\infty$ and hence $\sigma(w_0) \in S^\infty$.

(B) Construct the following epistemic model of game G: $W = S^\infty$ and, for every player i and every $s \in S^\infty$ let $R_i(s) = \{s' \in S^\infty : s'_i = s_i\}$. Then R_i is an equivalence relation (hence serial, transitive and euclidean). For all $s \in S^\infty$, let $\sigma_i(s) = s_i$. Fix an arbitrary $s \in S^\infty$ and an arbitrary player i. By definition of S^∞, it is not the case that there exists an $\hat{s}_i \in S_i$ such that, for all $s_{-i} \in S_{-i}^\infty$, $\pi_i(s_i, s_{-i}) < \pi_i(\hat{s}_i, s_{-i})$. Thus, since - by construction - for all $s' \in R_i(s)$, $\sigma_{-i}(s') \in S_{-i}^\infty$, $s \in RAT_i$ (see Definition 9.3). Since i was chosen arbitrarily, $s \in RAT$; hence, since $s \in S^\infty$ was chosen arbitrarily, $RAT = S^\infty$. It follows that $s \in \mathbb{CB}RAT$ for every $s \in S^\infty$. ⊣

9.4 Syntactic Characterization of Common Belief of Rationality

We now turn to the syntactic analysis of rationality in strategic-form games. In order to be able to describe a game syntactically, the set of propositional variables (or atoms) At will be taken to include:

- Strategy symbols s_i^1, s_i^2, ... The intended interpretation of s_i^k is "player i chooses her k^{th} strategy s_i^k".[8]
- Atoms of the form $s_i^\ell \succeq_i s_i^k$, whose intended interpretation is "strategy s_i^ℓ of player i is at least as good, for player i, as her strategy s_i^k", and atoms of the form $s_i^\ell \succ_i s_i^k$, whose intended interpretation is "for player i strategy s_i^ℓ is better than strategy s_i^k".

Fix a strategic-form game with ordinal payoffs $G = \left\langle \mathsf{Ag}, \{S_i, \pi_i\}_{i \in \mathsf{Ag}} \right\rangle$ and let $S_i = \{s_i^1, s_i^2, ..., s_i^{m_i}\}$ (thus the cardinality of S_i is m_i). We denote by $\mathbf{KD45}_G$ the **KD45** multi-agent logic *without a common belief operator* that satisfies the following additional axioms: for all $i \in \mathsf{Ag}$ and for all $k, \ell = 1, ..., m_i$, with $k \neq \ell$,

$$
\begin{array}{ll}
\left(s_i^1 \vee s_i^2 \vee ... \vee s_i^{m_i}\right) & (\mathbf{G1}) \\
\neg\left(s_i^k \wedge s_i^\ell\right) & (\mathbf{G2}) \\
s_i^k \to B_i s_i^k & (\mathbf{G3}) \\
\left(s_i^k \succeq_i s_i^\ell\right) \vee \left(s_i^\ell \succeq_i s_i^k\right) & (\mathbf{G4}) \\
\left(s_i^\ell \succ_i s_i^k\right) \leftrightarrow \left(\left(s_i^\ell \succeq_i s_i^k\right) \wedge \neg\left(s_i^k \succeq_i s_i^\ell\right)\right) & (\mathbf{G5})
\end{array}
$$

Axiom **G1** says that player i chooses at least one strategy, while axiom **G2** says that player i cannot choose more than one strategy. Thus **G1** and **G2** together imply that each player chooses exactly one strategy. Axiom **G3**, on the other hand, says that player i is conscious of his own choice: if he chooses strategy s_i^k then he believes that he chooses s_i^k. The remaining axioms state that the ordering of strategies is complete (**G4**) and that the corresponding strict ordering is defined as usual (**G5**).[9]

Proposition 9.2
The following is a theorem of logic $\mathbf{KD45}_G$: $B_i s_i^k \to s_i^k$. That is, every player has correct beliefs about her own choice of strategy.[10] $\dashv$

Proof In the following PL stands for 'Propositional Logic' and RK denotes the inference rule "from $\psi \to \chi$ infer $\Box\psi \to \Box\chi$", which is a derived rule of inference that applies to every modal operator $\Box$ that satisfies axiom **K** and the rule of Necessitation. Fix a player i and $k, \ell \in \{1, ..., m_i\}$ with $k \neq \ell$. Let φ_k denote the formula

[8] Thus, with slight abuse of notation, we use the symbol s_i^k to denote both an element of S_i, that is, a strategy of player i, and an element of At, that is, an atom whose intended interpretation is "player i chooses strategy s_i^k".

[9] We have not included the axiom corresponding to transitivity of the ordering, namely $\left(s_i^{k_1} \succeq_i s_i^{k_2}\right) \wedge \left(s_i^{k_2} \succeq_i s_i^{k_3}\right) \to \left(s_i^{k_1} \succeq_i s_i^{k_3}\right)$, because it is not needed in what follows.

[10] Note that, in general, logic $\mathbf{KD45}_G$ allows for incorrect beliefs. In particular, a player might have incorrect beliefs about the choices made by *other* players. By Proposition 9.2, however, a player cannot have mistaken beliefs about her own choice.

$$(s_i^1 \vee ... \vee s_i^{m_i}) \wedge \neg s_i^1 \wedge ... \wedge \neg s_i^{k-1} \wedge \neg s_i^{k+1} \wedge ... \wedge \neg s_i^{m_i}.$$

1.	$\varphi_k \to s_i^k$	tautology
2.	$\neg(s_i^k \wedge s_i^\ell)$	axiom **G2** (for $\ell \neq k$)
3.	$s_i^k \to \neg s_i^\ell$	2, PL
4.	$B_i s_i^k \to B_i \neg s_i^\ell$	3, rule RK
5.	$B_i \neg s_i^\ell \to \neg B_i s_i^\ell$	axiom $\mathbf{D}_i$
6.	$s_i^\ell \to B_i s_i^\ell$	axiom **G3**
7.	$\neg B_i s_i^\ell \to \neg s_i^\ell$	6, PL
8.	$B_i s_i^k \to \neg s_i^\ell$	4, 5, 7, PL (for $\ell \neq k$)
9.	$s_i^1 \vee ... \vee s_i^{m_i}$	axiom **G1**
10.	$B_i s_i^k \to (s_i^1 \vee ... \vee s_i^{m_i})$	9, PL
11.	$B_i s_i^k \to \varphi_k$	8 (for every $\ell \neq k$), 10, PL
12.	$B_i s_i^k \to s_i^k$	1, 11, PL.

⊣

Given a game G, let $\mathcal{F}_G$ denote the set of epistemic models of G (see Definition 9.2).

Definition 9.6
Given a game G and an epistemic model $F \in \mathcal{F}_G$ a *syntactic model of G based on F* is obtained by adding to F any propositional valuation $V : W \to (\mathsf{At} \to \{true, false\})$ that satisfies the following restrictions (we write $w \models p$ instead of $V(w)(p) = true$):

- $w \models s_i^h$ if and only if $\sigma_i(w) = s_i^h$,
- $w \models (s_i^k \succeq_i s_i^\ell)$ if and only if $\pi_i(s_i^k, \sigma_{-i}(w)) \geq \pi_i(s_i^\ell, \sigma_{-i}(w))$,
- $w \models s_i^k \succ_i s_i^\ell$ if and only if $\pi_i(s_i^k, \sigma_{-i}(w)) > \pi_i(s_i^\ell, \sigma_{-i}(w))$.

Thus, in a syntactic model of a game, at state w it is true that player i chooses strategy s_i^h if and only if the strategy of player i associated with w (in the semantic model on which the syntactic model is based) is s_i^h (that is, $\sigma_i(w) = s_i^h$) and it is true that strategy s_i^k is at least as good as (respectively, better than) strategy s_i^ℓ if and only if s_i^k in combination with $\sigma_{-i}(w)$ (the profile of strategies of players other than i associated with w) yields an outcome which player i considers at least as good as (respectively, better than) the outcome yielded by s_i^ℓ in combination with $\sigma_{-i}(w)$.

For example, a syntactic model of the game shown in Part a of Figure 9.1 based on the semantic model shown in Part b of Figure 9.1 satisfies the following formula at state w_1:

$B \wedge e \wedge (A \succ_1 B) \wedge (A \succ_1 C) \wedge (A \succ_1 D) \wedge (B \succeq_1 C) \wedge (C \succeq_1 B) \wedge (B \succ_1 D)$
$\wedge (C \succ_1 D) \wedge (e \succ_2 f) \wedge (e \succ_2 g) \wedge (e \succ_2 h) \wedge (f \succ_2 g) \wedge (f \succ_2 h) \wedge (g \succ_2 h)$.

Remark 9.5
Let $\mathcal{M}_G$ denote the set of all syntactic models of game G. It is straightforward to verify that logic $\mathbf{KD45}_G$ is sound with respect to $\mathcal{M}_G$.[11] $\dashv$

We now provide an axiom that, for every game, characterizes the output of the IDSDS procedure (see Definition 9.5), namely the set of strategy profiles S^∞. The following axiom says that if player i chooses strategy s_i^k then it is not the case that she believes that a different strategy s_i^ℓ is better for her:

$$s_i^k \rightarrow \neg B_i(s_i^\ell \succ_i s_i^k). \qquad (\mathbf{WR})$$

Proposition 9.3
Fix a strategic-form game with ordinal payoffs $G = \left\langle \mathsf{Ag}, \{S_i, \pi_i\}_{i \in \mathsf{Ag}} \right\rangle$. Then

(A) If $M = \langle W, \{R_i\}_{i\in\mathsf{Ag}}, \{\sigma_i\}_{i\in\mathsf{Ag}}, V\rangle$ is a syntactic model of G that validates axiom **WR**, then $\sigma(w) \in S^\infty$, for every state $w \in W$.

(B) There exists a syntactic model M of G that validates axiom **WR** and is such that (1) for every $s \in S^\infty$, there exists a state w such that $w \models s$, and (2) for every $s \in S$ and for every $w \in W$, if $w \models s$ then $\sigma(w) \in S^\infty$. $\dashv$

Proof (A) Fix a game and a syntactic model of it that validates axiom **WR**. Fix an arbitrary state w_0 and an arbitrary player i. By Axioms **G1** and **G2** (see Remark 9.5) $w_0 \models s_i^k$ for a unique strategy $s_i^k \in S_i$. Fix an arbitrary $s_i^\ell \in S_i$, with $s_i^\ell \neq s_i^k$. Since the model validates axiom **WR**, $w_0 \models \neg B_i(s_i^\ell \succ_i s_i^k)$, that is, there exists a $w_1 \in R_i(w_0)$, such that $w_1 \models \neg(s_i^\ell \succ_i s_i^k)$. Hence, by Definition 9.6, $\sigma_i(w_0) = s_i^k$ and $\pi_i(s_i^k, \sigma_{-i}(w_1)) \geq \pi_i(s_i^\ell, \sigma_{-i}(w_1))$, so that, by Definition 9.3, $w_0 \in RAT_i$. Since w_0 and i were chosen arbitrarily, $RAT = W$ and thus, $\mathbb{CB}RAT = W$, that is, for every $w \in W$, $w \in \mathbb{CB}RAT$. Hence, by Part A of Proposition 9.1, $\sigma(w) \in S^\infty$.

[11] It follows from the following observations: (1) axioms **G1** and **G2** are valid in every syntactic model because, for every state w, there is a unique strategy $s_i^k \in S_i$ such that $\sigma_i(w) = s_i^k$ and, by the validation rules (see Definition 9.6), $w \models s_i^k$ if and only if $\sigma_i(w) = s_i^k$; (2) axiom **G3** is an immediate consequence of the fact (see Definition 9.2) that if $w' \in R_i(w)$ then $\sigma_i(w') = \sigma_i(w)$; (3) axioms **G4** and **G5** are valid because, for every state w, there is a unique profile of strategies $\sigma_{-i}(w)$ of the players other than i and the payoff function π_i of player i restricted to the set $S_i \times \{\sigma_{-i}(w)\}$ induces a complete (and transitive) ordering of S_i.

(B) Let F be the semantic epistemic model constructed in the proof of Part B of Proposition 9.1 and let M be a syntactic model based on F that satisfies the validation rules of Definition 9.6. First we show that M validates axiom **WR**. Recall that, in F, $W = S^\infty$, $s' \in R_i(s)$ if and only if $s_i = s'_i$ and σ is the identity function. Fix an arbitrary player i and an arbitrary state $\hat{s}$. We need to show that, for every $s_i^\ell \in S_i$, $\hat{s} \models \neg B_i(s_i^\ell \succ_i \hat{s}_i)$. Suppose that, for some $s_i^\ell \in S_i$, $\hat{s} \models B_i(s_i^\ell \succ_i \hat{s}_i)$, that is, for every $s' \in R_i(\hat{s})$, $s' \models (s_i^\ell \succ_i \hat{s}_i)$. Then, by Definition 9.6, for every $s' \in R_i(\hat{s})$, $\pi_i(s_i^\ell, s'_{-i}) > \pi_i(\hat{s}_i, s'_{-i})$, so that, by Definition 9.3, $\hat{s} \notin RAT_i$. But, as shown in the proof of Proposition 9.1, $RAT = S^\infty$ so that, since $RAT \subseteq RAT_i \subseteq W = S^\infty$, $RAT_i = S^\infty$, yielding a contradiction. Thus M validates axiom **WR**. Now fix an arbitrary $s \in S^\infty$. Then, by Definition 9.6, $s \models s$; thus (1) holds; conversely, let $s \models s$; then, by construction of F, $\sigma(s) = s$ and $s \in S^\infty$. Thus (2) holds. $\dashv$

Remark 9.6
Since, by Proposition 9.1, the set of strategy-profiles S^∞ characterizes the semantic notion of common belief of rationality, it follows from Proposition 9.3 that axiom **WR** provides a syntactic characterization of common belief or rationality in strategic-form games with ordinal payoffs. $\dashv$

Remark 9.7
Note that axiom **WR** provides a syntactic characterization of common belief of rationality in a logic that does *not* contain a common belief operator. However, since **WR** expresses the notion that player i chooses rationally, by the Necessitation rule every player believes that player i is rational [that is, from **WR** we obtain that, for every player $j \in \mathsf{Ag}$, $B_j\left(s_i^k \to \neg B_i(s_i^\ell \succ_i s_i^k)\right)$ is a theorem], and every player believes this [from $B_j\left(s_i^k \to \neg B_i(s_i^\ell \succ_i s_i^k)\right)$, by Necessitation, we get that $B_r B_j\left(s_i^k \to \neg B_i(s_i^\ell \succ_i s_i^k)\right)$ is a theorem, for every player $r \in \mathsf{Ag}$] and so on, so that - essentially - the rationality of every player's choice is commonly believed. Indeed, if one adds the common belief operator CB to the logic, then, by Necessitation, $CB\left(s_i^k \to \neg B_i(s_i^\ell \succ_i s_i^k)\right)$ becomes a theorem.[12] $\dashv$

Remark 9.8
There appears to be an important difference between the result of Section 9.3 and the result of this section: Proposition 9.1 gives a *local* result, while

[12]Despite the fact that the intuitive definition of common belief involves an infinite conjunction (see Footnote 5), there is a finite axiomatization of common belief. For example, the following three axioms are sufficient (without any additional rule of inference: see Section 9.10 for a reference): (1) $CB\varphi \to B_i\varphi$, (2) $CB\varphi \to B_iCB\varphi$ and (3) $CB(\varphi \to B_1\varphi \wedge \cdots \wedge B_n\varphi) \to (B_1\varphi \wedge \cdots \wedge B_n\varphi \to CB\varphi)$.

Proposition 9.3 provides a *global* one. For example, Part A of Proposition 9.1 states that if *at a state* there is common belief of rationality, then the strategy profile played *at that state* belongs to S^∞, while Part A of Proposition 9.3 states that in a syntactic model that validates axiom **WR** the strategy profile played *at every state* belongs to S^∞. As a matter of fact, the result of Section 9.3 is also "global" in nature. To see this, fix an epistemic model and a state w_0 and suppose that $w_0 \in \mathbb{CB}RAT$. By transitivity of R^+ (see Footnote 6) $\mathbb{CB}RAT \subseteq \mathbb{CB}\,\mathbb{CB}RAT$. Thus, for every $w \in R^+(w_0)$, $w \in \mathbb{CB}RAT$. Hence, by Proposition 9.1, $\sigma(w) \in S^\infty$. That is, if at a state there is common belief of rationality, then at that state, *as well as at all states reachable from it by the common belief relation* R^+, it is true that the strategy profile played belongs to S^∞. This is essentially a global result, since from the point of view of a state w_0, the "global" space is precisely the set $R^+(w_0)$. ⊣

9.5 Common Belief versus Common Knowledge

In the previous two sections we studied the implications of common *belief* of rationality in strategic-form games. What distinguishes belief from knowledge is that belief may be erroneous, while knowledge is veridical: if I know that φ then φ is true, while it is possible for me to believe that φ when φ is in fact false. In a game a player might have erroneous beliefs about the choices of the other players or about their beliefs. Perhaps one might be able to draw sharper conclusions about what the players will do in a game if one rules out erroneous beliefs. Thus a natural question to ask is: If we replace belief with knowledge, what can we infer from the hypothesis that there is *common knowledge* of rationality? Is the set of strategy profiles that are compatible with common knowledge of rationality a proper subset of S^∞? The answer is negative as can be seen from the epistemic model constructed in the proof of Part B of Proposition 9.1: that model is one where each accessibility relation is an equivalence relation and thus the underlying frame is an $\mathcal{S}5$ frame. Hence the set of strategy profiles that are compatible with common knowledge of rationality coincides with the set of strategy profiles that are compatible with common belief of rationality, namely S^∞. However, it is possible to obtain sharper predictions by replacing belief with knowledge and, at the same time, introducing a mild strengthening of the notion of rationality. Given a strategic-form game with ordinal payoffs $G = \left\langle \mathsf{Ag}, \{S_i, \pi_i\}_{i\in\mathsf{Ag}} \right\rangle$ we will now consider epistemic models of G of the form $\langle W, \{\sim_i\}_{i\in\mathsf{Ag}}, \{\sigma_i\}_{i\in\mathsf{Ag}}\rangle$ where $\langle W, \{\sim_i\}_{i\in\mathsf{Ag}}\rangle$ is an $\mathcal{S}5$ Kripke frame, that is, the accessibility relation $\sim_i$ of each player $i \in \mathsf{Ag}$ is an *equivalence* relation. Since we are dealing with $\mathcal{S}5$ frames, instead of belief we will speak of knowledge and denote the semantic operators for individual knowledge and common knowledge by $\mathbb{K}_i$ and $\mathbb{CK}$, respectively. Thus $\mathbb{K}_i : 2^W \to 2^W$ and $\mathbb{CK} : 2^W \to 2^W$ are given by:

$$\begin{aligned} \mathbb{K}_i E &= \{w \in W : \ \sim_i (w) \subseteq E\}, \text{ and} \\ \mathbb{CK} E &= \{w \in W : \ \sim^* (w) \subseteq E\} \end{aligned} \tag{9.5}$$

where, as before, $\sim_i (w) = \{w' \in W : w \sim_i w'\}$ and $\sim^*$ is the transitive closure of $\bigcup_{i \in \mathsf{Ag}} \sim_i$.[13] $\mathbb{K}_i E$ is interpreted as the event that (that is, the set of states at which) player i knows event E and $\mathbb{CK}E$ as the event that E is commonly known.

We now consider a stronger notion of rationality than the one given in Definition 9.3, which we will call *s-rationality* ('s' stands for 'strong').

Definition 9.7

Fix a strategic-form game G and an $\mathcal{S}5$ epistemic model of G. At state w player i's strategy $\sigma_i(w)$ is *s-rational* if it is not the case that there is another strategy $s_i' \in S_i$ which (1) yields *at least as high* a payoff as $\sigma_i(w)$ against *all* the strategy profiles of the other players that player i considers possible and (2) a higher payoff than $\sigma_i(w)$ against *at least one* strategy profile of the other players that player i considers possible, that is, if

there is no strategy $s_i' \in S_i$ such that

(1) $\pi_i(s_i', \sigma_{-i}(w')) \geq \pi_i(\sigma_i(w), \sigma_{-i}(w'))$, $\forall w' \in \ \sim_i (w)$, and

(2) $\pi_i(s_i', \sigma_{-i}(\tilde{w})) > \pi_i(\sigma_i(w), \sigma_{-i}(\tilde{w}))$, for some $\tilde{w} \in \ \sim_i (w)$.

[recall that, by Definition 9.2, the function $\sigma_i(\cdot)$ is constant on the set $\sim_i (w)$]. Equivalently, $\sigma_i(w)$ is s-rational at state w if, for every $s_i' \in S_i$, whenever there is a $w' \in \ \sim_i (w)$ such that $\pi_i(s_i', \sigma_{-i}(w')) > \pi_i(\sigma_i(w), \sigma_{-i}(w'))$ then there is another state $w'' \in \ \sim_i (w)$ such that $\pi_i(\sigma_i(w), \sigma_{-i}(w'')) > \pi_i(s_i', \sigma_{-i}(w''))$. $\dashv$

Denote by $SRAT_i$ the event that (i.e. the set of states at which) player i's choice is s-rational and let $SRAT = \bigcap_{i \in \mathsf{Ag}} SRAT_i$. Then $SRAT$ is the event that the choice of every player is s-rational.

As we did in Section 9.3 for the weaker notion of rationality and for common belief, we will now determine, for every game G, the set of strategy profiles that are compatible with common knowledge of s-rationality. Also in this case, the answer is based on an iterated deletion procedure. However, unlike the IDSDS procedure given in Definition 9.5, the deletion procedure defined below operates not at the level of individual players' strategies but at the level of strategy profiles.

---

[13] Thus, in addition to the properties listed in Footnote 6, the operator $\mathbb{K}_i$ satisfies the veridicality property $\mathbb{K}_i E \subseteq E, \forall E \subseteq W$ (because of reflexivity of $\sim_i$). Since reflexivity is inherited by $\sim^*$, also the common knowledge operator satisfies the veridicality property: $\mathbb{CK}E \subseteq E$.

Definition 9.8
Given a strategic-form game with ordinal payoffs $G = \left\langle \mathsf{Ag}, \{S_i, \pi_i\}_{i\in\mathsf{Ag}} \right\rangle$, a subset of strategy profiles $X \subseteq S$ and a strategy profile $x \in X$, we say that x is *inferior relative to* X if there exists a player i and a strategy $s_i \in S_i$ of player i (thus s_i need not belong to the projection of X onto S_i) such that:

1. $\pi_i(s_i, x_{-i}) > \pi_i(x_i, x_{-i})$, and
2. for all $s_{-i} \in S_{-i}$, if $(x_i, s_{-i}) \in X$ then $\pi_i(s_i, s_{-i}) \geq \pi_i(x_i, s_{-i})$.

The *Iterated Deletion of Inferior Profiles* (IDIP) is defined as follows. For $m \in \mathbb{N}$ define $T^m \subseteq S$ recursively as follows: $T^0 = S$ and, for $m \geq 1$, $T^m = T^{m-1} \backslash I^{m-1}$, where $I^{m-1} \subseteq T^{m-1}$ is the set of strategy profiles that are inferior relative to T^{m-1}. Let $T^\infty = \bigcap\limits_{m\in\mathbb{N}} T^m$.[14] $\dashv$

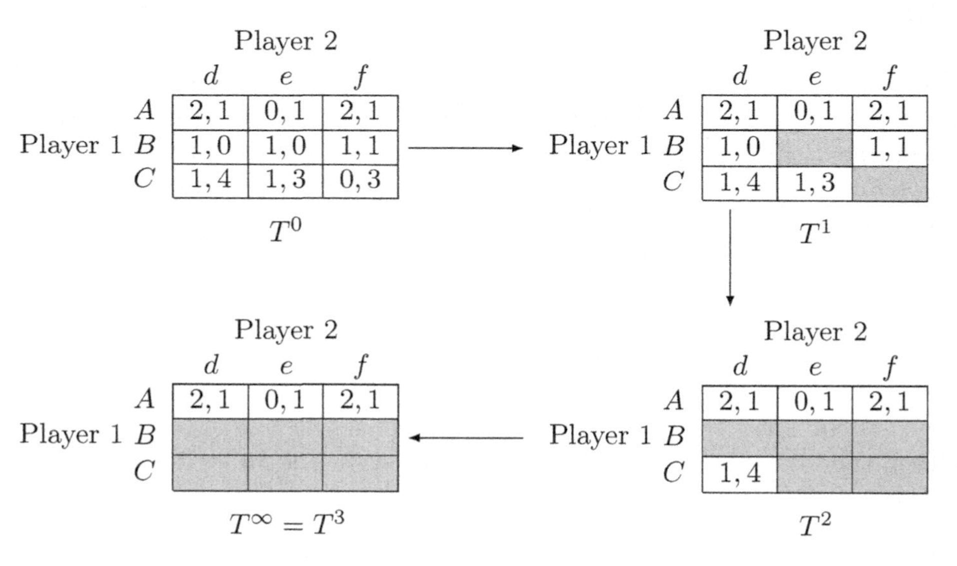

Figure 9.3: Illustration of the IDIP procedure

The IDIP procedure is illustrated in Figure 9.3, where

$T^0 = \{(A,d),(A,e),(A,f),(B,d),(B,e),(B,f),(C,d),(C,e),(C,f)\}$. This equals S.

$I^0 = \{(B,e),(C,f)\}$ (the elimination of (B,e) is done through Player 2 and strategy f, while the elimination of (C,f) is done through Player 1 and strategy B);

$T^1 = \{(A,d),(A,e),(A,f),(B,d),(B,f),(C,d),(C,e)\}$,

[14] Since the strategy sets are finite, there exists an integer r such that $T^\infty = T^r = T^{r+k}$ for every $k \in \mathbb{N}$.

$I^1 = \{(B,d),(B,f),(C,e)\}$ (the elimination of (B,d) and (B,f) is now done through Player 1 and strategy A, while the elimination of (C,e) is done through Player 2 and strategy d);

$T^2 = \{(A,d),(A,e),(A,f),(C,d)\}$,

$I^2 = \{(C,d)\}$ (the elimination of (C,d) is done through Player 1 and strategy A);

$T^3 = \{(A,d),(A,e),(A,f)\}$,

$I^3 = \varnothing$; thus

$T^\infty = T^3$.

The following Proposition is the counterpart to Proposition 9.1, when rationality is replaced with s-rationality, belief with knowledge and the IDSDS procedure with the IDIP procedure.

Proposition 9.4
Fix a strategic-form game with ordinal payoffs $G = \left\langle \mathsf{Ag}, \{S_i, \pi_i\}_{i \in \mathsf{Ag}} \right\rangle$ and let $T^\infty \subseteq S$ be the set of strategy profiles obtained by applying the IDIP procedure. Then:

(A) given an arbitrary $\mathcal{S}5$ epistemic model of G, if w is a state at which there is common knowledge of s-rationality, then the strategy profile chosen at w belongs to T^∞: if $w \in \mathbb{CK}SRAT$ then $\sigma(w) \in T^\infty$, and

(B) for every $s \in T^\infty$, there exists an $\mathcal{S}5$ epistemic model of G and a state w such that $\sigma(w) = s$ and $w \in \mathbb{CK}SRAT$. ⊣

Proof (A) Fix an $\mathcal{S}5$ epistemic model of G and a state w_0 and suppose that $w_0 \in \mathbb{CK}SRAT$. We want to show that $\sigma(w_0) \in T^\infty$.

First we prove by induction that

$$\forall w \in W \text{ such that } w_0 \sim^* w, \forall m \geq 0,\ \sigma(w) \notin I^m. \tag{9.6}$$

1. Base step ($m = 0$). Fix an arbitrary $w_1 \in W$ such that $w_0 \sim^* w_1$. If $\sigma(w_1) \in I^0$ (that is, $\sigma(w_1)$ is inferior relative to the entire set of strategy profiles S) then there exist a player i and a strategy $\hat{s}_i \in S_i$ such that, $\pi_i(\hat{s}_i, \sigma_{-i}(w_1)) > \pi_i(\sigma_i(w_1), \sigma_{-i}(w_1))$, and, for every $s_{-i} \in S_{-i}$, $\pi_i(\hat{s}_i, s_{-i}) \geq \pi_i(\sigma_i(w_1), s_{-i})$; thus, in particular, for all w' such that $w_1 \sim_i w'$, $\pi_i(\hat{s}_i, \sigma_{-i}(w')) \geq \pi_i(\sigma_i(w_1), \sigma_{-i}(w'))$. Furthermore, by reflexivity of $\sim_i$, $w_1 \sim_i w_1$. It follows from Definition 9.7 that $w_1 \notin SRAT_i$, so that, since $SRAT \subseteq SRAT_i$, $w_1 \notin SRAT$, contradicting the hypothesis that $w_0 \in \mathbb{CK}SRAT$ (since $w_0 \sim^* w_1$).

2. Inductive step: assume that (9.6) holds for all $k \leq m$; we want to show that it holds for $k = m + 1$. Suppose that $\forall w \in W$ such that $w_0 \sim^* w, \forall k \leq m$, $\sigma(w) \notin I^k$. Then

$$\forall w \in W \text{ such that } w_0 \sim^* w, \sigma(w) \in T^{m+1}. \tag{9.7}$$

Fix an arbitrary $w_1 \in W$ such that $w_0 \sim^* w_1$ and suppose that $\sigma(w_1) \in I^{m+1}$, that is, $\sigma(w_1)$ is inferior relative to T^{m+1}. Then, by definition of I^{m+1}, there exist a player i and a strategy $\hat{s}_i \in S_i$ such that, $\pi_i(\hat{s}_i, \sigma_{-i}(w_1)) > \pi_i(\sigma_i(w_1), \sigma_{-i}(w_1))$ and, for every $s_{-i} \in S_{-i}$, if $(\hat{s}_i, s_{-i}) \in T^{m+1}$ then $\pi_i(\hat{s}_i, s_{-i}) \geq \pi_i(\sigma_i(w_1), s_{-i})$. By Definition 9.2, for every w such that $w \sim_i w_1$, $\sigma_i(w) = \sigma_i(w_1)$ and by (9.7), for every w such that $w_0 \sim^* w$, we have that $(\sigma_i(w), \sigma_{-i}(w)) \in T^{m+1}$. Thus, since $\sim_i (w_1) \subseteq \; \sim^* (w_1) \subseteq \; \sim^* (w_0)$, we have that, for every w such that $w \sim_i w_1$, $(\sigma_i(w_1), \sigma_{-i}(w)) \in T^m$. By reflexivity of $\sim_i$, $w_1 \sim_i w_1$; hence, by Definition 9.7, $w_1 \notin SRAT_i$ and thus $w_1 \notin SRAT$ (since $SRAT \subseteq SRAT_i$). This, together with the fact that $w_0 \sim^* w_1$, contradicts the hypothesis that $w_0 \in \mathbb{CK}SRAT$.

Thus, we have shown by induction that, $\forall w \in W$ such that $w \sim^* w_0$, $\sigma(w) \in \bigcap_{m \in \mathbb{N}} T^m = T^\infty$. It only remains to establish that $\sigma(w_0) \in T^\infty$, but this follows from reflexivity of $\sim^*$.

(B) Construct the following epistemic model of game G: $W = T^\infty$ and, for every player i and every $s, s' \in T^\infty$ let $s \sim_i s'$ if and only if $s'_i = s_i$ Then $\sim_i$ is an equivalence relation and thus the frame is an $\mathcal{S}5$ frame. For all $s \in T^\infty$, let $\sigma(s) = s$. Fix an arbitrary $\tilde{s} \in T^\infty$ and an arbitrary player i. By definition of T^∞, it is not the case that there exists an $\hat{s}_i \in S_i$ such that $\pi_i(\hat{s}_i, \tilde{s}_{-i}) > \pi_i(\tilde{s}_i, \tilde{s}_{-i})$ and, for every $s'_{-i} \in S_{-i}$, if $(\hat{s}_i, s'_{-i}) \in T^\infty$ then $\pi_i(\hat{s}_i, s'_{-i}) \geq \pi_i(\tilde{s}_i, s'_{-i})$. Thus $\tilde{s} \in SRAT_i$; hence, since player i was chosen arbitrarily, $\tilde{s} \in SRAT$. Since $\tilde{s}$ was chosen arbitrarily, it follows that $SRAT = T^\infty$ and thus $\mathbb{CK}SRAT = T^\infty$. ⊣

We now turn to the syntactic analysis. Given a strategic-form game with ordinal payoffs $G = \left\langle \mathsf{Ag}, \{S_i, \pi_i\}_{i \in \mathsf{Ag}} \right\rangle$, let $\mathbf{S5}_G$ be the **S5** multi-agent logic *without a common knowledge operator* that satisfies axioms **G1-G5** of Section 9.4. Clearly, $\mathbf{S5}_G$ is an extension of $\mathbf{KD45}_G$. Let $\mathcal{M}_G^{S5}$ denote the set of all syntactic models of game G (see Definition 9.6) based on $\mathcal{S}5$ epistemic models of G. It is straightforward to verify that logic $\mathbf{S5}_G$ is sound with respect to $\mathcal{M}_G^{S5}$.

In parallel to the analysis of Section 9.4, we now provide an axiom that, for every game, characterizes the output of the IDIP procedure, namely the set of strategy profiles T^∞. The following axiom is a strengthening of axiom **WR** of Section 9.4: it says that if player i chooses strategy s_i^k then it is not the case that (1) she believes that a different strategy s_i^ℓ is at least as good for her as s_i^k and (2) she considers it possible that s_i^ℓ is better than s_i^k:

$$s_i^k \to \neg\left(B_i(s_i^\ell \succeq_i s_i^k) \wedge \neg B_i \neg(s_i^\ell \succ_i s_i^k)\right). \tag{SR}$$

The following proposition confirms that axiom **SR** is a strengthening of axiom **WR**: the latter is derivable in the logic obtained by adding **SR** to $\mathbf{KD45}_G$.

Proposition 9.5
Axiom **WR** is a theorem of $\mathbf{KD45}_G + \mathbf{SR}$. ⊣

Proof

1.	$s_i^k \to \neg\left(B_i(s_i^\ell \succeq_i s_i^k) \wedge \neg B_i \neg(s_i^\ell \succ_i s_i^k)\right)$	**SR**
2.	$(s_i^\ell \succ_i s_i^k) \leftrightarrow (s_i^\ell \succeq_i s_i^k) \wedge \neg(s_i^k \succeq_i s_i^\ell)$	**G5**
3.	$(s_i^\ell \succ_i s_i^k) \to (s_i^\ell \succeq_i s_i^k)$	2, PL
4.	$B_i(s_i^\ell \succ_i s_i^k) \to B_i(s_i^\ell \succeq_i s_i^k)$	3, RK
5.	$B_i(s_i^\ell \succ_i s_i^k) \to \neg B_i \neg(s_i^\ell \succ_i s_i^k)$	Axiom $\mathbf{D}_i$
6.	$B_i(s_i^\ell \succ_i s_i^k) \to \left(B_i(s_i^\ell \succeq_i s_i^k) \wedge \neg B_i \neg(s_i^\ell \succ_i s_i^k)\right)$	4, 5, PL
7.	$\neg\left(B_i(s_i^\ell \succeq_i s_i^k) \wedge \neg B_i \neg(s_i^\ell \succ_i s_i^k)\right) \to \neg B_i(s_i^\ell \succ_i s_i^k)$	6, PL
9.	$s_i^k \to \neg B_i(s_i^\ell \succ_i s_i^k)$	1, 7, PL

⊣

The following proposition is the counterpart to Proposition 9.3: it shows that - when belief is replaced with knowledge - axiom **SR** provides a syntactic characterization of the output of the IDIP procedure (namely, the set of strategy-profiles T^∞) and thus, by Proposition 9.4, provides a syntactic characterization of common knowledge of s-rationality in strategic-form games with ordinal payoffs.

Proposition 9.6
Fix a strategic-form game with ordinal payoffs $G = \left\langle \mathsf{Ag}, \{S_i, \pi_i\}_{i\in\mathsf{Ag}} \right\rangle$. Then

(A) If $M = \langle W, \{\sim_i\}_{i\in\mathsf{Ag}}, \{\sigma_i\}_{i\in\mathsf{Ag}}, V\rangle$ is an $\mathcal{S}5$ syntactic model of G that validates axiom **SR**, then $\sigma(w) \in T^\infty$, for every state $w \in W$.

(B) There exists an $\mathcal{S}5$ syntactic model M of G that validates axiom **SR** and is such that (1) for every $s \in T^\infty$, there exists a state w in M such that $w \models s$, and (2) for every $s \in S$ and for every $w \in W$, if $w \models s$ then $\sigma(w) \in T^\infty$. ⊣

Proof

To stress the fact that we are dealing with $\mathcal{S}5$ models, we shall use the operator K_i (knowledge) instead of B_i (belief).

(A) Fix a game and an $\mathcal{S}5$ syntactic model of it that validates axiom **SR**. Fix an arbitrary state w_0 and an arbitrary player i. By Axioms **G1** and **G2** (see Remark 9.5) $w_0 \models s_i^k$ for a unique strategy $s_i^k \in S_i$. Fix an arbitrary $s_i^\ell \in S_i$, with $s_i^\ell \neq s_i^k$. Since the model validates axiom **SR**, $w_0 \models \neg\left(K_i(s_i^\ell \succeq_i s_i^k) \wedge \neg K_i \neg(s_i^\ell \succ_i s_i^k)\right)$, that is (since the formula $\neg\left(K_i(s_i^\ell \succeq_i s_i^k) \wedge \neg K_i \neg(s_i^\ell \succ_i s_i^k)\right)$ is propositionally equivalent to $\neg K_i \neg(s_i^\ell \succ_i s_i^k) \to \neg K_i(s_i^\ell \succeq_i s_i^k)$),

$$w_0 \models \neg K_i \neg(s_i^\ell \succ_i s_i^k) \to \neg K_i(s_i^\ell \succeq_i s_i^k). \tag{9.8}$$

If, for every w such that $w_0 \sim_i w$, $\pi_i(s_i^k, \sigma_{-i}(w)) \geq \pi_i(s_i^\ell, \sigma_{-i}(w))$, then, by Definition 9.7, $w \in SRAT_i$. If, on the other hand, there is a w_1 such that $w_0 \sim_i w_1$ and $\pi_i(s_i^\ell, \sigma_{-i}(w_1)) > \pi_i(s_i^k, \sigma_{-i}(w_1))$, then, by Definition 9.6, $w_1 \models (s_i^\ell \succ_i s_i^k)$ and thus $w_0 \models \neg K_i \neg(s_i^\ell \succ_i s_i^k)$. Hence, by (9.8), $w_0 \models \neg K_i(s_i^\ell \succeq_i s_i^k)$, that is, there exists a w_2 such that $w_0 \sim_i w_2$ and $w_2 \models \neg(s_i^\ell \succeq_i s_i^k)$, so that, by Axioms **G4** and **G5**, $w_2 \models s_i^k \succ_i s_i^\ell$; that is, by Definition 9.6, $\pi_i(s_i^k, \sigma_{-i}(w_2)) > \pi_i(s_i^\ell, \sigma_{-i}(w_2))$. Hence, by Definition 9.7, $w \in SRAT_i$. Since w_0 and i were chosen arbitrarily, it follows that $SRAT = W$ and thus $\mathbb{CK}SRAT = W$. Hence, by Proposition 9.4, $\sigma(w) \in T^\infty$ for every $w \in W$.

(B) Let F be the $\mathcal{S}5$ epistemic model constructed in the proof of Part B of Proposition 9.4 and let M be a syntactic model based on F that satisfies the validation rules of Definition 9.6. First we show that M validates axiom **SR.** Recall that in F, $W = T^\infty$, $s' \in \sim_i (s)$ if and only if $s_i = s_i'$ and σ is the identity function. Fix an arbitrary player i and an arbitrary state $\hat{s}$. We need to show that, for every $s_i^\ell \in S_i$, $\hat{s} \models \neg\left(K_i(s_i^\ell \succeq_i \hat{s}_i) \wedge \neg K_i \neg(s_i^\ell \succ_i \hat{s}_i)\right)$. Suppose that, for some $s_i^\ell \in S_i$, $\hat{s} \models \left(K_i(s_i^\ell \succeq_i \hat{s}_i) \wedge \neg K_i \neg(s_i^\ell \succ_i \hat{s}_i)\right)$, that is, for every s such that $\hat{s} \sim_i s$ (recall that $\hat{s} \sim_i s$ if and only if $\hat{s}_i = s_i$), $s \models s_i^\ell \succeq_i \hat{s}_i$ and there exists an $\tilde{s}$ such that $\hat{s} \sim_i \tilde{s}$ (that is, $\hat{s}_i = \tilde{s}_i$) and $\tilde{s} \models s_i^\ell \succ_i \hat{s}_i$. Then, by Definition 9.6, for all s such that $\hat{s} \sim_i s$, $\pi_i(s_i^\ell, s_{-i}) \geq \pi_i(\hat{s}_i, s_{-i})$ and $\hat{s} \sim_i \tilde{s}$ and $\pi_i(s_i^\ell, \tilde{s}_{-i}) > \pi_i(\hat{s}_i, \tilde{s}_{-i})$. Then by Definition 9.7, $\hat{s} \notin SRAT_i$. But, as shown in the proof of Proposition 9.4, $SRAT = T^\infty$ so that, since $SRAT \subseteq SRAT_i \subseteq W = T^\infty$, $SRAT_i = T^\infty$, yielding a contradiction. Thus M validates axiom **SR**. Now fix an arbitrary $s \in T^\infty$. Then, by Definition 9.6, $s \models s$; thus (1) holds. Conversely, let $s \models s$; then, by construction of F, $\sigma(s) = s$ and $s \in T^\infty$. Thus (2) holds.

$\dashv$

As noted in Section 9.4 for the case of axiom **WR** (see Remark 9.7), axiom **SR** provides a syntactic characterization of common knowledge of s-rationality in a logic that does not include a common knowledge operator. However, since **SR** expresses the notion that player i chooses s-rationally, by the Necessitation rule every player knows that player i is s-rational and every player knows this, and so on, so that essentially the s-rationality of every player is commonly known. Indeed, if one adds the common knowledge operator CK to the logic, then, by Necessitation, $CK\left(s_i^k \to \neg\left(B_i(s_i^\ell \succeq_i s_i^k) \wedge \neg B_i \neg(s_i^\ell \succ_i s_i^k)\right)\right)$ becomes a theorem.

It is also worth repeating (see Remark 9.8), that the difference between the local character of Proposition 9.4 and the global character of Proposition 9.6 is only apparent: the characterization of Proposition 9.4 can in fact be viewed as a global characterization.

Remark 9.9
Note that neither Proposition 9.4 nor Proposition 9.6 is true if one replaces knowledge with belief, as illustrated in the game of Part a of Figure 9.4 and corresponding $\mathcal{KD}45$ frame of Part b. In the corresponding model we have that, according to the stronger notion of s-rationality (Definition 9.7), $SRAT = \{w_1, w_2\}$ so that $w_1 \in \mathbb{CB}SRAT$, despite the fact that $\sigma(w_1) = (b, d)$, which is an inferior strategy profile (relative to the entire game).[15] In other words, common *belief* of s-rationality is compatible with the players collectively choosing an inferior strategy profile. Thus, unlike the weaker notion expressed by axiom **WR**, with axiom **SR** there is a crucial difference between the implications of common *belief* and those of common *knowledge* of rationality. $\dashv$

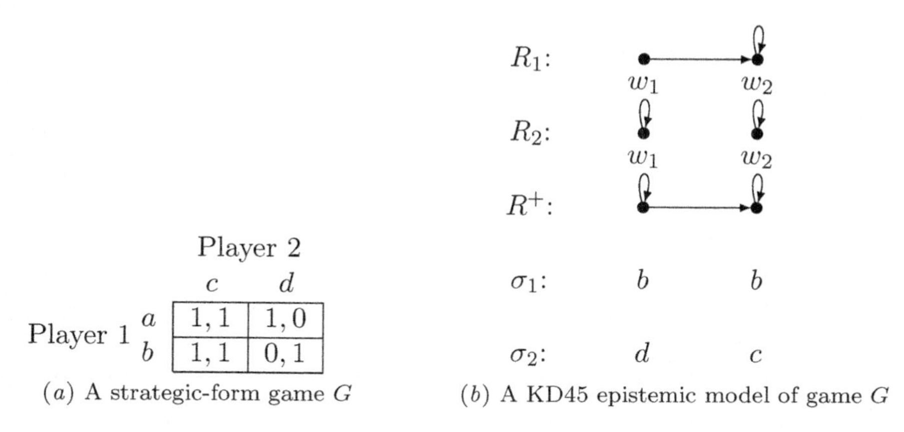

(a) A strategic-form game G (b) A KD45 epistemic model of game G

Figure 9.4: A model with common *belief* of s-rationality at every state

[15] In the game of Figure 9.4 we have that, while $S^\infty = S = \{(a,c),(a,d),(b,c),(b,d)\}$, $T^\infty = \{(a,c),(b,c)\}$.

9.6 Probabilistic Beliefs and von Neumann Morgenstern Payoffs

So far we have assumed that each player has an *ordinal* ranking of the possible outcomes; furthermore, we restricted attention to *qualitative* beliefs, represented by Kripke frames. In such a framework one can express the fact that, say, Player 1 is uncertain as to whether Player 2 will choose strategy c or strategy d but one cannot express graded forms of beliefs, such as "Player 1 believes that it is twice as likely that Player 2 will play c rather than d". The preponderant approach in the game-theoretic literature is to endow players with probabilistic beliefs and to assume that the players' preferences can be represented by a Bernoulli (also called von Neumann-Morgenstern) utility function. In this section we briefly describe this approach.

		Player 2	
		c	d
Player 1	A	o_1	o_2
	B	o_3	o_4

Figure 9.5: A strategic-form game-frame

Consider the strategic-form game-frame shown in Figure 9.5 (a game-frame is a game without the players' ranking of the outcomes), where o_1, o_2, o_3 and o_4 are the possible outcomes, and suppose that Player 1 assigns subjective probability $\frac{1}{3}$ to the possibility that Player 2 will choose c and probability $\frac{2}{3}$ to Player 2 choosing d. What choice should Player 1 make? If he chooses A, then the outcome will be o_1 with probability $\frac{1}{3}$ and o_2 with probability $\frac{2}{3}$; on the other hand, choosing B will yield outcome o_3 with probability $\frac{1}{3}$ and o_4 with probability $\frac{2}{3}$. Thus comparing A to B amounts to comparing the lottery $\begin{pmatrix} o_1 & o_2 \\ \frac{1}{3} & \frac{2}{3} \end{pmatrix}$ to the lottery $\begin{pmatrix} o_3 & o_4 \\ \frac{1}{3} & \frac{2}{3} \end{pmatrix}$. An ordinal ranking of the set of basic outcomes $\{o_1, o_2, o_3, o_4\}$ is no longer sufficient to determine what is rational for Player 1 to do (given the hypothesized beliefs). Thus we need to modify the models that we have been using so far in two ways: we need to enrich our structures so that we can express probabilistic beliefs and we need to go beyond ordinal rankings of the outcomes.

Definition 9.9

A *probabilistic frame* is a tuple $\langle W, \{R_i\}_{i\in\mathsf{Ag}}, \{p_i\}_{i\in\mathsf{Ag}}\rangle$ where $\langle W, \{R_i\}_{i\in\mathsf{Ag}}\rangle$ is a $\mathcal{KD}45$ Kripke frame and, for every agent $i \in \mathsf{Ag}$, $p_i : W \to \Delta(W)$

(where $\Delta(W)$ denotes the set of probability measures over W) is a function that satisfies the following properties (we use the notation $p_{i,w}$ instead of $p_i(w)$):[16] $\forall w, w' \in W$,
1. $supp(p_{i,w}) = R_i(w)$, and
2. if $w' \in R_i(w)$ then $p_{i,w'} = p_{i,w}$. ⊣

Thus $p_{i,w} \in \Delta(W)$ is agent i's subjective probability measure at state w. Condition 1 says that the agent assigns positive probability to all and only the states that she considers possible (according to her accessibility relation R_i) and Condition 2 says that the agent knows her own probabilistic beliefs, since she has the same probability measure at every state that she considers possible.

The semantic belief operator $\mathbb{B}_i : 2^W \to 2^W$ of player i (obtained from the doxastic accessibility relation R_i) is defined as in Section 9.2 (see 9.1) and so is the common belief operator $\mathbb{CB} : 2^W \to 2^W$. In this context, the interpretation of $\mathbb{B}_i E$ is "the event that player i assigns probability 1 to event E".

As noted above, the ordinal ranking of the set of outcomes O that we have postulated so far is not sufficient to determine whether one lottery is better than another. Traditionally, game theorists have assumed that every player has a complete ranking of all the lotteries over the set of basic outcomes O. The *theory of expected utility*, developed by the founders of game theory, namely John von Neumann and Oscar Morgenstern, provides a list of "rationality" or "consistency" axioms for how lotteries should be ranked and yields the following representation theorem. Given a finite set O of *basic outcomes*, we denote by $\Delta(O)$ the set of probability distributions or *lotteries* over O. A *von Neumann-Morgenstern ranking* of $\Delta(O)$ is a binary relation $\succsim^{vnm}$ on $\Delta(O)$ that satisfies a number of properties, known as the von Neumann-Morgenstern axioms or expected utility axioms.[17] If $L, L' \in \Delta(O)$, the interpretation of $L \succsim^{vnm} L'$ is that lottery L is considered to be at least as good as lottery L'.

Theorem 9.10

[von Neumann and Morgenstern (1944)]. Let $O = \{o_1, ..., o_m\}$ be a set of basic outcomes and $\succsim^{vnm}$ a von Neumann-Morgenstern ranking of $\Delta(O)$. Then there exists a function $U : O \to \mathbb{R}$, called a *Bernoulli* (or *von Neumann-Morgenstern*) *utility function* such that, given any two lotteries $L = \begin{pmatrix} o_1 & \cdots & o_m \\ p_1 & \cdots & p_m \end{pmatrix}$ and $L' = \begin{pmatrix} o_1 & \cdots & o_m \\ q_1 & \cdots & q_m \end{pmatrix}$, $L \succsim^{vnm} L'$ if and

[16] If μ is a probability measure over W, we denote by $supp(\mu)$ the support of μ, that is, the set of states to which μ assigns positive probability.

[17] Because of space limitations we shall not list those axioms. The interested reader is referred to Kreps (1988).

only if $\sum_{j=1}^{m} U(o_j)p_j \geq \sum_{j=1}^{m} U(o_j)q_j$. The number $\sum_{j=1}^{m} U(o_j)p_j$ is called the *expected utility of lottery L*.
Furthermore, if $U : O \to \mathbb{R}$ is a Bernoulli utility function that represents the ranking $\succsim^{vnm}$, then, for every pair of real numbers $a, b \in \mathbb{R}$ with $a > 0$, the function $V : O \to \mathbb{R}$ defined by $V(o) = aU(o) + b$ is also a Bernoulli utility function that represents $\succsim^{vnm}$. $\dashv$

Definition 9.10
A *finite strategic-form game with cardinal (or von Neumann Morgenstern) payoffs* is a quintuple $G = \left\langle \mathsf{Ag}, \{S_i\}_{i\in\mathsf{Ag}}, O, z, \{\succsim_i^{vnm}\}_{i\in\mathsf{Ag}} \right\rangle$, where Ag, S_i, O and z are as in Definition 9.1 and, for every player $i \in N$, $\succsim_i^{vnm}$ is a von Neumann-Morgenstern ranking of $\Delta(O)$. Such games are often represented in *reduced form* by replacing the triple $\left\langle O, z, \{\succsim_i^{vnm}\}_{i\in\mathsf{Ag}} \right\rangle$ with a set of *cardinal payoff functions* $\{\pi_i\}_{i\in\mathsf{Ag}}$ with $\pi_i : S \to \mathbb{R}$ defined by $\pi_i(s) = U_i(z(s))$, where $U_i : O \to \mathbb{R}$ is a Bernoulli utility function that represents the ranking $\succsim_i^{vnm}$ (whose existence is guaranteed by Theorem 9.10). $\dashv$

Going back to the above example based on Figure 9.5, where Player 1 assigns subjective probability $\frac{1}{3}$ to Player 2 choosing c and probability $\frac{2}{3}$ to Player 2 choosing d, if Player 1 has a von Neumann-Morgenstern ranking $\succsim_1^{vnm}$ of $\Delta(\{o_1, o_2, o_3, o_4\})$, then it is rational for him to choose A if and only if $\frac{1}{3}U_1(o_1) + \frac{2}{3}U_1(o_2) \geq \frac{1}{3}U_1(o_3) + \frac{2}{3}U_1(o_4)$, where U_1 is a Bernoulli utility function that represents $\succsim_1^{vnm}$.

It is worth stressing that the move from games where players have ordinal rankings of the basic outcomes to games where they have von Neumann-Morgenstern rankings of lotteries (over basic outcomes) is not an innocuous move. The reason is not only that much more is assumed about each individual player's preferences, but also that - since the game is implicitly assumed to be common knowledge among the players - each player is assumed to know the cardinal rankings of his opponents (how they rank all possible lotteries, what their attitude to risk is, etc.).

The definition of an epistemic model of a game (Definition 9.2) can be straightforwardly extended to games with von Neumann-Morgenstern payoffs.

Definition 9.11
Given a strategic-form game with von Neumann Morgenstern payoffs G of the form $G = \left\langle \mathsf{Ag}, \{S_i\}_{i\in\mathsf{Ag}}, \{\pi_i\}_{i\in\mathsf{Ag}} \right\rangle$, an *epistemic-probabilistic model* of G is a tuple
$\langle W, \{R_i\}_{i\in\mathsf{Ag}}, \{p_i\}_{i\in\mathsf{Ag}}, \{\sigma_i\}_{i\in\mathsf{Ag}} \rangle$ where $\langle W, \{R_i\}_{i\in\mathsf{Ag}}, \{p_i\}_{i\in\mathsf{Ag}} \rangle$ is a probabilistic frame (see Definition 9.9) and $\sigma_i : W \to S_i$ is - as before - a function

that associates, with every state, a strategy of player i, satisfying the property that if $w' \in R_i(w)$ then $\sigma_i(w') = \sigma_i(w)$. ⊣

As before, given a state w and a player i, we denote by $\sigma_{-i}(w)$ the strategy profile of the players other than i at state w. The definition of rationality (Definition 9.3) can now be sharpened, as follows.

Definition 9.12
Fix a strategic-form game with von Neumann Morgenstern payoffs G and an epistemic-probabilistic model of G. At state w player i's strategy $s_i = \sigma_i(w)$ is *rational* if it maximizes player i's payoff, given his beliefs at w, that is, if

$$\sum_{x \in R_i(w)} p_{i,w}(x)\ \pi_i(s_i, \sigma_{-i}(x)) \geq \sum_{x \in R_i(w)} p_{i,w}(x)\ \pi_i(s_i', \sigma_{-i}(x)), \quad \forall s_i' \in S_i.$$

[Recall that, by Definition 9.11, the function $\sigma_i(\cdot)$ is constant on the set $R_i(w)$]. ⊣

What are the implications of common belief of rationality in this framework? It turns out that a result similar to Proposition 9.1 holds in this case too: common belief of rationality is characterized by a strengthening of the IDSDS procedure (Definition 9.5).[18] Because of space limitations we omit the details. Similarly, a result along the lines of Proposition 9.4 holds in this case too for a strengthening of the IDIP procedure (see Stalnaker (1994)).

9.7 Dynamic Games with Perfect Information

So far we have restricted attention to strategic-form games, where the players make their choices simultaneously or in ignorance of the other players' choices. We now turn to dynamic games, where players make choices sequentially, having some information about the moves previously made by their opponents. If information is partial, the game is said to have *imperfect information*, while the case of full information is referred to as *perfect information*. Because of space limitations we shall restrict attention to perfect-information games.

[18] The modified procedure allows the deletion of pure strategies that are strictly dominated by a mixed strategy, that is, by a probability distribution over the set of pure strategies. This is because, as shown by Pearce (1984), a pure strategy s is strictly dominated by another, possibly mixed, strategy if and only if there is no (probabilistic) belief concerning the strategies chosen by the opponents that makes s a best reply, that is, there is no belief that makes s a rational choice.

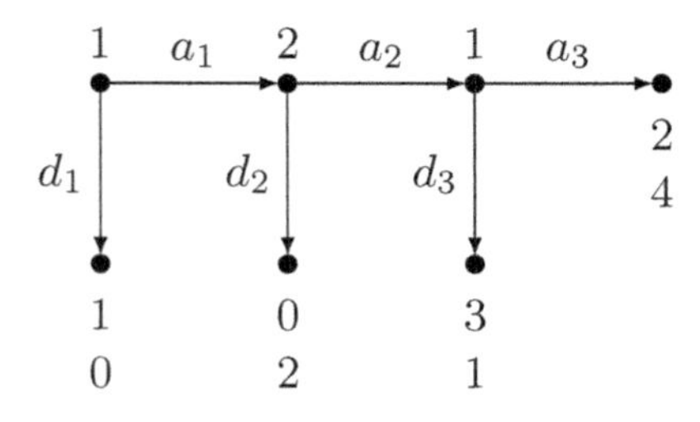

Figure 9.6: A dynamic game with perfect information

An example of a dynamic game with perfect information is shown in Figure 9.6 in the form of a tree. Each node in the tree represents a history of prior moves and is labeled with the player whose turn it is to move. For example, at history $a_1 a_2$ it is Player 1's turn to move (after his initial choice of a_1 followed by Player 2's choice of a_2) and he has to choose between two actions: a_3 and d_3. The terminal histories (the leaves of the tree) represent the possible outcomes and each player i is assumed to have an ordinal preference relation $\succsim_i$ over the set of terminal histories (in Figure 9.6 the players' preferences over the terminal histories have been represented by means of ordinal utility functions, as explained below).

The formal definition of a perfect-information game is as follows. If A is a set, we denote by A^* the set of finite sequences in A. If $h = \langle a_1, ..., a_k \rangle \in A^*$ and $1 \leq j \leq k$, the sequence $\langle a_1, ..., a_j \rangle$ is called a *prefix* of h. If $h = \langle a_1, ..., a_k \rangle \in A^*$ and $a \in A$, we denote the sequence $\langle a_1, ..., a_k, a \rangle \in A^*$ by ha.

Definition 9.13
A *finite extensive game with perfect information and ordinal payoffs* is a tuple $\left\langle A, H, \mathsf{Ag}, \iota, \{\succsim_i\}_{i \in \mathsf{Ag}} \right\rangle$ whose elements are:

- A finite set of actions A.
- A finite set of histories $H \subseteq A^*$ which is closed under prefixes (that is, if $h \in H$ and $h' \in A^*$ is a prefix of h, then $h' \in H$). The null history $\langle\rangle$, denoted by $\emptyset$, is an element of H and is a prefix of every history. A history $h \in H$ such that, for every $a \in A$, $ha \notin H$, is called a *terminal history*. The set of terminal histories is denoted by Z. $D = H \backslash Z$ denotes the set of non-terminal or *decision* histories. For every decision history $h \in D$, we denote by $A(h)$ the set of actions available at h, that is, $A(h) = \{a \in A : ha \in H\}$.
- A finite set Ag of players.

- A function $\iota : D \to \mathsf{Ag}$ that assigns a player to each decision history. Thus $\iota(h)$ is the player who moves at history h. For every $i \in \mathsf{Ag}$, let $D_i = \iota^{-1}(i)$ be the set of histories assigned to player i.

- For every player $i \in \mathsf{Ag}$, $\succsim_i$ is an ordinal ranking of the set Z of terminal histories. ⊣

The ordinal ranking of player i is normally represented by means of an ordinal *utility* (or *payoff*) *function* $U_i : Z \to \mathbb{R}$ satisfying the property that $U_i(z) \geq U_i(z')$ if and only if $z \succsim_i z'$. In the game of Figure 9.6, associated with every terminal history is a pair of numbers: the top number is the utility of Player 1 and the bottom number is the utility of Player 2.

Histories will be denoted more succinctly by listing the corresponding actions, without angled brackets and without commas; thus instead of writing for instance $\langle \emptyset, a_1, a_2, a_3, a_4 \rangle$ we simply write $a_1a_2a_3a_4$.

In their seminal book, von Neumann and Morgenstern (1944) showed that a dynamic game can be reduced to a strategic-form game by defining strategies as complete, contingent plans of action. In the case of perfect-information games a *strategy* for a player is a function that associates with every decision history assigned to that player one of the choices available there. For example, a possible strategy of Player 1 in the game of Figure 9.6 is (d_1, d_3). A profile of strategies (one for each player) determines a unique path from the null history (the root of the tree) to a terminal history (a leaf of the tree). Figure 9.7 shows the strategic-form corresponding to the extensive form of Figure 9.6.

		Player 2	
		a_2	d_2
Player 1	a_1a_3	2, 4	0, 2
	a_1d_3	3, 1	0, 2
	d_1a_3	1, 0	1, 0
	d_1d_3	1, 0	1, 0

Figure 9.7: The strategic-form of the game of Figure 9.6

How should a model of a dynamic game be constructed? One approach in the literature has been to consider models of the corresponding strategic-form (the type of models considered in Section 9.2). However, there are several conceptual issues that arise in this context. The interpretation of $s_i = \sigma_i(w)$ is that at state w player i "chooses" strategy s_i. Now consider a model of the game of Figure 9.6 and a state w where $\sigma_1(w) = (d_1, a_3)$. What does it mean to say that Player 1 "chooses" strategy (d_1, a_3)? The first part of the strategy, namely d_1, can be interpreted as a description of Player 1's actual choice to play d_1, but the second part of the strategy, namely a_3, has no such interpretation: if Player 1 in fact plays d_1 then he knows that he will not have to make any further choices and thus it is not clear what it means for him to "choose" to play a_3 in a situation that is made impossible by his decision to play d_1.[19] Thus it does not seem to make sense to interpret $\sigma_1(w) = (d_1, a_3)$ as 'at state w Player 1 chooses (d_1, a_3)'. Perhaps the correct interpretation is in terms of a more complex sentence such as 'Player 1 chooses to play d_1 and if - contrary to this - he were to play a_1 and Player 2 were to follow with a_2, then Player 1 would play a_3'. Thus while in a simultaneous game the association of a strategy of player i to a state can be interpreted as a description of player i's actual behavior at that state, in the case of dynamic games this interpretation is no longer valid, since one would end up describing not only the actual behavior of player i at that state but also his counterfactual behavior. Methodologically, this is not satisfactory: if it is considered to be necessary to specify what a player would do in situations that do not occur in the state under consideration, then one should model the counterfactual explicitly. But why should it be necessary to specify at state w (where Player 1 is playing d_1) what he would do at the counterfactual history a_1a_2? Perhaps what matters is not so much what Player 1 would actually do there but what Player 2 believes that Player 1 would do: after all, Player 2 might not know that Player 1 has decided to play d_1 and needs to consider what to do in the eventuality that Player 1 actually ends up playing a_1. So, perhaps, the strategy of Player 1 is to be interpreted as having two components: (1) a description of Player 1's behavior and (2) a conjecture in the mind of Player 2 about what Player 1 would do. If this is the correct interpretation, then one could - from a methodological point of view - object that it would be preferable to disentangle the two components and model them explicitly.[20]

An alternative - although less common - approach in the literature dispenses with strategies and considers models of games where (1) states are described in terms of players' *actual behavior* and (2) players' conjectures concerning the actions of their opponents (as well as their own actions) in various hypothetical situations are modeled by means of a generalization

[19] For this reason, some authors, instead of using strategies, use the weaker notion of "plan of action" introduced in Rubinstein (1991). A plan of action for a player only contains choices that are not ruled out by his earlier choices. For example, the possible plans of action for Player 1 in the game of Figure 9.6 are d_1, (a_1, a_3) and (a_1, d_3). However, most of the issues raised below apply also to plans of action. The reason for this is that a choice of player i at a later decision history of his may be counterfactual at a state because of the choices of *other* players (which prevent that history from being reached).

[20] For a more in-depth discussion of these issues see (Bonanno, 2014).

of the Kripke frames considered so far. The generalization is obtained by encoding not only the initial beliefs of the players (at each state) but also their *dispositions to revise those beliefs* under various hypothesis. These structures are reviewed in the next section.

9.8 The Semantics of Belief Revision

A $\mathcal{KD}45$ Kripke frame $\langle W, \{R_i\}_{i\in \mathsf{Ag}}\rangle$ represents the actual beliefs of the agents at every state w. In order to capture the agents' disposition to revise their beliefs under various hypotheses, we need to consider extensions of those frames.

Definition 9.14
A *belief revision frame* is a triple $\langle W, \{R_i\}_{i\in \mathsf{Ag}}, \{\mathcal{E}_i, f_i\}_{i\in \mathsf{Ag}}\rangle$, where the pair consisting of $\langle W, \{R_i\}_{i\in \mathsf{Ag}}\rangle$ is a $\mathcal{KD}45$ Kripke frame and, for every agent $i \in \mathsf{Ag}$, $\mathcal{E}_i \subseteq 2^W \backslash \varnothing$ is a set of admissible hypotheses (or potential items of information) and $f_i : W \times \mathcal{E}_i \to 2^W$ is a function that satisfies the following properties: $\forall w \in W$, $\forall E, F \in \mathcal{E}_i$,

1. $f_i(w, E) \neq \varnothing$,
2. $f_i(w, E) \subseteq E$,
3. if $R_i(w) \cap E \neq \varnothing$ then $f_i(w, E) = R_i(w) \cap E$,
4. if $E \subseteq F$ and $f_i(w, F) \cap E \neq \varnothing$ then $f_i(w, E) = f_i(w, F) \cap E$.

$\dashv$

The event $f_i(w, E)$ is interpreted as the set of states that player i would consider possible, at state w, under the supposition that (or if informed that) E is true. Condition 1 requires these suppositional beliefs to be consistent. Condition 2 requires that, under the supposition that E is true, E be indeed considered true. Condition 3 says that if E is compatible with the initial beliefs (given by $R_i(w)$) then the suppositional beliefs coincide with the initial beliefs conditioned on event E.[21] Condition 4 is an extension of Condition 3: if E implies F and E is compatible (not with player i's prior beliefs but) with the *posterior* beliefs that player i would have if she supposed (or learned) that F were the case (let's call these her posterior F-beliefs), then her beliefs under the supposition (or information) that E must coincide with her posterior F-beliefs conditioned on event E.

Thus the function f_i can be used to model the full epistemic attitude of player i at every state w: her prior (or initial) beliefs are given by the set

[21]Note that it follows from Condition 3 and seriality of R_i that, for every $w \in W$, $f_i(w, W) = R_i(w)$, so that one could simplify the definition of a belief revision frame by dropping the relations R_i and recovering the initial beliefs at state w from the set $f_i(w, W)$. We have chosen not to do so in order to maintain continuity in the exposition.

$R_i(w)$ and, for every event E, the set $f_i(w, E)$ captures how she is disposed to revise those beliefs under the supposition that E is true. In particular, the function f_i tells us how player i would revise her prior beliefs if she learned information that contradicted those beliefs.

Remark 9.11

If $\mathcal{E}_i = 2^W \backslash \varnothing$ then Conditions 1-4 of Definition 9.14 imply that, for every $w \in W$, there exists a "plausibility" relation Q_i^w on W which is complete ($\forall w_1, w_2 \in W$, either $w_1 Q_i^w w_2$ or $w_2 Q_i^w w_1$ or both) and transitive ($\forall w_1, w_2, w_3 \in W$, if $w_1 Q_i^w w_2$ and $w_2 Q_i^w w_3$ then $w_1 Q_i^w w_3$) and such that, for every $E \subseteq W$ with $E \neq \varnothing$, $f_i(w, E) = \{x \in E : x Q_i^w y, \ \forall y \in E\}$. The interpretation of $x Q_i^w y$ is that - at state w and according to player i - state x is at least as plausible as state y. Thus $f_i(w, E)$ is the set of most plausible states in E (according to player i at state w). If $\mathcal{E}_i \neq 2^W \backslash \varnothing$ then Conditions 1-4 in Definition 9.14 are necessary but not sufficient for the existence of such a plausibility relation. The existence of a plausibility relation that rationalizes the function $f_i(w, \cdot) : \mathcal{E}_i \to 2^W$ is necessary and sufficient for the belief revision policy encoded in $f_i(w, \cdot)$ to be compatible with the syntactic theory of belief revision introduced in Alchourrón, Gärdenfors, and Makinson (1985), known as the AGM theory.

One can associate with each function f_i a conditional belief operator $\overline{\mathbb{B}}_i : 2^W \times \mathcal{E}_i \to 2^W$ as follows, with $F \in 2^W$ and $E \in \mathcal{E}_i$:

$$\overline{\mathbb{B}}_i(F|E) = \{w \in W : f_i(w, E) \subseteq F\}. \tag{9.9}$$

Possible interpretations of the event $\overline{\mathbb{B}}_i(F|E)$ are "according to player i, if E were the case, then F would be true" or "if informed that E, player i would believe that F" or "under the supposition that E, player i would believe that F".

The unconditional belief operator $\mathbb{B}_i : 2^W \to 2^W$ remains as defined in Section 9.5 and represents the initial beliefs of agent i.[22] Similarly, the common belief operator $\mathbb{CB}$ remains as defined in Section 9.5 and captures what is *initially* common belief among the agents.

9.9 Common Belief of Rationality in Perfect-Information Games

We can now return to dynamic games with perfect information. First we define an algorithm, known as *backward induction*, which is meant to capture the "rational" way of playing these games and explore the possibility of providing an epistemic foundation for it.

[22]Note that, for every event F, $\mathbb{B}_i F = \overline{\mathbb{B}}_i(F|W)$.

The backward induction algorithm starts at the end of the game and proceeds backwards towards the root:

1. Start at a decision history h whose immediate successors are only terminal histories (e.g. history a_1a_2 in the game of Figure 9.6) and select a choice that maximizes the utility of player $\iota(h)$ (in the example of Figure 9.6, at a_1a_2 Player 1's optimal choice is d_3 (since it gives her a payoff of 3 rather than 2, which is the payoff that she would get if she played a_3). Delete the immediate successors of history h (that is, turn h into a terminal history) and assign to h the payoff vector associated with the selected choice.

2. Repeat Step 1 until all the decision histories have been exhausted.

For example, the choices selected by the backward-induction algorithm in the game of Figure 9.6 are d_3, d_2 and d_1.[23]

A question that has been studied extensively in the literature is whether *initial* common belief of rationality can provide an epistemic justification for the backward-induction solution. In order to answer this question we need to introduce the notion of an epistemic model of a perfect-information game.

Definition 9.15
Given a dynamic game with perfect information and ordinal payoffs $\Gamma = \left\langle A, H, \mathsf{Ag}, \iota, \{\precsim_i\}_{i\in\mathsf{Ag}}\right\rangle$, an *epistemic model* of Γ is a tuple $\langle W, \{R_i\}_{i\in\mathsf{Ag}}, \{\mathcal{E}_i, f_i\}_{i\in\mathsf{Ag}}, \zeta\rangle$ where $\langle W, \{R_i\}_{i\in\mathsf{Ag}}, \{\mathcal{E}_i, f_i\}_{i\in\mathsf{Ag}}\rangle$ is a belief revision frame (Definition 9.14) and $\zeta : W \to Z$ is a function that associates with every state a terminal history and satisfies the following property: $\forall w, w' \in W, \forall i \in \mathsf{Ag}, \forall h \in H, \forall a \in A$,

$$\begin{array}{l}\text{If } h \text{ is a decision history of player } i,\ a \text{ an action at } h\\ \text{and } ha \text{ a prefix of } \zeta(w) \text{ then, } \forall w' \in R_i(w),\\ \text{if } h \text{ is a prefix of } \zeta(w') \text{ then } ha \text{ is a prefix of } \zeta(w').\end{array} \tag{9.10}$$

⊣

The function ζ describes the *actual behavior* of the players at any given state. Thus we are not associating a strategy profile with a state but a sequence of actions leading from the null history to a terminal history. Condition (9.10) states that if at a state the play of the game reaches decision history h of player i, where she actually takes action a, then either player i initially believes that history h will not be reached or, if she considers it

[23]The backward induction algorithm may yield more than one solution. Multiplicity arises if there is at least one player who has more than one payoff-maximizing choice at a decision history of his.

possible that history h will indeed be reached, then she has correct beliefs about what action she will take (namely a) if h is reached.

Condition (9.10) can be stated more succinctly in terms of events. If E and F are two events, we denote by $E \to F$ the event $\neg E \cup F$. Thus $E \to F$ captures the material conditional. Given a history h in the game, we denote by $[h]$ the event that h is reached, that is, $[h] = \{w \in W : h$ is a prefix of $\zeta(w)\}$. Recall that D_i denotes the set of decision histories of player i and $A(h)$ the set of choices available at h. Then (9.10) can be stated as follows:[24]

$$\begin{aligned} &\forall h \in D_i, \forall a \in A(h), \\ &[ha] \subseteq \mathbb{B}_i([h] \to [ha]). \end{aligned} \tag{9.11}$$

In words: if, at a state, player i takes action a at her decision history h, then she believes that if h is reached then she takes action a.

Condition (9.11) rules out the possibility that a player may be uncertain about her own choice of action at decision histories of hers that are not ruled out by her initial beliefs. In general, a corresponding condition might not hold for *revised* beliefs. That is, suppose that at state w player i erroneously believes that her decision history h will not be reached ($w \in [h]$ but $w \in B_i\neg[h]$); suppose also that a is the action that she will choose at h ($w \in [ha]$). It may be the case that, according to her revised beliefs on the supposition that h is reached, she believes that she takes an action b different from the action that she actually takes, namely a. In order to rule this out we need to impose the following strengthening of (9.11):[25]

$$\begin{aligned} &\forall h \in D_i, \ \forall a \in A(h), \\ &[ha] \subseteq \overline{\mathbb{B}}_i([ha] | [h]). \end{aligned} \tag{9.12}$$

How can rationality be captured in the models that we are considering?

---

[24] Note that, if at state w player i believes that history h will *not* be reached ($\forall w' \in R_i(w)$, $w' \notin [h]$) then $R_i(w) \subseteq \neg[h] \subseteq [h] \to [ha]$, so that $w \in \mathbb{B}_i([h] \to [ha])$ and therefore (9.11) is satisfied even if $w \in [ha]$.

[25] (9.12) is implied by (9.11) whenever player i's initial beliefs do not rule out h. That is, if $w \in \neg\mathbb{B}_i\neg[h]$ (equivalently, $R_i(w) \cap [h] \neq \varnothing$) then, for every $a \in A(h)$,

$$\text{if } w \in [ha] \text{ then } w \in \overline{\mathbb{B}}_i([ha] | [h]). \quad \text{(F1)}$$

In fact, by Condition 3 of Definition 9.14 (since, by hypothesis, $R_i(w) \cap [h] \neq \varnothing$),

$$f_i(w, [h]) = R_i(w) \cap [h]. \quad \text{(F2)}$$

Let $a \in A(h)$ be such that $w \in [ha]$. Then, by (9.11), $w \in \mathbb{B}_i([h] \to [ha])$, that is, $R_i(w) \subseteq \neg[h] \cup [ha]$. Thus $R_i(w) \cap [h] \subseteq (\neg[h] \cap [h]) \cup ([ha] \cap [h]) = \varnothing \cup [ha] = [ha]$ (since $[ha] \subseteq [h]$) and therefore, by (F2), $f_i(w, [h]) \subseteq [ha]$, that is, $w \in \overline{\mathbb{B}}_i([ha] | [h])$.

Various definitions of rationality have been suggested in the literature, most notably *material rationality* and *substantive rationality* . The former notion is weaker in that a player can be found to be irrational only at decision histories of hers that are actually reached. The latter notion, on the other hand, is more stringent since a player can be judged to be irrational at a decision history h of hers even if she correctly believes that h will not be reached. We will focus on the weaker notion of material rationality. As before, we shall define a player's rationality as a proposition, that is, an event. Recall that Z denotes the set of terminal histories and $u_i : Z \to \mathbb{R}$ is player i's ordinal utility function (representing her preferences over the set Z). Define $\pi_i : W \to \mathbb{R}$ by $\pi_i(w) = u_i(\zeta(w))$. For every $x \in \mathbb{R}$, let $[\pi_i \leq x]$ be the event that player i's payoff is not greater than x, that is, $[\pi_i \leq x] = \{w \in W : \pi_i(w) \leq x\}$ and, similarly, let $[\pi_i > x] = \{w \in W : \pi_i(w) > x\}$. Then we say that player i is materially rational at a state if, for every decision history h of hers that is actually reached at that state and for every real number x, it is not the case that she believes that – under the supposition that h is reached – (1) her payoff from her actual choice would not be greater than x and (2) it would be greater than x if she were to take an action different from the one that she is actually taking (at that history in that state).[26]

Formally this can be stated as follows (recall that D_i denotes the set of decision histories of player i and $A(h)$ the set of actions available at h):

Player i is *materially rational* at $w \in W$ if, $\forall h \in D_i, \forall a \in A(h)$
if ha is a prefix of $\zeta(w)$ then, $\forall b \in A(h)$, $\forall x \in \mathbb{R}$, (9.13)
$\overline{\mathbb{B}}_i([\pi_i \leq x] \,|\, [ha]) \to \neg\overline{\mathbb{B}}_i([\pi_i > x] \,|\, [hb])$.

Note that, in general, we cannot replace the antecedent $\overline{\mathbb{B}}_i([\pi_i \leq x] \,|\, [ha])$ with $\mathbb{B}_i([ha] \to [\pi_i \leq x])$, because at state w player i might initially believe that h will not be reached, in which case it would be trivially true that $w \in \mathbb{B}_i([ha] \to [\pi_i \leq x])$; however, if decision history h is actually reached at w then player i will be surprised and will have to revise her beliefs. Thus her rationality is judged on the basis of her *revised* beliefs. Note, however, that if $w \in \neg\mathbb{B}_i\neg[h]$, that is, if at w she does not rule out the possibility that h will be reached and $a \in A(h)$ is the action that she actually takes at history h at state w ($w \in [ha]$), then, for every event F, $w \in \mathbb{B}_i([ha] \to F)$

[26] This definition is a "local" definition in that it only considers, for every decision history of player i, a change in player i's choice at that decision history and not also at later decision histories of hers. One could make the definition of rationality more stringent by simultaneously considering changes in the choices at a decision history and subsequent decision histories of the same player (if any).

if and only if $w \in \overline{\mathbb{B}}_i(F|[ha])$.[27]

Note also that, according to (9.13), a player is trivially rational at any state at which she does not take any actions.

Does initial common belief that all the players are materially rational (according to 9.13) imply backward induction in perfect-information games? The answer is negative.[28] To see this, consider the perfect-information game shown in Figure 9.6 and the model of it shown in Figure 9.8.[29]

First of all, note that the common belief relation R^+ is obtained by adding to R_2 the pair (w_2, w_2); thus, in particular, $R^+(w_2) = \{w_2, w_3\}$. We want to show that both players are materially rational at both states w_2 and w_3, so that at state w_2 it is initially common belief that both players are materially rational, despite that fact that the play of the game at w_2 is $a_1a_2d_3$, which is not the backward-induction play. Clearly, Player 1 is rational at state w_2 (since he obtains his largest possible payoff); he is also rational at state w_3 because he knows that he plays d_1, obtaining a payoff of 1, and believes that if he were to play a_1 then Player 2 would respond with d_2 and give him a payoff of zero: this belief is encoded in $f_1(w_3, [a_1]) = \{w_4\}$, where $[a_1] = \{w_1, w_2, w_4\}$, and $\zeta(w_4) = a_1d_2$. Player 2 is trivially rational at state w_3 since she does not take any actions there. Now consider state w_2. Player 2 initially erroneously believes that Player 1 will end the game by playing d_1: $R_2(w_2) = \{w_3\}$ and $\zeta(w_3) = d_1$. However,

---

[27] Proof. Suppose that $w \in [ha] \cap \neg\mathbb{B}_i\neg[h]$. As shown in Footnote 25 (see (F2)),

$$R_i(w) \cap [h] = f_i(w, [h]). \quad \text{(G1)}$$

Since $[ha] \subseteq [h]$,

$$R_i(w) \cap [h] \cap [ha] = R_i(w) \cap [ha]. \quad \text{(G2)}$$

As shown in Footnote 25, $f_i(w, [h]) \subseteq [ha]$ and, by Condition 1 of Definition 9.14, $f_i(w, [h]) \neq \varnothing$. Thus $f_i(w, [h]) \cap [ha] = f_i(w, [h]) \neq \varnothing$. Hence, by Condition 4 of Definition 9.14

$$f_i(w, [h]) \cap [ha] = f_i(w, [ha]). \quad \text{(G3)}$$

By intersecting both sides of (G1) with $[ha]$ and using (G2) and (G3) we get that $R_i(w) \cap [ha] = f_i(w, [ha])$.

[28] In fact, common belief of material rationality does not even imply a Nash equilibrium outcome. A Nash equilibrium is a strategy profile satisfying the property that no player can increase her payoff by unilaterally changing her strategy. A Nash equilibrium outcome is a terminal history associated with a Nash equilibrium. Note that a backward-induction solution of a perfect-information game can be expressed as a strategy profile and is always a Nash equilibrium.

[29] In Figure 9.8 we have only represented parts of the functions f_1 and f_2, namely the following: $f_1(w_3, \{w_1, w_2, w_4\}) = \{w_4\}$ and $f_2(w_2, \{w_1, w_2, w_4\}) = f_2(w_3, \{w_1, w_2, w_4\}) = \{w_1\}$. Note that $[a_1] = \{w_1, w_2, w_4\}$).

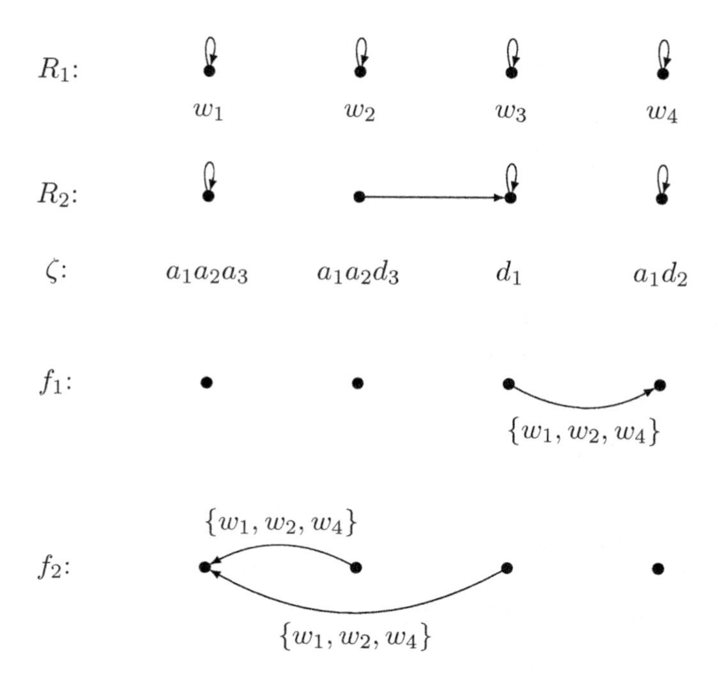

Figure 9.8: A (partial) model of the game of Figure 9.6

at state w_2, Player 1 is in fact playing a_1 and thus Player 2 will be surprised. Her initial disposition to revise her beliefs on the supposition that Player 1 plays a_1 is such that she would believe that she herself would play a_2 and Player 1 would follow with a_3, thus giving her the largest possible payoff: this belief is encoded in $f_2(w_2, [a_1]) = \{w_1\}$ (recall that $[a_1] = \{w_1, w_2, w_4\}$) and $\zeta(w_1) = a_1a_2a_3$. Hence she is rational at state w_2, according to (9.13).

In order to obtain the backward-induction solution, one needs to go beyond common initial belief of material rationality. Proposals in the literature include the notions of epistemic independence, strong belief, stable belief and substantive rationality. Space limitations prevent us from discussing these topics.

It is worth stressing that *in the models considered above, strategies do not play any role*: states are described in terms of the players' actual behavior along a play of the game. One could view a player's strategy as her (conditional) beliefs about what she would do under the supposition that each of her decision histories is reached. However, the models considered so far do not guarantee that a player's revised beliefs select a unique action at each of her decision histories. One could impose such a restriction on the players' dispositions to revise their beliefs.[30] However, in this setup strategies would then be cognitive constructs rather than objective counterfactuals about what a player would actually do at each of her decision histories.

9.10 Notes

In this section we point to the main references in the areas reviewed in this chapter, as well as references for related topics.

The birth of game theory The beginning of game theory is normally associated with the publication, in 1944, of the book *Theory of games and economic behavior* by von Neumann and Morgenstern (1944), although Cournot (1838) provided an analysis of simultaneous games among firms as early as 1838. Cournot's analysis of competition was later elaborated on by Bertrand (1883), von Stackelberg (1934) and by Hotelling (1929). Other notable precursors of the book by von Neumann and Morgenstern are an article by Zermelo (1913) (where he proved that in the game of chess either White has a strategy that guarantees him a win, or Black has a strategy that guarantees her a win, or both players have a strategy that guarantees a draw) and an article by von Neumann (1928) (where he proved the existence of a value in every finite zero-sum game). For a brief history of the first forty years of the development of game theory see the paper by Aumann (1987b).

The birth of the epistemic foundation program The origins of the literature on the epistemic foundations of solution concepts in non-cooperative games can be traced to two seminal papers by Bernheim (1984) and by Pearce (1984), both published in 1984. The purpose of these two articles was to capture the notion of "common recognition of rationality" in games. The analysis, however, was not developed explicitly in terms of epistemic notions: the idea of common belief of rationality was captured indirectly through the notion of rationalizability, which is an iterative procedure of elimination of strategies that are never a best response.

Another pioneering contribution was that by Aumann (1987a), providing an epistemic characterization of the notion of correlated equilibrium in terms of common knowledge of rationality when the players' beliefs share a common prior.

[30] The relevant restriction is as follows: $\forall h \in D_i, \forall a, b \in A(h), \forall w, w', w'' \in W$, if $w', w'' \in f_i(w, [h])$ and ha is a prefix of $\zeta(w')$ and hb is a prefix of $\zeta(w'')$ then $a = b$.

Extensive surveys of the literature on the epistemic foundation program are provided by Battigalli and Bonanno (1999), Dekel and Gul (1997) and by Perea (2012).

Epistemic models of strategic-form games There are two types of epistemic models of strategic-form games used in the game-theoretic literature: the "state-space" models and the "hierarchy of beliefs" models. The qualitative Kripke models considered in Sections 9.2 and 9.3 and their probabilistic counterparts considered in Section 9.6 are known in the game-theoretic literature as state-space models. Although, in the philosophy literature, Kripke frames date back to the work of Kripke (1963), in game theory state-space models first appeared in the work of Aumann (1976). Aumann (1987a) used a state-space model to obtain a characterization of the notion of correlated equilibrium using $\mathcal{S}5$ frames. Stalnaker (1994, 1996) provided the first systematic analysis of solution concepts in terms of $\mathcal{KD}45$ epistemic models of games.

The alternative approach in the game-theoretic literature uses the probabilistic hierarchy-of-belief models and type spaces that where introduced in the seminal papers of Harsanyi (1968), which started the literature on incomplete-information games. The first epistemic characterization of common belief of rationality in strategic-form games using these structures was provided by Tan and Werlang (1988). They showed that the (probabilistic version of) the iterative elimination of strictly dominated strategies identifies the strategy profiles that are compatible with common belief of rationality. The state-space formulation of this result is due to Stalnaker (1994), but it was implicit in Brandenburger and Dekel (1987). All these characterizations were for games with von Neumann-Morgenstern payoffs and for probabilistic beliefs. The stronger iterative elimination procedure (the stronger version of the IDIP algorithm given in Definition 9.8) and corresponding epistemic characterization is due to Stalnaker (1994) (with a correction by Bonanno and Nehring (1998)). The qualitative characterizations of Propositions 9.1 and 9.4 are based on work by Bonanno (2008).

Epistemic foundations of other strategic-form solution concepts Because of space limitations, we have restricted attention to the epistemic foundations of only some solution concepts. In the literature, epistemic conditions have been studied for additional solution concepts, such as for correlated equilibrium by Aumann (1987a) and Barelli (2009), for Nash equilibrium by Aumann and Brandenburger (1995), Bach and Tsakas (2014), Barelli (2009), Perea (2007b), and by Polak (1999), and for iterated admissibility by Brandenburger (1992), Barelli and Galanis (2013) Brandenburger,

Friedenberg, and Keisler (2008), Samuelson (1992), and by Stahl (1995). Surveys of the literature are given by Battigalli and Bonanno (1999), Dekel and Gul (1997) and by Perea (2012).

The use of logic in the analysis of games The literature on the epistemic foundation program is predominantly based on the semantic approach. The first to use formal logic in the analysis of games were Bacharach (1987) (who used first-order logic to investigate the notion of Nash equilibrium in strategic- form games) and Bonanno (1991) (who used propositional logic to investigate the notion of backward-induction in dynamic games with perfect information). There is now a sizeable literature that analyzes games using logic, in particular epistemic logic (see, for example, work by Board (2004), Bonanno (2001), Clausing (2003, 2004), de Bruin (2010) and by van Benthem (2011)). The analysis of Sections 9.4 and 9.5 is based on a paper by Bonanno (2008). The three axioms given in footnote 12 that provide a finite axiomatisation for common belief are also taken from a paper by Bonanno (1996).

Epistemic foundations of backward induction The issue of whether the backward-induction algorithm can be given an epistemic foundation has given rise to a large literature. The seminal paper was by Ben-Porath (1997). There are two strands in this literature. One group of papers uses epistemic models where states are described in terms of strategies (see, for example, work by Aumann (1995, 1998), Balkenborg and Winter (1997), Battigalli and Siniscalchi (2002) ,Halpern (2001), and by Stalnaker (1998)). A second group of papers (by, for example, Baltag, Smets, and Zvesper (2009), Battigalli, Di-Tillio, and Samet (2013), and Samet (1996)) uses the "behavioral" models discussed in Section 9.9, which were introduced by Samet (1996). There is a bewildering collection of claims in the literature concerning the implications of rationality in dynamic games with perfect information: Aumann (1995) proves that common *knowledge* of rationality implies the backward induction solution, Ben-Porath (1997) and Stalnaker (1998) prove that common *belief / certainty* of rationality is *not* sufficient for backward induction, Samet (1996) proves that what is needed for backward induction is common *hypothesis* of rationality, Feinberg (2005) shows that common *confidence* of rationality logically contradicts the knowledge implied by the structure of the game, etc. The sources of this wide variety of results are partly clarified in two recent surveys of this literature, by Brandenburger (2007) and by Perea (2007a).

It is worth noting that the models of dynamic games considered in Section 9.9 are not the only possibility. Instead of modeling the epistemic states of the players in terms of their prior beliefs and prior disposition to revise those beliefs in a static framework, one can model the actual beliefs that the players hold at the time at which they make their choices. In such a framework the players' initial belief revision policies (or dispositions to revise their initial beliefs) can be dispensed with: the analysis can be carried out entirely in terms of the actual beliefs at the time of choice. This alternative approach is put forward by Bonanno (2013), where an epistemic characterization of backward induction is provided that does not rely on (objective or subjective) counterfactuals.

Epistemic foundations of other extensive-form solution concepts Because of space constraints, for extensive-form games we have restricted attention to the epistemic foundations of only one solution concept, namely backward induction. In the literature, epistemic conditions have been studied for additional solution concepts, such as extensive-form rationalizability (by Battigalli (1997) and Pearce (1984)), forward induction (by Battigalli and Siniscalchi (2002)), and perfect Bayesian equilibrium (by Bonanno (2011)). An account of part of this literature can be found in a paper by Perea (2012).

Belief revision The semantics for belief revision described in Section 9.8 has its roots in the well-known AGM theory which was introduced by Alchourrón et al. (1985). The AGM theory is a syntactic theory, whose semantic counterpart was first explored by Grove (1988). There is a vast literature on AGM belief revision. For a recent overview see the special issue of the *Journal of Philosophical Logic* on *25 Years of AGM Theory* (Volume 40 (2), April 2012). The conditions under which there is a precise correspondence between the subjective counterfactual functions f_i described in Section 9.8 and the syntactic AGM theory are explored by Bonanno (2009).

Acknowledgements I am grateful to Elias Tsakas for helpful comments and suggestions.

References

Alchourrón, C., P. Gärdenfors, and D. Makinson (1985). On the logic of theory change: partial meet contraction and revision functions. *The Journal of Symbolic Logic 50*, 510–530.

Aumann, R. (1976). Agreeing to disagree. *The Annals of Statistics 4*, 1236–1239.

Aumann, R. (1987a). Correlated equilibrium as an expression of Bayesian rationality. *Econometrica 55*, 1–18.

Aumann, R. (1987b). Game theory. In J. Eatwell, M. Milgate, and P. Newman (Eds.), *The New Palgrave, a dictionary of economics*, Volume 2, pp. 460–482. London: Macmillan.

Aumann, R. (1995). Backward induction and common knowledge of rationality. *Games and Economic Behavior 8*, 6–19.

Aumann, R. (1998). On the centipede game. *Games and Economic Behavior 23*, 97–105.

Aumann, R. and A. Brandenburger (1995). Epistemic conditions for Nash equilibrium. *Econometrica 63*, 1161–1180.

Bach, C. and E. Tsakas (2014). Pairwise epistemic conditions for Nash equilibrium. *Games and Economic Behaviour 85*, 48–59.

Bacharach, M. (1987). A theory of rational decision in games. *Erkenntnis 27*, 17–55.

Balkenborg, D. and E. Winter (1997). A necessary and sufficient epistemic condition for playing backward induction. *Journal of Mathematical Economics 27*, 325–345.

Baltag, A., S. Smets, and J. Zvesper (2009). Keep hoping for rationality: a solution to the backward induction paradox. *Synthese 169*, 301–333.

Barelli, P. (2009). Consistency of beliefs and epistemic conditions for Nash and correlated equilibria. *Games and Economic Behavior 67*, 363–375.

Barelli, P. and S. Galanis (2013). Admissibility and event rationality. *Games and Economic Behavior 77*, 21–40.

Battigalli, P. (1997). On rationalizability in extensive games. *Journal of Economic Theory 74*, 40–61.

Battigalli, P. and G. Bonanno (1999). Recent results on belief, knowledge and the epistemic foundations of game theory. *Research in Economics 53*, 149–225.

Battigalli, P., A. Di-Tillio, and D. Samet (2013). Strategies and interactive beliefs in dynamic games. In D. Acemoglu, M. Arellano, and E. Dekel (Eds.), *Advances in Economics and Econometrics. Theory and Applications: Tenth World Congress.* Cambridge: Cambridge University Press.

Battigalli, P. and M. Siniscalchi (2002). Strong belief and forward induction reasoning. *Journal of Economic Theory 106*, 356–391.

Ben-Porath, E. (1997). Nash equilibrium and backwards induction in perfect information games. *Review of Economic Studies 64*, 23–46.

van Benthem, J. (2011). *Logical Dynamics of Information and Interaction.* Cambridge: Cambridge University Press.

Bernheim, D. (1984). Rationalizable strategic behavior. *Econometrica 52*, 1002–1028.

Bertrand, J. (1883). Théorie mathématique de la richesse sociale. *Journal des Savants 67*, 499–508.

Board, O. (2004). Dynamic interactive epistemology. *Games and Economic Behavior 49*, 49–80.

Bonanno, G. (1991). The logic of rational play in games of perfect information. *Economics and Philosophy 7*, 37–65.

Bonanno, G. (1996). On the logic of common belief. *Mathematical Logic Quarterly 42*, 305–311.

Bonanno, G. (2001). Branching time logic, perfect information games and backward induction. *Games and Economic Behavior 36*, 57–73.

Bonanno, G. (2008). A syntactic approach to rationality in games with ordinal payoffs. In G. Bonanno, W. van der Hoek, and M. Wooldridge (Eds.), *Logic and the Foundations of Game and Decision Theory (LOFT 7)*, Volume 3 of *Texts in Logic and Games*, pp. 59–86. Amsterdam University Press.

Bonanno, G. (2009). Rational choice and *AGM* belief revision. *Artificial Intelligence 173*, 1194–1203.

Bonanno, G. (2011). *AGM* belief revision in dynamic games. In K. Apt (Ed.), *Proceedings of the 13th Conference on Theoretical Aspects of Rationality and Knowledge*, TARK XIII, New York, pp. 37–45. ACM.

Bonanno, G. (2013). A dynamic epistemic characterization of backward induction without counterfactuals. *Games and Economics Behavior 78*, 31–43.

Bonanno, G. (2014). Reasoning about strategies and rational play in dynamic games. In J. van Benthem, S. Ghosh, and R. Verbrugge (Eds.), *Modeling strategic reasoning*, Texts in Logic and Games. Springer. forthcoming.

Bonanno, G. and K. Nehring (1998). On Stalnaker's notion of strong rationalizability and Nash equilibrium in perfect information games. *Theory and Decision 45*, 291–295.

Brandenburger, A. (1992). Lexicographic probabilities and iterated admissibility. In P. Dasgupta, D. Gale, O. Hart, and E. Maskin (Eds.), *Economic analysis of markets and games*, pp. 282–290. MIT Press.

Brandenburger, A. (2007). The power of paradox: some recent developments in interactive epistemology. *International Journal of Game Theory 35*, 465–492.

Brandenburger, A. and E. Dekel (1987). Rationalizability and correlated equilibria. *Econometrica 55*, 1391–1402.

Brandenburger, A., A. Friedenberg, and J. Keisler (2008). Admissibility in games. *Econometrica 76*, 307–352.

Clausing, T. (2003). Doxastic conditions for backward induction. *Theory and Decision 54*, 315–336.

Clausing, T. (2004). Belief revision in games of perfect information. *Economics and Philosophy 20*, 89–115.

Cournot, A. A. (1838). *Recherches sur les principes mathématiques de la théorie des richesses.* Paris: Hachette.

de Bruin, B. (2010). *Explaining games: the epistemic programme in game theory.* Springer.

Dekel, E. and F. Gul (1997). Rationality and knowledge in game theory. In D. Kreps and K. Wallis (Eds.), *Advances in economics and econometrics*, pp. 87–172. Cambridge University Press.

Feinberg, Y. (2005). Subjective reasoning - dynamic games. *Games and Economic Behavior 52*, 54–93.

Grove, A. (1988). Two modellings for theory change. *Journal of Philosophical Logic 17*, 157–170.

Halpern, J. (2001). Substantive rationality and backward induction. *Games and Economic Behavior 37*, 425–435.

Harsanyi, J. (1967-1968). Games with incomplete information played by "Bayesian players", Parts I-III. *Management Science 8*, 159–182, 320–334, 486–502.

Hotelling, H. (1929). Stability in competition. *Economic Journal 39*, 41–57.

Kreps, D. (1988). *Notes on the theory of choice.* Boulder: Westview Press.

Kripke, S. (1963). A semantical analysis of modal logic I: normal propositional calculi. *Zeitschrift für Mathematische Logik und Grundlagen der Mathematik 9*, 67–96.

von Neumann, J. (1928). Zur Theorie der Gesellschaftsspiele. *Mathematische Annalen 100*, 295–320.

von Neumann, J. and O. Morgenstern (1944). *Theory of games and economic behavior*. Princeton University Press.

Pearce, D. (1984). Rationalizable strategic behavior and the problem of perfection. *Econometrica 52*, 1029–1050.

Perea, A. (2007a). Epistemic foundations for backward induction: an overview. In J. van Benthem, D. Gabbay, and B. Löwe (Eds.), *Interactive logic. Proceedings of the 7th Augustus de Morgan Workshop*, Volume 1 of *Texts in Logic and Games*, pp. 159–193. Amsterdam University Press.

Perea, A. (2007b). A one-person doxastic characterization of Nash strategies. *Synthese 158*, 251–271.

Perea, A. (2012). *Epistemic game theory: reasoning and choice*. Cambridge: Cambridge University Press.

Polak, B. (1999). Epistemic conditions for Nash equilibrium, and common knowledge of rationality. *Econometrica 67*, 673–676.

Rubinstein, A. (1991). Comments on the interpretation of game theory. *Econometrica 59*, 909–924.

Samet, D. (1996). Hypothetical knowledge and games with perfect information. *Games and Economic Behavior 17*, 230–251.

Samuelson, L. (1992). Dominated strategies and common knowledge. *Games and Economic Behavior 4*, 284–313.

von Stackelberg, H. (1934). *Marktform und Gleichgewicht*. Vienna: Julius Springer.

Stahl, D. (1995). Lexicographic rationalizability and iterated admissibility. *Economics Letters 47*, 155–159.

Stalnaker, R. (1994). On the evaluation of solution concepts. *Theory and Decision 37*, 49–74.

Stalnaker, R. (1996). Knowledge, belief and counterfactual reasoning in games. *Economics and Philosophy 12*, 133–163.

Stalnaker, R. (1998). Belief revision in games: forward and backward induction. *Mathematical Social Sciences 36*, 31–56.

Tan, T. and S. Werlang (1988). The Bayesian foundation of solution concepts of games. *Journal of Economic Theory 45*, 370–391.

Zermelo, E. (1913). Über eine Anwendung der Mengenlehre auf die Theorie des Schachspiels. *Proceedings Fifth International Congress of Mathematicians 2*, 501–504.

Chapter 10

BDI Logics

John-Jules Ch. Meyer, Jan Broersen and Andreas Herzig

Contents

Abstract This paper presents an overview of so-called BDI logics, logics where the notion of Beliefs, Desires and Intentions play a central role. Starting out from the basic ideas about BDI by Bratman, we consider various formalizations in logic, such as the approach of Cohen and Levesque, slightly remodelled in dynamic logic, Rao & Georgeff's influential BDI logic based on the branching-time temporal logic CTL*, the KARO framework and BDI logic based on STIT (seeing to it that) logics.

10.1 Introduction

In this chapter we present an overview of so-called BDI (for Beliefs, Desires, and Intentions) logics, that is, logics that describe the mental attitudes of intelligent agents in terms of folk-psychological notions of beliefs, desires and intentions. This theory is based on the work of the philosopher Michael Bratman — as with all chapters in his book, references to the literature are

Chapter 10 of the *Handbook of Epistemic Logic*, H. van Ditmarsch, J.Y. Halpern, W. van der Hoek and B. Kooi (eds), College Publications, 2015, pp. 489–542.

provided in the final section. The chapter is organized as follows: we start with some of the basic ideas in Bratman's philosophy, which is about practical reasoning (the reasoning about performing actions) on the basis of the agent's beliefs, desires and, very importantly, intentions, which are special desires to which the agent is committed. Then a number of formalizations of BDI theory in logic is reviewed, the so-called BDI logics. Starting out with Cohen & Levesque's approach, slightly reworked in a dynamic logic setting by Herzig and colleagues. Then we look at Rao & Georgeff's BDI logic, based on the branching-time temporal logic CTL*. Next we discuss the KARO framework which is based on dynamic logic since its conception. We sketch as a small excursion how the KARO framework, which is devised to capture the behaviour of (rational) intelligent agents, also can be used for describing emotional behaviour of agents. We then present a relatively new approach to BDI, based on so-called STIT (seeing to it that) logic. We finally round off with a conclusion section and a section containing the pointers to all bibliographical references.

10.2 Bratman's theory of Belief-Desire-Intention

> "What happens to our conception of mind and rational agency when we take seriously future-directed intentions and plans and their roles as inputs into further practical reasoning? This question drives much of this book."

This is how the preface of Michael E. Bratman's famous book "Intention, Plans, and Practical Reason" starts. In this book the author lays down the foundations of what later would be called the BDI (Belief-Desire-Intention) theory of agency, a folk-psychological theory of how humans make decisions and take action (referred to as practical reasoning after Aristotle), and which would lead to a new computing paradigm, agent-oriented programming or agent technology more in general, when AI researchers started to apply it to the specification and implementation of artificial agents.

The main new ingredient in Bratman's theory is that of *intention.* Beliefs and desires were already known to be of importance in human behaviour. For instance, Daniel Dennett's intentional stance, the strategy of interpreting the behaviour of an entity by treating it as if it were a rational agent that governed its choice of action by a consideration of its beliefs and desires, already mentions the role of beliefs and desires in 1987. But Bratman claims that to fully understand the practical reasoning of humans also the notion of intention is needed. An intention is not just a mere desire but something the agent is committed to, that is, not given up too soon by the agent. For instance, if I have an intention to give a lecture in Amsterdam tomorrow, it is not a mere wish to do so, but I'm really taking measures (making plans, e.g., cancelling other plans or making sure my laptop will be in my bag) to do it and unless something happens that seriously interferes with my intention to give that lecture tomorrow, I really will do so. Thus Bratman takes intention to be a first-class citizen, and not something that can be reduced to beliefs and desires. In other words, a reduction of intention to a theory of only beliefs and desires is rejected by Bratman.

Another important notion in Bratman's theory is that of a *plan*. As he explains, rational agents need plans for two reasons. First of all, agents need to "allow deliberation and rational reflection to influence action beyond the present", since agents have only limited resources (including time) to deliberate at the time of action. Secondly agents need to coordinate their actions, both on an intrapersonal and a interpersonal level, and plans help agents with that, too. As to the relation between plans and intentions, Bratman says that 'our intentions concerning our future actions are typically elements in larger plans'. Bratman focuses on *future-oriented intentions*. Such intentions differ from present-directed intentions, alias intentions-in-action, which accompany an agent's actions (more precisely, an agent's intentional actions), and pertain to what to do beginning now.

To explain the differences between beliefs, desires and intentions Bratman introduces the notion of a *pro-attitude*. A pro-attitude is an agent's mental attitude directed toward an action under a certain description. It plays a motivational role. So desires and intentions are both pro-attitudes while beliefs typically are not. But although desires and intentions are both pro-attitudes they differ. Intentions are conduct-controlling pro-attitudes, while ordinary desires are merely potential influences of action. The '*volitional*' dimension of the commitment involved in future-directed intentions comes from the conduct-controlling nature of intentions: as a conduct-controlling pro-attitude an intention involves a special commitment to action that ordinary desires do not.

Besides identifying intentions as conduct-controlling pro-attitudes, Bratman argues that intentions also have other properties: they have inertia and they serve as inputs into further practical reasoning. By the former is meant that intentions resist reconsideration. This has to do again with the resource-boundedness of realistic cognitive agents. The agent normally simply lacks the time to compute, at any given time, the optimal plan of action given his beliefs and desires, so it has to form future-directed intentions and store them in his mental agenda and use them to avoid computing plans all the time. Once an intention has been formed (and a commitment to action has been made) the intention will normally remain intact until the time of action: it has a characteristic stability / inertia. By this Bratman means that intentions made influence further reasoning about (decisions about) action, where also refinements of intentions (intentions to do more concrete actions) may play a role. For example, if I have the intention to speak in Amsterdam tomorrow, I can form a more refined intention to take the car driving to Amsterdam in order to speak. As a consequence after the second intention it won't be rational anymore to consider time tables for trains going to Amsterdam, while it was so after the first intention. All this has led to seeing intentions as distinctive states of mind, distinct from beliefs and desires, and to a belief-desire-intention model rather than a desire-belief model of practical reasoning.

As we have seen, Bratman describes how prior intentions and plans provide a filter of admissibility on options. This is what later by Cohen & Levesque has been called a 'screen of admissibility'. The basis of this role of intentions in further practical reasoning is the need for consistency in one's web of intentions and beliefs, as Bratman calls it: other things being equal, it should be possible for me to do all that I intend in a world in which my beliefs are true. But as Bratman explains this is not as simple as it looks. In particular "not every option that is incompatible with what the agent already intends and believes is inadmissible." In short, it depends on whether beliefs can be forced to be changed by the new intention so that the inconsistency disappears or not. In the former case the new intention is admissible, in the latter it is not.

In Bratman's view there is an intrinsic relation between intentions and plans. Plans are intentions. They share the properties of intentions: they resist reconsideration and have inertia, they are conduct controllers and not merely conduct influencers, and they provide crucial inputs for further practical reasoning and planning. But they have increased complexity as compared to simple intentions: they are typically partial in the sense of incomplete (typically I have a partial plan to do something and fill in the details later) and have a hierarchical structure (plans concerning ends embed plans concerning means and preliminary steps, and more general intentions embed more specific ones).

To sum it up, according to Bratman, future-oriented intentions have the following characteristics:

- An intention is a *high-level plan.*
- An intention guides deliberation and triggers further planning: it typically leads to the *refinement* of a high-level plan into a more and more precise plan.
- An intention comes with the agent's *commitment* to achieve it.
- An agent abandons an intention only under the following conditions:

- the intention has been achieved;
- he believes it is impossible to achieve it;
- he abandons another intention for which it is instrumental.

Let us illustrate Bratman's future-oriented intentions by an example. Suppose we are in autumn and I desire to go to Paris next spring. Under certain conditions —such as the importance of that desire and my beliefs about its feasibility—, that desire will make me form an intention to travel to Paris next spring. This is a very high-level plan: I do not settle the exact dates, I do not decide by which means of transportation I am going to go to Paris, and I do not know where to stay yet. I am however committed to that plan: during the following months I will stick to my intention to go to Paris, unless I learn that it is impossible to go to Paris in spring (say because my wife wants to spend our spring holidays in Spain, or because I changed my mind due to an invitation to give a talk at an important conference). During the next months I am going to refine my high-level plan: I will decide to go on a particular weekend, I will decide to go by train and not by plane, and I will book a hotel for the weekend under concern. This more elaborated plan is going to be refined further as time goes by: I decide to take the 7am train and not the 9am train, and I decide to go to the train station by metro and not by taxi, etc. Finally, once I have spent that weekend in Paris I no longer pursue that goal and drop it.

Bratman's theory might be called semi-formal: while he isolates the fundamental concepts and relates them, he does not provide a formal semantics. This was both undertaken by Phil Cohen and Hector Levesque and, more or less at the same time, by Anand Rao and Michael Georgeff. In the next section we will go into the details of how they casted Bratman's theory into a logic of intention and we present some subsequent modifications and extensions of their original logic. This will be followed by a section on Rao and Georgeff's approach.

10.2.1 Cohen and Levesque's approach to intentions

We have seen that the concepts of belief, desire, time and action play an important role in Bratman's theory of intention. A logical analysis of that theory should involve combining a logic of belief, a logic of desire, a logic of time and a logic of action.

Belief, time, and action play a fundamental role in Cohen and Levesque's logic. However, the concept of desire is somewhat neglected: Cohen and Levesque rather base their logic on the concept of *realistic preference*. The latter can be viewed as a desire that has already been filtered by the agent's beliefs about its realisability. This is highlighted by the property that belief implies realistic preference: when I am convinced that φ is true then I also have to prefer that φ is true. (I might however prefer that φ be false at some point in the future.)

Cohen and Levesque's analysis amounts to a *reduction* of the concept of intention to those of belief, realistic preference, time and action: they define intention in terms of the latter four concepts. The reader may note that this is actually a surprising move, given that Bratman had strongly argued that intentions are independent of desires and cannot be reduced to them.

In the next two sections we present the four building blocks of Cohen and Levesque's logic, grouping together action and time on the one hand, and belief and realistic preference on the other.

Action and time

The basic building block of Cohen and Levesque's logic is a linear version of propositional dynamic logic PDL. The semantics of linear PDL allows to also interpret the temporal operators of linear-time temporal logic LTL.

Standard PDL Standard PDL is not about actions but about events. It has a set $\mathcal{A}$ of atomic event names. Cohen and Levesque add agents to the picture and provide an *agentive version* of PDL. Let us write i, j, etc. for agents from some set of individuals $\mathcal{I}$. Then atomic actions are elements of $\mathcal{I} \times \mathcal{A}$. We write them $i{:}\alpha$ where $\alpha \in \mathcal{A}$ is an atomic event and $i \in \mathcal{I}$. Formulas of language of PDL are built from atomic formulas and atomic actions by means of modal operators $\texttt{Poss}_\pi$, where π is an action. The formula $\texttt{Poss}_\pi\varphi$ reads "there is a possible execution of π after which φ is true".[1] This reading highlights that the standard version of PDL allows for several possible executions of π in order to account for indeterminism.

While $\texttt{Poss}_\pi$ quantifies existentially over the executions of π, the dual modal operator $\texttt{After}_\pi$ quantifies universally. It is definable from $\texttt{Poss}_\pi$ as follows:

$$\texttt{After}_\pi\varphi \stackrel{\text{def}}{=} \neg\texttt{Poss}_\pi\neg\varphi$$

Let us consider the case where φ is truth $\top$ or falsity $\bot$: $\texttt{Poss}_\pi\top$ has to be read "π is executable", while $\texttt{After}_\pi\bot$ has to be read "φ is inexecutable".

The semantics of PDL is based on *transition systems* where an atomic action $i{:}\alpha$ can be interpreted as a set of edges. Such a transition system is a pair $\langle W, R\rangle$ where W is a non-empty set of possible worlds and R maps every action π to an accessibility relation $R_\pi \subseteq W \times W$ relating possible

[1] The standard notation is $\langle\pi\rangle\varphi$; we here deviate in order to be able to distinguish actual action from potential action.

worlds. An edge from world w to world u that is labeled π means that it is possible to execute π in w and that u is a possible outcome world when π is executed. The set of all these π edges makes up the accessibility relation R_π interpreting the action π.

A PDL *model* is a transition system together with a valuation V mapping atomic formulas p from the set of propositional variables $\mathcal{P}$ to their extension $V(p) \subseteq W$, i.e., to the set of worlds $V(p)$ where p is true.

Models allow to give truth values to formulas. In particular, $\texttt{Poss}_\pi\varphi$ is true at a world w if there is a couple (w, w') in R_π such that φ is true at world w':

$$M, w \models \texttt{Poss}_\pi\varphi \quad \text{iff} \quad \text{there is a } u \in W \text{ such that } wR_\pi u \text{ and } M, u \models \varphi$$

The formula $\texttt{Poss}_\pi\varphi$ therefore expresses a weak notion of ability: the action π might occur and φ could be true afterwards.

Linear PDL Probably Cohen and Levesque were the first to adapt PDL in order to model actual agency. The modalities are interpreted in *linear* PDL models. In such models, for every possible world w there is at most one successor world u that is temporally related to w. The accessibility relation linking w to u may be labelled by several atomic actions. Formally, a transition system $\langle W, R\rangle$ is linear if for every world $w \in W$ such that $\langle w, u_1\rangle \in R_{\pi_1}$ and $\langle w, u_2\rangle \in R_{\pi_2}$ we have $u_1 = u_2$. An edge from world w to world u that is labeled π means that π is executed in w and that u will be the result. (The reader might note the difference with the above standard PDL.) This allows for the simultaneous performance of two different actions; they must however lead to the same outcome world. The models of linear PDL are the class of linear transition systems.

In order to distinguish the modal operators of actual action from the modal operators of possible action we write the former as $\texttt{Happ}_\pi\varphi$, read "π is going to be performed, and φ is true afterwards". Just as $\texttt{After}_\pi$ is the dual of $\texttt{Poss}_\pi$, we define a modal operator $\texttt{IfHapp}_\pi$ that is the dual of $\texttt{Happ}_\pi$ by stipulating:

$$\texttt{IfHapp}_\pi\varphi \stackrel{\text{def}}{=} \neg\texttt{Happ}_\pi\neg\varphi$$

$\texttt{Happ}_\pi\varphi$ and $\texttt{IfHapp}_\pi\varphi$ say different things: the first formula says that π is executable and that φ is true after it, while the second says that *if* π is executable then φ is true after it. The former should therefore imply the latter.

The truth condition for $\texttt{Happ}_\pi$ is:

$$M, w \models \texttt{Happ}_\pi\varphi \quad \text{iff} \quad \text{there is a } u \in W \text{ such that } wR_\pi u \text{ and } M, u \models \varphi$$

So it has exactly the same form as that for $\texttt{Poss}_\pi$. We changed the name of the modal operator in order to better suit the linearity of the models.

The following axiom schema characterises linear PDL models:

$$(\texttt{Happ}_{i:\alpha}\top \wedge \texttt{Happ}_{j:\alpha'}\varphi) \rightarrow \texttt{Happ}_{i:\alpha}\varphi \qquad (10.1)$$

Beyond atomic events, PDL also has complex events such as sequential and nondeterministic composition, test, and iteration. We will however not refer to them in our present introduction.

Cohen and Levesque's logic has the temporal operators "eventually" (noted $\texttt{F}$), "henceforth" (noted $\texttt{G}$), and "until" (noted $\texttt{U}$). These operators are interpreted in linear PDL models in the obvious way. Let us give the truth condition for the 'eventually" operator, for example:

$M, w \models \texttt{F}\varphi$ iff there is an integer n and there are $v_1, \ldots, v_n \in W$ such that $v_1 = w$, $\langle v_k, v_{k+1}\rangle \in R_{\pi_k}$ for some π_k, and $M, v_n \models \varphi$

Cohen and Levesque also need existential quantification over actions. We here present their account in terms of an operator fusing existential quantification $\exists$ over events α with the dynamic operator $\texttt{Happ}_{i:\alpha}$. Its truth condition is as follows:

$M, w \models \exists\alpha\texttt{Happ}_{i:\alpha}\varphi$ iff there are $\alpha \in \mathcal{A}, u \in W$ such that $\langle w, u\rangle \in R_{i:\alpha}$ and $M, u \models \varphi$

Belief and preference

Cohen and Levesque's account of belief is standard, while their account of preference is in terms of the somewhat unusual notion of strong realistic preference.

Belief Cohen and Levesque have modal operators of belief $\texttt{Bel}_i$, one per agent i. The modal logic of each of these operators is the standard logic of belief KD45 (see also Chapter 1 of this handbook). Such operators can be interpreted if we add accessibility relations B_i to the transition systems of linear PDL, one per agent i. The set of worlds $B_i(w) = \{u : \langle w, u\rangle \in B_i\}$ is the set of those worlds that are possible for agent i at world w: the set of worlds that are compatible with his beliefs at w.

In order to be an accessibility relation for KD45, each of these relations has to satisfy the following constraints:

- for every $w \in W$ there is at least one $u \in W$ such that $\langle w, u\rangle \in B_i$ (seriality);

- if $\langle w, u\rangle \in B_i$ and $\langle u, v\rangle \in B_i$ then $\langle w, v\rangle \in B_i$ (transitivity);
- if $\langle w, u\rangle \in B_i$ and $\langle w, v\rangle \in B_i$ then $\langle u, v\rangle \in B_i$ (Euclideaness).

These constraints make that the following implications become valid:

- $\mathtt{Bel}_i\varphi \to \neg\mathtt{Bel}_i\neg\varphi$ (consistency of belief, axiom D)
- $\mathtt{Bel}_i\varphi \to \mathtt{Bel}_i\mathtt{Bel}_i\varphi$ (positive introspection, axiom 4)
- $\neg\mathtt{Bel}_i\varphi \to \mathtt{Bel}_i\neg\mathtt{Bel}_i\varphi$ (negative introspection, axiom 5)

Preference For Cohen and Levesque, intentions are particular *strong realistic preferences.* The latter are true in a subset of the worlds that are doxastically possible for an agent. There is a modal operator $\mathtt{Pref}_i$, one per agent i, and the formula $\mathtt{Pref}_i\varphi$ reads "i chooses φ to be true".[2] Such a notion of preference is strongly realistic in the sense that belief logically implies preference. Semantically, strong realistic preference can be modelled by accessibility relations P_i, one per agent $i \in \mathcal{I}$, such that $P_i \subseteq B_i$. The latter constraint implements realism: a world that is compatible with agent i's preferences cannot be incompatible with i' beliefs. In other words, at world w agents have to select their preferred worlds among the worlds that are epistemically possible for them at w.

The logic of action, time, belief, and preference

Let us sum up Cohen and Levesque's semantics. A *frame* is a quadruple $M = \langle W, R, B, P\rangle$ where

- W is a non-empty set of possible worlds;
- $R : (\mathcal{I} \times \mathcal{A}) \longrightarrow (W \times W)$ maps actions π to accessibility relations R_π;
- $B : \mathcal{I} \longrightarrow (W \times W)$ maps agents i to accessibility relations B_i;
- $P : \mathcal{I} \longrightarrow (W \times W)$ maps agents i to accessibility relations P_i;

These frames have to satisfy the following constraints:

- $\langle W, R\rangle$ is a linear transition system;
- every B_i is serial, transitive and Euclidean;

[2] Cohen and Levesque's original notation is $\mathtt{Goal}_i$ instead of $\mathtt{Pref}_i$ (while they actually refer to it in their title as 'choice'). We moved to our notation in order to avoid confusion with the concept of choice in STIT theory.

- $P_i \subseteq B_i$, for every $i \in \mathcal{I}$.

Let us call $\mathcal{CL}$ that class of frames. As usual, a model is a frame together with a valuation $V : \mathcal{P} \longrightarrow 2^W$ mapping atomic formulas p to their extension $V(p) \subseteq W$. Validity and satisfiability in $\mathcal{CL}$ frames are defined as usual.

We are now ready to formulate Cohen and Levesque's reduction of intention.

Defining intention

Cohen and Levesque define a modal operator of intention by means of a cascade of definitions. We here reproduce them in a slightly simplified form. We then discuss them and finally comment on the modifications.

1. i has the *achievement goal* that φ if i prefers that φ is eventually true and believes that φ is currently false. Formally:

$$\texttt{AGoal}_i\varphi \stackrel{\text{def}}{=} \texttt{Pref}_i\texttt{F}\varphi \land \texttt{Bel}_i\neg\varphi$$

2. i has the *persistent goal* that φ if i has the achievement goal that φ and will keep that goal until it is either fulfilled or believed to be out of reach. Formally:

$$\texttt{PGoal}_i\varphi \stackrel{\text{def}}{=} \texttt{AGoal}_i\varphi \land (\texttt{AGoal}_i\varphi)\,\texttt{U}\,(\texttt{Bel}_i\varphi \lor \texttt{Bel}_i\,\texttt{G}\neg\varphi)$$

3. i has the *intention* that φ if i has the persistent goal that φ and believes he can achieve φ by an action of his. This requires to quantify over i's actions by means of the fused operator quantifying over events. Formally:

$$\texttt{Intend}_i\varphi \stackrel{\text{def}}{=} \texttt{PGoal}_i\varphi \land \texttt{Bel}_i\,\texttt{F}\,\exists\alpha\,\texttt{Happ}_{i:\alpha}\varphi$$

Some valid and invalid principles for intention The construction of Cohen and Levesque guarantees several desirable properties and avoids some that are unwanted. Here are two of them.

First, i's intention that φ logically implies i's belief that $\neg\varphi$. Formally this writes:

$$\texttt{Intend}_i\varphi \to \texttt{Bel}_i\neg\varphi$$

Second, the formula schema $\texttt{Bel}_i(\varphi \to \psi) \to (\texttt{Intend}_i\varphi \to \texttt{Intend}_i\psi)$ is invalid: i's intention that φ together with i's belief that φ implies ψ does not logically imply i's intention that ψ. This is crucial both for Bratman and for Cohen and Levesque. Here is a famous example illustrating why the principle should not be valid: if I intend to go to the dentist and believe that going to the dentist will cause pain then I do not necessarily intend to have pain. This is called the Side-Effect-Free Principle by Su *et al.* They have proposed a logic with a semantics in terms of linear neighbourhood structures (instead of accessibility relations) in order to interpret a modal operator of preference. Such structures allow to validate the principle of consistency of beliefs and intentions ($\mathtt{Bel}_i\varphi \to \neg\mathtt{Pref}_i\neg\varphi$) while guaranteeing the Side-Effect-Free Principle. The price for that is that preference no longer satisfy the monotony axiom $\mathtt{Pref}_i(\varphi \wedge \psi) \to (\mathtt{Pref}_i\varphi \wedge \mathtt{Pref}_i\psi)$.

Comments on the simplifications We have simplified the definition of a persistent goal. Cohen and Levesque's original definition allows agents to abandon a persistent goal for some other, superior reason. Their definition is

$$\mathtt{PGoal}_i\varphi \stackrel{\text{def}}{=} \mathtt{AGoal}_i\varphi \wedge (\mathtt{AGoal}_i\varphi)\,\mathtt{U}\,(\mathtt{Bel}_i\varphi \vee \mathtt{Bel}_i\mathtt{G}\neg\varphi \vee \psi)$$

where ψ is an unspecified condition accounting for that other reason.

This definition stipulates that a persistent goal φ is also abandoned if some other condition ψ becomes true. This leaves room for the abandonment of persistent goals that are instrumental for some goal that obtains without the agent's intervention. A classical example is a student i coming back late in the night to the dorms who forgot to take the key of the entrance door: his overall goal is to be able to get into the building and he starts to plan to climb over the wall (φ), but then some other student who also comes home late happens to pass just in front of him and opens the door (ψ), thus enabling i to drop φ. Hence, i abandons his persistent goal φ although it has neither been achieved nor turned out to be impossible.

Dealing with such a general condition ψ however makes it difficult to go beyond specific cases.

Variants and extensions

We now overview an extension of the basic logic, together with several alternatives to Cohen and Levesque's incremental definition of intention.

Introspection of intention Cohen and Levesque do not assume principles of positive and negative introspection of preference. However, they seem natural in the following form

- $\mathtt{Pref}_i\varphi \to \mathtt{Bel}_i\mathtt{Pref}_i\varphi$
- $\neg\mathtt{Pref}_i\varphi \to \mathtt{Bel}_i\neg\mathtt{Pref}_i\varphi$

They correspond to the following constraints on the accessibility relations for preference and belief:

- if $\langle w, u \rangle \in B_i$ and $\langle u, v \rangle \in P_i$ then $\langle w, v \rangle \in P_i$;
- if $\langle w, u \rangle \in B_i$ and $\langle w, v \rangle \in P_i$ then $\langle u, v \rangle \in P_i$.

This allows to prove principles of positive and negative introspection of goals, achievement goals, persistent goals, and intention. For instance, they validate $\texttt{Intend}_i\varphi \rightarrow \texttt{Bel}_i\texttt{Intend}_i\varphi$ and $\neg\texttt{Intend}_i\varphi \rightarrow \texttt{Bel}_i\neg\texttt{Intend}_i\varphi$. For example, both when I intend to go to Paris and when I don't intend to go to Paris then I am aware of this.

Weakly realistic preference Sadek has argues for a slightly different notion of realistic preference. The latter does not demand that all preference-accessible worlds be in the set of belief-accessible worlds, but only requires that they have a non-empty intersection. (This is sometimes called weak realism, as opposed to Cohen and Levesque's strong realism.) In frames that are weakened in this way the somewhat counterintuitive principle $\texttt{Bel}_i\varphi \rightarrow \texttt{Pref}_i\varphi$ is no longer valid. Instead, the weaker $\texttt{Bel}_i\varphi \rightarrow \neg\texttt{Pref}_i\neg\varphi$ is valid, which can be reformulated as

$$\neg(\texttt{Bel}_i\varphi \wedge \texttt{Pref}_i\neg\varphi)$$

It says that one cannot simultaneously believe that φ and prefer that $\neg\varphi$.

An epistemic version of achievement goals Herzig and Longin have advocated a different definition of an achievement goal. It is weaker than Cohen and Levesque's in that they only require that i does not believe that φ is currently true (instead of i's belief that φ is currently false). It is stronger in that they replace i's goal that φ will be true by i's goal that φ will be *believed.* This gives the following definition:

$$\texttt{AGoal}^{\texttt{W}}_i\varphi \stackrel{\text{def}}{=} \texttt{Pref}_i\,\texttt{F}\,\texttt{Bel}_i\varphi \wedge \neg\texttt{Bel}_i\varphi$$

They start by arguing for the replacement of $\texttt{Pref}_i\texttt{F}\varphi$ by $\texttt{Pref}_i\texttt{F}\texttt{Bel}_i\varphi$. As they point out, it is the *raison d'être* of an intention to be abandoned at some stage, and an agent can only do so if he believes that he has achieved that goal. So the agent's goal cannot just be that φ be true, but should be that he *believes* that φ is true. Using the same reasons they then argue for the replacement of $\texttt{Bel}_i\neg\varphi$ by $\neg\texttt{Bel}_i\varphi$: as long as φ is not believed to be true the agent should stick to his achievement goal φ, so $\neg\texttt{Bel}_i\varphi$ is better in line with this than $\texttt{Bel}_i\neg\varphi$. They illustrate the first replacement by a variant of the Byzantine generals example. Let r mean that a message of general i has been received by general j. Suppose i initially believes that j has not received the message yet, i.e., $\mathtt{Bel}_i \neg r$. Suppose moreover that i believes that he will actually *never* know whether j received the message or not, i.e., $\mathtt{Bel}_i \mathtt{G}(\neg \mathtt{Bel}_i r \wedge \neg \mathtt{Bel}_i \neg r)$. (This differs from the original Byzantine generals example, where it is possible that the messengers get through and where it is just possible for i that he will never know.) If we express i's achievement goal that r as $\mathtt{Pref}_i\, \mathsf{F}\, r$ then Cohen and Levesque make us conclude that $\mathtt{AGoal}_i r$, i.e., i has the achievement goal that φ although he believes that he will never be able to abandon that goal. In contrast, if we express i's achievement goal that r as $\mathtt{Pref}_i\, \mathsf{F}\, \mathtt{Bel}_i r$ then we have $\neg \mathtt{AGoal}_i^w r$: i cannot have the achievement goal that r.[3]

Weaker link between action and goal Sadek and Bretier point out that the definition of intention is too strong in particular in cooperative situations where agent i's action need not directly achieve his goal φ: it is enough that i triggers a subsequent action of another agent j which will achieve i's goal. Their modification can be formulated as follows:

$$\mathtt{Intend}_i \varphi \stackrel{\text{def}}{=} \mathtt{PGoal}_i \varphi \wedge \mathtt{Pref}_i \mathsf{F}(\exists \alpha \mathtt{Happ}_{i:\alpha} \mathsf{F} \varphi)$$

Stronger commitment Sadek and Bretier also discuss a stronger definition of intention where the agent is committed to do all he can to achieve his goal. They express this by a universal quantification over events.[4] We formulate their definition as follows:

$$\mathtt{Intend}_i \varphi \stackrel{\text{def}}{=} \mathtt{PGoal}_i \varphi \wedge \mathtt{Pref}_i \forall \alpha (\mathtt{Bel}_i\, \mathtt{Happ}_{i:\alpha}\, \mathsf{F} \varphi \rightarrow \mathtt{Pref}_i\, \mathsf{F}\, \mathtt{Happ}_{i:\alpha} \top)$$

That definition was criticised in the literature, in particular by Herzig and Long, as being too strong. Indeed, it postulates that agents want to achieve their intentions by all possible means, including illegal actions and actions with a huge cost for them. For example, it might commit me to steal a car if this is the only means to go to Paris on that spring weekend (say because there is a train strike).

Attempts Lorini and Herzig complement Cohen and Levesque's approach by integrating the concept of an *attempt* to perform an action. The motivation is that intentions typically make an agent *try* to perform an action,

[3] Our hypothesis $\mathtt{Bel}_i \neg \varphi$ implies the second condition $\neg \mathtt{Bel}_i \varphi$ because the logic of belief contains the D axiom, and $\mathtt{Bel}_i \mathtt{G}(\neg \mathtt{Bel}_i r \wedge \neg \mathtt{Bel}_i \neg r)$ implies $\neg \mathtt{Pref}_i \mathsf{F} \mathtt{Bel}_i r$, which is the negation of the first condition.

[4] This is therefore not a fused operator. In order to save space we do not give the the details of the semantics of that quantifier and rely on the reader's intuitions about it.

while the successful performance of that action is not guaranteed. The central principle there is "can and attempts implies does" : if i intends to (attempt to) perform α and α is feasible then α will indeed take place. This requires a logic with both modal operators of possible action $\texttt{Poss}_\pi$ and modal operators of actual action $\texttt{Happ}_\pi$.

Conclusion

Cohen and Levesque succeeded in providing a fine-grained analysis of intention by relating that concept to action, belief and realistic preference. A central point in Bratman's theory their logic does not account for is the refinement of intentions. According to Bratman, an agent starts by forming high-level intentions such as going to Paris in a month, and as time goes by he makes that intention more precise: he first starts to intend to go to Paris by train and not by plane; at a later stage he decomposes the intention to go to Paris by train into the intention to take a taxi to the train station (instead of a bus), then take the TGV to Paris, and then take the metro. It is probably an interesting direction of future research to integrate intention refinement mechanisms e.g. by resorting to dynamic epistemic logics (see Chapter 6).

10.2.2 Rao & Georgeff's BDI logic

As mentioned earlier, besides Cohen and Levesque, also Rao and Georgeff, more or less at the same time, published a formalisation of the groundbreaking work of Bratman on the philosophy of intelligent (human) agents. As we have seen, Bratman made a case for the notion of *intention* besides belief and desire, to describe the behaviour of rational agents. Intentions force the agent to commit to certain desires and to really 'go for them'. So focus of attention is an important aspect here, which also enables the agent to monitor how s/he is doing and take measures if things go wrong. Rao & Georgeff stress that in the case of resource-bounded agents it is imperative to focus on desires / goals and make choices. This was also observed by Cohen & Levesque, who try to formalize the notion of intention in a linear-time temporal logic (or, as we have seen in the previous section, a linear version of dynamic logic) in terms of the notion of a (persistent) goal.

Here we treat Rao & Georgeff's approach who base it on branching-time temporal logic framework CTL* to give a formal-logical account of BDI theory. The reader may also like to look at Chapter 5 of this book, the chapter that relates knowledge and time. Like Cohen & Levesque's approach, BDI logic has influenced many researchers (including Rao & Georgeff themselves) to think about architectures of agent-based systems in order to realize these systems. Rao & Georgeff's BDI logic is more liberal than that of Cohen & Levesque in the sense that they *a priori* regard each of the three attitudes of belief, desire and intention as primitive: they introduce separate modal operators for belief, desire and intention, and then study possible relations between them.

(The language of) BDI logic is constructed as follows. Two types of formulas are distinguished: state formulas and path formulas. We assume some given first-order signature. Furthermore, we assume a set E of event types with typical element e. The operators BEL, $GOAL$, $INTEND$ have as obvious intended reading the belief, goal and intention of an agent, respectively, while $\mathsf{U}, \diamond, \mathsf{O}$ are the usual temporal operators, viz. until, eventually and next, respectively.

Definition 10.1 (State and path formulas)

1. The set of *state formulas* is the smallest closed under:
   - any first-order formula w.r.t. the given signature is a state formula
   - if φ_1 and φ_2 are state formulas then also $\neg\varphi_1, \varphi_1 \vee \varphi_2, \exists x \varphi_1(x)$ are state formulas
   - if e is an event type, then $succeeded(e), failed(e)$ are state formulas
   - if φ is a state formula, then $BEL(\varphi), GOAL(\varphi), INTEND(\varphi)$ are state formulas
   - if ψ is a *path formula*, then $optional(\psi)$ is a state formula
2. The set of *path formulas* is the smallest set closed under:
   - any state formula is a path formula
   - if ψ_1, ψ_2 are path formulas, then $\neg\psi_1, \psi_1 \vee \psi_2, \psi_1 \mathsf{U} \psi_2, \diamond\psi_1, \mathsf{O}\psi_1$ are path formulas $\dashv$

State formulas are interpreted over a state, that is a (state of the) world at a particular point in time, while path formulas are interpreted over a path of a time tree (representing the evolution of a world). In the sequel we will see how this will be done formally. Here we just give the informal readings of the operators.

The operators *succeeded* and *failed* are used to express that events have (just) succeeded and failed, respectively. Next there are the modal operators for belief, goal and intend. (In the original version of BDI theory, desires are represented by goals, or rather a GOAL operator. In a later paper the $GOAL$ operator was replaced by DES for desire.) The optional operator states that there is a future (represented by a path) where the argument of the operator holds. Finally, there are the familiar (linear-time) temporal operators, such as the 'until', 'eventually' and 'nexttime', which are to be interpreted along a linear time path.

Definition 10.2
The following abbreviations are defined:

1. $\Box\psi = \neg \diamond \neg\psi$ (always)
2. $inevitable(\psi) = \neg optional(\neg\psi)$
3. $done(e) = succeeded(e) \vee failed(e)$
4. $succeeds(e) = inevitable\mathsf{O}(succeeded(e))$
5. $fails(e) = inevitable\mathsf{O}(failed(e))$
6. $does(e) = inevitable\mathsf{O}(done(e))$ $\dashv$

The 'always' operator is the familiar one from (linear-time) temporal logic. The 'inevitability' operator expresses that its argument holds along all possible futures (paths from the current time). The 'done' operator states that an event occurs (action is done) no matter whether it is succeeding or not. The final three operators state that an event succeeds, fails, or is done iff it is inevitable (i.e. in any possible future) it is the case that at the next instance the event has succeeded, failed, or has been done, respectively (note that this means that an event, succeeding or failing, is supposed to take one unit of time).

Definition 10.3 (Semantics)
The semantics is given w.r.t. models of the form $\mathcal{M} = \langle W, E, T, \prec, \mathcal{U}, B, G, I, \Phi\rangle$, where

- W is a set of possible worlds
- E is a set of primitive event types
- T is a set of time points
- $\prec$ is a binary relation on time points, which is serial, transitive and back-wards linear
- $\mathcal{U}$ is the universe of discourse
- Φ is a mapping of first-order entities to $\mathcal{U}$, for any world and time point

- $B, G, I \subseteq W \times T \times W$ are accessibility relations for $BEL, GOAL, INTEND$, respectively ⊣

The semantics of BDI logic, Rao & Georgeff-style, is rather complicated. Of course, we have possible worlds again, but as we will see below, these are not just unstructured elements, but they are each time trees, describing possible flows of time. So, we also need time points and an ordering on them. As BDI logic is based on branching time, the ordering need not be linear in the sense that all time points are related in this ordering. However, it is stipulated that the time ordering is serial (every time point has a successor in the time ordering), the ordering is transitive and backwards-linear, which means that every time point has only one direct predecessor. The accessibility relations for the 'BDI'-modalities are standard apart from the fact that they are also time-related, that is to say that worlds are (belief/goal/intend-)accessible with respect to a time point. Another way of viewing this is that – for all three modalities – for every time point there is a distinct accessibility relation between worlds.

In order to obtain reasonable properties for beliefs, desires and intentions, a number of constraints on the accessibility relations are stipulated. First of all, a *world / time point compatibility* requirement is assumed for all of the B, G, I accessibility relations: for $R = B, G, I$:

$$\textit{If } w' \in R(w,t) \textit{ then } t \in w \textit{ and } t \in w'$$

where $R(w,t) = \{w' \mid R(w,t,w')\}$ for $R = B, G, I$. This requirement is needed for the semantic clauses for the BEL, GOAL and INTEND modalities that we will give below to work. And next there are the usual requirements of the B accessibility relation to satisfy seriality, transitivity and Euclideaness in order to obtain the familiar KD45 properties of belief: beliefs are consistent, and satisfy positive and negative introspection. As to the G and I accessibility relations we require seriality in order to obtain the well-known KD property of consistent goals and intentions.

Next we elaborate on the structure of the possible worlds.

Definition 10.4 (Possible worlds)
Possible worlds in W are assumed to be *time trees*: an element $w \in W$ has the form $w = \langle T_w, A_w, S_w, F_w \rangle$ where

- $T_w \subseteq T$ is the set of time points in world w
- A_w is the restriction of the relation $\prec$ to T_w
- $S_w : T_w \times T_w \to E$ maps adjacent time points to (successful) events
- $F_w : T_w \times T_w \to E$ maps adjacent time points to (failing) events

- the domains of the functions S_w and F_w are disjoint ⊣

As announced before, a possible world itself is a time tree, a temporal structure representing possible flows of time. The definition above is just a technical one stating that the time relation within a possible world derives naturally from the *a priori* given relation on time points. Furthermore it is indicated by means of the functions S_w and F_w how events are associated with adjacent time points.

Now we come to the formal interpretation of formulas on the above models. Naturally we distinguish state formulas and path formulas, since the former should be interpreted on states whereas the latter are interpreted on paths. In the sequel we use the notion of a *fullpath*: a fullpath in a world w is an *infinite* sequence of time points such that, for all i, $(t_i, t_{i+1}) \in A_w$. We denote a fullpath in w by $(w_{t0}, w_{t1}, \ldots)$, and define $fullpaths(w)$ as the set of all fullpaths occurring in world w (i.e., all fullpaths that start somewhere in the time tree w).

Definition 10.5 (Interpretation of formulas)
Given a model $\mathcal{M} = \langle W, E, T, \prec, \mathcal{U}, B, G, I, \Phi \rangle$, the interpretation of formulas is now given by:

1. (state formulas)

   - $\mathcal{M}, v, w_t \models q(y_1, \ldots, y_n)$ iff $(v(y_1), \ldots, v(y_n)) \in \Phi(q, w, t)$
   - $\mathcal{M}, v, w_t \models \neg\varphi$ iff $\mathcal{M}, v, w_t \not\models \varphi$
   - $\mathcal{M}, v, w_t \models \varphi_1 \vee \varphi_2$ iff $\mathcal{M}, v, w_t \models \varphi_1$ or $\mathcal{M}, v, w_t \models \varphi_2$
   - $\mathcal{M}, v, w_t \models \exists x \varphi$ iff $\mathcal{M}, v\{d/x\}, w_t \models \varphi$ for some $d \in \mathcal{U}$
   - $\mathcal{M}, v, w_{t0} \models optional(\psi)$ iff exists fullpath $(w_{t0}, w_{t1}, \ldots)$ such that $\mathcal{M}, v, (w_{t0}, w_{t1}, \ldots) \models \psi$
   - $\mathcal{M}, v, w_t \models BEL(\varphi)$ iff for all $w' \in B(w, t) : \mathcal{M}, v, w'_t \models \varphi$
   - $\mathcal{M}, v, w_t \models GOAL(\varphi)$ iff for all $w' \in G(w, t) : \mathcal{M}, v, w'_t \models \varphi$
   - $\mathcal{M}, v, w_t \models INTEND(\varphi)$ iff for all $w' \in I(w, t) : \mathcal{M}, v, w'_t \models \varphi$
   - $\mathcal{M}, v, w_t \models succeeded(e)$ iff exists $t0$ such that $S_w(t0, t) = e$
   - $\mathcal{M}, v, w_t \models failed(e)$ iff exists $t0$ such that $F_w(t0, t) = e$ ⊣

   where $v\{d/x\}$ denotes the function v modified such that $v\{d/x\}(x) = d$. (Note that clauses for BEL, GOAL and INTEND are well-defined due to the world / time point compatibility requirement that we have assumed to hold.)

2. (path formulas)

- $\mathcal{M}, v, (w_{t0}, w_{t1}, \ldots) \models \varphi$ iff $\mathcal{M}, v, w_{t0} \models \varphi$, for φ state formula
- $\mathcal{M}, v, (w_{t0}, w_{t1}, \ldots) \models \mathsf{O}\varphi$ iff $\mathcal{M}, v, (w_{t1}, w_{t2}, \ldots) \models \varphi$
- $\mathcal{M}, v, (w_{t0}, w_{t1}, \ldots) \models \diamond\varphi$ iff $\mathcal{M}, v, (w_{tk}, \ldots) \models \varphi$ for some $k \geq 0$
- $\mathcal{M}, v, (w_{t0}, w_{t1}, \ldots) \models \psi_1 \mathsf{U} \psi_2$ iff
  either there exists $k \geq 0$ such that $\mathcal{M}, v, (w_{tk}, \ldots) \models \psi_2$ and for all $0 \leq j < k :\ \mathcal{M}, v, (w_{tj}, \ldots) \models \psi_1$, or
  for all $j \geq 0 :\ \mathcal{M}, v, (w_{tj}, \ldots) \models \psi_1$

Most of the above clauses should be clear, including those concerning the modal operators for belief, goal and intention. The clause for the 'optional' operator expresses exactly that optionally ψ is true if ψ is true in one of the possible futures represented by fullpaths starting at the present time point. The interpretation of the temporal operators is as usual.

Rao & Georgeff now discuss a number of properties that may be desirable to have as axioms. In the following we use α to denote so-called *O-formulas*, which are formulas that contain no positive occurrences of the '*inevitable*' operator (or negative occurrences of '*optional*") outside the scope of the modal operators $BEL, GOAL$ and $INTEND$.

1. $GOAL(\alpha) \rightarrow BEL(\alpha)$
2. $INTEND(\alpha) \rightarrow GOAL(\alpha)$
3. $INTEND(does(e)) \rightarrow does(e)$
4. $INTEND(\varphi) \rightarrow BEL(INTEND(\varphi))$
5. $GOAL(\varphi) \rightarrow BEL(GOAL(\varphi))$
6. $INTEND(\varphi) \rightarrow GOAL(INTEND(\varphi))$
7. $done(e) \rightarrow BEL(done(e))$
8. $INTEND(\varphi) \rightarrow inevitable \diamond (\neg INTEND(\varphi))$

In order to render these formulas validities further constraints should be put on the models, since in the general setting above these are not yet valid.

For reasons of space we only consider the first two. In order to define constraints on the models such that these two become valid, we introduce the relation $\lhd$ on worlds, as follows:

$w'' \lhd w' \Leftrightarrow fullpaths(w'') \subseteq fullpaths(w')$. So $w'' \lhd w'$ means that there the world (time tree) w'' represents less choices than w'.

Now we define the *B-G condition* as the property that the following holds:

$$\forall w' \in B(w,t) \exists w'' \in G(w,t) : w'' \lhd w'$$

Informally, this condition says that for any belief accessible world there is a goal accessible world that contains less choices. It is now easy to show the following proposition.

Proposition 10.1
Let $\mathcal{BG}$ be the class of models of the above form that satisfy the B-G condition. Then: $\mathcal{BG} \models GOAL(\alpha) \rightarrow BEL(\alpha)$ for O-formulas α. ⊣

Similarly one can define the *G-I condition* as

$$\forall w' \in G(w,t) \exists w'' \in I(w,t) : w'' \lhd w'$$

and obtain:

Proposition 10.2
Let $\mathcal{GI}$ be the class of models of the above form that satisfy the G-I condition. Then: $\mathcal{GI} \models INTEND(\alpha) \rightarrow GOAL(\alpha)$ for O-formulas α. ⊣

Let us now consider the properties deemed desirable by Rao & Georgeff again. The first formula describes Rao & Georgeff's notion of 'strong realism' and constitutes a kind of belief-goal compatibility: it says that the agent believes he can optionally achieve his goals. There is some controversy on this. Interestingly, but confusingly, Cohen & Levesque adhere to a form of realism that renders more or less the converse formula $BELp \rightarrow GOALp$. But we should be careful and realize that Cohen & Levesque have a different logic in which one cannot express options as in the branching-time framework of Rao & Georgeff. Furthermore, it seems that in the two frameworks there is a different understanding of goals (and beliefs) due to the very difference in ontologies of time employed: Cohen & Levesque's notion of time could be called 'epistemically nondeterministic' or 'epistemically branching', while 'real' time is linear: the agents envisage several future courses of time, each of them being a linear history, while in Rao & Georgeff's approach also 'real' time is branching, representing options that are available to the agent.

The second formula is a similar one to the first. This one is called goal-intention compatibility, and is defended by Rao & Georgeff by stating that if an optionality is intended it should also be wished (a goal in their terms). So, Rao & Georgeff have a kind of selection filter in mind: intentions (or rather intended options) are filtered / selected goals (or rather goal (wished) options), and goal options are selected believed options. If one views it this way, it looks rather close to Cohen & Levesque's 'Intention is choice (chosen

/ selected wishes) with commitment', or loosely, wishes that are committed to. Here the commitment acts as a filter.

The third one says that the agent really does the primitive actions that s/he intends to do. This means that if one adopts this as an axiom the agent is not allowed to do something else (first). (In our opinion this is rather strict on the agent, since it may well be that postponing its intention for a while is also an option.) On the other hand, as Rao & Georgeff say, the agent may also do things that are not intended since the converse does not hold. And also nothing is said about the intention to do complex actions.

The fourth, fifth and seventh express that the agent is conscious of its intentions, goals and what primitive action he has done in the sense that he believes what he intends, has as a goal and what primitive action he has just done.

The sixth one says something like that intentions are really wished for: if something is an intention then it is a goal that it is an intention.

The eighth formula states that intentions will inevitably (in every possible future) be dropped eventually, so there is no infinite deferral of its intentions. This leaves open, whether the intention will be fulfilled eventually, or will be given up for other reasons. Below we will discuss several possibilities of giving up intentions according to different types of commitment an agent may have.

BDI-logical expressions can be used to characterize different types of agents. Rao & Georgeff mention the following possibilities:

1. (blindly committed agent) $INTEND(\mathit{inevitable} \diamond \varphi) \to \mathit{inevitable}(INTEND(\mathit{inevitable} \diamond \varphi) \mathsf{U} BEL(\varphi))$

2. (single-minded committed agent) $INTEND(\mathit{inevitable} \diamond \varphi) \to \mathit{inevitable}(INTEND(\mathit{inevitable} \diamond \varphi) \mathsf{U} (BEL(\varphi) \vee \neg BEL(\mathit{optional} \diamond \varphi)))$

3. (open minded committed agent) $INTEND(\mathit{inevitable} \diamond \varphi) \to \mathit{inevitable}(INTEND(\mathit{inevitable} \diamond \varphi) \mathsf{U} (BEL(\varphi) \vee \neg GOAL(\mathit{optional} \diamond \varphi)))$

A blindly committed agent maintains his intentions to inevitably obtaining eventually something until he actually believes that that something has been fulfilled. A single-minded committed agent is somewhat more flexible: he maintains his intention until he believes he has achieved it *or he does not believe that it can be reached (i.e. that it is still an option in some future) anymore.* Finally, the open minded committed agent is even more flexible: he can also drop his intention if it is not a goal (desire) anymore.

Rao & Georgeff are then able to obtain results under which conditions the various types of committed agents will reach their intentions. For example, for a blindly committed agent it holds that under the assumption of the axioms we have discussed earlier plus an axiom that expresses no infinite deferral of intentions:

$$INTEND(\varphi) \rightarrow \mathit{inevitable} \diamond \neg INTEND(\varphi)$$

that

$$INTEND(\mathit{inevitable}(\diamond\varphi)) \rightarrow \mathit{inevitable}(\diamond BEL(\varphi))$$

expressing that if the agent intends to eventually obtain φ it will inevitably eventually believe that it has succeeded in achieving φ.

The branching-time setup of the approach as opposed to a linear-time one is much more expressive and is shown to solve problems such as the *Little Nell problem.* This is about a girl, Little Nell, that is in mortal peril, and a rescue agent that reasons like this: I intend to rescue Little Nell, and therefore I believe (because I'm confident that my actions will succeed) that she will be safe, but then I can drop my intention to rescue her just because she will be safe...! In a linear-time approach – if one is not very careful – this scenario results in a contradictory (or unintuitive) representation (basically because there is only one future in which apparently Little Nell will be safe), while in a branching-time approach such as Rao and Georgeff's this presents no problem at all. In fact in $\mathsf{CTL_{BDI}}$ the scenario comes down to something like (here φ stands for "Little Nell is safe")

$$\begin{aligned}&INTEND(\mathit{inevitable} \diamond \varphi) \rightarrow\\&\mathit{inevitable}(INTEND(\mathit{inevitable} \diamond \varphi)\mathsf{U}BEL(\mathit{optional} \diamond \varphi))\end{aligned}$$

informally saying that since the agent believes that there is a way (by performing its plan) to eventually reaching the goal φ, it may drop its intention to perform the plan to achieve eventually φ, which is definitely not valid in $\mathsf{CTL_{BDI}}$! Intuitively, this is the case, because there may be other branches along which Little Nell will not be safe, so that there is no reason to give up the intention to rescue her.

In the next section we will look at yet another approach, based on (non-linear) dynamic logic, which may perhaps be viewed as an amalgam of those of Cohen & Levesque (using dynamic logic) but allowing for non-linear, i.e. branching, structures.

10.3 KARO Logic

In this section we review the KARO formalism, in which *action*, together with knowledge / belief, is the primary concept, on which other agent notions are built. Historically, the KARO approach was the first approach truly based on dynamic logic, although as we have seen, in retrospect, we may view Cohen & Levesque's approach as being based on a linear variant of PDL (Propositional Dynamic Logic). There are differences, though. We will see that in KARO the fact that it is based on a logic of action is even more employed than in Cohen & Levesque: besides BDI-like notions such as knowledge, belief, desires, and goals that are operators that take formulas as arguments, and are quite similar in nature as the notions that are in Cohen & Levesque's approach, in KARO there are also operators taking actions as arguments such as ability and commitment, and operators that take both actions and formulas as arguments, such as a Can operator and a (possible) intention operator. All these operators are used to describe the mental state of the agent. But even more importantly, in the KARO framework (dedicated) actions are used to *change* the mental state of the agent. So there are revise, commit and uncommit actions to revise beliefs and update the agenda (the commitments) of the agent. In this sense KARO is related to dynamic epistemic logic, the topic of Chapter 6 in this handbook.

KARO logic for rational agents

The KARO formalism is an amalgam of dynamic logic and epistemic / doxastic logic, augmented with several additional (modal) operators in order to deal with the motivational aspects of agents. So, besides operators for knowledge ($\mathbf{K}$), belief ($\mathbf{B}$) and action ($[\alpha]$, "after performance of α it holds that"), there are additional operators for ability ($\mathbf{A}$) and desires ($\mathbf{D}$).

Assume a set $\mathcal{A}$ of atomic actions and a set $\mathcal{P}$ of atomic propositions.

Definition 10.6 (Language)
The language $\mathcal{L}_{\mathsf{KARO}}$ of KARO-formulas is given by the BNF grammar:

$$\begin{array}{rcl}\varphi & ::= & p(\in \mathcal{P}) \mid \neg\varphi \mid \varphi_1 \wedge \varphi_2 \mid \ldots \\ & & \mathbf{K}\varphi \mid \mathbf{B}\varphi \mid \mathbf{D}\varphi \mid [\alpha]\varphi \mid \mathbf{A}\alpha\end{array}$$

$$\alpha \quad ::= \quad a(\in \mathcal{A}) \mid \varphi? \mid \alpha_1;\alpha_2 \mid \alpha_1 + \alpha_2 \mid \alpha^* \qquad \dashv$$

Here the formulas generated by the second (α) part are referred to as actions (or rather action expressions). We use the abbreviations $\mathtt{tt} \equiv p \vee \neg p$ (for some fixed $p \in \mathcal{P}$) and $\mathtt{ff} \equiv \neg\mathtt{tt}$. Conditional and while-action are introduced by the usual abbreviations: $\mathtt{if}\ \varphi\ \mathtt{then}\ \alpha_1\ \mathtt{else}\ \alpha_2\ \mathtt{fi} \equiv (\varphi?;\alpha_1) + (\neg\varphi?;\alpha_2)$ and $\mathtt{while}\ \varphi\ \mathtt{do}\ \alpha\ \mathtt{od} \equiv (\varphi?;\alpha)^*;\neg\varphi?$.

Thus formulas are built by means of the familiar propositional connectives and the modal operators for knowledge, belief, desire, action and ability. Actions are the familiar ones from imperative programming: atomic ones, tests, sequential composition, (nondeterministic) choice and repetition.

Definition 10.7 (KARO models)

1. The semantics of the knowledge, belief and desires operators is given by means of Kripke structures of the form $\mathcal{M} = \langle W, \vartheta, R_K, R_B, R_D \rangle$, where

   - W is a non-empty set of states (or worlds)
   - ϑ is a truth assignment function per state
   - R_K, R_B, R_D are accessibility relations for interpreting the modal operators $\mathbf{K}, \mathbf{B}, \mathbf{D}$. The relation R_K is assumed to be an equivalence relation, while the relation R_B is assumed to be euclidean, transitive and serial. Furthermore we assume that $R_B \subseteq R_K$. No special constraints are assumed for the relations R_D.

2. The semantics of actions is given by means of structures of type $\langle \Sigma, \{R_a \mid a \in \mathcal{A}\}, \mathcal{C}, Ag \rangle$, where

   - Σ is the set of possible model/state pairs (i.e. models of the above form, together with a state appearing in that model)
   - R_a $(a \in \mathcal{A})$ are relations on Σ encoding the behaviour of atomic actions
   - $\mathcal{C}$ is a function that gives the set of actions that the agent is able to do per model/state pair
   - Ag is a function that yields the set of actions that the agent is committed to (the agent's 'agenda') per model/state pair. ⊣

We have elements in the structures for interpreting the operators for knowledge, belief, and desire. Actions are modelled as model/state pair transformers to emphasize their influence on the mental state (that is, the complex of knowledge, belief and desires) of the agent rather than just the state of the world. Both (cap)abilities and commitments are given by functions that yield the relevant information per model / state pair.

Definition 10.8 (Interpretation of formulas in KARO)
In order to determine whether a formula $\varphi \in \mathcal{L}$ is true in a model/state pair (M, w) (if so, we write $(M, w) \models \varphi$), we stipulate:

- $\mathcal{M}, w \models p$ iff $\vartheta(w)(p) = \mathit{true}$, for $p \in \mathcal{P}$

- The logical connectives are interpreted as usual.
- $\mathcal{M}, w \models \mathbf{K}\varphi$ iff $\mathcal{M}, w' \models \varphi$ for all w' with $R_K(w, w')$
- $\mathcal{M}, w \models \mathbf{B}\varphi$ iff $\mathcal{M}, w' \models \varphi$ for all w' with $R_B(w, w')$
- $\mathcal{M}, w \models \mathbf{D}\varphi$ iff $\mathcal{M}, w' \models \varphi$ for all w' with $R_D(w, w')$
- $\mathcal{M}, w \models [\alpha]\varphi$ iff $\mathcal{M}', w' \models \varphi$ for all M', w' with $R_\alpha((\mathcal{M}, w), (\mathcal{M}', w'))$
- $\mathcal{M}, w \models \mathbf{A}\alpha$ iff $\alpha \in \mathcal{C}(\mathcal{M}, w)$[5]
- $\mathcal{M}, w \models \mathbf{Com}(\alpha)$ iff $\alpha \in Ag(\mathcal{M}, w)$[6] ⊣

Here R_α is defined as usual in dynamic logic by induction from the basic case R_a, but now on model/state pairs rather than just states. So, e.g. $R_{\alpha_1+\alpha_2} = R_{\alpha_1} \cup R_{\alpha_2}$, $R_{\alpha^*} = R^*_\alpha$, the reflective transitive closure of R_α, and $R_{\alpha_1;\alpha_2}$ is the relational product of R_{α_1} and R_{α_2}. Likewise the function $\mathcal{C}$ is lifted to complex actions. We call an action α *deterministic* if $\#\{w' \mid R_\alpha(w, w')\} \leqslant 1$ for any $w \in W$, and *strongly deterministic* if $\#\{w' \mid R_\alpha(w, w')\} = 1$. (Here # stands for cardinality.)

We have clauses for knowledge, belief and desire. The action modality gets a similar interpretation: something (necessarily) holds after the performance / execution of action α if it holds in all the situations that are accessible from the current one by doing the action α. The only thing which is slightly nonstandard is that, as stated above, a situation is characterised here as a model / state pair. The interpretations of the ability and commitment operators are rather trivial in this setting (but see the footnotes): an action is enabled (or rather: the agent is able to do the action) if it is indicated so by the function $\mathcal{C}$, and, likewise, an agent is committed to an action α if it is recorded so in the agent's agenda.

Furthermore, we will make use of the following syntactic abbreviations serving as auxiliary operators:

Definition 10.9

- (dual) $\langle\alpha\rangle\varphi = \neg[\alpha]\neg\varphi$, expressing that the agent has the opportunity to perform α resulting in a state where φ holds.
- (opportunity) $\mathbf{O}\alpha = \langle\alpha\rangle\mathtt{tt}$, i.e., an agent has the opportunity to do an action iff there is a successor state w.r.t. the R_α-relation;

[5] In fact, the ability operator can alternatively defined by means of a second accessibility relation for actions, in a way analogous to the opportunity operator below.

[6] The agenda is assumed to be closed under certain conditions such as taking 'prefixes' of actions (representing initial computations). Details are omitted here, but see Section 10.6 for references.

- (practical possibility) $\mathbf{P}(\alpha, \varphi) = \mathbf{A}\alpha \wedge \mathbf{O}\alpha \wedge \langle\alpha\rangle\varphi$, i.e., an agent has the practical possibility to do an action with result φ iff it is both able and has the opportunity to do that action and the result of actually doing that action leads to a state where φ holds;

- (can) $\mathbf{Can}(\alpha, \varphi) = \mathbf{KP}(\alpha, \varphi)$, i.e., an agent can do an action with a certain result iff it knows it has the practical possibility to do so;

- (realisability) $\Diamond\varphi = \exists a_1, \ldots, a_n \mathbf{P}(a_1; \ldots; a_n, \varphi)$[7], i.e., a state property φ is realisable iff there is a finite sequence of atomic actions of which the agent has the practical possibility to perform it with the result φ;

- (goal) $\mathbf{G}\varphi = \neg\varphi \wedge \mathbf{D}\varphi \wedge \Diamond\varphi$, i.e., a goal is a formula that is not (yet) satisfied, but desired and realisable.[8]

- (possible intend) $\mathbf{I}(\alpha, \varphi) = \mathbf{Can}(\alpha, \varphi) \wedge \mathbf{KG}\varphi$, i.e., an agent (possibly) intends an action with a certain result iff the agent can do the action with that result and it moreover knows that this result is one of its goals. $\dashv$

Remark 10.1

- The dual of the (box-type) action modality expresses that there is at least a resulting state where a formula φ holds. It is important to note that in the context of *deterministic* actions, i.e. actions that have at most one successor state, this means that the *only* state satisfies φ, and is thus in this particular case a stronger assertion than its dual formula $[\alpha]\varphi$, which merely states that if there are any successor states they will (all) satisfy φ. Note also that if atomic actions are assumed to be deterministic all actions including the complex ones will be deterministic.

- Opportunity to do an action is modelled by having at least one successor state according to the accessibility relation associated with the action.

---

[7] We abuse our language here slightly, since strictly speaking we do not have quantification in our object language. See our references to KARO in Section 10.6 for a proper definition.

[8] In fact, here we simplify matters slightly. One might stipulate that a goal should be explicitly selected somehow from the desires it has, which could be modelled by means of an additional modal operator. Here we leave this out for simplicity's sake.

- Practical possibility to to an action with a certain result is modelled as having both ability and opportunity to do the action with the appropriate result. Note that $\mathbf{O}\alpha$ in the formula $\mathbf{A}\alpha \wedge \mathbf{O}\alpha \wedge \langle\alpha\rangle\varphi$ is actually redundant since it already follows from $\langle\alpha\rangle\varphi$. However, to stress the opportunity aspect it is added.

- The Can predicate applied to an action and formula expresses that the agent is 'conscious' of its practical possibility to do the action resulting in a state where the formula holds.

- A formula φ is realisable if there is a 'plan' consisting of (a sequence of) atomic actions of which the agent has the practical possibility to do them with φ as a result.

- A formula φ is a goal in the KARO framework if it is not true yet, but desired and realisable in the above meaning, that is, there is a plan of which the agent has the practical possibility to realise it with φ as a result.

- An agent is said to (possibly) intend an action α with result φ if it 'Can' do this (knows that it has the practical possibility to do so), and, moreover, knows that φ is a goal. $\dashv$

In order to manipulate both knowledge / belief and motivational matters special actions `revise`, `commit` and `uncommit` are added to the language. (We assume that we cannot nest these operators. So, for instance $\texttt{commit}(\texttt{uncommit}\alpha)$ is not a well-formed action expression.) The semantics of these are again given as model/state transformers (We only do this here in a very abstract manner, viewing the accessibility relations associated with these actions as functions. For further details we refer the reader to inspect some of the KARO references mentioned in Section 10.6.

Definition 10.10 (Accessibility of revise, commit and uncommit)

1. $R_{\texttt{revise}\varphi}(\mathcal{M}, w) = \mathit{update\_belief}(\varphi, (\mathcal{M}, w))$.

2. $R_{\texttt{commit}\alpha}(\mathcal{M}, w) = \mathit{update\_agenda}^{+}(\alpha, (\mathcal{M}, w))$, if $\mathcal{M}, w \models \mathbf{I}(\alpha, \varphi)$ for some φ, otherwise $R_{\texttt{commit}\alpha}(\mathcal{M}, w) = \emptyset$ (indicating failure of the commit action).

3. $R_{\texttt{uncommit}\alpha}(\mathcal{M}, w) = \mathit{update\_agenda}^{-}(\alpha, (\mathcal{M}, w))$, if it holds that $\mathcal{M}, w \models \mathbf{Com}(\alpha)$, otherwise $R_{\texttt{uncommit}\alpha}(\mathcal{M}, w) = \emptyset$ (indicating failure of the uncommit action);

4. $\texttt{uncommit}\alpha \in \mathcal{C}(\mathcal{M}, w)$ iff $\mathcal{M}, w \models \neg\mathbf{I}(\alpha, \varphi)$ for all formulas φ, that is, an agent is able to uncommit to an action if it is not intended to do it (any longer) for any purpose. $\dashv$

Here *update_belief*, *update_agenda*$^+$ and *update_agenda*$^-$ are functions that update the agent's belief and agenda (by adding or removing an action), respectively. Details are omitted here, but essentially these actions are model/state transformers again, representing a change of the mental state of the agent (regarding beliefs and commitments, respectively). The *update_belief*$(\varphi, (\mathcal{M}, w))$ function changes the model $\mathcal{M}$ in such a way that the agent's belief is updated with φ, while *update_agenda*$^+(\alpha, (\mathcal{M}, w))$ changes the model $\mathcal{M}$ such that α is added to the agenda, and likewise for the *update_agenda*$^-$ function, but now with respect to removing an action from the agenda. The `revise` operator can be used to cater for revisions due to observations and communication with other agents, which we will not go into further here.

The interpretation of formulas containing revise and (un)commit actions is now done using the accessibility relations above. One can now define validity as usual with respect to the KARO-models. One then obtains the following validities (of course, in order to be able to verify these one should use the proper model and not the abstraction / simplification we have presented here.) Typical properties of this framework, called the KARO logic, include:

Proposition 10.3

1. $\models \Box(\varphi \to \psi) \to (\Box\varphi \to \Box\psi)$, for $\Box \in \{\mathbf{K}, \mathbf{B}, \mathbf{D}, [\alpha]\}$
2. $\models \langle\alpha\rangle\varphi \to [\alpha]\varphi$, for deterministic α
3. $\models \Box\varphi \to \Box\Box\varphi$, for $\Box \in \{\mathbf{K}, \mathbf{B}\}$
4. $\models \neg\Box\varphi \to \Box\neg\Box\varphi$, for $\Box \in \{\mathbf{K}, \mathbf{B}\}$
5. $\models \mathbf{K}\varphi \to \varphi$
6. $\models \neg\mathbf{B}\mathtt{ff}$
7. $\models \mathbf{O}(\alpha;\beta) \leftrightarrow \langle\alpha\rangle\mathbf{O}\beta$
8. $\models \mathbf{Can}(\alpha;\beta,\varphi) \leftrightarrow \mathbf{Can}(\alpha, \mathbf{P}(\beta,\varphi))$
9. $\models \mathbf{I}(\alpha,\varphi) \to \mathbf{K}\langle\alpha\rangle\varphi$
10. $\models \mathbf{I}(\alpha,\varphi) \to \langle\mathtt{commit}\alpha\rangle\mathbf{Com}(\alpha)$
11. $\models \mathbf{I}(\alpha,\varphi) \to \neg\mathbf{A}\mathtt{uncommit}(\alpha)$
12. $\models \mathbf{Com}(\alpha) \to \langle\mathtt{uncommit}(\alpha)\rangle\neg\mathbf{Com}(\alpha)$

13. $\models \mathbf{Com}(\alpha) \wedge \neg\mathbf{Can}(\alpha, \top) \rightarrow \mathbf{Can}(\mathtt{uncommit}(\alpha), \neg\mathbf{Com}(\alpha))$

14. $\models \mathbf{Com}(\alpha) \rightarrow \mathbf{KCom}(\alpha)$

15. $\models \mathbf{Com}(\alpha_1; \alpha_2) \rightarrow \mathbf{Com}(\alpha_1) \wedge \mathbf{K}[\alpha_1]\mathbf{Com}(\alpha_2)$

16. $\models \mathbf{Com}(\mathtt{if}\ \varphi\ \mathtt{then}\ \alpha_1\ \mathtt{else}\ \alpha_2\ \mathtt{fi}) \wedge \mathbf{K}\varphi \rightarrow \mathbf{Com}(\varphi?; \alpha_1)$

17. $\models \mathbf{Com}(\mathtt{if}\ \varphi\ \mathtt{then}\ \alpha_1\ \mathtt{else}\ \alpha_2\ \mathtt{fi}) \wedge \mathbf{K}\neg\varphi \rightarrow \mathbf{Com}(\neg\varphi?; \alpha_2)$

18. $\models \mathbf{Com}(\mathtt{while}\ \varphi\ \mathtt{do}\ \alpha\ \mathtt{od}) \wedge \mathbf{K}\varphi \rightarrow \mathbf{Com}((\varphi?; \alpha); \mathtt{while}\ \varphi\ \mathtt{do}\ \alpha\ \mathtt{od})$
$\dashv$

The first of these properties says that all the modalities mentioned are 'normal' in the sense that they are closed under implication. The second states that the dual operator $\langle\alpha\rangle$ is stronger than the operator $[\alpha]$ in case the action α is deterministic: if there is at most one successor state after performing α and we know that there is at least one successor state satisfying φ then *all* successor states satisfy φ. The third and fourth properties are the so-called introspection properties for knowledge and belief. The fifth property says that knowledge is true, while the sixth states that belief (may not be true but) is not inconsistent. The seventh property states that having the opportunity to do a sequential composition of two actions amounts to having the opportunity of doing the first action first and then having the opportunity to do the second. The eighth states that an agent that *can* do a sequential composition of two actions with result φ iff the agent can do the first actions resulting in a state where it has the practical possibility to do the second with φ as result. The ninth states that if one possibly intends to do α with result φ then one knows that there is a possibility of performing α resulting in a state where φ holds. The tenth asserts that if an agent possibly intends to do α with some result φ, it has the opportunity to commit to α with result that it is committed to α (i.e. α is put into its agenda). The eleventh says that if an agent intends to do α with a certain purpose, then it is unable to uncommit to it (so, if it is committed to α it has to persevere with it). This is the way persistence of commitment is represented in KARO. Note that this is much more 'concrete' (also in the sense of computability) than the persistence notions in the other approaches we have seen, where temporal operators pertaining to a possibly infinite future were employed to capture them...! We think it is no coincidence that Hindriks *et al.* established an almost perfect match in the sense of a correspondence between the agent programming language GOAL and Cohen & Levesque's logic of intention, the main difference being the inability of GOAL to express the persistence properties of intentions in this logic...!) In KARO we have the advantage of having dedicated actions in the action language dealing with the change of commitment that can be used to express persistence without referring to the (infinite) future, rendering the notion of persistence much 'more computable'. The twelfth property says that if an agent is committed to an action and it has the opportunity to uncommit to it, as result then indeed the commitment is removed. The thirteenth says that whenever an agent is committed to an action that is no longer known to be practically possible, it knows that it can undo this impossible commitment. The fourteenth property states that commitments are known to the agent. The last four properties have to do with commitments to complex actions. For instance, the fifteenth says that if an agent is committed to a sequential composition of two actions then it is committed to the first one, and it knows that after doing the first action it will be committed to the second action.

KARO logic for emotional agents

In this subsection we look at a recent application of BDI logic that deals with agent behaviour that is strictly beyond the scope of the original aim of BDI logic, viz. describing the behaviour of rational agents. We will sketch how the KARO framework can be used for describing emotional agents. Although it is perhaps a bit paradoxical to describe emotions and emotional behaviour with logic, one should bear in mind that we are dealing with behaviour here, and this can be described in logic, especially a logic that deals with actions such as the KARO framework. Furthermore, as we shall see, emotional behaviour will turn out to be complimentary rather than opposed to rational behaviour of agents, something that is also acknowledged by recent work in cognitive science through the work of for instance Damasio. Our presentation here is inspired by two psychological theories: that of Oatley & Jenkins and that of OCC. Since the latter is much more involved (treating 22 emotions, while the former only treats 4 basic emotions), we here mainly follow the ideas of Oatley & Jenkins, and say a few words on modelling OCC later.

According to Oatley & Jenkins, the 4 basic emotions, happiness, sadness, anger and fear, have the following characteristics:

- Happiness results when in the process of trying to achieve a goal, things go 'right', as expected, i.e., subgoals are achieved thus far.
- Sadness results when in the process of trying to achieve a goal, things go 'wrong', i.e., not as expected, i.e., subgoals are not being achieved.
- Anger is the result of frustration about not being able to execute the current plan, and makes the agent try harder to execute the plan.

- Fear results when a 'maintenance goal' is threatened, so that the agent will make sure that this maintenance goal is restored before going on with other activities.

It is directly obvious from these descriptions that these emotions are BDI-related notions! So it is not so strange to use a BDI-logic like KARO to describe them, which is what Meyer did.

Let us take sadness as an example. For simplicity, assume that plans consist of sequences of atomic actions. In KARO we can then express the trigger condition for sadness as follows:

$$\mathbf{I}(\pi, \varphi) \wedge \mathbf{Com}(\pi) \wedge \mathbf{B}([\alpha]\psi) \rightarrow$$

$$[\alpha]((\mathbf{B}\neg\psi \wedge \mathbf{Com}(\pi\backslash\alpha)) \rightarrow \mathit{sad}(\pi\backslash\alpha, \varphi))$$

where α is a prefix of plan π. Intuitively, this says that if the agent has the (possible) intention to perform plan π with goal φ, it is committed to π (so it has a true intention to do π), and it believes that after doing the initial fragment α of the plan π it holds that ψ, then after doing α if it believes that ψ does not hold while it is still committed to the rest of the plan, it is sad (with respect to the rest of the plan and goal φ). In a similar way the trigger conditions of the other emotions can be formalised.

Also, together with Steunebrink and Dastani, Meyer looked at modeling OCC. Particularly, they show how to formalise the (trigger conditions of) emotions in OCC in three steps: first by presenting a more general logical structure of the emotions, which are later refined in terms of doxastic logic and finally in the full-blown BDI logic KARO again. The way emotions get a semantics based on BDI models is quite intricate and beyond the scope of this chapter, but one of the properties that can be proven valid in this approach is the following, using KARO's (possible) intend operator (here parametrized by an agent):

$$\mathbf{I}_i(\alpha, \varphi) \rightarrow [i:\alpha](\mathit{Pride}^T_j(i:\alpha) \wedge \mathit{Joy}^T_i(\varphi) \wedge \mathit{Gratification}^T_i(i:\alpha, \varphi))$$

(Here $i : \alpha$ in the dynamic logic box refers to the action of i performing α, and the superscript ^{T} placed at the emotion operators pertains to the idea that we are considering triggering / elicitation forms of the emotions concerned.) Informally this reads that if the agent i has the possible intention to do α with goal φ, then if he has performed α he is proud (triggered pride) of his action, has (triggered) joy about the achievement of the goal φ and has (triggered) gratification with respect to action $i : \alpha$ and goal φ.

Finally, let us also mention here the strongly related work by Adam *et al.* Also this work is devoted to a formalisation of OCC emotions in BDI terms. There are differences with the work of Steunebrink *et al.*, though.

For instance, Adam simply defines joy as a conjunction of belief and desire: $Joy\ \varphi =_{def} Bel\ \varphi \wedge Des\ \varphi$. This seems to express a 'state of joy' (experience) rather than a trigger for joy. This raises a confusion of emotion elicitation (triggering) and experience, which is kept separately in the approach of Steunebrink. This confusion also appears at other places in the work of Adam, e.g. where she defines gratification as the conjunction of pride (which pertains to triggering) and joy (which is about experience as we saw earlier). In later work by Adam, this issue is improved upon and it is explained that the above definition of Joy is solely about the triggering of joy, not the experience. However, the confusion of triggering versus experience is still not resolved completely since it is still present in the introspective properties of emotional awareness $Emotion\varphi \leftrightarrow BelEmotion\varphi$ and $\neg Emotion\varphi \leftrightarrow Bel\neg Emotion\varphi$, which hold in Adam's framework, for any Emotion. This is counterintuitive if Emotion should capture the triggering of the associated emotion, since an agent may not be aware of this triggering.

10.4 BDI-modalities in STIT logic

The principles of BDI logics reflect rationality postulates for agent modalities. In particular, the BDI principles model how B, D and I modalities interact with each other over time (well known are the so called 'commitment strategies' of Cohen & Levesque, stating under which belief and desire conditions intentions have to be dropped, see Section 10.2.1). BDI logics are not meant for knowledge representation but for agent specification: ideally concretely built agents will some day be verified against the logic principles of BDI-logics (how exactly this could ever be done is a question we set aside here).

An essential component of any BDI logic is then its dynamic part. Traditionally, either the dynamic part is formed by a dynamic logic fragment (Cohen & Levesque, KARO) or a temporal fragment (Rao & Georgeff). Recently a third alternative has been considered: STIT (seeing to it that) logic. STIT logics can be said to be in between dynamic logic and temporal logic. Where dynamic logic sees actions as the steps of a program, and temporal logic leaves actions entirely out of the picture, STIT logic sees action as a relation between agents and the effects they can see to. STIT logic achieves this by generalizing temporal structures to choice structures. The most distinguishing feature of STIT logic is that truth of formulas often expresses information about the dynamics of the world. For instance, the STIT logic formula $[ag\ \texttt{stit}]X(at\_station))$ says that agent ag currently sees to it that next it is at the station. But, it does *not* say that the agent *can* see to it that next it is at the station (this is however a logical consequence). Abilities are truths about static conditions, and not about dynamic conditions.

In the present section we will discuss how in recent years several authors have aimed to combine STIT logic and BDI notions. There are two parts. In the first part, Section 10.4.1, we focus on classical instantaneous STIT logics and the BDI extensions that have been suggested for them. In the second part, Section 10.4.2, we consider dynamic variants of the BDI modalities and discuss the notion of 'knowingly doing' within a version of STIT where effects of actions take effect in next states: XSTIT.

There is a strong connection, both conceptually and technically, between the family of logics STIT and *Alternating-time Temporal Logic*, which is at the center of attention in Chapter 11.

10.4.1 BDI modalities in instantaneous stit

Traditionally STIT logics encompass operators for agency that assume that an agentive choice performance is something that takes no time. So, an instantaneous stit operator $[ag\ \texttt{stit}]\varphi$ typically obeys the success axiom $[ag\ \texttt{stit}]\varphi \rightarrow \varphi$ to capture the intuition concerning instantaneity saying that if *ag now* sees to it that φ holds, then φ must be true *now*. Before putting forward an alternative to this view, where the central agency operator has a built-in step to a next moment in time, we give the formal definition of standard (Chellas) instantaneous STIT logic and discuss its logical properties. We will here use a slightly different syntax and semantics than used by Chellas himself and also different from that of Horty, but, the logic is the same.

CSTIT

Definition 10.11

Given a countable set of propositions P and $p \in P$, and given a finite set Ags of agent names, and $ag \in Ags$, the formal language $\mathcal{L}_{\mathsf{CSTIT}}$ is:

$$\varphi \ := \ p \mid \neg\varphi \mid \varphi \wedge \varphi \mid \Box\varphi \mid [ag\ \texttt{Cstit}]\varphi \qquad \dashv$$

Besides the usual propositional connectives, the syntax of CSTIT comprises two modal operators. The operator $\Box\varphi$ expresses 'historical necessity', and plays the same role as the well-known path quantifiers in logics such as CTL and CTL*. Another way of talking about this operator is to say that it expresses that φ is 'settled'. We abbreviate $\neg\Box\neg\varphi$ by $\Diamond\varphi$. The operator $[ag\ \texttt{Cstit}]\varphi$ stands for 'agent ag sees to it that φ' (the 'C' referring to Chellas). $\langle ag\ \texttt{Cstit}\rangle\varphi$ abbreviates $\neg[ag\ \texttt{Cstit}]\neg\varphi$.

The semantics given in Definition 10.12 below is an alternative to the semantics given by Belnap and colleagues in terms of BT+AC structures (Branching Time + Agentive Choice structures). The differences are not essential. Where the branching of time in BT + AC structures is represented by tree-like orderings of moments, the structures below use 'bundles' of linearly ordered sets of moments. We use the latter to be uniform in the semantic structures across different STIT formalisms in this section.

Definition 10.12

A CSTIT-frame is a tuple $\langle S, H, R_{ag}\rangle$ such that[9]:

1. S is a non-empty set of static states. Elements of S are denoted s, s', etc.

2. H is a non-empty set of system histories $\ldots s_{-2}, s_{-1}, s_0, s_1, s_2, \ldots$ with $s_x \in S$ for $x \in \mathbb{Z}$. Elements of H are denoted h, h', etc.

3. Dynamic states are tuples $\langle s, h\rangle$, where $s \in S$, $h \in H$ and s appears on h. Now the relations R_{ag} are 'effectivity' equivalence classes over dynamic states such that $\langle s, h\rangle R_{ag}\langle s', h'\rangle$ only if $s = s'$. For any state s and agent ag, the relation R_{ag} defines a partition of the dynamic states built with s. The partition models the possible choices $C^s_{ag}, C'^s_{ag}, C''^s_{ag}, \ldots$ of ag in s. A choice profile $\langle C^s_{ag_1}, C^s_{ag_2} \ldots C^s_{ag_n}\rangle$ at s is a particular combination of choices $C^s_{ag_i}$ at s, one for each agent ag_i in the system. For any s the intersection of choices in any choice profile is non-empty: $\bigcap_{ag_i \in Ags} C^s_{ag_i} \neq \emptyset$

In Definition 10.12 above, we refer to the states s as 'static states'. This is to distinguish them from 'dynamic states', which are combinations $\langle s, h\rangle$ of a static state and a history. Dynamic states function as the elementary units of evaluation of the logic. This means that the basic notion of 'truth' in the semantics of this logic is about dynamic conditions concerning choice performances.

We now define models by adding a valuation of propositional atoms to the frames of Definition 10.12.

Definition 10.13

A frame $\mathcal{F} = \langle S, H, R_{ag}\rangle$ is extended to a model $\mathcal{M} = \langle S, H, R_{ag}, V\rangle$ by adding a valuation V of atomic propositions:

- V is a valuation function $V : P \longrightarrow 2^{S\times H}$ assigning to each atomic proposition the set of state history pairs relative to which they are true.

[9] In the meta-language we use the same symbols both as constant names and as variable names, and we assume universal quantification of unbound meta-variables.

We evaluate truth with respect to dynamic states.

Definition 10.14
Relative to a model $\mathcal{M} = \langle S, H, R_{ag}, V \rangle$, truth $\langle s, h \rangle \models \varphi$ of a formula φ in a dynamic state $\langle s, h \rangle$, with $s \in h$, is defined as:

$$\begin{array}{lll}
\langle s,h\rangle \models p & \Leftrightarrow & \langle s,h\rangle \in V(p) \\
\langle s,h\rangle \models \neg\varphi & \Leftrightarrow & \text{not } \langle s,h\rangle \models \varphi \\
\langle s,h\rangle \models \varphi \wedge \psi & \Leftrightarrow & \langle s,h\rangle \models \varphi \text{ and } \langle s,h\rangle \models \psi \\
\langle s,h\rangle \models \Box\varphi & \Leftrightarrow & \forall h' : \text{if } s \in h' \text{ then } \langle s,h'\rangle \models \varphi \\
\langle s,h\rangle \models [ag\ \texttt{Cstit}]\varphi & \Leftrightarrow & \forall h' : \text{if } \langle s,h\rangle R_{ag} \langle s,h'\rangle \text{ then } \langle s,h'\rangle \models \varphi
\end{array} \quad \dashv$$

Satisfiability, validity on a frame and general validity are defined as usual.

Now we proceed with the axiomatization.

Theorem 10.2
The following axiom schemes, in combination with a standard axiomatization for propositional logic, and the standard rules (like necessitation) for the normal modal operators, define a complete Hilbert system for CSTIT:

$$\begin{array}{ll}
 & \text{The S5 axioms for } \Box \\
 & \text{For each } ag \text{ the S5 axioms for } [ag\ \texttt{Cstit}] \\
(SettC) & \Box\varphi \rightarrow [ag\ \texttt{Cstit}]\varphi \\
(Indep) & \Diamond[ag_1\ \texttt{Cstit}]\varphi \wedge \ldots \wedge \Diamond[ag_n\ \texttt{Cstit}]\psi \rightarrow \\
 & \Diamond([ag_1\ \texttt{Cstit}]\varphi \wedge \ldots \wedge [ag_n\ \texttt{Cstit}]\psi) \\
 & \text{for } Ags = \{ag_1, \ldots, ag_n\}
\end{array} \quad \dashv$$

Balbiani *et al.* propose an alternative axiomatization and a semantics whose units of evaluation are not two dimensional pairs $\langle s, h \rangle$ but one dimensional worlds w. Here we have chosen to give a two-dimensional semantics to emphasize the relation with the XSTIT semantics in section 10.4.2.

BDI-stit

Semmling and Wansing add BDI modalities to a basic Chellas stit logic as the one just defined. Their BDI-stit formalism extends the syntax as follows (we take the liberty of using our own notation for the BDI operators and to define an alternative but equivalent semantics).

Definition 10.15
Given a countable set of propositions P and $p \in P$, and given a finite set Ags of agent names, and $ag \in Ags$, the formal language $\mathcal{L}_{\textsf{bdi-stit}}$ is:

$$\varphi \;:=\; p \mid \neg\varphi \mid \varphi \wedge \varphi \mid \Box\varphi \mid [ag\ \texttt{Cstit}]\varphi \mid \langle[ag\ \texttt{bel}]\rangle\varphi \mid \langle[ag\ \texttt{des}]\rangle\varphi \mid \langle[ag\ \texttt{int}]\rangle\varphi$$

⊣

To emphasize their weak modal character, we denote the introduced belief, desire, intention and possibility operators with a combination of sharp and square brackets. This alludes to the combination of first order existential and universal quantifications that is present in any first order simulation of a weak modal operator. The reading of the operators speaks for itself; they express belief, desire and intention concerning a proposition φ.

Definition 10.16
A bdi-stit-frame is a tuple $\langle S, H, R_{ag}, N_b, N_d, N_i\rangle$ such that:

1. $\langle S, H, R_{ag}\rangle$ is a CSTIT-frame
2. N_b N_d and N_i are neighborhood functions of the form $N : S \times H \times Ags \mapsto 2^{2^{S\times H}}$ mapping any combination of a dynamic state $\langle s, h\rangle$ and an agent ag to a set of neighborhoods of $\langle s, h\rangle$. Semmling & Wansing then impose constraints on neighborhood frames that are equivalent to:
   - **a.** All three functions N_b, N_d and N_i obey $\emptyset \notin N(s, h, ag)$
   - **b.** All three functions N_b, N_d and N_i obey that if $N \in N(s, h, ag)$ and $N \subset N'$ then $N' \in N(s, h, ag)$
   - **c.** $N \in N_i(s, h, ag)$ and $N' \in N_i(s, h, ag)$ implies $N \cap N' \neq \emptyset$

The intuition underlying neighborhood functions is that $N_b(s, h, ag)$ gives for agent ag in situation $\langle s, h\rangle$ the clusters of possible worlds (situations / dynamic states) the joint possibility of which it believes in. Since clusters and propositions correspond to each other one-to-one (modulo logic equivalence of the propositions), it will also be convenient to look at the clusters or neighborhoods as propositions and to say that if $N \in N_b(s, h, ag)$ the agent ag believes the proposition (modulo logical equivalence) corresponding to N, that $N \in N_d(s, h, ag)$ holds if ag desires the proposition and that $N \in N_i(s, h, ag)$ holds if ag intends the proposition.

Now **a.** says that there is no belief, desire or intention for impossible states of affairs. **b.** says that belief, desire and intention are closed under weakening of the propositions believed, desired or intended. **c.** says that intentions are consistent in the sense that it is not possible to hold at the same time an intention for a proposition and for its negation.

Definition 10.17 (Truth conditions BDI operators)
Relative to a model $\langle S, H, R_{ag}, N_b, N_d, N_i, V\rangle$, truth of belief, desire and intention operators is defined as *($[\![\varphi]\!]$ is the truth set of φ, that is, the subset of all dynamic elements in $S \times H$ satisfying φ)*:

$$\langle s,h\rangle \models \langle[ag\ \mathtt{bel}]\rangle\varphi \Leftrightarrow [\![\varphi]\!] \in N_b(s,h,ag)$$
$$\langle s,h\rangle \models \langle[ag\ \mathtt{des}]\rangle\varphi \Leftrightarrow [\![\varphi]\!] \in N_d(s,h,ag)$$
$$\langle s,h\rangle \models \langle[ag\ \mathtt{int}]\rangle\varphi \Leftrightarrow [\![\varphi]\!] \in N_i(s,h,ag) \qquad \dashv$$

An axiomatization of a probabilistic epistemic logic is obtained by formulating axioms corresponding to the conditions on neighborhood functions.

Theorem 10.3 (Hilbert system BDI operators)
Relative to the semantics following from definitions 10.16 and 10.17 we define the following Hilbert system. We assume the standard derivation rules for the weak modalities, like closure under logical equivalence.

(BelPos)	$\neg\langle[ag\ \mathtt{bel}]\rangle\bot$	*(BelWk)*	$\langle[ag\ \mathtt{bel}]\rangle\varphi \rightarrow \langle[ag\ \mathtt{bel}]\rangle(\varphi \vee \psi)$
(DesPos)	$\neg\langle[ag\ \mathtt{des}]\rangle\bot$	*(DesWk)*	$\langle[ag\ \mathtt{des}]\rangle\varphi \rightarrow \langle[ag\ \mathtt{des}]\rangle(\varphi \vee \psi)$
(IntPos)	$\neg\langle[ag\ \mathtt{int}]\rangle\bot$	*(IntWk)*	$\langle[ag\ \mathtt{int}]\rangle\varphi \rightarrow \langle[ag\ \mathtt{int}]\rangle(\varphi \vee \psi)$
		(IntD)	$\langle[ag\ \mathtt{int}]\rangle\varphi \rightarrow \neg\langle[ag\ \mathtt{int}]\rangle\neg\varphi$

$\dashv$

Relative to their own version of the semantics, Semmling & Wansing prove completeness of their logic. Here the completeness of the axiomatization relative to the frames of Definition 10.16 follows from general results in neighborhood semantics and monotonic modal logic. One can check that the conditions on the frames correspond one-to-one with the axioms in the axiomatization.

As can be seen from the axioms and conditions we have shown above, Semmling and Wansing chose to make their BDI-stit logic rather weak, trying to commit only to a minimum of logical properties. But even with this minimalistic approach there is room for debate. For instance, the condition on intentions only looks at pairwise consistency of intentions, but conflicts are still possible in case there is a combination of three intentions: $\{\langle[ag\ \mathtt{int}]\rangle\varphi, \langle[ag\ \mathtt{int}]\rangle(\varphi \rightarrow \psi), \langle[ag\ \mathtt{int}]\rangle\neg\psi\}$ is satisfiable. If we do not want that, it is straightforward to adapt the condition on neighborhood functions (demand that any finite number of neighborhoods has a state in common), but it is unclear how to axiomatise it.

Even though the STIT framework's notion of truth refers to the dynamics of a system of agents (which is why we talk about 'dynamic states'), Semmling and Wansing do not focus on dynamic interpretations of the BDI attitudes. A formula like $\langle[ag\ \texttt{bel}]\rangle[ag\ \texttt{Cstit}]\varphi$ must express something like "ag believes that it sees to it that φ", but the inherent dynamic aspect of this notion is not analyzed. In particular, no interactions between STIT and BDI modalities are studied. This is likely to be due to the fact that explicit dynamic temporal operators are absent in the logic and because agency is instantaneous. In Section 10.4.2 we will report on the study of the inherent dynamic aspect of such combinations of operators.

BDI, STIT and regret

Lorini and Schwarzentruber use STIT logic as the basis for investigations into what they call counterfactual emotions[10]. The typical counterfactual emotion is 'regret'. Regret can be described as a discrepancy between what actually occurs and what could have happened. Based on this they argue that for a definition of regret in the STIT framework, it needs to be extended with modalities for knowledge and desire. They consider three different STIT formalisms to base their definitions on, but here it will suffice to discuss their ideas using the Chellas STIT logic given earlier.

Knowledge is added to their STIT framework in a straightforward way: a normal S5 knowledge operator $[ag_i\ \texttt{kno}]$ is added for every agent ag_i in the system. The interpretation is in terms of equivalence classes over the basic units of evaluation (for the stit language we consider here: dynamic states). So, for knowledge we get the truth condition $\langle s,h\rangle \models [ag_i\ \texttt{kno}]\varphi$ iff for all s',h' such that $\langle s,h\rangle \sim \langle s',h'\rangle$ we have $\langle s',h'\rangle \models \varphi$. The second operator we take from their system is an operator for desire, which is defined using propositional constants.

Definition 10.18
Let $good_{ag_i}$ denote a propositional constant, one for each agent ag_i in the system, whose truth expresses that a state is good for that agent. Now the modal operators $[ag_i\ \textsf{good}]\varphi$ and $[ag_i\ \textsf{des}]\varphi$ are defined by:

$$[ag_i\ \textsf{good}]\varphi \equiv_{def} \Box(good_{ag_i} \to \varphi)$$
$$[ag_i\ \textsf{des}]\varphi \equiv_{def} [ag_i\ \texttt{kno}][ag_i\ \textsf{good}]\varphi \qquad \dashv$$

The counterfactual aspect is introduced by the definition of the notion of "could have prevented" (CHP). For definitions of such counterfactual properties, the STIT framework is more suited than dynamic logic or situation calculus frameworks, since in STIT we reason about actual performances of actions which also makes it possible to reason about choices that are not (f)actual. Lorini and Schwarzentruber define their notion of CHP in a group

[10] However, it is not the emotions that are counterfactual; the theory is about factual emotions based on beliefs about counterfactual conditions

stit framework. Here we have only defined individual agency. Therefore we will only consider the two agent case, for which there is no essential difference between group operators and operators for individual agents. The two agent case will enable us to discuss the ideas properly.

In the two agent setting, Lorini and Schwarzentruber's intuition can be described as follows: agent ag_1 could have prevented φ if and only if (1) φ is currently true, and (2) agent ag_2 does not see to it that φ. They reformulate the second condition as 'provided that agent ag_2 sticks to its choice, for agent ag_1 there is the possibility to act otherwise in a way that does not guarantee the outcome φ'[11].

Definition 10.19
$\langle[ag_1 \texttt{ CHP}]\rangle\varphi \equiv_{def} \varphi \land \neg[ag_2 \texttt{ Cstit}]\varphi$

CHP is not a normal modality (which is why we use the combination of sharp and square brackets, as explained before). Lorini and Schwarzentruber argue that their notion of CHP obeys agglomeration '$\langle[ag \texttt{ CHP}]\rangle\varphi \land \langle[ag \texttt{ CHP}]\rangle\psi \to \langle[ag \texttt{ CHP}]\rangle(\varphi \land \psi)$', but not weakening '$\langle[ag \texttt{ CHP}]\rangle\varphi \to \langle[ag \texttt{ CHP}]\rangle(\varphi \lor \psi)$'. But, agglomeration is not necessarily an intuitive property for a notion of 'could have prevented': if the guard could have prevented prisoner 1 to escape through exit 1 and if the guard could have prevented prisoner 2 to escape through exit 2, it does not follow that the guard could have prevented both prisoner 1 to escape though exit 1 and prisoner 2 to escape through exit 2. We believe alternative definitions of the notion of 'could have prevented' are possible. In particular it seems promising to look for notions where the counterfactual conditions concerning alternatives are made explicit using the $\Diamond$ operator.

With the right concepts in place, finally we are able to define the notion of regret for a proposition φ as the desire for $\neg\varphi$ in conjunction with knowledge about the fact that φ could have been prevented.

Definition 10.20
$\langle[ag_i \texttt{ rgt}]\rangle\varphi \equiv_{def} [ag_i \texttt{ des}]\neg\varphi \land [ag_i \texttt{ kno}]\langle[ag_i \texttt{ CHP}]\rangle\varphi$

Definitions like the one above for regret hinge on the possibility to express that certain actions or choices are actually performed. This type of expressivity is not provided by many other action formalisms.

[11]The reformulation is equivalent, but suggests a meaning of group action that is not standard: if ag_1 only has an alternative under the provision that the choice of ag_2 is kept fixed, then a standard interpretation would be that ag_1 and ag_2 have an alternative in *cooperation*. But, then the possibility to prevent also relies on cooperation. What seems needed for a notion of 'could have prevented' for individual agents, is that such agents have alternatives *individually*.

10.4.2 BDI modalities in XSTIT: dynamic attitudes

In the systems discussed in Section 10.4.1, BDI notions were combined with STIT operators. However, in both approaches there was no special attention for a dynamic interpretation of the combination of BDI and STIT operators. Yet this interpretation strongly suggests itself. If an operator like $[ag\ \texttt{Cstit}]\varphi$ expresses that agent ag now exercises his choice to ensure that φ, and a knowledge operator like $[ag\ \texttt{kno}]\varphi$ also has a dynamic reading (note that the truth condition of knowledge is not with respect to static states s but with respect to dynamic states $\langle s, h\rangle$, cf. the clause for $[ag\ \texttt{kno}]\varphi$ just above Definition 10.18), then a natural interpretation of a combination like $[ag\ \texttt{kno}][ag\ \texttt{Cstit}]\varphi$ is that agent ag "knowingly sees to it that φ". We suspect that this dynamic reading was not suggested by the authors of the discussed systems because these systems do not contain temporal modalities and because the used version of STIT has instantaneous effects. In this section we report on work studying the notions of knowingly and intentionally doing that is based on a version of STIT where agency inherently involves a move to some next state: XSTIT.

XSTIT

We give here the basic definitions for XSTIT. XSTIT enriches the CSTIT language of Section 10.4.1 with a temporal next operator and replaces the operator $[ag\ \texttt{Cstit}]\varphi$ by the operator $[ag\ \texttt{xstit}]\varphi$ where the effect φ is not instantaneous but occurs in a next state. Further explanations and motivations can be found elsewhere (see Section 10.6).

Definition 10.21
Given a countable set of propositions P and $p \in P$, and given a finite set Ags of agent names, and $ag \in Ags$, the formal language $\mathcal{L}_{\mathsf{XSTIT}}$ is:

$$\varphi \ := \ p \mid \neg\varphi \mid \varphi \wedge \varphi \mid \Box\varphi \mid [ag\ \texttt{xstit}]\varphi \mid X\varphi \qquad \dashv$$

The operator $[ag\ \texttt{xstit}]\varphi$ stands for 'agent ag sees to it that φ in the next state'. The operator $X\varphi$ is a standard next time operator. $\Box$ is again the operator for historical necessity (settledness).

Definition 10.22
An XSTIT-frame is a tuple $\langle S, H, E\rangle$ such that:

1. S is a non-empty set of static states. Elements of S are denoted s, s', etc.

2. H is a non-empty set of system histories $\ldots s_{-2}, s_{-1}, s_0, s_1, s_2, \ldots$ with $s_x \in S$ for $x \in \mathbb{Z}$. Elements of H are denoted h, h', etc. We denote

that s' succeeds s on the history h by $s' = succ(s, h)$ and by $s = prec(s', h)$. We have the following bundling constraint on the set H:

a. if $s \in h$ and $s' \in h'$ and $s = s'$ then $prec(s, h) = prec(s', h')$

3. $E : S \times H \times Ags \mapsto 2^S$ is an h-effectivity function yielding for an agent ag the set of next static states allowed by the choice performed by the agent relative to a history. We have the following constraints on h-effectivity functions:

a. if $s \notin h$ then $E(s, h, ag) = \emptyset$

b. if $s' \in E(s, h, ag)$ then $\exists h' : s' = succ(s, h')$

c. if $s' = succ(s, h')$ and $s' \in h$ then $s' \in E(s, h, ag)$

d. $E(s, h, ag_1) \cap E(s, h', ag_2) \neq \emptyset$ for $ag_1 \neq ag_2$

Condition **2.a** is the 'backwards-linear' requirement that we have seen in Section 10.2.2. Condition **3.b** ensures that next state effectivity as seen from a current state s does not contain states s' that are not reachable from the current state through some history. Condition **3.c** expresses the STIT condition of 'no choice between undivided histories'. Condition **3.d** above states that simultaneous choices of different agents never have an empty intersection. This is the central condition of 'independence of agency'. It reflects that a choice performance of one agent can never have as a consequence that some other agent is limited in the choices it can exercise simultaneously.

Again, we evaluate truth with respect to dynamic states.

Definition 10.23
Relative to a model $\mathcal{M} = \langle S, H, E, V \rangle$, truth $\langle s, h \rangle \models \varphi$ of a formula φ in a dynamic state $\langle s, h \rangle$, with $s \in h$, is defined as:

$$
\begin{array}{lcl}
\langle s, h \rangle \models p & \Leftrightarrow & s \in V(p) \\
\langle s, h \rangle \models \neg\varphi & \Leftrightarrow & \text{not } \langle s, h \rangle \models \varphi \\
\langle s, h \rangle \models \varphi \wedge \psi & \Leftrightarrow & \langle s, h \rangle \models \varphi \text{ and } \langle s, h \rangle \models \psi \\
\langle s, h \rangle \models \Box\varphi & \Leftrightarrow & \forall h' : \text{ if } s \in h' \text{ then } \langle s, h' \rangle \models \varphi \\
\langle s, h \rangle \models X\varphi & \Leftrightarrow & \text{if } s' = succ(s, h) \text{ then } \langle s', h \rangle \models \varphi \\
\langle s, h \rangle \models [ag \texttt{ xstit}]\varphi & \Leftrightarrow & \forall s', h' : \text{if } s' \in E(s, h, ag) \text{ and} \\
 & & s' \in h' \text{ then } \langle s', h' \rangle \models \varphi
\end{array}
$$
⊣

Satisfiability, validity on a frame and general validity are defined as usual.

Now we proceed with the axiomatization.

Theorem 10.4
The following axiom schemes, in combination with a standard axiomatization for propositional logic, and the standard rules (like necessitation) for the normal modal operators, define a complete Hilbert system for XSTIT:

$$\begin{array}{ll} & \text{S5 for } \Box \\ & \text{For each } ag\text{, KD for } [ag\ \mathtt{xstit}] \\ \textit{(Lin)} & \neg X \neg \varphi \leftrightarrow X\varphi \\ \textit{(Sett)} & \Box X\varphi \rightarrow [ag\ \mathtt{xstit}]\varphi \\ \textit{(XSett)} & [ag\ \mathtt{xstit}]\varphi \rightarrow X\Box\varphi \\ \textit{(Indep)} & \Diamond[ag_1\ \mathtt{xstit}]\varphi \wedge \ldots \wedge \Diamond[ag_n\ \mathtt{xstit}]\psi \rightarrow \\ & \Diamond([ag_1\ \mathtt{xstit}]\varphi \wedge \ldots \wedge [ag_n\ \mathtt{xstit}]\psi) \\ & \text{for } Ags = \{ag_1, \ldots, ag_n\} \end{array}$$

⊣

Knowingly doing

To study the notion of knowingly doing, like before, in Section 10.4.1, a normal S5 knowledge operator $[ag_i\ \mathtt{kno}]\varphi$ is added for every agent ag_i in the system. The equivalence classes, or 'information sets' of these operators contain state-history pairs, which means knowledge concerns information about the dynamics of the system of agents. Now knowingly doing is suitably modeled by the combination of operators $[ag_i\ \mathtt{kno}][ag_i\ \mathtt{xstit}]\varphi$. For the logic of this notion we can consider several interactions between the contributing modalities. Here we only briefly mention some of the possibilities.

It is a fundamental property of agency that an agent cannot know what other agents choose simultaneously. This is expressed by the following axiom.

Definition 10.24
The property of ignorance about concurrent choices of others is defined by the axiom IgnCC, which reads as follows:

$$[ag_1\ \mathtt{kno}][ag_2\ \mathtt{xstit}]\varphi \rightarrow [ag_1\ \mathtt{kno}]\Box[ag_2\ \mathtt{xstit}]\varphi \text{ for } ag_1 \neq ag_2$$

⊣

(IgnCC) expresses that if an agent knows that something results from the choice of another agent, it can only be that the agent knows it is settled that that something results from a choice of the other agent.

Definition 10.25
The property of knowledge about the next state is defined by the axiom:

$$\textit{(XK)} \qquad [ag\ \mathtt{kno}]X\varphi \rightarrow [ag\ \mathtt{kno}][ag\ \mathtt{xstit}]\varphi$$

⊣

The (XK) property expresses that the only things an agent can know about the next state are the things it knows to be seeing to it itself.

Definition 10.26
The property of effect recollection is defined by the axiom:

$$(\textit{Rec-Eff}) \qquad [ag\ \texttt{kno}][ag\ \texttt{xstit}]\varphi \rightarrow [ag\ \texttt{xstit}][ag\ \texttt{kno}]\varphi \qquad \dashv$$

(Rec-Eff) expresses that if agents knowingly see to something, then they know that something is the case in the resulting state.

The above three properties for knowingly doing just exemplify some of the possibilities. More properties have been studied. Also the theory on these dynamic attitudes has been extended to beliefs and to intentions. The case of intentional action is particularly interesting because there is an extensive philosophical literature on this notion. One of the philosophical scenario's discussed by Broersen for instance is the side effect problem. Here we can only point to the fact that XSTIT, after addition of the right BDI modalities, seems to be a suitable base logic for the study of such notions.

10.5 Conclusion

In this chapter we have reviewed the use of epistemic logic, extended with other modalities for motivational attitudes such as desires and intentions, for describing (the behaviour of) intelligent agents. What is immediately clear is that although all logical approaches are based on and inspired by Bratman's seminal work on BDI theory for practical reasoning, the formalizations themselves are quite different in nature. They also enjoy different (and sometimes even on first sight contradictory) properties. In our view this means that although Bratman did his uttermost to present a clear philosophy, as is often the case when formalizing this kind of philosophical theories, there is still a lot of ambiguity or, put more positively, freedom to formalise these matters. We have even seen that quite different base logics may be used, such as (branching-time) temporal logic, dynamic logic and also stit logic. This makes the formal logics in themselves hard to compare. We think it depends on the purpose of the formalisation (is it used for better understanding, or does it serve as a basis for computational and executable frameworks) which of the BDI logics will be most appropriate. The latter is especially important for designers and programmers of agent systems.

10.6 Notes

Pointers to the theory of Bratman and colleagues include the work on practical reasoning by Bratman (1987) and by Bratman, Israel, and Pollack

(1988), as well as the paper on intentions by Bratman (1990). The concept of pro-attitude is also introduced by Bratman (1987). The term 'Practical Reasoning' dates back to Aristotle: For further reading and a contemporary reference list the reader is also advised to read the short introduction 'Practical Reason' in the Stanford Encyclopedia of Philosophy (2013), or to consult work on 'Practical Syllogism'. The term 'screen of admissibility' to describe Bratman's theory of how intentions filter admissible options was coined by Cohen and Levesque (1990). For more on the subtle issue of admissible intentions and inconsistent beliefs and an elaborate example, see pp. 41–42 of the paper by Bratman (1987). Daniel Dennett set out his work on the intentional stance (based on the notions of desire and belief) in (Dennett, 1987). A reduction of intention to beliefs and desires alone is rejected (pp. 6–9) by Bratman (1987).

A pioneering paper on intelligent agents is by Wooldridge and Jennings (1995), and also by Wooldridge (1999). The latter for instance addresses the issue of how logics like those studied in this chapter may eventually be *realised* in agent-based systems. The area of agents and multi-agent systems is now a mature and still active area: a good start for further orientation is provided by the website of IFAAMAS (IFAAMAS), which hosts proceedings of the Autonomous Agents and Multi-Agent Systems conference since 2007. As mentioned in the text, Bratman's semi-formal theory on intentions was further formalised by Cohen and Levesque (1990), and, more or less at the same time, by Rao and Georgeff (1991). In fact their work won, respectively, the 2006 and 2007 IFAAMAS Awards for Influential Papers in Autonomous Agents and Multiagent Systems.

Cohen and Levesque (1990) were among the first to adapt PDL to model agency. Many responses, extensions and variations of this logic have been proposed. For instance, the Side-Effect-Free Principle was coined by Su, Sattar, Lin, and Reynolds (2007), while the study of introspective properties of preferences was undertaken by Herzig and Longin (2004), Lorini and Demolombe (2008), and by Herzig, Lorini, Hübner, and Vercouter (2010). The first of those three references also put forward a definition of epistemic achievement goal. Weakening the link between action and goals, and at the same time allowing for a stronger commitment (a notion criticised by Herzig and Longin (2004)) was undertaken by Sadek (2000) and by Bretier (1995). Sadek (1992) moreover argues for a notion of realistic preference, also called weak realism. The notion of *attempting* to do an action was formalised by Lorini and Herzig (2008).

Our treatment of Rao and Georgeff's BDI logic is mainly based on (Rao and Georgeff, 1991) (which includes the GOAL-operator) and (Rao and Georgeff, 1998) (where this operator is replaced by an operator DES for desire). Their approach, in turn, is for its formal model heavily inspired by the framework on branching time logic put forward by Emerson (1990). The abbreviation CTL stands for Computation-Tree Logic, in which one can quantify over branches, and, given a branch, over its points. CTL* is an extension of CTL where some of CTL's restrictions on the occurrences of branching and linear time quantifiers, are lifted. The semantic constraints that make some of the desirable axioms of Section 10.2.2 valid, are discussed by Rao and Georgeff (1991, 1998) and by Wooldridge (2000). In particular, the world/time point compatibility requirement is taken from Wooldridge (2000). We have used the little Nell problem as an advocate to use branching time logic, rather than linear time logic. This problem was discussed by McDermott (1982), and also addressed by the papers of Cohen & Levesque and Rao & Georgeff. It is interesting to note that in a series of talks and papers however, Moshe Vardi has argue that from a computer science perspective, it is far from clear that branching time provides a superior model over that of linear time (see for instance the paper by Vardi (2001)).

KARO stands for Knowledge, Ability, Result and Opportunity. The KARO framework was developed in a number of papers by van Linder, van der Hoek, and Meyer (1995, 1997), van der Hoek, van Linder, and Meyer (1998) and Meyer, van der Hoek, and van Linder (1999) and the thesis by van Linder (1996). All the basic operators of KARO can be interpreted as modal operators, cf. also footnote 5, which was proven by van der Hoek, Meyer, and van Schagen (2000). Adding a notion of agenda to KARO, in order to deal with commitments, was proposed and studied by Meyer et al. (1999) (see also footnote 6). Readers interested in knowing the details of our simplification as mentioned in footnote 8 are referred to the paper by Meyer et al. (1999). One of KARO's foundations is dynamic logic, for which the book by Harel (1984) is a standard reference. An early reference to the agent programming language GOAL (Goal-Oriented Agent Language) is provided by Hindriks, Boer, Hoek, and Meyer (2001), while de Boer, Hindriks, van der Hoek, and Meyer (2007) provide, rather than the language for declarative goals itself, both a programming framework and a programming logic for such computational agents. The connection with intention logic of Cohen & Levesque was given by Hindriks, van der Hoek, and Meyer (2012). A starting point to get familiar with the software platform for GOAL is (2013). As an example of how logic frameworks can be beneficial for designers and programmers of agent systems, the KARO framework played an important role in devising the agent programming language 3APL by Hindriks, de Boer, van der Hoek, and Meyer (1999). Moreover, by formalising the relations between the KARO framework and the 3APL language, Hindriks and Meyer (2007) and Meyer (2007) were able to propose a verification logic for 3APL and its derivatives. As a proper treatment of this aspect of BDI logics goes beyond the scope of this chapter we refer to the literature on this issue, e.g., the work by van der Hoek and Wooldridge (2003), Dastani, van Riemsdijk, and Meyer (2007), Alechina, Dastani, Logan, and Meyer (2007), Dastani, Hindriks, and Meyer (2010), and by Hindriks et al. (2012).

Adding emotions to the mix of agents' attitudes is by now a well-respected avenue in agent-based modeling. Emotions are now regarded to complement rationality, rather than being in tension with it. This was acknowledged in for instance cognitive science through the work of Damasio (1994), and is comparable with the rise of 'behavioural economics' complementing 'classical game theory' (for the latter, see also Chapter 9 of this book). See for instance 'Emotions and Economic Theory' by Elster (1998). The work that we presented on emotions was inspired by two psychological theories: that of Oatley and Jenkins (1996) and that of Ortony, Clore, and Collins (1988) (also referred to as OCC). More specifically, KARO logic for emotional agents as based on Oatley & Jenkins' framework is worked out by Meyer (2006), while the formalisation of OCC theories is done by Steunebrink, Dastani, and Meyer (2007, 2012). References to the work of Adam *et al.* on emotions include the papers by Adam (2007) and by Adam, Herzig, and Longin (2009). Their introspective properties for emotion presented at the end of Section 10.3 are taken from Theorem 13 of the latter reference.

STIT (Seeing To It That) logics are an attempt of treating agency as a modality in an approach originating in a series of papers, beginning with the paper by Belnap and Perloff (1988). Other important contributions in this field are by Chellas (1995) and by Horty (2001). A semantics for STIT based on worlds (rather than state-history pairs) was given by Balbiani, Herzig, and Troquard (2007). The work of Semmling and Wansing (2008) adds BDI modalities to STIT logic. Lorini and Schwarzentruber (2011) use STIT logic to analyse emotions. For further details and motivation for next-state STIT logic XSTIT we refer to work by Broersen (2009, 2011a). The latter, together with Broersen (2011b) (which, among other things, discusses the side effect problem), is the place to further read on knowingly doing. Monotonic modal logic is presented in Hansen (2003). Theorem 10.2 is taken from Xu (1998), and Theorem 10.4 from Broersen (2011a).

There have been a number of further developments, which can be seen as consequences of BDI logics. First of all we mention that on the basis of a BDI logic, Shoham (1993) initiated the field of agent-oriented programming (AOP), which is basically a way of programming the (BDI-like) mental states of agents as a new paradigm of programming. It could be viewed as a successor (or even refinement) of the well-known Object-Oriented (OO) Programming paradigm. Momentarily there is a host of such dedicated agent cognitive (BDI) programming languages (see the work by Bordini, Dastani, Dix, and Seghrouchni (2005, 2009)), including 2APL//3APL by

Hindriks et al. (1999) and Dastani (2008) which is inspired by KARO.

Another extension that is made in the literature is a combination with reasoning with uncertainty. Casali, l. Godo, and Sierra (2011) propose a so-called Graded BDI logic. This means that beliefs, desires and intentions are not crisp anymore, but fuzzy. This is done by using a multi-valued base logic. In particular, a fuzzy modal operator D^+ is introduced with as reading "φ is positively desired". Its truth degree corresponds to the agent's level of satisfaction would φ become true. It satisfies the principle $D^+(\varphi \vee \psi) \leftrightarrow (D^+\varphi \wedge D^+\psi)$. There is also an analogous operator D^-, read "φ is negatively desired". Intention is then defined from this. Their logic combines Rational Pavelka Logic (an extension of Lukasiewicz's many-valued logic with infinitely many truth values by rational truth degree constants) as described by Hájek (1998) with a multi-context logic of Giunchiglia and Serafini (1994) in order to account for the agents' mental attitudes. The approach allows for a much more fine-grained deliberation process. In particular, the process of deriving intentions can now be formally expressed as follows: the intention degree to reach a desire φ by means of a plan α is taken as a trade-off between the benefit of reaching this desire and the cost of the plan, weighted by the belief degree r.

Lorini and Demolombe (2008) and Lorini (2011) propose a similar graded approach to BDI, but now based on a sense of plausibility (non-exceptionality) as proposed by Spohn (1998). In this logic both the belief and the goal attitude are graded. Beyond modal operators of knowledge $\mathtt{Know}_i$, the logic of Lorini (2011) has special atoms exc_h indicating that the degree of exceptionality of the current state is h, where h ranges over a finite set of natural numbers Num. Then graded belief is defined as

$$\mathtt{Bel}_i^{\geq h}\varphi \stackrel{\text{def}}{=} \bigvee_{k\geq h} \Big(\neg\mathtt{Know}_i(exc_k \to \varphi) \wedge \bigwedge_{l<k} \mathtt{Know}_i(exc_l \to \varphi)\Big)$$

Graded goals $\mathtt{Goal}_i^{\geq h}\varphi$ are defined in a similar way from special atoms des_h indicating that the degree of desirability of the current state is h. This theory is next applied to the modelling of expectation-based emotions such as hope, fear, disappointment and relief, and their intensity.

Finally, in analogy to the area of belief revision, also a lot of work has appeared on the revision of the other mental states in BDI, notably intentions. In fact, intention reconsideration is already part of the first implementation of Bratman's theory as set out by Bratman et al. (1988) and is present in the BDI architecture of Rao and Georgeff (1992) via the so-called BDI control loop. It also is present in the logical theory KARO of van Linder et al. (1995) and Meyer et al. (1999) that we have seen earlier. Using a dynamic logic it is very natural to also incorporate actions in KARO that are mental state-revising such as belief revision as proposed by van Linder et al. (1995) and changes in commitments as studied by Meyer et al. (1999). But recently there have appeared more fundamental theories of intention revision, such as by van der Hoek, Jamroga, and Wooldridge (2007), where a logic of intention dynamics is developed based on dynamic logic and a dynamic update operator of the form $[\Omega]\varphi$, meaning after the agent has updated on the basis of observations Ω, it must be the case that φ. The approach of beliefs, desires and intentions is rather syntactical (set of sentences without modalities for these mental attitudes interpreted by model structures). Contrary to this, van Ditmarsch, de Lima, and Lorini (2010) present a model-theoretic approach of intentions and intention dynamics. It uses modal operators for time, belief and choice as basis, and intention as a derived operator, and for intention change it employs a dynamic operator called local assignment. This is an operation on the model that changes the truth value of atomic formulae at specific time points. Shoham (2009) discusses the interaction of belief revision and intention revision. This is done more in a traditional AGM sense as put forward by Alchourrón, Gärdenfors, and Makinson (1985), than in a dynamic epistemic modal sense.

Acknowledgements

Thanks to Emiliano Lorini for his useful comments on an earlier version of this chapter.

References

Adam, C. (2007). *Emotions: from psychological theories to logical formalization and implementation in a BDI agent.* Ph. D. thesis, Institut National Polytechnique de Toulouse, Toulouse.

Adam, C., A. Herzig, and D. Longin (2009). A logical formalization of the OCC theory of emotion. *Synthese 168*(2), 201–248.

Alchourrón, C. E., P. Gärdenfors, and D. Makinson (1985). On the logic of theory change: Partial meet contractions and revision functions. *Journal of Symbolic Logic 50*(2), 510–530.

Alechina, N., M. Dastani, B. Logan, and J.-J. Ch. Meyer (2007). A logic of agent programs. In R. C. Holte and A. E. Howe (Eds.), *Proceedings of AAAI-07*, Vancouver, Canada, pp. 795–800. AAAI Press.

Balbiani, P., A. Herzig, and N. Troquard (2007). Alternative axiomatics and complexity of deliberative stit theories. *Journal of Philosophical Logic 37*, 387–406.

Belnap, N. and M. Perloff (1988). Seeing to it that: a canonical form for gentiles. *Theoria 54*, 175–199.

Bordini, R. H., M. Dastani, J. Dix, and A. E. F. Seghrouchni (Eds.) (2005). *Multi-Agent Programming: Languages, Platforms and Applications*, Volume 15 of *Multiagent Systems, Artificial Societies, and Simulated Organizations.* New York: Springer.

Bordini, R. H., M. Dastani, J. Dix, and A. E. F. Seghrouchni (Eds.) (2009). *Multi-Agent Programming (Languages, Tools and Applications).* Dordrecht Heidelberg: Springer.

Bratman, M. E. (1987). *Intentions, Plans, and Practical Reason.* Massachusetts: Harvard University Press.

Bratman, M. E. (1990). What is intention? In P. R. Cohen, J. Morgan, and M. E. Pollack (Eds.), *Intentions in Communication*, Chapter 2, pp. 15–31. Cambridge, Massachusetts: MIT Press.

Bratman, M. E., D. J. Israel, and M. E. Pollack (1988). Plans and resource-bounded practical reasoning. *Computational Intelligence 4*, 349–355.

Bretier, P. (1995). *La communication orale coopérative: contribution à la modélisation logique et à la mise en oeuvre d'un agent rationnel dialoguant.* Ph. D. thesis, Université Paris Nord, Paris, France.

Broersen, J. M. (2009). A complete stit logic for knowledge and action, and some of its application. In M. Baldoni, T. C. Son, M. B. van Riemsdijk, and M. Winikoff (Eds.), *Declarative Agent Languages and Technologies VI (DALT 2008)*, Volume 5397 of *LNCS*, Berlin, pp. 47–59. Springer.

Broersen, J. M. (2011a). Deontic epistemic *stit* logic distinguishing modes of 'Mens Rea'. *Journal of Applied Logic 9*(2), 127–152.

Broersen, J. M. (2011b). Making a start with the stit logic analysis of intentional action. *Journal of Philosophical Logic 40*, 399–420.

Casali, A., L. l. Godo, and C. Sierra (2011). A graded BDI agent model to represent and reason about preferences. *Artificial Intelligence 175*(7-8), 1468–1478.

Chellas, B. F. (1995). On bringing it about. *Journal of Philosophical Logic 24*, 563–571.

Cohen, P. R. and H. J. Levesque (1990). Intention is choice with commitment. *Artificial Intelligence 42*(3), 213–261.

Damasio, A. (1994). *Descartes' Error: Emotion, Reason, and the Human Brain.* New York: Grosset/Putnam Press.

Dastani, M. (2008). 2APL: a practical agent programming language. *Autonomous Agents and Multi-Agent Systems 16*(3), 214–248.

Dastani, M., K. V. Hindriks, and J.-J. Ch. Meyer (Eds.) (2010). *Specification and Verification of Multi-Agent Systems.* New York/Dordrecht/Heidelberg/London: Springer.

Dastani, M., B. van Riemsdijk, and J.-J. Ch. Meyer (2007). A grounded specification language for agent programs. In M. Huhns, O. Shehory, E. H. Durfee, and M. Yokoo (Eds.), *Proceedings of the 6th International Joint Conference On Autonomous Agents and Multi-Agent Systems (AAMAS2007*, pp. 578–585.

de Boer, F., K. Hindriks, W. van der Hoek, and J.-J. Meyer (2007). A verification framework for agent programming with declarative goals. *Journal of Applied Logic 5*(2), 277 – 302.

Dennett, D. C. (1987). *The Intentional Stance.* Cambridge, Massachusetts: MIT Press.

van Ditmarsch, H., T. de Lima, and E. Lorini (2010). Intention change via local assignments. In *Proceedings of LADS 2010*, Volume 6822 of *LNCS*, Berlin/Heidelberg, pp. 136–151. Springer-Verlag.

Elster, J. (1998). Emotions and economic theory. *Journal of Economic Literature 36*(1), 47–74.

Emerson, E. A. (1990). Temporal and modal logic. In J. van Leeuwen (Ed.), *Handbook of Theoretical Computer Science*, Volume B: Formal Models and Semantics, Chapter 14, pp. 996–1072. Amsterdam: Elsevier Science.

Giunchiglia, F. and L. Serafini (1994). Multilanguage hierarchical logics (or: how we can do without modal logics). *Artificial Intelligence 65*, 29–70.

Goal (2013). The software platform for GOAL. `http://ii.tudelft.nl/trac/goal`, retrieved July 2013.

Hájek, P. (1998). *Metamathematics of Fuzzy Logic*, Volume 4 of *Trends in Logic*. Dordrecht: Kluwer Academic Publishers.

Hansen, H. H. (2003). Monotonic modal logics. Master's thesis, ILLC, Amsterdam.

Harel, D. (1984). Dynamic logic. In D. Gabbay and F. Guenther (Eds.), *Handbook of Philosophical Logic*, Volume II, pp. 497–604. Dordrecht/Boston: Reidel.

Herzig, A. and D. Longin (2004). C&L intention revisited. In D. Dubois, C. Welty, and M.-A. Williams (Eds.), *Proceedings of the 9th International Conference on Principles on Principles of Knowledge Representation and Reasoning (KR2004)*, pp. 527–535. AAAI Press.

Herzig, A., E. Lorini, J. F. Hübner, and L. Vercouter (2010). A logic of trust and reputation. *Logic Journal of the IGPL 18*(1), 214–244.

Hindriks, K. V., F. S. d. Boer, W. v. d. Hoek, and J.-J. Ch. Meyer (2001). Agent programming with declarative goals. In *Proceedings of the 7th International Workshop on Intelligent Agents VII. Agent Theories Architectures and Languages*, ATAL '00, London, UK, UK, pp. 228–243. Springer-Verlag.

Hindriks, K. V., F. S. de Boer, W. van der Hoek, and J.-J. Ch. Meyer (1999). Agent programming in 3APL. *International Journal of Autonomous Agents and Multi-Agent Systems 2*(4), 357–401.

Hindriks, K. V. and J.-J. Ch. Meyer (2007). Agent logics as program logics: grounding KARO. In C. Freksa, M. Kohlhase, and K. Schill (Eds.), *29th Annual German Conference on AI, KI 2006*, Volume 4314 of *LNAI*, pp. 404–418. Springer.

Hindriks, K. V., W. van der Hoek, and J.-J. Ch. Meyer (2012). GOAL agents instantiate intention logic. In A. Artikis, R. Craven, N. K. Çiçekli, B. Sadighi, and K. Stathis (Eds.), *Logic Programs, Norms and Action (Sergot Festschrift)*, Volume 7360 of *LNAI*, Heidelberg, pp. 196–219. Springer.

van der Hoek, W., W. Jamroga, and M. Wooldridge (2007). Towards a theory of intention revision. *Synthese 155*(2), 265–290.

van der Hoek, W., J.-J. Ch. Meyer, and J. W. van Schagen (2000). Formalizing potential of agents: the KARO framework revisited,. In M. Faller, S. Kaufmann, and M. Pauly (Eds.), *Formalizing the Dynamics of Information*, Volume 91 of *CSLI Lecture Notes*, pp. 51–67. Stanford: CSLI Publications.

van der Hoek, W., B. van Linder, and J.-J. Ch. Meyer (1998). An integrated modal approach to rational agents. In M. Wooldridge and A. Rao (Eds.), *Foundations of Rational Agency*, Volume 14 of *Applied Logic Series*, pp. 133–168. Dordrecht: Kluwer.

van der Hoek, W. and M. Wooldridge (2003). Towards a logic of rational agency. *Logic Journal of the IGPL 11*(2), 133–157.

Horty, J. F. (2001). *Agency and Deontic Logic.* Oxford University Press.

IFAAMAS. `http://www.ifaamas.org`, retrieved July 2013.

van Linder, B. (1996). *Modal Logics for Rational Agents.* Ph. D. thesis, Utrecht University.

van Linder, B., W. van der Hoek, and J.-J. Ch. Meyer (1995). Actions that make you change your mind: belief revision in an agent-oriented setting. In A. Laux and H. Wansing (Eds.), *Knowledge and Belief in Philosophy and Artificial Intelligence*, pp. 103–146. Berlin: Akademie Verlag.

van Linder, B., W. van der Hoek, and J.-J. Ch. Meyer (1997). Seeing is believing (and so are hearing and jumping),. *Journal of Logic, Language and Information 6*, 33–61.

Lorini, E. (2011). A dynamic logic of knowledge, graded beliefs and graded goals and its application to emotion modelling. In *Proceedings of LORI 2011*, Volume 6953 of *LNCS*, Berlin / Heidelberg, pp. 165–178. Springer-Verlag.

Lorini, E. and R. Demolombe (2008). Trust and norms in the context of computer security: toward a logical formalization. In R. V. der Meyden and L. V. der Torre (Eds.), *Proceedings of the International Workshop on Deontic Logic in Computer Science (DEON 2008)*, Volume 5076 of *LNCS*, Berlin/Heidelberg, pp. 50–64. Springer-Verlag.

Lorini, E. and A. Herzig (2008). A logic of intention and attempt. *Synthese KRA 163*(1), 45–77.

Lorini, E. and F. Schwarzentruber (2011). A logic for reasoning about counterfactual emotions. *Artificial Intelligence 175*, 814–847.

McDermott, D. V. (1982). A temporal logic for reasoning about processes and plans. *Cognitive Science 6*, 101–155.

Meyer, J.-J. Ch. (2006). Reasoning about emotional agents. *International Journal of Intelligent Systems 21*(6), 601–619.

Meyer, J.-J. Ch. (2007). Our quest fot the holy grail of agent verification. In N. Olivetti (Ed.), *Proceedings of TABLEAUX 2007*, Volume 4548 of *LNAI*, Berlin/Heidelberg, pp. 2–9. Springer.

Meyer, J.-J. Ch., W. van der Hoek, and B. van Linder (1999). A logical approach to the dynamics of commitments. *Artificial Intelligence 113*, 1–40.

Oatley, K. and J. M. Jenkins (1996). *Understanding Emotions.* Blackwell Publishing.

Ortony, A., G. L. Clore, and A. Collins (1988). *The Cognitive Structure of Emotions.* Cambridge: Cambridge University Press.

Rao, A. and M. P. Georgeff (1992). An abstract architecture for rational agents. In B. Nebel, C. Rich, and W. Swartout (Eds.), *Proceedings of the Third International Conference on Principles of Knowledge Representation and Reasoning (KR'92)*, San Mateo, Califorinia, pp. 439–449. Morgan Kaufmann.

Rao, A. S. and M. P. Georgeff (1991). Modeling rational agents within a BDI-architecture. In R. F. J. Allen and E. Sandewall (Eds.), *Proceedings of the Second International Conference on Principles of Knowledge Representation and Reasoning (KR'91)*, San Mateo, California, pp. 473–484. Morgan Kaufmann.

Rao, A. S. and M. P. Georgeff (1998). Decision procedures for BDI logics. *Journal of Logic and Computation 8*(3), 293–344.

Sadek, M. D. (1992). A study in the logic of intention. In B. Nebel, C. Rich, and W. Swartout (Eds.), *Proceedings of the Third International Conference on Principles of Knowledge Representation and Reasoning (KR'92)*, San Mateo, Califorinia, pp. 462–473. Morgan Kaufmann.

Sadek, M. D. (2000). Dialogue acts are rational plans. In M. Taylor, F. Nel, and D. Bouwhuis (Eds.), *The structure of mutimodal dialogue*, Philadelphia/Amsterdam,, pp. 167–188. From ESCA/ETRW, Workshop on The Structure of Multimodal Dialogue (Venaco II), 1991.

Semmling, C. and H. Wansing (2008). From BDI and *stit* to *bdi-stit* logic. *Logic and Logical Philosophy 17*(1-2), 185–207.

Shoham, Y. (1993). Agent-oriented programming. *Artificial Intelligence 60*(1), 51–92.

Shoham, Y. (2009). Logical theories of intention and the database perspective. *Journal of Philosophical Logic 38*(6), 633–647.

Spohn, W. (1998). Ordinal conditional functions: a dynamic theory of epistemic states. In *Causation in Decision, Belief Change and Statistics*, pp. 105–134. Dordrecht: Kluwer.

Steunebrink, B., M. Dastani, and J.-J. Ch. Meyer (2007). A logic of emotions for intelligent agents. In R. C. Holte and A. E. Howe (Eds.), *Proceedings of AAAI-07*, Vancouver, Canada, pp. 142–147. AAAI Press.

Steunebrink, B. R., M. Dastani, and J.-J. Ch. Meyer (2012). A formal model of emotion triggers for BDI agents with achievement goals. *Synthese/KRA 185*(1), 83–129.

Su, K., A. Sattar, H. Lin, and M. Reynolds (2007). A modal logic for beliefs and pro attitudes. In *Proceedings of the 22nd AAAI Conference on Artificial Intelligence (AAAI 2007)*, pp. 496–501.

Vardi, M. Y. (2001). Branching vs. linear time: Final showdown. In *Tools and Algorithms for the Construction and Analysis of Systems*, pp. 1–22. Springer.

Wallace, R. J. (2013). http://en.wikipedia.org/wiki/Practical_reason, retrieved July 2013.

Wooldridge, M. J. (1999). Intelligent agents. In G. Weis (Ed.), *Multiagent Systems*, pp. 27–77. Cambridge, Massachusetts: The MIT Press.

Wooldridge, M. J. (2000). *Reasoning about Rational Agents.* Cambridge, Massachusetts: MIT Press.

Wooldridge, M. J. and N. R. Jennings (Eds.) (1995). *Intelligent Agents.* Berlin: Springer.

Xu, M. (1998). Axioms for deliberative stit. *Journal of Philosophical Logic 27*(5), 505–552.

Chapter 11

Knowledge and Ability

Thomas Ågotnes, Valentin Goranko, Wojciech Jamroga and Michael Wooldridge

Contents

Abstract In this chapter we relate epistemic logics with logics for strategic ability developed and studied in computer science, artificial intelligence and multi-agent systems. We discuss approaches from philosophy and artificial intelligence to modelling the interaction of agents' knowledge and abilities and then focus on concurrent game models and the alternating-time temporal logic ATL. We show how ATL enables reasoning about agents' coalitional abilities to achieve qualitative objectives in concurrent game models, first assuming complete information and then under incomplete information and uncertainty about the structure of the game model. We then discuss epistemic extensions of ATL enabling explicit reasoning about the interaction of knowledge and strategic abilities on different epistemic levels, leading inter alia to the notion of constructive knowledge.

Chapter 11 of the *Handbook of Epistemic Logic*, H. van Ditmarsch, J.Y. Halpern, W. van der Hoek and B. Kooi (eds), College Publications, 2015, pp. 543–589.

Our aim in this chapter is to survey logics that attempt to capture the interplay between knowledge and *ability*. The term "ability", in the sense we mean it in this chapter, corresponds fairly closely to its everyday usage. That is, ability means the capability to do things, and to bring about states of affairs. There are several reasons why ability is worth studying in a formal setting:

- First, the concept of ability is surely worth studying from the perspective of philosophy, and in particular the philosophy of language. In this context, the point is to try to gain an understanding of what people mean when they make statements like "I can X", and in particular to understand what such a claim implies about the mental state of the speaker.
- Second, and more recently, researchers in computer science and artificial intelligence are interested in the notion of *what machines can achieve*. For example, imagine an artificial intelligence to whom we can issue instructions in natural English. Then, if we give the machine an instruction "X" (where X might be "make me a cup of tea" or, famously, "bring me a beer"), then the question of what this instruction means in relation to the abilities of the machine becomes important. An artificial intelligence that is presented with an instruction to X surely needs to understand whether or not it actually can in fact X before it proceeds; this implies some model or theory of ability, and the ability of the artificial intelligence to reason with this model or theory.

The remainder of this chapter is structured as follows. We begin, in the following section, with a discussion on the way that philosophers have considered knowledge and ability. In section 11.2, we move on to discuss how the concept of ability has been considered and formalised in artificial intelligence; we focus in particular on the seminal and enormously influential work of Robert Moore on the relationship between knowledge and ability. We then go on to review more recent contributions to the logic of ability, arising from the computer aided verification community: specifically, Alternating-time Temporal Logic (ATL) and the issues that arise when attempting to integrate ATL with a theory of knowledge.

11.1 Philosophical Treatments of Knowledge and Ability

We will begin by reviewing the way that the concept of ability has been considered in philosophy. We will start with the work of the philosopher

Gilbert Ryle (1900–1976). Ryle was greatly interested in the concept of mind, and one of the questions to which he addressed himself was the distinction between *knowing how* and *knowing that*. Crudely, "knowing that" is the concept of knowledge with which this handbook is largely concerned: we think of "knowing that" as a kind of relation between agents and true propositions, and in this book we write $K_i\varphi$ to mean that agent i knows that φ, where φ is some proposition. The concept of "knowing how" seems related, but is clearly different: it is concerned with the knowledge of how to achieve things.

One question in particular that Ryle considered was whether know-how could be reduced to know-that: that is, whether know-how was in fact just a type of know-that. Ryle argued that such a reduction was not possible. His argument against such a reduction has since become quite celebrated (it is commonly known as "Ryle's regress", and has been applied in various other settings). Broadly, the argument goes as follows:

> If know-how were a species of know-that, then, to engage in any action, one would have to contemplate a proposition. But, the contemplation of a proposition is itself an action, which presumably would itself have to be accompanied by a distinct contemplation of a proposition. If the thesis that knowledge-how is a species of knowledge-that required each manifestation of knowledge-how to be accompanied by a distinct action of contemplating a proposition, which was itself a manifestation of knowledge-how, then no knowledge-how could ever be manifested.

In other words, know-how does not reduce to know-that because if we had to know-that a proposition every time we exercised know-how, we would get an infinite regress of know-thats.

There are other arguments against a reduction of know-how to know-that. For example, consider the following sentence:

$$\text{Michael knows how to ski.} \qquad (*)$$

If know-how reduces to know-that, then there is some proposition, call it φ^*, such that knowing how to ski is equivalent to knowing φ^*. Presumably, telling somebody the proposition φ^* would then be enough to convey the ability to ski, assuming of course the hearer of the message was able to process and understand it. We are trivialising the arguments at stake, of course, but nevertheless, anybody who has learned to ski will recognise that such a simplistic reduction of know-how to know-that is implausible.

The exact relationship of know-how to know-that remains the subject of some debate, with arguments on either side; see, e.g., Stanley and Williamson for a contemporary discussion.

Despite the philosophical objections of Ryle and others, researchers in artificial intelligence have largely adopted the view that know-how can be reduced to know-that. In the section that follows, we will see how researchers in artificial intelligence have attempted to formalise such reductions. Interestingly, with a sufficiently rich logical formalism, the two notions turn out distinct again, as we will argue in Section 11.5.3.

There have been several other early philosophical approaches to developing logics of agency and ability, including works of Georg Henrik von Wright (1916–2003), Stig Kanger (1924–88), Brian Chellas, Mark Brown and Nuel Belnap and Michael Perloff. In particular, Brown proposes a modal logic with non-normal modalities formalising the idea that the modality for ability has a more complex, existential-universal meaning (the agent has *some* action or choice, such that *every* outcome from executing that action (making that choice) achieves the aim), underlying all further approaches to formalizing agents' ability that will be presented here. At about the same time Belnap and Perloff developed the basics of their theory of *seeing to it that*, usually abbreviated to "STIT". The STIT family of logics were formulated in an attempt to be able to give a formal semantics to the concept of agency, where an agent is loosely understood to be an entity that makes deliberate purposeful actions. Although accounts differ on details, the general aim is to be able to make sense of statements such as "the agent i brings it about that φ". We will denote that STIT operator by $\nabla_i \varphi$. The semantics of Belnap and Perloff's operator (technically, known as the *achievement stit* operator), which makes the idea behind Brown's ability operators explicit, was as follows. Formulae of the form $\nabla_i \varphi$ are interpreted in a branching-time model containing histories of the world, each of which contains various "moments". Branching in the model occurs as a result of choices made by agents. Intuitively, in making a particular choice, an agent *constrains* the possible future histories of the world. Roughly speaking, we then say that $\nabla_i \varphi$ is true in a world/moment pair if there was some earlier moment in the history at which point the agent i made a choice such that:

- φ is true in all histories consistent with that choice;
- at the point at which the choice was made, the status of φ was not settled, i.e., there were histories arising from the choice moment in which φ was false.

Expressed differently, $\nabla_i \varphi$ means that the agent i made a choice such that φ was a necessary consequence of this choice, while φ would not necessarily have been true had the agent not made this choice.

Although we have not presented the full formal semantics for Belnap and Perloff's operator, it should be obvious that the operator is semantically rather complex, at least compared to conventional tense and dynamic operators (the semantics constitute a "formidable definition" in the words of Belnap, Perloff, and Xu). The main source of difficulty is that an expression $\nabla_i \varphi$ seems to refer to both the past (the moment at which the relevant choice was made) and the present/future (the moments at which φ is true). Several subsequent proposals have been made to refine and simplify the logic of STIT, with the aim of dealing with some of their counterintuitive properties, and contemporary formulations are much simpler and more intuitive. Here is a list of simple candidate axioms and deduction rules for a logic of agency, due to Troquard:

(M) $\nabla_i(\varphi \wedge \psi) \rightarrow (\nabla_i \varphi \wedge \nabla_i \psi)$

(C) $(\nabla_i \varphi \wedge \nabla_i \psi) \rightarrow \nabla_i(\varphi \wedge \psi)$

(N) $\nabla_i \top$

(No) $\neg \nabla_i \top$

(T) $\nabla_i \varphi \rightarrow \varphi$

(RE) If $\varphi \leftrightarrow \psi$ then $\nabla_i \varphi \rightarrow \nabla_i \psi$.

Taken together, axioms (M) and (C) essentially state that bringing it about is compositional with respect to conjunction: i brings it about that $\varphi \wedge \psi$ iff i brings it about that φ and i it about that ψ. Axioms (N) and (No) are clearly contradictory, and so one can accept at most one of them. Axiom (N) says that agent i brings about anything that is inevitably true, while (No) says that an agent does not bring about necessary truths. Most treatments of STIT reject axiom (N), preferring instead the axiom (No). Axiom (T) essentially states that agents are successful: if i brings it about that φ then φ indeed holds.

STIT logics are closely connected with other logics of ability in multi-agent systems, known as 'alternating-time temporal logics' (ATL), which we will discuss in the remainder of this chapter. In particular, as shown by Broersen, Herzig, and Troquard, STIT logics can be used to encode the logic ATL, as we discuss in section 11.3. However, since STIT logics are somewhat tangential to the main thrust of this article, we will say no more about them here, and simply refer the reader to the bibliographic notes for references to the literature.

11.2 Ability in Artificial Intelligence

In this section, we survey attempts within the artificial intelligence (AI) community to develop logics of ability. Any survey of such work must surely begin with the 1969 paper *Some Philosophical Problems from the Standpoint of Artificial Intelligence* by John McCarthy and Pat Hayes. McCarthy and Hayes start from the following observation:

> We want a computer program that decides what to do by inferring in a formal language [i.e., a logic] that a certain strategy will achieve a certain goal. This requires formalizing concepts of causality, ability, and knowledge.

Much of their article is concerned with speculating about the features that would be required in a logic to be used for decision-making of the type they describe. To illustrate their discussion, they introduced some ideas that have subsequently become more-or-less standard in the AI literature: in particular, the idea of what we now call the *situation calculus* ("sit calc"), a formalism for reasoning about actions in which world states (a.k.a. situations) are introduced as terms into the object language that the program uses to reason about the world. The situation calculus is one of the cornerstone formalisms of AI, and has been developed significantly since McCarthy and Hayes' original work. Although they did not present a formalisation of ability, they did speculate on what such a formalism might look like. They suggested three possible interpretations of what it means for a computer program π to be able to achieve a state of affairs φ:

1. There is a sub-program σ and room for it in memory which would achieve φ if it were in memory, and control were transferred to π. No assertion is made that π knows σ or even knows that σ exists.
2. σ exists as above and that σ will achieve φ follows from information in memory according to a proof that π is capable of checking.
3. π's standard problem-solving procedure will find σ if achieving φ is ever accepted as a subgoal.

The distinction between cases (1) and (2) is that, in the first, ability is seen from the standpoint of an omniscient external observer, who can see that there is some action or procedure such that, if π executed the action or followed the procedure, then the achievement of φ would result, while in the second, there is some potential awareness of this on the part of the program. The third conception implies *practical* ability: not only does the possibility for the program to achieve φ exist, but the program would actually be able to compute and execute an appropriate strategy σ. Considerations such as these are reflected in most subsequent logical treatments of ability.

After the seminal work of McCarthy and Hayes, probably the best-known and most influential studies of ability in AI was Robert Moore's analysis of the relationship between knowledge and ability in a dynamic variant of epistemic logic. Although the intuitions underpinning Moore's account of knowledge and action are intuitively appealing and deeply insightful, the technical details of the actual formalism he used are rather involved. Moore's aim was to develop a formalism that could be used in a classical first-order theorem proving system, of the type that was widely studied in artificial intelligence research at the time. However, he realised the value of using a modal language to express dynamic epistemic properties. His solution was to axiomatise a quantified modal logic using a first-order logic meta-language, in effect, allowing statements in the quantified modal language to be translated into the first-order language. The technical details are along the lines of the "standard translation" of modal logic into first-order logic. The resulting framework is rich and powerful; but it is technically very involved, and not intuitive for people to read. For these reasons, we here present a greatly simplified version of Moore's formalism, which is not intended to be a fully-fledged logic, but a notation for representing the key insights of the original work.

The basic components of the framework are a set of possible worlds (essentially, system states), which we denote by St, and set of actions, which we denote by Act. To keep things simple, we will assume there is just one agent in the system. We write $q \models \varphi$ to mean that the proposition φ is true in state q (we won't worry about the full formal definition of the satisfaction relation $\models$).

Epistemic properties of the system are captured in an epistemic accessibility relation $\sim \subseteq \mathsf{St} \times \mathsf{St}$, with the usual meaning: $q \sim q'$ iff q' and q are indistinguishable to the agent. A unary knowledge modality K is defined in the standard way:

$$q \models K\varphi \text{ iff } \forall q' : q \sim q' \text{ implies } q' \models \varphi.$$

As an aside, readers familiar with S5 treatments of knowledge might be interested to hear that Moore only required epistemic accessibility relations $\sim$ to be reflexive and transitive, giving an epistemic logic KT4 = S4. In this respect, his treatment of knowledge was slightly unorthodox, in that he rejected the negative introspection axiom. However, the omission of negative introspection plays no significant part in his formalisation of ability.

To capture the effects of actions, Moore used a ternary relation $R \subseteq \mathsf{Act} \times \mathsf{St} \times \mathsf{St}$, where $R(a, q, q')$ means that q' results from performing a in when in state q. Moore assumed actions were deterministic, so that only one state could result from the performance of an action in a given state.

Moore then introduced a modal operator $(\mathsf{Res}\ a\ \varphi)$ to mean that *after action a is performed, φ will be true* – that is, φ is a "result" of performing action a. The truth condition for this operator is as follows:

$$q \models (\mathsf{Res}\ a\ \varphi) \quad \text{iff} \quad \exists q' \cdot R(a, q, q') \text{ and } q' \models \varphi$$

Thus, Moore's expression $(\mathsf{Res}\ a\ \varphi)$ is similar to the dynamic logic expression $\langle a \rangle \varphi$; and since Moore assumed actions are deterministic, it is in turn closely related to the dynamic logic expression $[a]\varphi$.

Quantification plays an important role in Moore's logic of ability, and in particular, the theory relies on some important properties of *quantifying in* to modal knowledge contexts. To illustrate the issues, we assume that we have in our logic the technical machinery of first-order quantification. Now, consider the distinction between the following two quantified epistemic formulae, where $\mathit{Murderer}(x)$ is intended to mean that the individual denoted by x is a murderer:

$$K(\exists x : \mathit{Murderer}(x)) \tag{11.1}$$

$$\exists x : (K\ \mathit{Murderer}(x)) \tag{11.2}$$

The first formula is said to be a *de dicto* formula, while the second is a *de re* formula. These two formulae do not express the same properties. The first formula asserts that, in every world consistent with the agent's knowledge, the formula $\exists x : \mathit{Murderer}(x)$ is true. Observe that, in one such world, the individual x could be Alice, while in another such world, the individual x could be Bob. In this case, the identity of the murderer is different in different epistemic alternatives for the agent. Thus, (11.1) asserts that the agent knows somebody is a Murderer, *but does not imply that the agent knows the identity of the individual in question.*

In contrast, (11.2) asserts something stronger. It says that there is some individual x such that in all epistemic alternatives for the agent, x is a Murderer. This time, the value of x is fixed across all epistemic alternatives for the agent, and so *the agent knows the identity of the individual in question.*

This distinction – between knowing that some x has a property, and knowing the identity of the x that has the property – is, as we will see, important in Moore's theory of ability.

With these concepts in place, we can now turn to Moore's formalisation of ability. He was concerned with developing a theory of ability that would capture the following two aspects of the interaction between knowledge and action:

1. As a result of performing an action, an agent can gain knowledge, and in particular, agents can perform "test" actions, in order to find things out.

2. In order to perform some actions, an agent needs knowledge: these are *knowledge pre-conditions.* For example, in order to open a safe, it is necessary to know the combination.

We will develop Moore's model in two stages. The aim is ultimately to define a unary operator $(\mathsf{Can}\ \varphi)$, which is intended to mean that the agent has the ability to achieve φ. First, we will see a definition that captures some important intuitions, but which fails to capture some other of our intuitions about ability. Then, we will adapt our definition to rectify these problems. Here is our first attempt to define Can:

$$(\mathsf{Can}\ \varphi) \quad \leftrightarrow \quad (\exists a : K(\mathsf{Res}\ a\ \varphi))$$

This definition says that the agent can φ if there exists some action a such that the agent knows that φ will result from the performance of a. Notice that the variable a denoting the relevant action is quantified *de re*, and so this definition implies that the agent *knows the identity of the action* that results in the achievement of φ. To see why this is important, suppose Can had instead been defined as follows:

$$(\mathsf{Can}\ \varphi) \quad \leftrightarrow \quad K(\exists a : (\mathsf{Res}\ a\ \varphi))$$

In this *de dicto* definition of ability, in every epistemic alternative there is some action that will result in φ, but the identity of the action may be different in different epistemic alternatives: but then, since the agent is uncertain about which of its epistemic alternatives is the actual world, which of these actions should it perform? While the *de dicto* definition of ability tells us something, it does not seem strong enough to serve as a practical definition of ability.

So, according to our first definition, the agent can achieve φ if there exists some action a, such that the agent knows the identity of a, and that the result of performing a is φ. Moore pointed out, however, that while this definition certainly embodies some intuition of value in understanding ability, it suffers from one major drawback: the agent is required to know the identity of the whole of the action required to achieve the desired state. In everyday usage, this seems much too strong a requirement. For example, we might say to a friend "I can be at your house at 8pm", being entirely comfortable about the truth of this statement, without knowing in advance exactly what action will be used or required to achieve it. Moore therefore proposed an adapted version of the definition, as follows:

$$
\begin{aligned}
&(\mathsf{Can}\ \varphi) \leftrightarrow \\
&\qquad \exists a.K(\mathsf{Res}\ a\ \varphi) \vee \\
&\qquad \exists a.K(\mathsf{Res}\ a\ (\mathsf{Can}\ \varphi))
\end{aligned}
$$

Thus, an agent has the ability to achieve φ if either:

1. it knows the identity of an action such that it knows that after this action is performed, φ will hold; or

2. it knows the identity of an action such that it knows that after this action is performed, $(\mathsf{Can}\ \varphi)$ will hold.

Thus, the second case allows for the possibility of performing an action that will furnish the agent with the capability to achieve φ.

Moore's formalism was enormously influential in the AI community. For example, Douglas Appelt used an enriched version of Moore's formalism directly in a theorem-proving system for planning natural language utterances. Many other researchers used ideas from Moore's work to formalise ability; of particular lasting significance has been the idea of an agent requiring *de re* knowledge of an action. In this chapter, we focus on a somewhat more complex notion of ability, that looks at *strategies*, i.e., conditional plans rather than simple actions. Still, the issue to what extent the agent (or group of agents) know the right strategy, and the difference between knowing a good strategy *de re* vs. *de dicto* is a central one. We demonstrate it in Section 11.5.

11.3 Logics for Abilities of Agents and Coalitions

In this section we present semantic models of game-like interaction in multi-agent systems, as well as two basic logics for reasoning about such models, namely *Coalition Logic* (CL) and *Alternating-time Temporal Logic* (ATL). We begin with an example leading to the concept of concurrent game model.

Example 11.1 (Shared file updates)
Two agents share a file in the cyberspace. The file can be updated only at designated moments (e.g., only on the hour). Agents can try to *update* the file (action U) or *do nothing* (action N). However, updating is disabled in some states. Both agents act simultaneously. Initially (at state E), both agents are enabled to apply U. If at any moment both agents try to update simultaneously, there is a conflict, and as a consequence the file is locked forever. If one agent tries to update while the other applies N, the file gets updated by the updating agent i (resulting in state U_i). Also, as a very

simple form of fairness, we assume that no agent can update the file twice in a row. That is, if agent i applies again U before the other has applied U, then the process goes to a state D_i where i is disabled from updating, and can only apply N. At that state the other agent can enable i to apply further updates by applying N, too, as the procedure then goes back to state U_i, for $i = 1, 2$; alternatively, she can decide to apply U herself. If both agents apply N at any state but D_i, then nothing happens. After, if ever, the other agent applies U at state U_i or D_i, while agent i applies N the procedure goes to state P ("the file has been *processed*") where both agents are disabled from making further updates. From that state, either nothing happens, if both agents apply N, or agent 1 (and only agent 1) can reset the process to state E, by applying action R (*reset*). ⊣

The example above describes a procedure formalisable as a transition system. At each state of that system every agent chooses one of its available actions, and the resulting tuple of actions is executed synchronously. The combination of actions determines the transition to a successor state, and so on, ad infinitum. Below we define formally such multi-agent transition systems.

11.3.1 Concurrent Game Models

Definition 11.1 (Concurrent game structures and models)
A *concurrent game structure* (CGS) is a tuple $\mathcal{S} = (\mathsf{Ag}, \mathsf{St}, \mathsf{Act}, \mathsf{act}, \mathsf{out})$ which consists of:

- a finite, non-empty set of *players* or *agents*[1] $\mathsf{Ag} = \{1, \ldots, \mathsf{k}\}$; the subsets of Ag are often called *coalitions*;

- a non-empty set of *states* (*game positions*) St;

- a non-empty set of *actions*, or *moves* Act;

- an *action manager* function $\mathsf{act} : \mathsf{Ag} \times \mathsf{St} \rightarrow \mathcal{P}(\mathsf{Act})$ assigning to every player a and a state q a non-empty set of actions available for execution by a at the state q.
  An *action profile* is a tuple of actions $\alpha = \langle \alpha_1, \ldots, \alpha_{\mathsf{k}} \rangle \in \mathsf{Act}^k$. The action profile is *executable at the state* q if $\alpha_{\mathsf{i}} \in \mathsf{act}(\mathsf{i}, q)$ for every $\mathsf{i} \in \mathsf{Ag}$. We denote by $\mathsf{act}(q)$ the subset of $\prod_{a \in \mathsf{Ag}} \mathsf{act}(a, q)$ consisting of of all action profiles executable at the state q.

- a *transition function* out that assigns a unique *outcome state* $\mathsf{out}(q, \alpha)$ to every state q and every action profile α which is executable at q.

[1] We use the terms 'agent' and 'player' as synonyms.

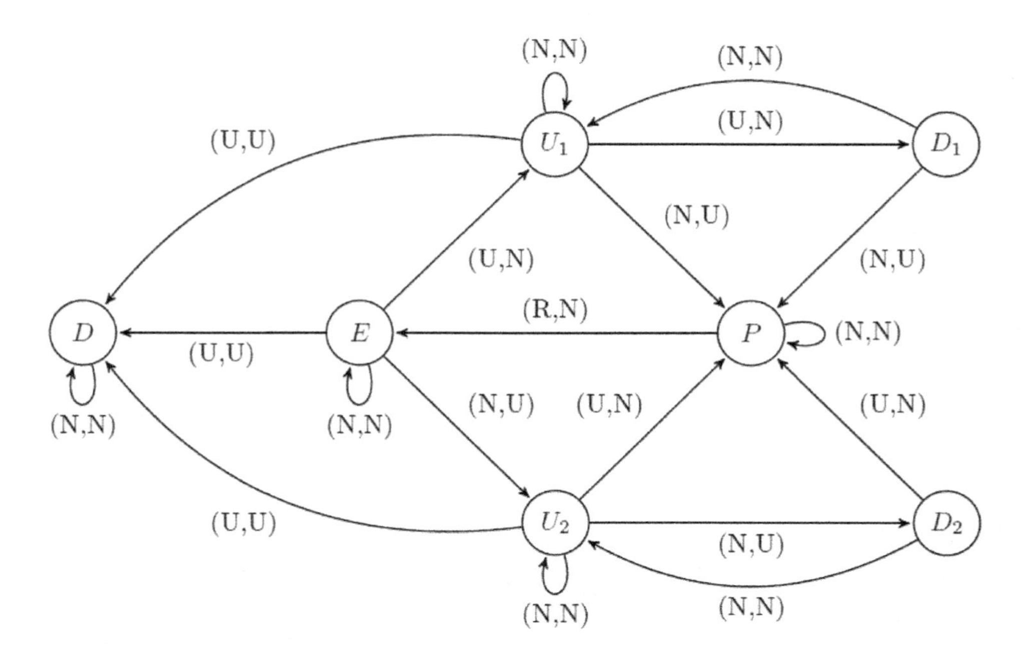

Figure 11.1: A concurrent game model encoding the shared file updating story.

A *concurrent game model* (CGM) is a CGS endowed with a labeling $L : \mathsf{St} \to \mathcal{P}(\text{PROP})$ of the states with sets of atomic propositions from a fixed set PROP. As usual, that labeling describes which atomic propositions are true at a given state. ⊣

Example 11.2 (Shared file updates as a concurrent game model)
The procedure described in Example 11.1 is encoded in the concurrent game model on Figure 11.1. Each transition is labelled by a pair of actions (the action of agent 1 and the action of agent 2), and the atomic propositions are the same as the names of states where they hold. ⊣

11.3.2 Plays and strategies

A *play* in a CGS/CGM $\mathcal{M}$ is just a path in $\mathcal{M}$, that is, an infinite sequence of states that can result from subsequent transitions in $\mathcal{M}$ obtained by joint moves of all players. The formal definitions follow below.

Definition 11.2 (Plays and histories)
We fix a concurrent game model $\mathcal{S} = (\mathsf{Ag}, \mathsf{St}, \mathsf{Act}, \mathsf{act}, \mathsf{out}, L)$. For a play λ and positions $i, j \geq 0$, we use $\lambda[i]$, $\lambda[j, i]$, and $\lambda[j, \infty)$ to denote the ith state

of λ, the finite segment $q_j, q_{j+1} \dots, q_i$, and the tail subsequence $q_j, q_{j+1}, \dots$ of λ, respectively. A play with $\lambda[0] = q$ will be called an *q-play*.

The finite initial segments $\lambda[0, i]$ of plays λ will be called *histories* in $\mathcal{S}$; a typical history will be denoted by h and its length by $| h |$. $\dashv$

Definition 11.3 (Strategies)
A *positional (aka. memoryless) strategy* in $\mathcal{S}$ for a player $\mathsf{a} \in \mathsf{Ag}$ is a function $s_{\mathsf{a}} : \mathsf{St} \to \mathsf{Act}$, such that $s_{\mathsf{a}}(q) \in \mathsf{act}(\mathsf{a}, q)$. A positional strategy in $\mathcal{S}$ for a coalition $C \subseteq \mathsf{Ag}$ is a tuple of positional strategies, one for each player in C.

A *perfect recall strategy* in $\mathcal{S}$ for the player a is a function $s_{\mathsf{a}} : \mathsf{St}^+ \to \mathsf{Act}$ such that $s_{\mathsf{a}}(\langle \dots, q \rangle) \in \mathsf{act}(\mathsf{a}, q)$, where St^+ is the set of all histories in $\mathcal{S}$. A perfect recall strategy in $\mathcal{S}$ for the coalition C is a tuple of perfect recall strategies, one for each player in C. $\dashv$

Clearly, every positional strategy can be seen as a perfect recall strategy.

Definition 11.4 (Coalitional actions)
Let $q \in \mathsf{St}$ and let $C \subseteq \mathsf{Ag}$. We denote the complement $\mathsf{Ag} \setminus C$ by $\overline{C}$.

- A *C-action* at the state q is a tuple α_C such that $\alpha_C(\mathsf{i}) \in \mathsf{act}(\mathsf{i}, q)$ for every $\mathsf{i} \in C$ and $\alpha_C(\mathsf{j}) = \sharp_{\mathsf{j}}$ for every $\mathsf{j} \notin C$, where $\sharp_{\mathsf{j}}$ is a fixed symbol used as a placeholder for an arbitrary action of player j.

  We denote by $\mathsf{act}(C, q)$ the set of all C-actions at state q.

  Alternatively, C-actions at q can be defined as equivalence classes on the set of all action profiles at q, where each equivalence class is determined by the choices of actions of the players in C.

- An *action profile* $\alpha \in \mathsf{act}(q)$ *extends a C-action* α_C, denoted $\alpha_C \sqsubseteq \alpha$, if $\alpha(\mathsf{i}) = \alpha_C(\mathsf{i})$ for every $\mathsf{i} \in C$.

- Given a C-action $\alpha_C \in \mathsf{act}(C, q)$ and a $\overline{C}$-action $\alpha_{\overline{C}} \in \mathsf{act}(\overline{C}, q)$, we denote by $\alpha_C \oplus \alpha_{\overline{C}}$ the unique action profile $\alpha \in \mathsf{act}(q)$ such that both $\alpha_C \sqsubseteq \alpha$ and $\alpha_{\overline{C}} \sqsubseteq \alpha$. $\dashv$

Definition 11.5 (Outcome sets)
Let $q \in \mathsf{St}$, $C \subseteq \mathsf{Ag}$, and $\alpha_C \in \mathsf{act}(C, q)$.

The *outcome set* of the C-action α_C at q is the set of states

$$\mathsf{out\_set}(q, \alpha_C) := \{\mathsf{out}(q, \alpha) \mid \alpha \in \mathsf{act}(q) \text{ and } \alpha_C \sqsubseteq \alpha\} \qquad \dashv$$

The outcome set function $\mathsf{out\_set}$ can be naturally extended to act not just on joint actions but on all joint strategies applied at a given state (resp., history) in a given CGM. Then $\mathsf{out\_set}(q, s_C)$ (resp., $\mathsf{out\_set}(h, s_C)$) returns the set of all possible successor states that can result from applying a given positional (resp., perfect recall) joint strategy s_C of the coalition C at the state q (resp., the history h).

Further, we extend the function out_set to out_plays that returns the set of all *plays* which can be realized when the players in C follow the strategy s_C from a given state q in $\mathcal{S}$ onward. Formally, it is defined as:

For memoryless strategies,

$$\mathsf{out\_plays}(q, s_C) := \\ \{\lambda \in \mathsf{St}^\omega \mid \lambda[0] = q \text{ and } \lambda[j+1] \in \mathsf{out\_set}(\lambda[j], s_C) \text{ for each } j \in \mathbb{N}\}$$

Respectively, for perfect recall strategies,

$$\mathsf{out\_plays}(q, s_C) := \\ \{\lambda \in \mathsf{St}^\omega \mid \lambda[0] = q \text{ and } \lambda[j+1] \in \mathsf{out\_set}(\lambda[0, j], s_C) \text{ for each } j \in \mathbb{N}\}$$

A fundamental question about a multi-player game is: *What can a given player or coalition achieve in the game?* So far the objectives of players and coalitions have not been formally specified. A typical objective would be to reach a state satisfying a given property expressed in terms of the atomic propositions, e.g. a *winning* state. Generally, an objective is a property of plays, e.g. one can talk about winning or losing plays for the given player or coalition. More precisely, if the current state of the game is q, we say that a coalition of players C can bring about (or enforce) an objective φ from that state if there is a joint strategy s_C for C such that every play from $\mathsf{out\_plays}(q, s_C)$ satisfies the objective φ. We will now introduce suitable logics that can be used to formally specify strategic objectives of players and coalitions and to formally determine their ability to bring about such objectives.

11.3.3 Expressing Local Coalitional Powers: Coalition Logic

The power of a coalition to enforce the outcome to be in one or another designated outcome set naturally corresponds to a *non-normal modal operator with monotone neighbourhood semantics.* This observation was formalized by the *Coalition Logic (CL)* introduced by Marc Pauly as a multi-modal logic capturing coalitional abilities in strategic games. It extends classical propositional logic with a family of modal operators $\langle\!\langle C \rangle\!\rangle$ indexed with the subsets (coalitions) of the set of agents Ag. Formulas of CL are defined recursively as follows:

$$\varphi := p \mid \neg\varphi \mid \varphi \vee \psi \mid \langle\!\langle A \rangle\!\rangle \varphi.$$

where $p \in \Pi$ is a proposition, and $A \subseteq \mathsf{Ag}$ is a set (coalition) of agents.

Intuitively, the formula $\langle\langle C\rangle\rangle\varphi$ says that *the coalition C has, at the given game state, the power to guarantee an outcome satisfying φ*. Formally, one can interpret the operator $\langle\langle C\rangle\rangle$ in a state q of concurrent game model $\mathcal{M}$ as follows:

$$\mathcal{M}, q \models \langle\langle C\rangle\rangle\varphi \text{ iff } \mathsf{out\_set}(q, \alpha_C) \subseteq [\![\varphi]\!]_{\mathcal{M}} \text{ for some } \alpha_C \in \mathsf{act}(C, q),$$

where $[\![\varphi]\!]_{\mathcal{M}} := \{q \in \mathsf{St} \mid \mathcal{M}, q \models \varphi\}$.

Example 11.3 (Expressing properties in CL)
Here are some properties expressed in CL. The reader might ponder each of them whether it should be satisfiable, valid, or neither.

1. *"If Player 1 has an action to guarantee a winning successor state, then Player 2 cannot prevent reaching a winning successor state."*

$$\langle\langle 1\rangle\rangle\, \mathsf{win}_1 \rightarrow \neg\langle\langle 2\rangle\rangle\neg\mathsf{win}_1$$

2. *"Player* a *has an action to guarantee a successor state where he is rich, and has an action to guarantee a successor state where he is happy, but does not have an action to guarantee a successor state where he is both rich and happy."*

$$\langle\langle \mathsf{a}\rangle\rangle\mathsf{rich} \wedge \langle\langle \mathsf{a}\rangle\rangle\mathsf{happy} \wedge \neg\langle\langle \mathsf{a}\rangle\rangle(\mathsf{rich} \wedge \mathsf{happy})$$

3. *"None of the players 1 and 2 has an action ensuring an outcome state satisfying* goal*; but they both have a collective action ensuring such an outcome state ."*

$$\neg\langle\langle 1\rangle\rangle\mathsf{goal} \wedge \neg\langle\langle 2\rangle\rangle\mathsf{goal} \wedge \langle\langle 1, 2\rangle\rangle\mathsf{goal} \qquad \dashv$$

Pauly has shown that the following is a sound and complete axiomatization of CL over the classical propositional logic:

1. $\langle\langle \mathsf{Ag}\rangle\rangle\top$
2. $\neg\langle\langle C\rangle\rangle\bot$ for any coalition $C \subseteq \mathsf{Ag}$
3. $\neg\langle\langle \emptyset\rangle\rangle\varphi \rightarrow \langle\langle \mathsf{Ag}\rangle\rangle\neg\varphi$
4. $\langle\langle C\rangle\rangle\varphi \wedge \langle\langle D\rangle\rangle\psi \rightarrow \langle\langle C \cup D\rangle\rangle(\varphi \wedge \psi)$ for any disjoint $C, D \subseteq \mathsf{Ag}$,

plus the inference rule Monotonicity:

$$\frac{\varphi \rightarrow \psi}{\langle\langle C\rangle\rangle\varphi \rightarrow \langle\langle C\rangle\rangle\psi}$$

11.3.4 Alternating-time temporal logics

Coalition logic is a very natural language to express *local* strategic abilities of players and coalitions, that is, their powers to guarantee desired properties in the *successor states*. However, it is not expressive enough for reasoning about *long term* strategic abilities. A suitable logic for that purpose is Alternating-time Temporal Logic ATL*, widely used for specifying and reasoning about temporal properties of plays in multiagent systems that different players and coalitions can enforce by adopting suitable strategies.

Syntax of ATL* and ATL

The language of ATL* extends that of the Computation Tree Logic CTL* with *strategic path quantifiers* indexed by all subsets of a fixed finite non-empty set Ag of *agents* (or *players*).

Definition 11.6 (ATL*-formulae)
The formulae of ATL* are defined for the fixed set of agents Ag and a fixed countably infinite set PROP of atomic propositions, recursively as follows:

$$\varphi := p \mid \neg\varphi \mid (\varphi \wedge \varphi) \mid \mathsf{X}\varphi \mid \mathsf{G}\varphi \mid (\varphi \mathsf{U} \varphi) \mid \langle\!\langle C \rangle\!\rangle \varphi,$$

where p ranges over PROP and C ranges over $\mathcal{P}(\mathsf{Ag})$. ⊣

The intuitive interpretation of $\langle\!\langle C \rangle\!\rangle \varphi$ is: "*The coalition C has a joint strategy that ensures that every path enabled by that strategy satisfies φ*", and the temporal operators have the usual interpretation: X stands for "next", G for "always from now on", and U for "until". The other boolean connectives and the propositional constants $\top$ ("truth") and $\bot$ ("falsum") are defined as usual. Just like in LTL and CTL*, $\mathsf{F}\varphi$ ("eventually") can be defined as $\top\mathsf{U}\varphi$. We will also use dual strategic path quantifiers $[\![C]\!]$ defined as

$$[\![C]\!]\,\varphi := \neg\langle\!\langle C \rangle\!\rangle \neg\varphi,$$

with the intended interpretation *"the coalition C cannot prevent an outcome satisfying φ"*.

As in CTL*, the formulae of ATL* can be classified in two sorts: *state formulae*, that are evaluated at game states, and *path formulae*, that are evaluated on game plays. These are respectively defined by the following grammars, where $C \subseteq \mathsf{Ag}, p \in \mathrm{PROP}$:

State formulae: $\varphi ::= p \mid \neg\varphi \mid (\varphi \wedge \varphi) \mid \langle\!\langle C \rangle\!\rangle \gamma$,
Path formulae: $\gamma ::= \varphi \mid \neg\gamma \mid (\gamma \wedge \gamma) \mid \mathsf{X}\gamma \mid \mathsf{G}\gamma \mid (\gamma \mathsf{U} \gamma)$.

By analogy with CTL, we can now define the fragment ATL consisting of the 'purely-state-formulae' of ATL*, i.e., those ATL* formulae where every temporal operator is in the immediate scope of a strategic path quantifier. Formally:

Definition 11.7
The formulae of ATL are defined recursively as follows:

$$\varphi := p \mid \neg\varphi \mid (\varphi \wedge \varphi) \mid \langle\langle C\rangle\rangle \mathsf{X}\varphi \mid \langle\langle C\rangle\rangle \mathsf{G}\varphi \mid \langle\langle C\rangle\rangle (\varphi \mathsf{U} \varphi),$$

where p ranges over PROP and C ranges over $\mathcal{P}(\mathsf{Ag})$. ⊣

ATL operators can be intuitively interpreted as follows:

- $\langle\langle C\rangle\rangle \mathsf{X}\varphi$: 'The coalition C *has a joint action that ensures* φ *in the next moment (state)*';
- $\langle\langle C\rangle\rangle \mathsf{G}\varphi$: 'The coalition C *has a joint strategy to maintain forever outcomes satisfying* φ';
- $\langle\langle C\rangle\rangle \psi\mathcal{U}\varphi$: 'The coalition C *has a joint strategy to eventually reach an outcome satisfying* φ*, while meanwhile maintaining the truth of* ψ'.

Example 11.4 (Expressing properties in ATL*)
Here are some properties expressed in ATL* (actually, all but the last one, are formulae in ATL).

- *If the system has a strategy to eventually reach a safe state, then the environment cannot prevent it from reaching a safe state*:

$$\langle\langle system\rangle\rangle \mathsf{F}\,\mathsf{safe} \rightarrow [[env]]\,\mathsf{F}\,\mathsf{safe}$$

- *If the system has a strategy to stay in a safe state forever and has a strategy to eventually achieve its goal, then it has a strategy to stay in a safe state until it achieves its goal*:

$$(\langle\langle system\rangle\rangle \mathsf{G}\,\mathsf{safe} \wedge \langle\langle system\rangle\rangle \mathsf{F}\,\mathsf{goal}) \rightarrow \langle\langle system\rangle\rangle (\mathsf{safe}\,\mathsf{U}\,\mathsf{goal})$$

- *The coalition C has a joint action to ensure that the coalition B cannot prevent C from eventually winning*:

$$\langle\langle C\rangle\rangle\, [\![B]\!]\, \mathsf{F}\,\mathsf{win}_C.$$

⊣

Semantics of ATL*

Definition 11.8 (Semantics of ATL* on concurrent game models) Let $\mathcal{M} = (\mathsf{Ag}, \mathsf{St}, \mathsf{Act}, \mathsf{act}, \mathsf{out}, L)$ be a concurrent game model over a fixed set of atomic propositions PROP. The truth of ATL*-formulae is defined by mutual induction on state and path formulae as follows:

1. For state formulae, at a state $q \in \mathsf{St}$:
   - $\mathcal{M}, q \models p$ iff $p \in L(q)$, for all $p \in \text{PROP}$;
   - $\mathcal{M}, q \models \neg\varphi$ iff $\mathcal{M}, q \not\models \varphi$;
   - $\mathcal{M}, q \models \varphi \wedge \psi$ iff $\mathcal{M}, q \models \varphi$ and $\mathcal{M}, q \models \psi$;
   - $\mathcal{M}, q \models \langle\!\langle C \rangle\!\rangle\varphi$ iff there exists a perfect recall C-strategy s_C such that $\mathcal{M}, \lambda \models \varphi$ holds for all $\lambda \in \mathsf{out\_plays}(q, s_C)$.
2. For path formulae, at a path $\lambda \in \mathsf{St}^\omega$:
   - $\mathcal{M}, \lambda \models \varphi$ iff $\mathcal{M}, \lambda[0] \models \varphi$, for every state formula φ;
   - $\mathcal{M}, \lambda \models \neg\gamma$ iff $\mathcal{M}, \lambda \not\models \gamma$;
   - $\mathcal{M}, q \models \gamma_1 \wedge \gamma_2$ iff $\mathcal{M}, q \models \gamma_1$ and $\mathcal{M}, q \models \gamma_2$;
   - $\mathcal{M}, \lambda \models \mathsf{X}\gamma$ iff $\mathcal{M}, \lambda[1, \infty) \models \gamma$;
   - $\mathcal{M}, \lambda \models \mathsf{G}\gamma$ iff $\mathcal{M}, \lambda[i, \infty) \models \gamma$ holds for all positions $i \geq 0$;
   - $\mathcal{M}, q \models \gamma_1 \mathsf{U} \gamma_2$ iff for some position $i \geq 0$ both $\mathcal{M}, \lambda[i, \infty) \models \gamma_2$ and $\mathcal{M}, \lambda[j, \infty) \models \gamma_1$ hold for all positions $0 \leq j < i$. ⊣

Because ATL only involves state formulae, the memory used in the strategies required to satisfy $\langle\!\langle C \rangle\!\rangle$-formulae turns out redundant and these can be replaced with positional strategies. Formally, let us denote by $\models^p$ the truth definition for ATL*-formulae given above, where perfect recall strategies are replaced by positional strategies throughout. Then, the following holds:

> For every concurrent game model $\mathcal{M}$, ATL-formula φ and a state $q \in \mathcal{M}$, we have that $\mathcal{M}, q \models \varphi$ iff $\mathcal{M}, q \models^p \varphi$.[2]

Thus, we have the following equivalent, but more explicit truth definition for ATL-formulae.

[2] Note that the analogous claim does *not* hold for the broader language of ATL*. To see this, consider the last formula in Example 11.5, which holds according to the perfect recall semantics $\models$ but not according to the positional semantics $\models^p$.

Definition 11.9 (Truth of ATL-formulae)
Let $\mathcal{M} = (\mathsf{Ag}, \mathsf{St}, \mathsf{Act}, \mathsf{act}, \mathsf{out}, L)$ be a concurrent game model over a fixed set of atomic propositions PROP. The truth of ATL-formulae at a state $q \in \mathsf{St}$ is defined inductively as follows:

- $\mathcal{M}, q \models p$ iff $p \in L(q)$, for all $p \in \text{PROP}$;
- $\mathcal{M}, q \models \neg\varphi$ iff $\mathcal{M}, q \not\models \varphi$;
- $\mathcal{M}, q \models \varphi \wedge \psi$ iff $\mathcal{M}, q \models \varphi$ and $\mathcal{M}, q \models \psi$;
- $\mathcal{M}, q \models \langle\!\langle C \rangle\!\rangle \mathsf{X}\varphi$ iff there exists a C-action $\alpha_C \in \mathsf{act}(C, q)$ such that $\mathcal{M}, q' \models \varphi$ for all $q' \in \mathsf{out\_set}(q, \alpha_C)$;
- $\mathcal{M}, q \models \langle\!\langle C \rangle\!\rangle \mathsf{G}\varphi$ iff there exists a positional C-strategy s_C such that, for all $\lambda \in \mathsf{out\_plays}(q, s_C)$, $\mathcal{M}, \lambda[i] \models \varphi$ holds for all positions $i \geq 0$;
- $\mathcal{M}, q \models \langle\!\langle C \rangle\!\rangle \varphi \mathsf{U} \psi$ iff there exists a positional C-strategy s_C such that, for all $\lambda \in \mathsf{out\_plays}(q, s_C)$, there exists a position $i \geq 0$ such that $\mathcal{M}, \lambda[i] \models \psi$ and $\mathcal{M}, \lambda[j] \models \varphi$ holds for all positions $0 \leq j < i$. ⊣

While ATL and CL were developed independently, it is easy to see that CL is equivalent to the next-time fragment of ATL. That is, we have $\mathcal{M}, q \models_{CL} \langle\!\langle C \rangle\!\rangle \varphi$ iff $\mathcal{M}, q \models_{ATL} \langle\!\langle C \rangle\!\rangle \mathsf{X}\varphi$, for every $\mathcal{M}, q, C$, and φ.

Example 11.5 (Properties of shared file updates)
Consider the shared file updates in Example 11.1. Denote the model by $\mathcal{M}_1$. With a slight abuse we use the same notation for states and for the atomic propositions (nominals) that identify them completely. We leave to the reader to verify the following.

- $\mathcal{M}_1, E \models \neg\langle\!\langle 1 \rangle\!\rangle \mathsf{X} D \wedge \langle\!\langle 1 \rangle\!\rangle \mathsf{X} \neg D \wedge \langle\!\langle 1 \rangle\!\rangle \mathsf{X}(E \vee U_2) \wedge \neg\langle\!\langle 2 \rangle\!\rangle \mathsf{X}(E \vee U_2)$
- $\mathcal{M}_1, E \models \neg\langle\!\langle 1 \rangle\!\rangle \mathsf{F} D \wedge \neg\langle\!\langle 2 \rangle\!\rangle \mathsf{F} D \wedge \langle\!\langle 1 \rangle\!\rangle \mathsf{G} \neg D \wedge \langle\!\langle 2 \rangle\!\rangle \mathsf{G} \neg D$
- $\mathcal{M}_1, U_1 \models \neg\langle\!\langle 1 \rangle\!\rangle \mathsf{F} P \wedge \langle\!\langle 1, 2 \rangle\!\rangle \mathsf{GF} P$
- $\mathcal{M}_1, U_2 \models \langle\!\langle 1, 2 \rangle\!\rangle (\neg D_2)\, \mathsf{U}\, U_1 \wedge \neg\langle\!\langle 1 \rangle\!\rangle (\neg D_2)\, \mathsf{U}\, U_1$
- $\mathcal{M}_1, E \models \langle\!\langle 1, 2 \rangle\!\rangle \mathsf{F}(P \wedge \mathsf{X}(E \wedge \mathsf{G}\neg P))$. NB: the truth of this formula requires a strategy for 1 and 2 using memory. ⊣

Validity and Satisfiability in ATL

Definition 11.10
Let θ be an ATL-formula and Γ be a set of ATL-formulae.

- θ is *satisfiable* if $\mathcal{M}, q \models \theta$ for some CGM $\mathcal{M}$ and a state $q \in \mathcal{M}$;
- θ is *valid* if $\mathcal{M}, q \models \theta$ for every CGM $\mathcal{M}$ and every state $q \in \mathcal{M}$.

Satisfiability and validity of Γ are defined similarly. ⊣

Pauly's axioms for Coalition Logic, that we presented in Section 11.3.3, can be rephrased in ATL to provide validities of one-step ability:

($\top$) $\langle\langle \mathsf{Ag} \rangle\rangle \mathsf{X} \top$

($\bot$) $\neg \langle\langle C \rangle\rangle \mathsf{X} \bot$

(Ag) $\neg \langle\langle \emptyset \rangle\rangle \mathsf{X} \varphi \rightarrow \langle\langle \mathsf{Ag} \rangle\rangle \mathsf{X} \neg \varphi$

(Sup) $\langle\langle C \rangle\rangle \mathsf{X} \varphi \wedge \langle\langle D \rangle\rangle \mathsf{X} \psi \rightarrow \langle\langle C \cup D \rangle\rangle \mathsf{X}(\varphi \wedge \psi)$ for any disjoint $C, D \subseteq \mathsf{Ag}$.

Long-term abilities satisfy the following validities in ATL that define operators $\langle\langle C \rangle\rangle \mathsf{X}$, $\langle\langle C \rangle\rangle \mathsf{G}$, $\langle\langle C \rangle\rangle \mathsf{U}$ recursively as fixpoints of certain monotone operators:

(FP$_\mathsf{G}$) $\langle\langle C \rangle\rangle \mathsf{G} \varphi \leftrightarrow \varphi \wedge \langle\langle C \rangle\rangle \mathsf{X} \langle\langle C \rangle\rangle \mathsf{G} \varphi$

(GFP$_\mathsf{G}$) $\langle\langle \emptyset \rangle\rangle \mathsf{G}(\theta \rightarrow (\varphi \wedge \langle\langle C \rangle\rangle \mathsf{X} \theta)) \rightarrow \langle\langle \emptyset \rangle\rangle \mathsf{G}(\theta \rightarrow \langle\langle C \rangle\rangle \mathsf{G} \varphi)$

(FP$_\mathcal{U}$) $\langle\langle C \rangle\rangle \psi \mathsf{U} \varphi \leftrightarrow \varphi \vee (\psi \wedge \langle\langle C \rangle\rangle \mathsf{X} \langle\langle C \rangle\rangle \psi \mathsf{U} \varphi)$

(LFP$_\mathcal{U}$) $\langle\langle \emptyset \rangle\rangle \mathsf{G}((\varphi \vee (\psi \wedge \langle\langle C \rangle\rangle \mathsf{X} \theta)) \rightarrow \theta) \rightarrow \langle\langle \emptyset \rangle\rangle \mathsf{G}(\langle\langle C \rangle\rangle \psi \mathsf{U} \varphi \rightarrow \theta)$.

Axioms ($\top$)–**(LFP$_\mathcal{U}$)** plus the inference rules of $\langle\langle C \rangle\rangle \mathsf{X}$-*Monotonicity* and $\langle\langle \emptyset \rangle\rangle \mathsf{G}$-*Necessitation:*

$$\frac{\varphi \rightarrow \psi}{\langle\langle C \rangle\rangle \mathsf{X} \varphi \rightarrow \langle\langle C \rangle\rangle \mathsf{X} \psi} \qquad \qquad \frac{\varphi}{\langle\langle \emptyset \rangle\rangle \mathsf{G} \varphi}$$

provide a sound and complete axiomatization for the validities of ATL. Furthermore, the satisfiability in ATL is decidable and EXPTIME-complete. The whole ATL* is 2EXPTIME-complete, but no explicit complete axiomatization for it is known yet.

11.4 Abilities under Incomplete Information

Usually agents have *incomplete knowledge* about the environment where they act, as well as about the current course of affairs, including the current states of the other agents, the actions they take, etc. That, of course, affects essentially their abilities to achieve individual or collective objectives. In this section, we combine concurrent game models and epistemic models in order to give semantics to a language expressing strategic ability under imperfect/incomplete information. We do not involve epistemic operators in the object language yet (this will be done in Section 11.5). Thus, knowledge is reflected only in the models, through agents' epistemic relations (aka indistinguishability relations), but not explicitly referred in the language. As we will show, taking knowledge into account on purely semantic level is already sufficient to make a strong impact on the meaning of strategic operators, and the patterns of ability that can emerge.

In game theory, two different terms are traditionally used to indicate lack of information: "incomplete" and "imperfect" information. Usually, the former refers to uncertainties about the game structure and rules, while the latter refers to uncertainties about the history, current state, etc. of the specific *play of the game*. The models that we use allow for representing both types of uncertainty in a uniform way. We take the liberty to use the term "imperfect information" in the stricter game-theoretic sense, whereas we will use "incomplete information" more loosely, to indicate any possible relevant lack of information.

11.4.1 Incomplete Information and Uniform Strategies

The decision making and abilities of strategically reasoning players are strongly influenced by the knowledge they possess about the world, the other players, past actions, etc. So far we have considered structures of *complete and (almost) perfect information* in the sense that players were completely aware of the rules and structure of the game system and of the current state of the play, and the only information they lack is the choice of actions of the other players at the current state. However, in reality this is seldom the case: usually players have only partial information, both about the structure and the rules of the game, and about the precise history and the current state of a specific play of the game. In the following we are concerned with the following question: *What can players achieve in a game if they are not completely informed about its structure and the current play?*

We represent players' incomplete information about the world by *indistinguishability relations* $\sim \subseteq \mathsf{St} \times \mathsf{St}$ on the state space, as discussed in Section 11.2. We write $q \sim_{\mathsf{a}} q'$ to describe player a's inability to discern between states q and q'. Both states appear identical from a's perspective. The indistinguishability relations are traditionally assumed to be *equivalence relations.* The knowledge of a player is then determined as follows: a player *knows* that property φ holds in a state q if φ is the case in all states indistinguishable from q for that player.

How does the incomplete information modelled by an indistinguishability relation affect the strategic abilities of the player? In the case of ATL with complete information, abilities were derived from strategies defined on states or their sequences, i.e., memoryless or perfect recall strategies. For incomplete information, the picture is similar. However, the two notions of a strategy (as well as the definition of outcome) must take into account some constraints due to uncertainty of the players.

Formally, a concurrent game with incomplete information can be modelled by a *concurrent epistemic game structure* (CEGS) is a tuple

$$\mathcal{S} = (\mathsf{Ag}, \mathsf{St}, \{\sim_{\mathsf{a}} \mid \mathsf{a} \in \mathsf{Ag}\}, \mathsf{Act}, \mathsf{act}, \mathsf{out})$$

where $(\mathsf{Ag}, \mathsf{St}, \mathsf{Act}, \mathsf{act}, \mathsf{out})$ is a CGS and $\sim_{\mathsf{a}}$ are indistinguishability relations over St, one per agent in Ag. A basic assumption in the case of incomplete information is that *players have the same choices of actions in indistinguishable states*, for otherwise, they would have a way to discern between these states. That is, we require that if $q \sim_{\mathsf{a}} q'$ then $\mathsf{act}_{\mathsf{a}}(q) = \mathsf{act}_{\mathsf{a}}(q')$.

Just like CGM, a *concurrent epistemic game model* (CEGM) is defined by adding to a CEGS a labeling of game states with sets of atomic propositions.

The notion of strategy must be refined in the incomplete information setting in order to be realistic. In such setting an "executable" strategy has to assign the same choices to indistinguishable situations. Such strategies are called *uniform.*

Definition 11.11 (Uniform strategies)
A memoryless strategy s_{a} is *uniform* if the following condition is satisfied:

$$\text{for all states } q, q' \in \mathsf{St}, \text{ if } q \sim_{\mathsf{a}} q' \text{ then } s_{\mathsf{a}}(q) = s_{\mathsf{a}}(q').$$

For perfect recall strategies we first lift the indistinguishability between states to indistinguishability between sequences of states. Two histories $h = q_0q_1 \dots q_n$ and $h' = q'_0q'_1 \dots q'_{n'}$ are *indistinguishable* for agent a, denoted by $h \approx_{\mathsf{a}} h'$, if and only if $n = n'$ and $q_i \sim_{\mathsf{a}} q'_i$ for $i = 1, \dots, n$. Now, a perfect recall strategy s_{a} is *uniform* if the following condition holds:

$$\text{for all histories } h, h' \in \mathsf{St}^+, \text{ if } h \approx_{\mathsf{a}} h' \text{ then } s_{\mathsf{a}}(h) = s_{\mathsf{a}}(h').$$

Uniform joint strategies are defined as tuples of uniform individual strategies. Note that this definition presumes no communication between

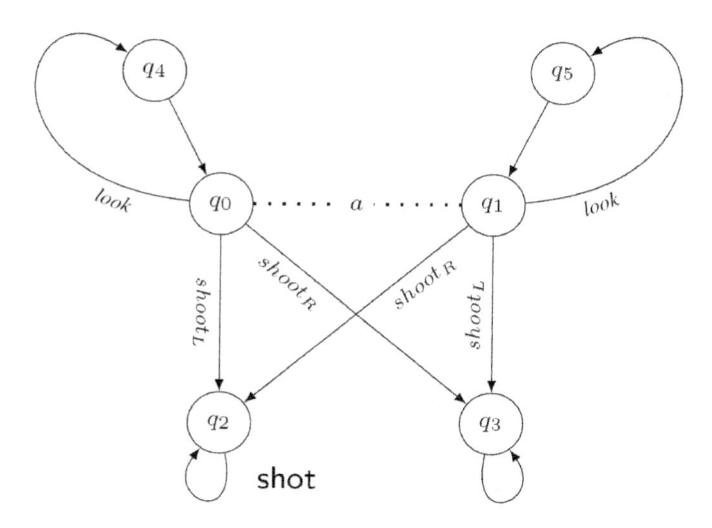

Figure 11.2: "Poor duck model" $\mathcal{M}_2$ with one player (a) and transitions labeled with a's actions. Dotted lines depict the indistinguishability relations. Automatic transitions (i.e., such that there is only one possible transition from the starting state) are left unlabeled.

the agents that could reduce their uncertainties. We will come back to this issue again later. ⊣

Example 11.6 (Poor duck)
Consider model $\mathcal{M}_2$ from Figure 11.2, with the following story. A man wants to shoot down a yellow rubber duck in a shooting gallery. The man knows that the duck is in one of the two cells in front of him, but he does not know in which one. He can either decide to shoot to the left (action $shoot_L$), or to the right (action $shoot_R$), or reach out to the cells and look what is in (action *look*). Note that only one of the shooting actions can be successful – which one it depends on the starting state of the game.[3]

Intuitively, the man does not have a strategy to ensure shooting down the duck in one step, at least not from his subjective point of view. On the other hand, he should be able to ensure it in multiple steps if he has a perfect recall of his observations. We will formalize the intuitions in the next subsection. ⊣

[3] For more seriously minded readers, we propose an alternative story: agent a is a doctor who can apply two different treatments to a patient with symptoms that fit two different diseases. Additionally, the doctor can order a blood test to identify the disease precisely.

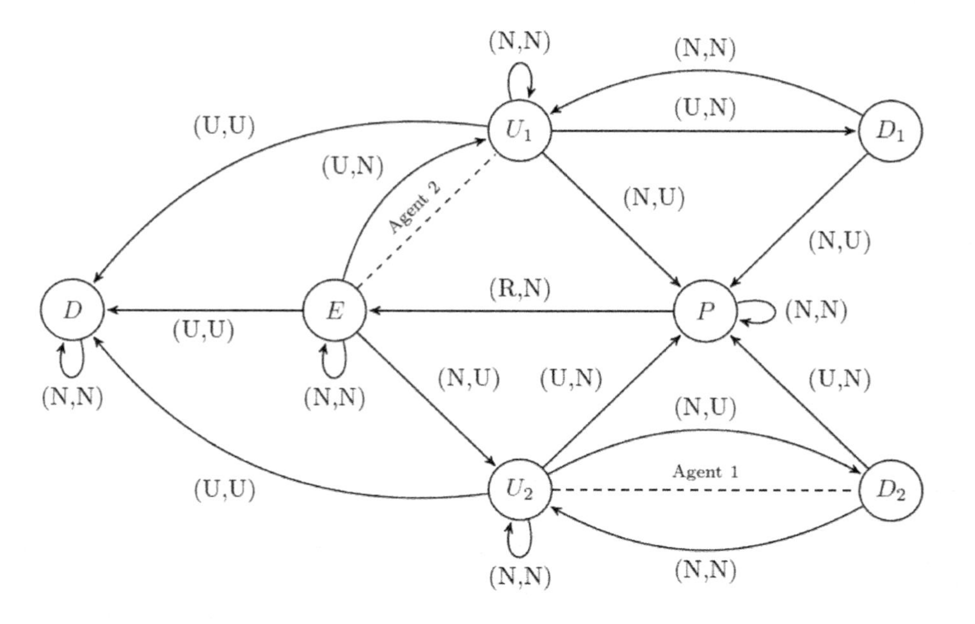

Figure 11.3: A concurrent game model with incomplete information encoding the shared file updating story.

Example 11.7 (Shared file updates with incomplete information) Consider again the shared file updating procedure in Example 11.1 and now suppose that Agent 2 cannot distinguish states E and U_1 (e.g., because he cannot observe the action of Agent 1 at state E) and, likewise, Agent 1 cannot distinguish states U_2 and D_2. The resulting concurrent game model with incomplete information in given in Figure 11.3. ⊣

11.4.2 Reasoning about Abilities under Uncertainty

Agents' incomplete information and use of memory can be incorporated into ATL* in different ways. One possible approach is not to change the logical language but to consider variations of the semantics, reflecting these by suitably varying the notion of strategy employed in the truth definition of the strategic operators. Combining perfect recall and memoryless strategies, on the one hand, with complete and incomplete information, on the other hand, gives rise to 4 natural variants of the semantics for ATL* (as originally discussed by Schobbens):

$\models_{IR}$: complete information and perfect recall strategies;

$\models_{Ir}$: complete information and memoryless strategies;

$\models_{iR}$: incomplete information and perfect recall strategies;

$\models_{ir}$: incomplete information and memoryless strategies.

The 4 semantic variants are obtained by different updating of the main semantic clause from Section 11.3. Let $\mathcal{M}$ be a CEGM, and let $\sim_C := \bigcup_{\mathsf{a}\in C} \sim_{\mathsf{a}}$ be the epistemic relation corresponding to the *group knowledge of coalition* C (i.e., what everybody in C knows). Moreover, let $[q]_{\mathsf{a}} = \{q' \mid q \sim_{\mathsf{a}} q'\}$ denote the information set of agent a, and analogously for coalitions C. The truth definitions for $\models_{xy}$, where $x \in \{I, i\}, y \in \{R, r\}$, read as follows:

$\mathcal{M}, q \models_{IR} \langle\langle C\rangle\rangle\gamma$ iff there is a joint perfect recall strategy s_C for C such that $\mathcal{M}, \lambda \models \gamma$ for every play $\lambda \in \mathsf{out\_plays}(q, s_C)$

$\mathcal{M}, q \models_{Ir} \langle\langle C\rangle\rangle\gamma$ iff there is a joint memoryless strategy s_C for C such that $\mathcal{M}, \lambda \models \gamma$ for every play $\lambda \in \mathsf{out\_plays}(q, s_C)$

$\mathcal{M}, q \models_{iR} \langle\langle C\rangle\rangle\gamma$ iff there is a uniform joint perfect recall strategy s_C for C such that $\mathcal{M}, \lambda \models \gamma$ for every play $\lambda \in \bigcup_{q'\in[q]_C} \mathsf{out\_plays}(q', s_C)$

$\mathcal{M}, q \models_{ir} \langle\langle C\rangle\rangle\gamma$ iff there is a uniform joint memoryless strategy s_C for C such that $\mathcal{M}, \lambda \models \gamma$ for every play $\lambda \in \bigcup_{q'\in[q]_C} \mathsf{out\_plays}(q', s_C)$.

It is easy to see that the standard semantic relation $\models$ introduced in Section 11.3 corresponds to $\models_{IR}$, and the positional semantics $\models^p$ is the same as $\models_{Ir}$.

Another option is to extend the *syntax* of cooperation modalities with subscripts: $\langle\langle A\rangle\rangle_{xy}$ where x indicates the use of memory in the strategies (memoryless if $x = r$ and perfect recall if $x = R$) and y indicates the information setting (complete if $y = I$ and incomplete information if $y = i$).

Example 11.8 (Shooting the duck in variants of ATL)
Consider the "poor duck" model $\mathcal{M}_2$ from Example 11.6. There is no good strategy for the man to shoot down the duck in one step regardless of the kind of memory that the man possesses – formally, $\mathcal{M}_2, q_0 \models_{iR} \neg\langle\langle a\rangle\rangle\mathsf{X}\mathsf{shot}$ and $\mathcal{M}_2, q_0 \models_{ir} \neg\langle\langle a\rangle\rangle\mathsf{X}\mathsf{shot}$. However, he should be able to achieve it in multiple steps if he can remember and use his observations, i.e., $\mathcal{M}_2, q_0 \models_{iR} \langle\langle a\rangle\rangle\mathsf{F}\mathsf{shot}$.

On the other hand, suppose that it has been a long party, and the man is very tired, so he is only capable of using memoryless strategies at the moment. Does he have a memoryless strategy which he knows will achieve the goal? No. In each of these cases the man risks that he will fail (at least from his subjective point of view). In consequence, $\mathcal{M}_2, q_0 \models_{ir} \neg\langle\langle a\rangle\rangle \mathsf{F}\mathsf{shot}$. Interestingly enough, the man *can* identify an opening strategy that will guarantee his knowing how to shoot the duck in the next moment: $\mathcal{M}_2, q_0 \models_{ir} \langle\langle a\rangle\rangle \mathsf{X} \langle\langle a\rangle\rangle \mathsf{F}\mathsf{shot}$. The opening strategy is to look; if the system proceeds to q_4 then the second strategy is to shoot to the left, otherwise the second strategy is to shoot to the right. ⊣

Example 11.9 (Properties of shared file under incomplete informati
Consider Example 11.7 about shared file updates with incomplete information. Denote the model on Figure 11.3 by $\mathcal{M}_3$. Again, we use the same notation for states and for the atomic propositions identifying them. The following hold:

- $\mathcal{M}_3, E \models_{ir} \langle\langle 1,2\rangle\rangle \mathsf{X} U_1$. Indeed, there is a joint uniform strategy for $\{1,2\}$ that enforces $\mathsf{X} U_1$ from all states q such that $E \sim_{\{1,2\}} q$, i.e., states E, U_1. The strategy is agent 1 playing U at E and N at U_1; agent 2 playing N at E, U_1 (and anything at the other states).

- On the other hand, $\mathcal{M}_3, E \models_{ir} \neg\langle\langle 1,2\rangle\rangle \mathsf{X} U_2$. Although there is a joint uniform strategy that *objectively* enforces $\mathsf{X} U_2$ (e.g., 1 playing N and 2 playing U at E, U_1, and anything at the other states), the strategy does not guarantee success in the subjective judgment of agent 2, because it does not achieve the goal from state U_1 (which player 2 thinks might be the case).

- $\mathcal{M}_3, E \models_{ir} \langle\langle 1,2\rangle\rangle ((\neg D_1 \wedge \neg D_2) \,\mathsf{U}\, P)$. A joint uniform strategy that enforces $(\neg D_1 \wedge \neg D_2) \,\mathsf{U}\, P$ prescribes agent 1 to play N at E and U_1, U at U_2 and anything at all other states, while it requires agent 2 to play U at states E and U_1, N at U_2 and anything at all other states. Note that the requirement for agent 1 to play N at U_1 is irrelevant for the *objective* outcome of the strategy, but it is needed to assure the subjective judgment of agent 2 that the joint strategy will also work from state U_1.

- $\mathcal{M}_3, U_2 \models_{ir} \neg\langle\langle 1,2\rangle\rangle (\neg D_2 \,\mathsf{U}\, P)$. Again, the goal $(\neg D_2 \,\mathsf{U}\, P)$ can be achieved objectively from state U_2 by a uniform strategy, but not subjectively, because for all that agent 1 knows at U_2 the formula may be already falsified if the current state were D_2.

- Still, $\mathcal{M}_3, E \models_{ir} \langle\langle 1,2\rangle\rangle (\neg D_2 \,\mathsf{U}\, P)$. Indeed, the uncertainty of agent 2 between states U_2 and D_2 no longer matters, because *being at state* E agent 1 knows for sure that the joint action (N, U) would take the system to state U_2 and not to D_2 and therefore playing (U, N) at that state must succeed. All he needs to do is follow blindly that strategy when/if the system gets in his information set $\{U_2, D_2\}$. Thus, the following strategy enforces $\neg D_2 \,\mathsf{U}\, P$ from E, U_1: agent 1 plays N at state E and U at U_2, D_2; agent 2 plays U at states E, U_1 and N at U_2.

- $\mathcal{M}_3, E \models_{ir} \neg\langle\langle 1,2\rangle\rangle \mathsf{F}(D_1 \wedge \mathsf{F}(U_1 \wedge \mathsf{F}P))$, because any successful joint memoryless strategy would have to prescribe joint actions (U, N) at E and U_1 and (N, N) at D_1 and therefore would never reach P thereafter.

- However, $\mathcal{M}_3, \langle E\rangle \models_{iR} \langle\langle 1,2\rangle\rangle \mathsf{F}(D_1 \wedge \mathsf{F}(U_1 \wedge \mathsf{F}P))$, where $\langle E\rangle$ is the one state history starting at state E. Indeed, a joint uniform memory-based strategy prescribing a play beginning with $E, U_1, D_1, U_1, P, \ldots$ would succeed. Note that this strategy only requires only 1 extra memory cell per agent, needed to record the passing through state D_1. We leave the details to the reader.

Some remarks:

1. Ability under incomplete information can additionally be classified into *objective* and *subjective*. The former refers to existence of a strategy that is guaranteed to succeed from the perspective of an external observer with complete information about the model, while the latter requires that strategy to be guaranteed to succeed from the perspective of the player/coalition executing it. The definition above is based on *subjective* ability. As it requires that the executing players must be able to identify the right strategy on their own (within the limits of their incomplete information), it imposes stronger requirements on the strategy than the objective ability. Technically, this is because in the evaluation of a state formula $\langle\langle C\rangle\rangle\gamma$ in state q, when judging the suitability of the selected uniform strategy for C in terms of the coalition's subjective ability all epistemic alternatives of q wrt $\sim_C$ are to be taken into account, whereas objectively it suffices to check that the strategy only succeeds from q.

2. The distinction between *objective* and *subjective* ability is closely related to the first two interpretations of ability by McCarthy and Hayes, that were discussed in Section 11.2. Objective ability to enforce φ means that there exists a strategy (a "subprogram" σ in McCarthy and Hayes' terminology) that, if executed, will guarantee φ. Subjective ability means that that the decision maker(s) (the "main program" π for McCarthy and Hayes) has enough information to verify that σ enforces φ. We note that the third, strongest level of ability from

McCarthy and Hayes cannot be expressed in the logics presented in this chapter, as they embed no notion of problem-solving procedure *inside* their semantics.

3. Alternative semantics, where common or distributed knowledge of C are taken into account when judging the ability of the coalition to identify the right strategy, have also been considered. We will return to this topic in Section 11.5 where knowledge operators are added explicitly to the language.

4. Most modal logics of strategic ability agree that *executable* strategies are exactly the ones that obey the uniformity constraints. For a joint strategy, this means that each member of the coalition must be able to carry out his part of the joint plan individually. An alternative approach has been presented by Guelev and colleagues, where the notion of uniform joint strategy for a coalition is redefined, based on a suitable *group indistinguishability relation* for the coalition. The most interesting case is when the "distributed knowledge" relation is used. Conceptually, this amounts to assuming that members of the coalition have unlimited communicating capabilities, and freely share relevant information *during the execution of the strategy.*

11.4.3 Impact of Knowledge on Strategic Ability

Clearly, knowledge and abilities are not independent: the more one knows, the more one can achieve by choosing a better suited strategy. We have seen that limited information influences the range of available strategies that agents and coalitions can choose. Respectively, it also affects the semantics of claims about an agent (or a coalition) being able to enforce a given outcome of the game. How big is the impact? As it turns out, adding agents' knowledge (or, dually, uncertainty) to the semantics of ATL changes a lot. We will review the impact *very* briefly here. The interested reader is referred to the bibliographic notes for details.

Valid sentences

Traditionally, a logic is identified with the set of sentences that are true in the logic; a semantics is just a possible way of defining the set, alternative to an axiomatic inference system. Thus, by comparing the sets of validities we compare the respective logics in the traditional sense. As it turns out, all semantic variants of ATL* discussed here are different on the level of general properties they induce. The pattern is as follows: properties of perfect information strategies refine properties of imperfect information strategies,

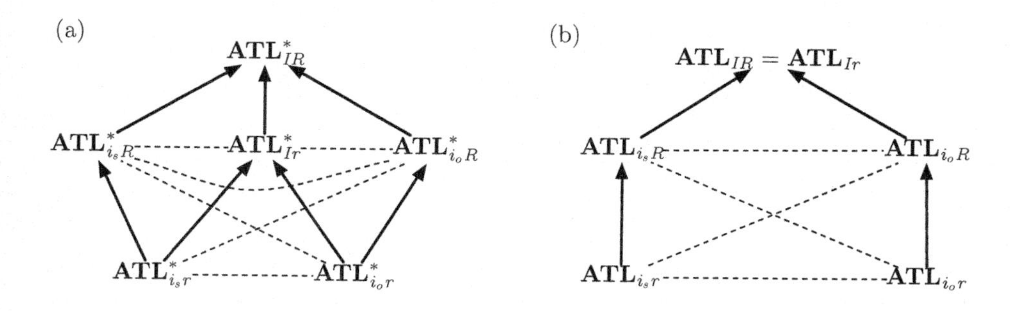

Figure 11.4: Comparison of the sets of validities induced by various semantics of (a) ATL*, and (b) ATL. Arrows depict strict subsumption of the sets of validities, e.g., "$\mathrm{ATL}^*_{Ir} \rightarrow \mathrm{ATL}^*_{IR}$" means that $Validities(\mathrm{ATL}^*_{Ir}) \subsetneq Validities(\mathrm{ATL}^*_{IR})$. Dotted lines connect semantic variants with incomparable sets of validities. We do not include links that follow from transitivity of the subsumption relation. The subscript i_s refers to "subjective" ability under imperfect information, that is, the standard "i" semantics presented in Section 11.4.2. Subscript i_o refers to the "objective" ability under imperfect information, where we only look at the paths starting from the actual initial state of the game (see the remark at the end of Section 11.4.2).

perfect recall games are special case of memoryless games, and properties of objective and subjective abilities of agents are incomparable (see Figure 11.4 for a graphical summary of the relationships).

Thus, assuming imperfect information changes the set of properties that are universally true, i.e., ones that hold in every model and every state. Moreover, the change is essential in the sense that some fundamental validities of standard ATL (with perfect information) do not hold anymore under imperfect information. We give three examples of such formulae here:

$$\langle\langle A\rangle\rangle \mathsf{G}\varphi \leftrightarrow \varphi \wedge \langle\langle A\rangle\rangle \mathsf{X} \langle\langle A\rangle\rangle \mathsf{G}\varphi \tag{11.3}$$

$$\langle\langle A\rangle\rangle \varphi_1 \mathsf{U} \varphi_2 \leftrightarrow \varphi_2 \vee \varphi_1 \wedge \langle\langle A\rangle\rangle \mathsf{X} \langle\langle A\rangle\rangle \varphi_1 \mathsf{U} \varphi_2 \tag{11.4}$$

$$\langle\langle \mathsf{Ag}\rangle\rangle \mathsf{F}\varphi \leftrightarrow \neg\langle\langle \emptyset\rangle\rangle \mathsf{G}\neg\varphi \tag{11.5}$$

Formulae (11.3) and (11.4) provide fixpoint characterizations of strategic-temporal modalities. Formula (11.5) addresses the duality between necessary and obtainable outcomes in a game. All the three sentences are validities of ATL_{IR} and ATL_{Ir}, but they are *not* valid in ATL_{iR} and ATL_{ir}.[4]

[4] Interestingly, (11.5) becomes valid again with the "objective" interpretation of ability under imperfect information.

	CL	ATL	ATL*
Ir	*P*-complete	*P*-complete	*PSPACE*-complete
IR	*P*-complete	*P*-complete	*2EXPTIME*-complete
ir	*P*-complete	Δ_2^P-complete	*PSPACE*-complete
iR	*P*-complete	*undecidable*	*undecidable*

Figure 11.5: Overview of model checking complexity for variants of strategic logics. Each cell represents the logic over the set of formulae given in the column using the semantics given in the row. $\Delta_2^P = NP^{NP}$ is the class of problems solvable by a deterministic Turing machine making adaptive calls to an oracle of a problem in *NP*.

Expressivity

Assuming imperfect information in the semantics of ATL-related logics changes also their relative expressivity. It is well known that, under the perfect information semantics: (i) ATL* is strictly less expressive than the *alternating μ-calculus*, and (ii) ATL is strictly less expressive than *alternation-free alternating μ-calculus*. Alternating μ-calculus (AMC) is the logic obtained by adding the *least fixpoint operator* to Coalition Logic. Thus, characterization of long-term abilities under perfect information can be reduced to one-step abilities and fixpoint reasoning.

In contrast, when imperfect information semantics is used, the languages of ATL and alternation-free alternating μ-calculus become incomparable with respect to their expressive power. In particular, the simple ATL formula $\langle\langle A\rangle\rangle \mathsf{F} \mathsf{p}$ cannot be equivalently translated into alternation-free AMC. Thus, long-term abilities cannot be equivalently expressed with one-step abilities and fixpoints anymore. This has an interesting consequence. Recall Moore's notion of ability that we presented in Section 11.2. It is clearly based on fixpoint reasoning. Thus, under imperfect information, alternating μ-calculus seems a better logic than ATL to formalize Moore's concept of ability.

Computational complexity

Limited information of players in a game has also impact on the complexity of "solving" the game. In logical terms, this means that the complexity of model checking for a strategic logic is usually worse when imperfect information is involved. That is because the fixpoint characterizations of strategic-temporal modalities do not hold anymore under imperfect information, and hence model checking cannot be done with the standard fixpoint algorithm.

In essence, synthesizing strategies for imperfect information games cannot be done incrementally. Figure 11.5 summarizes how adding epistemic relations to the semantics of strategic operators increases the complexity of model checking for CL, ATL, and ATL*. Interestingly, it turns out that for the restricted language of ATL the type of information has a bigger impact on the complexity than the type of recall. For the broader language of ATL*, it is the other way around. The table indicates also some really bad news: for imperfect information *and* perfect recall, model checking becomes undecidable for any logic that can express long-term abilities.

Few results are known for the other decision problems. Satisfiability checking (and hence also validity checking) under perfect information is known to be *PSPACE*-complete for Coalition Logic, and *2EXPTIME*-complete for ATL. To the best of our knowledge, there are no complexity results for satisfiability and validity checking of imperfect information variants of ATL. It is not even known if the problems are decidable.

11.5 Reasoning about Strategies and Knowledge

So far, we have taken into account agents' knowledge only semantically, by introducing epistemic indistinguishability relations in the models to capture agents' uncertainty. However, we did not refer to knowledge in the object language through epistemic operators. In this section, we add knowledge operators to the language of strategic logic and show how interplay between knowledge and abilities can be specified.

11.5.1 Towards an Epistemic Extension of ATL

A natural extension of CL and ATL is to add epistemic operators to the languages, in order to formalise reasoning about the interaction of knowledge and abilities of agents and coalitions. A straightforward combination of ATL and the multi-agent epistemic logic MAEL, called *Alternating-time Temporal Epistemic Logic* (ATEL). ATEL enables specification of various modes and nuances of interaction between knowledge and strategic abilities, e.g.:

- $\langle\langle A\rangle\rangle\varphi \rightarrow \mathrm{E}_A\langle\langle A\rangle\rangle\varphi$ (if group A can bring about φ then everybody in A knows that they can);
- $\mathrm{E}_A\langle\langle A\rangle\rangle\varphi \wedge \neg\mathrm{C}_A\langle\langle A\rangle\rangle\varphi$ (the agents in A have mutual knowledge but not common knowledge that they can enforce φ);
- $\langle\langle i\rangle\rangle\varphi \rightarrow \mathrm{K}_i\neg\langle\langle \mathbb{A}\mathrm{gt} \setminus \{i\}\rangle\rangle\neg\varphi$ (if agent i can bring about φ then she knows that the rest of agents cannot prevent it).

The combined language can be interpreted in *concurrent epistemic game models* (CEGM)

$$\mathcal{S} = (\mathsf{Ag}, \mathsf{St}, \{\sim_{\mathsf{a}} | \ \mathsf{a} \in \mathsf{Ag}\}, \mathsf{Act}, \mathsf{act}, \mathsf{out}),$$

combining the CGM-based models for ATL and the multi-agent epistemic models. That is, we interpret ATEL in the same structures as those used in the variants of ATL for incomplete information, presented in Section 11.4.2. The semantics of ATEL is given by union of the semantic clauses for epistemic logic and those for ATL. This straightforward combination gives a logic that is an independent fusion of the two parts. For example, the next-time fragment of ATL (i.e., coalition logic) combined with epistemic logic is completely axiomatised simply by adding the standard S5 axioms for knowledge to the axiomatisation of coalition logic described in Section 11.3.3. If common and/or distributed knowledge operators are included, adding standard axioms for group knowledge again results in a complete axiomatisation. Analogous axiomatisations can be obtained for the full fusion of ATL with MAEL, i.e. ATEL.

The independent product of ATL and epistemic logic is useful and allows capturing meaningful properties such as the ones mentioned above. Some interesting properties, however, turn out to be difficult to express, and some formulae turn out to have counterintuitive meaning. This is particularly true for the interplay between epistemic properties and long-term abilities of agents and coalitions, expressible in the full language of ATEL.

Importantly, one would expect that an agent's ability to achieve property φ should imply that the agent has enough control and knowledge to *identify* and *execute* a strategy that enforces φ. A number of more sophisticated approaches to combining strategic and epistemic operators have been proposed in order to be able to capture such issues. Most of the solutions agree that only uniform strategies are really executable, cf. our exposition of ATL variants for incomplete information in Section 11.4.2. However, in order to identify a successful strategy, the agents must consider not only the courses of action, starting from the current state of the system, but also from states that are indistinguishable from the current one. There are many cases here, especially when group epistemics is concerned: the agents may have common, ordinary, or distributed knowledge about a strategy being successful, or they may be hinted the right strategy by a distinguished member (the "leader"), a subgroup ("headquarters committee") or even another group of agents ("consulting company"). We discuss the most interesting cases in the subsequent paragraphs.

11.5.2 Epistemic Levels of Strategic Ability

There are several possible interpretations of A's ability to bring about property γ, formalized by formula $\langle\langle A\rangle\rangle\gamma$, under imperfect information:

1. There exists a specification of A's behavior σ_A (not necessarily executable!) such that, for every execution of σ_A, γ holds.

2. There is a uniform strategy s_A such that, for every execution of s_A, γ holds *(A have objective ability to enforce γ).*

3. A know (in one sense or another, see below) that there is a uniform s_A such that, for every execution of s_A, γ holds *(A have a strategy "de dicto" to enforce γ).*

4. There is a uniform s_A such that A know that, for every execution of s_A, γ holds *(A have a strategy "de re" to enforce γ).*

The above interpretations form a sequence of increasingly stronger levels of ability – each next one implies the previous ones. We observe that the case (1) above corresponds to formula $\langle\langle A\rangle\rangle\gamma$ interpreted in the original semantics of ATL (or, alternatively, in ATEL). The other cases, however, are not expressible in straightforward combinations of ATL and epistemic logic such as ATEL, not even for the next-time fragment (i.e., Coalition Logic) and not even when there is only a single agent. We will show in Section 11.5.3 how they can be formally characterized with a suitable combination of strategic and epistemic modalities.

Note also that cases (2)-(4) come close to various philosophical notions of ability that we discussed in Sections 11.1 and 11.2. Cases (3) and (4) closely resemble Ryle's distinction between "knowing that" and "knowing how". In (3), *A know that* they can enforce γ. In (4), they *know how* to do it. Case (2) corresponds to the first level of ability according to McCarthy and Hayes (i.e., objective existence of a subprogram that achieves γ). Case (4) is analogous to the second level of McCarthy and Hayes (the subprogram exists, and there is enough information to realize it) as well as Moore's formalization of ability. Finally, case (3) can be seen as Moore's notion of ability with knowledge *de re* (of the right strategy to play) replaced with knowledge *de dicto* (that such a strategy exists).

Out of the four levels of ability, "knowing how to play" (4) is arguably most interesting. However, the statement "A know that every execution of s_A satisfies γ" is precise only if A consists of a single agent a. Then, we take into account the paths starting from states indistinguishable from the current one according to a (i.e., $\bigcup_{q'\in \mathrm{img}(q,\sim_a)} \mathsf{out\_plays}(q', s_a)$). In case of multiple agents, there are several different "modes" in which they can know the right strategy. That is, given strategy s_A, coalition A can have:

- *Common knowledge* that s_A enforces γ. This requires the least amount of additional communication when coordinating a joint strategy (agents from A may agree upon a total order over their collective strategies at the beginning of the game and that they will always choose the maximal successful strategy with respect to this order).
- *Mutual knowledge* that s_A enforces γ: everybody in A knows that s_A brings about γ.
- *Distributed knowledge* that s_A enforces γ: if the agents share their knowledge at the current state, they can identify the strategy as successful.
- *"Leader"*: the strategy can be identified by an agent $a \in A$;
- *"Headquarters committee"*: s_A can be identified by a subgroup $A' \subseteq A$.
- *"Consulting company"*: s_A can be identified by another group B.
- Other variations are possible, too.

11.5.3 Expressing Epistemic Levels of Ability and Constructive Knowledge

The issue of expressing various knowledge-related levels of ability through a suitable combination of strategic and epistemic logics has attracted significant attention. Most extensions (or refinements) of ATL, proposed as solutions, cover only some of the possibilities, albeit in an elegant way. Others offer a more general treatment of the problem at the expense of an unnecessarily complex logical language. One of the main problems is how to capture the interplay between epistemic, strategic, and temporal aspects of play with the kind of quantifiers offered by ATL on hand, and epistemic logic on the other. For example, "A have distributed knowledge about how to play to enforce γ" can be rephrased as "there is a strategy for A such that, for every state that is indistinguishable from the current one for all agents in A, and for every path from that state, possibly resulting from execution of the strategy, γ must hold on the path." This, however, cannot be directly expressed in ATL which combines quantification over strategies and paths within a single operator $\langle\langle A\rangle\rangle$. One way out is to use separate modal operators to quantify over strategies and paths, the way it is done e.g. in *Strategic STIT* and *Strategy Logic*. Indeed, "knowing how to play" in the simpler case of single-step games can be expressed using a combination of Chellas STIT and epistemic logic. One can speculate that a straightforward combination of epistemic operators with Strategic STIT (or Strategy Logic) should work equally well for long-term abilities. However, such combinations have not been investigated yet.

Another solution is to define the success of a strategy *from a set of states*, instead of a single global state. This idea was used in *Constructive Strategic Logic* (*CSL*) which extends ATL with so called *constructive knowledge* operators. In CSL, each formula is interpreted in a *set* of states (rather than a single state). We write $\mathcal{M}, Q \models \langle\langle A \rangle\rangle\varphi$ to express the fact that A must have a strategy which is successful for all "opening" states from Q. The new epistemic operators $\mathbb{K}_i, \mathbb{E}_A, \mathbb{C}_A, \mathbb{D}_A$ for "practical" or "constructive" knowledge yield the set of states for which a single evidence (i.e., a successful strategy) should be presented (instead of checking if the required property holds in each of the states separately, like standard epistemic operators do).

Formally, the semantics of CSL over concurrent epistemic game models is defined by the following clauses:

$\mathcal{M}, Q \models p$ iff $p \in L(q)$ for every $q \in Q$;

$\mathcal{M}, Q \models \neg\varphi$ iff $\mathcal{M}, Q \not\models \varphi$;

$\mathcal{M}, Q \models \varphi \wedge \psi$ iff $\mathcal{M}, Q \models \varphi$ and $\mathcal{M}, Q \models \psi$;

$\mathcal{M}, Q \models \langle\langle A \rangle\rangle\varphi$ iff there is a uniform strategy s_A such that $\mathcal{M}, \lambda \models \varphi$ for every $\lambda \in \bigcup_{q \in Q} \mathsf{out\_plays}(q, s_A)$;

$\mathcal{M}, Q \models \mathbb{K}_i\varphi$ iff $\mathcal{M}, \{q' \mid \exists_{q \in Q}\, q \sim_i q'\} \models \varphi$;

$\mathcal{M}, Q \models \mathbb{C}_A\varphi$ iff $\mathcal{M}, \{q' \mid \exists_{q \in Q}\, q \sim_A^C q'\} \models \varphi$;

$\mathcal{M}, Q \models \mathbb{E}_A\varphi$ iff $\mathcal{M}, \{q' \mid \exists_{q \in Q}\, q \sim_A^E q'\} \models \varphi$;

$\mathcal{M}, Q \models \mathbb{D}_A\varphi$ iff $\mathcal{M}, \{q' \mid \exists_{q \in Q}\, q \sim_A^D q'\} \models \varphi$.

The semantic clauses for temporal operators are exactly as in ATL*. Additionally, we define that $\mathcal{M}, q \models \varphi$ iff $\mathcal{M}, \{q\} \models \varphi$.

A nice feature of CSL is that standard knowledge operators can be defined using constructive knowledge, see Section 11.5.4 for details. Thus, one can use formulae of CSL to express the following:

1. $\langle\langle a \rangle\rangle\varphi$ expresses that agent a has a uniform strategy to enforce φ *from the current state* (but may not know about it);

2. $K_a\langle\langle a \rangle\rangle\varphi$ refers to agent a having a strategy *"de dicto"* to enforce φ (i.e. knowing only that *some* successful uniform strategy is available);

3. $\mathbb{K}_a\langle\langle a \rangle\rangle\varphi$ refers to agent a having a strategy *"de re"* to enforce φ (i.e. having a successful uniform strategy and knowing the strategy);

Again, this connects neatly to the fundamental work on ability in philosophy and AI that we discussed in Sections 11.1 and 11.2. It can be argued that standard knowledge operators K_a capture Ryle's notion of know-that, whereas constructive knowledge operators $\mathbb{K}_a$ refer to the notion of know-how. Also, formalisations (1) and (3) above roughly correspond to McCarthy and Hayes's levels (1) and (3) of program ability. Finally, formulae $\mathbb{K}_a\langle\langle a\rangle\rangle\varphi$ and $K_a\langle\langle a\rangle\rangle\varphi$ capture formally Moore's distinction between ability *de re* and *de dicto* for long-term strategies. This extends naturally to abilities of coalitions, with $\mathbb{C}_A\langle\langle A\rangle\rangle\varphi, \mathbb{E}_A\langle\langle A\rangle\rangle\varphi, \mathbb{D}_A\langle\langle A\rangle\rangle\varphi$ formalizing common, mutual, and distributed knowledge how to play, $\mathbb{K}_a\langle\langle A\rangle\rangle\varphi$ capturing the "leader" scenario, and so on (and similarly for different levels of knowledge "de dicto"). We conclude the topic with the following example.

Example 11.10 (Onion Soup Robbery)
A virtual safe contains the recipe for the best onion soup in the world. The safe can only be opened by a k-digit binary code, where each digit c_i is sent from a prescribed location i $(1 \leq i \leq k)$. To open the safe and download the recipe it is enough that at least $n \leq k$ correct digits are sent at the same moment. However, if a *wrong* value is sent from one of the locations, or if an insufficient number (i.e., between 1 and $n-1$) of digits is submitted, then the safe locks up and activates an alarm.

k agents are connected at the right locations; each of them can send 0, send 1, or do nothing (*nop*). Moreover, individual agents have only partial information about the code. To make the example more concrete, we assume that agent i (connected to location i) knows the values of $c_{i-1}\,\mathrm{XOR}\,c_i$ and $c_i\,\mathrm{XOR}\,c_{i+1}$ (we take $c_0 = c_{k+1} = 0$). This implies that only agents 1 and k know the values of "their" digits. Still, every agent knows whether his neighbors' digits are the same as his.[5]

Formally, the concurrent epistemic game model $Attack_k^n$ is constructed as follows:

- $\mathsf{Ag} = \{1, \ldots, k\}$;
- $\mathsf{St} = Q \cup S$, where states in $Q = \{(c_1, \ldots, c_k) \in \{0,1\}^k\}$ identify possible codes for the (closed) safe, and states in $S = \{open, alarm\}$ represent the situations when the safe has been opened, or when the alarm has been activated;
- $\mathrm{PROP} = \{\mathsf{open}\}$; $\quad L(\mathsf{open}) = \{open\}$; $\quad \mathsf{Act} = \{0, 1, nop\}$;
- $\mathsf{act}(i,q) = \{0, 1, nop\}$ for $q \in Q$, and $\mathsf{act}(i,q) = \{nop\}$ for $q \in S$;

[5] For the more seriously minded readers, we observe that the story is just a variant of coordinated attack.

- For all $x \in \mathsf{St}$: $\mathsf{out}(x, nop, \ldots, nop) = x$. For $q \in Q$, and at least one $\alpha_i \neq nop$: $\mathsf{out}(q, \alpha_1, \ldots, \alpha_k) = open$ if $\alpha_j = c_j$ for at least n agents j and $\alpha_i \notin \{c_i, nop\}$ for no i; else, $\mathsf{out}(q, \alpha_1, \ldots, \alpha_k) = alarm$.

- $q \sim_i q'$ iff $q[i-1] \,\mathrm{XOR}\, q[i] = q'[i-1] \,\mathrm{XOR}\, q'[i]$ and $q[i] \,\mathrm{XOR}\, q[i+1] = q'[i] \,\mathrm{XOR}\, q'[i+1]$.

The following CSL formulae hold in every state $q \in Q$ of model $Attack_k^n$, assuming that $k \geq 3$:

- $\langle\!\langle \mathsf{Ag} \rangle\!\rangle \mathsf{F}\mathsf{open} \wedge \neg \mathbb{E}_{\mathsf{Ag}} \langle\!\langle \mathsf{Ag} \rangle\!\rangle \mathsf{F}\mathsf{open}$: there is an executable strategy for the agents, which guarantees a win, but not all of them can identify it (in fact, none of them can in this case);

- $\mathbb{D}_{\mathsf{Ag}} \langle\!\langle \mathsf{Ag} \rangle\!\rangle \mathsf{F}\mathsf{open}$: if the agents share information they can recognize who should send what;

- $\mathbb{D}_{\{1,\ldots,n-1\}} \langle\!\langle \mathsf{Ag} \rangle\!\rangle \mathsf{F}\mathsf{open}$: it is enough that the first $n-1$ agents devise the strategy. Note that the same holds for the last $n-1$ agents, i.e., the subteam $\{k-n+2, \ldots, k\}$;

- Still, $\neg \mathbb{D}_{\{1,\ldots,n-1\}} \langle\!\langle 1, \ldots, n-1 \rangle\!\rangle \mathsf{F}\mathsf{open}$: all agents are necessary to *execute* the strategy.

We observe that constructive knowledge operators allow to approximate the amount of communication that is needed to establish a winning strategy in scenarios where explicit modeling of communication is impossible or too expensive. For instance, formula $\mathbb{D}_{\mathsf{Ag}} \langle\!\langle \mathsf{Ag} \rangle\!\rangle \mathsf{F}\mathsf{open}$ says that if the agents in Ag share their information they will be able to determine a strategy that opens the safe. Of course, the model does not include a possibility of such "sharing", at least not explicitly. That is, there is no transition that leads to a state in which the epistemic relations of agents have been combined via intersection. Still, $D_A\varphi$ indicates that there is *epistemic potential* for agents in A to realize/infer φ; what might be missing is means of exploiting the potential (e.g., by communication). In the same way, $\mathbb{D}_A \langle\!\langle A \rangle\!\rangle \mathsf{F}\varphi$ says that the epistemic potential for A to determine the right strategy for $\mathsf{F}\varphi$ is there, too. So, it might be profitable to design efficient communication mechanisms to make the most of it. ⊣

11.5.4 Closer Look at Constructive Knowledge

In order to "constructively know" that φ, agents A must be able to find (or "construct") a mathematical object that supports φ. This is relevant when $\varphi \equiv \langle\!\langle B \rangle\!\rangle \psi$; in that case, the mathematical object in question is a strategy for B which guarantees achieving ψ. The semantic role of *constructive knowledge operators* is to produce sets of states that will appear on the left hand side of the satisfaction relation. In a way, these modalities "aggregate" states into sets, and sets into bigger sets.

Note that in CSL we can use two different notions of validity. We say that a formula is *weakly valid* (or simply *valid*) if it is satisfied individually by *each state* in every model, i.e.,, if $\mathcal{M}, q \models \varphi$ for all models $\mathcal{M}$ and states q in $\mathcal{M}$. It is *strongly valid* if it is satisfied by all non-empty *sets* in all models; i.e.,, if for each $\mathcal{M}$ and every non-empty set of states Q it is the case that $\mathcal{M}, Q \models \varphi$. We are ultimately interested in the former. The importance of strong validity, however, lies in the fact that strong validity of $\varphi \leftrightarrow \psi$ makes φ and ψ completely interchangeable. That is, if $\varphi_1 \leftrightarrow \varphi_2$ is strongly valid, and ψ' is obtained from ψ through replacing an occurrence of φ_1 by φ_2, then $\mathcal{M}, Q \models \psi$ iff $\mathcal{M}, Q \models \psi'$. It is not difficult to see that the same is not true for weak validity.

Clearly, strong validity implies validity, but not vice versa.

Defining Standard Knowledge from Constructive Knowledge

In the semantics of CSL, formulae are interpreted in sets of states; in order for φ to hold in $\mathcal{M}, Q$, the formula must be "globally" satisfied in all states from Q at once (i.e., with a single strategy). Notice, however, that $\mathcal{M}, Q \models \langle\langle\emptyset\rangle\rangle\varphi \mathsf{U} \varphi$ iff $\mathcal{M}, q \models \varphi$ for every $q \in Q$. This can be used as a technical trick to evaluate φ "locally" (i.e., in every state of Q separately). In particular, we can use it to define standard knowledge from constructive knowledge as:

$$K_a\varphi \;\equiv\; \mathbb{K}_a\langle\langle\emptyset\rangle\rangle\varphi \mathsf{U} \varphi,$$

and analogously for group knowledge operators. It is not difficult to see that $\mathcal{M}, q \models \mathbb{K}_a\langle\langle\emptyset\rangle\rangle\varphi \mathsf{U} \varphi$ iff $\mathcal{M}, q' \models \varphi$ for every q' such that $q \sim_a q'$. More generally, the following formula of CSL is strongly valid:

$$\mathcal{K}_A\varphi \;\leftrightarrow\; \hat{\mathcal{K}}_A\langle\langle\emptyset\rangle\rangle\varphi \mathsf{U} \varphi,$$

where $\mathcal{K} = C, E, D$ and $\hat{\mathcal{K}} = \mathbb{C}, \mathbb{E}, \mathbb{D}$. In consequence, we obtain that *standard knowledge can be seen as a special case of constructive knowledge.*

Properties of Constructive Knowledge

We believe that operators $\mathbb{C}_A$, $\mathbb{E}_A$, $\mathbb{D}_A$ and $\mathbb{K}_a$ do capture a special kind of knowledge of agents. An interesting question is: do these notions of knowledge have the properties usually associated with knowledge? In particular, do postulates $\mathbf{K}, \mathbf{D}, \mathbf{T}, \mathbf{4}, \mathbf{5}$ of epistemic logic hold for constructive knowledge? Below, we list the constructive knowledge versions of the S5 axioms of individual knowledge. "Yes" means that the schema is strongly valid; "No" means that it is not even weakly valid (keep in mind that strong validity implies validity). Incidentally, none of the properties turns out to be weakly but not strongly valid.

K	$\mathbb{K}_a(\varphi \to \psi) \to (\mathbb{K}_a\varphi \to \mathbb{K}_a\psi)$	Yes
D	$\neg\mathbb{K}_a\mathsf{false}$	Yes
T	$\mathbb{K}_a\varphi \to \varphi$	No
4	$\mathbb{K}_a\varphi \to \mathbb{K}_a\mathbb{K}_a\varphi$	Yes
5	$\neg\mathbb{K}_a\varphi \to \mathbb{K}_a\neg\mathbb{K}_a\varphi$	Yes

Thus, in general, the answer is *no*; particularly, the truth axiom does not hold. Note, however, that the axiom does hold in the restricted case when constructive knowledge is applied to positive strategic formulae, i.e., $\mathbb{K}_a\langle\langle B\rangle\rangle\varphi \to \langle\langle B\rangle\rangle\varphi$ *is* strongly valid in CSL. Moreover, the above results show that $\mathbb{K}_a$ satisfies the standard postulates for beliefs: logical closure, consistency, and positive as well as negative introspection. This suggests that "knowing how to play" sits somewhere in between the realms of doxastic and epistemic logic: it is stronger than belief but not quite as strong as standard knowledge.

11.5.5 Public Announcements

The combination of ATL modalities and epistemic modalities can be seen as extending ATL to take imperfect and incomplete information into account, but it can also be seen as a way of making epistemic logic dynamic: to allow us to reason about how knowledge changes. The "dynamic epistemic logics", dealing with how different types of actions and other events change knowledge, were discussed in Chapter 6. A difference between these logics and epistemic variants of ATL is, in addition to the important difference that the ATL language uses strategic/coalitional modalities, that the latter use a more abstract model of actions. But what if we restrict the actions considered in ATL to be one of the specific types of actions considered in dynamic epistemic logic? Here we will look at the case where actions are restricted to be *public announcements.*

For simplicity, we will only consider the language of Coalition logic CL (i.e., the next-step fragment of the ATL language), and we will consider a $\langle G\rangle$ modality with a simpler semantics: we let $\langle G\rangle\varphi$ mean that "there exists a joint public announcement by group G, such that φ will be true". The discussion that follows extend directly also to the cases where we allow the full ATL language, as well as the combined exists-forall interpretation "there exists a joint public announcement by group G such that no matter what the other agents announce, φ will be true".

In the case that actions are public announcements we can interpret the language simply in a standard epistemic model, which indeed models both which actions (public announcements) are available to an agent, and what the consequences of a group making a joint action (joint public announcement) are. Regarding the former, an agent can truthfully announce something she knows. Regarding the latter, the consequence of an announcement by a single agent i is the model update by a formula of the form $K_i\varphi$, assuming it is common knowledge that public announcements are truthful. The consequence of a joint announcement of a group G will then be a model update with a formula of the form $\bigwedge_{i\in G} K_i\varphi_i$.

Formally, let $\mathcal{M} = (\mathsf{St}, \sim_1, \ldots, \sim_n, V)$ be an epistemic model, where St is a set of states, $\sim_i \subseteq \mathsf{St} \times \mathsf{St}$ is an epistemic indistinguishability relation and is assumed to be an equivalence relation for each agent $i \in N$, and $V : \Theta \to 2^{\mathsf{St}}$ assigns primitive propositions to the states in which they are true. When q is a state in $\mathcal{M}$, we let:

$$\begin{aligned}\mathcal{M}, q \models \quad & \langle G\rangle\varphi \text{ iff } \exists\{\psi_i : i \in G\} \subseteq \mathcal{L}_{el} \text{ such that} \\ & \mathcal{M}, q \models \textstyle\bigwedge_{i\in G} K_i\psi_i \text{ and } \mathcal{M}|\textstyle\bigwedge_{i\in G} K_i\psi_i, q \models \varphi\end{aligned}$$

where $\mathcal{L}_{el}$ is the standard epistemic language and $\mathcal{M}|\psi = (\mathsf{St}', \sim'_1, \ldots, \sim'_n, V')$ is the update of $\mathcal{M}$ by ψ: $\mathsf{St}' = \{q' \in \mathsf{St} : \mathcal{M}, q' \models \psi\}$; $\sim'_i = \sim_i \cap (\mathsf{St}' \times \mathsf{St}')$; $V'(p) = V(p) \cap \mathsf{St}'$.

The consequences of restricting actions to truthful public announcements are (perhaps surprisingly) significant. This is due to the fact that now there is a very intimate relationship between knowledge and ability, since action is knowledge (the set of available actions *is* the set of known formulas). Let us revisit one of the key relationships between knowledge and ability we have considered in this chapter, in the case that actions are public announcements: the *de dicto*/*de re* distinction.

It is easy to see that

$$K_i\langle i\rangle\varphi \tag{11.6}$$

expresses the fact that agent i knows *de dicto* that she can enforce φ, i.e., that $\mathcal{M}, q \models K_i\langle i\rangle\varphi$ iff for any t such that $q \sim_i t$, there exists a $\psi \in \mathcal{L}_{el}$ such that $\mathcal{M}, t \models K_i\psi$ and $\mathcal{M}|K_i\psi, t \models \varphi$.

Consider now the formula

$$\langle i\rangle K_i\varphi \tag{11.7}$$

obtained by swapping the two modalities. Intuitively, this formula says that agent i can perform some action such that she will know φ afterwards. Note that in coalition logic (or ATL), the corresponding formula does *not* express the fact that agent i knows *de re* that she can enforce φ (this is true also for the single agent case, when the semantics of the $\langle\langle i \rangle\rangle$ operator of coalition logic is "there exists an action by i such that φ will be true, like the $\langle i \rangle$ operator introduced here). In fact, that claim is not expressible in ATL at all, as discussed before. However, it turns out that it *is* in fact expressed by (11.7) in the special case of public announcements.

Proposition 11.1
$\mathcal{M}, q \models \langle i \rangle K_i \varphi$ iff there exists a $\psi \in \mathcal{L}_{el}$ such that for any t such that $q \sim_i t$, $\mathcal{M}, t \models K_i \psi$ and $\mathcal{M}|K_i\psi, t \models \varphi$. $\dashv$

Again, this fact is due to the intimate relationship between knowledge and ability in the case that actions are public announcements, and does not hold in general.

11.6 Concluding remarks

The development of a logical theory that adequately captures the subtle interactions between knowledge and ability has been a long-standing goal in philosophy and, more recently, artificial intelligence. While it seems relatively straightforward to develop a theory that captures some meaningful notion of ability, it seems surprisingly hard to develop a theory that explains the interaction between knowledge and ability. In this article, we have surveyed work on this problem in philosophy, artificial intelligence, and mainstream computer science.

11.7 Notes

The seminal work of Robert Moore on the relationship between knowledge and ability can be found in (Moore, 1977, 1990). Much of the discussion in Section 11.1 is inspired by (and modelled on) Stanley and Williamson (2001). Ryle's argument that a reduction from know-how to know-that is not possible is found in (Ryle, 1949). Some other early philosophical approaches to logics of agency and ability are due to Chellas (1969), Brown (1988), and Belnap and Perloff (1988). See (Belnap, Perloff, and Xu, 2001, p.36) for comments on the semantics of Belnap and Perloff's operator. For a simpler variant of STIT, see, e.g., the work of Broersen, Herzig, and Troquard (2006). Troquard (2007) surveys STIT axioms and logics. We refer the reader to this contribution of Troquard (2007) for a discussion of STIT logics in artificial intelligence, and to a paper by Belnap et al. (2001) for a detailed review from the perspective of philosophical logic. Broersen et al. (2006) show how STIT logics can be used to encode the logic ATL.

The seminal work on ability in AI is by McCarthy and Hayes (1969). See, e.g., the work of Reiter (2001), for further developments in situation calculus. For details on the "standard translation" of modal logic into first-order logic mentioned in Section 11.2, we refer to Blackburn, de Rijke, and Venema (2001). An overview of dynamic logic is provided by Harel, Kozen, and Tiuryn (2000) while, e.g., Hughes and Cresswell (1968, pp. 83–91) present a discussion of *quantifying in* in the context of modal logic generally, Konolige (1986, pp. 38–42) discusses quantifying in for modal logics of knowledge and belief, and Fagin, Halpern, Moses, and Vardi (1995, pp. 80-83) present a semantics of first-order quantification in modal epistemic logic. Douglas Appelt's theorem-proving system is described by Appelt (1982, 1985). Other work using ideas from Moore's work to formalise ability include papers by Morgenstern (1986, 1987), Morgenstern and Stein (1988) Singh (1994), Wooldridge (2000), Jamroga and van der Hoek (2004) and by Jamroga and Ågotnes (2007).

Alternating-time Temporal Logic was introduced by Alur, Henzinger, and Kuperman (1997, 2002), and Coalition Logic by Pauly (2001, 2002). See the book of Chellas (1980) for an introduction to neighbourhood semantics. Soundness and completeness of the ATL axioms and inference rules shown in Section 11.3.4 was proved by Goranko and van Drimmelen (2006). EXPTIME-completeness of ATL was shown by van Drimmelen (2003) and by Goranko and van Drimmelen (2006) using alternation tree automata, and a practically efficient tableau-based decision procedure for it was developed by Goranko and Shkatov (2009). The whole ATL* was proved decidable and 2EXPTIME-complete by Schewe (2008) using alternating tree automata.

Different ways of incorporating agents' incomplete information and use of memory into ATL* have been studied by van der Hoek and Wooldridge (2003), Alur et al. (2002), Schobbens (2004), and by Jamroga and Ågotnes (2007). The four variants of the semantics of ATL* with incomplete information and the i/I/r/R notation is due to Schobbens (2004). The classification of ability under incomplete information as *objective* and *subjective* is due to Jamroga and Bulling (2011). The alternative approach to modeling executable strategies, mentioned at the end of Section 11.4.2, is studied by Guelev, Dima, and Enea (2011) and by Diaconu and Dima (2012).

See the work of Bulling and Jamroga (2013, 2011) and of Bulling, Dix, and Jamroga (2010) for further details on the impact of knowledge on strategic ability. Readers interested in modal μ-calculus and fixpoint operators can consult, e.g., the work of Bradfield and Stirling (2006). An introduction to alternation-free μ-calculus is given by Alur et al. (2002), while Bulling et al. (2010) provide more details on how adding epistemic relations to the semantics of strategic operators increases the complexity of model checking for CL, ATL, and ATL*. ATEL was first studied by van der Hoek and Wooldridge (2002, 2003). Completeness of coalition logic extended with epistemic axioms was shown by Ågotnes and Alechina (2012). The different possible interpretations of "bringing about" under imperfect information were discussed by Jamroga and van der Hoek (2004). Variants of ATL that cover only some of the possible ways strategic and epistemic operators can interact include approaches by Schobbens (2004), van Otterloo and Jonker (2004), and by Ågotnes (2006), while a more general treatment is proposed by Jamroga and van der Hoek (2004). Strategic STIT is studied by Horty (2001), and by Broersen et al. (2006), and Strategy Logic by Mogavero, Murano, and Vardi (2010). The paper by Herzig and Troquard (2006) shows how "knowing how to play" in the simpler case of single-step games can be expressed using a combination of Chellas STIT and epistemic logic. Constructive Strategic Logic (CSL) was introduced and studied by Jamroga and Ågotnes (2007).

Interpretations of coalition operators where actions are restricted to public announcements, leading, e.g., to *Group announcement logic*, are further studied by Ågotnes and van Ditmarsch (2008) and by Ågotnes, Balbiani, van Ditmarsch, and Seban (2010).

Acknowledgements The authors would like to thank Jan Broersen, who provided extensive comments on earlier drafts of this chapter, which helped to improve it considerably. Wojciech Jamroga acknowledges the support of the National Research Fund (FNR) Luxembourg under the project GALOT (INTER/DFG/12/06), as well as the support of the 7th Framework Programme of the European Union under the Marie Curie IEF project ReVINK (PIEF-GA-2012-626398). Michael Wooldridge would like to acknowledge the support of the ERC under Advanced Grant 291528 ("RACE").

References

Ågotnes, T. (2006). Action and knowledge in alternating-time temporal logic. *Synthese 149*(2), 377–409.

Ågotnes, T. and N. Alechina (2012, June). Epistemic coalition logic: Completeness and complexity. In *Proceedings of the Eleventh International Conference on Autonomous Agents and Multiagent Systems (AAMAS 2012)*, Valencia, Spain, pp. 1099–1106. IFAMAAS.

Ågotnes, T., P. Balbiani, H. van Ditmarsch, and P. Seban (2010). Group announcement logic. *Journal of Applied Logic 8*(1), 62–81.

Ågotnes, T. and H. van Ditmarsch (2008, May). Coalitions and announcements. In L. Padgham, D. Parkes, J. Muller, and S. Parsons (Eds.), *Proceedings of the Seventh International Conference on Autonomous Agents and Multiagent Systems (AAMAS 2008)*, Estoril, Portugal, pp. 673–680. IFAMAAS/ACM DL.

Alur, R., T. A. Henzinger, and O. Kuperman (1997, October). Alternating-time temporal logic. In *Proceedings of the 38th IEEE Symposium on Foundations of Computer Science*, pp. 100–109.

Alur, R., T. A. Henzinger, and O. Kuperman (2002). Alternating-time temporal logic. *Journal of the ACM 49*(5), 672–713.

Appelt, D. E. (1982). Planning natural language utterances. In *Proceedings of the Second National Conference on Artificial Intelligence (AAAI-82)*, Pittsburgh, PA, pp. 59–62.

Appelt, D. E. (1985). *Planning English Sentences.* Cambridge University Press: Cambridge, England.

Belnap, N. and M. Perloff (1988). Seeing to it that: a canonical form for agentives. *Theoria 54*, 175–199.

Belnap, N., M. Perloff, and M. Xu (2001). *Facing the Future: Agents and Choices in Our Indeterminist World.* Oxford University Press: Oxford, England.

Blackburn, P., M. de Rijke, and Y. Venema (2001). *Modal Logic.* Cambridge University Press: Cambridge, England.

Bradfield, J. and C. Stirling (2006). Modal mu-calculi. In P. Blackburn, J. van Benthem, and F. Wolter (Eds.), *Handbook of Modal Logic*, pp. 721–756. Elsevier.

Broersen, J., A. Herzig, and N. Troquard (2006). Embedding alternating-time temporal logic in strategic stit logic of agency. *Journal of Logic and Computation 16*(5), 559–578.

Brown, M. A. (1988). On the logic of ability. *Journal of Philosophical Logic 17*, 1–26.

Bulling, N., J. Dix, and W. Jamroga (2010). Model checking logics of strategic ability: Complexity. In M. Dastani, K. Hindriks, and J.-J. Meyer (Eds.), *Specification and Verification of Multi-Agent Systems*, pp. 125–159. Springer.

Bulling, N. and W. Jamroga (2011). Alternating epistemic mu-calculus. In *Proceedings of IJCAI-11*, pp. 109–114.

Bulling, N. and W. Jamroga (2013). Comparing variants of strategic ability. *Journal of Autonomous Agents and Multi-Agent Systems 26*(2), 288–314.

Chellas, B. (1980). *Modal Logic: an Introduction.* Cambridge University Press.

Chellas, B. F. (1969). *The Logical Form of Imperatives.* Ph. D. thesis, Stanford University.

Diaconu, R. and C. Dima (2012). Model-checking alternating-time temporal logic with strategies based on common knowledge is undecidable. *Applied Artificial Intelligence 26*(4), 331–348.

van Drimmelen (2003). Satisfiability in alternating-time temporal logic. In *Proceedings of 18th IEEE Symposium on Logic in Computer Science (LICS)*, pp. 208–217.

Fagin, R., J. Y. Halpern, Y. Moses, and M. Y. Vardi (1995). *Reasoning About Knowledge.* The MIT Press: Cambridge, MA.

Goranko, V. and D. Shkatov (2009). Tableau-based decision procedures for logics of strategic ability in multiagent systems. *ACM Trans. Comput. Log. 11*(1), 1–48.

Goranko, V. and G. van Drimmelen (2006). Complete axiomatization and decidablity of Alternating-time temporal logic. *Theor. Comp. Sci. 353*, 93–117.

Guelev, D., C. Dima, and C. Enea (2011). An alternating-time temporal logic with knowledge, perfect recall and past: axiomatisation and model-checking. *Journal of Applied Non-Classical Logics 21*(1), 93–131.

Harel, D., D. Kozen, and J. Tiuryn (2000). *Dynamic Logic.* The MIT Press: Cambridge, MA.

Herzig, A. and N. Troquard (2006). Knowing how to play: Uniform choices in logics of agency. In *Proceedings of AAMAS'06*, pp. 209–216.

van der Hoek, W. and M. Wooldridge (2002). Tractable multiagent planning for epistemic goals. In C. Castelfranchi and W. Johnson (Eds.), *Proceedings of the First International Joint Conference on Autonomous Agents and Multi-Agent Systems (AAMAS-02)*, pp. 1167–1174. ACM Press, New York.

van der Hoek, W. and M. Wooldridge (2003). Cooperation, knowledge and time: Alternating-time Temporal Epistemic Logic and its applications. *Studia Logica 75*(1), 125–157.

Horty, J. (2001). *Agency and Deontic Logic.* Oxford University Press.

Hughes, G. E. and M. J. Cresswell (1968). *Introduction to Modal Logic.* Methuen and Co., Ltd.

Jamroga, W. and T. Ågotnes (2007). Constructive knowledge: What agents can achieve under incomplete information. *Journal of Applied Non-Classical Logics 17*(4), 423–475.

Jamroga, W. and N. Bulling (2011, July). Comparing variants of strategic ability. In *Proceedings of the 22nd International Joint Conference on Artificial Intelligence (IJCAI)*, Barcelona, Spain, pp. 252–257.

Jamroga, W. and W. van der Hoek (2004). Agents that know how to play. *Fundamenta Informaticae 63*(2–3), 185–219.

Konolige, K. (1986). *A Deduction Model of Belief.* Pitman Publishing: London and Morgan Kaufmann: San Mateo, CA.

McCarthy, J. and P. J. Hayes (1969). Some philosophical problems from the standpoint of artificial intelligence. In B. Meltzer and D. Michie (Eds.), *Machine Intelligence 4*, pp. 463–502. Edinburgh University Press.

Mogavero, F., A. Murano, and M. Vardi (2010). Reasoning about strategies. In *Proceedings of FSTTCS*, pp. 133–144.

Moore, R. C. (1977). Reasoning about knowledge and action. In *Proceedings of the Fifth International Joint Conference on Artificial Intelligence (IJCAI-77)*, Cambridge, MA.

Moore, R. C. (1990). A formal theory of knowledge and action. In J. F. Allen, J. Hendler, and A. Tate (Eds.), *Readings in Planning*, pp. 480–519. Morgan Kaufmann Publishers: San Mateo, CA.

Morgenstern, L. (1986). A first-order theory of planning, knowledge, and action. In J. Y. Halpern (Ed.), *Proceedings of the 1986 Conference on Theoretical Aspects of Reasoning About Knowledge*, pp. 99–114. Morgan Kaufmann Publishers: San Mateo, CA.

Morgenstern, L. (1987). Knowledge preconditions for actions and plans. In *Proceedings of the Tenth International Joint Conference on Artificial Intelligence (IJCAI-87)*, Milan, Italy, pp. 867–874.

Morgenstern, L. and L. Stein (1988). Why things go wrong: A formal theory of causal reasoning. In *Proceedings of the Fifth National Conference on Artificial Intelligence (AAAI-86)*, Philadelphia, PA.

van Otterloo, S. and G. Jonker (2004). On Epistemic Temporal Strategic Logic. *Electronic Notes in Theoretical Computer Science 126*, 77–92.

Pauly, M. (2001). *Logic for Social Software.* Ph. D. thesis, University of Amsterdam.

Pauly, M. (2002). A modal logic for coalitional power in games. *Journal of Logic and Computation 12*(1), 149–166.

Reiter, R. (2001). *Knowledge in Action.* The MIT Press: Cambridge, MA.

Ryle, G. (1949). *The Concept of Mind.* Chicago University Press.

Schewe, S. (2008). ATL* satisfiability is 2ExpTime-complete. In *Proceedings of the 35th International Colloquium on Automata, Languages and Programming, Part II (ICALP 2008), 6–13 July, Reykjavik, Iceland,* Volume 5126 of *Lecture Notes in Computer Science*, pp. 373–385. Springer-Verlag.

Schobbens, P.-Y. (2004). Alternating-time logic with imperfect recall. *Electronic Notes in Theoretical Computer Science 85*(2), 82–93.

Singh, M. P. (1994). *Multiagent Systems: A Theoretical Framework for Intentions, Know-How, and Communications (LNAI Volume 799).* Springer-Verlag: Berlin, Germany.

Stanley, J. and T. Williamson (2001). Knowing how. *The Journal of Philosophy 98*(8), 411–444.

Troquard, N. (2007). *Independent Agents in Branching Time.* Ph. D. thesis, Université Paul Sabatier.

Wooldridge, M. (2000). *Reasoning about Rational Agents.* The MIT Press: Cambridge, MA.

Chapter 12

Knowledge and Security

Riccardo Pucella

Contents

Security: the state of being free from danger or threat.

(New Oxford American Dictionary)

Abstract A persistent intuition in the field of computer security says that epistemic logic, and more generally epistemic concepts, are relevant to the formalization of security properties. Confidentiality, integrity, authentication, anonymity, non-repudiation, all can be expressed as epistemic properties. This survey illustrates the use of epistemic concepts and epistemic logic to formalize a specific security property, confidentiality, in two large domains of application: cryptographic protocol analysis, and multi-level security systems.

12.1 Introduction

A persistent intuition in some quarters of the security research community says that epistemic logic and, more generally, epistemic concepts are useful

Chapter 12 of the *Handbook of Epistemic Logic*, H. van Ditmarsch, J.Y. Halpern, W. van der Hoek and B. Kooi (eds), College Publications, 2015, pp. 591–655.

for reasoning about the security of systems. What grounds this intuition is that much work in the field is based on epistemic concepts—sometimes explicitly, but more often implicitly, by and large reinventing possible-worlds semantics for knowledge and belief.[1]

Reasoning about the security of systems in practice amounts to establishing that those systems satisfy various security properties. A security property, roughly speaking, is a property of a system stating that the system is not vulnerable to a particular threat. Threats, in this context, are generally taken to be attacks by agents intent on subverting the system.

While what might be considered a threat—and therefore what security properties are meant to protect against—is in the eye of the beholder, several properties have historically been treated as security properties:

- **Data Confidentiality:** only authorized agents should have access to a piece of data; more generally, only authorized agents should be able to infer any information about a piece of data.
- **Data Integrity:** only authorized agents should have access to alter a piece of data.
- **Agent Authentication:** an agent should be able to prove her identity to another agent.
- **Data Authentication:** an agent should be able to determine the source of a piece of data.
- **Anonymity:** the identity of an agent or the source of a piece of data should be kept hidden except from authorized agents.
- **Message Non-repudiation:** the sender of a message should not be able to deny having sent the message.

These properties may seem intuitive on a first reading but they are vague and depend on terms that require clarification: *secret*, *authorized*, *access to a piece of data*, *source*, *identity*.

Epistemic concepts come into play when defining many of the terms that appear in the statements of security properties. Indeed, those terms can often be usefully understood in terms of knowledge: confidentiality can be read as *no agent except for authorized agents can know a piece of information*; authentication as *an agent knows the identity of the agent with*

[1]This chapter assumes from the reader a basic knowledge of epistemic logic and its Hintikka-style possible-worlds semantics; see §12.6 for references. Furthermore, to simplify the exposition, the term *epistemic* is used to refer both to knowledge and to belief throughout.

whom she is interacting, or *an agent knows the identity of the agent who sent the information*; anonymity as *no one knows the identity of the agent who performed a particular action*; and so on. While it is not the case that every security property can be read as an epistemic property, enough of them can to justify studying them *as* epistemic properties.

Epistemic logic and epistemic concepts play two roles in security research:

- **Definitional**: they are used to formalize security properties and concepts, and provide a clear semantic grounding for them. Epistemic logic may be explicitly used as an explanatory and definitional language for properties of interest.

- **Practical**: they are used to derive verification and enforcement techniques for security properties, that is, to either establish that a security property is true in a system, or to force a security property to be true in a system.

It is fair to say that after nearly three decades of research, epistemic logic has had several successes on the definitional front and somewhat fewer on the practical front. This is perhaps not surprising. While epistemic logic and epistemic concepts are well suited for definitions and for describing semantic models, verification of epistemic properties tends to be expensive, and tools for the verification of security properties in practice often approximate epistemic properties using properties that are easier to check, such as safety properties.[2]

This chapter illustrates the use of epistemic logic and epistemic concepts for reasoning about security through the study of a specific security property, confidentiality. Not only is confidentiality a prime example of the use of knowledge to make a security property precise, but it has also been studied extensively from several perspectives. Moreover, many of the issues arising while studying confidentiality also arise for other security properties with an epistemic flavor.

Confidentiality is explored in two contexts: cryptographic protocols in §12.2, and multi-level security systems in §12.3. Cryptographic protocols are communication protocols that use cryptography to protect information exchanged between agents in a system. While it may seem simple enough for Alice to send a confidential message to Betty by encrypting it, Alice

[2] A safety property is a property of the form *a bad state is never reached in any execution of the system*, for some definition of *bad state*. A safety property can be checked by examining every possible execution independently of any other; in contrast, checking an epistemic property requires examining every possible execution in the context of all other possible executions.

and Betty need to share a common key for this to work. How is such a key distributed before communication may take place? Most cryptographic protocols involve key creation and distribution, and these are notoriously difficult to get right. Key distribution also forces the consideration of authentication as an additional security property. Cryptographic protocol analysis is the one field of security research that has explicitly and extensively used epistemic logic, and the bulk of this chapter is dedicated to that topic.

In multi-level security systems, one is generally interested in confidentiality guarantees even when information is used or released within the system during a computation. The standard example is that of a centralized system where agents have different security clearances and interact with data with different security classifications; the desired confidentiality guarantees ensure that classified data, no matter how it is manipulated by agents with an appropriate security clearance, never flows to an agent that does not have an appropriate security clearance. Most of the work in this field of security research uses epistemic concepts implicitly— the models use possible-worlds definitions of knowledge, but no epistemic logic is introduced. All reasoning is semantic reasoning in the models.

Security properties other than confidentiality are briefly discussed in §12.4. The chapter concludes in §12.5 with some personal views on the use of epistemic logic and epistemic concepts in security research. My observations should not be particularly controversial, but my main conclusion remains that progress beyond the current state of the art in security research—at least in security research that benefits from epistemic logic—will require a deeper understanding of resource-bounded knowledge, which is itself an active research area in epistemic logic.

All bibliographic references are postponed to §12.6, where full references and additional details are given for topics covered in the main body of the chapter. It is worth noting that the literature on reasoning about security draws from several fields besides logic. For instance, much of the research on cryptographic protocols derives from earlier work in distributed computing. Similarly, recent research both on cryptographic protocol analysis and on confidentiality in multi-level systems is based on work in programming language semantics and static analysis. The interested reader is invited to follow the references given in §12.6 for details.

12.2 Cryptographic Protocols

Cryptographic protocols are communication protocols—rules for exchanging messages between agents—that use cryptography to achieve a security goal such as authenticating one agent to another, or exchanging confidential messages.

Cryptographic protocols are a popular object of study for several reasons. First, they are concrete—they correspond to actual artifacts implemented and used in practice. Second, their theory extends that of distributed protocols and network protocols in general, which are themselves thoroughly studied.

Cryptographic protocols have characteristics that distinguish them from more general communication protocols. In particular, they

(1) enforce security properties;

(2) rely on cryptography;

(3) execute in the presence of attackers that might attempt to subvert them.

Protocols can be analyzed concretely or symbolically. The concrete perspective views protocols as exchanging messages consisting of sequences of bits and subject to formatting requirements, which is the perspective used in most network protocols research. The symbolic perspective views protocols as exchanging messages consisting of symbols in some formal language, which is the perspective used in most distributed protocols research. The focus of this section is on symbolic cryptographic protocols analysis.

12.2.1 Protocols

A common notation for protocols is to list the sequence of messages exchanged between the parties involved in the protocol, since the kinds of protocols studied rarely involve complex control flow.

A simple protocol between Alice and Betty (represented by A and B) in which Alice sends message m_1 to Betty and Betty responds by sending message m_2 to Alice would be described by:

$$\begin{array}{llcll} 1. & A & \longrightarrow & B & : & m_1 \\ 2. & B & \longrightarrow & A & : & m_2 \end{array} \qquad (12.1)$$

The message sequence notation takes a global view of the protocol, describing the protocol from the outside, so to speak. An alternate way to describe a protocol is to specify the roles of the parties involved in the protocol. For protocol (12.1), for instance, there are two roles: the initiator role, who sends message m_1 to the receiver and waits for a response message, and the receiver role, who waits for a message to arrive from the initiator and responds with m_2.

A protocol executes in an environment, which details anything relevant to the execution of said protocol, such as the agents participating in the protocol, whether other instances of the protocol are also executed concurrently, the possible attackers and their capabilities. The result of executing a protocol in a given environment can be modeled by a set of traces, where a trace corresponds to a possible execution of the protocol. A trace is a sequence of global states. A global state records the local state of every agent involved in the protocol, as well as the state of the environment. This general description is compatible with most representations in the literature, and can be viewed as a Kripke structure by defining a suitable accessibility relation over the states of the system.

To illustrate protocols in general, and initiate the study of confidentiality, here are two simple protocols that achieve a specific form of confidentiality without requiring cryptography. One lesson to be drawn from these examples is that confidentiality in some cases can be achieved without complex operations.

The first protocol solves an instance of the following problem: how two agents may exchange secret information in the open, without an eavesdropping third agent learning about the information. The instance of the problem, called the Russian Cards problem, is pleasantly concrete and can be explained to children: Alice and Betty each draw three cards from a pack of seven cards, and Eve (the eavesdropper) gets the remaining card. Can players Alice and Betty learn each other's cards without revealing that information to Eve? The restriction is that Alice and Betty can only make public announcements that Eve can hear.

Several protocols for solving the Russian Cards problem have been proposed; a fairly simple solution is the *Seven Hands protocol.* Recall that there are seven cards: three are dealt to Alice, three are dealt to Betty, and the last card is dealt to Eve. Call the cards dealt to Alice a_1, a_2, a_3, and the cards dealt to Betty b_1, b_2, b_3. The card dealt to Eve is e.

The Seven Hands protocol is a two-step protocol that Alice can use to tell her cards to Betty and learn Betty's cards in response:

$$\begin{array}{llll} 1. & A \longrightarrow & B & : SH_A \\ 2. & B \longrightarrow & A & : SH_B \end{array} \qquad (12.2)$$

where SH_A and SH_B are the following specific messages:

(1) Message SH_A is constructed by Alice as follows. Alice first chooses a random renaming W, X, Y, Z of the elements in $\{b_1, b_2, b_3, e\}$, that is, a random permutation of the four cards not in her hand. Message SH_A then consists of the following seven subsets of cards, in some

arbitrary order:

$$\begin{array}{cc} \multicolumn{2}{c}{\{a_1, a_2, a_3\}} \\ \{a_1, W, X\} & \{a_1, Y, Z\} \\ \{a_2, W, Y\} & \{a_2, X, Z\} \\ \{a_3, W, Z\} & \{a_3, X, Y\} \end{array}$$

These subsets are carefully chosen: for every possible hand of Betty, that is, for every possible subset S of size three of $\{W, X, Y, Z\}$, there is exactly one set in SH_A with which S has an empty intersection, and that set is Alice's hand $\{a_1, a_2, a_3\}$. Thus, upon receiving SH_A, Betty can identify Alice's hand by examining the sets Alice sent and picking the one with which her own hand has an empty intersection, and in the process Betty learns Alice's hand, and by elimination, Eve's card.

(2) Message SH_B, Betty's response, is simply Eve's card. Alice, upon receiving SH_B, knows her own hand and Eve's card, and therefore can infer by elimination Betty's hand.

At the end of the exchange, Alice knows Betty's hand and Betty knows Alice's hand, as required.

What about Eve? She does not learn anything about the cards in Alice or Betty's hand. Indeed, after seeing Alice's message, Eve has no information about Alice's hand, since every card appears in exactly three of the sets Alice sent. There is no way for Eve to isolate which of those cards might be one of Alice's. Furthermore, after seeing Betty's message, all she has learned is her own card, which she already knew.

If we define $c \in i$ to be the primitive proposition *card c is in player i's hand* (where A, B, E represent Alice, Betty, and Eve, respectively) then we expect that the following epistemic formula holds after the first message is received by Betty:

$$K_B(a_1 \in A) \wedge K_B(a_2 \in A) \wedge K_B(a_3 \in A),$$

that the following epistemic formula holds after the second message is received by Alice:

$$K_A(b_1 \in B) \wedge K_A(b_2 \in B) \wedge K_B(b_3 \in B),$$

and that the following formula holds after either of the messages is received:

$$\neg K_E(a_1 \in A) \wedge \neg K_E(a_2 \in A) \wedge \neg K_E(a_3 \in A) \\ \wedge \neg K_E(b_1 \in B) \wedge \neg K_E(b_2 \in B) \wedge \neg K_E(b_3 \in B).$$

It is an easy exercise to construct the Kripke structures describing this scenario.

The Seven Hands protocol is ideally suited for epistemic reasoning via a possible-worlds semantics for knowledge, as it relies on combinatorial analysis. Its applicability, however, is limited.

The second protocol is a protocol to ensure anonymity, which is a form of confidentiality (see §12.4). It does not rely on combinatorial analysis but rather on properties of the XOR operation.[3] The Dining Cryptographers protocol was originally developed to solve the following problem. Suppose that Alice, Betty, and Charlene are three cryptographers having dinner at their favorite restaurant. Their waiter informs them that arrangements have been made for the bill to be paid anonymously by one party. That payer might be one of the cryptographers, but it might also be U.S. National Security Agency. The three cryptographers respect each other's right to make an anonymous payment, but they would like to know whether the NSA is paying.

The following protocol can be used to satisfy the cryptographers' curiosity and allow each of them to determine whether the NSA or one of her colleagues is paying, without revealing the identity of the payer in the latter case.

(1) Every cryptographer i flips a fair coin privately with her neighbor j on her right: the Boolean result $T_{\{i,j\}}$ is *true* if the coin lands tails, and *false* if the coin lands heads. Thus, the cryptographers produce the Boolean results $T_{\{A,B\}}$, $T_{\{A,C\}}$, $T_{\{B,C\}}$; Alice sees $T_{\{A,B\}}$ and $T_{\{A,C\}}$; Betty sees $T_{\{A,B\}}$ and $T_{\{B,C\}}$; Charlene sees $T_{\{A,C\}}$ and $T_{\{B,C\}}$.

(2) Every cryptographer i computes a private Boolean value Df_i as *true* if the two coin tosses she has witnessed are different, and *false* if they are the same. Thus, $Df_A = T_{\{A,B\}} \oplus T_{\{A,C\}}$, $Df_B = T_{\{A,B\}} \oplus T_{\{B,C\}}$, and $Df_C = T_{\{A,C\}} \oplus T_{\{B,C\}}$.

(3) Every cryptographer i publicly announces Df_i, except for the paying cryptographer (if there is one) who announces $\neg Df_i$, the negation of Df_i.

Once the protocol is executed, any curious cryptographer interested in determining who paid for dinner simply has to take the XOR of all the announcements: if the result is *false*, then the NSA paid, and if the result is *true*, then one of the cryptographers paid.

---

[3] XOR (exclusive or) is a binary Boolean operation $\oplus$ defined by taking $b_1 \oplus b_2$ to be true if and only if exactly one of b_1 or b_2 is true. It is associative and commutative.

To see why this is the case, consider the two possible scenarios. Suppose the NSA paid. Then the XOR of all the announcements is:

$$\begin{aligned}
&Df_A \oplus Df_B \oplus Df_C \\
&\quad = (T_{\{A,B\}} \oplus T_{\{A,C\}}) \oplus (T_{\{B,C\}} \oplus T_{\{A,B\}}) \oplus (T_{\{A,C\}} \oplus T_{\{B,C\}}) \\
&\quad = (T_{\{A,B\}} \oplus T_{\{A,B\}}) \oplus (T_{\{B,C\}} \oplus T_{\{B,C\}}) \oplus (T_{\{A,C\}} \oplus T_{\{A,C\}}) \\
&\quad = \mathit{false} \oplus \mathit{false} \oplus \mathit{false} \\
&\quad = \mathit{false}
\end{aligned}$$

whereas if one of the cryptographers paid (without loss of generality, suppose it is Alice), then the XOR of all the announcements is:

$$\begin{aligned}
&\neg Df_A \oplus Df_B \oplus Df_C \\
&\quad = \neg\,(T_{\{A,B\}} \oplus T_{\{A,C\}}) \oplus (T_{\{B,C\}} \oplus T_{\{A,B\}}) \oplus (T_{\{A,C\}} \oplus T_{\{B,C\}}) \\
&\quad = (\neg T_{\{A,B\}} \oplus T_{\{A,C\}}) \oplus (T_{\{B,C\}} \oplus T_{\{A,B\}}) \oplus (T_{\{A,C\}} \oplus T_{\{B,C\}}) \\
&\quad = (\neg T_{\{A,B\}} \oplus T_{\{A,B\}}) \oplus (T_{\{B,C\}} \oplus T_{\{B,C\}}) \oplus (T_{\{A,C\}} \oplus T_{\{A,C\}}) \\
&\quad = \mathit{true} \oplus \mathit{false} \oplus \mathit{false} \\
&\quad = \mathit{true}
\end{aligned}$$

If one of the cryptographers paid, neither of the two other cryptographers will know which of her colleagues paid, since either possibility is compatible with what they can observe. Again, it is an easy exercise to construct the Kripke structures capturing these scenarios.

12.2.2 Cryptography

While protocols such as the Seven Hands protocol and the Dining Cryptographers protocol enforce confidentiality by carefully constructing specific messages meant to convey specific information in a specific context, most cryptographic protocols rely on cryptography for confidentiality.

Cryptography seems a natural approach for confidentiality. After all, the whole point of cryptography is to hide information in such a way that only agents with a suitable key can access the information. And indeed, if the goal is for Alice to send message m to Betty when Alice and Betty alone share a key to encrypt and decrypt messages, then the simplest protocol for confidential message exchange is simply for Alice to encrypt m and send it to Betty. But how do Alice and Betty come to share a key in the first place? Distributing keys is tricky, because keys have to be sent to the right agents, in such a way that no other agent can get them.

Before addressing those problems, let us review the basics of cryptography. The reader is assumed to have been exposed to at least informal descriptions of cryptography. An encryption scheme is defined by a set of sourcetexts, a set of ciphertexts, a set of keys, and for every key k an injective encryption function e_k producing a ciphertext from a sourcetext and a decryption function d_k producing a sourcetext from a ciphertext, with the property that $d_k(e_k(x)) = x$ for all sourcetexts x. We often assume that ciphertexts and keys are included in sourcetexts to allow for nested encryption and encrypted keys.[4]

There are two broad classes of encryption schemes, which differ in how keys are used for decryption. *Shared-key encryption schemes* require an agent to have a full key to both encrypt and decrypt a message. They tend to be efficient, and can often be implemented directly in hardware. *Public-key encryption schemes*, on the other hand, are set up so that an agent only needs to know part of a key (called the public key) to encrypt a message, while needing the full key to decrypt a message. The full key cannot be easily recovered from knowing only the public key. Public keys are generally made public (hence the name), so that any agent can encrypt a message intended for, say, Alice, by looking up and using Alice's public key. Since only Alice has the full key, only she can decrypt that message. DES and AES are concrete examples of shared-key encryption schemes, while RSA and elliptic-curve encryption schemes are concrete examples of public-key encryption schemes.

Cryptographic protocols are needed with shared-key encryption schemes because agents need to share a key in order to exchange encrypted messages. How is such a shared key distributed? And how can agents make sure they are not tricked into sharing those keys with attackers? An additional difficulty is that when the same shared key is reused for every interaction between two agents, the content of all those interactions becomes available to an attacker that manages to learn that key. To minimize the impact of a key compromise, many systems create a fresh session key for any two agents that want to communicate, which exacerbates the key distribution problem.

Public-key encryption simplifies key distribution, since public keys can simply be published. Any agent wanting to send a confidential message to Alice has only to look up Alice's public key and use it to encrypt her message. The problem, from Alice's perspective, is that anyone can encrypt a message and send it to her, which means that if Alice wants to make sure that the encrypted message she received came from Betty, some sort of authentication mechanism is needed. Furthermore, all known public-key encryption schemes are computationally expensive, so a common approach

[4]This section considers deterministic encryption schemes only, ignoring probabilistic encryption schemes.

is to have agents that want to exchange messages in a session first use public-key encryption to generate a session key for a shared-key encryption scheme that they use for their exchanged messages. Such a scenario requires authentication to ensure that agents are not tricked into communicating with an attacker.

Sample Cryptographic Protocols Most classical cryptographic protocols are designed to solve the problem of key distribution for shared-key encryption schemes, and of authentication for public-key encryption schemes. In these contexts, confidentiality and authentication are the key properties: confidentiality to enforce that distributed keys remain secret from attackers, and authentication to ensure that agents can establish the identity of the other agents involved in a message exchange.

This section presents two protocols, each illustrating different problems that can arise and highlighting vulnerabilities that attackers can exploit. (Attackers will be introduced more carefully in the next section.) The first protocol distributes session keys for a shared-key encryption scheme, while the second protocol aims at achieving mutual authentication for public-key encryption schemes. Not all of the problems illustrated will occur in every protocol, of course, nor are vulnerabilities in one context necessarily vulnerabilities in another context.

For the first protocol, consider the following situation. Suppose Alice wants to communicate with Betty, and there is a trusted server Serena who will generate a shared session key (for some shared-key encryption scheme) for them to use. Assume that every registered user of the system shares a distinct key with the trusted server in some shared-key encryption scheme; these keys for Alice and Betty are denoted k_{AS} and k_{BS}.

The idea is for Serena to generate a fresh key and send it to both Alice and Betty. Sending it in the clear, however, would allow an eavesdropping attacker to read it and then use it to decrypt messages between Alice and Betty. Since Alice and Betty both share a key with Serena, one solution might be to use those keys to encrypt the session key sent to Alice and Betty, but this turns out to be difficult to implement in practice. Here is the problem. Alice, wanting to communicate with Betty, sends a message to Serena asking her to generate a session key, and Serena sends it to both Alice and Betty. As far as Betty is concerned, she receives a key with an indication that Alice will use it to send her messages. Betty now has to store the key and wait for Alice to send her messages encrypted with that key. If Alice wants to set up several concurrent communications with Betty, then Betty will have to match each incoming communication with the appropriate key, which is annoying at best and inefficient at worst. It turns out to be more efficient for Serena to send the fresh session key k_{sess} to Alice, and for Alice to forward the key to Betty in her first message. This observation leads to the following protocol:

$$\begin{array}{llcll} 1. & A \longrightarrow & S & : & B \\ 2. & S \longrightarrow & A & : & B, \{k_{sess}\}_{k_{AS}}, \{k_{sess}\}_{k_{BS}} \\ 3. & A \longrightarrow & B & : & A, \{k_{sess}\}_{k_{BS}} \end{array} \qquad (12.3)$$

Both Alice and Betty learn key k_{sess}, which is kept secret from eavesdroppers.

While this protocol might seem sufficient to distribute a key to both Alice and Betty, several things can go wrong in the presence of an insider attacker, that is, an attacker that is also a registered user of the system and has control over the network (i.e., can intercept and forge messages; see §12.2.3).

The insider attacker, Isabel, can initiate a communication with trusted server Serena via her shared key k_{IS} (Isabel is assumed to have such a key because she is a registered user of the system), and use the key to pose as Alice to Betty. Here is a sequence of messages exemplifying the attack, where the notation $I[A]$ denotes I posing as A:[5]

$$\begin{array}{rcccl} I & \longrightarrow & S & : & B \\ S & \longrightarrow & I & : & B, \{k_{sess}\}_{k_{IS}}, \{k_{sess}\}_{k_{BS}} \\ I[A] & \longrightarrow & B & : & A, \{k_{sess}\}_{k_{BS}} \end{array} \qquad (12.4)$$

Betty believes that she is sharing key k_{sess} with Alice, while she is in fact sharing it with Isabel. This is a failure of authentication—the protocol does not authenticate the initiator to the responder.

Isabel can also trick Alice into believing she is talking to Betty, by posing as the server and intercepting messages between Alice and the server, as the following sequence of messages exemplifies:

$$\begin{array}{rcccl} A & \longrightarrow & I[S] & : & B \\ I[A] & \longrightarrow & S & : & I \\ S & \longrightarrow & I[A] & : & I, \{k_{sess}\}_{k_{AS}}, \{k_{sess}\}_{k_{IS}} \\ I[S] & \longrightarrow & A & : & B, \{k_{sess}\}_{k_{AS}}, \{k_{sess}\}_{k_{IS}} \\ A & \longrightarrow & I[B] & : & A, \{k_{sess}\}_{k_{IS}} \end{array} \qquad (12.5)$$

This form of attack is commonly known as a man-in-the-middle attack. Isabel intercepts Alice's message to the server, and turns around and sends a different message to the server posing as Alice. The response from the

[5]We can assume that every message has a *from* and *to* field—think email—and that these can be forged. Isabel posing as Alice means that Isabel sends a message and forges the *from* field of the message to hold Alice's name.

server is intercepted by Isabel, who crafts a suitable response back to Alice. Alice takes that response (which she believes is coming from the server) and sends it to Betty, but that message is intercepted by Isabel as well. Now, as far as Alice is concerned, she has successfully completed the protocol, and holds a key k_{IS} that she believes she can use to communicate confidentially with Betty, while she is really communicating with Isabel.

How can we correct these vulnerabilities? One feature on which these attacks rely is that the identity of the intended parties for the keys in the protocol are easily forged by the attacker. So one fix is to bind the intended parties to the appropriate copies of the key. Here is an amended version of the protocol:

$$\begin{array}{llcl} 1. & A \longrightarrow S & : & B \\ 2. & S \longrightarrow A & : & B, \{B, k_{sess}\}_{k_{AS}}, \{A, k_{sess}\}_{k_{BS}} \\ 3. & A \longrightarrow B & : & \{A, k_{sess}\}_{k_{BS}} \end{array} \quad (12.6)$$

When Alice receives her response from the server and she decrypts her message $\{B, k_{sess}\}_{k_{AS}}$, she can verify that the key she meant Serena to create to communicate with Betty is in fact a key meant to communicate with Betty. This suffices to foil Isabel in attack (12.5). Similarly, when Betty receives her message from Alice containing $\{A, k_{sess}\}_{k_{BS}}$, she can verify that the key is meant to communicate with Alice. This suffices to foil attack (12.4).

Protocol (12.6) now seems to work as intended. It does suffer from another potential vulnerability, though, one that is less directly threatening, but can still cause problems: it is susceptible to a replay attack. Here is the scenario. Suppose that Isabel eavesdrops on messages as Alice gets a session key k_O from the trusted server to communicate with Betty, and holds on to messages $\{B, k_O\}_{k_{AS}}$ and $\{A, k_O\}_{k_{BS}}$. Suppose further that after a long delay Isabel manages to somehow obtain key k_O, perhaps by breaking into Alice's or Betty's computer, or by expending several months' worth of effort to crack the encryption. Once Isabel has k_O, she can subvert an attempt by Alice to get a session key for communicating with Betty by simply intercepting the messages from Alice to Serena, and replaying the messages $\{B, k_O\}_{k_{AS}}$ and $\{A, k_O\}_{k_{BS}}$ she intercepted in the past. The following sequence of messages exemplifies this attack:

$$\begin{array}{rcccl} A & \longrightarrow & I[S] & : & B \\ I[S] & \longrightarrow & A & : & B, \{B, k_O\}_{k_{AS}}, \{A, k_O\}_{k_{BS}} \\ A & \longrightarrow & B & : & \{A, k_O\}_{k_{BS}} \end{array} \quad (12.7)$$

The main point here is that Alice and Betty after this protocol interaction end up using key k_O as their session key, but that key is one that Isabel knows, meaning that Isabel can decrypt every single message that Alice and Betty exchange in that session. So even though she does not have the shared keys k_{AS} and k_{BS}, she has managed to trick Alice and Betty into using a key she knows.

Preventing this kind of replay attack requires ensuring that messages from earlier executions of the protocol cannot be used in later executions. One way to do that is to have every agent record every message they have ever sent and received, but that is too expensive to be practical. The common alternative is to use timestamps, or nonces. A nonce is a large random number, meant to be unpredictable and essentially unique—the likelihood that the same nonce occurs twice within two different sessions should be negligible. To fix protocol (12.6) and prevent replay attacks, it suffices for Alice and Betty to generate nonces n_A and n_B, respectively, and send them to trusted server Serena so that she can include them in her responses:

$$\begin{array}{lllll} 1. & A & \longrightarrow & B & : & n_A \\ 2. & B & \longrightarrow & S & : & A, n_A, n_B \\ 3. & S & \longrightarrow & A & : & \{B, k_{sess}, n_A\}_{k_{AS}}, \{A, k_{sess}, n_B\}_{k_{BS}} \\ 4. & A & \longrightarrow & B & : & \{A, k_{sess}, n_B\}_{k_{BS}} \end{array} \tag{12.8}$$

As long as Alice and Betty, when each receives her encrypted message containing the session key, both check that the nonce in the encrypted message is the one that they generated, then they can be confident that the encrypted messages have not been reused from earlier sessions.

Protocol (12.8) now seems to work as intended and is not vulnerable to replay attacks. But it does not actually guarantee mutual authentication; that is, it does not guarantee to Alice that she is in fact talking to Betty when she believes she is, and to Betty that she is in fact talking to Alice when she believes she is. Consider the following attack, in which attacker Trudy is not an insider—she is not a registered user of the system—but has control of the network and thus can intercept and forge messages. Trudy poses as Betty by intercepting messages from Alice and forging responses:

$$\begin{array}{ccccl} A & \longrightarrow & T[B] & : & n_A \\ T[B] & \longrightarrow & S & : & A, n_A, n_T \\ S & \longrightarrow & A & : & \{B, k_{sess}, n_A\}_{k_{AS}}, \{A, k_{sess}, n_T\}_{k_{BS}} \\ A & \longrightarrow & T[B] & : & \{A, k_{sess}, n_T\}_{k_{BS}} \end{array} \tag{12.9}$$

From Alice's perspective, she has completed the protocol by exchanging messages with Betty, and holds a session key for sending confidential messages to Betty. But of course Alice has been communicating with Trudy, and Betty is not even aware of the exchange. Trudy cannot actually read the messages sent by Alice, so there is no breach of confidentiality, but Trudy has managed to trick Alice into believing she shares a session key with Betty. In terms of knowledge, Alice knows the session key, but she does not know that Betty does.

There is also a way for Trudy to trick Betty into believing she shares a session key with Alice, by posing as Alice:

$$\begin{array}{rcccl} T[A] & \longrightarrow & B & : & n_T \\ B & \longrightarrow & S & : & A, n_T, n_B \\ S & \longrightarrow & T[A] & : & \{B, k_{sess}, n_T\}_{k_{AS}}, \{A, k_{sess}, n_B\}_{k_{BS}} \\ T[A] & \longrightarrow & B & : & \{A, k_{sess}, n_B\}_{k_{BS}} \end{array} \tag{12.10}$$

From Betty's perspective, she has completed the protocol by exchanging messages with Alice, and holds a session key for sending confidential messages to Alice. But of course Betty has been communicating with Trudy, and Alice is not even aware of the exchange. In terms of knowledge, Betty knows the session key, but she does not know that Alice does.

Mutual authentication is achieved through an additional nonce exchange at the end of the protocol which uses the newly created session key:

$$\begin{array}{rrcccl} 1. & A & \longrightarrow & B & : & n_A \\ 2. & B & \longrightarrow & S & : & A, n_A, n_B, n'_B \\ 3. & S & \longrightarrow & A & : & n'_B, \{B, k_{sess}, n_A\}_{k_{AS}}, \{A, k_{sess}, n_B\}_{k_{BS}} \\ 4. & A & \longrightarrow & B & : & n'_A, \{A, k_{sess}, n_B\}_{k_{BS}}, \{A, n'_B\}_{k_{sess}} \\ 5. & B & \longrightarrow & A & : & \{B, n'_A\}_{k_{sess}} \end{array} \tag{12.11}$$

This protocol now seems to work as intended without being vulnerable to replay attacks or authentication failures. How can this be guaranteed?

Intuitively, protocol (12.11) is not susceptible to replay attacks because of the use of nonces: Alice can deduce that the first encrypted component of the third message was not reused from an earlier protocol execution, and Betty can deduce that the first encrypted component of the fourth message was not reused from an earlier protocol execution.

Similarly, mutual authentication in protocol (12.11) follows from the use of shared keys: Alice can deduce that Serena created the first encrypted component of the third message; and Betty can deduce that Serena created the first encrypted component of the fourth message. Moreover, if Betty believes that k_{sess} is a key known only to Alice and herself, then she can deduce that Alice created the second encrypted component in the fourth message, and similarly, if Alice believes that k_{sess} is a key known only to Betty and herself, then she can deduce that Betty created the encrypted component in the fifth message.

The confidentiality of the session key requires the assumption that trusted server Serena is indeed trustworthy and creates keys that have not previously been used and distributed to other parties. If so, then Alice can deduce that the session key she receives in the third message is a confidential key for communicating with Betty, since Alice can also deduce that she has been executing the protocol with Betty. Similarly, Betty can deduce that the session key she receives in the fourth message is a confidential key for communicating with Alice, since Betty can deduce that she has been executing the protocol with Alice

In a precise sense, the goal of cryptographic protocol analysis is to prove these kind of properties formally, and many techniques have been developed which are surveyed below in §12.2.5.

The second cryptographic protocol uses a public-key encryption scheme to achieve mutual authentication:

$$\begin{array}{llcll} 1. & A & \longrightarrow & B & : & \{A, n_A\}_{pk_B} \\ 2. & B & \longrightarrow & A & : & \{n_A, n_B\}_{pk_A} \\ 3. & A & \longrightarrow & B & : & \{n_B\}_{pk_B} \end{array} \tag{12.12}$$

where pk_A and pk_B are the public keys of Alice and Betty, respectively, and n_A and n_B are nonces.

Intuitively, when Alice receives her nonce n_A back in the second message, she knows that Betty must have decrypted her first message at some point during the execution of the protocol (because only Betty could have decrypted the message that contained it), and similarly, when Betty receives her nonce n_B back, she knows that Alice must have decrypted her second message. Note that n_A and n_B are kept confidential throughout the protocol, and that mutual authentication relies on that confidentiality.

Protocol (12.12), known as the Needham-Schroeder protocol, achieves mutual authentication even in the presence of an attacker that has control of the network and can intercept and forge messages. It is however vulnerable to insider attackers that are registered users of the system and have control of the network. For example, insider attacker Isabel can use an attempt by Alice to initiate an authentication session with her to trick unsuspecting Betty into believing that Alice is initiating an authentication session with her:

$$\begin{array}{ccccl} A & \longrightarrow & I & : & \{A, n_A\}_{pk_I} \\ I[A] & \longrightarrow & B & : & \{A, n_A\}_{pk_B} \\ B & \longrightarrow & I[A] & : & \{n_A, n_B\}_{pk_A} \\ I & \longrightarrow & A & : & \{n_A, n_B\}_{pk_A} \\ A & \longrightarrow & I & : & \{n_B\}_{pk_I} \\ I[A] & \longrightarrow & B & : & \{n_B\}_{pk_B} \end{array} \tag{12.13}$$

From Alice's perspective, she has managed to complete a mutual authentication session with Isabel, which was her goal all along. But Isabel also managed to complete an authentication session with Betty, tricking Betty into believing she is interacting with Alice.

There is a simple fix that eliminates that vulnerability:

$$\begin{array}{llllll} 1. & A & \longrightarrow & B & : & \{A, n_A\}_{pk_B} \\ 2. & B & \longrightarrow & A & : & \{B, n_A, n_B\}_{pk_A} \\ 3. & A & \longrightarrow & B & : & \{n_B\}_{pk_B} \end{array} \tag{12.14}$$

It is interesting to see how the fix works: if Alice, during her mutual authentication attempt with Isabel, notices that the response message she receives from Isabel names a different agent than Isabel, then she can deduce that her authentication attempt is being subverted to try to confound another agent, and she can abort the authentication attempt at that point.

12.2.3 Attackers

A distinguishing feature of cryptographic protocols, besides the use of cryptography, is that they are deployed in potentially hostile environments in which attackers may attempt to subvert the operations of the protocol.

Reasoning about cryptographic protocols, therefore, requires a threat model, describing the kind of attackers against which the cryptographic protocol should protect. Attackers commonly considered in the literature include:

- **Eavesdropping attackers:** assumed to be able to read all messages exchanged between agents. Eavesdropping attackers do not affect communication in any way, however, and remain hidden from other agents.

- **Active attackers:** assumed to have complete control over communications between agents, that is, able to read all messages as well as intercept them and forge new messages. They remain hidden from other agents, and thus no agent will intentionally attempt to communicate with an active attacker.

- **Insider attackers:** assumed to have complete control over communications between agents just like active attackers, but also considered legitimate registered users in their own right. They can therefore initiate interactions with other agents as themselves, and other agents can intentionally initiate interactions with them.[6]

[6] A fourth class of attackers, less commonly considered, shares characteristics with both eavesdropping attackers and insider attackers: dishonest agents are assumed not to have control over the network but may attempt to subvert the protocol while acting within the limits imposed on legitimate users.

The class of insider attackers includes the class of active attackers, which itself includes the class of eavesdropping attackers. Thus, in that sense, an insider attacker is stronger than an active attacker which is stronger than an eavesdropping attacker. In practice, this means that a cryptographic protocol that is deemed secure in the presence of an insider attacker will remain so in the presence of active and eavesdropping attackers, and so on.

We saw several examples of attacks in §12.2.2, performed by different kind of attackers. Most of the protocols in §12.2.2 achieve their goals in the presence of eavesdropping attackers, while some also achieve their goals in the presence of active attackers but fail in the presence of insider attackers. The Needham-Schroeder protocol (12.12), for instance, can be shown to satisfy mutual authentication in the presence of active attackers, but not in the presence of insider attackers—as exemplified by attack (12.13)—while the variant protocol (12.14) achieves mutual authentication even in the presence of insider attackers.

The attacks described in §12.2.2 took place at the level of the protocols themselves, and not at the level of the encryption schemes used by the protocols. But vulnerabilities in encryption schemes are also relevant: an attacker cracking an encrypted message from the trusted server to the agents in protocol (12.11) will learn the session key, which will invalidate any confidentiality guarantees claimed for the protocol. Despite this, cryptographic protocol are typically analyzed independently from the details of the encryption scheme. The main reason is that it abstracts away from vulnerabilities specific to the encryption scheme used, leaving only those relating to the cryptographic protocol. Vulnerabilities in encryption schemes are usually independent of the cryptographic protocols that use them, and can be investigated separately. A vulnerability in the protocol will be a vulnerability no matter what encryption scheme is used, and requires a change in the protocol to correct the flaw.

The standard way to analyze cryptographic protocols independently of any encryption scheme is to use a *formal model of cryptography* that assumes perfect encryption leaking no information about encrypted content. It can be defined as the following symbolic encryption scheme. If P is a set of plaintexts and K is a set of keys, then the set of sourcetexts is taken to be the smallest set S of symbolic terms containing P and K such that $(x, y) \in S$ and $\{x\}_k \in S$ when $x, y \in S$ and $k \in K$. Intuitively, (x, y) represents the concatenation of x and y, and $\{x\}_k$ represents the encryption of x with key k. The ciphertexts are all sourcetexts of the form $\{x\}_k$. The symbolic encryption function $e_k(x)$ simply returns $\{x\}_k$, and the symbolic decryption function $d_k(x)$ returns y if x is $\{y\}_k$, and some special token **fail** otherwise.[7]

[7]The symbolic decryption function embodies an assumption that encrypted messages

In the context of analyzing protocols with a formal model of cryptography, attackers are usually modeled using *Dolev-Yao capabilities.* These capabilities go hand in hand with the symbolic aspect of formal models of cryptography. Intuitively, eavesdropping Dolev-Yao attackers can split up concatenated messages and decrypt them if they know the decryption key; active Dolev-Yao attackers can additionally create new messages by concatenating existing messages and encrypting them with known keys. Dolev-Yao attackers do not have the capability of cracking encryptions, nor can they access messages at the level of their component bits.

12.2.4 Modeling Knowledge

The analyses in §12.2.2 show that various notions of knowledge arise rather naturally when reasoning informally about properties of cryptographic protocols. There are essentially two main kinds of knowledge described in the literature. In some frameworks, both kinds of knowledge are used.

Message Knowledge The first kind of knowledge, the most common and in some sense the most straightforward, tries to capture the notion of *knowing a message.*

There are several equivalent approaches to modeling this kind of knowledge, at least in a formal model of cryptography with Dolev-Yao capabilities. Intuitively, the idea is a constructive one: an attacker knows a message if she can construct that message from other messages she has received or intercepted. (Message knowledge in the context of confidentiality properties is often presented from the perspective of an attacker, since confidentiality is breached when the attacker comes to know a particular message.) In such a context, knowing a message is sometimes called *having a message*, *possessing a message*, or *seeing a message.*

Message knowledge may be described via the following sets. Let H be a set of messages that the attacker has received or intercepted. The set $\mathit{Parts}(H)$, the set of all components of messages from H, is defined inductively by the following inference rules:

$$\frac{m \in H}{m \in \mathit{Parts}(H)} \qquad \frac{\{m\}_k \in \mathit{Parts}(H)}{m \in \mathit{Parts}(H)}$$

$$\frac{(m_1, m_2) \in \mathit{Parts}(H)}{m_1 \in \mathit{Parts}(H)} \qquad \frac{(m_1, m_2) \in \mathit{Parts}(H)}{m_2 \in \mathit{Parts}(H)}$$

We see that the content of all encrypted messages in H is included in $\mathit{Parts}(H)$, even those that the attacker cannot decrypt. In a sense,

have enough redundancy for an agent to determine when decryption is successful.

$\mathit{Parts}(H)$ is an upper bound on messages the attacker can know. In contrast, the set $\mathit{Analyzed}(H)$ of messages that the attacker can actually see is more restricted:

$$\frac{m \in H}{m \in \mathit{Analyzed}(H)}$$

$$\frac{(m_1, m_2) \in \mathit{Analyzed}(H)}{m_1 \in \mathit{Analyzed}(H)} \qquad \frac{(m_1, m_2) \in \mathit{Analyzed}(H)}{m_2 \in \mathit{Analyzed}(H)}$$

$$\frac{\{m\}_k \in \mathit{Analyzed}(H) \quad k \in \mathit{Analyzed}(H)}{m \in \mathit{Analyzed}(H)}$$

Clearly, $\mathit{Analyzed}(H) \subseteq \mathit{Parts}(H)$. One definition of message knowledge is to say that an attacker knows message m in a state where she has received or intercepted a set H of messages if $m \in \mathit{Analyzed}(H)$. This is the *attacker knows what she can see* interpretation of message knowledge.

The best way to understand this concept of knowledge is to use a physical analogy: we can think of plaintext messages as stones, and encrypted messages as locked boxes. Encrypting a message means putting it in a box and locking it. A message is known if it can be held in one's hands. An encrypted message is known because the box can be held. The content of an encrypted message is known only if the box can be opened (decrypted) and the content (a stone or another box) taken and held.

This form of message knowledge can be captured fairly easily in any logic without using heavy technical machinery, since the data required to define message knowledge is purely local. If we let $\mathit{Messages}_i(s)$ be the set of messages received or intercepted by agent i in state s of the system, then we can capture message knowledge via a proposition $\mathit{knows}_i(m)$, where i is an agent and m is a message, defined to be true at state s if and only if $m \in \mathit{Analyzed}(\mathit{Messages}_i(s))$.

Rather than using a dedicated proposition, another approach relies on a dedicated modal operator to capture message knowledge. Message knowledge as defined above can be seen as a form of *explicit knowledge*, often represented by a modal operator $X_i\varphi$, read *agent i explicitly knows φ*. (Explicit knowledge is to be contrasted with the implicit knowledge captured by the possible-worlds definition of knowledge.) One form of explicit knowledge, *algorithmic knowledge*, uses a local algorithm stored in the local state of an agent to determine if φ is explicitly known to that agent. Thus, $X_i\varphi$ is true at a state s if the local algorithm of agent i says that the agent knows φ in state s. If we let proposition $\mathit{part}_i(m)$ be true at a state s when $m \in \mathit{Parts}(\mathit{Messages}_i(s))$, then it is a simple matter to define a local algorithm to check if $m \in \mathit{Analyzed}(\mathit{Messages}_i(s))$ and capture knowledge of message m via $X_i(\mathit{part}_i(m))$: *agent i explicitly knows that message m*

is part of the messages she has received. Thus, $part_i(m)$ may be true at a state while $X_i(part_i(m))$ is false at that state if the message is encrypted with a key that agent i does not know.

A variant of the *can see* interpretation of message knowledge is to consider instead the messages that an attacker can create. Define the set $\mathit{Synthesized}(H)$ of messages that the attacker can create from a set H of messages inductively by the following inference rules:

$$\frac{m \in H}{m \in \mathit{Synthesized}(H)}$$

$$\frac{m_1 \in \mathit{Synthesized}(H) \quad m_2 \in \mathit{Synthesized}(H)}{(m_1, m_2) \in \mathit{Synthesized}(H)}$$

$$\frac{m \in \mathit{Synthesized}(H) \quad k \in \mathit{Synthesized}(H)}{\{m\}_k \in \mathit{Synthesized}(H)}$$

An alternative interpretation of message knowledge, the *attacker knows what she can send* interpretation, can be defined as: an attacker knows message m in a state where she has received or intercepted a set H of messages if $m \in \mathit{Synthesized}(\mathit{Analyzed}(H))$. Since it holds that $\mathit{Analyzed}(H) \subseteq \mathit{Synthesized}(\mathit{Analyzed}(H))$, everything an attacker can see she can also send.

The *can send* interpretation of message knowledge is tricky, because clearly any agent can send any plaintext and any key—this is akin to anyone being able to send any password—and it is easy to inadvertently define nondeterministic attackers that can synthesize any message. The intent is for attackers to be able to send only messages based on those she has received or intercepted, but that is a restriction that can be difficult to justify. This suggests some subtleties in choosing the right definition of message knowledge.

Message knowledge, whether under the *can see* or *can send* interpretation, is severely constrained. It is knowledge of terms, as opposed to knowledge of facts—although terms can be facts, facts are more general than terms. Message knowledge is conducive to formal verification using a variety of techniques, mostly because it does not require anything but looking at the local state of an agent. Indeed, message knowledge is inherently local.

Possible-Worlds Knowledge The other kind of knowledge that arises in the study of cryptographic protocols is the standard possible-worlds definition of knowledge via an accessibility relation over the states of a structure. The Kripke structures interpreting knowledge are usually sets of traces of the protocol and the accessibility relation for agent i is an equivalence relation over the states of the system that relates two states in which agent i

has the same local state (including having received or intercepted the same messages).

In the presence of cryptography, the standard accessibility relation, meant to capture when two states are indistinguishable to an agent, seems inappropriate. After all, the whole point of cryptography is to hide information—and in particular, most cryptographic definitions say that if an agent receives message m_1 encrypted with a key k_1 that she does not know and message m_2 encrypted with key k_2 that she also does not know, then that agent should be unable to distinguish the two messages, in the sense of being able to identify which is which. Thus, goes the argument, a state where an agent has received $\{m_1\}_{k_1}$ and a state where that agent has received $\{m_2\}_{k_2}$ instead should be indistinguishable if k_1 and k_2 are not known.

To capture a more appropriate definition of state indistinguishability, one approach is to filter the local states through a function that replaces all messages encrypted with an unknown key by a special token $\Box$. More precisely, if H is a set of messages, we write $[H] = \{[m]^H : m \in H\}$, where $[m]^H$ is inductively defined as follows:

$$\begin{aligned} [m]^H &= m \qquad \text{if } m \text{ is a plaintext} \\ [(m_1, m_2)]^H &= ([m_1]^H, [m_2]^H) \\ [\{m\}_k]^H &= \begin{cases} \{[m]^H\}_k & \text{if } k \in \mathit{Analyzed}(H) \\ \Box & \text{otherwise} \end{cases} \end{aligned}$$

The revised equivalence relations $\sim_i^\Box$ through which knowledge is interpreted can now be defined to be $s \sim_i^\Box t$ if and only if $[\mathit{Messages}_i(s)] = [\mathit{Messages}_i(t)]$.[8]

The definition $[-]^H$ above, which is typical, uses $\mathit{Analyzed}(-)$ to extract the keys that the agent knows. Alternate definitions can be given, from a simpler definition that looks for keys appearing directly in the local state, to a more complex recursive definition defined using possible-worlds knowledge.

Possible-worlds knowledge interpreted via an $\sim_i^\Box$ accessibility relation is general enough to express message knowledge. If we assume a class of propositions $\mathit{part}_i(m)$ as before, true at a state s when $m \in \mathit{Parts}(\mathit{Messages}_i(s))$, then formula $K_i(\mathit{part}_i(m))$ says that agent i knows message m—intuitively, she knows that m is part of some message in her local state, and has access to it.

[8]This definition does not account for the possibility that an agent, even if she does not know the content of an encrypted message, may still recognize that she has already seen that encrypted message. (This is an issue when encryption is deterministic, so that encrypting m with key k always yields the same string of bits.) One approach is to refine the definition so that every encryption $\{m\}_k$ is replaced by a unique token $\Box_{m,k}$.

To see that $K_i(part_i(m))$ corresponds to message knowledge as defined above, we can relate it to the definition of message knowledge in terms of a local $knows_i(m)$ proposition, using the *can see* interpretation of message knowledge. It is not difficult to show that if $knows_i(m)$ is true at state s, then $K_i(part_i(m))$ must also be true at state s. The converse direction requires a suitable richness condition that guarantees that there are enough encrypted messages to compare: for every message $\{m\}_k$ received or intercepted where k is not known to the agent, there should exist another state in which the agent has received $\{m'\}_{k'}$ for a different m' and a different k'.[9] Under such a richness condition, if $K_i(part_i(m))$ is true at a state s, then $knows_i(m)$ is true at that same state s.

Thus, possible-worlds knowledge can be used to express message knowledge, and can also capture higher-order knowledge, that is, knowledge about general facts, including other agents' knowledge. The informal analyses of §12.2.2 show that it makes sense to state that Alice may know that Betty knows the key. While Betty's knowledge here is message knowledge (knowledge of the key) and therefore can be modeled with any of the approaches above, Alice's knowledge is higher-order knowledge, knowledge about knowledge of another agent. Logics that allow reasoning about Alice's knowledge of Betty's knowledge of the key tend to rely on possible-worlds definitions of knowledge.

12.2.5 Reasoning about Cryptographic Protocols

Several approaches have been developed for reasoning about cryptographic protocols. Most are not based on epistemic logic, but extend a classical propositional or first-order logic—possibly with temporal operators—with a simple form of message knowledge in the spirit of $knows_i(m)$. This allows them to leverage well-understood techniques for system analysis from the formal verification community and from the programming language community. Other approaches are explicitly epistemic in nature.

Techniques for reasoning about cryptographic protocols roughly split along two axes, each corresponding to a way of using logic to reason about protocols in general.

(1) Reasoning can be performed either deductively using the proof theory of the logic (e.g., through deductions in a theorem prover), or semantically, using the models of the logic (e.g., through model checking).

[9] To see the need for a richness condition, if there is a single state in which agent i has received an encrypted message, then $K_i(part_i(m))$ holds vacuously when m is the content of the encrypted message.

(2) Reasoning can be performed either directly on the description of the protocol—either taken as a sequence of messages or a program for each role in the protocol—or indirectly on the set of traces generated by protocol executions.

Comparing reasoning methods across these axes is difficult, as each have their advantages and their disadvantages.

The Inductive Method A good example of a deductive approach for reasoning about security protocols is the Inductive Method, based on inductive definitions in higher-order logic (a generalization of first-order predicate logic allowing quantification over arbitrary relations). These inductive definitions admit powerful induction principles which become the main proof technique used to establish confidentiality and authentication properties.

The Inductive Method is fairly characteristic of many deductive approaches to cryptographic protocol analysis: the deductive system is embedded in a powerful logic such as higher-order logic, and does not use epistemic concepts beyond a local definition of message knowledge equivalent to the use of a $knows_i(m)$ proposition.

The Inductive Method proper is based on defining a theory—a set of logical rules—for analyzing a given protocol. The theory for a protocol describes how to generate the protocol execution traces, where a trace is a sequence of events such as *A sends m to B*, represented by the predicate $\mathsf{Say}(A, B, m)$. Rules state which events can possibly follow a given sequence of events, thereby describing traces inductively. In general, there is a rule in the theory for every message in the protocol description. Rules inductively define a set Prot of traces representing all the possible traces of the protocol.

If we consider a theory for Protocol (12.14), a rule for message (1) would say:

$$\begin{aligned} &tr \in \mathsf{Prot} \\ &\qquad \Rightarrow \langle tr, \mathsf{Say}(A, B, \{A, n_A\}_{pk_B})\rangle \in \mathsf{Prot} \end{aligned}$$

where $\langle tr, e\rangle$ adds event e to trace tr. That is, if tr represents a valid trace of the protocol, then that trace can be extended with the first message of a new protocol execution. Similarly, a rule for message (2) would say:

$$\begin{aligned} &tr \in \mathsf{Prot} \\ &\wedge \mathsf{Say}(A', B, \{A, n_A\}_{pk_B}) \in tr \\ &\qquad \Rightarrow \langle tr, \mathsf{Say}(B, A, \{B, n_A, n_B\}_{pk_A})\rangle \in \mathsf{Prot} \end{aligned}$$

That is, if tr is a valid trace of the protocol in which an agent has received the first message of a protocol execution, that agent can respond appropriately with the second message of the protocol execution.[10] Rules are simply implications and conjunctions over a vocabulary of events.

The attacker S is also defined by rules; these rules describe how attacker actions can extend traces with new events. For a Dolev-Yao attacker, these rules define a nondeterministic process that can intercept any message, decompose it into parts and decrypt it if the correct key is known, and that can create new messages from other messages it has observed. The theory includes inductive definitions for the *Analyzed* and *Synthesized* sets given in §12.2.4, as well as rules of the form

$$\begin{aligned} & tr \in \mathsf{Prot} \\ & \wedge\, m \in \mathit{Synthesized}(\mathit{Analyzed}(\mathsf{Spied}(tr))) \\ & \wedge\, B \in \mathsf{Agents} \\ & \qquad \Rightarrow \langle tr, \mathsf{Say}(S, B, m)\rangle \in \mathsf{Prot} \end{aligned}$$

that states that if m can be synthesized from the messages the attacker observed on trace tr (captured by an inductively-defined set $\mathsf{Spied}(tr)$), then the attacker can add an event $\mathsf{Say}(S, B, m)$ for any agent B to the trace.

The Inductive Method is geared for proving safety properties: for every state in every trace, that state is not a bad state. A protocol is proved correct by induction on the length of the traces: choosing the shortest sequence to a bad state, assuming all states earlier on the trace are good, then deriving a contradiction by showing that any state following these good states must be good itself.

A confidentiality property such as *the attacker never learns message* m is established by making sure that the attacker is unable to ever send message m, by proving the following formula:

$$(\forall tr \in \mathsf{Prot})\,(\forall B \in \mathsf{Agents})\,\mathsf{Say}(S, B, m) \notin tr$$

This is a *can send* interpretation of message knowledge. Indeed, according to the rules for the attacker, if the attacker knows message m at any point during a trace, then there exists a extension of that trace where the attacker sends message m. Thus, showing that the attacker never learns message m amounts to showing that there is no trace in which an event $\mathsf{Say}(S, B, m)$ appears, for any agent B.

Abstracting away from the details of the approach, the Inductive Method essentially relies on rules to describe the evolution of a protocol execution, and verifying a confidentiality property is reduced to verifying that a

[10] These rules are simplifications. Actual rules would contain appropriate quantification and additional side conditions to ensure that A and B are different agents, that nonces do not clash, and so on.

certain bad state is not reachable. Other approaches to cryptographic protocol analysis share this methodology, many of them using a logic programming language rather than higher-order logic to express protocol evolution rules; see §12.6.

BAN Logic The Inductive Method relies on encoding rules for generating protocol execution traces in an expressive general logic suitable for automating inductive proofs. In contrast, the next approach, BAN Logic, is a logic tailored for reasoning about cryptographic protocols described as a sequence of message exchanges. It has the additional feature of including a higher-order *belief* operator as a primitive.

BAN Logic is a logic in the tradition of Hoare Logic, in that it advocates an axiomatic approach for reasoning about cryptographic protocols. BAN Logic tracks the evolution of beliefs during the execution of cryptographic protocol, and is described by a set of inference rules for deriving new beliefs from old. BAN Logic includes primitive formulas stating that k is a shared key known only to A and B ($A \stackrel{k}{\leftrightarrow} B$), that m is a secret between A and B ($A \stackrel{m}{\rightleftharpoons} B$), that agent A believes formula F (A $\mathsf{believes}$ F), that agent A controls the truth of formula F (A $\mathsf{controls}$ F), that agent A sent a message meaning F (A said F), that agent A received and understood a message meaning F (A sees F), and that a message meaning F was created during the current protocol execution ($\mathsf{fresh}(F)$). The precise semantics of these formulas is given indirectly through inference rules, some of which are presented below.

BAN Logic assumes that agents can recognize when an encrypted message is one they have created themselves; encryption is in consequence written $\{F\}^i_k$, where i denotes the agent who encrypted a message meaning F with key k. (This also highlights another characteristic of BAN Logic: messages are formulas.)

Here are some of the inference rules of BAN Logic:

$$\text{(R1)} \qquad \frac{A \text{ believes } B \stackrel{k}{\leftrightarrow} A \quad A \text{ sees } \{F\}^i_k \quad i \neq A}{A \text{ believes } B \text{ said } F}$$

$$\text{(R2)} \qquad \frac{A \text{ believes } B \text{ said } (F, F')}{A \text{ believes } B \text{ said } F}$$

$$\text{(R3)} \qquad \frac{A \text{ believes fresh}(F) \quad A \text{ believes } (B \text{ said } F)}{A \text{ believes } B \text{ believes } F}$$

$$\text{(R4)} \qquad \frac{A \text{ believes } B \text{ controls } F \quad A \text{ believes } B \text{ believes } F}{A \text{ believes } F}$$

$$\text{(R5)} \qquad \frac{A \text{ sees } (F, F')}{A \text{ sees } F}$$

(R6) $$\frac{A \text{ believes } B \overset{k}{\leftrightarrow} A \quad A \text{ sees } \{F\}_k^i \quad i \neq A}{A \text{ sees } F}$$

(R7) $$\frac{A \text{ believes fresh}(F)}{A \text{ believes fresh}((F', F))}$$

(R8) $$\frac{A \text{ believes } B \text{ believes } (F, F')}{A \text{ believes } B \text{ believes } F}$$

Rule (R1), for instance, says that if agent A believes that k is shared only between B and herself, and she receives a message encrypted with key k that she did not encrypt herself, then she believes that B sent the original message. Rule (R3) is an honesty rule: it says that agents send messages meaning F only when they believe F. There are commutative variants of rules (R2), (R5), (R7), and (R8), as well as variants for more general tuples; there are also variants of (R8) for any level of nested belief.

BAN Logic does not attempt to model protocol execution traces. Reasoning is done directly on the sequence of messages in the description of the protocol. Because sequences of messages do not carry enough information to permit this kind of reasoning, a transformation known as *idealization* must be applied to the protocol. Roughly speaking, idealization consists of replacing the messages in the protocol by formulas of BAN Logic that capture the intent of each message. For instance, if agent A sends key k to agent B with the intention of sharing a key that is known only to A, then a suitable idealization would have A send the formula $A \overset{k}{\leftrightarrow} B$ to B. Idealization is an annotation mechanism, and as such is somewhat subjective.

To illustrate reasoning in BAN Logic, consider the following simple protocol in which Alice sends a secret value m_0 to Betty encrypted with their shared key k_{AB}, along with a nonce exchange to convince B that the message is not a replay of a message in a previous execution of the protocol (see §12.2.2):

$$\begin{array}{llcll} 1. & A & \longrightarrow & B & : \; A \\ 2. & B & \longrightarrow & A & : \; n_B \\ 3. & A & \longrightarrow & B & : \; A, \{m_0, n_B\}_{k_{AB}} \end{array} \tag{12.15}$$

A possible idealization of protocol (12.15) would be:

$$3'. \quad A \longrightarrow B \; : \; \{A \overset{m_0}{\rightleftharpoons} B, n_B\}_{k_{AB}} \tag{12.16}$$

The first two messages in protocol (12.15) carry information that BAN Logic does not use, so they are not present in the idealized protocol. The third message is idealized to A sending formula $A \overset{m_0}{\rightleftharpoons} B$ to B along with the nonce n_B, indicating that A considers m_0 to be a secret at that point.

Reasoning about an idealized protocol consists in laying out the initial beliefs of the agents, and deriving new beliefs from those and from the messages exchanged between the agents, using the inference rules of the logic. For protocol (12.16), initial beliefs include that both parties believe that key k_{AB} has not been compromised, that nonce n_B has not already been used, and that message m_0 that A wants to send to B is initially secret. These initial beliefs are captured by the following formulas:

$$A \text{ believes } A \overset{k_{AB}}{\leftrightarrow} B$$

$$B \text{ believes } A \overset{k_{AB}}{\leftrightarrow} B$$

$$B \text{ believes } \mathsf{fresh}(n_B)$$

$$A \text{ believes } A \overset{m_0}{\rightleftharpoons} B$$

We can derive new formulas from these initial beliefs in combination with the messages exchanged by the agents. The idea is to update this set of formulas after each protocol step: after an idealized step $A \rightarrow B : F$, which says that B receives a message meaning F, we can add formula B sees F to the set of formulas, and we use the inference rules to derive additional formulas to add to the set.

For example, in idealized protocol (12.16), after message (3'), we add formula

$$B \text{ sees } \{A \overset{m_0}{\rightleftharpoons} B, n_B\}_{k_{AB}} \tag{12.17}$$

to the set of initial beliefs. Along with the initial belief B believes $A \overset{k_{AB}}{\leftrightarrow} B$, formula (12.17) allows us to apply inference rule (R1) to derive:

$$B \text{ believes } A \text{ said } (A \overset{m_0}{\rightleftharpoons} B, n_B). \tag{12.18}$$

From the initial belief B believes $\mathsf{fresh}(n_B)$, inference rule (R7) lets us derive that any message combined with n_B must be fresh, and thus we can derive:

$$B \text{ believes } \mathsf{fresh}(A \overset{m_0}{\rightleftharpoons} B, n_B). \tag{12.19}$$

Formula (12.19) together with (12.18) give us, via inference rule (R3):

$$B \text{ believes } A \text{ believes } (A \overset{m_0}{\rightleftharpoons} B, n_B).$$

Via inference rule (R8), this yields:

$$B \text{ believes } A \text{ believes } A \overset{m_0}{\rightleftharpoons} B. \tag{12.20}$$

Thus, after the messages of the idealized protocol have been exchanged, B believes that A believes that m_0 is a secret between A and B. This is about as much as we can expect.

We can say more if we are willing to assume that B believes that the secrecy of m_0 is in fact controlled by A. If so, we can add the following formula to the set of initial beliefs:

$$B \text{ believes } A \text{ controls } A \overset{m_0}{\rightleftharpoons} B \tag{12.21}$$

and formulas (12.21) and (12.20) combine via inference rule (R4) to yield the stronger conclusion:

$$B \text{ believes } A \overset{m_0}{\rightleftharpoons} B.$$

In other words, if B believes that A controls the secrecy of m_0 and also that A believes m_0 to be secret, then after the protocol executes B also believes that m_0 is a secret shared only with A.

Attackers are not explicit in BAN Logic. In a sense, an active Dolev-Yao attacker is implicitly encoded within the inference rules of the logic, but the focus of BAN Logic is reasoning about the belief of agents in the presence of an active attacker, as opposed to reasoning about the knowledge of an attacker. A successful attack in BAN Logic shows up as a failure to establish a desired belief for one of the agents following a protocol execution.

Temporal and Epistemic Temporal Logics Another class of approaches for reasoning about cryptographic protocols rely on a form of temporal logic to express desired properties of the protocol and show that they are true of a model representing the protocol—generally through a suitable representation of its traces. This is done through model-checking techniques to determine algorithmically whether a formula is true in the models representing the protocol. These model-checking techniques vary in terms of how the models are described: these can be either directly expressed by finite state machines, or through domain-specific languages.

The simplest approaches to cryptographic protocol analysis via temporal logics merely extend existing temporal-logic verification techniques. At least two challenges arise in these cases: modeling attackers, and expressing message knowledge. For attackers, while eavesdropping attackers do not affect the execution of protocols and therefore are comparatively easy to handle in standard temporal-logic verification frameworks, active attackers require work. In some cases, it is possible to simply encode an active attacker within the model using the tools of the framework. Message knowledge is usually dealt with by introducing a variant of a $\mathit{knows}_i(m)$ proposition.

In general, the logics themselves are completely straightforward: they are standard propositional or first-order temporal logics extended with a message knowledge predicate. All the action is in the interpretation of the message knowledge predicate, and in the construction of the models to account for the actions of active attackers. There is not much to say about those approaches as far as pertains to epistemic concepts, but they are popular in practice.

More interesting from an epistemic perspective are those frameworks relying on a temporal epistemic logic, that is, a logic with both temporal and epistemic operators. The MCK model-checker is an example of a verification framework that uses a linear-time temporal logic with epistemic operators to verify protocols that do not use cryptography, such as the Seven Hands or the Dining Cryptographers protocols of §12.2.1. Protocols are described via finite state machines, and formulas express properties of paths through that finite state machine, each such path corresponding to a possible execution of the protocol.

As an example, consider the Dining Cryptographers protocol, which translates well to a finite state machine. States can be described using three agent-indexed Boolean variables $\mathit{paid}[i]$, $\mathit{chan}[i]$, and $\mathit{df}[i]$, where variable $\mathit{paid}[i]$ records whether agent i paid;, variable $\mathit{chan}[i]$ is a communication channel used by agent i to send the result of its coin toss to her right neighbor, and variable $\mathit{df}[i]$ records the announcement of Df_i by agent i at the end of the protocol. The initial states are all the states satisfying:

$$\begin{aligned}
&(\neg \mathit{paid}[1] \wedge \neg \mathit{paid}[2] \wedge \neg \mathit{paid}[3]) \\
&\quad \vee (\mathit{paid}[1] \wedge \neg \mathit{paid}[2] \wedge \neg \mathit{paid}[3]) \\
&\quad \vee (\neg \mathit{paid}[1] \wedge \mathit{paid}[2] \wedge \neg \mathit{paid}[3]) \\
&\quad \vee (\neg \mathit{paid}[1] \wedge \neg \mathit{paid}[2] \wedge \mathit{paid}[3])
\end{aligned}$$

Every agent executes the following program, where a single step of the program for each agent is executed in a transition of the state machine:

```
protocol diningcrypto (paid : observable Bool,
                       chan_left, chan_right : Bool,
                       df : observable Bool[])

  coin_left, coin_right : observable Bool

  begin
    if    True -> coin_right := True
       [] True -> coin_right := False
    fi;
    chan_right.send(coin_right);
    coin_left := chan_left.recv();
    df[self] := coin_left xor coin_right xor paid;
  end
```

Program `diningcrypto`[11] is instantiated for every agent with suitable variables for the parameters:

agent 1 executes `diningcrypto` $(paid[1], chan[1], chan[2], df)$
agent 2 executes `diningcrypto` $(paid[2], chan[2], chan[3], df)$
agent 3 executes `diningcrypto` $(paid[3], chan[3], chan[1], df)$

At the first state transition, every agent nondeterministically chooses a value for their coin toss into local variable `coin_right`; at the second state transition, the result of the coin toss is sent on the channel given as the `chan_right` parameter; at the third state transition, the local variable `coin_left` for each agent is updated to reflect the result of the coin toss received from the agent's left neighbor; at the fourth state transition, variable *df* is updated for every agent.

Given such a state machine, a formula expressing the anonymity of the payer from agent 1's perspective can be written as:

$$\begin{aligned}&\mathbf{X}^4\,(\neg paid[1]\\&\qquad\Rightarrow (K_1(\neg paid[1] \wedge \neg paid[2] \wedge \neg paid[3]))\\&\qquad\quad\vee (K_1(paid[2] \vee paid[3]) \wedge \neg K_1 paid[2] \wedge \neg K_1 paid[3]))\end{aligned}$$

where $\mathbf{X}^4$ is a temporal operator meaning *after four rounds*. This formula, which is true or false of an initial state, says that after the protocol terminates, if cryptographer 1 did not pay, then she either knows that no cryptographer paid, or she knows that one of the other two cryptographers paid but does not know which. (Formulas expressing anonymity from agent 2 and agent 3's perspectives are similar.)

MCK has no built-in support for active attackers, so it cannot easily deal with cryptographic protocols even if we were to add a message knowledge primitive to the language that can deal with encrypted messages. Of course, it is possible to encode some attackers within the language that MCK provides for describing models, but the effect on the efficiency of model checking is unclear.

The theoretical underpinnings of model checking for temporal epistemic logic are fairly well understood, even though the problem has not been studied nearly as much as model checking for temporal logics. Message knowledge does not particularly complicate matters, once the choice of how

[11]The `observable` annotation is used to derive the indistinguishability relation: two states are indistinguishable to agent i if the observable variables of the program executed by agent i have the same value in both states. The `if ... [] ... fi` construct nondeterministically executes one of its branches with an associated condition that evaluates to true. Variable `self` is assigned the name of the agent executing the program.

to interpret message knowledge is made. Accounting for active attackers is more of an issue, since active attackers introduce additional actions into the model, increasing its size.

The main difficulty with model checking epistemic temporal logic is its inherent complexity. While model checking a standard epistemic logic such as S5 takes time polynomial in the size of the model, adding temporal operators and interpreting the logic over the possibly infinite paths in a finite state machine increases that complexity. For example, in the presence of perfect recall (when agents remember their full history) and synchrony (when agents have access to a global clock), the model-checking problem has non-elementary complexity if the logic includes an *until* temporal operator, and is PSPACE-complete otherwise. The problem tends to become PSPACE-complete when perfect recall is dropped. Progress has been made to control the complexity of model-checking epistemic temporal logics by a careful analysis of the complexity of specific classes of formulas that, while restricted, are still sufficiently expressive to capture interesting security properties, but much work remains to be done to make the resulting techniques efficient.

12.3 Information Flow in Multi-Level Systems

Confidentiality in cryptographic protocols is mainly viewed through the lens of access control: some privilege (a key) is required in order to access the confidential data (the content of an encrypted message). An agent who has the key can access the content, an agent without the key cannot. Those access restrictions can control the release of information, but once that information is released there is nothing stopping it from being propagated by agents or by the system through error or malice, or because the released information is needed for the purpose of computations. For systems in which confidentiality is paramount, it is not sufficient to simply ensure that access to confidential data is controlled, there also needs to be a guarantee that even when the confidential data is released it does not land in unauthorized hands. These sorts of confidentiality guarantees require understanding the flow of information in a system.

Confidentiality in the presence of released information is usually studied in the context of systems in which all data are classified with a security level, and where agents have security clearances allowing them to access data at their security level or lower. For simplicity, only scenarios with security levels *high* and *low* (think *classified* and *unclassified* in military settings) will be considered. Intuitively, a high-security agent should be allowed to read both high- and low-security data, and a low-security agent should be allowed to read only low-security data. This is an example of security policy, which describes the forms of information flows that are allowed and those that are disallowed. Information flows that are disallowed capture the desired confidentiality guarantee.

As an example, imagine a commercial system such as a bank mainframe, where agents perform transactions via credit cards or online accounts. In such a system, credit card numbers and bank account numbers might be considered high-security data, and low-security agents should be prevented from accessing them. However, what about the last four digits of a credit card number? Even that information is often considered sensitive. What about a single digit? What about the digits frequency in any given credit card number? Because it is in general difficult to characterize exactly what kind of information about high-security data should not be leaked to low-security agents, it is often easier to prevent any kind of partial information disclosure.

The problem of preventing information disclosure is made more interesting, and more complicated, by the fact that information may not only flow directly from one point to another (e.g, by an agent sending a message to another, or by information being posted, or by updating an observable memory location) but may also flow indirectly from one point to another. Suppose that the commercial system described above sends an email to a central location whenever a transfer of more than one million dollars into a given account A occurs. Anyone observing email traffic can see those emails being sent and learn that account A now contains at least a million dollars. This is an extreme example, but it illustrates indirect information flow: information is gained not by directly observing an event, but by correlating an observation with the event.

Epistemic concepts arise naturally in this setting—a security policy saying that there is no flow of information from high-security data to low-security agents can be expressed as *low-security agents do not learn anything about high-security data.* Moreover, the definitions used in the literature essentially rely on a possible-worlds definition of knowledge within a specific class of models.

Two distinct models of information flow will be described. Both of these models are observational models: they define the kind of observations that agents can make about the system and about the activity of other agents. These observations form the basis of agents' knowledge.

The first model considered takes a fairly abstract view of a system, as sequences of events such as inputs from agents, outputs to agents, internal computation, and so on. These events are the observations that agents can make. In such a setting, security policies regulate information about the occurrence of events. The second model considered is more concrete, and stems from practical work on defining verification techniques for information flow at the level of the source code implementing a system. In that model, observations take the form of content of memory locations that programs can manipulate.

12.3.1 Information Flow in Event Systems

The first model of information flow uses sets of traces corresponding to the possible executions of the system. Every trace is a sequence of events; some of those events are high-security events (and only observable by high-security agents), and some of those events are low-security events (and observable by both high-security and low-security agents). The intuition is that a low-security agent, observing only the low-security events in a trace, should not be able to infer any information about the high-security events in a trace.

How can a low-security agent infer information? If we assume that the full set of traces of the system is known to all agents, then a low-security agent, upon observing a particular sequence of low-security events, can narrow down a set of possible traces that could be the actual trace by considering all the traces that are compatible with her view of the low-security events. By looking at those possible traces, she may infer information about high-security events. For instance, maybe a particular high-security event e appears in every such possible trace, and thus she learns that high-security event e has occurred. In the most extreme case, there may be a single trace compatible with her view of the low-security events, and therefore that low-security agent learns exactly which high-security events have occurred.

The model can be formalized using event systems. An event system is a tuple $S = (E, I, O, Tr)$ where E is a set of events, $I \subseteq E$ a set of input events, $O \subseteq E$ a set of output events, and $Tr \subseteq E^*$ a set of finite traces representing the possible executions of the system. Given a trace $\tau \in E^*$ and a subset $E' \subseteq E$ of events, we write $\tau|_{E'}$ for the subtrace of τ consisting of events from E' only.

We assign a security level to events in E by partitioning them into low-security events L and high-security events H: events in $I \cap L$ are low-security input events, events in $O \cap L$ are low-security output events, and so on.

A naive attempt at defining information flow in this setting might be to say that there is information flowing from high-security events to low-security agents if a low-security agent's view of $\tau|_L$ implies that at least one high-security event subsequence is not possible. In other words, seeing a particular sequence of low-security events rules out one possible high-security event subsequence. Formally, if we write $Tr|_H$ for $\{\tau|_H : \tau \in Tr\}$, information flows from high-security events to low-security agents if there is a trace $\tau \in Tr$ such that $\{\tau'|_H : \tau'|_L = \tau|_L\} \neq Tr|_H$.

Such a definition turns out to be too strong—it is equivalent to separability described below—because it pinpoints information flows where there are none: since low-security events may influence high-security events, a particular subsequence of high-security events may be ruled out due to the influence of low-security events, and in that case there should be no information flow since the low-security agent could have already predicted that the high-security subsequence would have been ruled out. Intuitively, there is information flow when one high-security event subsequence that should be possible as far as the low-security agent expects is not in fact possible. This argument gives an inkling as to why the definition of information flow is not entirely trivial.

Security policies in event systems are often defined as closure properties of the set of traces. Security policies that historically were deemed interesting for the purpose of formalizing existing multi-level systems include the following:

- **Separability:** no nontrivial interaction between high-security events and low-security agents is possible because for any such interaction there is a trace with the same high-security events but different low-security events, and a trace with the same low-security events but different high-security events. Formally, for every pair of traces $\tau_1, \tau_2 \in Tr$, there is a trace $\tau \in Tr$ such that $\tau|_L = \tau_1|_L$ and $\tau|_H = \tau_2|_H$.

- **Noninference:** a low-security agent cannot learn about the occurrence of high-security events because any trace, as far as the low-security agent can tell, could be a trace where there are no high-security events at all. Formally, for all traces $\tau \in Tr$, there is a trace $\tau' \in Tr$ such that $\tau|_L = \tau'|_L$ and $\tau'|_H = \langle\,\rangle$.

- **Generalized Noninference:** a more lenient form of noninference, where a low-security agent cannot learn about the occurrence of high-security input events because any trace could be a trace where there are no high-security input events at all. Formally, for all traces $\tau \in Tr$, there is a trace $\tau' \in Tr$ such that $\tau|_L = \tau'|_L$ and $\tau'|_{(H \cap I)} = \langle\,\rangle$.

- **Generalized Noninterference:** a low-security agent cannot learn about high-security input events, and high-security input events cannot influence low-security events. Formally, for all traces $\tau \in Tr$ and all traces $\tau' \in \mathit{interleave}((H \cap I)^*, \{\tau|_L\})$, there is a $\tau'' \in Tr$ such that $\tau''|_L = \tau|_L$ and $\tau''|_{(L \cup (H \cap I))} = \tau'$. (Function $\mathit{interleave}(T, U)$ returns every possible interleaving of every trace from T with every trace from U.)

A closure property says that if some traces are in the model, then other variations on these traces must also be in the model. This is clearly an epistemic property. Under a possible-worlds definition of knowledge, an agent knows a formula if that formula is true at all traces that the agent considers possible given her view of the system. In general, the fewer possible traces there are, the more facts can be known, since it it easier for a fact to be true at all possible traces if there are few of them. The closure properties ensure that there are enough possible traces from the perspective of a low-security agent to prevent a specific of class facts from being known.[12]

Closure conditions on sets of traces are therefore just a way to enforce lack of knowledge, given a possible-worlds definition of knowledge. We can make this precise by viewing event systems as Kripke frames.

The accessibility relation of each agent depends on the agent's security level. In the case of interest, a low-security agent is assigned an accessibility relation $\sim_L$ defined as $\tau_1 \sim_L \tau_2$ if and only if $\tau_1|_L = \tau_2|_L$.

We identify a proposition with a set of traces, intuitively, those traces in which the proposition is true. As usual, conjunction is intersection of propositions, disjunction is union of propositions, negation is complementation of propositions with respect to the full set of traces in the event system, and implication is subset inclusion. To define the proposition *the low-security agent knows* P, we first define the low-security agent's knowledge set of a trace τ as the set of all traces $\sim_L$-equivalent to τ, $\mathcal{K}_L(\tau) = \{\tau' : \tau \sim_L \tau'\}$. The proposition *the low-security agent knows* P can be defined in the usual way, as:

$$\mathsf{K}_L(P) = \{\tau : \mathcal{K}_L(\tau) \subseteq P\}$$

It is easy to see that K_L satisfies the usual S5 axioms, suitably modified to account for propositions being sets:

$$\begin{array}{ll} \text{(D)} & \mathsf{K}_L(P) \cap \mathsf{K}_L(Q) = \mathsf{K}_L(P \cap Q) \\ \text{(K)} & \mathsf{K}_L(P) \subseteq P \\ \text{(PI)} & \mathsf{K}_L(P) \subseteq \mathsf{K}_L(\mathsf{K}_L(P)) \\ \text{(NI)} & \neg\mathsf{K}_L(P) \subseteq \mathsf{K}_L(\neg\mathsf{K}_L(P)) \end{array}$$

These properties are the set-theoretic analogues of *Distribution*, *Knowledge*, *Positive Introspection*, and *Negative Introspection*, respectively.

As an example, consider an event system (E, I, O, Tr) that satisfies generalized noninterference, and the proposition *high-security input event* e *has occurred*. This proposition is represented by the set P_e of all traces in Tr in which e occurs. The proposition *the low-security agent knows that* e *has occurred* corresponds to the set of traces $\mathsf{K}_L(P_e)$. It is easy to check that

[12] As in §12.2, this does not take probabilistic information into account.

because the system satisfies generalized noninterference, the set $\mathsf{K}_L(P_e)$ is empty, meaning that there is no trace on which the low-security agent ever knows P_e, that is, that e has occurred. By way of contradiction, suppose that $\tau \in \mathsf{K}_L(P_e)$. By definition, $\tau \in \mathsf{K}_L(P_e)$ if and only if $\mathcal{K}_L(\tau) \subseteq P_e$. But the closure condition for generalized noninterference implies that there must exist a trace $\tau' \sim_L \tau$, that is, a trace in $\mathcal{K}_L(\tau)$, such that e does not occur in τ'. Thus, there is a trace in $\mathcal{K}_L(\tau)$ which is not in P_e, and $\mathcal{K}_L(\tau) \not\subseteq P_e$. Thus, $\tau \notin \mathsf{K}_L(P_e)$, a contradiction.

This is a somewhat roundabout way to see that there is an implicit epistemic logic lurking which explains the notions of information flow security policies in event systems. It is certainly possible to make such a logic explicit by introducing a syntax and adding an interpretation to event systems, and study information flow in event systems from such a perspective.

The key point here is that event-system models of information flow and the expression of security policies in those models intrinsically use epistemic concepts, and all reasoning is essentially classical epistemic reasoning performed directly on the models.

12.3.2 Language-Based Noninterference

A more concrete model for information flow is obtained by moving away from trace-based models of systems and relying instead on the program code implementing those systems.

Defining information flow at the level of programs has several advantages: the system is described in detail, information can be defined in terms of the data explicitly manipulated by the program, and enforcement can be automated; the latter turns out to be especially important given the complexity of modern computing systems which makes manual analysis often infeasible.

The observational model used by most language-based information-flow security research is not event-based, although it is still broadly concerned with input and output. The focus here is on information flow in imperative programs, which operates by changing the state of the environment as a program executes. The state of the environment is represented by a store holding values associated with variables. Variables can be read and written by programs. Every variable is tagged with a security level, describing the security level of the data it contains. A low-security agent can observe all low-security variables, but not the high-security ones. Inputs to programs are modeled as initial values of variables, while outputs are modeled as final value of variables: low-security inputs are initial value of low-security variables, and so on. The basic security policy generally considered is a form of noninterference: that low-security outputs do not reveal anything about high-security inputs, and that high-security inputs do not influence the value of low-security outputs.

Consider the following short programs:

(P1) $h := l + 1$
(P2) $l := h + 1$
(P3) $\mathsf{if}\ l = 0\ \mathsf{then}\ h := h + 1\ \mathsf{else}\ l := l + 1$
(P4) $\mathsf{if}\ h = 0\ \mathsf{then}\ h := h + 1\ \mathsf{else}\ l := l + 1$

In all of these programs, variable h is a high-security variable, and variable l is a low-security variable.

A program executes in a store assigning initial values to variables, and execution steps modify the store until the program terminates in a final store. Several simplifying assumption are made: programs are deterministic, and programs always terminate. This is purely to keep the discussion and the technical machinery light. These restrictions can be lifted easily. Moreover, the programming language under consideration will not be described in detail; the sample programs should be intuitive enough.

How do we formalize noninterference in this setting? A store σ is a mapping from variables x to values $\sigma(x)$. We assume every variable x is tagged with a security level $sec(x) \in \{L, H\}$. Let Σ be the set of all possible stores. We model execution of a program C using a function $[\![C]\!] : \Sigma \longrightarrow \Sigma$ from initial stores to final stores. Thus, executing program C in store σ yields a final store $[\![C]\!](\sigma)$. For example, executing program (P1) in store $\langle l \mapsto 5, h \mapsto 10\rangle$ yields store $\langle l \mapsto 5, h \mapsto 6\rangle$, and executing program (P3) in store $\langle l \mapsto 5, h \mapsto 10\rangle$ yields store $\langle l \mapsto 6, h \mapsto 10\rangle$.

Two stores σ_1 and σ_2 are L-equivalent, written $\sigma_1 \approx_L \sigma_2$, if they assign the same values to the same low-security variables: $\sigma_1 \approx_L \sigma_2$ if and only if for all variables x with $sec(x) = L$, $\sigma_1(x) = \sigma_2(x)$. A program C satisfies noninterference if executing C in two L-equivalent states (that is, in two states that a low-security agent cannot distinguish) yields two L-equivalent states: for all σ_1 and σ_2, if $\sigma_1 \approx_L \sigma_2$, then $[\![C]\!](\sigma_1) \approx_L [\![C]\!](\sigma_2)$.[13]

How do programs (P1–4) fare under this definition of noninterference? Program (P1) clearly satisfies noninterference, since the final value of low-security variable l does not depend on the value of any high-level variable, while program (P2) clearly does not. The other two programs are more interesting. The final value of low-security variable l in program (P3) only

[13] Another way of understanding this definition is that it requires the relation on stores induced by program execution to be a refinement of L-equivalence $\approx_L$. If we define $[\![C]\!]_{\approx_L}$ as the relation $\{([\![C]\!](\sigma_1), [\![C]\!](\sigma_2)) : \sigma_1 \approx_L \sigma_2\}$, then the noninterference condition can be rephrased as $[\![C]\!]_{\approx_L} \subseteq\ \approx_L$.

depends on the initial value of l, and thus we expect (P3) to satisfy noninterference, and it does. Program (P4), however, does not, as we can see by executing the program in stores $\langle l \mapsto 0, h \mapsto 0\rangle$ and $\langle l \mapsto 0, h \mapsto 1\rangle$, both $\approx_L$-equivalent, but which yield stores $\langle l \mapsto 0, h \mapsto 1\rangle$ and $\langle l \mapsto 1, h \mapsto 1\rangle$, respectively, two stores that cannot be L-equivalent since they differ in the value they assign to variable l. And indeed, observing the final value of l reveals information about the initial value of h.

Noninterference is usually established by a static analysis of the program code, which approximates the flow of information through a program before execution. While the details of the static analyses are interesting in their own right, they have little to do with epistemic logic beyond providing an approach to verifying a specific kind of epistemic property in a specific context.

Recent work on language-based information-flow security has highlighted the practical importance of declassification, that is, the controlled release of high-security data to low-security agents. The problem of password-based authentication illustrates the need for such release: when a low-security agent tries to authenticate herself as a high-security agent, she may be presented with a login screen asking for the password of the high-security agent. That password should of course be considered high-security information. However, the login screen leaks information, since entering an incorrect password will reveal that the attempted password is not the right password, thereby leaking a small amount of information about the correct password. The leak is small, but it exists, and because of it the login screen does not satisfy the above definition of noninterference. Defining a suitable notion of security policy that allows such small release of information while still preventing more important information flow is a complex problem.

While the concepts underlying information-flow security are clearly epistemic in nature—taking stores as possible worlds and L-equivalence as an accessibility relation for low-security agents—there is no real demand for an explicit epistemic logic in which to describe policies. One reason is that it is in general difficult to precisely nail down, in a given system, what high-security information should be kept from low-security agents. It is simply easier to ask that no information be leaked to low-security agents. This *no information* condition is easier to state semantically than through an explicit logical language—not learning any information in the sense of noninterference can be stated straightforwardly as a relationship between equivalence relations, while if we were to use an epistemic logic, we would have to say something along the lines of *for all formulas φ that do not depend only on the state of the low-security agent,* $\neg K_L\varphi$ where K_L expresses the knowledge of that low-security agent. The latter is patently clunkier to work with. It may be the case that an explicit epistemic logic would be more useful in the context of declassification, where not all information needs to be kept from low-security agents.

12.4 Beyond Confidentiality

The focus of this chapter has been on confidentiality, because it is by far the most studied security property. It is not only important, it also underpins several other security properties. Other related properties are also relevant.

Anonymity A specific form of confidentiality is anonymity, where the information to be kept secret is the association between actions and agents who perform them. Anonymity has been studied using epistemic logic, and several related definitions have been proposed and debated.

To discuss anonymity, we need to be able to talk about actions and agents who perform them. Let $\delta(i, a)$ be a proposition interpreted as *agent* i *performed action* a.

The simplest definition of anonymity is lack of knowledge: action a performed by agent i is minimally anonymous with respect to agent j if agent j does not know that agent i performed a. This can be captured by the formula

$$\neg K_j \delta(i, a).$$

Minimal anonymity is, well, minimal. It does not rule out that agent j may narrow down the list of possible agents that performed a to agent i and one other agent. Stronger forms of anonymity can be defined: action a performed by agent i is totally anonymous with respect to agent j if, as far as agent j is concerned, action a could have been performed by any agent in the system (except for agent j). This can be captured by the formula

$$\delta(i, a) \Rightarrow \bigwedge_{i' \neq j} P_j \delta(i', a)$$

where $P_i \varphi$ is the usual dual to knowledge, $\neg K_i(\neg \varphi)$, read as *agent* i *considers* φ *possible.*

Total anonymity is at the other extreme on the spectrum from minimal anonymity; it is a very strong requirement. Intermediate definitions can be obtained by requiring that actions be anonymous only up to a given set of agents—sometimes called an anonymity set: action a performed by agent i is anonymous up to I with respect to agent j if, as far as agent j is concerned, action a could have been performed by any agent in I. This can be captured by the formula:

$$\delta(i, a) \Rightarrow \bigwedge_{i' \in I} P_j \delta(i', a).$$

As an example of this last definition of anonymity, note that it can be used to describe the anonymity provided by the Dining Cryptographers protocol from §12.2.1. Recall that if one of the cryptographers paid, the Dining Cryptographers protocol guarantees that each of the non-paying cryptographers think it possible that any of the cryptographers but herself paid. In other words, if $C = \{Alice, Betty, Charlene\}$ are the cryptographers and if cryptographer i paid, then the protocol guarantees that the paying action is anonymous up to $C \setminus \{j\}$ with respect to cryptographer j, as long as $j \neq i$.

Coercion Resistance Voting protocols are protocols in which anonymity plays an important role. Voting protocols furthermore satisfy other interesting security properties. Aside from secrecy of votes (that every voter's choice should be private, and observers should not be able to figure out who voted how), other properties include fairness (voters do not have any knowledge of the distribution of votes until the final tallies are announced), verifiability (every voter should be able to check whether her vote was counted properly), and receipt freeness (no voter has the means to prove to another that she has voted in a particular manner).

This last property, receipt freeness, is particularly interesting in terms of epistemic content. Roughly speaking, receipt freeness says that a voter Alice cannot prove to a potential coercer Corinna that she voted in a particular way. This is the case even if Alice wishes to cooperate with Corinna; receipt freeness guarantees that such cooperation cannot lead to anything because it will be impossible for Corinna to be certain how Alice voted. In that sense, receipt freeness goes further than secrecy of votes. Even if Alice tells Corinna that she voted a certain way, Corinna has no way to verify Alice's assertion, and Alice has no way to convince her.

Coercion resistance is closely related to receipt freeness but is slightly stronger. Intuitively, a voting protocol is coercion resistant if it prevents voter coercion and vote buying even by active coercers: a coercer should not be able to influence the behavior of a voter. Coercion resistance can be modeled epistemically, although the details of the modeling is subtle, and many important details will be skipped in the description below. Part of the difficulty and subtlety is that the idea of coercion means changing how a voter behaves based on a coercer's desired outcome or goal, which needs to be modeled somehow.

One formalization of coercion resistance uses a model of voting protocols based on traces, where some specific agents are highlighted: a voter that the coercer tries to influence (called the coerced voter), the coercer, and the remaining agents and authorities, assumed to be honest. Every voter in the system votes according to a voting strategy, which in the case of honest voters is the strategy corresponding to the voting protocol.

The formalization assumes that every voter has a specific voting goal, formally captured by the set of traces in which that voter successfully votes according to her desired voting goal. The coercer, however, is intent on affecting the coerced voter—for instance, to coerce a vote for a given candidate, or perhaps to coerce a vote away from a given candidate. To coerce a voter, the coercer hands the coerced voter a particular strategy that will fulfill the coercer's goals instead of the coerced voter's. For instance, the coercer's strategy may simply be one that forwards all messages to and from the coercer, effectively making the coerced voter a proxy for the coercer.

Let V be the space of possible strategies that voters and coercers can follow. Coercion resistance can be defined by saying that for every possible strategy $v \in V$, there is another strategy $v' \in V$ that the coerced voter can use instead of v with the property that: (1) the voter always achieves her goal by using v', and (2) the coercer does not know whether the coerced voter used strategy v or v'. In other words, in every trace in which the coerced voter uses strategy v, the coercer considers it possible, given her view of the trace, that the coerced voter is using strategy v' instead. Conversely, in every trace in which the coerced voter uses strategy v', the coercer considers it possible that the coerced voter is using strategy v. So, the coercer cannot know whether the coerced voter followed the coercer's instructions (i.e., used v) or tried to achieve her own goal (i.e., used v'). As in the case of information flow in event systems in §12.3.1, the definition of coercion resistance is a form of closure property on traces, which corresponds to lack of knowledge in the expected way, where knowledge is captured by an indistinguishability relation on states based on the coercer's observations.

Zero Knowledge The property *an agent does not learn anything about something*, as embodied in information-flow security policies and other forms of confidentiality, is generally modeled using an indistinguishability relation over states and enforced by making sure that there are enough states to prevent the confidential information from being known by unauthorized agents.

Another approach to modeling and enforcing this lack of learning is demonstrated by *zero knowledge interactive proof systems*. An interactive proof system for a string language L is a two-party system (P, V) in which a prover P tries to convince a verifier V that some string x is in L through a sequence of message exchanges amounting to an interactive proof of $x \in L$. Classically, the prover is assumed to be infinitely powerful, while the verifier is assumed to be a probabilistic polynomial-time Turing machine.

An interactive proof system has the property that if $x \in L$, the conversation between P and V will show $x \in L$ with high probability, and if $x \notin L$, the conversation between *any* prover and V will show $x \in L$ with low probability. (The details for why the second condition refers to any prover rather than just P is beyond the scope of this discussion.)

An interactive proof system for L is zero knowledge if whenever $x \in L$ holds the verifier is able to generate *on its own* the conversations it would have had with the prover during an interactive proof of $x \in L$. The intuition here is that the verifier does not learn anything from a conversation with the prover (other than $x \in L$) if it can learn exactly the same thing by generating that whole conversation itself. Thus, the only knowledge gained by the verifier is that which the prover initially set out to prove.

Zero knowledge interactive proof systems rely on indistinguishability, but not indistinguishability among a large set of states. Rather, it is indistinguishability between two scenarios: a scenario where the verifier interacts with the prover, and a scenario where the verifier does not interact with the prover but instead simulates a complete interaction with the prover. This simulation paradigm, a core notion in modern theoretical computer science, says roughly that an agent does not gain any knowledge from interacting with the outside world if she can achieve the same results without interacting with the outside world.

To give a sense of the kinds of definitions that arise in this context, here is one formal definition of perfect zero knowledge:

> Let (P, V) be an interactive proof system for L, where P (the prover) is an interactive Turing machine and V (the verifier) is a probabilistic polynomial-time interactive Turing machine. System (P, V) is *perfect zero-knowledge* if for every probabilistic polynomial-time interactive Turing machine V^* there is a probabilistic polynomial-time Turing machine M^* (the simulator) such that for every $x \in L$ the following two random variables are identically distributed:
>
> (i) the output of V^* interacting with P on common input x;
>
> (ii) the output of machine M^* on input x.

While the details are beyond the scope of this chapter, the intuition behind this definition is to have, for every possible verifier V^* (and not only V) interacting with P, a machine M^* that can simulate the interaction of V^* and P even though it does not have access to the prover P. The existence of such simulators implies that V^* does not gain any knowledge from P.

This gives a different epistemic foundation for confidentiality, one that is intimately tied to computation and its complexity. The relationship with classical epistemic logic is essentially unexplored.

12.5 Perspectives

The preceding sections illustrate how extensively epistemic concepts, explicitly framed as an epistemic logic or not, have been applied to security research. Whether the application of these concepts has been successful is a more subjective question.

In a certain sense, the problems described in this chapter are solved problems by now. Confidentiality and authentication in cryptographic protocol analysis under a formal model of cryptography and Dolev-Yao attackers, for example, can be checked quite efficiently with a vast array of methods, at least for common security properties, and the definitions used approximate the epistemic definitions quite closely.

So what are the remaining challenges in cryptographic protocol analysis, and has epistemic logic a role to play? The most challenging aspect of cryptographic protocol analysis is to move beyond Dolev-Yao attackers and beyond formal models of cryptography, towards more concrete models of cryptography.

Moving beyond a Dolev-Yao attacker requires shifting the notion of message knowledge to use richer algebras of message with more operations. Directions that have been explored include providing attackers with the ability to perform offline dictionary attacks, working with an XOR operation, or even number-based operations such as exponentiation. One problem is that when the algebra of messages is subject to too many algebraic properties, determining whether an attacker knows a message may quickly become undecidable. Even when message knowledge for an attacker is decidable, it may still be too complex for efficient reasoning. It is not entirely clear how epistemic concepts can help solve problems in that arena.

Moving from a formal model of cryptography to a concrete model, one that reflects real encryption schemes more accurately using sequences of bits and computational indistinguishability, requires completely shifting the approach to cryptographic protocol analysis.

Formal models of cryptography work by abstracting away the *one-way security* property of encryption schemes—that it is computationally hard to recover the sourcetext from a ciphertext without knowing the encryption key. More concrete models of cryptography rely on stronger properties than one-way security, properties such as *semantic security*, which intuitively says that if any information about a message m can be computed by an efficient algorithm given the ciphertext $e_k(m)$ for a random k and m chosen according to an arbitrary probability distribution, that same information can be computed without knowing the ciphertext. In other words, the ciphertext $e_k(m)$ offers no advantage in computing information about some message m chosen from an arbitrary probability distribution.

The definition of semantic security is reminiscent of the definition of zero knowledge interactive proof systems in §12.4, and it is no accident, as they both rely on a simulation paradigm to express the fact that no knowledge is gained. As in the case of zero knowledge interactive proof systems, there is a clear epistemic component to the definition of semantic security, one to which classical epistemic logic has not been applied.

The main difficulty with applying classical epistemic logic to concrete models of cryptography is that these models take attackers to be probabilistic polynomial-time Turing machines, and take security properties to be probabilistic properties relative to those probabilistic polynomial-time Turing machines. This means that an epistemic approach to concrete models of cryptography needs to be probabilistic as well as computationally bounded. The former is not a problem, since probabilistic reasoning shares much of the same foundations as epistemic reasoning. But the latter is more complicated. Concrete models of cryptography are not based on impossibility, but on computational hardness. And while possible-worlds definitions of knowledge are well suited to talking about impossible versus possible outcomes, they fare less well at talking about difficult versus easy outcomes.

The trouble that possible-worlds definitions of knowledge run into when trying to incorporate a notion of computational difficulty is really the problem of logical omniscience in epistemic logic under a different guise. Agents, under standard possible-worlds definitions of knowledge, know all tautologies, and know all logical consequences of their knowledge: if $K\varphi$ is true and $\varphi \Rightarrow \psi$ is valid, then $K\psi$ is also true. Any normal epistemic operator will satisfy these properties, and in particular, any epistemic logic based on Kripke structures will satisfy these properties. Normality does not deal well with computational difficulty, because while it may be computationally difficult to establish that $\varphi \Rightarrow \psi$ is valid, a normal modal logic will happily derive all knowledge-based consequences of that valid formula. It would seem that giving a satisfactory epistemic account of concrete models of cryptography requires a non-normal epistemic logic, one that supports a form of resource-bounded knowledge. Resource-bounded knowledge is not well understood, and logics for resource-bounded knowledge still feel too immature to form a solid basis for reasoning about concrete models of cryptography.

Leaving aside concrete models of cryptography, it is almost impossible to discuss epistemic logic in the context of cryptographic protocols without addressing the issue of BAN Logic. BAN Logic is an interesting and original use of logic, developed to prove cryptographic protocol properties manually by paralleling informal arguments for protocol correctness.

BAN Logic has spilled a lot of virtual ink. Aside from its technical limitations—it requires a protocol idealization step that remains outside the purview of the logic but affects the results of analysis—the logic is considered somewhat *passé*. Other approaches we saw in §12.2.5 operate in the same space, namely analyzing cryptographic protocols under a formal model of cryptography in the presence of Dolev-Yao attackers, and most are less limited and more easily automated. Other approaches, such as Protocol Composition Logic, even advocate Hoare-style reasoning about the protocol text from within the logic, just like BAN Logic.

My perspective on BAN Logic is that it tried to do something which has not really been tried since, something that remains a sort of litmus test for our understanding of security in cryptographic protocols: identifying high-level primitives that capture relevant concepts for security, high-level primitives that match our intuitive understanding of security properties, those same intuitions that guide our design of cryptographic protocols in the first place. We do not have such high-level primitives in any other framework, all of which tend to work at much lower levels of abstraction. The primitives in BAN Logic are intuitively attractive, but poorly understood. The continuing conversation on BAN Logic is a reminder that we still do not completely understand the basic concepts and basic terms needed to discuss cryptographic protocols, and I think BAN Logic remains relevant, if only as a nagging voice telling us that we have not quite gotten it right yet.

Many of the issues that arise when trying to push cryptographic protocol analysis from a formal model of cryptography to a more concrete model also come up in the context of information-flow security. As mentioned in §12.3.2, recent work has turned to the question of declassification, or controlled release of information. The reason for this is purely pragmatic: most applications need to release some kind of information in order to do any useful work, even under a lax interpretation of noninterference.

But it does not take long to see that even a controlled release of information can lead to unwanted release of information in the aggregate. Returning to the password-login problem from §12.3.2, it is clear that every wrong attempt at entering a password leaks some information, something that needs to happen if the login screen is to operate properly. But of course, repeated attempts at checking the password will eventually lead to the correct password as the only remaining possibility, which is a severe undesirable release of information. Security policies controlling declassification therefore seem to require a way to account for more quantitative notions of leakage which aggregate over time, something that symbolic approaches to information-flow security have difficulty handling well.

Modeling information flow quantitatively can be seen as a move from reasoning about information as a monolithic unit to reasoning about information as a resource. Once we make that leap, other resources affecting information flow start suggesting themselves. For example, execution time can leak information. Consider the simple program:

if *(high-security Boolean variable)*
then *fast code*
else *slow code*

By observing the execution time of the program, we can determine the value of the high-security Boolean variable. This example is rather silly, of course, but it illustrates the point that information leakage can occur based on observations of other resources than simply the state of memory.

What about the combination of information flow and cryptography? After all, in practice, systems do use cryptography internally to help keep data confidential. Encrypted data can presumably be written on shared storage (which might be easier to manage than storage segregated into high-security and low-security storage) or moved online, or in general given to low-security agents without information being released, as long as they do not have the key or the resources to decrypt. Accounting for cryptography in information-flow security raises questions similar to those in cryptographic protocol analysis concerning what models of cryptography to use and how to account for the cryptographic capabilities of attackers. It also raises difficulties similar to those in cryptographic protocol analysis when trying to move from a formal model of cryptography to a concrete model, including how to provide an epistemic foundation for information flow using a resource-bounded definition of knowledge.

Conclusion Epistemic concepts are central to many aspects of reasoning about security. In some cases, these epistemic concepts may even naturally take expression in a *bona fide* epistemic logic that can be used to formalize the reasoning. But whether an epistemic logic is used or not, the underlying concepts are clearly epistemic. In particular, the notion of truth at all possible worlds reappears in many different guises throughout the literature.

Research in security analysis has reached a sort of convergence point around the use of symbolic methods. The challenge seems to be to move beyond this convergence point, and such a move requires taking resources seriously: realistic definitions of security rely on the notion that exploiting a vulnerability should require more resources (time, power, information) than are realistically available to an attacker. In epistemic terms, what is needed is a reasonably well-behaved definition of resource-bounded knowledge, itself an active area of research in epistemic logic. It would appear, then, that advances in epistemic logic may well help increase our ability to reason about security in direct ways.

12.6 Notes

For the basics of epistemic logic, both the syntax and the semantics, the reader is referred to the introductory chapter (Chapter 1) of the current volume. For the sake of making this chapter as self-contained as possible, most of the background material can be usefully obtained from the textbooks of Fagin, Halpern, Moses, and Vardi (1995) and of Meyer and van der Hoek (1995). The possible-worlds definition of knowledge used throughout this chapter is simply the view that knowledge is truth at all worlds that an agent considers as possible alternatives to the current world, a view which goes back to Hintikka (1962).

Cryptographic Protocols While the focus of the section is on symbolic cryptographic protocol analysis, cryptographic protocols can also be studied from the perspective of more computationally-driven cryptography, of the kind described in §12.5; see Goldreich (2004). The Russian Cards problem, which was first presented at the Moscow Mathematic Olympiads in 2000, is described formally and studied from an epistemic perspective by van Ditmarsch (2003). The problem has been used as a benchmark for several epistemic logic model checkers (van Ditmarsch, van der Hoek, van der Meyden, and Ruan, 2006). The Dining Cryptographers problem and the corresponding protocol is described by Chaum (1988). It was proved correct in an epistemic temporal logic model checker by van der Meyden and Su (2004).

For a good overview of classical cryptography along with some perspectives on protocols, see Stinson (1995) and Schneier (1996); both volumes contain descriptions of DES, AES, RSA, and elliptic-curve cryptography. Goldreich (2001, 2004) is also introductory, but from the perspective of modern computational cryptography.

For a good high-level survey of the kind of problems surrounding the design and deployment of cryptographic protocols, see Anderson and Needham (1995), then follow up with the prudent engineering practices by Abadi and Needham (1996). The key distribution protocol used as the first example in §12.2.2 is related to the Yahalom protocol described by Burrows, Abadi, and Needham (1990a). The Needham-Schroeder protocol was first described by Needham and Schroeder (1978). The man-in-the-middle attack on the Needham-Schroeder protocol in the presence of an insider attacker was pointed out by Lowe (1995), and the fix was analyzed by Lowe (1996).

The Dolev-Yao model of the attacker given in §12.2.3 is due to Dolev and Yao (1983).

The formal definition of message knowledge via *Analyzed* and *Synthesized* sets is taken from Paulson (1998). Equivalent definitions are given in nearly every formal system for reasoning about cryptographic protocols in a formal model of cryptography. Message knowledge can be defined using a local deductive system, which makes it fit nicely within the deductive knowledge framework of Konolige (1986)—see also Pucella (2006). More generally, message knowledge is a form of algorithmic knowledge (Halpern, Moses, and Vardi, 1994), that is, a local form of knowledge that relies on an algorithm to compute what an agent knows based on the local state of the agent. In the case of a Dolev-Yao attacker, this local algorithm simply computes the sets of analyzed and synthesized messages (Halpern and Pucella, 2012).

Another way of defining message knowledge is the hidden automorphism model, due to Merritt (1983), which is a form of possible-worlds knowledge. While it never gained much traction, it has been used in later work by Toussaint and Wolper (1989) and also in the logic of Bieber (1990). It uses algebraic presentations of encryption schemes called cryptoalgebras. There is a unique surjective cryptoalgebra homomorphism from the free cryptoalgebra over a set of plaintexts and keys to any cryptoalgebra over the same plaintexts and keys which acts as the identity on plaintexts and keys. Message knowledge in a given cryptoalgebra C is knowledge of the structure of messages as given by that surjective homomorphism from the free cryptoalgebra to C. A revealed reduct is a subset of C that the agent has seen. A state of knowledge with respect to revealed reduct R is a set of of mappings f from the free cryptoalgebra to C that are homomorphisms on $f^{-1}(R)$. In this context, an agent knows message m if the agent knows the representation of message m, meaning that m is the image of the same free cryptoalgebra term under every mapping in the state of knowledge of the agent. Thus, if an agent receives the messages $\{m_1\}_k$ and $\{m_2\}_k$ but does not receive k, then only $\{m_1\}_k$ and $\{m_2\}_k$ are in the revealed reduct; the agent may consider any distinct messages m_1' and m_2' to map to $\{m_1\}_k$ and $\{m_2\}_k$ after encryption with k, since any such mapping will act as a homomorphism on the pre-image of the revealed reduct.

Possible-worlds definitions of knowledge in the presence of cryptography are problematic because cryptography affects what agents can observe, and this impacts the definition of the accessibility relation between worlds. The idea of replacing encrypted messages in the local state of agents by a token goes back to the semantics for BAN Logic by Abadi and Tuttle (1991). Treating encrypted messages as tokens while still allowing agents to distinguish different encrypted messages is less common, but has been used at least by Askarov and Sabelfeld (2007) and by Askarov, Hedin, and Sabelfeld (2008) in the context of information flow.

There are several frameworks for formally reasoning about cryptographic protocols, and I shall not list them all here. But I hope to provide enough pointers to the literature to ensure that the important ones are covered. For an early survey on the state of the art in formal reasoning about cryptographic protocols until 1995, see Meadows (1995).

The Inductive Method described in §12.2.5 is due to Paulson (1998), and is built atop the Isabelle logical framework (Paulson, 1994), a framework for higher-order logic. BAN Logic is introduced by Burrows et al. (1990a), who use it to perform an analysis of several existing protocols in the literature. The logic courted controversy pretty much right from the start (Nessett, 1990; Burrows, Abadi, and Needham, 1990b). Probably the most talked-about problem with BAN Logic is the lack of an independently-motivated semantics which would ensure that statements of the logic match operational intuition. Without such a semantics, it is difficult to argue for the reasonableness of the result of a BAN Logic analysis, except for the pragmatic observation that failure to prove a statement in BAN Logic often indicates a problem with the cryptographic protocol. Abadi and Tuttle (1991) attempt to remedy the situation by defining a semantics for BAN Logic. Successor logics extending or modifying BAN Logic generally start with a variant of the Abadi-Tuttle semantics (Syverson, 1990; Gong, Needham, and Yahalom, 1990; van Oorschot, 1993; Syverson and van Oorschot, 1994; Wedel and Kessler, 1996; Stubblebine and Wright, 1996). Contemporary epistemic logic alternatives to BAN Logic were also developed, using a semantics in terms of protocol execution, but they never really took hold (Bieber, 1990; Moser, 1990).

The model checker MCK is described by Gammie and van der Meyden (2004), and was used to analyze the Dining Cryptographers protocol (van der Meyden and Su, 2004) as well as the Seven Hands protocol for the Russian Cards problem (van Ditmarsch et al., 2006). TDL is an alternative epistemic temporal logic for reasoning about cryptographic protocols with a model checker developed by Lomuscio and Woźna (2006), based on a earlier model checker (Raimondi and Lomuscio, 2004). TDL is a branching-time temporal epistemic logic extended with a message knowledge primitive in addition to standard possible-worlds knowledge for expressing higher-order knowledge, and does not provide explicit support for attackers in its modeling language. The model-checking complexity results mentioned are due to van der Meyden and Shilov (1999); see also Engelhardt, Gammie, and van der Meyden (2007), and Huang and van der Meyden (2010).

Another epistemic logic which forms the basis for reasoning about cryptographic protocol is Dynamic Epistemic Logic (DEL) (Gerbrandy, 1999). DEL is an epistemic logic of broadcast announcements which includes formulas of the form $[\rho]_i\varphi$, read φ *holds after agent* i *broadcasts formula* ρ, where ρ is a formula in a propositional epistemic sublanguage. (The actual syntax of DEL is slightly different.) Agents may broadcast that they know a fact, and this broadcast affects the knowledge of other agents. Kripke structures are used to capture the state of knowledge of agents at a point in time, and agent i announcing ρ will change Kripke structure M representing the current state of knowledge of all agents into a Kripke structure $M^{\rho,i}$ representing the new state of knowledge that obtains. Dynamic Epistemic Logic has been used to analyze the Seven Hands protocol in great detail (van Ditmarsch, 2003). Extensions to handle cryptography are described by Hommersom, Meyer, and de Vink (2004), as well as by van Eijck and Orzan (2007).

Process calculi, starting with the spi calculus (Abadi and Gordon, 1999) and later the applied pi calculus (Abadi and Fournet, 2001), have been particularly successful tools for reasoning about cryptographic protocols. These use either observational equivalence to show that a process implementation of the protocol is equivalent to another process that clearly satisfies the required properties, or static analysis such as type checking to check the properties (Gordon and Jeffrey, 2003). Epistemic logics defined against models obtained from processes are given by Chadha, Delaune, and Kremer (2009) and by Toninho and Caires (2009). Another process calculus, CSP, has also proved popular as a foundation for reasoning about cryptographic protocols (Lowe, 1998; Ryan and Schneider, 2000).

Finally, other approaches rely on logic programming ideas: the NRL protocol analyzer (Meadows, 1996), Multiset Rewriting (Cervesato, Durgin, Lincoln, Mitchell, and Scedrov, 1999), and ProVerif (Blanchet, 2001). Thayer, Herzog, and Guttman (1999) introduce a distinct semantic model for protocols, strand spaces, which has some advantages over traces. Syverson (1999) develops an authentication logic on top of strand spaces, while Halpern and Pucella (2003) investigate the suitability of strand spaces as a basis for epistemic reasoning.

Information Flow Bell and LaPadula (1973) and LaPadula and Bell (1973) were among the first to develop mandatory access control, and introducing the idea of attaching security levels to data to enforce confidentiality.

Early work on information flow security mostly focused on event traces, and tried to describe both closure conditions on traces, as well as unwinding conditions that would allow one to check that a set of event traces satisfies the security condition. Separability was defined by McLean (1994), noninference by O'Halloran (1990), generalized noninference by McLean (1994), and generalized noninterference by McCullough (1987, 1988) following the work of Goguen and Meseguer (1982, 1984). Other definitions of information-flow security for event systems are given by Sutherland (1986) and by Wittbold and Johnson (1990). A modern approach to information-flow security in event systems is described by Mantel (2000). The set-theoretic description of the knowledge operator is taken from a paper by Halpern (1999), but appears in various guises in the economics literature (Aumann, 1989). Halpern and O'Neill (2008) layer an explicit epistemic language on top of the event models re-expressed as Kripke structures, and show that the resulting logic can capture common definitions of confidentiality in event systems.

Denning and Denning (1977) first pointed out that programming languages are a useful setting for reasoning about information flow by observing that static analysis can be used to identify and control information flow. Most recent work on information-flow security from a programming language perspective goes back to the Secure Lambda Calculus by Heintze and Riecke (1998) in a functional language setting, and to Smith and Volpano (1998) in an imperative language setting. Honda, Vasconcelos, and Yoshida (2000) give a similar development in the context of a process calculus. Sabelfeld and Myers (2003) give a survey and overview of the state of the field up to 2003. Balliu, Dam, and Le Guernic (2011, 2012) offer a rare use of an explicit epistemic temporal logic to reason about information-flow security. Sabelfeld and Sands (2009) give a good overview of the issues involved in declassification for language-based information flow. Askarov and Sabelfeld (2007) use an epistemic logic in the context of declassification. Chong (2010) uses a form of algorithmic knowledge to model information release requirements.

Beyond Confidentiality Protocols for anonymous communication generally rely on a cloud of intermediaries that prevent information about the identity of the original sender to be isolated; Crowds is an example of such a protocol (Reiter and Rubin, 1998). Anonymity has been well studied as an instance of confidentiality (Hughes and Shmatikov, 2004; Garcia, Hasuo, Pieters, and van Rossum, 2005). The explicit connection with epistemic logic was made by Halpern and O'Neill (2005), which is the source of the definitions in §12.4. An early analysis of anonymity via epistemic logic is given by Syverson and Stubblebine (1999).

Anonymity is an important component of voting protocols. Van Eijck and Orzan (2007) prove anonymity for a specific voting protocol using epistemic logic. More general analyses of voting protocols with epistemic logic have also been attempted (Baskar, Ramanujam, and Suresh, 2007; Küsters and Truderung, 2009). The model of coercion resistance in §12.4 is from Küsters and Truderung (2009).

Zero knowledge interactive proof systems were introduced by Goldwasser, Micali, and Rackoff (1989) and have become an important tool in theoretical computer science. A good overview is given by Goldreich (2001). Halpern, Moses, and Tuttle (1988) give an epistemically-motivated analysis of zero knowledge interactive proof systems using a computationally-bounded definition of knowledge devised by Moses (1988).

Another context in which epistemic concepts—or perhaps more accurately, epistemic vocabulary—appear is that of authorization and trust management. Credential-based authorization policies can be used to control access to resources by requiring agents to present appropriate credentials (such as certificates) proving that they are allowed access. Because systems that rely on credential-based authorization policies are often decentralized systems, meaning that there is no central clearinghouse for determining for every authorization request whether an agent has the appropriate credentials, the entire approach relies on a web of trust between agents and credentials. Since in many such systems credentials can be delegated—an agent may allow another agent to act on her behalf—not only can credential checking become complicated, but authorization policies themselves become nontrivial to analyze to determine contradictions (an action being both allowed and forbidden by the policy under certain conditions) or coverage (a class of actions remaining unregulated by the policy under certain conditions). Where do epistemic concepts come up in such a scenario? Authorization logics from the one introduced by Abadi, Burrows, Lampson, and Plotkin (1993) to the recent NAL (Schneider, Walsh, and Sirer, 2011) have been described as logics of belief, and are somewhat reminiscent of BAN Logic. One of their basic primitives is a formula $A \mathsf{\ says\ } F$, which as a credential means that A believes and is accountable for the truth of F. Delegation, for example, is captured by a formula $(A \mathsf{\ says\ } F) \Rightarrow (B \mathsf{\ says\ } G)$. This form of belief is entirely axiomatic, just like belief in BAN Logic.

Perspectives Ryan and Schneider (1998) have extended the Dolev-Yao model of attackers with an XOR operation; Millen and Shmatikov (2003) with products and enough exponentiation to model the Diffie-Hellman key-establishment protocol of Diffie and Hellman (1976); and Lowe (2002) and later Corin, Doumen, and Etalle (2005) and Baudet (2005) with the ability to mount offline dictionary attacks. As described by Halpern and Pucella (2012), many of these can be expressed using algorithmic knowledge, at least in the context of eavesdropping attackers. More generally, extending

Dolev-Yao with additional operations can best be studied using equational theories, that is, equations induced by looking at the algebra of the additional operations; see for example Abadi and Cortier (2004), and Chevalier and Rusinowitch (2008).

While it would be distracting to discuss the back and forth over BAN Logic in the decades since its inception, I will point out that recent work by Cohen and Dam has taken a serious look at the logic with modern eyes, and highlighted both interesting interpretations as well as subtleties (Cohen and Dam, 2005b,a). The protocol composition logic PCL of Datta, Derek, Mitchell, and Roy (2007), which builds on earlier work by Durgin, Mitchell, and Pavlovic (2003), is a modern attempt at devising a logic for Hoare-style reasoning about cryptographic protocols.

A good overview of concrete models of cryptography is given by Goldreich (2001). Semantic security, among others, is studied by Bellare, Chor, Goldreich, and Schnorr (1998). The relationship between formal models of cryptography and concrete models—how well does the former approximate the latter?—has been explored by Abadi and Rogaway (2002), and later extended by Micciancio and Warinschi (2004), among others. Backes, Hofheinz, and Unruh (2009) provide a good overview.

Approaches to analyze cryptographic protocols in a concrete model of cryptography have been developed (Lincoln, Mitchell, Mitchell, and Scedrov, 1998; Mitchell, Ramanathan, Scedrov, and Teague, 2001). In recent years some of the approaches for analyzing cryptographic protocols in a formal model of cryptography have been modified to work with a concrete model of cryptography, such as PCL (Datta, Derek, Mitchell, Shmatikov, and Turuani, 2005) and ProVerif (Blanchet, 2008). In some cases, cryptographic protocol analysis in a concrete model relies on extending indistinguishability over states to indistinguishability over the whole protocol (Datta, Küsters, Mitchell, Ramanathan, and Shmatikov, 2004).

Defining a notion of resource-bounded knowledge that does not suffer from the logical omniscience problem is an ongoing research project in the epistemic logic community, and various approaches have been advocated, each with its advantages and its deficiencies: algorithmic knowledge (Halpern et al., 1994), impossible possible worlds (Hintikka, 1975), awareness (Fagin and Halpern, 1988). A comparison between the approaches in terms of expressiveness and pragmatics appears in (Halpern and Pucella, 2011).

Information flow in probabilistic programs was first investigated by Gray and Syverson (1998) using probabilistic multiagent systems (Halpern and Tuttle, 1993). Backes and Pfitzmann (2002) study it in a more computational setting. Smith (2009) presents some of the tools that need to be considered to analyze the kind of partial information leakage occurring in the password-checking example. Preliminary work on information flow in the presence of cryptography includes Laud (2003), Hutter and Schairer (2004), and Askarov et al. (2008).

Acknowledgments Thanks to Aslan Askarov, Philippe Balbiani, Stephen Chong, and Vicky Weissman for comments on an early draft of this chapter.

References

Abadi, M., M. Burrows, B. Lampson, and G. Plotkin (1993). A calculus for access control in distributed systems. *ACM Transactions on Programming Languages and Systems 15*(4), 706–734.

Abadi, M. and V. Cortier (2004). Deciding knowledge in security protocols under equational theories. In *Proc. 31st Colloquium on Automata, Languages, and Programming (ICALP'04)*, Volume 3142 of *Lecture Notes in Computer Science*.

Abadi, M. and C. Fournet (2001). Mobile values, new names, and secure communication. In *Proc. 28th Annual ACM Symposium on Principles of Programming Languages (POPL'01)*, pp. 104–115.

Abadi, M. and A. D. Gordon (1999). A calculus for cryptographic protocols: The spi calculus. *Information and Computation 148*(1), 1–70.

Abadi, M. and R. Needham (1996). Prudent engineering practice for cryptographic protocols. *IEEE Transactions on Software Engineering 22*(1), 6–15.

Abadi, M. and P. Rogaway (2002). Reconciling two views of cryptography (the computational soundness of formal encryption). *Journal of Cryptology 15*(2), 103–127.

Abadi, M. and M. R. Tuttle (1991). A semantics for a logic of authentication. In *Proc. 10th ACM Symposium on Principles of Distributed Computing (PODC'91)*, pp. 201–216.

Anderson, R. and R. Needham (1995). Programming Satan's computer. In J. van Leeuwen (Ed.), *Computer Science Today: Recent Trends and Developments*, Volume 1000 of *Lecture Notes in Computer Science*, pp. 426–440. Springer-Verlag.

Askarov, A., D. Hedin, and A. Sabelfeld (2008). Cryptographically-masked flows. *Theoretical Computer Science 402*(2–3), 82–101.

Askarov, A. and A. Sabelfeld (2007). Gradual release: Unifying declassification, encryption and key release policies. In *Proc. 2007 IEEE Symposium on Security and Privacy*, pp. 207–221. IEEE Computer Society Press.

Aumann, R. J. (1989). Notes on interactive epistemology. Cowles Foundation for Research in Economics working paper.

Backes, M., D. Hofheinz, and D. Unruh (2009). CoSP: A general framework for computational soundness proofs. In *Proc. 16th ACM Conference on Computer and Communications Security (CCS'09)*, pp. 66–78. ACM Press.

Backes, M. and B. Pfitzmann (2002). Computational probabilistic non-interference. In *Proc. 7th European Symposium on Research in Computer Security (ESORICS'02)*, Volume 2502 of *Lecture Notes in Computer Science*, pp. 1–23. Springer-Verlag.

Balliu, M., M. Dam, and G. Le Guernic (2011). Epistemic temporal logic for information flow security. In *Proc. ACM SIGPLAN 6th Workshop on Programming Languages and Analysis for Security (PLAS'11)*. ACM Press.

Balliu, M., M. Dam, and G. Le Guernic (2012). ENCOVER: Symbolic exploration for information flow security. In *Proc. 25th IEEE Computer Security Foundations Symposium (CSF'12)*, pp. 30–44. IEEE Computer Society Press.

Baskar, A., R. Ramanujam, and S. P. Suresh (2007). Knowledge-based modelling of voting protocols. In *Proc. 11th Conference on Theoretical Aspects of Rationality and Knowledge (TARK'07)*, pp. 62–71. ACM Press.

Baudet, M. (2005). Deciding security of protocols against off-line guessing attacks. In *Proc. 12th ACM Conference on Computer and Communications Security (CCS'05)*, pp. 16–25. ACM Press.

Bell, D. E. and L. J. LaPadula (1973). Secure computer systems: Mathematical foundations. Technical Report MTR-2547, Volume 1, MITRE Corporation.

Bellare, M., B. Chor, O. Goldreich, and C. Schnorr (1998). Relations among notions of security for public-key encryption schemes. In *Proc. 18th Annual International Cryptology Conference (CRYPTO'98)*, Volume 1462 of *Lecture Notes in Computer Science*, pp. 26–45. Springer-Verlag.

Bieber, P. (1990). A logic of communication in hostile environment. In *Proc. 3rd IEEE Computer Security Foundations Workshop (CSFW'90)*, pp. 14–22. IEEE Computer Society Press.

Blanchet, B. (2001). An efficient cryptographic protocol verifier based on Prolog rules. In *Proc. 14th IEEE Computer Security Foundations Workshop (CSFW'01)*, pp. 82–96. IEEE Computer Society Press.

Blanchet, B. (2008). A computationally sound mechanized prover for security protocols. *IEEE Transactions on Dependable and Secure Computing 5*(4), 193–207.

Burrows, M., M. Abadi, and R. Needham (1990a). A logic of authentication. *ACM Transactions on Computer Systems 8*(1), 18–36.

Burrows, M., M. Abadi, and R. Needham (1990b). Rejoinder to Nessett. *ACM Operating Systems Review 24*(2), 39–40.

Cervesato, I., N. Durgin, P. Lincoln, J. Mitchell, and A. Scedrov (1999). A metanotation for protocol analysis. In *Proc. 12th IEEE Computer Security Foundations Workshop (CSFW'99)*, pp. 55–69. IEEE Computer Society Press.

Chadha, R., S. Delaune, and S. Kremer (2009). Epistemic logic for the applied pi calculus. In *Proc. IFIP International Conference on Formal Techniques for Distributed Systems*, Volume 5522 of *Lecture Notes in Computer Science*, pp. 182–197. Springer-Verlag.

Chaum, D. (1988). The dining cryptographers problem: Unconditional sender and recipient untraceability. *Journal of Cryptology 1*(1), 65–75.

Chevalier, Y. and M. Rusinowitch (2008). Hierarchical combination of intruder theories. *Information and Computation 206*(2–4), 352–377.

Chong, S. (2010). Required information release. In *Proc. 23rd IEEE Computer Security Foundations Symposium (CSF'10)*, pp. 215–227. IEEE Computer Society Press.

Cohen, M. and M. Dam (2005a). A completeness result for BAN logics. In *Proc. Methods for Modalities (M4M)*, pp. 202–219.

Cohen, M. and M. Dam (2005b). Logical omniscience in the semantics of BAN logic. In *Proc. Workshop on Foundations of Computer Security (FCS'05)*, pp. 121–132.

Corin, R., J. Doumen, and S. Etalle (2005). Analysing password protocol security against off-line dictionary attacks. In *Proc. 2nd International Workshop on Security Issues with Petri Nets and other Computational Models (WISP'04)*, Volume 121 of *Electronic Notes in Theoretical Computer Science*, pp. 47–63. Elsevier Science Publishers.

Datta, A., A. Derek, J. C. Mitchell, and A. Roy (2007). Protocol Composition Logic (PCL). *Electronic Notes in Theoretical Computer Science 172*(1), 311–358.

Datta, A., A. Derek, J. C. Mitchell, V. Shmatikov, and M. Turuani (2005). Probabilistic polynomial-time semantics for a protocol security logic. In *Proc. 32nd Colloquium on Automata, Languages, and Programming (ICALP'05)*, pp. 16–29.

Datta, A., R. Küsters, J. C. Mitchell, A. Ramanathan, and V. Shmatikov (2004). Unifying equivalence-based definitions of protocol security. In *Proc. Workshop on Issues in the Theory of Security (WITS'04)*.

Denning, D. E. and P. J. Denning (1977). Certification of programs for secure information flow. *Communications of the ACM 20*(7), 504–513.

Diffie, W. and M. E. Hellman (1976). New directions in cryptography. *IEEE Transactions on Information Theory 22*, 664–654.

van Ditmarsch, H. P. (2003). The Russian Cards problem. *Studia Logica 75*, 31–62.

van Ditmarsch, H. P., W. van der Hoek, R. van der Meyden, and J. Ruan (2006). Model checking Russian cards. *Electronic Notes in Theoretical Computer Science 149*(2), 105–123.

Dolev, D. and A. C. Yao (1983). On the security of public key protocols. *IEEE Transactions on Information Theory 29*(2), 198–208.

Durgin, N. A., J. C. Mitchell, and D. Pavlovic (2003). A compositional logic for proving security properties of protocols. *Journal of Computer Security 11*(4), 677–722.

van Eijck, J. and S. Orzan (2007). Epistemic verification of anonymity. *Electronic Notes in Theoretical Computer Science 168*, 159–174.

Engelhardt, K., P. Gammie, and R. van der Meyden (2007). Model checking knowledge and linear time: PSPACE cases. In *Proc. Symposium on Logical Foundations of Computer Science*, Volume 4514 of *Lecture Notes in Computer Science*, pp. 195–211. Springer-Verlag.

Fagin, R. and J. Y. Halpern (1988). Belief, awareness, and limited reasoning. *Artificial Intelligence 34*, 39–76.

Fagin, R., J. Y. Halpern, Y. Moses, and M. Y. Vardi (1995). *Reasoning about Knowledge.* MIT Press.

Gammie, P. and R. van der Meyden (2004). MCK: Model checking the logic of knowledge. In *Proc. 16th International Conference on Computer Aided Verification (CAV'04)*, Lecture Notes in Computer Science, pp. 479–483. Springer-Verlag.

Garcia, F. D., I. Hasuo, W. Pieters, and P. van Rossum (2005). Provable anonymity. In *Proc. 3rd ACM Workshop on Formal Methods in Security Engineering (FMSE 2005)*, pp. 63–72. ACM Press.

Gerbrandy, J. (1999). Dynamic epistemic logic. In *Logic, Language, and Information*, Volume 2. CSLI Publication.

Goguen, J. A. and J. Meseguer (1982). Security policies and security models. In *Proc. 1982 IEEE Symposium on Security and Privacy*, pp. 11–20. IEEE Computer Society Press.

Goguen, J. A. and J. Meseguer (1984). Unwinding and inference control. In *Proc. 1984 IEEE Symposium on Security and Privacy*, pp. 75–86. IEEE Computer Society Press.

Goldreich, O. (2001). *Foundations of Cryptography: Volume 1, Basic Tools.* Cambridge University Press.

Goldreich, O. (2004). *Foundations of Cryptography: Volume 2, Basic Applications.* Cambridge University Press.

Goldwasser, S., S. Micali, and C. Rackoff (1989). The knowledge complexity of interactive proof systems. *SIAM Journal on Computing 18*(1), 186–208.

Gong, L., R. Needham, and R. Yahalom (1990). Reasoning about belief in cryptographic protocols. In *Proc. 1990 IEEE Symposium on Security and Privacy*, pp. 234–248. IEEE Computer Society Press.

Gordon, A. D. and A. Jeffrey (2003). Authenticity by typing for security protocols. *Journal of Computer Security 11*(4), 451–520.

Gray, III, J. W. and P. F. Syverson (1998). A logical approach to multilevel security of probabilistic systems. *Distributed Computing 11*(2), 73–90.

Halpern, J. Y. (1999). Set-theoretic completeness for epistemic and conditional logic. *Annals of Mathematics and Artificial Intelligence 26*, 1–27.

Halpern, J. Y., Y. Moses, and M. R. Tuttle (1988). A knowledge-based analysis of zero knowledge. In *Proc. 20th Annual ACM Symposium on the Theory of Computing (STOC'88)*, pp. 132–147.

Halpern, J. Y., Y. Moses, and M. Y. Vardi (1994). Algorithmic knowledge. In *Proc. 5th Conference on Theoretical Aspects of Reasoning about Knowledge (TARK'94)*, pp. 255–266. Morgan Kaufmann.

Halpern, J. Y. and K. O'Neill (2005). Anonymity and information hiding in multiagent systems. *Journal of Computer Security 13*(3), 483–514.

Halpern, J. Y. and K. O'Neill (2008). Secrecy in multiagent systems. *ACM Transactions on Information and System Security 12*(1), 5:1–5:47.

Halpern, J. Y. and R. Pucella (2003). On the relationship between strand spaces and multi-agent systems. *ACM Transactions on Information and System Security 6*(1), 43–70.

Halpern, J. Y. and R. Pucella (2011). Dealing with logical omniscience: Expressiveness and pragmatics. *Artificial Intelligence 175*(1), 220–235.

Halpern, J. Y. and R. Pucella (2012). Modeling adversaries in a logic for security protocol analysis. *Logical Methods in Computer Science 8*(1:21), 1–26.

Halpern, J. Y. and M. R. Tuttle (1993). Knowledge, probability, and adversaries. *Journal of the ACM 40*(4), 917–962.

Heintze, N. and J. G. Riecke (1998). The SLam calculus: Programming with secrecy and integrity. In *Proc. 25th Annual ACM Symposium on Principles of Programming Languages (POPL'98)*, pp. 365–377. ACM Press.

Hintikka, J. (1962). *Knowledge and Belief.* Cornell University Press.

Hintikka, J. (1975). Impossible possible worlds vindicated. *Journal of Philosophical Logic 4*, 475–484.

Hommersom, A., J.-J. Meyer, and E. de Vink (2004). Update semantics of security protocols. *Synthese 142*(2), 229–267.

Honda, K., V. Vasconcelos, and N. Yoshida (2000). Secure information flow as typed process behaviour. In *European Symposium on Programming*, Volume 1782 of *Lecture Notes in Computer Science*, pp. 180–199. Springer-Verlag.

Huang, X. and R. van der Meyden (2010). The complexity of epistemic model checking: Clock semantics and branching time. In *Proc. 19th European Conference on Artificial Intelligence (ECAI'10)*, pp. 549–554. IOS Press.

Hughes, D. and V. Shmatikov (2004). Information hiding, anonymity and privacy: A modular approach. *Journal of Computer Security 12*(1), 3–36.

Hutter, D. and A. Schairer (2004). Possibilistic information flow control in the presence of encrypted communication. In *Proc. 9th European Symposium on Research in Computer Security (ESORICS'04)*, Volume 3193 of *Lecture Notes in Computer Science*, pp. 209–224. Springer-Verlag.

Konolige, K. (1986). *A Deduction Model of Belief.* Morgan Kaufmann.

Küsters, R. and T. Truderung (2009). An epistemic approach to coercion-resistance for electronic voting protocols. In *Proc. 2009 IEEE Symposium on Security and Privacy*, pp. 251–266. IEEE Computer Society Press.

LaPadula, L. J. and D. E. Bell (1973). Secure computer systems: A mathematical model. Technical Report MTR-2547, Volume 2, MITRE Corporation.

Laud, P. (2003). Handling encryption in an analysis for secure information flow. In *Proc. 12th European Symposium on Programming*, Volume 2618 of *Lecture Notes in Computer Science*. Springer-Verlag.

Lincoln, P., J. C. Mitchell, M. Mitchell, and A. Scedrov (1998). A probabilistic poly-time framework for protocol analysis. In *Proc. 5th ACM Conference on Computer and Communications Security (CCS'98)*, pp. 112–121.

Lomuscio, A. and B. Woźna (2006). A complete and decidable security-specialised logic and its application to the tesla protocol. In *Proc. 5th International Joint Conference on Autonomous Agents and Multiagent Systems (AAMAS'06)*, pp. 145–152.

Lowe, G. (1995). An attack on the Needham-Schroeder public-key authentication protocol. *Information Processing Letters 56*, 131–133.

Lowe, G. (1996). Breaking and fixing the Needham-Schroeder public-key protocol using FDR. In *Proc. 2nd International Workshop on Tools and Algorithms for the Construction and Analysis of Systems (TACAS'96)*, Volume 1055, pp. 147–166. Springer-Verlag.

Lowe, G. (1998). Casper: A compiler for the analysis of security protocols. *Journal of Computer Security 6*, 53–84.

Lowe, G. (2002). Analysing protocols subject to guessing attacks. In *Proc. Workshop on Issues in the Theory of Security (WITS'02)*.

Mantel, H. (2000). Possibilistic definitions of security — an assembly kit. In *Proc. 13th IEEE Computer Security Foundations Workshop (CSFW'00)*, pp. 185–199. IEEE Computer Society Press.

McCullough, D. (1987). Specifications for multi-level security and a hook-up property. In *Proc. 1987 IEEE Symposium on Security and Privacy*, pp. 161–166. IEEE Computer Society Press.

McCullough, D. (1988). Noninterference and the composability of security properties. In *Proc. 1988 IEEE Symposium on Security and Privacy*, pp. 177–186. IEEE Computer Society Press.

McLean, J. (1994). A general theory of composition for trace sets closed under selective interleaving functions. In *Proc. 1994 IEEE Symposium on Security and Privacy*, pp. 79–93.

Meadows, C. (1995). Formal verification of cryptographic protocols: A survey. In *Advances in Cryptology (ASIACRYPT'94)*, Volume 917 of *Lecture Notes in Computer Science*, pp. 133–150. Springer-Verlag.

Meadows, C. (1996). The NRL protocol analyzer: An overview. *Journal of Logic Programming 26*(2), 113–131.

Merritt, M. J. (1983). *Cryptographic Protocols.* Ph. D. thesis, Georgia Institute of Technology.

van der Meyden, R. and N. V. Shilov (1999). Model checking knowledge and time in systems with perfect recall (extended abstract). In *Proc. Conference on Foundations of Software Technology and Theoretical Computer Science*, Volume 1738 of *Lecture Notes in Computer Science*, pp. 432–445. Springer-Verlag.

van der Meyden, R. and K. Su (2004). Symbolic model checking the knowledge of the dining cryptographers. In *Proc. 17th IEEE Computer Security Foundations Workshop (CSFW'04)*, pp. 280–291. IEEE Computer Society Press.

Meyer, J.-J. C. and W. van der Hoek (1995). *Epistemic Logic for AI and Computer Science*, Volume 41 of *Cambridge Tracts in Theoretical Computer Science.* Cambridge University Press.

Micciancio, D. and B. Warinschi (2004). Soundness of formal encryption in the presence of active adversaries. In *Proc. Theory of Cryptography Conference (TCC'04)*, Volume 2951 of *Lecture Notes in Computer Science*, pp. 133–151. Springer-Verlag.

Millen, J. and V. Shmatikov (2003). Symbolic protocol analysis with products and Diffie-Hellman exponentiation. In *Proc. 16th IEEE Computer Security Foundations Workshop (CSFW'03)*, pp. 47–61. IEEE Computer Society Press.

Mitchell, J., A. Ramanathan, A. Scedrov, and V. Teague (2001). A probabilistic polynomial-time calculus for analysis of cryptographic protocols. In *Proc. 17th Annual Conference on the Mathematical Foundations of Programming Semantics*, Volume 45 of *Electronic Notes in Theoretical Computer Science.* Elsevier Science Publishers.

Moser, L. E. (1990). A logic of knowledge and belief for reasoning about computer security. In *Proc. 3rd IEEE Computer Security Foundations Workshop (CSFW'90)*, pp. 57–63. IEEE Computer Society Press.

Moses, Y. (1988). Resource-bounded knowledge. In *Proc. 2nd Conference on Theoretical Aspects of Reasoning about Knowledge (TARK'88)*, pp. 261–276. Morgan Kaufmann.

Needham, R. M. and M. D. Schroeder (1978). Using encryption for authentication in large networks of computers. *Communications of the ACM 21*(12), 993–999.

Nessett, D. M. (1990). A critique of the Burrows, Abadi and Needham logic. *ACM Operating Systems Review 24*(2), 35–38.

O'Halloran, C. (1990). A calculus of information flow. In *Proc. European Symposium on Research in Computer Security.*

van Oorschot, P. C. (1993). Extending cryptographic logics of belief to key agreement protocols. In *Proc. 1st ACM Conference on Computer and Communications Security (CCS'93)*, pp. 232–243. ACM Press.

Paulson, L. C. (1994). *Isabelle, A Generic Theorem Prover*, Volume 828 of *Lecture Notes in Computer Science.* Springer-Verlag.

Paulson, L. C. (1998). The inductive approach to verifying cryptographic protocols. *Journal of Computer Security 6*(1/2), 85–128.

Pucella, R. (2006). Deductive algorithmic knowledge. *Journal of Logic and Computation 16*(2), 287–309.

Raimondi, F. and A. Lomuscio (2004). Verification of multiagent systems via ordered binary decision diagrams. In *Proc. AAMAS'04*, pp. 630–637. IEEE Computer Society Press.

Reiter, M. K. and A. D. Rubin (1998). Crowds: Anonymity for web transactions. *ACM Transactions on Information and System Security 1*(1), 66–92.

Ryan, P. and S. Schneider (2000). *Modelling and Analysis of Security Protocols.* Addison Wesley.

Ryan, P. Y. A. and S. A. Schneider (1998). An attack on a recursive authentication protocol: A cautionary tale. *Information Processing Letters 65*(1), 7–10.

Sabelfeld, A. and A. C. Myers (2003). Language-based information-flow security. *IEEE Journal on Selected Areas in Communications 21*(1), 1–15.

Sabelfeld, A. and D. Sands (2009). Declassification: Dimensions and principles. *Journal of Computer Security 17*(5), 517–548.

Schneider, F. B., K. Walsh, and E. G. Sirer (2011). Nexus Authorization Logic (NAL): Design rationale and applications. *ACM Transactions on Information and System Security 14*(1), 1–28.

Schneier, B. (1996). *Applied Cryptography* (Second ed.). John Wiley & Sons.

Smith, G. (2009). On the foundations of quantitative information flow. In *Proc. 12th International Conference on Foundations of Software Science and Computation Structures (FOSSACS'09)*, Volume 5504 of *Lecture Notes in Computer Science*, pp. 288–302. Springer-Verlag.

Smith, G. and D. Volpano (1998). Secure information flow in a multi-threaded imperative language. In *Proc. 25th Annual ACM Symposium on Principles of Programming Languages (POPL'98)*, pp. 355–364. ACM Press.

Stinson, D. R. (1995). *Cryptography: Theory and Practice.* CRC Press.

Stubblebine, S. and R. Wright (1996). An authentication logic supporting synchronization, revocation, and recency. In *Proc. 3rd ACM Conference on Computer and Communications Security (CCS'96).* ACM Press.

Sutherland, D. (1986). A model of information. In *Proc. 9th National Computer Security Conference*, pp. 175–183.

Syverson, P. (1990). A logic for the analysis of cryptographic protocols. NRL Report 9305, Naval Research Laboratory.

Syverson, P. (1999). Towards a strand semantics for authentication logic. In *Proc. 15th Annual Conference on the Mathematical Foundations of Programming Semantics*, Volume 20 of *Electronic Notes in Theoretical Computer Science.* Elsevier Science Publishers.

Syverson, P. F. and S. G. Stubblebine (1999). Group principals and the formalization of anonymity. In *Proc. World Congress on Formal Methods in the Development of Computing Systems*, Volume 1708 of *Lecture Notes in Computer Science*, pp. 814–833.

Syverson, P. F. and P. C. van Oorschot (1994). On unifying some cryptographic protocol logics. In *Proc. 1994 IEEE Symposium on Security and Privacy*, pp. 14–28. IEEE Computer Society Press.

Thayer, F. J., J. C. Herzog, and J. D. Guttman (1999). Strand spaces: Proving security protocols correct. *Journal of Computer Security* 7(2/3), 191–230.

Toninho, B. and L. Caires (2009). A spatial-epistemic logic and tool for reasoning about security protocols. Technical report, Departamento de Informática, FCT/UNL.

Toussaint, M.-J. and P. Wolper (1989). Reasoning about cryptographic protocols. In J. Feigenbaum and M. Merritt (Eds.), *Distributed Computing and Cryptography*, Volume 2 of *DIMACS Series in Discrete Mathematics and Theoretical Computer Science*, pp. 245–262. American Mathematical Society.

Wedel, G. and V. Kessler (1996). Formal semantics for authentication logics. In *Proc. 4th European Symposium on Research in Computer Security (ESORICS'96)*, Volume 1146 of *Lecture Notes in Computer Science*, pp. 219–241. Springer-Verlag.

Wittbold, J. T. and D. Johnson (1990). Information flow in nondeterministic systems. In *Proc. 1990 IEEE Symposium on Security and Privacy*. IEEE Computer Society Press.